Predictive and Optimised Life Cycle Management

Predictive and Optimised Life Cycle Management

Buildings and infrastructure

Edited by
Asko Sarja

Routledge
Taylor & Francis Group

LONDON AND NEW YORK

First published 2006
by Routledge
2 Park Square, Milton Park, Abingdon, Oxon OX14 4RN

Simultaneously published in the USA and Canada
by Routledge
52 Vanderbilt Avenue, New York, NY 10017

*Routledge is an imprint of the Taylor & Francis Group,
an informa business*

First issued in paperback 2020

Copyright © 2006 Taylor & Francis

Typeset in Sabon by
Integra Software Services Pvt. Ltd, Pondicherry, India

British Library Cataloguing in Publication Data
A catalogue record for this book is available from the British Library

Library of Congress Cataloging in Publication Data
Predictive and optimised life cycle management: buildings and infrastructure /
   edited by Asko Sarja.
     p.   cm.
   ISBN 0-415-35393-9 (hardback: alk. paper)   1. Production engineering.
   2. Product life cycle.   3. Production management.
   I. Sarja, Asko.
   TS176.P697 2006
   658.5—dc22   2005020355

ISBN 13: 978-0-367-57785-8 (pbk)
ISBN 13: 978-0-415-35393-9 (hbk)

Contents

Preface

All over the world societies have identified and accepted the goal of sustainable development to reach a stable social and economic development in harmony with nature. Against all these aspects – social, economic, cultural and ecological, the construction industry is a major player. In ecological aspects the most dictating is energy efficiency. Building sector is responsible for a major share of all the influences mentioned above. Hence, the goal of a sustainable building sector is a big challenge for civil engineering.

These changes are creating new challenges for the construction industry, which can be fulfilled successfully only through active and innovative changes in the design, products, manufacturing methods and management. Introducing the latest advancements of new technologies in tackling the challenges, a significant development can be achieved.

Lifetime engineering is a theory and practice of *predictive, optimising and integrated* long-term investment planning, design, construction, management in use, MRR&M (Maintenance, Repair, Rehabilitation and Modernisation) end-of-life management of assets. With the aid of lifetime engineering, we can *control and optimise the lifetime properties of built assets with design and management* corresponding to the objectives of owners, users and society.

Principles, methodologies and methods of *Integrated Life Cycle Design* have been presented in the year 2002 in the book: [Sarja, Asko 2002, *Integrated Life Cycle Design of Structures*. 142pp. Spon Press, London. ISBN 0-415-25235-0.]. This book is focused on the planning and design phase of the life cycle of buildings and civil infrastructures.

This book, *Predictive and Optimised Life Cycle Management*, is focused on the phase of management of facilities in use, which is concretised as principles, methodologies, procedures and methods of maintenance, repair and rehabilitation planning (MR&R). The idea of this presentation is an *integrated, predictive and optimising lifetime management* of buildings and civil infrastructures. The term *integrated* means integration between different phases of the life cycle and integration of all generic classes of the requirements of sustainable building: human (usability) requirements, economy, ecology and culture. The term *predicitve* means that all requirements and properties are modelled and calculated for the entire design lifetime period. The term *optimising* means that all functional and performance properties of a facility and its parts are optimised in relation to lifetime quality of the facility. Most of the detailed methods are the same in planning, design and management, but they are applied differently and in different processes.

Chapter I

Introduction

Asko Sarja

Building and civil engineering has an important role in the service of the societies of our time in:

- providing safe, healthy and convenient conditions for the living and working of people;
- equipping the societal infrastructure for industrial production, logistics and transport on all hierarchical levels of the global economy (local, national, regional and global); and
- participating in preserving the balance of the ecosphere (biosphere and geosphere).

The importance of this mission has been rising with the increasing development of society, and with the increasing level of the standard of living. In order to control the balance for a sustainable development, we need more than the traditional synthetic principles and processes, methodologies, procedures and methods for making rational and balanced optimisations and decisions in all levels and phases of product and production development, design, manufacturing, construction, MR&R, demolition, recovery, reuse and rehabilitation of buildings, civil and industrial infrastructures.

A schedule of the process for innovatively solving the challenge of the mission in the societal context is presented in Figure 1.1 [1]. The generic requirements are demanded by society, owners and users, and are then interpreted as architectural and technical requirements in different levels and time spans:

- for long span in the strategic development of products, organisations, management, production, manufacturing, MR&R (Maintenance, Repair, Rehabilitation and Renewal) and even in *End of Life* (demolition, recovery, reuse and recycling processes); and
- for short span on the project level in briefing, design, manufacturing, construction and management while in use and in end-of-life processing.

Our current society is based on scientific knowledge, which includes natural sciences (physics, chemistry and biology). In this long- and short-term processing, the applied technical sciences serve as practical tools for technology, and architecture gets support from arts. The objectives and targets of construction are defined in the general politics of each society in connection to environmental, sociological and economic objectives.

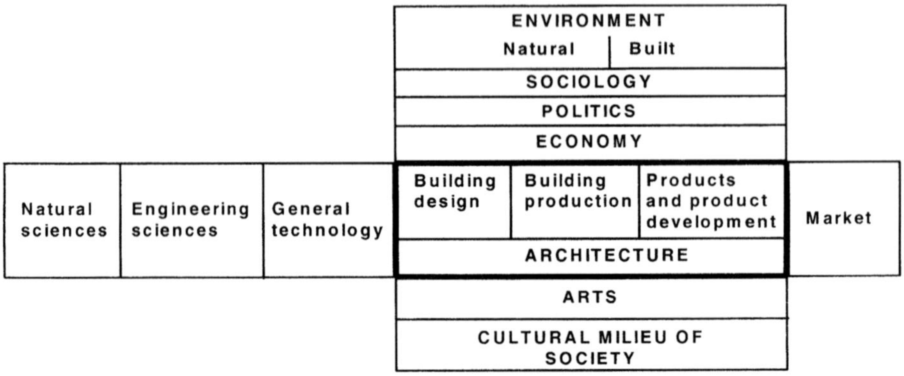

			ENVIRONMENT			
			Natural | Built			
			SOCIOLOGY			
			POLITICS			
			ECONOMY			
Natural sciences	Engineering sciences	General technology	Building design	Building production	Products and product development	Market
			ARCHITECTURE			
			ARTS			
			CULTURAL MILIEU OF SOCIETY			

Figure 1.1 Development environment of building technology [1].

In the background of the development, there exists the heritage of general culture, arts and architecture of each region and society. Knowledge push is served by natural sciences, engineering sciences and general technology. This knowledge is transferred into technology through design, building production and building products. The entire evolution is strongly controlled by market forces and businesses of the companies [1].

All over the world, societies have identified and accepted the goal of sustainable development to reach a stable social and economic development in harmony with nature and cultural heritage. In social, economic, cultural and ecological aspects, the construction branch is a major player. In ecological aspects, the most dictating is energy efficiency. The building sector is responsible for a major share of all the influences mentioned above. Hence, the goal of sustainable building is a big challenge for civil engineering.

These changes are creating new challenges for construction, which can be tackled successfully only through active and innovative changes in design, products, manufacturing methods and management. Introducing the latest advancements of the new technologies into the solving of the challenges, a major development can be achieved. In the next 10–15 years, we can witness an entirely new generation of building technology. New solutions will be developed, or in fact already are partly existing in prototypes and in limited applications. The advanced technology will then penetrate into more wider applications until it becomes common practice. This evolution phase has started in the second part of 1980s and will be implemented into common practice until 2010–2015.

The construction branch includes building and civil engineering and all their life cycle phases: construction, operation, maintenance, repair, renewal, demolition and recycling. For example, in Europe the construction branch share is 10% of GNP, 15% of employment and 40% of raw material consumption, energy consumption and waste generation. Building and civil engineering structures are responsible for a major share of all the influences mentioned above. Building and civil engineering structures are the long-lasting products in societies. Typically, the real service-life of structures lies between fifty years and several hundreds of years. We know, and this is especially known in Greece, that some of the most valued historic structures currently have even reached an age of several thousand years. This is the reason why sustainable engineering in the field of buildings and civil infrastructures is especially challenging in comparison to all other areas of technology.

In order to reach these objectives, we have to make changes even the paradigm, and especially the frameworks, processes and methods of engineering in all phases of the life cycle: investment planning and decisions, design, construction, use and facility management, demolition, reuse, recycling and disposal. Hence, it is apt to discuss on lifetime engineering. The lifetime (also called *whole life* or *life cycle*) principle has recently been introduced into design and management of structures and this development process is getting increasing attention in the practice of structural engineers [2–5].

This new paradigm can be called as *Lifetime Engineering*, defined as a theory and practice of *predictive, optimising and integrated* long-term investment planning, design, construction, management in use, MRR&M (Maintenance, Repair, Rehabilitation and Modernisation) end-of-life management of assets. With the aid of lifetime engineering we can *control and optimise* the *lifetime properties* of built assets with *design and management* corresponding to the objectives of owners, users and society.

Lifetime engineering includes:

- lifetime investment planning and decision-making;
- integrated lifetime design;
- lifetime procurement and contracting: Integrated lifetime construction;
- integrated lifetime management and MRR&M planning and actions; and
- end-of-life management: Recovery, reuse, recycling and disposal.

The integrated lifetime engineering methodology concerns the development and use of technical performance parameters to optimise and guarantee the *lifetime quality* of structures in relation to the requirements arising from human conditions, economical, cultural and ecological considerations. The *lifetime quality* is the capability of the whole network or an object to fulfil the requirements of users, owners and society over its entire life, which means in the practice the planning period (usually 50 to 100 years). The survival life usually includes several service (working) life periods, which follow after a refurbishment (or redevelopment) between the working periods.

Lifetime investment planning and decision-making is a basic issue not only in starting the construction, but also in lifetime asset management. Generally, the investment planning and decision-making is used in evaluating project alternatives, either in planning of a construction project or in MR&R planning. The procedure includes the consideration of characteristics or attributes which decision makers regard as important, but which are not readily expressed in monetary terms. Examples of such attributes in case of buildings are: location, accessibility, site security, maintainability and image.

Integrated lifetime design includes a framework, a description of the design process and its phases, special lifetime design methods with regard to different aspects: human conditions, economy, cultural compatibility and ecology. These aspects will be treated with parameters of technical performance and economy, in harmony with cultural and social requirements, and with relevant calculation models and methods [2]. The optimising planning and design procedure thus includes consideration of characteristics of attributes which decision makers regard as important, but which are not readily expressed in monetary terms. Examples of such attributes in case of buildings are: location, accessibility, site security, maintainability, and image.

Integrated lifetime management and maintenance planning includes continuous condition assessment, predictive modelling of performance, durability and reliability of the facility, maintenance and repair planning and decision-making procedure regarding alternative maintenance and repair actions.

End-of-life management is the last phase of the life cycle, including the selective demolition, recovery and reuse of components and modules, recycling of materials, and disposal of non-usable modules, components and materials.

The role of lifetime engineering as a link between normative practice and the targets of sustainable building is presented in Figure 1.2 [6].

When looking at the statistics of demolitions of buildings and structures, we can notice the following reasons for refurbishment or demolition. Degradation is the main reason for refurbishment of buildings in 17% [7] and in 26 (steel) to 27 (concrete) of demolition of bridges [8]. In individual cases, degradation is a dominating reason for refurbishment or demolition of the structures, which are working in highly degrading environment. Obsolescence is the cause of refurbishment of buildings in 26% [7] and the reason of demolition of bridges in 74% of demolition cases [8]. In the case of modules or component renewals of facilities, the share of obsolescence is still higher. This means that the obsolescence is the dominating reason for demolitions in cases, where structures are working in non-degrading environment.

A conclusion of this, and a challenge for structural engineering, is that we have to consider the degradation and obsolescence criteria in the design, as well as into the MR&R planning of structures [9]. Hence, we need new methodology and methods

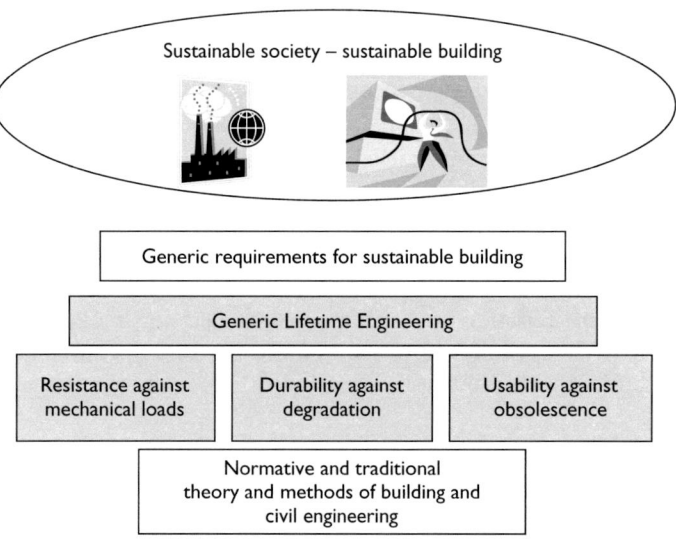

Figure 1.2 Generic Lifetime Engineering as a link between normative practice and the targets and requirements of sustainable building [6].

for analysis, optimisation and decision-making. This methodology and corresponding detailed methods are included in this book.

This book is mainly based on results of a European research project in the fifth framework program, GROWTH: 'Life Cycle Management of Concrete infrastructures, Lifecon', which was carried out during 2001–2003 [10]. Totally eighteen partner organisations from six countries participated in the project. Selected parts of these results were produced as deliveries of the project. The authors of these selected parts of Lifecon results are also the contributors of corresponding chapters of this book.

References

[1] Sarja, A. 1987. Towards the advanced industrialized construction technique of the future. *Betonwerk + Fertigteil-Technik*, 236–239.

[2] Sarja, A. 2002. *Integrated life cycle design of structures*. London: Spon Press, 142 pp. ISBN 0-415-25235-0.

[3] Sarja, A., and E. Vesikari, eds. 1996. *Durability design of concrete structures*. RILEM Report of TC 130-CSL. RILEM Report Series 14. London: E&FN Spon, Chapman & Hall, 165 pp.

[4] Sarja, A., ed. 1998. *Open and industrialised building*. London: E&FN Spon, 228 pp.

[5] Sarja, A. 2002. Reliability based life cycle design and maintenance planning. Workshop on reliability based code calibration, Swiss Federal Institute of Technology, *ETH* Zürich, Switzerland, March 21–22. http://www.jcss.ethz.

[6] Sarja, A. 2004. Generalised lifetime limit state design of structures. *Proceedings of the 2nd international conference, Lifetime-oriented design concepts, ICDLOC*, pp. 51–60. Ruhr-Universität Bochum, Germany. ISBN 3-00-013257-0.

[7] Aikivuori, A. 1994. Classsification of demand for refurbishment Projects. *Acta Universitatis Ouluensis, Series C, Technica 77*. University of Oulu, Department of Civil Engineering. ISBN 951-42-3737-4.

[8] Iizuka, H. 1988. A statistical study of life time of bridges. *Structural engineering/ earthquake engineering* 5(1): 51–60, April. Japan Society of Civil Engineers.

[9] Iselin, D. G., and A. C. Lemer, eds. 1993. The Fourth Dimension. *Building: Strategies for minimizing obsolescence*. National Research Council, Building Research Board. Washington, D.C.: National Academy Press.

[10] Sarja, A. (Co-ordinator) *et al.* 2004. *Life cycle managemnt of concrete infrastuctures*, Lifecon, Deliverables D1-D15. http://www.vtt.fi/rte/strat/projects/lifecon/.

Chapter 2

Theory, systematics and methods

Asko Sarja

2.1 Generic requirements for sustainability

The prevailing goal and trend in all areas of mechanical industry as well as in building and civil engineering is *Lifetime Engineering* (also called 'Life Cycle Engineering'). The integrated lifetime engineering methodology aims at regulating optimisation and guaranteeing the life cycle requirements with technical performance parameters. With the aid of lifetime engineering, we can thus control and optimise the human conditions (functionality, safety, health and comfort), the monetary (financial) economy and the economy of nature (ecology), also taking into account the cultural and social needs (Figure 2.1).

The integrated lifetime engineering methodology concerns the development and use of technical performance parameters to guarantee that the structures fulfil through the life cycle the requirements arising from human conditions, economic, cultural and ecological considerations.

1. Human requirements
• Functionality in use
• Safety
• Health
• Comfort
	2. Economic requirements
	• Investment economy
	• Construction economy
	• Lifetime economy in:
	   – Operation
	   – Maintenance
	   – Repair
	   – Rehabilitation
	   – Renewal
	   – Demolition
	   – Recovery and reuse
	   – Recycling of materials
	   – Disposal

3. Cultural requirements • Building traditions • Life style • Business culture • Aesthetics • Architectural styles and trends • Image	4. Ecological requirements • Raw materials economy • Energy economy • Environmental burdens economy • Waste economy • Biodiversity and geodiversity

Figure 2.1 Generic requirements as components of the criteria for a sustainable lifetime quality [1, 2, 3].

For the life cycle design, the analysis and design are expanded to two economical levels: monetary economy and ecology. The life cycle expenses are calculated to the present value or annual costs after discounting the expenses from manufacture, construction, maintenance, repair, changes, modernisation, reuse, recycling and disposal. The monetary costs are treated as usual in current value calculations. The expenses of nature are the use of non-renewable natural resources (materials and energy) and the management of the generation of air, water and soil pollutants and the solid waste. Consequences of air pollution are health problems, inconvenience for people, ozone depletion and global climatic change. The goal is to limit the natural expenses under the allowed values and to minimise them.

2.2 Criteria of lifetime quality

In the level of practical engineering, the generic requirements can be treated with a number of technical factors which are presented in Figure 2.2. These can be used in analysing the indicators of the lifetime quality.

The central life cycle quality indicators of a structural system are:

- functional usability in the targeted use
- adaptability in use
- changeability during use
- reliable safety
- healthy
- durability, which means resistance against degratation loads
- resistance against obsolescence
- ecological efficiency.

In buildings, the compatibility and easy changeability between load-bearing structures, partition structures and building service systems is important. Regarding the life cycle ecology of buildings, the energy efficiency of the building is a dictating factor. Envelope structures are responsible for most of the energy consumption, and therefore the envelope must be durable and have an effective thermal insulation and safe static and hygro-thermal behaviour. The internal walls have a more moderate length of service life length, but they have the requirement of coping with relatively high degrees of change, and must therefore possess good changeability and reuseability. In the production phase it is important to ensure the effective recycling of the production wastes in factories and on site. Finally, the requirement is to recycle the components and materials after demolition. Obsolescence of buildings is either technical or functional, sometimes even aesthetic in nature. Technical and functional obsolescence is usually related to the primary lifetime quality factors of structures. Aesthetic obsolescence is usually architectural in nature.

Civil engineering structures like harbours, bridges, dams, offshore structures, towers, cooling towers, etc. are often massive and their target service life is long. Their repair works under use are difficult. Therefore their life cycle quality is tied to high durability and easy maintainability during use, saving of materials and selection of environmentally friendly raw materials, minimising and recycling of construction wastes, and finally recycling of the materials and components after demolition. Some

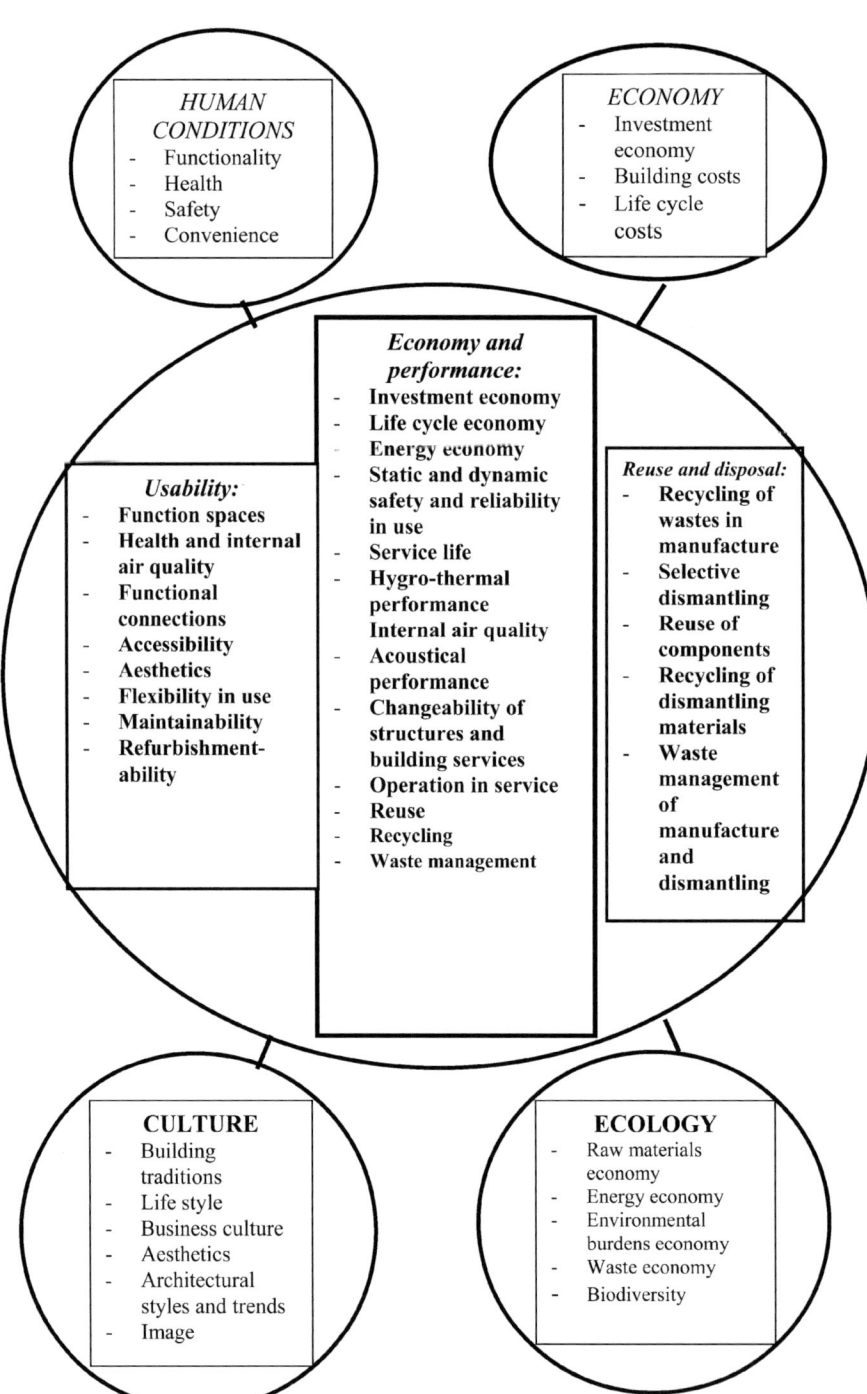

Figure 2.2 Transforming the generic requirements into technical factors [1].

parts of the civil engineering structures like waterproof membranes and railings have a short or moderate service life and therefore the aspects of easy reassembly and recycling are most important. Technical or performance-related obsolescence is the dominant reason for demolition of civil engineering structures, which raise the need for careful planning of the whole civil engineering system, e.g. the traffic system, and for selection of relevant and future-oriented design criteria.

2.3   Lifetime engineering for lifetime quality

2.3.1   Definition

Lifetime engineering is a theory and practice of *predictive, optimising and integrated* long-term investment planning, design, construction, management in use, MRR&M (Maintenance, Repair, Rehabilitation and Modernisation) end-of-life management of assets.

2.3.2   Content and framework

Lifetime engineering includes:

- lifetime investment planning and decision-making
- integrated lifetime design
- integrated lifetime construction
- integrated lifetime management and MR&R (Maintenance, Repair and Rehabilitation) planning
- modernisation
- dismantling, reuse, recycling and disposal.

Integrated lifetime design includes a framework, a description of the design process and its phases, special lifetime design methods with regard to different aspects: human conditions, economy, cultural compatibility and ecology. These aspects will be treated with parameters of technical performance and economy, in harmony with cultural and social requirements, and with relevant calculation models and methods.

Integrated lifetime management and MRR&M planning and actions include continuous condition assessment, predictive modelling of performance, durability and reliability of the facility, maintenance and repair planning and the decision-making procedure regarding alternative maintenance and repair actions. Modernisation includes actions for modification and improvements to an existing asset or structure to bring it up to an acceptable condition, which meets the changed generic and specific requirements.

End-of-life management Dismantling, reuse, recycling and disposal includes comprehensive recycling system together with logistic principles and solutions, consisting of selective demolition, the selection of site wastes, treatment and quality control of treated wastes for use as raw materials in buildings, roads and trafficked areas. The system is operated by several partners: cities as client and firms involved in the construction process as technical actors.

The entire context of Lifetime Engineering is presented in Figure 2.3.

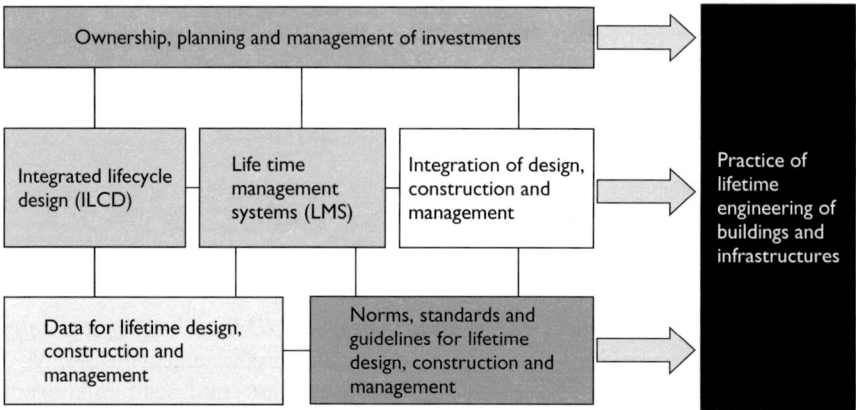

Figure 2.3 Context of the praxis of lifetime engineering.

2.3.3   *Lifetime management as a part of lifetime engineering*

The *objective* of the integrated and predictive lifetime management is to achieve optimised and controlled *lifetime quality* of buildings or civil infrastructures in relation to the generic requirements. The lifetime quality means the *capability of an asset to fulfil the requirements of users, owners and society on an optimised way during the entire design life of the asset*. This objective can be achieved with a *performance-based methodology*, applying *generic limit state approach*. This means, that the *generic requirements have to be modelled with technical and economic numerical parameters* into quantitative models and procedures, and with semi-numerical or non-numerical ranking lists, classifications and descriptions into qualitative procedures. This methodology can be described in a schedule, which is presented in Figure 2.4 [1]. The generic requirements are listed in Figure 2.1.

The lifetime performance modelling (Figure 2.4) and the limit state approach build an essential core of the lifetime management – MR&R planning. Performance-based modelling includes the following classes:

- modelling of the behaviour under mechanical (static, dynamic and fatigue) loads
- modelling of the behaviour under physical, chemical and biological loads
- modelling of the usability and functional behaviour.

The mechanical modelling has been traditionally developed on the limit state principles, which were started in the 1930s and introduced into common practice in the 1970s. The newest specific standard for reliability of structures is Eurocode EN 1990:2000 [5]. The mechanical behaviour (safety and serviceability), besides the other categories mentioned above, has to be checked in the several phases of the management process. This is especially important in condition assessment and MR&R planning. It is sometimes possible to combine the mechanical calculations with degradation and service life calculations, but often it is better to keep these separated. Because the models and calculation methods of mechanical behaviour are

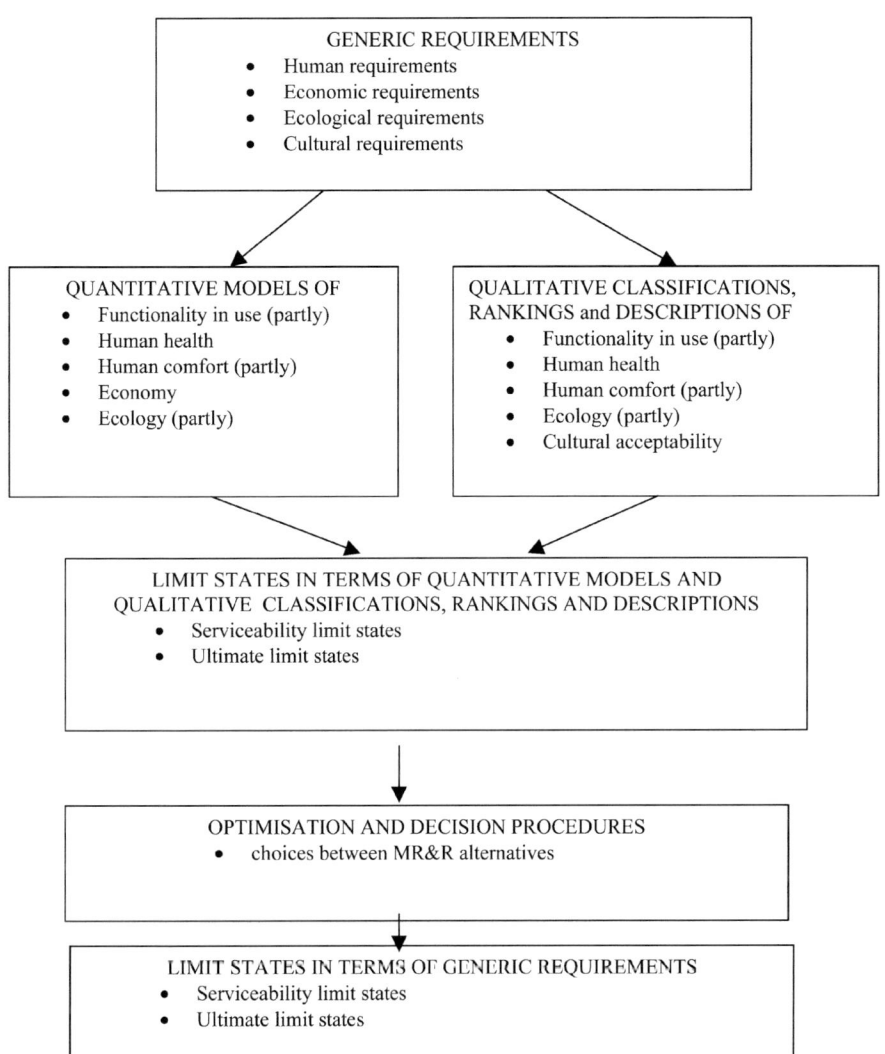

GENERIC REQUIREMENTS
- Human requirements
- Economic requirements
- Ecological requirements
- Cultural requirements

QUANTITATIVE MODELS OF
- Functionality in use (partly)
- Human health
- Human comfort (partly)
- Economy
- Ecology (partly)

QUALITATIVE CLASSIFICATIONS, RANKINGS and DESCRIPTIONS OF
- Functionality in use (partly)
- Human health
- Human comfort (partly)
- Ecology (partly)
- Cultural acceptability

LIMIT STATES IN TERMS OF QUANTITATIVE MODELS AND QUALITATIVE CLASSIFICATIONS, RANKINGS AND DESCRIPTIONS
- Serviceability limit states
- Ultimate limit states

OPTIMISATION AND DECISION PROCEDURES
- choices between MR&R alternatives

LIMIT STATES IN TERMS OF GENERIC REQUIREMENTS
- Serviceability limit states
- Ultimate limit states

Figure 2.4 Schedule of the generic procedure of reliability management in Lifecon LMS [1].

very traditional and included in normative documents of limit state design, this issue is not discussed in this book which is focused on durability limit state design and obsolescence limit state design.

Modelling for physical, chemical and biological loads includes a large variety of thermal behaviour, behaviour under fire conditions, moisture behaviour and behaviour under biological impacts and biological phenomena (e.g. mould and decay). These are connected with several phenomena and properties of structures in use, and in this context this section is devoted to the different procedures of reliability assessment. Traditional analysis of thermal, fire and moisture behaviour are not discussed in this book.

Modelling of usability and functionality aims at the management of obsolescence in the life cycle management system. Obsolescence is the inability to satisfy changing functional (human), economic, cultural or ecological requirements. Obsolescence can affect the entire building or civil infrastructural asset, or just some of its modules or components. Obsolescence is the cause for the demolition of buildings or infrastructures, which is about 50% of all demolition cases. Therefore, this issue is very central in developing asset management for sustainability, which is the aim of Lifecon LMS.

The main health issue during the MR&R works arises from the use of unhealthy materials [Lifecon D5.1] [4]. During the use of assets (especially in closed spaces such as buildings or tunnels), moisture in structures and on finished surfaces should be avoided because it can cause mould; and it is essential to check that no materials used can cause emissions or radiation which are dangerous for health and comfort of the users. In some areas, radiation from the ground must be also be eliminated through insulation and ventilation of the foundations. Thus, the main tools for health management are: selection of materials (especially finishing materials), eliminating risks of moisture in structures (through waterproofing, drying during construction and ventilation) and elimination of possible radioactive ground radiation with airproofing and ventilation of ground structure. Health requirements can follow the guidelines of national and international codes, standards and guides. The modelling of the health issues thus focuses on calculating comparable indicators on the health properties mentioned above, and on comparing these between alternatives in the optimisation and decision-making procedures. These can usually be calculated numerically, and thus they are mainly quantitative variables and indicators, which can be compared in the optimisation and decision-making procedures.

Comfort properties are related to the functionality and performance of assets. Following are the comfort properties:

- acoustic comfort, including noise level during MR&R works or in the use of closed spaces like tunnels and buildings
- insulation of airborne sound between spaces
- comfortable internal climate of closed spaces like tunnels and buildings
- aesthetic comfort externally and in functions of use in all kinds of assets
- vibrations of structures.

These are calculated with special rules and calculation methods, which are also traditional and therefore will not be discussed in this book. Mainly quantitative (exact numerical or classified) values can be used for these properties.

Ecology can be linked to the environmental expenditures: consumption of energy, consumption of raw materials, production of environmental burdens into air, soil and water and loss of biodiversity. Most of these can be calculated numerically, and thus are quantitative variables and indicators. These can also be compared quantitatively in the optimisation and decision-making procedures. In buildings, energy consumption mostly dictates environmental properties. For this reason, the thermal insulation of the envelope is important. Finally, the reuse and recycling of the components and materials after demolition belong to the ecological indicators. Engineering structures such as bridges, dams, towers, cooling towers are often massive and their material consumption is an important factor. Their environmental efficiency depends on the selection of environmental-friendly local raw materials, high durability and easy maintainability of the structures during use, recycling of construction wastes and finally recycling of the components and materials after demolition. Some parts of engineering structures, such as waterproofing membranes and railings, have a short or moderate service life and consequently easy reassembly and recycling are most important to minimise the annual material consumption property. During MR&R works, it is important to apply effective recycling of production wastes. This leads to calculations of waste amounts as quantitative variables of this component of ecology. Some ecological properties, like loss of biodiversity, are difficult to calculate numerically, and they often can be only described qualitatively. This qualitative description can then be used in comparing alternatives during optimisation and decision-making procedures.

The functionality of civil infrastructures means the capability to serve the main targets of an asset, e.g. in the case of tunnels and bridges, the capability to transmit traffic. This can be modelled numerically using variables and indicators, suitable geometric dimensions and load-bearing capacity, etc. The functionality of buildings is very much related to the flexibility for changes of spaces and often also on the loading capacity of floors. Also, the changeability of building service systems is important. Internal walls have a moderate requirement of service life and hence a quite high need to accommodate changes. These demand the capability of a building to enable changes in the functions during the lifetime management. For this reason, internal walls must have good changeability and recycleability. An additional property is good and flexible compatibility with the building services system because the services system is the part of the building that is most often changed.

To avoid repeating of traditional and well-known issues, the generalised and reliability-based life cycle management approach can be focused and formulated into following three categories:

1. static and dynamic (mechanical) modelling and design
2. degradation-based durability and service life modelling and design
3. obsolescence-based performance and service life modelling and design.

In Lifecon LMS system, the transformation of generic requirements into functional and performance property definitions and further into technical specifications and performance models will be realised with the following methods [2, 4, 5].

1. Requirements analysis and performance specifications: Quality Function Deployment (QFD) method: Deliverable D2.3
2. Service Life Estimation:

   • Probabilistic service life models: Deliverable D3.2
   • RILEM TC 130 CSL models: Deliverable D2.1
   • Reference structure method: Deliverable D2.2

3. Condition matrix: Markovian Chain method: Deliverable D2.2, and Condition assessment protocol: Deliverable D3.1

4. Total and systematic reliability based methodology: Deliverable D2.1
5. Risk analysis: Deliverable D2.3.

2.4 Generalised limit state methodology

2.4.1 Performance-based limit state approach

The lifetime performance modelling and the limit state approach, which are based on the integrated and generic requirements as shown in Figure 2.1, are working as an essential core of the lifetime design and management of structures.

Performance-based modelling includes the following three classes:

1. static and dynamic (mechanical) modelling and design
2. degradation-based durability and service-life modelling and design
3. obsolescence-based performance and service-life modelling and design.

When looking at the statistics of demolitions of buildings and structures we notice the following reasons for demolition:

- Obsolescence is the reason for demolition of buildings or infrastructures in about 50% of all demolition cases.
- Degradation is a dominating reason for demolition of the structures, which are working in highly degrading environment. This is most often the case in demolition of some civil infrastructures.
- Failures caused by mechanical loading of structures without durability problems are very rare, but consequences can be very high.

A conclusion of this, and a challenge for structural designers, is that we have to include the durability and obsolescence criteria into the design, as well as into the MR&R planning of structures.

2.4.2 Design life

Design life is a specified time period, which is used in calculations. The classification of design life of EN 1990:2002 is presented in Table 2.1 [5].

Table 2.1 Classification of EN 1990:2002 for design life of structures [5]

Class 1: 1–5 years	Special case temporary buildings
Class 2: 25 years	Temporary buildings, e.g. stores buildings, accommodation barracks
Class 3: 50 years	Ordinary buildings
Class 4: 100 years	Special buildings, bridges and other infrastructure buildings or where more accurate calculations are needed, for example, for safety reasons
Class 5: over 100 years	Special buildings e.g. monuments, very important infrastructure buildings

2.4.3 Generic limit states

The origination classes of generic limit states are as follows:

- static and dynamic
- degradations
- obsolescence.

The serviceability limit states and ultimate limit states of concrete structures in relation to this classification are presented in Table 2.2, together with some comparisons of terms in Table 2.3 [2, 6].

The generic durability limit states and their application in specific cases can be described with numerical models and treated with numerical methodology, which are quite analogous to the models and methodologies of the mechanical (static and dynamic) limit states design.

There are no international or national normative standards concerning obsolescence and no exactly defined limit states of obsolescence. Some remarks have been presented, e.g. in ISO standards [7], but real analysis methods have not been presented.

Table 2.2 Generic mechanical, degradation and obsolescence limit states of concrete structures

Classes of the limit states	Limit states		
	Mechanical (static and dynamic) limit states	Degradation limit states	Obsolescence limit states
Serviceability limit states	Deflection limit state Cracking limit state	Surface faults causing aesthetic harm (colour faults, pollution, splitting, minor spalling) Surface faults causing reduced service life (cracking, major spalling, major splitting) Carbonation of the concrete cover (grade 1: one-third of the cover carbonated, grade 2: half of the cover carbonated, grade 3: entire cover carbonated)	Reduced usability and functionality, but still usable The safety level does not allow the requested increased loads Reduced healthy, but still usable Reduced comfort, but still usable
Ultimate limit states	Insufficient safety against failure under loading	Insufficient safety due to indirect effects of degradation: • heavy spalling • heavy cracking causing insufficient anchorage of reinforcement • corrosion of the reinforcement causing insufficient safety	Serious obsolescence causing total loss of usability through: • loss of functionality in use • safety of use • health • comfort • economy in use • MR&R costs • ecology • cultural acceptance

Table 2.3 Comparison of static and dynamic (mechanical) limit state method, the degradation limit
state method and obsolescence limit state

S. No	Mechanical limit state design	Degradation limit state design	Obsolescency limit state design
1.	Strength class	Service life class	Service life class
2.	Target strength	Target service life	Target service life
3.	Characteristic strength (5% fractile)	Characteristic service life (5% fractile)	Characteristic service life (5% fractile)
4.	Design strength	Design life	Design life
5.	Partial safety factors of materials strength	Partial safety factors of service life	Partial safety factors of service life
6.	Static or dynamic loading onto structure	Environmental degradating loads onto structure	Obsolescence loading onto structure
7.	Partial safety factors of static loads	Partial safety factors of environmental loads	Partial safety factors of obsolescence loading
8.	Service limit state (SLS) and ultimate limit state (ULS)	Serviceability and ultimate limit states	Serviceability and ultimate limit states related to obsolescence in relation to the basic requirements

Because the limit states of obsolescence often cannot be described in quantitative
means, we often have to apply qualitative descriptions, criteria and methods.

2.4.4   Statistical basis of reliability

The simplest mathematical model for describing the 'failure' event comprises a load
variable $S(t)$ and a response variable $R(t)$ [3, 8, 9]. This means that both the resistance R and the load S are time-dependent, and the same equations can be used for
static reliability and durability. Usually, the time is neglected as a variable in static
and dynamic calculations; they are included only in fatigue reliability.

In durability-related limit states and service life calculations, the time is always
included as a variable of $R(t)$ and $S(t)$. In principle, the variables $S(t)$ and $R(t)$ can
be any quantities and expressed in any units. The only requirement is that they are
commensurable. Thus, for example, S can be a weathering effect and R the capability
of the surface to resist the weathering effect.

If R and S are independent of time, the 'failure' event can be expressed as follows:

$$\{\text{failure}\} = \{R(t) < S(t)\} \tag{2.1}$$

The failure probability P_f is now defined as the probability of that 'failure':

$$P_f = P\{R < S\} \tag{2.2}$$

Either the resistance R or the load S or both can be time-dependent quantities. Thus
the failure probability is also a time-dependent quantity. Considering $R(\tau)$ and $S(\tau)$

are instantaneous physical values of the resistance and the load at the moment, τ the failure probability in a lifetime t could be defined as:

$$P_f(t) = P\{R(\tau) < S(\tau)\} \text{ for all } \tau \leq t \tag{2.3a}$$

The determination of the function $P_f(t)$ according to the Equation (2.3a) is mathematically difficult. That is why R and S are considered to be stochastic quantities with time-dependent or constant density distributions. This means the failure probability can usually be defined as:

$$P_f(t) = P\{R(t) < S(t)\} \tag{2.3b}$$

According to the Equation (2.3b), the failure probability increases continuously with time as schematically presented in Figure 2.5 [8].

Considering continuous distributions, the failure probability P_f at a certain moment of time can be determined using the convolution integral:

$$P_f(t) = \int F_R(s, t) f_S(s, t) ds \tag{2.4}$$

where $F_R(s)$ is the cumulative distribution function of R, $f_S(s)$ is the probability density function of S and s the common quantity or measure of R and S.

The integral can be approximately solved by numerical methods. In static and dynamic calculations the time is not a variable, but the reliability is calculated at the moment $t = 0$. In durability calculations the time is a variable of the resistance $R(t)$, but usually the environmental degradation load $S(t)$ is considered to be constant. The value of S depends on the environmental exposure conditions and actual design life of the structure. The environmental loads are classified in the standards, for concrete structures the standard EN 206 can be used [10]. Mathematical formulation for applied statistical degradation methods are presented in the Model Code of JCSS (Joint Committee on Structural Safety).

The statistical reliability calculations serve an important basis for applied safety factor methods, which are now in common use. The statistical method is used in special cases, when the reliability has to be analysed in very individual terms. In such a case the material parameters and dimensions have to be determined in high

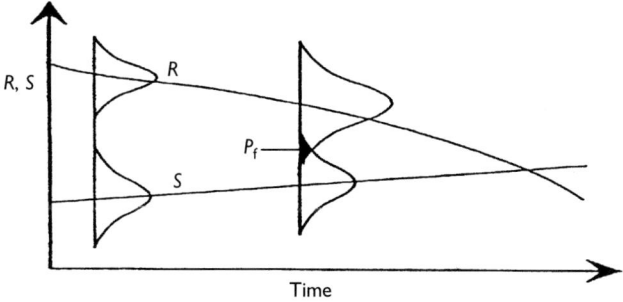

Figure 2.5 The increase of failure probability. Illustrative presentation [8].

number of samples, so that statistical values (mean value and standard deviation) can be calculated. In ordinary design or condition assessment this is not possible, and the safety factor method is applied.

The reliability index and the corresponding probability of failure can be calculated analytically only in some special cases. Usually, the equations are solved with suitable numerical methods of partial differential equations or with simulations.

2.4.4.1 Statistical reliability requirements

The statistical reliability methodology and requirements are defined in the European standard EN 1990. This standard is based on partial safety factor method, but the reliability requirements are expressed also in terms of statistical reliability index $\tilde{\beta}$. The general definition of the reliability index β of standard normal distribution is defined as a factor, which fulfils the equation:

$$P_f = \Phi(-\beta) \tag{2.5}$$

where Φ is the cumulative distribution function of the standardised Normal distribution.

The requirements of the standard EN 1990 for the reliability index β are shown in Table 2.4 for the design of new structures, as well as for the safety of existing structures [5].

Because these are European normative requirements, it is extremely important to use these reliability requirements as bases for all statistical and deterministic limit

Table 2.4 Recommended minimum values for reliability index β (Equation (2.5)) in ultimate limit states and in serviceability limit states, according to EN 1990:2002 [5]

Reliability class	Minimum values for reliability index β			
	1 year period		50 years period	
	Ultimate limit states	Serviceability limit states	Ultimate limit states	Serviceability limit states
RC3/CC3: High consequence for loss of human life, or economic, social or environmental consequences very great	5.2	No general recommendation	4.3	No general recommendation
RC2/CC2: Medium consequence for loss of human life, or economic, social or environmental consequences considerable	4.7	2.9	4.7	1.5
			Fatigue 1.5–3.8[1]	
RC1/CC1: Low consequence for loss of human life, or economic, social or environmental consequences small or negligible	4.2	No general recommendation	3.3	No general recommendation

Note
[1] Depends on degree of inspectability, reparability and damage tolerance.

states methods, which are used for reliability control of mechanical, durability and obsolescence reliability of assets and structures in Lifecon LMS. In degradation management direct statistical calculations, the values of safety index can be applied directly [Lifecon D3.2]. In deterministic limit state calculations, which will be discussed in this book, the lifetime safety factor is calculated with this statistical base, and then applied deterministically in practice. In usability management with obsolescence methodology the risk analysis method is applied, statistically applying these safety index values, or it can be calculated deterministically applying the QFD method or MADA method together with lifetime safety factor method as a deterministic limit state method.

2.4.5 Deterministic safety factor methods

2.4.5.1 Safety factor method for static, fatigue and dynamic loading

The partial safety factor has already been in common European codes and use for about three decades. The latest update of this methodology is presented in EN 1990 [5] and there is no need to present this methodology in this book.

2.4.5.2 Lifetime safety factor method for durability

In practice it is reasonable to apply the lifetime safety factor method in the design procedure for durability, which was for the first time presented in the report of RILEM TC 130 CSL [3, 8, 9]. The lifetime safety factor method is analogous with the static limit state design. The durability design using the lifetime safety factor method is related to controlling the risk of falling below the target service, while static limit state design is related to controlling the reliability of the structure against failure under external mechanical loading.

The durability design with lifetime safety factor method is always combined with static or dynamic design and aims to control the serviceability and service life of a new or existing structure, while static and dynamic design controls the loading capacity.

2.4.5.3 Durability limit states

The lifetime safety factor design procedure is somewhat different for structures consisting of different materials although the basic design procedure is the same for all kinds of materials and structures. Limit states can be the same as in static design, but some generalised limit states, including, e.g. visual or functional limit states, can be defined. In this way the principle of multiple requirements, which is essential for integrated life cycle design, can be introduced.

Limit states are divided into two main categories:

1. performance limit states
2. functionality limit states.

The performance limit states affect the technical serviceability or safety of structures, and the functional limit states affect the usability of structures. Both of these, but especially the latter, are often connected to obsolescence.

The performance limit states can be handled numerically, but the functional limit states cannot always be handled numerically – only qualitatively.

Investigations in practice have shown that about 50% of all demolished buildings or civil infrastructures have been demolished because of obsolescence, and the same amount because of insufficient technical performance or safety. A short summary of the parameters of durability limit states is presented in Table 2.1.

2.4.5.4 Design life

Design life is a specified time period, which is used in calculations. Ordinary design life is 50 years (EN 1990) for buildings and 100 years for civil engineering structures. In special cases, even longer design life cycles can be used. However, after 50 years the effect of increased design life cycle is quite small and it can be estimated as the residual value at the end of the calculation of life cycle. Temporary structures are designed for a shorter design life, which will be specified in each individual case. The classification of design life of EN 1990:2002 is presented in Table 2.1.

2.4.5.5 Reliability calculations

The design service life is determined by formula ([8], modified: [3, 9]):

$$t_{Ld} = t_{Lk}/\gamma_{tk} = t_g \tag{2.6}$$

Where t_{Ld} is the design service life, t_{Lk} the characteristic service life, γ_{tk} the lifetime safety factor and t_g the target service life.

Using the lifetime safety factor, the requirement of target service life (corresponding to a maximum allowable failure probability) is converted to the requirement of mean service life. The mean service life is approximated by service life models that show the crossing point of the degradation curve with the limit state of durability (Figure 2.6). The mean service life evaluated by the service life model divided by the central lifetime safety factor is *design life*, which must be greater than or equal to the requirement for the design life (also called *target service life*).

$$t_{Ld} = \mu(t_L)/\gamma_{t_0} \tag{2.7a}$$

$$t_{Ld} \geq t_g \tag{2.7b}$$

Where t_{Ld} is the design service life, t_L is the mean service life and γ_{t_0} the central safety factor.

When using ordinary characteristic values, the equations get the following formulations:

$$t_{Ld} = t_{Lk}/\gamma_{tk} = \text{required design life (target service life)} \tag{2.8a}$$

$$t_{Ld} = \mu(t_L)/\gamma_{t_0} = \text{required design life (target service life)} \tag{2.8b}$$

The lifetime safety factor depends on the maximum allowable failure probability. The lifetime safety factor also depends on the form of service life distribution. Figure 2.6

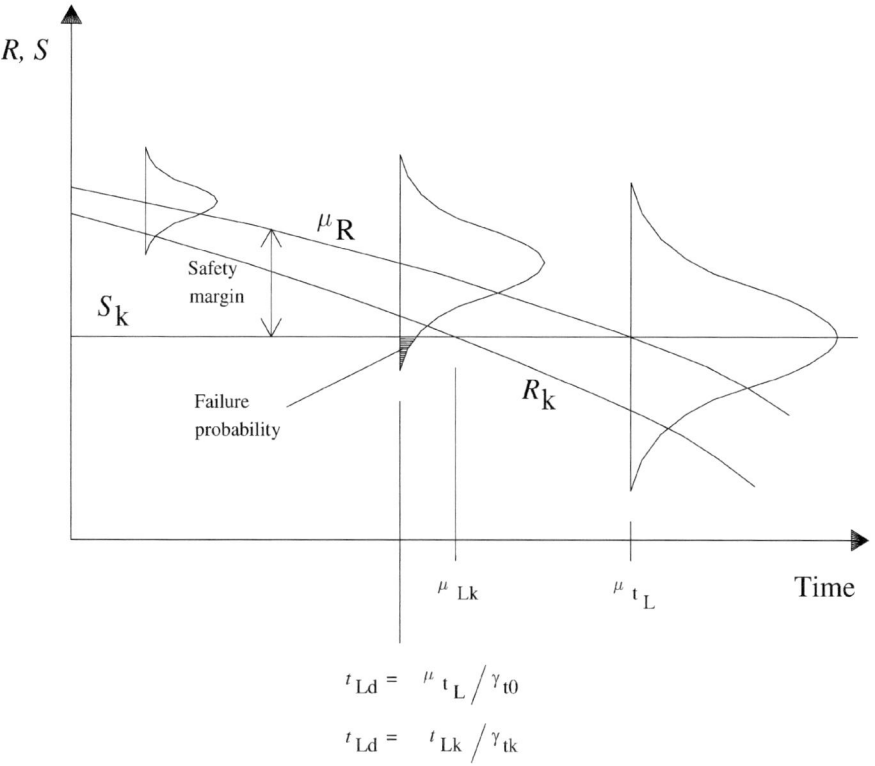

$$t_{Ld} = {}^{\mu}t_L \big/ {}^{\gamma}t_0$$

$$t_{Ld} = {}^{t}Lk \big/ {}^{\gamma}tk$$

Figure 2.6 The meaning of lifetime safety factor in a performance problem.

illustrates the meaning of lifetime safety factor when the design is done according to the performance principle. The function $R(t) - S$ is called the safety margin. Detailed description of the durability design is presented in Section 3.3.

2.4.6 Reliability requirements of existing structures

2.4.6.1 Design life

In MR&R planning, the design life periods of EN 1990:2002 can be basically applied. However, the total design life has to be prolonged in case of old structures. This leads to a new term: 'Residual design life'. The residual design life can be decided case by case, but it is usually the same or shorter than the design life of new structures. The residual design life can be optimised using MADA procedure. Proposed values of design life for MR&R planning are presented in Table 2.6.

2.4.6.2 Reliability requirements for service life

The reliability requirements for service life are different from the requirements for structural safety in mechanical limit states. Therefore, for mechanical safety it is

Table 2.5 Functional level usability limit states of obsolescence of structures

Reason of limit state	Serviceability limit state	Ultimate limit state
1. Human requirements		
Functional usability	Weakened functional usability	Total loss of functional usability
Convenience of use	Weakened convenience	
Healthiness of use	Minor health problems in use	Severe health problems in use
Safety of operation	Weakened safety of operation	Severe problems in safety of operation
2. Economic requirements		
Economy of operation	Weakened economy in operation	Total loss of economy in operation
Economy of MR&R	Weakened economy in MR&R	Total loss of economy in MR&R
3. Cultural requirements		
Cultural requirements of the society	Minor problems in meeting cultural requirements	Severe problems in meeting defined cultural requirements
4. Ecological requirements		
Requirements on the economy of nature: • Consumption of raw materials, energy and water • Pollution of air, waters and soil • Waste production • Loss of biodiversity	• Minor problems in meeting requirements of owners, users and society • Minor environmental problems	• Total loss of meeting the most severe requirements of society • Severe environmental problems

recommended to use the safety indexes of EN 1990:2002 as presented in Table 2.4, and for corresponding reliability classes the safety indexes of service life in durability limit states and in obsolescence limit states as presented in Table 2.4.

It is important to notice in each case the durability and obsolescence limit states, in addition the safety of mechanical (static, dynamic and fatigue) limit states has to be checked separately. Because durability works in interaction with structural mechanical safety, the recommended reliability indexes of durability service life are close to the level of requirements for mechanical safety.

Obsolescence does not usually have direct interaction with structural mechanical safety, hence the safety index recommendations are lower. The mechanical safety requirements of Table 2.5 have to be checked separately always in cases when obsolescence is caused by insufficient mechanical safety level in comparison to increased loading requirements or increased safety level requirements.

2.5 Performance under obsolescence loading

2.5.1 Principles

Obsolescence is the inability to satisfy changing functional (human), economic, cultural or ecological requirements. Obsolescence can affect the entire building or civil infrastructure facility or just some of its modules or components [1, 2, 7]. Obsolescence analysis and control aims to guarantee the ability of the buildings and civil infrastructures to meet all current and changing requirements with minor changes of the facilities. Lifetime design aims at minimising the need of early renewal or demolition.

Obsolescence is a real world-wide problem, which comes from the everyday world of events and ideas, and may be perceived differently by different people. Often these cannot be constructed by the investigators as the laboratory problems (degradation or static and dynamic stability) can be. As there is no direct threat to human life resulting from obsolescence, there are no set limits for obsolescence. Neither is there international or national normative standards concerning the issue. The responsibility to deal with obsolescence is thus left to the owners of the facilities. Consequently, when there are no standards or norms to follow, the decisions (corporate, strategic, MR&R, etc.) are readily made on economic grounds only, which too often lead to premature demolishing of sound facilities. It has been estimated that about 50% of all demolishing cases concerning buildings and civil infrastructures are due to obsolescence. In case of modules or component renewals the share of obsolescence is still higher.

Although analogy between the limit states of statics and dynamics, degradation and obsolescence can be found (see Table 2.3), the nature of obsolescence problem is *philosophically* different from the two others. While in the first two cases the limit states are reached because some real loads (e.g. environmental loads, live loads, etc.) are acting on the structure, in the obsolescence case there are no *actual tangible* loads causing the crossing of limit states. Instead, the obsolescence loading can be defined as the development of the society around the still-standing structure. This development that causes obsolescence includes human requirements, functional, economic, ecological and cultural changes. Behind these changes is the entire social, economic, technological and cultural change of the society. Some examples of different types of obsolescence are listed below.

- Functional obsolescence is due to changes in function and use of the building or its modules. This can even be when the location of the building becomes unsuitable. More common are changes in use that require changes in functional spaces or building services systems. This raises the need for flexible structural systems, usually requiring long spans and minimum numbers of vertical load-bearing structures. Partition walls and building services systems that are easy to change are also required.
- Technological obsolescence is typical for building service systems, but also the structure can be a cause when new products providing better performance become available. Typical examples are more efficient heating and ventilation systems and their control systems, new information and communication systems such as computer networks, better sound and impact insulation for floorings, and more accurate and efficient thermal insulation of windows or walls. Health and comfort of internal climate is the requirement that has increased in importance. The risk of technological obsolescence can be avoided or reduced by estimating future technical development when selecting products. The effects of technical obsolescence can also be reduced through proper design of structural and building service systems to allow easy change, renewal and recycling.
- Economic obsolescence means that operation and maintenance costs are too high in comparison to new systems and products. This can partly be avoided

in design by minimising the lifetime costs by selecting materials, structures and equipment that need minimum costs for maintenance and operation. Often this means simple and safe products that are not sensitive to defects and/or their effects. For example, monolith external walls are safer than layered walls.

- Cultural obsolescence is related to the local cultural traditions, ways of living and working, aesthetic and architectural styles and trends, and image of the owners and users.
- Ecological obsolescence happens often in cases of large infrastructure projects. In large projects this is often related to high waste and pollution production or loss of biodiversity. In the case of buildings we can foresee problems in the future, especially in the use of heating and cooling energy because heating and cooling is producing, for example, in northern and central Europe about 80–90% of all CO_2 pollution and acid substances into air.

The final objective of obsolescence analysis and optimisation is to reduce the demolishing of facilities that have not reached their mechanical (static or dynamic) or durability ultimate limit states, and thus to promote sustainable development.

2.5.2 Elements of obsolescence analysis

The obsolescence analysis can be divided into three elements:

1. Meaning of obsolescence
2. Factors and causes of obsolescence
3. Strategies and decisions on actions against obsolescence

2.5.2.1 Meaning of obsolescence

In this part of the analysis – when kept on general level – the owner should ask himself/herself, 'What does the obsolescence really mean with the type of facility in question (bridge, tunnel, wharf, lighthouse, cooling tower, etc.)?' Before the obsolescence can be made the subject of a deeper study, it must be clearly defined. The task can be facilitated with appropriate questions like:

- How do the different types of obsolescence (functional, technical, social...) show? What are the problems caused by obsolescence? Who suffers (and how) because of obsolescence (users, owner, environment)?
- Are there commonly accepted limit states for these different types of obsolescence? If not, how is obsolescence defined? Is the definition a result of a cost-benefit study? Or is the pressure from the public or authorities pushing hard and setting limits? What should the obsolescence limit states be for the facility type in question, and what other viewpoints than just the economic ones should be taken into consideration when defining obsolescence limit states? Who defines the obsolescence limit states? What are the obsolescence indicators?
- Is there data from the past available? What kind of data banks, sources of information or resources are there available for a deeper analysis? Does the decision

maker (facility manager, management team, etc.) have a *comprehensive picture* (also including societal approach, not just technical) of the obsolescence problem?

Of course this part of the analysis is a lot easier if the owner has documented examples of obsolescence cases in his/her facility stock. In any case, the previous task and its results should be duly documented.

2.5.2.2   *Factors and causes of obsolescence*

In this part of the obsolescence analysis, the possible causes for the different obsolescence types are sought after. This part follows straightforwardly the risk analysis procedure presented in deliverable D2.3 [11], where the causes of adverse incidents, i.e. so called *top events* are revealed, using fault tree analysis. In the obsolescence analysis these top events mean the obsolescence indicators of different obsolescence types. The reader is referred to the deliverable D2.3 for detailed description of the procedure.

The factors and causes of obsolescence can be physical needs, e.g. increased traffic on the route where the bridge is located, new type of ships that cannot dock to the existing wharf, etc. Many times the obsolescence causes can be traced to promulgation of new standards (that require, for example, stricter sound insulation in floors, etc. Although normally the existing facilities are exempted of these requirements, there will be pressure to follow the new standards). The factors can be fashion-originated: the existing façade of a building looks grim, the building is not located in 'the right part of the city', etc.

Although it is obvious that the top-level cause of obsolescence is the general development of society (technological, cultural, etc.), in this part of the analysis it must be studied at a deeper level. In the ideal case, the facility owner would become aware of the reasons behind trends, new norms and standards, migration, employment policy and all possible societal causes that have effect on the use of the facilities. After having these factors on hand, it is much easier for the facility owner to estimate the direction of the general development and plan the future actions for the facility. But as mentioned earlier, this requires quite a comprehensive touch to the whole process of facility management, and resources may be scarce in many organisations.

2.5.2.3   *Strategies and decisions on actions against obsolescence*

When the obsolescence indicators of possible obsolescence types and their causes for the facilities are identified, the owner should try to find actions to avoid or defer obsolescence. These actions generally have the purpose of minimising the impacts of obsolescence by anticipating change or accommodating changes that cause obsolescence before the costs of obsolescence become substantial.

Although obsolescence is best fought *before* entering the operations and maintenance phase in the life cycle of a facility, something can be done to minimise obsolescence costs also when dealing with existing structures. Good maintenance practices have the same effect in the maintenance phase as quality assurance in the construction phase, enhancing the likelihood that performance will indeed conform to design intent. Training of maintenance staff, preparation and updating of maintenance manuals and use of appropriate materials in maintenance activities contribute to avoiding the costs of obsolescence. Existing and new computer-assisted facility management systems that support condition monitoring, document management and maintenance scheduling, should be able to provide useful information that can help the facility manager to detect problems that could presage obsolescence. An idea of multidimensional 'obsolescence index' has been presented as a target for research, but so far this issue has stayed on the theoretical level.

The obsolescence studies and discussions have concentrated on buildings and on the business inside the building, like schools, hospitals, and office or industrial buildings. In these cases the location, inner spaces, etc. have great impact on the possible obsolescence, as the use of a building can change radically when the tenant or owner changes. The possible strategies include post-occupancy evaluation and report cards to achieve performance approaching the optimum of the facility, adaptive reuse, shorter terms for leasing and cost recovery calculations, etc. Often the strategy with obsolescence is 'making-do', which means finding low-cost ways to supplement performance that is no longer adequate. Normally, making-do is a short-term strategy with high user costs, leading eventually (after high complaint levels, loss of revenue, loss of tenants, etc.) to refurbishment of the facility.

However, with infrastructure facilities – on which Lifecon is focusing, like bridges, tunnels, wharves, lighthouses, etc. – the situation is not the same, as these facilities normally are already located in the most optimal place to serve that one certain business they were built for. Normally, this business (for example port activities, passing traffic through or over obstacles, etc.) cannot be totally halted, so the demolishing of obsolete – but otherwise sound – facility and construction of a new one is not a common or wise solution. One traditional solution (especially with bridges) has been to build a new facility near the old one and keep the old one for lesser service.

2.5.3   Limit states of obsolescence

In order to make possible the analysis of obsolescence, the obsolescence itself must be defined. For that definition limit states are needed. While in stability and durability analyses of structures there appear clear signs *in* the facility when limit states are reached (ruptures, cracks, spalling, corrosion, deflections, vibration, etc.), with obsolescence the case is not that simple. The signs about obsolescence are normally found *outside* of the facility (loss of revenue, complaints from users, traffic jams, increased maintenance costs, etc.). The decision when those *obsolescence indicators* have increased excessively, meaning that the limit states have been reached, is difficult and in most cases organisation specific. However, some qualitative limit states of obsolescence can be defined on a generic level. These are presented in Table 2.5.

As can be seen in Table 2.6, the difference between service limit state and ultimate limit state in obsolescence analysis is a question of interpretation. For example, there exists no standardised definition for 'minor problems' or 'severe problems', but they are organisation-specific matters. The obsolescence indicators are the same for service and ultimate limit states, but in ultimate limit state they are just stronger than in service limit state. Using an analogy with the traditional static and dynamic limit states definitions, one can come to the conclusion that the *ultimate limit state in obsolescence means that there is no recovery from that state without heavy measures while in service limit state minor actions can return the situation to the pre-obsolete state.* In the traditional static and dynamic analysis reaching the ultimate limit state

Table 2.6 Obsolescence indicators for different obsolescence types

	Functional and human	Economic	Ecological	Cultural
Bridge	• service capability of the bridge or network of bridges in the actual global, regional or local logistic system not adequate • weak capability to transmit the current traffic • weak bearing capacity for present traffic loads • low height for under-going road or water-borne traffic • heavy noise from traffic on bridge • heavy degradations cause uneasiness for users	• high costs for users because of traffic jams • high operation costs (e.g. bascule bridge) • high MR&R (Maintenance, Repair, Rehabilitation and Renewal) costs	• high production of environmental burdens because of traffic jams • high production of environmental burdens because of need for the use of by-pass roads • high production of environmental burdens because of highly increasing MR&R works • robust intermediate piers and long approach embankments impede free flow of water	• the image of the bridge does not meet the local image goals • the bridge is preserved as a cultural monument without adequate possibilities for changes • heavy abutments and intermediate piers block the free view of the under-going roadway users
Building	• the changeability of spaces not enough for the actual or future needs • the accessibility not adequate • not adaptable for modern installations • the quality of internal air does not meet actual health requirements • the emissions from materials cause danger for health • lighting does not meet the requirements of living or working • the living or working comfort does not meet present day requirements	• too high energy costs • too high operation costs • potential residual service life too short in comparison to required repair or rehabilitation cost	• the energy efficiency does not meet the current requirements of owners, users or society • high production of environmental burdens because of highly increasing MR&R works	• the spaces are not adaptable for the current ways of living or working • the architectural quality does not meet the local actual require-ments • building does not reflect the image that the user wants to give

means permanent deformations in the structure, while in the service limit state the deformations are not permanent.

To proceed in the obsolescence analysis, the generic level limit states must be converted into more specific and tangible descriptions. In this conversion, the facility type has a decisive role because the specific obsolescence indicators and their reasons vary a lot depending on the facility type (for example, traffic jam is obviously a bridge-related obsolescence indicator, but cannot be used for lighthouses). In Table 2.6 some obsolescence indicators for two different facility types are listed, categorising also the obsolescence type.

2.5.4   Methods for obsolescence analysis and decision-making

Although obsolescence is increasing in importance, no standards addressing obsolescence of civil infrastructure or building facilities have been enacted so far. Principled strategies and guidelines for dealing with obsolescence have been presented [7, 11] but the real analysis methods have not been applied. As obsolescence progress of a facility depends on the development of local conditions, as well as on the general development of society during the service life (or residual service life) of a facility, there is lot of uncertainty involved in obsolescence analyses. Like in any uncertainty-filled problem, in obsolescence situation also the case must be structured down to smaller parts, which can be consistently handled. It must be noted that *obsolescence avoidance thought* should be present in all life cycles of the facility: planning and programming; design; construction; operations, maintenance and renewal; retrofitting and reuse. The obsolescence analysis should be performed before the onset of obsolescence, as a part of the facility owning and management strategy.

The following methods can be applied in obsolescence analysis:

- Quality Function Deployment (QFD) method [Lifecon Deliverables D2.3 and D5.1]
- Life Cycle Costing method (LCC) [Lifecon Deliverable D5.3]
- Multiple Attribute Decision Aid (MADA) [Lifecon Deliverable D2.3]
- Risk Analysis (RA) [Lifecon Deliverable D2.3].

2.5.5   QFD in obsolescence analysis and decision-making

Quality Function Deployment method (QFD) can be used for interpreting any 'Requirements' into 'Specifications', which can be either 'Performance Properties' or 'Technical Specifications' [Lifecon Deliverable D2.3].

Thus QFD can serve as an optimising or selective linking tool between:

- changing 'Requirements'
- actual and predicted future 'Performance Properties'
- actual and predicted future 'Technical Specifications'

In the obsolescence issues QFD can be used for optimising the 'Technical Specifications' and/or 'Performance Properties' in comparison to changing 'Requirements' and their changing ranking and weights. These results can be used for selection between different design, operation and MR&R alternatives for avoiding the obsolescence.

Simply, the QFD method means building of a matrix between requirements (=What's) and Performance Properties or Technical Specifications (=How's). Usually, the Performance Properties are serving only as a link between Requirements and Technical Specifications, sometimes this link can be eliminated, and direct correlations between Requirements and Technical Specifications analysed. In practical planning and design, the application shall be limited to few key Requirements and key Specifications to maintain good control of variables and not to spend too many efforts for secondary factors.

The following procedure can be applied in LIFECON LMS when using QFD for analysis of functional requirements against owner's and user's needs, technical specifications against functional requirements, and design alternatives or products against technical specifications:

1. Identify and list factors for 'What' and 'How'
2. Aggregate the factors into Primary Requirements
3. Evaluate and list priorities or weighting factors of 'What's'
4. Evaluate correlation between 'What's' and 'How's'
5. Calculate the factor: correlation times weight for each 'How'.

Normalise the factor 'correlation times weight' of each 'How' for use as a priority factor or weighting factor of each 'How' at the next steps.

The obsolescence analysis and decision-making procedure includes six steps:

1. Define the individual 'Requirements' corresponding to alternative obsolescence assumptions
2. Aggregate the individual 'Requirements' into 'Primary Requirements'
3. Define the priorities of 'Primary Requirements' of the Object for alternative obsolescence assumptions
4. Define the ranking of alternative solutions for avoiding the obsolescence. One of these solutions is the demolition
5. Select between these alternatives using the priorities from step 1
6. Decide between the alternative solutions for avoiding the obsolescence or demolishing the facility.

The QFD method is described in more details in Lifecon Deliverable D2.3 [11], and applied into MR&R planning in Lifecon Deliverable D5.1 [12].

2.5.6 LCC in obsolescence analysis and decision-making

Life cycle costing can be effectively used in obsolescence analysis and decision-making between alternative obsolescence avoidance strategies and actions. It can be either alone, focusing on economic obsolescence options, or one part of the multiple analysis and decision-making, connected to other methods such as QFD, MADA or FTA.

The methodology of LCC in this connection is the same as presented for general MR&R planning and decision-making in Lifecon Deliverable D5.3. In obsolescence issues, the alternatives are different obsolescence options and alternative strategies and actions for avoiding the economic obsolescence.

Because economic obsolescence usually is only one of several categories of obsolescence, besides LCC other methods include QFD, MADA or FTA.

2.5.7 MADA in obsolescence analysis and decision-making

Multiple Attribute Decision Aid method is described in detail in Lifecon Deliverable D2.3. In order to 'measure' the influence of obsolescence factors and options on the ranking and choice between alternative strategies and actions for avoiding obsolescence, the method of sensitivity analysis of MADA can be applied.

Sensitivity analysis with Monte-Carlo simulation consists of four steps (Figure 2.7):

1. Random assessment of the weights or alternatives assessments simulating small variations (e.g. $\pm 5\%, \pm 10\% \ldots$)
2. Application of MADA methodology
3. Ranking of alternatives
4. Statistical analysis of the various rankings.

A simulated weight/alternative assessment is obtained by multiplying the initial weight/alternative assessment (given by the user) by a multiplicative factor (variation) modelling small variations (Figure 2.8).

For instance, an initial weight $W = 30$, subjected to small variations $[-10\%, +10\%]$, will vary in the range $[30 \times 0.9, -30 \times 1.1]$, i.e.

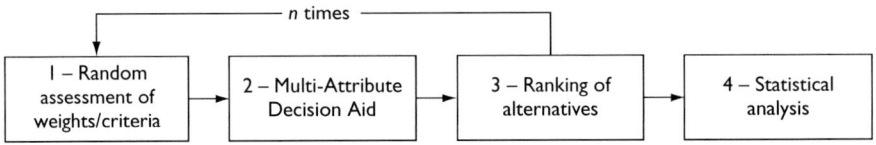

Figure 2.7 Monte-Carlo simulation in sensitivity analysis of MADA [Lifecon Deliverable D2.3].

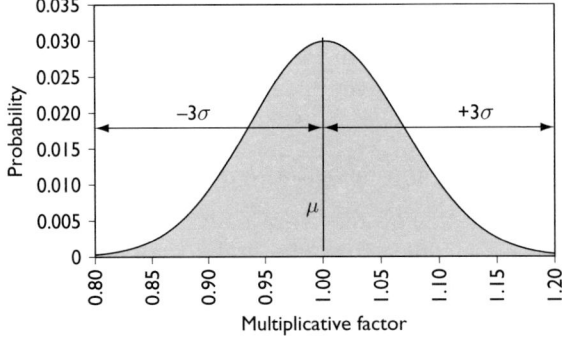

Figure 2.8 Example of multiplicative factor (Variation 20%) [Lifecon Deliverable D2.3].

These small variations can be calculated by means of a bounded Gaussian distribution defined with:

$$\begin{cases} \text{Mean: } \mu = 1 \\ \text{Standard deviation: } \sigma = \dfrac{\text{variation}}{3} \end{cases}$$

It is then bounded in lower values and upper values respectively by $(1 - \text{variation})$ and $(1 + \text{variation})$. The bounds and standard deviation are chosen in such a way to include 99.7% of the values (i.e. 99.7% of a Gaussian distribution is included between $(\mu - 3\sigma)$ and $(\mu + 3\sigma)$).

After n simulations, the various ranking of alternatives of strategies and actions and analysis of the variations will be carried out.

2.5.8   FTA in obsolescence analysis and decision-making

The use of Fault Tree Analysis (FTA) is explained with some examples of different cases.

2.5.8.1   Case 1: Bridge

In this illustrative example the top event is an obsolescence indicator which means the 'service capability of the bridge or network of bridges in the actual global, regional or local logistic system is not adequate':

2.5.8.1.1   TOP EVENT CLARIFICATION

Primary function of a bridge is to transmit traffic over an obstacle (another route, railway, ravine, etc.) *and* at the same time to make possible the transit under the bridge. So the service capacity refers both to the over-going and the under-passing traffic. Primary parameters of traffic are volume and weight, the corresponding counterparts of the bridge being free space (horizontal and vertical) and load-bearing capacity respectively. This leads to the conclusion:

The top event happens when the *dimensions* or the *load-bearing capacity* of the bridge does not meet the demands anymore. Two cases must be identified, i.e. traffic over the bridge and traffic under the bridge. For the under-passing traffic (vessels, trains, vehicles), the only important parameter of the bridge is free space as the traffic does not have contact with the bridge. For the over-passing traffic the load-bearing capacity of the bridge also is very important.

Note: Of course there are also other requirements that the bridge has to fulfil, like aesthetics, MR&R economy, ecological demands, etc. and consequently the bridge can be obsolete regarding those issues. However, in this example only the service capacity was of concern.

After this short reasoning, at the latest, the scope of the analysis should be defined: Is the analysis going to be carried out for the whole stock of bridges, for the bridges on some certain area or route, or for just one certain bridge. Logically, the more general the scope, the more branches the fault tree will have. In this illustrative example the obsolescence problem will be studied on the 'whole stock of bridges' (i.e. network) level. The fault tree for an individual bridge would of course be much smaller, because useless branches can be cut off immediately from the tree.

The resulting fault tree is shown in the Figure 2.9. First the whole tree is displayed to illustrate the possible extent of the analysis, and then it is shown in more detailed pieces to make the texts readable.

After finding out the primary reasons of obsolescence (circles in Figure 2.9), decisions can be made about countermeasures. There exist no thumb rules 'do this, avoid that', but the decisions are case- and organisation-specific.

2.5.8.2 Case 2: Building

Another example of short obsolescence analysis relates to the last cultural obsolescence indicator of Table 2.6: 'Building does not reflect the image that the user wants to give.'

This example is more difficult to analyse, but eventually can be handled with the same procedure as the bridge example above. The idea is again to split the problem into 'smaller pieces' (or parameters) in a structured way, and to find out the possible causes as to why the value of those parameters and their sub-parameters do not fit into user's image.

The splitting of the top event into smaller pieces could follow the following reasoning: The parameters of the building that have effect on the image of the user are mainly

- location
- outlook
- internal spaces, surfaces, decorations, hallways, etc.
- comfort feeling generally: inside and outside the building.

Each of those four main contributors can be further divided, for example the outlook of the building can be further split into the following five sub-contributors:

1. style of the building (castle, storehouse, box . . .)
2. colour of the building (colourful, trendy, old-fashioned, grim . . .)
3. dimensions of the building (overall size of the building, doors/windows, height, width . . .)
4. materials of the building (stone, brick, concrete, steel . . .)
5. condition of the building (brand new, worn, near to collapse . . .)

This way the analysis goes on until the fundamental level is reached. After finishing the fault tree it can be seen which basic factors contribute to the contradiction between the present building and the image expected by the user. Depending on the source data the relative importance of the basic factors can be estimated and consequently countermeasures launched. All the time those factors must be studied with image-oriented approach, i.e. throughout the analysis how the identified parameters affect the image of the user must be studied. Parameters that have no effect on the image will be excluded from this image-related obsolescence analysis, although these excluded parameters might have considerable effect on the overall business of the user. These

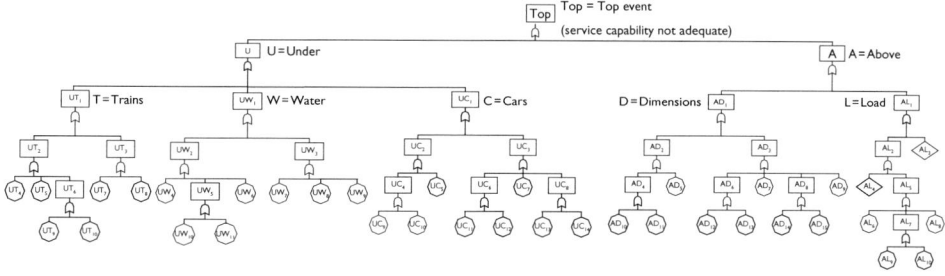

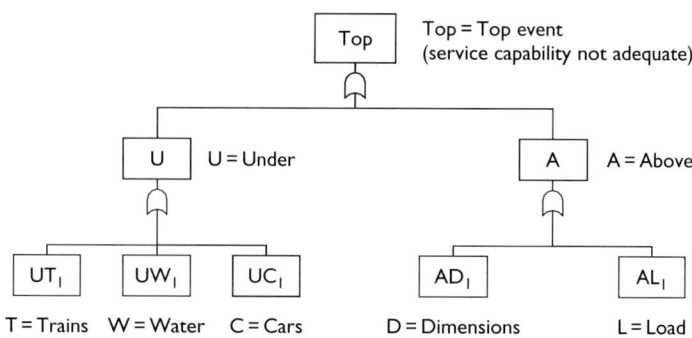

Abbr.	Explication of the event
U	Service capability not adequate for the traffic **U**nder the bridge
A	Service capability not adequate for the traffic **A**bove the bridge
UT_1	Service capability not adequate **U**nder the bridge for railway traffic (**T**rains)
UW_1	Service capability not adequate **U**nder the bridge for **W**ater-borne traffic
UC_1	Service capability not adequate **U**nder the bridge for road traffic (**C**ars)
AD_1	Service capability not adequate **A**bove the bridge due to **D**imension-related causes
AL_1	Service capability not adequate **A**bove the bridge due to **L**oad-related causes

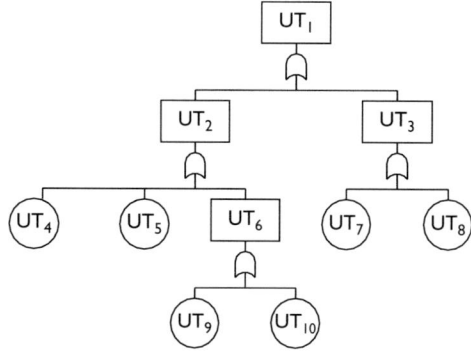

Figure 2.9 Fault tree in obsolescence analysis.

Abbr.	Explication of the event
UT$_1$	Service capability not adequate **U**nder the bridge for railway traffic (**T**rains)
UT$_2$	Vertical clearance for railway traffic limited
UT$_3$	Horizontal clearance for railway traffic limited
UT$_4$	Special cargo track (e.g. harbour activities) needs higher clearance
UT$_5$	Electrification problem: No room for installations (wires etc.) under the bridge
UT$_6$	Railway norms concerning vertical clearance are to be changed
UT$_7$	More tracks wanted but horizontal clearance does not allow that
UT$_8$	Wider clearance needed for special cargo tracks (e.g. harbour activities)
UT$_9$	Railway norms to be changed on international level
UT$_{10}$	Railway norms to be changed on national level

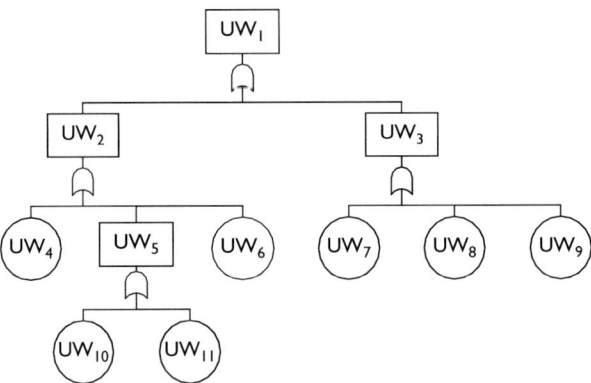

Abbr.	Explication of the event
UW$_1$	Service capability not adequate **U**nder the bridge for **W**ater-borne traffic
UW$_2$	Vertical clearance for water-borne traffic limited
UW$_3$	Horizontal clearance for water-borne traffic limited
UW$_4$	New water-level regulation policy keeps the water level very high
UW$_5$	Commercial water traffic needs higher clearance than what is the current situation
UW$_6$	Recreational yachting increases, with higher motor and sailing boats
UW$_7$	New route for seagoing ships requires wider navigation channel
UW$_8$	Intermediate piers badly situated in the middle of the watercourse
UW$_9$	Narrow navigation channels between abutments and piers cause difficult currents (e.g. for slow towboats, log floating, etc.)
UW$_{10}$	Deep-water channel to be opened, higher ships to be expected on watercourse
UW$_{11}$	Log floating to be commenced, towboats need higher clearance

Figure 2.9 (Continued).

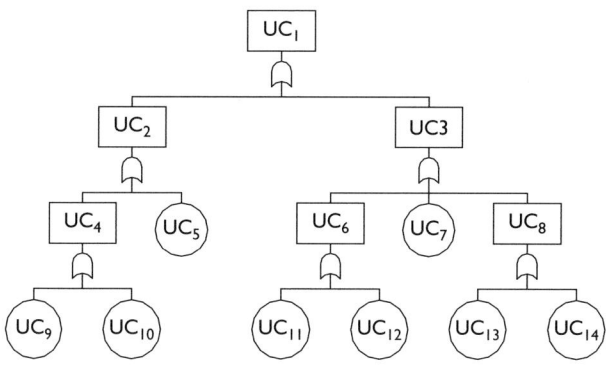

Abbr.	Explication of the event
UC_1	Service capability not adequate **U**nder the bridge for road traffic (**C**ars)
UC_2	Vertical clearance for under-passing road traffic limited
UC_3	Horizontal clearance for under-passing road traffic limited
UC_4	Road traffic norms concerning vertical clearance on normal roads are to be changed
UC_5	Special loads route (e.g. minimum height 7.2 m) network to be extended, including the under-passing road in question; need for higher clearance
UC_6	Stricter safety standards call for wider clearance between columns and abutments
UC_7	Change of the existing under-passing road into a 'wide lane road', but the clearance between columns is too narrow for that
UC_8	Change of the existing two-lane under-passing road into multilane road
UC_9	Road traffic norms to be changed on international level
UC_{10}	Road traffic norms to be changed on national level
UC_{11}	Standard to be changed on international level
UC_{12}	Standard to be changed on national level
UC_{13}	Too much traffic for two-lane road, more lanes needed
UC_{14}	Change from normal road to motorway

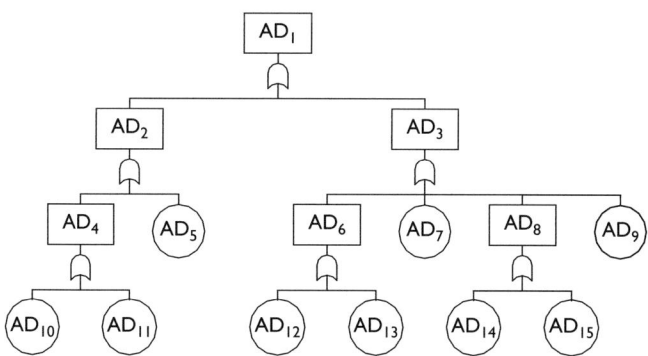

Abbr.	Explication of the event
AD_1	Service capability not adequate **A**bove the bridge due to **D**imension-related causes
AD_2	Vertical clearance on bridge limited
AD_3	Horizontal clearance on bridge limited
AD_4	Road traffic norms concerning vertical clearance on normal roads are to be changed
AD_5	Special loads route (e.g. height 7.2 m) network to be extended, need for higher clearance on the (truss) bridge in question
AD_6	New standard calls for wider lanes
AD_7	Change of the existing road into a 'wide lane road', but the horizontal clearance between railings is too narrow for that
AD_8	Change of the existing two-lane road into multilane road
AD_9	Pedestrians need a lane of their own, separated (e.g. elevated) from traffic lanes
AD_{10}	Road traffic norms to be changed on international level
AD_{11}	Road traffic norms to be changed on national level
AD_{12}	Standard to be changed on international level
AD_{13}	Standard to be changed on national level
AD_{14}	Too much traffic for two-lane road, more lanes needed
AD_{15}	Change of the road from normal road to motorway

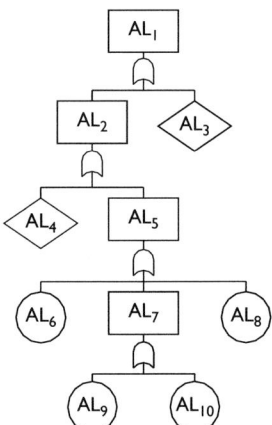

Abbr.	Explication of the event
AL_1	Service capability not adequate **A**bove the bridge due to **L**oad-related causes
AL_2	Loads increased
AL_3	Load bearing capacity decreased
AL_4	Overloads increased
AL_5	Legal loads increased
AL_6	Road class change from lower to higher
AL_7	Change of standards for normal road traffic loads
AL_8	Special loads (harbour, mine, foundry, factory)
AL_9	Standard to be changed on international level
AL_{10}	Standard to be changed on national level

Figure 2.9 (Continued).

contradictions must be taken into account in other analyses (e.g. in MADA) on corporate strategy level. An example of this kind of contradiction might be the following:

The company wants to give the image that they are open and very accessible to customers, and consequently have decided to have very large windows in the facade and open-plan office. However, the workers feel uncomfortable working close to the windows, where all passers-by can see them through the window, there is a nasty draft especially during cold days near the windows, and the open-plan office causes a lot of interruptions to work. If the company has not deemed workers' satisfaction as an image factor, it will be excluded from the image analysis, although it surely has an effect on the business of the company.

2.6 Methods for optimisation and decision-making

2.6.1 Principles of optimisation and decision-making

The *objective* of the integrated and predictive lifetime management is to achieve optimised and controlled lifetime quality of buildings or civil infrastructures in relation to the generic requirements. The *lifetime quality* means the *capability of an asset to fulfil the requirements of users, owners and society in an optimised way during the entire design life of the asset*. This objective can be achieved with a *performance-based methodology*, applying *generic limit state approach*. This means that the generic requirements have to be modelled with technical and economic numerical parameters into quantitative models and procedures, and with semi-numerical or non-numerical ranking lists, classifications and descriptions into qualitative procedures. This methodology can be described in a schedule, which is presented in Figure 2.4. The generic requirements are listed in Table 2.1.

The lifetime performance modelling (Figure 2.4) and the limit state approach are building an essential core of the lifetime management, MR&R planning. Performance-based modelling includes the following classes:

- modelling of the behaviour under mechanical (static, dynamic and fatigue) loads
- modelling of the behaviour under physical, chemical and biological loads
- modelling of the usability and functional behaviour.

The mechanical modelling has been traditionally developed on the limit state principles already started in 1930s, and introduced into common practice in the 1970s. The newest specific standard for reliability of structures is Eurocode EN 1990:2000 (Table 2.4). The mechanical behaviour (safety and serviceability), besides the other categories mentioned above, has to be checked in several phases of the management process. This is especially important in condition assessment and in MR&R planning. It is sometimes possible to combine the mechanical calculations with degradation and service life calculations, but often it is better to keep these separated. Because the models and calculation methods of mechanical behaviour are very traditional and included in normative documents of limit state design, this issue is not treated here, which is focused on durability limit state design and obsolescence limit state design.

Modelling for physical, chemical and biological loads includes a large variety of thermal behaviour, moisture behaviour and behaviour under fire conditions, and biological impacts, and biological phenomena (e.g. mould and decay). These are connected with several phenomena and properties of structures in use, and in this context this section is distributed into different procedures of the reliability assessment. Traditional analysis of thermal, fire and moisture behaviour are not treated in this book.

In life cycle management system, modelling for usability and functionality means management of obsolescence. *Obsolescence* means the inability to satisfy changing functional (human), economic, cultural or ecological requirements. Obsolescence can affect the entire building or civil infrastructural asset, or just some of its modules or components. Obsolescence is the cause of demolition of buildings or infrastructures in about 50% of all demolition cases. Therefore this issue is very central in developing asset management for sustainability, which is the aim of Lifecon LMS.

A main issue of healthiness during the MR&R works is to avoid unhealthy materials (Lifecon D5.1). During the use of assets (especially in closed spaces such as buildings or tunnels), it is important to avoid moisture in structures and on finishing surfaces, because it can facilitate mould growth, and to check that no materials used cause emissions or radiation, which are dangerous for health and can be uncomfortable for users. In some areas radiation from the ground must be also be eliminated through insulation and ventilation of the foundations. Thus the main tools for health management are: selection of materials (especially finishing materials), eliminating risks of moisture in structures (through waterproofing, drying during construction and ventilation), and elimination of possible radioactive ground radiation with airproofing and ventilation of ground structure. Health requirements can follow the guidelines of national and international codes, standards and guides. The modelling of the health issues thus focuses on calculating comparable indicators on the health properties mentioned above and on comparing these between alternatives in the optimisation and decision-making procedures. These can usually be calculated numerically, and they thus are mainly quantitative variables and indicators, which can be compared in the optimisation and decision-making procedures.

Comfort properties are related to the functionality and performance of asset, having for example the following properties:

- acoustic comfort, including noise level during MR&R works or in the use of closed spaces like tunnels and buildings
- insulation from airborne sound between spaces
- comfortable internal climate of closed spaces like tunnels and buildings
- aesthetic comfort externally and in functions of use in all kinds of assets; and vibrations of structures.

These are calculated with special rules and calculation methods, which are also traditional and therefore will not be treated in this book. Mainly quantitative (exact numerical or classified) values can be used for these properties.

Ecology can be linked to the environmental expenditures: consumption of energy, consumption of raw materials, release of environmental pollutants into air, soil and water, and loss of biodiversity. Most of these can be calculated numerically, and thus are quantitative variables and indicators. These can also be compared quantitatively in the optimisation and decision-making procedures. In buildings, energy consumption mostly dictates environmental properties. For this reason the thermal insulation of the envelope is important. Finally the reuse and recycling of the components and materials after the demolition belong to the ecological indicators. Engineering structures such as bridges, dams, towers and cooling towers are often very massive and their material consumption is an important factor. Their environmental efficiency depends on the selection of environmentally friendly local raw materials, high durability and easy maintainability of the structures during use, recycling of construction wastes and finally recycling of the components and materials after demolition. Some parts of engineering structures, such as waterproofing membranes and railings, have a short or moderate service life and consequently easy re-assembly and recycling are most important in order to minimise the annual material consumption property. During MR&R works it is important to apply effective recycling of production wastes. This leads to calculations of waste amounts as quantitative variables of this component of ecology. Some ecological properties, like loss of biodiversity, are difficult to calculate numerically, and they often can be only described qualitatively. This qualitative description can then be used in comparing alternatives during optimisation and decision-making procedures.

The functionality of civil infrastructures means the capability to serve for the main targets of an asset, e.g. in case of tunnels and bridges the capability to transmit traffic. This can be modelled numerically using, as variables and indicators, suitable geometric dimensions and load-bearing capacity and so on. The functionality of buildings is very much related to the flexibility for changes of spaces, and often also on the loading capacity of floors. Also the changeability of building service systems is important. Internal walls have a moderate requirement of service life and a relatively high need to accommodate changes. These dictate the capability of a building to enable changes in the functions during the lifetime management. For this reason internal walls must have good changeability and recycleability. An additional property is good and flexible compatibility with the building services system, because the services system is the part of the building that is most often changed.

To avoid the repetition of traditional and well-known issues, the generalised and reliability-based life cycle management approach can be focused and formulated into the following three categories:

- static and dynamic (mechanical) modelling and design
- degradation-based durability and service life modelling and design
- obsolescence-based performance and service life modelling and design.

In Lifecon LMS system the transformation of generic requirements into functional and performance property definitions, and further into technical specifications and performance models, will be realised with the following methods [11]:

- requirements analysis and performance specifications: QFD method: Deliverable D2.3
- service life estimation:

  - probabilistic service life models: Deliverable D3.2
  - RILEM TC 130 CSL models: Deliverable D2.1
  - Reference structure method: Deliverable D2.2

- condition matrix: Markov Chain method: Deliverable D2.2, and Condition Assessment Protocol: Deliverable D3.1
- total and systematic reliability-based methodology: Deliverable D2.1
- risk analysis: Deliverable D2.3.

2.6.2   Multiple Attribute Decision Aid (MADA)

2.6.2.1   An introductory example

As an example, we propose attributes and corresponding criteria that could be taken into account for the choice between cars (Figure 2.10):

- attribute 'economy' (purchasing cost and maintenance cost)
- attribute 'human requirements' (maximum speed, comfort, noise)
- attribute 'environment' (gas consumption, impact on air pollution, recycleability of materials).

For each criterion, a utility function is defined:

- 'Cost' is a quantitative criterion, in the range €0 − 100 000.
- 'Maintenance cost' is a qualitative criterion, in the domain low, medium or high.
- 'Noise' is a qualitative numerical criterion, with values of 0, 1, 2, 3, 4 or 5.

2.6.2.2   Terms of the MADA methodology

In MADA methodology the following specific terms are used [13].

Aggregation
Aggregation is a process leading from information on the preferences by criteria to information on a global preference between alternatives.

Comparison method

- Method 1 (Choice) – We clarify the decision by choosing a subset of alternatives (as small as possible) in order to choose the final alternative. This subset contains optimum and sufficient satisfying alternatives.

	Attributes/criteria							
	Economy		Human requirements			Environment		
Alternatives	Purchasing cost (€)	Maintenance cost	Maximum speed (km h^{-1})	Comfort	Noise	Average gas consumption	Impact on air pollution	Recycleability
Car 1	12 000	Low	165	Medium	4	8 l/100 km	High	Medium
Car 2	16 000	Medium	195	Good	2	6 l/100 km	Low	Good

Figure 2.10 Multi-attribute decision example: car selection.

- Method 2 (Problematic β – Sorting) – We assign the alternatives to categories. Each alternative is assigned to a single category (independent of other categories).
- Method 3 (Ranking) – We rank all or some of the alternatives (the most satisfying ones), by assigning a rank of ordering, which allows a total or partial ranking.
- Method 4 (Description) – The problem is correctly stated by describing alternatives and their consequences.

Core
The core is a subset of alternatives fulfilling the following conditions:

- Any alternative not belonging to the core is outclassed by at least one alternative of the core
- No alternative belonging to the core is outclassed by another alternative of the core.

Utility function
The utility function U_j describes the criterion j ($U_j(a)$ and $U_j(b)$ are the values corresponding to alternative a and b for the considered criterion j), with A being the set of alternatives.

Strict preference (P)
Let $p(U_j)$ be a strict preference threshold associated to criterion j.
Then the strict preference relation P is defined by

$$\forall a, b \in A, \text{`}a\,P\,b\text{'} \text{ if } U_j(a) > U_j(b) + p(U_j)$$

Example
Let the preference threshold for the criterion 'maximum speed' be $p(U_{\text{Speed}}) = 20\,\text{km}\,\text{h}^{-1}$.

Then U_{Speed} (Car 2) $> U_{\text{Speed}}$(Car 1)$ + p(U_{\text{Speed}})$
since $195 > 165 + 20 \rightarrow$ Car 2 is strictly preferred to Car 1.

Indifference (I)
Let $q(U_j)$ be an indifference threshold associated to criterion j.
Then the indifference relation I is defined by

$$\forall a, b \in A, \text{`}a\,I\,b\text{'} \text{ if } U_j(b) - q(U_j) \leq U_j(a) \leq U_j(b) + q(U_j)$$

Example
Let the indifference threshold for the criterion 'maximum speed' be $q(U_{\text{Speed}}) = 40\,\text{km}\,\text{h}^{-1}$.

Then U_{Speed} (Car 2) $\leq U_{\text{Speed}}$ (Car 1)$ + q(U_{\text{Speed}})$
since $165 - 40 \leq 195 \leq 165 + 40 \Rightarrow$ Choice between Car 1 and Car 2 is indifferent (we cannot prefer one to the other).

Weak preference (Q)
The weak preference relation Q is defined by

$$\forall a, b \in A, \text{'}a \; Q \; b\text{'} \; if \; U_j(b) + q(U_j) \leq U_j(a) \leq U_j(b) + p(U_j)$$

Example
Let the preference threshold for the criterion 'maximum speed' be $p(U_{Speed}) = 40 \, \text{km} \, \text{h}^{-1}$ and the indifference threshold be $q(U_{Speed}) = 20 \, \text{km} \, \text{h}^{-1}$.

> Then $U_{Speed}(\text{Car 1}) + q(U_{Speed}) \leq U_{Speed}(\text{Car 2}) \leq U_{Speed}(\text{Car 1}) + p(U_{Speed})$
> since $165 + 20 \leq 195 \leq 165 + 40 \Rightarrow$ Car 1 is weakly preferred to Car 2 (*but* not strictly preferred (we do not have $U_{Speed}(\text{Car 2}) > U_{Speed}(\text{Car 1}) + p(U_{Speed}))$)

Outclassing (S)
An outclassing relation S expresses the fact that alternative *a* is not 'strictly worse' than alternative *b* with:

$$\forall a, b \in A, \text{'}a \; S \; b\text{'} \; if \; U_j(a) \geq U_j(b) + q(U_j(b))$$

'Outclassing' means '*A* strictly preferred to *B*' *or* '*A* weakly preferred to *B*'.

Example
Let the indifference threshold for the criterion 'maximum speed' be $q(U_{Speed}) = 20 \, \text{km} \, \text{h}^{-1}$.

> Then $U_{Speed}(\text{Car 2}) \geq U_{Speed}(\text{Car 1}) + q(U_{Speed})$
> since $195 \geq 165 + 20$ Car 2 outclasses Car 1 (Weak preference).

Example
Let the indifference threshold for the criterion 'maximum speed' be $q(U_{Speed}) = 20 \, \text{km} \, \text{h}^{-1}$ and the preference threshold be $p(U_{Speed}) = 40 \, \text{km} \, \text{h}^{-1}$; and $U_{Speed}(\text{Car 2}) = 210 \, \text{km} \, \text{h}^{-1}$ (instead of $195 \, \text{km} \, \text{h}^{-1}$).

> Then $U_{Speed}(\text{Car 2}) \geq U_{Speed}(\text{Car 1}) + q(U_{Speed})$ *and*
> $U_{Speed}(\text{Car 2}) \geq U_{Speed}(\text{Car 1}) + p(U_{Speed})$
> since $210 \geq 165 + 20$ *and* $210 \geq 165 + 40 \Rightarrow$ Car 2 outclasses Car 1
> (strict preference).

Veto
A veto threshold $>^v$ is also defined. $a >^v b$ means that the difference between *a* and *b* for criterion *j* is such that *a* is definitely unacceptable in comparison with *b* (the outclassing of *a* compared to *b* is rejected even if *a* outclasses *b* concerning all the other criteria).

Example
Let the veto threshold for the criterion 'maximum speed' be $v(U_{Speed}) = 20 \, \text{km} \, \text{h}^{-1}$.

> Then $U_{Speed}(\text{Car 1}) + v(U_{Speed}) < U_{Speed}(\text{Car 2})$
> since $165 + 20 < 195 \Rightarrow$ Car 2 is definitely preferred to Car 1 (whatever the other criteria are).

These definitions are not valid for some criteria. Indeed, according to the criteria, the less could be the worst as for quality, speed, ... or the best as for noise, cost, consumption. This characteristic is called *minimisation*.

2.6.2.3 Types of criteria

Various types of criteria could be defined according to the quantity of considered thresholds: true, quasi, pseudo and pre-criteria. When comparing the values 'a' and 'b' of two alternatives for one criterion, various potential decisions are possible according to the type of criterion (Figure 2.11).

Example
Let A and B be two alternatives, 'a' and 'b' their respective value for the considered criterion.
Then for the *true criteria*:

- if $a > b$ then 'A outclasses B'
- if $a = b$ then there is 'indifference between A and B'
- if $a < b$ then 'A is outclassed by B'.

True criteria are too limited: it is a 'white or black decision', i.e. preference of one alternative except if alternative values are equal.

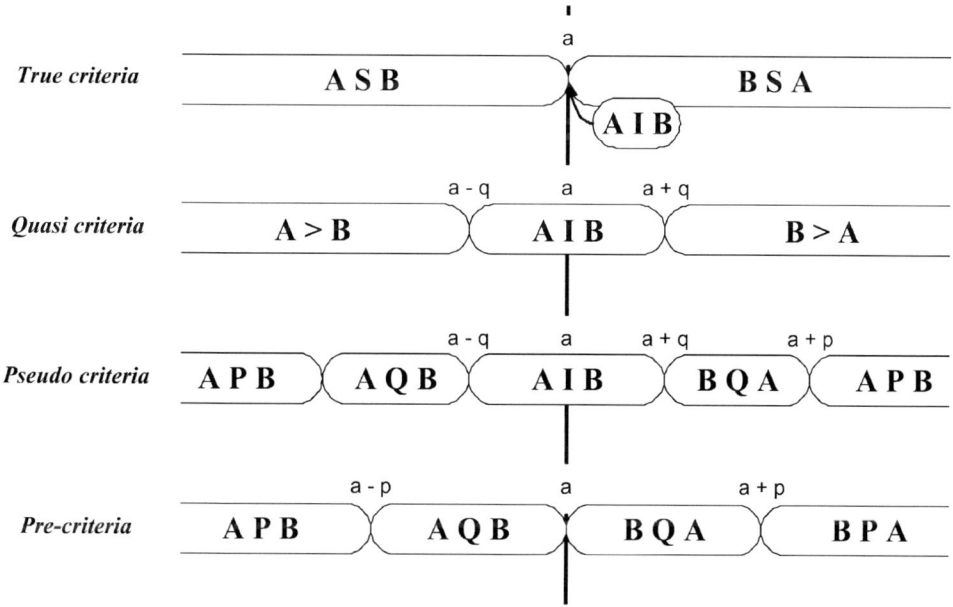

Figure 2.11 Criteria definition.

Pseudo-criteria include a gradation in preference (strict, weak preference or indifference). However, they require more information, such as the definition of the strict preference (p) and weak preference (q) thresholds.
Then:

- if $b < (a - p)$ then 'A strictly preferred to B'
- if $(a - p) < b < (a - q)$ then 'A weakly preferred to B'
- if $(a - q) < b < (a + q)$ then 'indifference between A and B'
- if $(a + q) < b$ then 'B weakly preferred to A'
- if $(a + p) < b$ then 'B strictly preferred to A'.

We will use pseudo-criteria, which are the most complete. They enable 'fuzzy comparisons' instead of 'white or black decisions'. Figures 2.12 and 2.13 illustrate this principle. When comparing two alternatives A and B, the decision is:

- based on the comparison of the values of A and B for true criteria
- based on the comparison of the values of A and B *and* the value of the difference $(A - B)$ for pseudo-criteria.

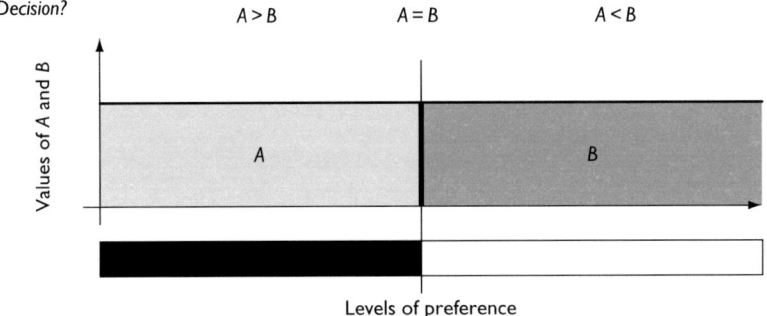

Figure 2.12 'White or black decisions'.

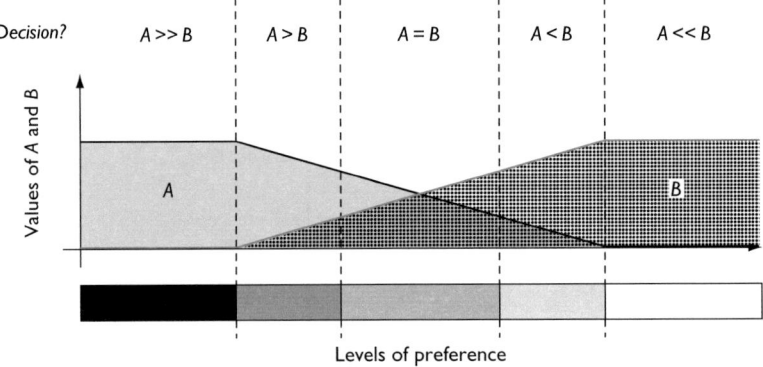

Figure 2.13 'Fuzzy decisions'.

2.6.2.4 Weights (of criteria)

A weight expresses the importance given by the user to a criterion. They have to be assessed in order to represent the decision-maker preferences: the higher the weight is, the higher the preference is.

It is difficult to assess the relative importance of each criterion. In order to help the user who does not know the criteria weights, we propose some guidelines [1, 14–18].

Several methodologies for weight determination are available:

- entropy methodology
- simple ranking methodology
- simple cardinal assessment methodology
- successive comparisons methodology
- eigen values methodology
- Analytical Hierarchy Process (AHP): ASTM E 1765–95 [14].

We will not give details of each methodology, but only the entropy methodology, the 'successive comparisons methodology' ('Revised Churchman Ackoff Technique') and the 'Analytical Hierarchy Process' [19]. The last one is programmed (Lifecon software).

2.6.2.4.1 ENTROPY METHODOLOGY

The relative importance of a criterion j, measured by a weight w_j, is proportional to the quantity of information supplied by this criterion. The larger the range of values for a criterion (i.e. the more easy is it to rank the alternatives), the higher is the weight of the particular criterion. The procedure is as follows.

Normalisation of alternatives evaluations (a_{ij}): dividing by the sum (for each criterion), entropy (E) assessment for each criterion (j) with $E_j = -(1/\ln(m)) \cdot \sum_i a_{ij} \cdot \ln(a_{ij})$ (m number of alternatives), assessment of the scattering measure $D_j = 1 - E_j$, assessment of weights with $w_j = \frac{D_j}{\sum_j D_j}$.

Example

Let us assess the weights of the following criteria, for the following set of alternatives:

	Criteria					
	Safety	Health	Investment costs	Future costs	Environmental impacts	Aesthetic
Alternative 1	4	1	100	20	4	16
Alternative 2	3	1	80	30	1	14
Alternative 3	2	1	85	30	5	6
Alternative 4	4	2	130	25	3	10
Alternative 5	1	1	30	35	6	2
Alternative 6	1	1	90	30	2	12
Alternative 7	2	1	88	20	7	8

(1) The first step consists in normalisation. The sum of alternatives for each column gives:

Criteria	Safety	Health	Investment costs	Future costs	Environmental impacts	Aesthetic
Sum	17	8	603	190	28	68

The new table, obtained after normalisation is:

| | Criteria | | | | | |
	Safety	Health	Investment costs	Future costs	Environmental impacts	Aesthetic
Alternative 1	0.235	0.125	0.166	0.105	0.143	0.235
Alternative 2	0.176	0.125	0.133	0.158	0.036	0.206
Alternative 3	0.118	0.125	0.141	0.158	0.179	0.088
Alternative 4	0.235	0.250	0.216	0.132	0.107	0.147
Alternative 5	0.059	0.125	0.050	0.184	0.214	0.029
Alternative 6	0.059	0.125	0.149	0.158	0.071	0.176
Alternative 7	0.118	0.125	0.146	0.105	0.250	0.118

(2) We then calculate the entropy for each criterion by means of $E_j = -\dfrac{1}{\ln(m)} \cdot \sum_i a_{ij} \cdot \ln(a_{ij})$ (with $m = 7$):

Criteria	Safety	Health	Investment costs	Future costs	Environmental impacts	Aesthetic
E_j	0.937	0.980	0.970	0.990	0.930	0.937

(3) We assess the scattering measure D_j for each criterion:

Criteria	Safety	Health	Investment costs	Future costs	Environmental impacts	Aesthetic
D_j	0.063	0.020	0.030	0.010	0.070	0.063

(4) We finally assess the weights by means of $w_j = \dfrac{D_j}{\sum\limits_j D_j}$, with $\sum\limits_j D_j = 0.2564$

Criteria	Safety	Health	Investment costs	Future costs	Environmental impacts	Aesthetic
w_j	0.245	0.080	0.118	0.038	0.274	0.245

This methodology is totally objective. This 'neutral' aspect is interesting in conflicting context or when it becomes difficult to determine weights. Nevertheless, decision-makers can intervene by multiplying each weight by a factor taking into account their preferences. Weights then gather objectivity of scattering measure as well as decision-makers' subjective preferences.

2.6.2.4.2 SUCCESSIVE COMPARISONS METHODOLOGY

The successive steps are:

1. ranking of criteria according to the importance
2. assessment of criteria according to a cardinal scale
3. systematic comparison of each criteria to the union of the following ones (comparisons between criteria and coalition of criteria)
4. checking of the consistency between the cardinal ranking (step 2) and the comparisons (step 3): possible modification of the value in case of conflict with the relations obtained by comparisons
5. normalisation of the obtained values.

Example

1. Ranking of criteria according to importance: 1st Performance (P), 2nd Cost (C), 3rd/4th (placed equal) Energy consumption (EC) and Service life (SL), 5th Waste (W).
2. Assessment of criteria according to a cardinal scale:

Criteria	P	C	EC	SL	W
Weights	5	4	2.5*	2.5*	I

* Share of the points between the criteria for which the alternatives are placed equal (2.5 instead of 3 and 2).

3. Systematic comparison of each criteria to the union of the following ones:

$$a - P \text{ compared with } C + EC + SL + W$$
$$b - P \text{ compared with } C + EC + SL$$
$$c - P \text{ compared with } C + EC$$
$$d - C \text{ compared with } EC + SL + W$$
$$e - C \text{ compared with } EC + SL$$
$$f - EC \text{ compared with } SL + W.$$

Starting with the first set of criteria (first column above), the user has to go down till the left criterion (in the example here: P) is considered as less important than the right coalition $(C + EC + SL + W, C + EC + SL, \dots)$. The user proceeds similarly for the two other columns.

Let us assume that the user answers as follows:

(i) $P > C + EC$ but $P < C + EC + SL$
(ii) $C < EC + SL$
(iii) $EC = SL$ (already known by the ranking).

4. Checking of the coherence between the cardinal ranking (step 2) and the comparisons (step 3): possible modification of the value in case of conflict with the relations obtained by comparisons.
   Comparing with the cardinal assessment:

(iii) $EC = SL \rightarrow 2.5 = 2.5$ Correct
(ii) $C < EC + SL \rightarrow 4 < 2.5 + 2.5$ Correct
(i) $P > C + EC \rightarrow 5 > 4 + 2.5$ Incorrect.

Weights have to be modified to achieve a correct comparison.
5. Normalisation of the obtained weights.

Criteria	P	C	EC	SL	W
Weights	0.44	0.23	0.14	0.14	0.06

2.6.2.4.3 ANALYTICAL HIERARCHY PROCESS (AHP)

The AHP has attracted the interest of many researchers mainly because the required input data are rather easy to obtain. The AHP is a decision support tool, which can be used to solve complex decision-making problems. This method is standardised for practical use in ASTM standard E 1765–98 [14].

The relevant information is derived by using a set of pairwise comparisons. These comparisons are used to obtain the weights of importance of the decision criteria. A Consistency Index indicates whether the pairwise comparisons are consistent.

The principle of this methodology is the assessment of the relative importance of each criterion over the others. In Figure 2.14, the values of the pairwise comparisons are members of the set {9, 8, 7, 6, 5, 4, 3, 2, 1, 1/2, 1/3, 1/4, 1/5, 1/6, 1/7, 1/8, 1/9}. The meaning of these values is detailed in Figure 2.15. If the preference of criterion i over criterion j is valued x, then the preference of criterion j over criterion i is valued $1/x$. The user just has to fill in the lower part of the matrix (white cells). Intermediate values 2, 4, 6 and 8 and their inverse values 1/2, 1/4, 1/6 and 1/8 are used to introduce more latitude in the comparison.

'Given a judgement matrix with pairwise comparisons, the corresponding maximum left eigen vector is approximated by using the geometric mean of each row' (that is to say, the elements in each row are multiplied with each other and then the nth root is taken, with n being the number of criteria). Next, the numbers are normalised by dividing them by their sum. We obtain a vector called *vector of priorities*.

Perfect consistency rarely occurs in practice. 'In the AHP the pairwise comparisons in a judgement matrix are considered to be adequately consistent if the corresponding *Consistency Ratio* (CR) is less than 10%.'

		1 Safety	2 Health	3 Investment costs	4 Future costs	5 Environmental impacts	6 Aesthetic						
1	Safety	1	5	2	3	7	1	1	1	1	1	1	1
2	Health	1/5	1	1/4	1/3	1	1/7	1	1	1	1	1	1
3	Investment costs	1/2	4	1	3	6	1/3	1	1	1	1	1	1
4	Future costs	1/3	3	1/3	1	5	1/2	1	1	1	1	1	1
5	Environmental impacts	1/7	1	1/6	1/5	1	1/6	1	1	1	1	1	1
6	Aesthetic	1	7	3	2	6	1	1	1	1	1	1	1
								1	1	1	1	1	1
									1	1	1	1	1
										1	1	1	1
											1	1	1
												1	1
													1

Consistency ratio = 0.04108

Figure 2.14 Pairwise comparisons.

Preference of criterion i over criterion j	Definition	Preference of criterion j over criterion i
1	Equal importance	1
2		1/2
3	Weak importance of one over another	1/3
4		1/4
5	Essential or strong importance	1/5
6		1/6
7	Demonstrated importance	1/7
8		1/8
9	Absolute importance	1/9

Figure 2.15 Levels of preference.

First, the *Consistency Index* (CI) is assessed. 'This is done by adding up the columns in the judgement matrix and multiplying the resulting vector by the vector of priorities.' We thus obtain δ_{max}. With this, we calculate $CI = \dfrac{\delta_{max} - n}{n - 1}$.

'The concept of *Random Consistency Index* (RCI) was also introduced by Saaty in order to establish an upper limit on how much inconsistency may be tolerated in a decision process.' The RCI values for different *n* values are given in Figure 2.16.

Then $CR = \dfrac{CI}{RCI}$. If the CR value is greater than 0.10, then a re-evaluation of the pairwise comparisons is recommended.

n	1	2	3	4	5	6	7	8	9	10	11	12	13	14	15
RCI	0	0	0.58	0.9	1.12	1.24	1.32	1.41	1.45	1.49	1.51	1.48	1.56	1.57	1.59

Figure 2.16 Random Consistency Index (function of the number of criteria).

		1 Purchasing cost	2 Maintenance cost	3 Maximum speed	4 Comfort	5 Noise	6 Gas consumption	7 Impact on air pollution	8 Recycle-ability
1	Purchasing cost	1	5	9	1	7	3	9	9
2	Maintenance cost	1/5	1	5	3	3	1	5	9
3	Maximum speed	1/9	1/5	1	1/7	1	1/5	1	3
4	Comfort	1	1/3	7	1	3	1	7	9
5	Noise	1/7	1/3	1	1/3	1	1/3	3	5
6	Gas consumption	1/3	1	5	1	3	1	5	7
7	Impact on air pollution	1/9	1/5	1	1/7	1/3	1/5	1	3
8	Recycle-ability	1/9	1/9	1/3	1/9	1/5	1/7	1/3	1

Figure 2.17 Pairwise comparisons.

The whole methodology has been programmed in order to simplify the use of the AHP method. We suggest the user to build a pre-ranking of criteria before establishing pairwise comparisons in order to reach more easily the 10% limit of the consistency ratio.

Example
Let us assess the weights of the criteria for the selection of cars (the user just have to fill in the yellow cells of the table).

The pre-ranking of criteria by the user is the following:

- 'Purchasing cost', 'comfort', 'gas consumption' and 'maintenance cost' are the four major criteria.
- 'Noise' and 'maximum speed' are secondary criteria.
- Finally, 'impact on air pollution' and 'recycleability' are minor criteria.

The user then fills in the pairwise comparisons table (horizontally) with the rate of preference, on the basis of the levels of preference (Figure 2.17). The easiest way may be:

- to start with the second criteria 'maintenance cost' and compare it with 'purchasing cost' (first criteria), for instance 'purchasing cost' has an 'essential or strong importance' relatively to 'maintenance cost' (the value is lower than 1 because the considered criteria 'maintenance cost' is less important than the criteria 'purchasing cost'
- to do the same with the third criteria (maximum speed)
- to go on with the other criteria.

For each row of the table, we calculate the product of values to the power '1/number of criteria':

The first value corresponding to the first row is: $V(1) = (1 \times 5 \times 9 \times 1 \times 7 \times 3 \times 9 \times 9)^{1/8}$

We obtain the following vector of priorities: $V = (4.078, 2.118, 0.457, 2.141,$ $0.729, 1.907, 0.398, 0.214)$, which is normalised by dividing each term by the sum of its terms.

The sum is: $4.078 + 2.118 + 0.457 + 2.141 + 0.729 + 1.907 + 0.398 + 0.214 = 12.042$.

The normalised vector of priorities is: $V = (0.339, 0.176, 0.038, 0.178, 0.060,$ $0.158, 0.033, 0.018)$

The consistency index is then calculated.

The sum of the columns in Figure 2.17 gives: $S = (3.009, 8.178, 29.333, 6.730,$ $18.533, 6.876, 31.333, 46)$

S and V are multiplied to obtain δ_{max}: $\delta_{max} = S(1) \times V(1) + S(2) \times V(2) + \cdots + S(8) \times V(8) = 8.831$.

The Consistency Index (CI) is: $\mathrm{CI} = \dfrac{\delta_{max} - n}{n-1} = \dfrac{8.831 - 8}{8 - 1} = 0.1187$.

With RCI $= 1.41$ (corresponding to eight criteria), we obtain: $\mathrm{CR} = \dfrac{\mathrm{CI}}{\mathrm{RCI}} = \dfrac{0.1187}{1.41} = 0.084 < 0.1$.

The value CR being lower than 0.1, the pairwise comparisons are consistent and we can use the weights produced with this methodology. If not, the user just has to refine the pairwise comparison, identifying the ones that are not consistent with the obtained ranking of criteria. Successive refinements will lead to a consistent result.

2.6.2.5 Further definitions

Compensation Compensation between alternatives in the decision process means that an alternative with a very negative assessment on a criterion can be counterbalanced by other positive assessments, and thus becomes equal or better than an alternative that has medium values for all the criteria.

For instance (Figure 2.18), we compare the two alternatives A_1 and A_2 by means of five criteria (in the range $\mathrm{U} = [0, 15]$). The 'mean value' leads to 'A_1 is equal to A_2', even though A_1 has a bad assessment for C_5. The 'mean value' involves compensation.

Incomparability Incomparability between alternatives means that we are not able to choose one of them.

Alternatives	Criteria					Mean
	C_1	C_2	C_3	C_4	C_5	
	$U = [0, 15]$	$U = [0, 15]$	$U = [0, 15]$	$U = [0, 15]$	$U = [0, 15]$	
A_1	12	12	12	12	2	10
A_2	10	10	10	10	10	10

Figure 2.18 Compensation.

Alternatives	Criteria	
	C_1	C_2
	$U = [0, 15]$	$U = [0, 15]$
	$W_1 = 1$	$W_2 = 1$
A_1	10	5
A_2	5	10

Figure 2.19 Incomparability.

Note
Incomparability also appears in the decision-making process when there are uncertainties in the available information. Two alternatives could be considered as incomparable even if there is a slight difference in the assessment (uncertainty in measurements for instance).

For instance (Figure 2.19), when comparing the alternatives A_1 and A_2 by means of two criteria C_1 and C_2 (in the range [0, 15]), with the same weights W_1 and W_2, we are not able to decide which one is preferable.

Properties of a set of criteria   A set of criteria should have three properties:

• *Exhaustive* – For a set of criteria, we must not have two equal alternatives *A* and *B* for the considered set of criteria if we can say '*A* is preferred to *B*' or '*B* is preferred to *A*'.
• *Consistent* – If two alternatives *A* and *B* are equal for a set of criteria, then the increase of the value *A* for one criterion and/or the decrease of *B* for another criterion must involve '*A* is preferred to *B*'.
• *Non-redundant* – Removing one criterion leads to the loss of the exhaustivity and consistency properties.

Independence of criteria   We must be able to rank the alternatives for a given criteria, without knowing the values of these alternatives for other criteria. For instance, when studying 'raw material depletion', we usually consider [1, 14–20]:

• consumed quantities
• available resources
• renewability.

Dependence is for instance: 'We couldn't judge the impact of the consumption of 50 kg of one material in comparison with 100 kg of another one if we don't know their available resources and their renewability.'

Minimisation   For some criteria, the less could be the worst as for quality ('Low quality' = 'Bad alternative') or the best as for cost ('Low cost' = 'Good alternative'). In the second case, criteria are minimised before any calculation (transformation of the utility function so that the less means the worst).

2.6.2.6 Selection between alternatives

This section describes the most common methods for selection between alternatives. Three types of methods have been collected in the referred literature (Figure 2.20):

- weighted methods such as additive weighting and weighted product (set of methodologies making use of the relative importance of criteria thanks to weights and leading to an aggregated results, i.e. a mark)
- outclassing methods such as ELECTRE and PROMETHEE methodologies (set of methodologies making use of *outclassing* concept; these methodologies compare alternatives two by two, criterion by criterion and lead to concordance and discordance indexes)
- ordinal methods (set of methodologies for which the result only depends on the initial ordinal ranking).

Type of method	Method	Selection procedure	Type of criteria	Thresholds[1]	
Aggregation method	Additive weighting	Ranking	True	NO	
	Weighting product	Ranking	True	NO	
Outclassing method	ELECTRE IS	Choice	Pseudo	YES	P, I, V
	ELECTRE III	Ranking	Pseudo	YES	P, I, V
	ELECTRE IV	Ranking	Pseudo	YES	P, I, V
	ELECTRE TRI	Sorting	Pseudo	YES	P, I, V
	PROMETHEE I	Choice	All	YES	P, I
	PROMETHEE II	Ranking	All	YES	P, I
	EXPROM I	Choice	All	YES	P, I
	EXPROM II	Ranking	All	YES	P, I
Ordinal method	MELCHIOR	Choice	Pseudo	YES	P, V
	BORDA	Ranking	True	NO	
	COPELAND	Ranking	True	NO	
	ORESTE	Choice	Pseudo	YES	P, V

Figure 2.20 MADA methodologies.

Note
[1] These methodologies compare alternatives two by two, criterion by criterion and lead to concordance and discordance indexes.

In the discussion following let us consider:

- m alternatives: $a_1, \ldots, a_i, \ldots, a_m$
- n criteria: $C_1, \ldots, C_j, \ldots, C_n$ and
- the weights of the n criteria: $w_1, \ldots, w_i, \ldots, w_n$.

2.6.2.6.1   ADDITIVE WEIGHTING METHOD

Comparison method: Ranking
Criteria: True criteria
Thresholds: No thresholds
Principle: A simple well-known method based on aggregation (i.e. result is a mark) but too easily influenced by arbitrary choices (normalisation)
Description: For the normalisation step (data preparation), four normalisation procedures are available (Figure 2.21). Also:

- normalisation of weights (division by the sum)
- assessment of ranking value $R(a_i) = \sum_j w_j \cdot a_{ij}$ for each alternative

- ranking of alternatives (the best alternative is one having the highest $R(a_i)$).

Note:

1. Criteria must be independent.
2. Method is subject to compensation (an alternative with a very negative assessment on a criteria can be counterbalanced by other positive assessments).

2.6.2.6.2   WEIGHTED PRODUCT METHOD

Comparison method: Ranking
Criteria: True criteria
Thresholds: No thresholds
Principle: A similar methodology to additive weighting method, but we multiply instead of adding up the values. It is used to avoid the influence of the normalisation method on the final results (additive weighting).

Normalisation procedures	Procedure 1	Procedure 2	Procedure 3	Procedure 4
Definition	$V_i = \frac{a_i}{\max(a_i)} \bullet 100$	$V_i = \frac{a_i - \min(a_i)}{\max(a_i) - \min(a_i)} \bullet 100$	$V_i = \frac{a_i}{\sum a_i} \bullet 100$	$V_i = \frac{a_i}{\sqrt{\sum a_i^2}}$
Interpretation	% of the maximum of a_i	% of the range $(\max(a_i) - \min a_i)$	% of the total $\sum_i a_i$	Component 1 of the unit vector

Figure 2.21 Normalisation procedures.

Description:

- normalisation of weights (division by the sum)
- assessment of $P(a_i) = \prod_j a_{ij}^{wj}$ for each alternative

- ranking of alternatives (the best alternative is one with the highest $P(a_i)$).

Note:

1. The main drawback of this methodology is the fact that it gives advantage/disadvantage to the utility that is far from the mean.
2. Normalisation of criteria not needed.

2.6.2.6.3   ANALYTICAL HIERARCHY PROCESS (ASTM STANDARD: E 1765–98)

This standard deals with 'Standard Practice for Applying Analytical Hierarchy Process (AHP) to Multiattribute Decision Analysis of Investments Related to Buildings and Building Systems'.

The AHP is one of a set of MADA methods that considers non-monetary attributes (qualitative and quantitative) in addition to common economic evaluation measures (such as life cycle costing or net benefits) when evaluating project alternatives.

The principles are mainly similar or the same as presented above in connection to the French methods. Because the method is presented in details in the ASTM standard E 1765–98, only a short general presentation of the method is described in this book. Each user can directly apply the standard in all calculations using the Lifecon classification of attributes and criteria. The procedure of this method is as follows (this mathematical procedure is presented in detail in the ASTM Standard E 1765–98):

1. Identify the elements of your problem to confirm that a MADA analysis is appropriate. Three elements are common to MADA problems:

    (i) MADA problems involve analysis of a finite and generally small set of discrete and predetermined options or alternatives.
    (ii) In MADA problems no single alternative is dominant, that is, no alternative exhibits the most preferred value or performance in all attributes.
    (iii) The attributes in a MADA problem are not all measurable in the same unit.

2. Identify the goal of the analysis, the attributes to be considered, and the alternatives to evaluate. Display the goal and attributes in a *hierarchy*.

    - A set of attributes refers to a complete group of attributes in the *hierarchy*, which is located under another attribute or under the goal of the problem.
    - A *leaf attribute* is an attribute that has no attribute below in the hierarchy.

3. Construct a decision matrix with data on the performance of each alternative for *each leaf attribute*.

4. Compare *in pairwise fashion* each alternative against every other alternative as to how much better one is than the other with respect to each leaf attribute. Repeat this process for each leaf attribute in the hierarchy.
5. Make *pairwise comparison* of the relative importance of each attribute in a given set, starting with sets at the bottom of the hierarchy, with respect to the attribute or goal immediately above that set.
6. Compute the final, overall desirability score for each alternative.

The ASTM Standard: E 1765–98 also includes examples, which help in understanding and applying the method for different types of buildings. This standard refers also to several other ASTM standards that support the applications on different fields, for example:

- E 1670 Classification of the serviceability of an office facility for management of operations and maintenance
- E 1701 Classification of serviceability of an office facility for manageability
- E 917 Practice for measuring life cycle costs of buildings and building systems
- E 1480 Terminology of facility management (building related)
- ASTM Adjunct: Computer program and user's guide to building maintenance, repair and replacement database for life cycle cost analysis; Adjunct to practices E 917, E 964, E 1057, E 1074 and E 1121
- ASTM Software Product: AHP/Expert Choice for ASTM building evaluation; Software to support practice E 1765.

2.6.2.6.4 ELECTRE METHODS (*ELIMINATION ET CHOIX TRADUISANT LA REALITÉ*)

The principle of this type of methodology is *outclassing*. Alternatives are compared two by two, criterion by criterion.
Let us note:

$P^+(a_i, a_k) = \sum P_j$, $j \in J^+(a_i, a_k)$, i.e. the sum of the weights of criteria for which a_i is better than a_k.
$P^=(a_i, a_k) = \sum P_j$, $j \in J^=(a_i, a_k)$, i.e. the sum of the weights of criteria for which a_i is equal to a_k.
$P^-(a_i, a_k) = \sum P_j$, $j \in J^-(a_i, a_k)$, i.e. the sum of the weights of criteria for which a_i is worse than a_k.

Concordance expresses how much the criteria support the hypothesis between alternatives *a* and *b*: '*a* outclasses *b*'.

$$\text{Concordance index is } C_{ik} = \frac{P^+(a_i, a_k) + P^=(a_i, a_k)}{P}, \text{ with } P = P^+ + P^= + P^-.$$

Discordance (opposite of concordance) measures the opposition to the hypothesis: '*a* outclasses *b*' expressed by discordant criteria.

$$\text{Discordance index is } D_{ik} = \begin{cases} 0 \text{ if } J^-(a_i, a_k) = \varnothing \ (\varnothing \text{ being the empty set}) \\ \frac{1}{\delta_i} \max\{g_j(a_k) - g_j(a_i)\}, \quad j \in J^-(a_i, a_k) \end{cases}$$

$g_i(a_i)$ is the value of alternative a_i for criteria j and δ_j is the amplitude of the criteria j scale, criteria for which we have the maximum of discordance (i.e. alternative b is 'better' than a).

Outclassing results from these two definitions: 'a outclasses b' means that

- concordance test is satisfied (a is at least as good as b for most of the criteria, concordance index is greater than a defined threshold c) and
- the remaining criteria do not involve a too strong opposition to this proposition 'a outclasses b' (discordance index is lower than a defined threshold d).

The main differences between the various ELECTRE methods are the different use of concordance index and the different types of criteria (but we present only methods using pseudo-criteria).

2.6.2.6.5 CHOICE METHOD: ELECTRE IS (S STANDS FOR *SEUIL* IN FRENCH, MEANING 'THRESHOLD')

Comparison method: Choice
Criteria: Pseudo-criteria
Thresholds: Strict preference, indifference, veto (one for each criterion); global concordance
Description:

- construction of concordance matrix for each criterion ($c_j(a_i, a_k)$)
- gathering of results in a global concordance matrix (C_{ik})
- construction of discordance matrix for each criterion ($d_j(a_i, a_k)$)
- gathering of results in a global discordance matrix (D_{ik})
- from concordance and discordance matrices, construction of outclassing matrix ($S(a_i, a_k)$)
- results expressed with outclassing graph and search of the core.

Note:

1. The method is easier than ELECTRE III
2. The method is suited also for management of incomparability and indifference.

2.6.2.6.6 RANKING METHOD: ELECTRE III

Comparison method: Ranking
Criteria: Pseudo-criteria
Thresholds: Strict preference, indifference, veto (one for each criterion); discrimination (refer to list of terms, definitions and symbols).
Description:

- construction of concordance matrix for each criterion ($c_j(a_i, a_k)$)
- results gathering in a global concordance matrix (C_{ik})
- construction of discordance matrix for each criterion ($d_j(a_i, a_k)$)

- gathering of results in a global discordance matrix (D_{ik})
- from global concordance matrix and discordance matrices, construction of belief matrix (δ_{ik}),
- ranking algorithm (downward and upward distillations, i.e. calculations ranking first from the worst alternative, second from the best one)
- ranking of alternatives according to their ranks in each distillation.

Note:

1. use of 'fuzzy outclassing' concept
2. management of incomparability and indifference
3. complex methodology taking into account minor differences in the assessments.

2.6.2.6.7 RANKING METHOD: ELECTRE IV

Comparison method: Ranking
Criteria: Pseudo-criteria
Thresholds: Strict preference, indifference, veto (one for each criterion); discrimination
Description:

- comparison of each couple of alternatives towards each criterion
- for each couple of alternatives (a, b), search of $a\ S_q\ b, a\ S_c\ b, a\ S_p\ b$ or $a\ S_v\ b$ relations
- affectation of a belief value to each outclassing relation
- construction of a matrix of belief degrees
- ranking algorithm (downward and upward distillations)
- ranking of alternatives according to their ranks in each distillation.

Note:

1. method without weights
2. use of 'fuzzy outclassing' concept
3. management of incomparability and indifference
4. complex methodology taking into account minor differences in the assessments.

2.6.2.6.8 SORTING METHOD: ELECTRE TRI

Comparison method: Sorting
Criteria: Pseudo-criteria
Thresholds: Strict preference, indifference, veto (one for each criterion); cut threshold λ
Description:

- definition of 'reference alternatives', either without any consideration of potential alternatives (e.g. use of standards) or in order to sort alternatives by groups
- assessment of concordance matrix by criterion using comparison of each alternative to a reference alternative
- assessment of global concordance index

- assessment of discordance matrix by criterion
- construction of a belief degree matrix
- implementation of outclassing relations (from belief degrees and cut threshold λ)
- allocation of each alternative to the various categories.

Note:

1. judgement of each alternative independently of other alternatives (less sensitive than γ methodologies concerning alternatives with similar assessments)
2. definition of one or several reference values (standards, etc.) for alternatives acceptation.

2.6.2.6.9 PROMETHEE

In PROMETHEE (Preference Ranking Organisation METHod for Enrichment Evaluation), alternatives are compared two by two, criterion by criterion. PROMETHEE methods are based on preference information. EXPROM methods are extensions of PROMETHEE methods and allow the distinction between strict and weak preferences.

PROMETHEE I

Comparison method: Choice
Criteria: All types
Thresholds: Strict preference, indifference (no thresholds for true criteria)
Description:

- construction of a preference matrix for each criterion $(S_j(a_i, a_k))$
- normalisation of weights
- results gathering in a global preference matrix $(C_{ik} = \sum_j w_j \cdot S_j(a_i, a_k))$
- assessment of input and output flows (respectively $\phi_{i+} = \sum_k C_{ik}$ and $\phi_{i-} = \sum_k C_{ki}$)
- identification of the outclassing relations between alternatives, on the basis of the following rule: the alternative a_i outclasses a_k if:
$$\begin{cases} \phi_{i+} > \phi_{k+} \text{ and } \phi_{i-} < \phi_{k-} \\ \text{or } \phi_{i+} > \phi_{k+} \text{ and } \phi_{i-} = \phi_{k-} \\ \text{or } \phi_{i+} = \phi_{k+} \text{ and } \phi_{i-} < \phi_{k-} \end{cases}$$

Note:

1. methodology less sensitive to the variations of the values of pseudo-criteria thresholds
2. management of indifference.

PROMETHEE II

Comparison method: Ranking
Criteria: All types
Thresholds: Strict preference, indifference (no thresholds for true criteria)

Description:

- construction of a preference matrix for each criterion $(S_j(a_i, a_k))$
- normalisation of weights
- results gathering in a global preference matrix $(C_{ik} = \sum_j w_j \cdot S_j(a_i, a_k))$
- assessment of input and output flows
- from input and output flows (similar to PROMETHEE I), assessment of net flows $(\phi_i = \phi_{i+} - \phi_{i-})$ and
- identification of the outclassing relations between alternatives and ranking (the higher the net flow is, the best is the alternative).

Note:

1. methodology less sensitive to the variations of the values of pseudo-criteria thresholds
2. management of indifference.

EXPROM I (Extension of PROMETHEE)

Comparison method: Choice
Criteria: All types
Thresholds: Strict preference, indifference (no thresholds for true criteria)
Description:

- normalisation of weights
- construction of a preference matrix for each criterion
- results gathering in a global weak preference matrix
- construction of a strict preference matrix for each criterion
- results gathering in a global strict preference matrix
- construction of a global preference matrix from weak and strict preference matrices
- assessment of input and output flows
- identification of the outclassing relations between alternatives.

Note:

1. management of indifference.

EXPROM II

Comparison method: Ranking
Criteria: All types
Thresholds: Strict preference, indifference (no thresholds for true criteria)
Description:

- normalisation of weights
- construction of a preference matrix for each criterion
- results gathering in a global weak preference matrix
- construction of a strict preference matrix for each criterion
- results gathering in a global strict preference matrix

- construction of a global preference matrix from weak and strict preference matrices
- assessment of input and output flows
- from input and output flows, assessment of net flows
- identification of the outclassing relations between alternatives and ranking.

Note:

1. management of indifference.

MELCHIOR

Comparison method: Choice
Criteria: Pseudo-criteria
Thresholds: Strict preference, veto
Principle: The principle of MELCHIOR (*Methode d'ELimination et de CHoix Incluant les relations d'ORdre*) methodology is outclassing. Alternatives are compared two by two, criterion by criterion.
Description:

- for each couple of alternatives (a, b), exclusion of couples for which veto thresholds are exceeded
- for each couple of alternatives (a, b), search of criteria supporting and not supporting the statement '*a* outclass *b*'
- identification of the outclassing relations between alternatives (masking concept: belief in supporting or non-supporting criteria).

Note:

1. management of indifference and incomparability
2. time-consuming method (limitation of the number of alternatives).

BORDA

Comparison method: Ranking
Criteria: True criteria
Thresholds: No thresholds
Principle: BORDA is an ordinal methodology (methodology only based on the initial ordinal ranking).
Description:

- for each criteria, attribution of n points to the best alternative towards the considered criterion, $m(m < n)$ points to the second, etc. (these points are called *Borda coefficients*)
- then, sum of the points obtained by each alternative for all the criteria and ranking of alternatives (the best is the one with the maximum of points).

Note:

1. compensation between alternatives
2. results depending on the insertion/suppression of alternatives.

COPELAND

Comparison method: Ranking
Criteria: True criteria
Thresholds: No thresholds
Principle: COPELAND is an ordinal methodology (methodology only based on the initial ordinal ranking).
Description:

- for each couple of alternatives (a, b), identification of preference relation (based on the number of criteria favourable to alternatives)
- then, sum of the coefficient obtained by each alternative and ranking of alternatives (the best is the one with the maximum of points).

Note:

1. compensation between alternatives.

ORESTE

Comparison method: Choice
Criteria: Pseudo-criteria
Thresholds: Strict preference, veto
Principle: ORESTE (*Organisation, Rangement Et Synthèse de données relaTionElles*) is an ordinal methodology i.e. methodology only based on the initial ordinal ranking.
Description:

- arrangement in the order of alternatives according to the decision-maker's ranking
- determinination of outclassing relations between alternatives.

Note:

1. no use of weights on criteria
2. management of indifference.

2.6.2.6.10 SUMMARY

The following graph (Figure 2.22) summarises the characteristics of each methodology, in order to help the choice for the best methodology for our objectives. For this purpose, various decision parameters have been selected:

- type of criteria
- weights determination
- problematic or not
- management of indifference and incomparability and
- risk of compensation existence.

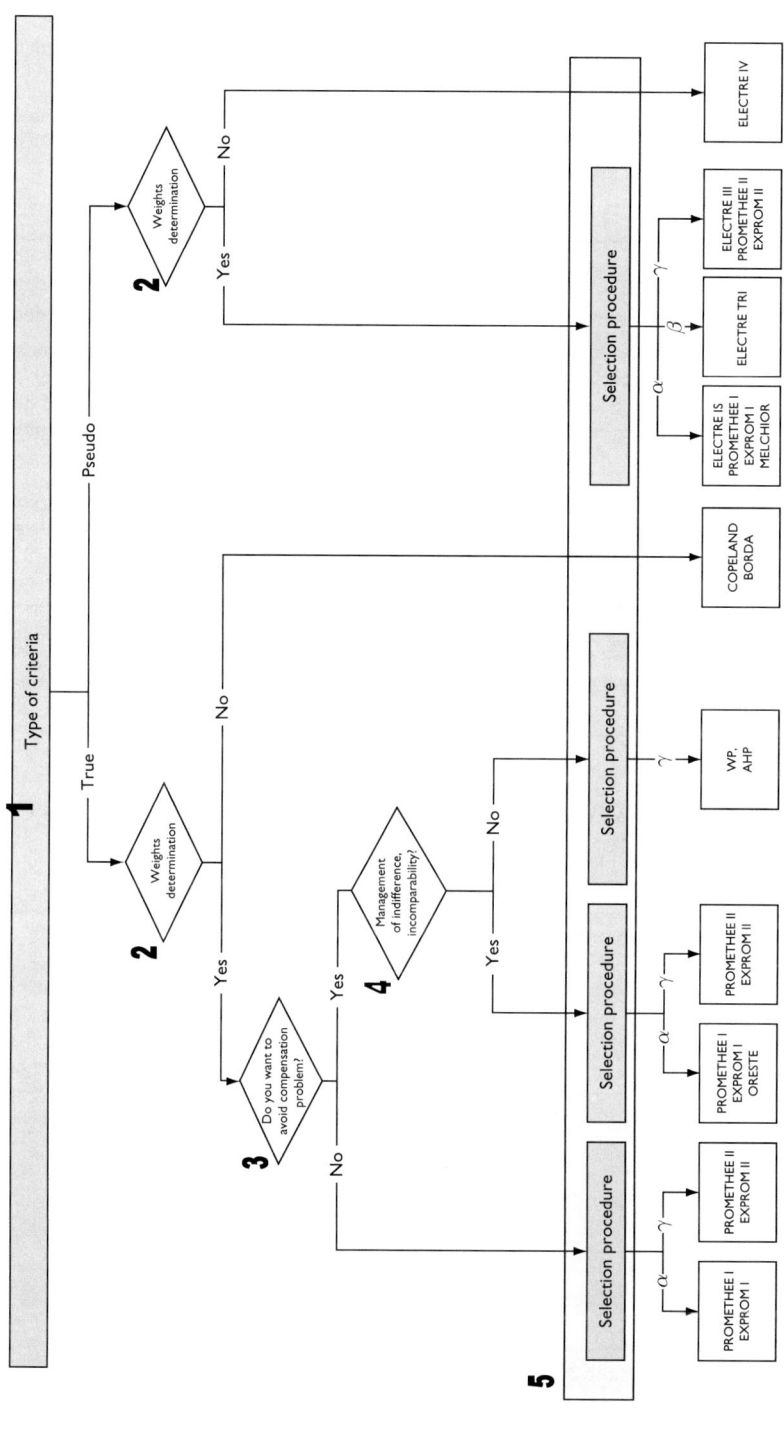

Figure 2.22 Choice of MADA methodology.

Other characteristics can be added to refine the choice:

- subjectivity of the thresholds values
- methodology being time-consuming

1. The first decision concerns the type of criteria. The question is: 'Do you want to use true criteria or pseudo-criteria?'

True criteria	Pseudo-criteria
You will only have a 'white or black decision'.	You will be able to include 'grey decisions'.
Result: for the considered criteria, preference of one alternative except if alternatives values are equal: $A \gg B; A = B$ or $A \ll B$	*Result*: For the considered criteria, strict or weak preferences, indifference: $A \gg B; A > B, A = B, A < B$ or $A \ll B$

2. The second decision concerns the weights. The question is: 'Do you want to assess the weights?'

YES (*Assessment of weights*)	NO (*No assessment*)
The user assess the weight according to his preferences.	The weights are not required (based on ordinal relations)

3. The third question concerns the alternatives. The question is: 'Do you want to avoid compensation?'

YES (*Avoid compensation problems*)	NO (*Let compensation be possible*)
The method will limit the effect of compensation	An alternative with a very negative assessment on a criterion is counterbalanced by other positive assessments, and can become equal or better than an alternative which has medium values for all the criteria.

4. The fourth question concerns indifference and incomparability. The question is: 'Do you want to manage indifference and incomparability?'

YES (*Manage indifference and incomparability*)	NO (*Do not manage indifference and incomparability*)
The method will limit the effect of compensation	An alternative with a very negative assessment on a criterion is counterbalanced by other positive assessments, and can become equal or better than an alternative which has medium values for all the criteria.

5. The fifth question concerns the problematic. The question is: 'What is your problematic?'

Choice	Sorting	Ranking	Description
We clarify the decision by choosing a subset of alternatives (as small as possible) in order to choose the final alternative. This subset contains optimum and sufficient satisfying alternatives.	We assign the alternatives by categories. Each alternative is assigned to a single category (independent of other categories).	We rank all or some of the alternatives (the most satisfying ones), by assigning a rank of ordering which allows a total or partial ranking.	The problem is correctly stated by describing alternatives and their consequences.

Among all these methodologies, three have been chosen. This choice according to the level of requirements fits Lifecon objectives: several methods are suggested, the most useful for the user will be chosen. When applying the methodology, the decision-maker will then have several solutions:

1. ADDITIVE WEIGHTING, a very simple method that does not avoid compensation and is not able to take into account indifference
2. COPELAND, a very simple method that does not require weights definition and
3. ELECTRE III, a more complex but more powerful method than the previous two.

We will thus have a software with various levels of complexity, a software that could be further developed and adapted according to users' needs. Obviously, the more degrees of freedom we leave to users, the more they need information and know-how.
  We will assume the following whenever possible (availability of information):

• We process pseudo-criteria in order to keep a gradation in preference, to manage indifference and incomparability (in some way to take into account the uncertainties on alternatives assessment; we can not decide with certainty if alternative 1 (service life is 70 years) is better than alternative 2 (service life is 80 years) because of the uncertainty on service life assessment).
• We do not accept compensation.
• We prefer outclassing methodologies to weighted methodologies.

2.6.3 Quality Function Deployment (QFD) method

2.6.3.1 Principles of QFD method

Quality Function Deployment (QFD) method is a tool for optimisation and decision-making, which has a strong numerical character thus serving especially the following functions:

• analysis and weighting of the requirements
• optimisation of solutions with a choice between different modifications of the solution and
• choice between alternatives of plans, designs, methods or products.

QFD method is related to methods of linear programming which have been developed in the 1950s and were widely used in the 1960s in the product development of industry. There was a need to modify the basic linear programming methodology for specific needs of each application field. Results of these applications are all the methods mentioned above: QFD, MADA and RAMS.

In the current formulation, QFD method was developed in Japan and it was first used in 1972 by Kobe Shipyard of Mitsubishi heavy industries. QFD has been increasingly used in Japan and since the 1980s also in USA, Europe and worldwide. Until now the use has mainly been in mechanical and electronics industry, but applications exist also in construction sector. In industrial engineering, manufacturing companies have successfully applied QFD to determine customers' needs for the features of the product into design at its early stages of development, to integrate design of products and their related processes, and to consider all elements of the product life cycle. Customer-oriented 'champion products' may also be priced higher than their competitors, and still become as market leaders. QFD has been little applied in construction. Examples have been reported for example from Japan, United States, Finland, Sweden and Chile which show its potential also in building design, construction planning and asset management.

2.6.3.2   General use of QFD method

QFD can be applied both on strategic level and on operational level of construction and asset management organisations. The strategic development of the owner, user, construction or management organisation may have the following focuses:

- strategic planning of the organisation and
- product development: product can be an entity (house, office, road, bridge, tunnel, etc.) or a more detailed product (module, component or material).

In practical construction or repair process QFD has to be applied in four stages:

1. analysis of the requirements of the client and their weights of importance
2. choice of the properties of the product (e.g. a house, office, bridge, railway or tunnel) based on the requirements and their weights, which have been resulted from the first stage
3. analysis of the requirements of the product for the production process
4. analysis of the requirements of the production process for the product.

This means that interactions between all phases of the planning, design and production are analysed and optimised with the QFD method. As results of the first stage are the weights of *requirements*. The result of the second stage is a list of the *properties* of the *product*, and the weights of these *properties*. The third stage results in a list of *requirements* of the product and their *weights* for a fluent production process. This stage consists of analysis of correlations between focused phases of the production process and the properties that the product requires from these phases. The fourth stage includes analysis on what the production process and its phases require from the product. This leads into an iterative optimisation between the properties of the product and the properties of the production process. The requirements from the first stage serve as constraints in this optimisation but they may have to be slightly modified if the iteration does not converge otherwise.

An example of this procedure is presented in the repair planning examples of Life-con deliverable D5.1, where QFD is used in combination with reliability, availability, maintainability, safety (RAMS) methodology. The components in RAMS include the requirements of the properties of the product and manufacturing process for the alternative repair technologies or materials. This is a mix of points 3 and 4 above (Deliverable D5.1).

QFD can be used in planning and design on quite different ways, for example:

- for interpreting any requirements into specifications, which can be either performance properties or technical specifications
- as an optimising or selective linking tool between requirements, performance properties and technical specifications
- at product development, at design of individual civil infrastructures or buildings, and at maintenance and repair planning.

Simply, the QFD method means building a matrix between requirements and performance properties or technical specifications. Usually the performance properties serve only as a link between requirements and technical specifications, the reason why the performance properties sometimes are not treated with QFD method. Additionally weighting factors of requirements and technical specifications as well as correlations between requirements and technical specifications are identified and determined numerically. As a computer tool, Excel is very suited for this calculation, as it has been used in examples (Deliverable D5.1).

2.6.3.3 Generic description of QFD method

In practical planning and design, the application is limited to a few key requirements and specifications in order to maintain good control of variables and not to spend too much effort on secondary factors. At product development some more detailed application can be used. A model 'house of quality' is presented in Figure 2.23.

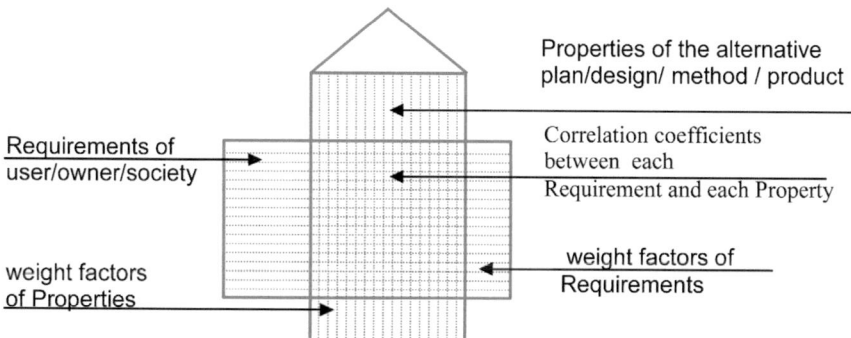

Figure 2.23 House of quality [1–5].

QFD provides an empty matrix called 'house of quality' (Figure 2.23). This matrix will be filled with requirements and their weighting factors in the rows along the left hand side and performance properties of the actual alternative in the columns along the top portion. The centre describes the matrix relationship of requirements and corresponding solutions. The importance measures (weight factors) are at the bottom, and the right hand side of the box shows the evaluation of competing alternatives.

The correlations and weights cannot usually be estimated with exact calculations, but they must be estimated with expertise knowledge, client questionnaires, long-term experiences and expectations on the future trends. The weights can be expressed in different scales, for example on the range of 0 (no importance) to 10 (extremely important). As final results of the matrix calculations, the weight factors of requirements and properties are normalised, as shown in the examples in the appendix below.

Example
[Lifecon Deliverable D5.1]

2.6.3.4 Alternative applications of QFD in LMS

QFD method basically means only handling the requirements and properties, analysing their interrelations and correlations as well as their weights, and, finally, optimising the Life Cycle Quality (LCQ) properties and selecting between alternative solutions of asset management strategies or MR&R plans, designs, methods and products. Thus, QFD can be applied in many variables, depending on the characteristic aims and contents of each application.

In Lifecon LMS, QFD can be used for following purposes:

- identifying functional requirements of owner, user and society
- interpreting and aggregating functional requirements into primary performance properties
- interpreting the performance properties into technical specifications of the actual object
- optimising the performance properties and technical specifications in comparison to requirements
- selecting between different design and repair alternatives and
- selecting between different products

QFD can be used on all levels of an LMS system:

- network level: prioritising the requirements of users, owners and society, strategic optimisation and decision-making between alternative MR&R strategies
- object level: ranking of priorities between objects, optimising and decision-making between MR&R alternatives, technologies, methods and products
- module, component and detail/materials levels: refined optimising and decision-making between MR&R alternatives, technologies, methods and products.

2.6.3.5 Phases of the QFD procedure

QFD procedure usually has three main phases, as presented in the application examples in the appendix below and in Deliverable D5.1:

- selecting the primary requirements and their weight factors from a set of numerous detailed requirements with the aid of QFD matrix
- moving the primary requirements and weight factors into second QFD matrix for selection between the alternatives of plans, designs, methods or products
- sensitivity analysis with simulation of variances of primary requirements and properties (Deliverable D5.1).

The following detailed procedure can be applied in Lifecon LMS when using QFD for analysis of functional requirements against owners' and users' needs, technical specifications against functional requirements, and design alternatives or products against technical specifications:

- Identify and list factors for requirements and properties.
- Aggregate and select the requirements into primary requirements.
- Evaluate and list priorities or weighting factors of primary requirements.
- Evaluate the correlation between requirements and properties.
- Normalise the factor 'correlation times weight' of each property for use as a priority factor or weighting factor of each property at the next steps.

2.6.3.6 Appendix to Section 2.6.3

Examples on the QFD procedure

QFD procedure is strictly guided with the standard formats. Therefore the steps of this procedure are easy to learn. A difficulty is to know how the requested correlations are really evaluated in each specific case. This cannot be given as a procedure, but it is a matter of expertise.

There are two kinds of problems related to the relations and correlation between requirements and properties: the so-called laboratory problems or real world problems [Lifecon Deliverable D2.1]. The laboratory problems can be modelled numerically, usually on the basis of natural sciences or simulations, etc. The real world problems and variables cannot be presented numerically as models, but they must be evaluated with qualitative descriptions first, and then presented subjectively numerically, if possible. Ranking and numerical classifications can be used as a help. These kinds of procedures are presented in the following examples.

1. *Design objectives for a housing development project* [4]

QFD was experimented in an afternoon brainstorming session to set design guidelines for a prototype building to be constructed for Tuusula Housing Fair 2000. The 'house of quality' matrices were formed to judge how well the original design criteria met customer requirements and to judge how well the technical solutions met customers' requirements.

The exercise was conducted together with 10 experts of different backgrounds. The following objectives were set for the working session:

- To share common understanding of the performance-based objectives of the end product (a building to be designed and constructed).
- To prioritise the project objectives.
- To strive for innovative design solutions that meet these objectives.

The first matrix (Figure 2.24) shows the selected main objectives of a housing project (adaptability, indoor conditions, economy, environmental friendliness, constructability and architecture) taken as a basis for building design. The second matrix (Figure 2.25) shows the structured approach in the design process based on the selection made in phase 1.

2. Energy-efficient design concepts for office refurbishment [4]

The second case study was done in an IEA (International Energy Agency) task 23 workshop together with Danish, Dutch, Japanese, Norwegian and Finnish experts. The group consisted of practitioners and researchers, architects and engineers. The session was structured and main decisions were documented using QFD (Figure 2.26). The selected design concepts (daylighting system, new windows, new construction, energy management system, double facades and solar walls) were taken as a basis for building design.

	Requirements	adaptability	resale value	indoor conditions	attractiveness	economy	autonomy	friendliness to the environment	futurity	habitability	respond to the environment	good indoor climate	constructability	identity	total ecology	architecture	simple user interfaces	recyclable fair house	transferability	dismountability		Importance factor
functionality	Utilisability	9	9	9	9	3	9	3	0	9	0	9	0	1	1	0	9	3	1	0		5
	Adaptability	9	3	0	9	3	1	9	3	9	0	0	1	1	9	0	1	9	9	9		2
	Maintainability	3	3	3	3	9	9	9	0	9	0	3	0	0	9	1	3	1	1	1		2
environmental loading	Operation	9	3	9	3	9	9	1	1	9	9	0	0	9	0	0	9	0	0	0		4
	Construction	0	0	0	3	3	0	9	0	0	0	0	9	1	9	1	0	9	9	9		2
resource use	Energy	9	3	9	3	9	9	9	9	0	9	9	0	3	9	0	0	1	1	1		5
	Water	9	1	0	1	3	9	9	3	1	0	0	0	0	3	0	3	1	0	0		1
	Materials	3	9	3	9	1	9	9	9	9	0	9	9	9	3	0	9	9	9	9		1
life cycle cost	Investment cost	9	9	3	3	9	3	0	0	0	3	3	9	1	0	0	1	3	3	3		3
	Operating cost	9	9	1	3	9	9	9	3	0	3	1	0	3	3	9	9	3	3	3		4
	Maintenance cost	9	9	3	9	9	9	9	9	0	9	3	0	3	3	9	3	3	3	3		2
indoor quality	Acoustic comfort	9	9	9	9	0	0	0	9	9	0	0	3	3	0	9	0	0	0	0		2
	Thermal comfort	9	9	9	9	0	0	3	9	9	9	9	3	3	0	9	3	0	0	0		3
	Lighting	9	9	9	9	3	9	3	9	9	0	3	9	1	9	1	0	0	0	0		4
	Indoor climate	3	9	9	9	0	0	3	9	9	9	9	9	9	1	0	0	0	0	0		5
architecture	Architecture	9	9	9	9	9	3	0	9	9	3	0	9	9	0	9	1	3	3	3		3
	Weight factor	393	355	322	307	285	273	258	250	248	246	241	182	180	179	169	118	112	102	97	0	4317
	Weight factor %	9 %	8 %	7 %	7 %	7 %	6 %	6 %	6 %	6 %	6 %	6 %	4 %	4 %	4 %	4 %	3 %	3 %	2 %	2 %	0 %	100 %
	Votes	4	1	3		2	1	3			1		2			4	4	1	1			
	Selected	x		x		x		x					x			x						

Figure 2.24 Design objectives for a housing project, phase 1.

PHASE 2 Properties / Requirements	SPACE	PROCESS	STRUCTURES	MATERIALS	ENERGY	EQUIPMENT	Importance factor (P1)
adaptability, simple interfaces, re-usable fair house	9	9	9	3	3	1	3
indoor conditions, responds to the environment	9	9	9	9	9	9	4
economy, resale value	9	9	9	9	9	9	1
environmental,autonomy, total ecology	9	3	9	9	9	9	5
constructability	1	9	3	1	1	1	3
architecture	9	9	3	9	1	0	2
Weight factor (P1)	138	134	133	120	104	95	724
Weight factor %	19 %	19 %	18 %	17 %	14 %	13 %	100 %

Figure 2.25 Design objectives for a housing project, phase 2.

	Requirements	Daylighting system	New windows	Big atrium	New construction	Nat ventil & heat rec.	Extra insulation	Energy mgmt system	Demolition	Underground space	New lighting system	Window renovation	Decentr. wat. heating system	Double facades	Shading	Opening facades	Roof extension	Solar walls PV	New office concept	Archive basement	Importance/Weight factor (P1)
Functionality	Flexibility	3	0	3	9	1	0	0	1	0	0	0	0	0	0	3	0	9	3		3
Functionality	Public spaces: access	0	0	3	9	0	0	0	3	3	0	0	0	0	0	0	0	0	0	0	4
Functionality	Public spaces: character	3	0	9	3	0	0	0	3	3	3	0	0	0	3	0	0	0	0		3
Longevity	Comfort	9	9	9	3	3	9	9	3	0	9	9	9	3	3	3	1	0	1	0	5
Longevity	200 years for the bldg	9	3	1	0	3	9	1	0	3	3	3	1	3	3	0	0	0	0	0	3
Longevity	20 years for the first user	9	9	3	0	9	9	9	0	3	3	9	9	3	3	1	0	3	3	0	5
Energy efficiency	–60% energy use	3	9	3	0	9	9	9	0	1	3	3	3	9	3	1	0	3	3	0	4
Energy efficiency	Daylight	9	0	9	3	0	0	0	9	0	9	0	0	0	0	9	0	0	0	0	4
Energy efficiency	Natural ventilation	0	9	3	9	9	0	3	9	0	0	3	0	3	3	3	0	0	0	0	3
Others	Architecture?	9	3	9	9	0	0	0	9	9	1	1	0	3	3	3	9	3	3	0	4
Others	Facades	0	3	0	9	0	0	0	0	1	0	3	0	9	3	0	0	9	1	0	2
Others	800 m² extra for public	0	0	3	9	0	0	0	1	9	0	0	0	0	0	0	0	9	0	0	5
Others	Environmentally friendly	9	9	3	0	9	9	9	0	9	9	3	9	3	9	1	0	9	3	3	4
	Weight factor (P1)	255	243	240	234	198	189	183	170	168	166	151	141	135	123	103	95	93	85	66	3038
	Weight factor %	8 %	8 %	8 %	8 %	7 %	6 %	6 %	6 %	6 %	5 %	5 %	5 %	4 %	4 %	3 %	3 %	3 %	3 %	2 %	100 % (0 %)
	Selected	x	x		x		x						x						x		

Figure 2.26 Design concepts for office refurbishment.

3. *Design priorities in an environmental friendly nursery school* [4]

The third QFD example was to set the project objectives with a view to the building's users' needs and requirements, and to show how the chosen criteria and the users' view affect the results. A QFD matrix was used to capture, record and verify the clients' requirements and to test the dependency between the requirements and the properties of the introduced building concept.

The project used in the test is construction of a nursery school for about 100 children, to be built in 2000. The design process of the building was to be finished towards the end of 1999, based on an architectural competition. The nursery school Merituuli will be built in a new suburban housing area, a former industrial area, where the basic infrastructure has already been developed (streets, access to main roads, district heating net, etc.). The location of the area is very close to the city of Helsinki with a good public access to the city, a fact that has made the area very popular especially among young families. This has also grown to be a design feature for the nursery school building and its connection to the surrounding housing area.

The building will serve as a nursery school daytime, and in the evening as a meeting point for local inhabitant activities. The total building area is $1\,260\,\text{m}^2$, one storey. The owner of the building is the City of Helsinki, and the building is constructed by the Construction Management Division of the City of Helsinki (HKR).

In a number of development sessions, arranged between HKR and VTT (Technical Research Centre of Finland) at the beginning of the project and later on between the designers, project management and VTT, the project goals and limits were discussed and the requirements were set. The design briefing tool ECOProP was used as a guideline for the sessions and to document the results and decisions reached.

The decision-making in the project was tested against the main criteria adopted from the IEA task 23 framework. The results of the design briefing sessions were used while building owner defined sub-requirements in compiling the QFD matrix (Figure 2.27).

	Requirements	district heat	bicycle access to site	cleanable ducts	multi-use playrooms	low energy envelope	mechanical ventilation + HR	changeable duct components	separated service space	super windows	floor heating	solar control	yard facing South	stimulating spaces, child scale	L-form	separated public evening use	ordinary windows	traditional envelope	radiators			Importance/Weight factor (P1)
LCC	low investment cost	9	1	0	9	0	0	0	0	0	0	0	0	3	0	1	3	0				5
	low service cost	9	1	9	3	9	0	0	3	9	3	3	0	3	0	0	0	0				4
	low maintenance cost	9	1	9	3	0	0	9	9	0	1	0	0	1	0	0	0	3				1
resource use	low electricity consumption	9	0	3	3	1	0	0	0	0	0	3	9	0	0	1	0	0				4
	low water consumption	0	0	0	0	0	0	0	0	0	0	0	0	0	0	0	0	0				4
	long service life	0	0	3	0	0	0	9	3	0	0	3	0	0	0	0	0	1				3
environmental loading	low CO2, NOx, SO2 emissions	9	0	0	0	9	9	0	0	9	0	9	0	0	0	0	0	0				5
	particles	0	0	9	0	3	9	9	0	0	0	0	0	0	3	0	0	0				5
	existing infrastructure	9	3	0	0	0	0	0	0	0	0	0	0	0	0	0	0	0				1
archit. quality	home-like	0	9	0	9	0	0	0	1	0	3	0	3	3	1	0	0	1				3
	attractive to children	0	3	0	9	0	0	0	0	0	9	0	3	9	1	0	0	0				4
	public service building	3	9	3	0	1	3	3	1	1	0	0	0	1	9	0	0	0				1
indoor quality	air purity + emissions	0	9	9	0	0	0	3	3	0	9	9	9	0	0	0	0	0				5
	high thermal quality	0	0	0	0	9	9	0	0	9	9	9	0	0	0	0	0	1				3
	illumination	0	0	0	0	0	0	0	0	0	0	3	0	3	0	3	0	0				5
	echoing	0	0	0	0	0	0	0	0	0	0	0	0	0	0	0	0	0				1
	low HVAC noise	0	0	1	0	3	0	0	9	0	1	0	0	0	0	0	0	0				2
functionality	user access to site	0	9	0	0	0	0	0	0	0	0	0	0	3	0	0	0	0				4
	service access	0	0	0	0	0	0	0	9	0	0	0	0	3	0	0	0	0				3
	safety in use	3	3	0	0	0	0	3	9	0	3	0	3	9	3	0	0	0				5
	evening use	0	9	0	0	0	0	0	0	0	0	0	0	0	9	0	0	0				1
	high adaptability	0	0	1	9	1	0	9	3	1	3	1	0	1	9	0	0	0				2
	Weight factor (P1)	207	169	163	153	136	135	132	130	111	108	107	90	90	89	42	24	15	12	0	0	1913
	Weight factor %	11%	9%	9%	8%	7%	7%	7%	7%	6%	6%	6%	5%	5%	5%	2%	1%	1%	1%	0%	0%	100%
	Selected	x	x	x	x			x	x	x			x									

Figure 2.27 Design priorities for a nursery school.

Properties for a nursery school building were discussed and selected keeping in mind the most important requirements for the building:

- low investment and service costs
- low environmental impacts in use
- good indoor climate
- existing infrastructure
- safety in use
- being attractive to children.

A set of building concepts was developed for evaluation purposes. The energy performance and environmental impacts of the concepts, ranging from a typical nursery school building in Helsinki to a low energy building utilising solar energy, were analysed using the results of energy analysis as a starting point. The environmental impacts were compared with the requirements set in the pre-design phase.

These evaluations show that the environmental targets of the project can be fulfilled with a typical building type used in the construction of nurseries and nursery schools. However, there is conflict between the environmental goals and life cycle costs, in terms of low investment and service costs. The extra building costs of a low energy building are in the order of magnitude €50–100 m^2.

Technical properties of the building alternatives, corresponding to the abovementioned criteria were documented as properties in the QFD matrix (Figure 2.27). The dependencies between the given requirements and properties were checked. According to the QFD results, the main properties of the nursery school building corresponding to the given requirements are:

- district heat
- bicycle access to the site
- cleansable ventilation ductwork
- multi-use playrooms for children
- low energy building envelope.

According to the QFD results, the requirements dealing with functionality or air quality in a nursery school are dominating the pre-design process. The present energy (district heat) price is so low that the extra costs of energy saving are difficult to argue.

2.6.4 Risk analysis

2.6.4.1 Aim and role of risk assessment and control in LMS

The aim of this deliverable is to cope with lifetime risks of concrete facility management keeping in mind the four principal viewpoints of LMS – human conditions, culture, economy and ecology. The main objectives of risk assessment and control are:

1. to make facility owners aware of the risks in relation to the generic requirements
2. to form a solid framework and base for risk-based decision-making
3. to give guidelines on how to use the risk approach in decision-making process.

Risk is a subject that has normally been interlinked with highly complex and complicated systems, like the operation of power plants, the processing industry, pipelines, oil rigs, the space industry and so on. Risk analysis techniques have also for long been a part of project management when economical issues have been treated.

In the construction industry, risks have traditionally been treated just in the context of structural safety. Of course, that is the main concern and target of the designer: how to design and maintain a structure in such a way that it satisfies the structural safety limits set by the authorities but at the same time would not be too conservatively designed and maintained.

Differing from the processing industry, in the construction sector the facilities can be in a poor condition and still 'satisfy' the basic need. In the processing industry for example cracks in the pipelines cannot be accepted, because they would be fatal for the system. In concrete facilities cracks are unwanted but unfortunately rather common phenomenon, but unlike in the processing industry the cracks do not cause immediate fatal threat to the safety of the system.

Fortunately, the present societal trends in the construction industry promote sustainable development and customer orientation and satisfaction, which all work in favour of better-maintained concrete facilities. Gradually, limit states are becoming more and stricter. With increasing national and global wealth, more emphasis is placed on environmental, human and cultural issues and not just on minimising construction and maintenance costs. At the same time with the development of the computational potency of modern computers, better and more accurate decision-making and risk analysis methods have been and are being developed. Authorities, stakeholders and funding partners know this. Consequently, the decisions as well as the explanations for allocation of expenses must be better optimised and argued. As an answer to these new challenges, risk analysis techniques have been proposed. They are flexible and can be applied to help in decision-making very widely. However, the construction industry in general is a very traditional discipline with old role models, and implementing new ideas and methods takes time. But sooner or later risk analysis methods will also be routine in the construction sector.

Dealing with risks should not be a separate item that is introduced only when there are emergencies. Instead, it should be a part of any management: cost management, time management, etc., and it should have a logical structure. A possible structure for risk management is shown in Figure 2.28.

As can be seen in Figure 2.28, risk analysis is an essential part of risk management, but on the other hand, just doing a risk analysis is not enough, it must not be excluded from the bigger context. Traditionally 'risks' have been treated in two different ways: *diagnostic risk analysis* and *risk-based decision-making*. The former concentrates on identification of the main contributors of risks, while the latter goes much further, trying to use the information of diagnostic risk analysis and then quantify the risks. Theories and methods for quantification exist, but risk analysis is not much practised in the construction sector.

2.6.4.2 Qualitative approach vs. quantitative approach

Qualitative approach in risk analysis is quite simple. More than anything it is logical thinking, structuring down the problem into smaller pieces, which can then be dealt

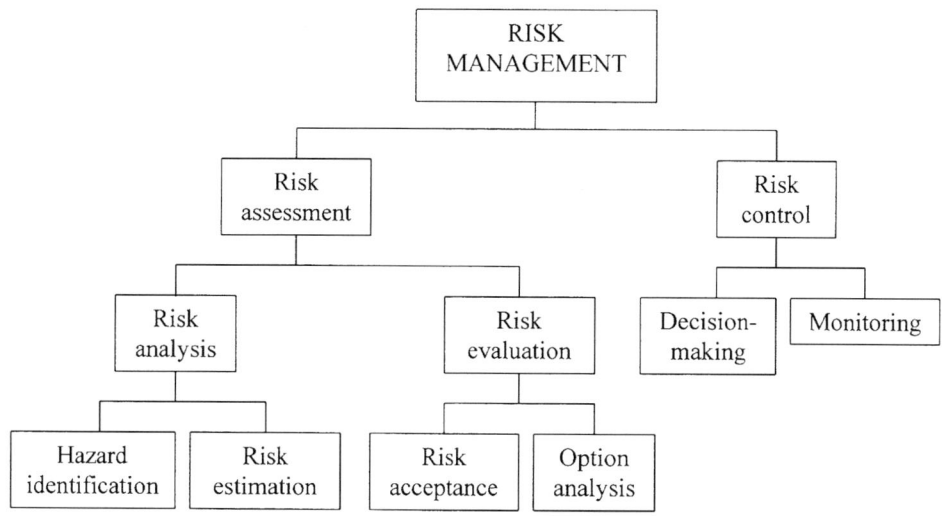

Figure 2.28 Structure of risk management [21].

with, one by one. An experienced engineer can produce rough estimations for failure frequencies and consequences, and a brainstorm session of many experts can make the estimations even better. If relative measures are used the quality of results normally maintains a good level, but if exact numbers are wanted, the situation is not the same. In some discussions it has been estimated that even the best risk calculations should be regarded as accurate to only within one or two orders of magnitude, when it comes to small probabilities [22].

In structural safety matters some limits are gaining consensus, namely 10^{-3} for the service limit state and 10^{-6} for the ultimate limit state, when new structures are concerned [23]. However, the situation is different when old structures are concerned, and the consensus about the probabilities of failure is no more complete. Numbers between 10^{-2} and 6×10^{-4} have been suggested but even then the discussion has been considering only the ultimate limit state [24]. In the Lifecon context this structural safety issue is only one part of the human viewpoint, and there are three more whole viewpoints (economical, ecological and cultural) without any established number-based limits [25]. Of course, there are some legislation about these issues also, but the regulations and restrictions are qualitative.

With fatal accidents, a principle of 'as low as reasonably practicable' (ALARP) is gaining popularity. The idea of ALARP is that if the probability of death is low enough (the frequency of death for an individual is for example 10^{-6}/year), the situation is acceptable. But if the frequency is greater than, say, 10^{-3}/year, the situation is unacceptable and improvements for the safety of the individual must be made immediately. In between these two limits the ALARP principle is applied: the probability of death is reduced to as low as reasonably practicable, meaning that if the costs of reducing the probability of death exceed the benefits or improvements gained, then the original risk is accepted. The question is once again of qualitative form, what does one mean by 'reasonably practicable'. And what are the ALARP limits for cultural, ecological and economical risks?

One problem in quantification is the use of deterministic values instead of statistical distributions. It is true that stakeholders (practical engineers, decision-makers, facility owners, etc.) are more familiar with exact numbers than distributions, but if a numerical estimation for risk is required, then using distributions in calculations gives better results. By using a characteristic value and a safety factor, it is possible to check if some condition for the risk is fulfilled, but the actual value of risk is not obtained. The variation and uncertainty of variables are best described with either standard mathematical or experimental distributions. The simulation methods will eliminate the problem of the difficult analytical integration. Most commercial Quantified Risk Analysis (QRA) softwares use simulation techniques. By using distributions in calculations the results of analyses will also be distributions which tell a lot more than a single value.

Unfortunately, finding out the source distributions for different variables is not an easy task. In the processing industry (where the risk analysis methods were developed) the situation is easier. Although the whole process may seem highly complex, it can be split into discrete phases, where the successful operation of that phase is a function of just a few variables. The high degree of automation has reduced the possibility of a human error, the operation conditions are always the same, and access to the area is restricted. All this has enabled consistent gathering of relevant information from the functioning of the process. With concrete civil infrastructures, the situation is different. The facilities stand in various open environments, access is quite easy for everybody, and construction and maintenance require a lot of manpower. The multi-dependent nature of a construction or maintenance project is not easy to handle or model. A characteristic feature in the construction industry compared with the processing industry is the lack of consistent source data and information, which causes problems in quantifying the uncertainty and risks.

One more problem in quantitative analysis is created by the mathematical definition of risk. Risk, being a product of two uncertain factors (the probability of occurrence of a scenario and the consequences of that scenario), can mislead the decision-maker, if it is introduced as one number only. This is illustrated in Table 2.7. The two cases have the same yearly risk (the numbers are more or less arbitrarily chosen for illustration purpose only), but for the facility owner the second case is disastrous, while the first one can be handled. The second case is not as probable as the first one, but the consequences are huge and will bankrupt the owner if the scenario comes true. But still the *risk* is the same in both cases, namely € 150/year.

Table 2.7 Illustration of the shortcoming of defining risk with only one number

Case (= adverse scenario)	Probability of occurrence	Consequences	Risk
Power failure silences the skyscraper for two hours	0.0001/year	€ 1 500 000	€ 150/year
Aeroplane crashes the skyscraper	0.00000001/year	€ 15 000 000 000	€ 150/year

The example in Table 2.7 clarifies the problem when using only one number for risk. On the other hand, this example highlights one more unsolved problem of risk analysis, namely 'the low probability–high consequences' problem. These scenarios cannot be included in normal risk analysis models, but somehow they should be taken into account in decision-making.

2.6.4.3   Risk analysis methods

The risk analysis methods were developed within the processing industry, where the systems and procedures are automatic and the role of human activity is not decisive. However, the principles of the methods are applicable also in other sectors of industry. For example, many routines in concrete facility management can be thought of as discrete processes with a logical structure, so in evaluating uncertainties the general risk analysis methods can be applied.

The risk analysis methods as such are simple logical chains of thinking, there is no higher mathematics included in the principle. As mentioned before, the methods were first used in the processing industry, and because that sector had the early head start there exist many detailed and case-tailored risk analysis methods in the processing industry, while in the construction sector more general methods are used. But the three basic questions to be answered remain the same, regardless of the method:

- What can go wrong?
- How likely is it to go wrong?
- What are the consequences?

The choice of analysis method depends on many variables like source data, resources, expertise, risk category, phase of the project, and especially the nature of the problem. In every method the basic structure of 'dealing with risks and uncertainty' is a logical, phased process that is roughly divided into five steps:

1. identification of the possible adverse incidents (hazards, mishaps, accidents)
2. identification of the causes and consequences of the adverse incidents, and building of structured causal relationships between them
3. estimation of the likelihood of causes and consequences, as well as the severity of the consequences
4. evaluation and quantification of the risks
5. decisions and actions to deal with the risks.

The first two steps are an essential part of any risk analysis (being a part of qualitative diagnostic risk analysis); the next two are necessary only if some quantitative values are needed; and the last step is clearly fundamental. Apart from the logic of risk analysis procedure, another fact binds the different risk analysis methods: strong expertise is needed and the results depend to a great extent on how rigorously the analyses are performed. No shortcuts should be taken if real benefits are wanted. It should be remembered that a huge part of the accidents, failures and unintended events happen due to negligence, not ignorance. All risk analysis methods (when pertinently carried out) include brainstorming and prioritisation processes performed by a *multi-discipline team* consisting of members from different stakeholder groups. These people who give 'raw material' (data, opinions, estimations, etc.) for risk analyses must be experts with solid experience in their business. These are, for example, maintenance engineers, facility owners, statisticians, inspectors, material suppliers and so on.

Those who conduct a risk analysis, on the other hand, do not have to be experts in facility management, repair methods or materials. Instead, they must have other skills, for example:

- experience in the risk management process
- experience with risk management tools
- neutrality in the project (e.g. having no partnership with contractors)
- an analytical way of thinking
- superior facilitation skills
- excellent communication skills.

The most common risk analysis methods are briefly presented in the following chapters, with some guidelines about their normal use and applications, and notes about their benefits and shortcomings [26–31].

2.6.4.3.1 PRELIMINARY HAZARD ANALYSIS (PHA)

This method is an initial effort to identify potential problem areas. It is a basic qualitative study considering larger operational components, not detailed interactions. The main benefit of PHA is the awareness of the hazards it creates. Depending on the depth of the analysis, the time to complete PHA is normally relatively short. PHA is not the most systematic or established method, and, for example, in the literature the results of PHA vary from presentiments of possible hazards to the evaluation of the risks, but it gives a good starting point to further analysis. The normal output of PHA is a list of possible hazards, classified, for example, by the phases of the process or system, or by the targets (personnel, product, environment, structure, reputation, etc.). A very thorough PHA output could include the following information:

- hazard description (source–mechanism–outcome)
- mission/system/project/process phases covered
- targets (meaning the potential 'hazard victims')
- probability interval
- subjective assessment of severity of consequences (for each target)
- subjective assessment of probability of occurrence (for each target)
- assessment of risk (product of the previous two)
- countermeasures, safeguards, actions.

In some contexts, the PHA and Hazard and Operability Study (HAZOP) have been used in an alternative sense, but of these two, HAZOP is a real risk analysis method while PHA has a more unofficial reputation.

2.6.4.3.2 HAZARD AND OPERABILITY STUDY (HAZOP)

The HAZOP method is mainly used in the processing industry to find out hazards, but in a wider context some routine phases of construction can be thought as processes, and in the very preliminary stage this method can be used. The idea of HAZOP is to study what kind of consequences can occur when there are little deviations from the intended use or operation of the process.

In the HAZOP method the process is first described completely and then it is divided into phases (called *nodes*) and the deviations are addressed at those nodes. The brainstorming team will consider one node at a time and the results will be recorded for each node in columnar form, under the following headings:

* deviation
* causes
* consequences
* safeguards (the existing protective methods/devices to prevent the cause or to safeguard against adverse consequence)
* action (to be taken in case of too serious consequences, for example applying the rule of the three Rs: remove the hazard, reduce the hazard, remedy the hazard).

HAZOP method consumes a lot of time and resources, but it is very easy to learn. It has long been used in the operational sector.

2.6.4.3.3 FAILURE MODE AND EFFECT ANALYSIS (FMEA)

This analysis describes potential failure modes in a system and identifies the possible effects on the performance of the system. In the five-step process, FMEA mostly deals with the steps 1 and 2, although some semi-quantitative estimations can be given. The product of an FMEA is a table of information that summarises the analysis of all possible failure modes. Traditionally FMEA has been used for concrete processes or structures, where the system can be divided into smaller parts, modules, components, etc., but theoretically it can be used in more abstract projects also.

First, the system must be described in such a way that the operation, interrelationship and interdependency of the functional entities of the system become clear to all parties involved. Then the FMEA starts by identifying the possible failure modes – meaning the ways in which a component or system failure occurs – for all the components. Theoretically, there are innumerable failure modes for each component (and no limit to the depth one can go), but practically there is a point after which the additional costs exceed the benefits.

After finding out the failure modes, the failure mechanisms must be identified. In this phase the question to be answered is: 'How could the component or system fail in this failure mode?' A very simple illustration of this is obtained from the durability of concrete: the erosion. The failure mode is the erosion and the failure mechanism is the flowing water acting on concrete. One failure mode can, of course, have more than just one failure mechanism. The failure modes of the components are normally known, but the failure mechanisms (the causes) are sometimes more difficult to identify.

The FMEA continues with the identification of the failure effects. The consequences of each failure mode must be carefully examined. In the concrete erosion example, the obvious effect is surface deterioration but it can have worse effects, too, like reduced bearing capacity and finally a collapse. The effects of the component failure should be studied on all the abstraction levels of the system (from the component level to the system level).

Once the failure modes are identified, the failure detection features for each failure mode should be described. Also, at each abstraction level (component, module, system) provisions that alleviate the effect of failure should be identified.

The above analysis is carried out in a qualitative way, but it is possible to add some semi-quantitative features in FMEA. When some estimations of the likelihood of occurrence of the failure modes and severity of the failure effects are given, the FMEA method is called FMECA. The letter C stands for 'criticality', which is the combination of the likelihood and severity. The criticality indicates the importance of that particular failure in risk analysis.

The results of FMEA are presented in table form. The practical minimum for the number of columns is four – the element (component), the failure mode, the failure mechanism and the failure effect. However, normally the columns are tailored for the case in question, and can have, for example, following column heads:

- event identification
- name of the element
- concise description of the function of the element
- modes of failure of the element
- causes of failure and operational conditions under which it can occur
- consequences of the failure on the system (locally and globally)
- means of detecting a failure of the element
- means of preventing the appearance of failure (redundancies, alarms, etc.)
- classification of severity
- comments and remarks
- probability of occurrence (estimate) in FMECA
- criticality (calculation) in FMECA.

Like any other risk analysis method, the FMEA should also be introduced into the project from the very beginning. Being more qualitative than being quantitative, the FMEA is never pointless. It reduces uncertainty in decision-making even if exact numbers are not required. Depending on the need, the FMEA can be anything from very rough to very detailed. For years, FMEA has been an integral part of engineering design and has grown to be one of the most powerful and practical process control and reliability tools in manufacturing environments. Especially, FMEA is a tool for identifying reliability, safety, compliance, and product non-conformities *in the design stage* rather than during the production process. A shortcoming in FMEA is that it is performed for only one failure at a time. So it may not be adequate for systems in which multiple failure modes can occur at the same time. Deductive methods are better in identifying these kinds of failures. FMEA does not include a human action interface, system interaction nor common cause failures. FMEA generally provides basic information for FTA.

2.6.4.3.4 EVENT TREE ANALYSIS (ETA)

If the successful operation of a system (or project, process, etc.) consists of chronological but discrete operation of its units, components or sub-systems, then ETA is
a very useful method to analyse the possible risks of the case. ETA is an inductive
method, it starts with a real or hypothetical event (the *initial event*) and proceeds
with forward analysis to identify all the possible consequences and final outcomes of
that initial event. The driving question in ETA is: 'What happens, if . . . ?'

No specific symbols are used in ETA (as is the case with FTA), but just simple
logic. Normally event trees are developed in binary format, meaning that the possible
events either occur or do not occur. This is quite logical, for example, in the case
of some accident in the processing industry, where some initial event should wake
the safeguard operation. This operation then occurs or does not occur (success or
failure), and then comes the next event and so on, until the final event is reached.

However, in the general case the initial or subsequent event can, of course, have
more than just two outcomes. In such a situation the events stemming from the node
(the inception of the subsequent event) are so chosen that they are mutually exclusive.
This means that no simultaneous occurrence of two or more subsequent events is
possible and as a consequence, the sum of the probabilities at a node is always equal
to one. The general case is presented in Figure 2.29.

The more common binary format use of ETA is illustrated in Figures 2.30 and
2.31. As in the general case, the construction of the event tree starts from the left.
The proceeding events (normally the safety systems and operations) are listed in
chronological order on the upper edge of the figure, on the right are mentioned the
final outcomes, consequences and calculated frequencies.

The event tree is very effective in determining how various initiating events can
result in accidents. On the other hand, the sequence of the events is analysed only for
that initiating event. So to get an exhaustive risk analysis, the selection of the initiating
events is a crucial task. Another limitation of ETA is the assumed independence of the
events. In reality there are always some subtle dependencies (for example common
components, operators) that may be overlooked in ETA. One more shortcoming of
ETA is the 'fail–not fail' dictonomy, because systems often degrade without sudden
failure.

Although ETA is mostly used to analyse accident scenarios, it can be applied to
almost any type of risk assessment, especially when it is used together with PHA,
HAZOP and FTA. One special application of ETA is Human Reliability Analysis.
Using this analysis (gross) human errors can be avoided by using ETA.

2.6.4.3.5 FAULT TREE ANALYSIS (FTA)

Fault tree analysis is one of the best and most used risk analysis methods. It is a
deductive method, trying to answer to the question: 'What causes . . . ?' The idea of
FTA is to go backwards from the failure or accident (so-called *top event*) and trace
all the possible events that can cause that top event, and then go on to the lower
levels until the final level is reached and the basic causes are found.

Like any other risk analysis method, FTA starts with the description of the system
(or project, process, etc.), where the fault tree is going to be applied. The bounds of
the system and the level of complexity must be clearly defined.

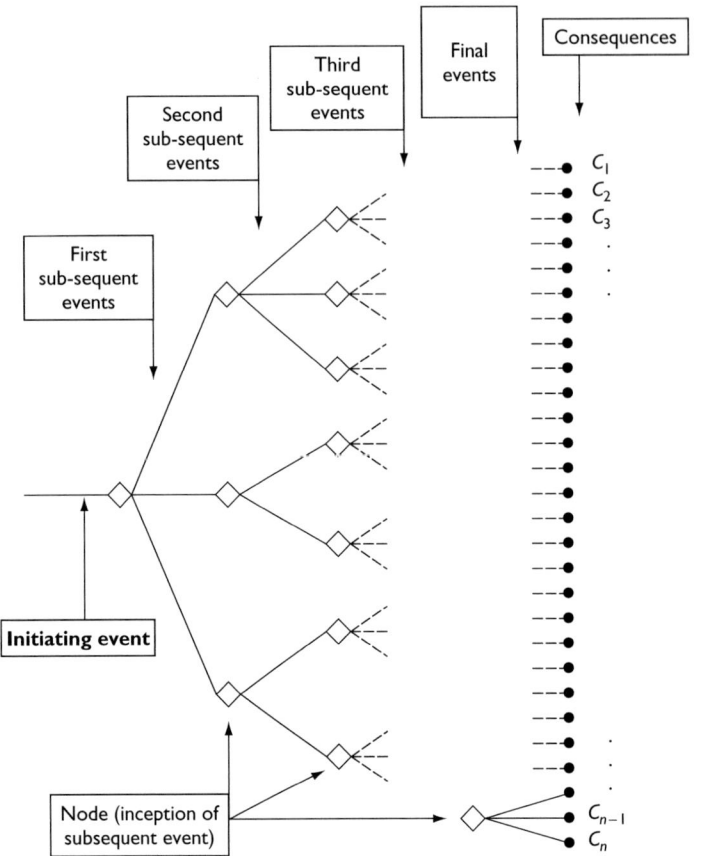

Figure 2.29 The event tree (general case).

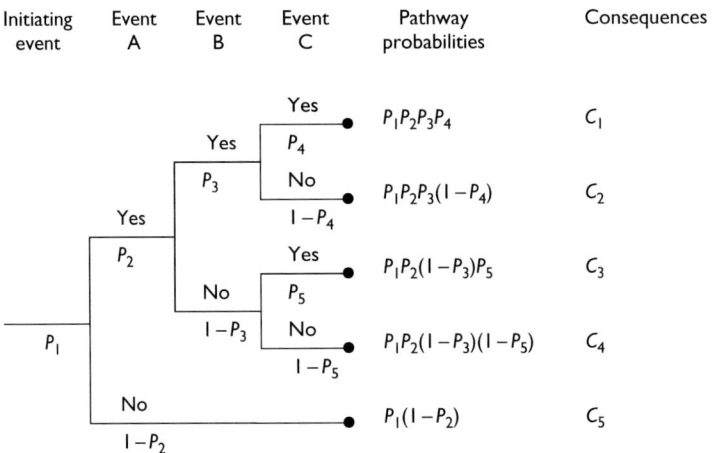

Figure 2.30 The event tree (binary case, general illustration).

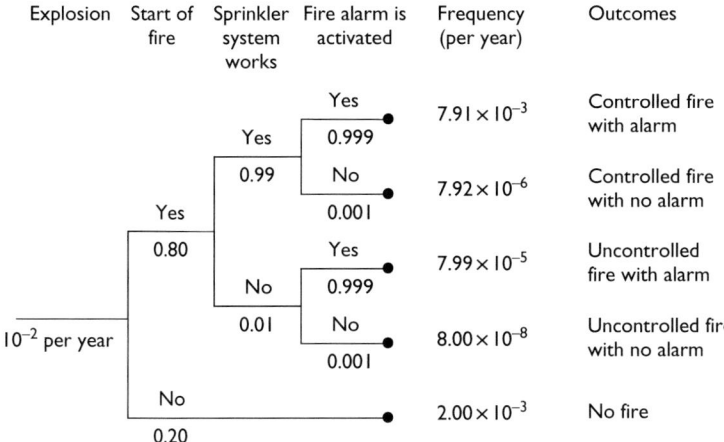

Explosion	Start of fire	Sprinkler system works	Fire alarm is activated	Frequency (per year)	Outcomes
			Yes	7.91×10^{-3}	Controlled fire with alarm
		Yes 0.99	0.999		
			No	7.92×10^{-6}	Controlled fire with no alarm
	Yes 0.80		0.001		
			Yes	7.99×10^{-5}	Uncontrolled fire with alarm
		No 0.01	0.999		
10^{-2} per year			No	8.00×10^{-8}	Uncontrolled fire with no alarm
			0.001		
	No 0.20			2.00×10^{-3}	No fire

Figure 2.31 The event tree (binary case, explosion example).

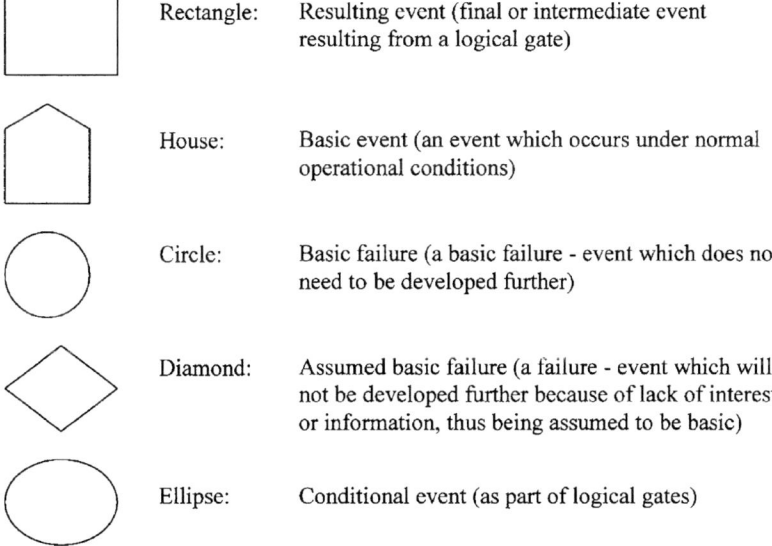

Rectangle: Resulting event (final or intermediate event resulting from a logical gate)

House: Basic event (an event which occurs under normal operational conditions)

Circle: Basic failure (a basic failure - event which does not need to be developed further)

Diamond: Assumed basic failure (a failure - event which will not be developed further because of lack of interest or information, thus being assumed to be basic)

Ellipse: Conditional event (as part of logical gates)

Figure 2.32 The basic symbols for events in fault tree analysis.

The fault tree is constructed by using standard logical symbols. The most used symbols and their meanings are presented in Figures 2.32 and 2.33. Although many more symbols exist, most fault tree analyses can be carried out using just four symbols (rectangle, circle, AND-gate and OR-gate).

The identification of the top event starts the construction of the fault tree. The top event is normally some undesired occurrence, for example a fire in a tunnel, falling

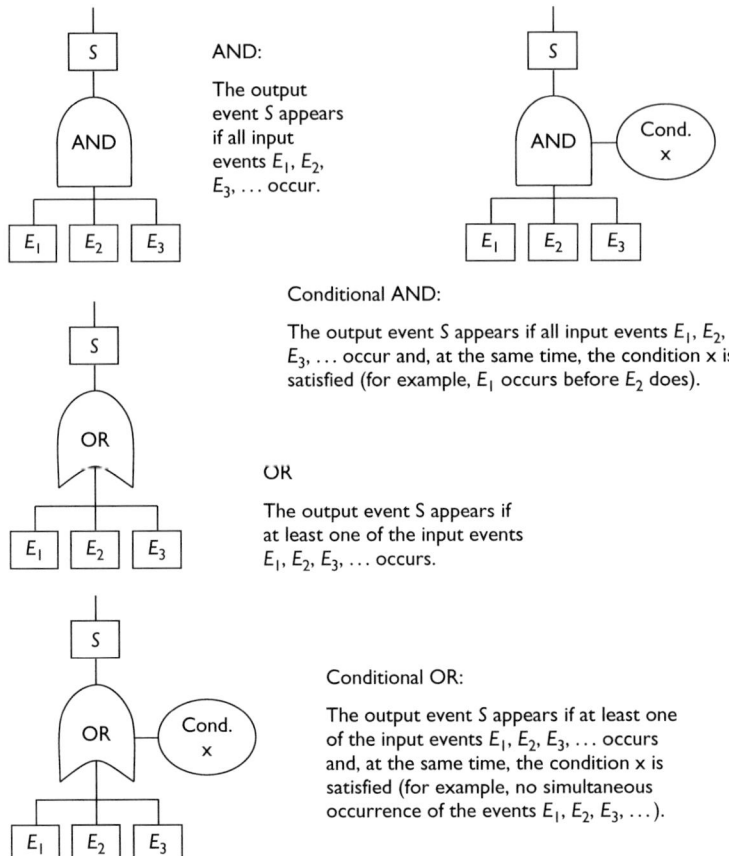

Figure 2.33 The basic symbols for logical gates in fault tree analysis.

of a worker from scaffolding, exposure to asbestos, or a cracking of an abutment. There are no strict rules for the definition of the top event, but the identification of the top event sets the framework as to how elaborate the analysis will be. The process of constructing a fault tree is explained in Figure 2.34.

Fault tree analysis is very useful because it can take into account not just internal causes of the system, but also external factors like human carelessness, natural disasters and so on. FTA can be used qualitatively or quantitatively. For most cases the qualitative part of the FTA is enough, because the construction of the fault tree forces 'the risk team' to improve their understanding of the system characteristics, and most of the errors and hazards can be removed or reduced to an acceptable level already in that phase.

In the quantitative phase of FTA the target is to find the probability for the occurrence of the top event. The probabilities for the other events of the fault tree are evaluated, and using the *minimal cut sets* (the smallest combination of basic events which,

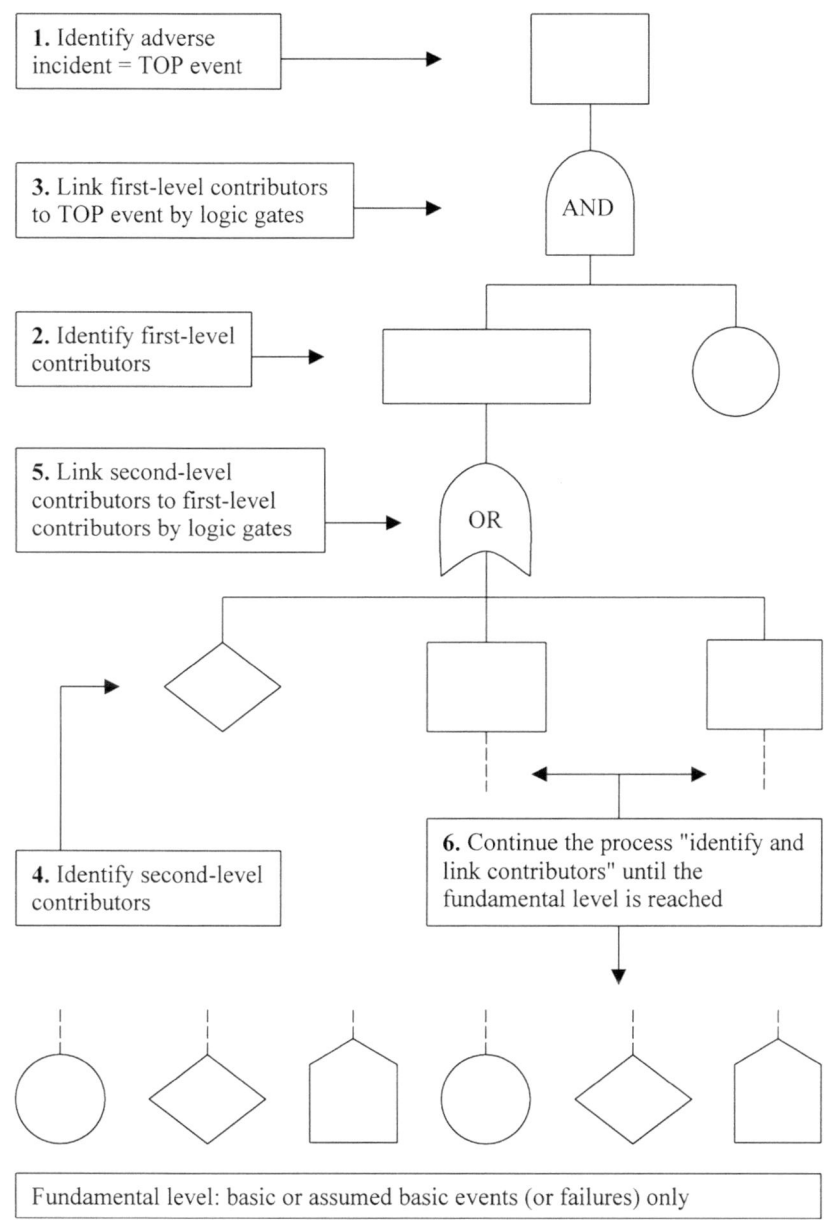

1. Identify adverse incident = TOP event

3. Link first-level contributors to TOP event by logic gates

AND

2. Identify first-level contributors

5. Link second-level contributors to first-level contributors by logic gates

OR

4. Identify second-level contributors

6. Continue the process "identify and link contributors" until the fundamental level is reached

Fundamental level: basic or assumed basic events (or failures) only

Figure 2.34 'Step by step' construction of the fault tree. Note the order (numbers) of the steps.

if they all occur, will cause the top event to occur) the probability for the top event can be calculated very easily. The OR-gate represents *union* and the AND-gate represents *intersection*, and the probabilities are obtained by *summing* and *multiplying*, respectively. The mathematical expression of union and intersection is explained in Figure 2.35.

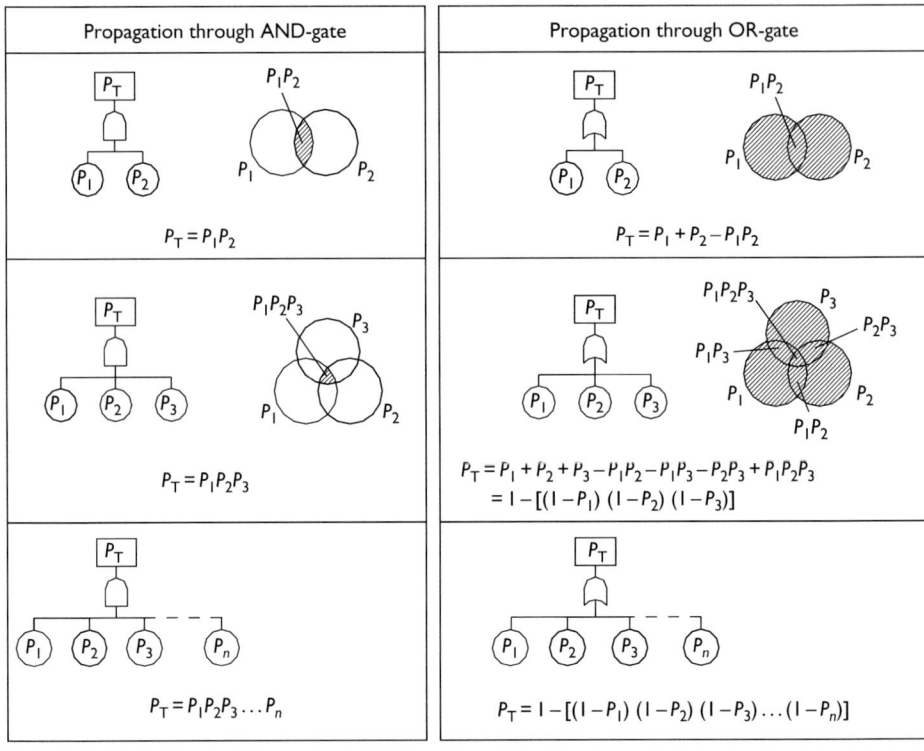

Figure 2.35 Mathematical expression of intersection (AND-gate) and union (OR-gate) in FTA.

As an illustration of the procedure from fundamental level to top event, a fictitious fault tree is constructed in Figure 2.36, with fictitious probabilities of the basic (or assumed basic) events. The probability of the top event of this fictitious fault tree is calculated in Figure 2.37.

FTA can be used for almost every type of risk assessment application, but it is used most effectively to find out the fundamental causes of specific accidents, where a complex combination of events is present. FTA has (like any other risk analysis method) some limitations. It examines only one specific accident at a time, and to analyse the next one another fault tree must be created. This is expensive and time consuming. FTA is also very dependent on the experience of the analyst. Two analysts with the same technical experience will probably get different fault trees. The third drawback is the same as with all the other risk analysis methods, namely the quantification problem. It needs a lot of expertise, knowledge, effort, data and patience.

However, carried out properly, FTA is extremely 'readable' and it makes the causes and interrelationship very visible. As a consequence, the actions and corrections are easily channelled to where they are most needed.

Probabilities of
the basic events:

$P(E_1) = 0.001$
$P(E_2) = 0.3$
$P(E_3) = 0.2$
$P(E_4) = 0.4$
$P(E_5) = 0.07$
$P(E_6) = 0.25$
$P(E_7) = 0.1$
$P(E_8) = 0.15$

Probabilities to be
calculated:

$P(E_9)\ldots P(E_{12})$
and the probability
$P(E)$ of the TOP
event E

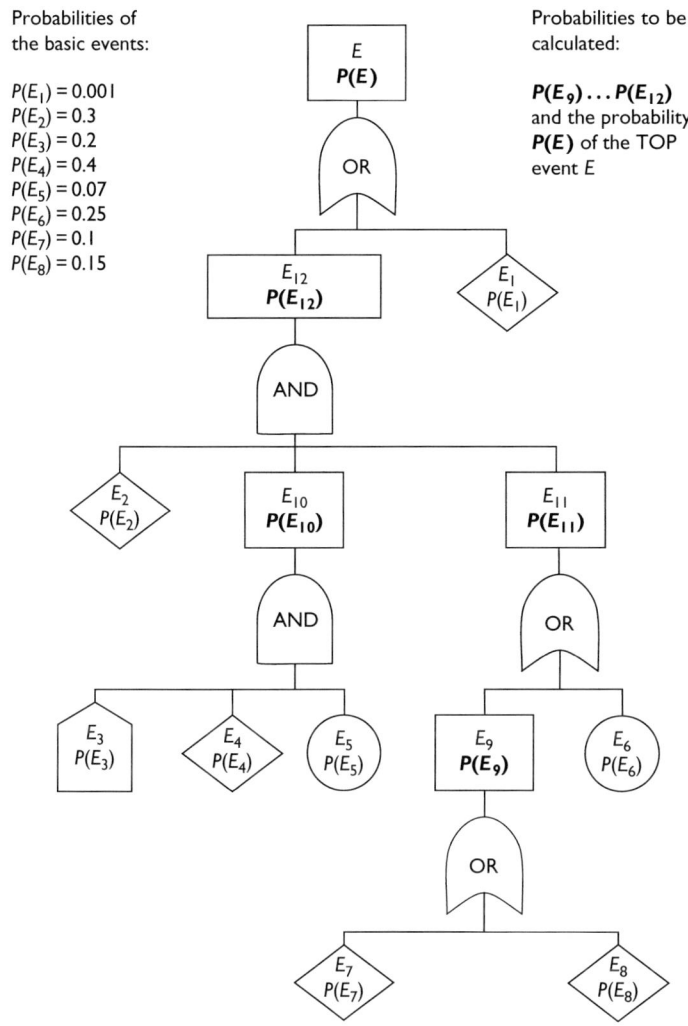

Figure 2.36 Illustrative example of a fault tree, with fictitious events (E_i) and the probabilities of their occurrence ($P(E_i)$).

2.6.4.4 Risk assessment and control procedure

2.6.4.4.1 PRINCIPLES

In LMS the subject 'risk' is treated on a very wide scale [25, 32]. Not just structural risks, but also environmental, ecological, cultural and human risks are taken into account. The idea is to make the facility owner *aware* of different risks throughout the whole lifetime of the facility, and to offer logical and easy procedure to deal consistently with them. Traditionally risk analysis has been carried out only in big

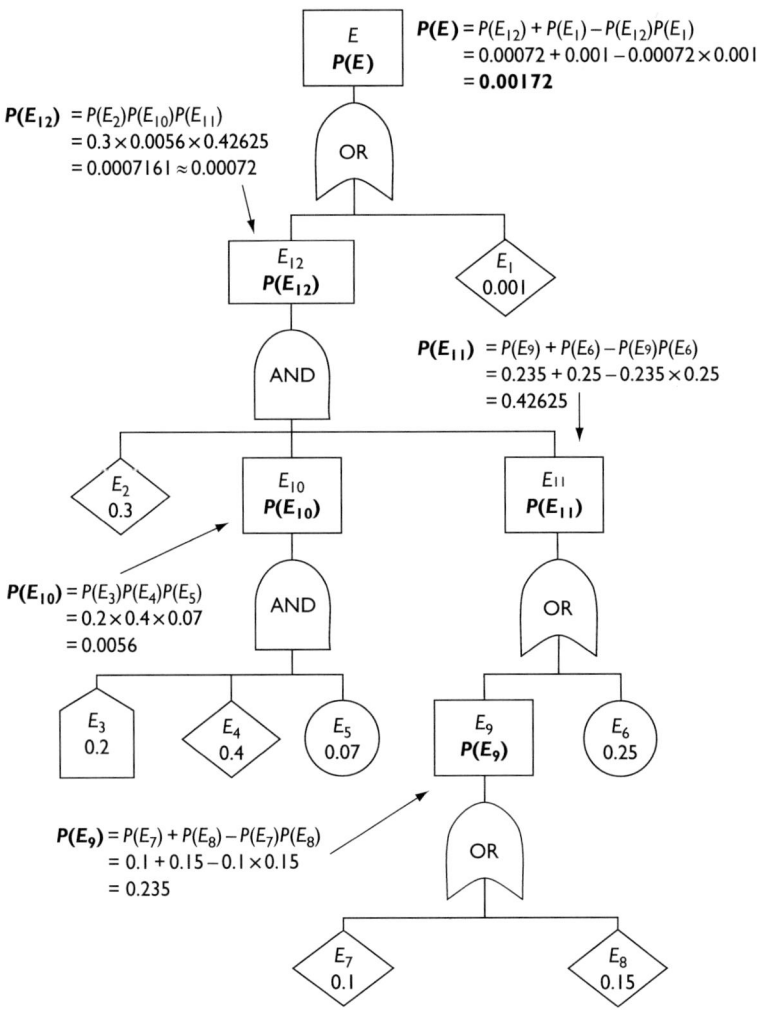

$$P(E_{12}) = P(E_2)P(E_{10})P(E_{11})$$
$$= 0.3 \times 0.0056 \times 0.42625$$
$$= 0.0007161 \approx 0.00072$$

$$P(E) = P(E_{12}) + P(E_1) - P(E_{12})P(E_1)$$
$$= 0.00072 + 0.001 - 0.00072 \times 0.001$$
$$= \mathbf{0.00172}$$

$$P(E_{11}) = P(E_9) + P(E_6) - P(E_9)P(E_6)$$
$$= 0.235 + 0.25 - 0.235 \times 0.25$$
$$= 0.42625$$

$$P(E_{10}) = P(E_3)P(E_4)P(E_5)$$
$$= 0.2 \times 0.4 \times 0.07$$
$$= 0.0056$$

$$P(E_9) = P(E_7) + P(E_8) - P(E_7)P(E_8)$$
$$= 0.1 + 0.15 - 0.1 \times 0.15$$
$$= 0.235$$

Figure 2.37 Calculation of the probability of occurrence of the top event (it is assumed, that the events (E_i) are independent from one another) of the fault tree of Figure 2.36.

construction or repair projects, but within Lifecon it is going to be an integrated information and optimisation tool in a predictive concrete facility management system. Change from a traditional point-in-time effort to continuous process sets some requirements for the risk analysis module of the LMS. The module must:

- have an informative role (instead of checking up)
- have a well-documented, updatable database structure
- be powerful and extendable enough for future challenges
- be compatible with other Lifecon decision-making tools.

Lifecon risk control proposal respects existing management systems. It does not demand abandonment of the old systems in order for it to work, but more likely offers a parallel system to be used with the old system. The biggest challenge for this risk control proposal is the implanting of new ways of thinking that it brings [1]. While there do not exist normative limits in all generic requirements (Table 2.1), it is up to the end user to decide which parts of the generic risk control proposal to exploit, to which extent and in which phases of the management process.

2.6.4.4.2 THE STEPS OF THE PROCEDURE

In short, the Lifecon risk assessment and control procedure follows the four steps which are explained below:

1. identification of adverse incidents
2. analysis of the identified adverse incidents

    (a) deductively (downwards), in order to find causes
    (b) inductively (upwards), in order to find consequences

3. quantitative risk analysis
4. risk-based decision-making (and continuous updating of risk database).

Steps 1, 2 and 4 are always performed if risk analysis is used, forming qualitative risk analysis. Step 3 is only performed if qualitative risk analysis is not enough for decision-making *and* if quantification is possible.

A very important feature in the procedure is the continuance. Management of concrete infrastructures is a continuous process and new experience is gained every day. The same applies to risk management. The steps described above form Lifecon risk management loop, which is continuously maintained and updated, with strict documentation.

2.6.4.4.2.1 Identification of adverse incidents

The risk analysis starts with the identification and listing of adverse incidents (threats, fears, unwanted happenings, mishaps), with regard to the whole lifetime of a facility or stock of facilities. Adverse incident means the same as top event in FTA or initiating event in event tree analysis. For easy follow-up and updating, the identified adverse incidents should be logically labelled and stored into the database. 'The whole lifetime of a facility' is too big a category, so smaller categories must be created. The lifetime of a facility is built up of a few functionally different but chronologically overlapping or coinciding phases. While identifying adverse incidents, the phase – the moment when the adverse incident can happen – is automatically identified. A logical categorisation of adverse incidents follows those functional phases, which are normally:

- everyday use
- inspection and condition assessment
- MR&R actions
- extremities (high floods, exceptional snow loads, collisions, high overloads, etc.).

Of course facility owners can categorise the identified adverse incidents differently, according to their own preferences. In theory, there is no limit for the number of categories, but the database easily becomes cumbersome if the number of categories increases too much.

Not only is it the facility owner's task to identify adverse incidents, but those who deal with them in their everyday work are the best people to identify incidents. For example, a contractor can help in identifying adverse incidents connected with MR&R actions, inspectors are suitable persons to identify the mishaps at inspection work and so on. In addition to instinct and experience, information about possible adverse incidents are gathered from statistics, research, expert opinions, accident reports, failure logs, MR&R data, monitoring data, material tests, material producers, future studies and so on.

The importance of rationality in identification of adverse incidents cannot be overemphasised. The idea is not to create horror scenarios, but to answer reasonably to the first question of risk analysis: 'What can go wrong?'. In Lifecon this means 'What can go wrong in the management of a concrete facility, during its whole lifetime?'. Possible adverse incidents to be identified in this first step could be for example (functional phase in brackets):

- an inspector hit by a car (inspection and condition assessment)
- falling from a road bridge in everyday use (everyday use)
- exceeding limit state in spite of LMS system (everyday use) or
- exceeding the MR&R budget (MR&R actions).

2.6.4.4.2.2 Analysis of the identified adverse incidents

After the adverse incidents are identified, they are analysed further. The goal of this second step of Lifecon risk control procedure is twofold: first to find the underlying *causes* of the adverse incidents, and second to find the *consequences* of the adverse incidents. The result – an unbroken nexus of events from causes to consequences – forms a structured skeleton that helps decision-makers to perceive causalities and logic of the risk problem at hand. This step is the most important in the whole risk analysis process and hence it should be carried out very carefully. The sources of information for construction of the skeleton are the same as in step 1. It must be noted that risk analysis is not 'one man's show', but requires multi-discipline expertise.

The downward analysis – to find causes for the identified adverse incidents – is made using fault tree analysis (FTA). The primary factors that lead up to a top event (an adverse incident) are looked for. The intermediate events are linked with corresponding logic gates, until the desired fundamental level is reached. The desired fundamental level depends on the end user. For example, for one end user it can be enough to know that there is approximately one severe car accident on a certain bridge every year, while another one wants to go further and find out why the accident frequency is so high.

The structure of a fault tree is illustrated in Figure 2.38. The top event refers to the adverse incident example from step 1, namely 'falling from road bridge'. The fault tree of Figure 2.38 is presented only for illustrative purpose. The idea is to show how a fault tree looks like and how it can be used. The depth of the analysis is stopped to a level that satisfies the fictitious decision-maker. At first glance the leftmost branch in the figure may seem strange. Why should a facility owner worry about intentional falling? The answer is that if the number of falling accidents is relatively high, the authorities may require some explanations. Consequently, if it is revealed that the bridge for some reason tempts people to climb on the railings, the authorities may demand immediate action to impede climbing. For example, in high rise buildings, lighthouses, etc., access to the top is normally controlled, whereas with bridges the access is (logically) free.

In Figure 2.38, the two branches on the right are not developed further, because the fictitious decision-maker is not interested in traffic accident–induced fallings or fallings during MR&R works, but wants to focus only on falling under normal circumstances, in everyday use.

The upward analysis – to find consequences for the identified adverse incidents – is made using event tree analysis (ETA). The goal of ETA is to find consequences and final outcomes for initiating event (adverse incident). In Lifecon the consequences are divided into four main categories:

- human conditions
- culture
- economy
- ecology.

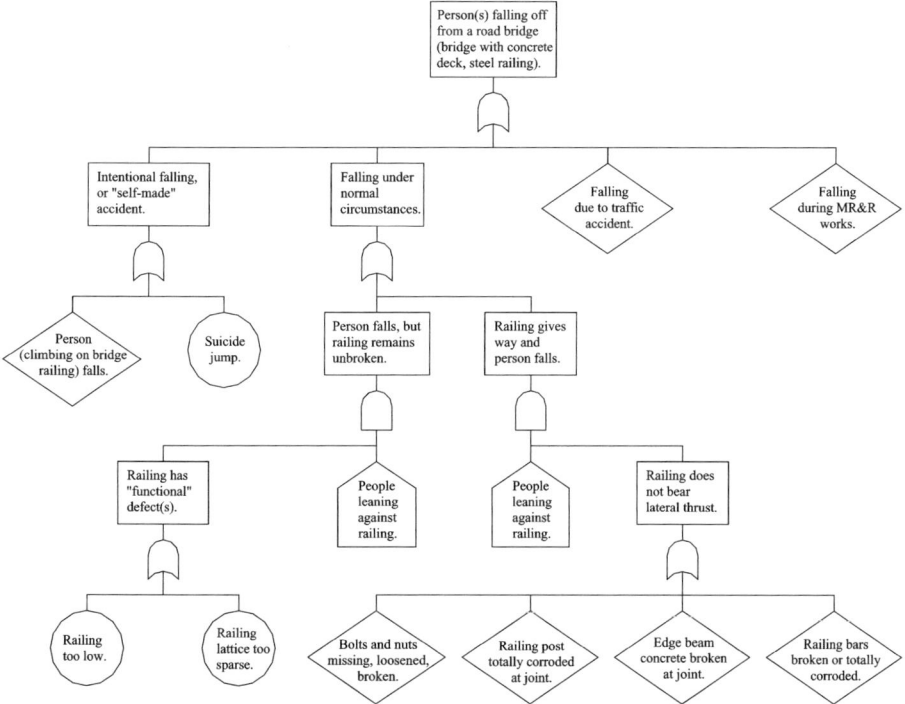

Figure 2.38 Illustration of a fault tree.

The four main categories of generic requirements (Table 2.1) are further divided into sub-categories. In risk control procedure, all these categories are examined (one by one) when finding out consequences for the identified adverse incidents. Once again it is up to the facility owner to decide how strictly the generic categorisation is complied with, when looking for consequences. For example, one facility owner may be interested only in direct economic consequences, whereas a more conscious facility owner takes into account also the consequences for culture. Of course, all adverse incidents do not necessarily have consequences in all generic categories.

The ETA is not as exhaustive as FTA described above. Normally after one or two nodes, the final consequences can be reached. Sometimes the identified adverse incidents are incidents that must not happen (collapse of main girder, pollution of ground water, fire in tunnel, etc.). In such cases the FTA is enough, revealing the causes of the incident, and if the top event probability is too high, a decision must be made to lower the probability.

An illustration of a possible event tree is presented in Figure 2.39. The consequences of falling come mainly into the safety category because the repair costs of a railing are almost nil compared to possible compensations in case of death or permanent injury. Falling from a bridge can have consequences also in the culture category, for example the 100-year-old decorated railing is found to be the cause of the fall and consequently authorities demand that the old railing must be replaced by a modern standard railing.

The first two steps described in Sections 2.6.4.4.2.1 and 2.6.4.4.2.2 are enough if risks are treated qualitatively only. With the aid of a visual, logical causes-consequences structure a facility owner can in most cases estimate the risk and make a consistent decision, even if no numbers are present in the analysis.

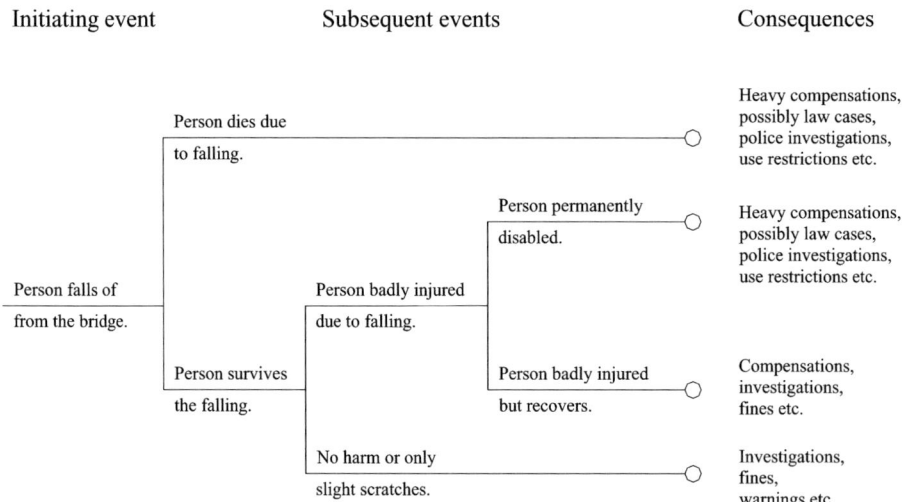

Figure 2.39 Illustration of an event tree.

2.6.4.4.2.3 Quantitative risk analysis

If the qualitative risk analysis is not enough, a quantitative risk analysis must be performed. The quantitative risk analysis utilises the same fault and event tree skeletons that were created in step two above.

In this quantitative phase, estimations about probabilities of basic events (or assumed basic events or failures, see Figures 2.35 and 2.37) are added to the fault tree part of the analysis. Likewise, in the event tree part of the analysis, estimations about the probabilities of the subsequent events (see Figure 2.36 are added to the event tree skeleton. The initiating event probability of ETA is the same numerical value that is obtained as a result from the fault tree analysis, i.e. the top event probability of FTA.

Because risk is defined as the product of probability and consequence, mere estimation and calculation of probabilities is not enough in calculating risk. Also the consequences must be evaluated numerically. Consequences of very different risks are taken into account, so there is no commensurate unit for all these different consequences. However, in practice the *very final* consequences are always calculated using some monetary unit. In this quantitative phase of risk analysis, all the ETA consequences generated in the qualitative phase are estimated in euros. In estimation of probabilities and consequences, the same sources of information are of help as in qualitative analysis, i.e. statistical data, experience and subjective opinions of experts. It must be noted that if quantification is not possible, quantitative risk analysis should not be requested at all.

In literature, the quantification is usually presented using deterministic values for probabilities. However, in reality it is impossible to give exact numerical values for uncertain probabilities and consequences. For that reason the use of distributions and simulation is preferred in this quantitative part of generic risk procedure. When giving estimates for probabilities and consequences, it is much easier to find a range of possible values instead of one consensual value. In FTA, the basic probabilities are expressed with appropriate distributions and after that the top event can be calculated using simulation. Likewise, in ETA, the numeric values for subsequent events and consequences are expressed with distributions. Then, using top event probability of FTA as the initiating event probability of ETA, the risk can be calculated with the aid of simulation. The result is of course a distribution, as all the input parameters are distributions.

2.6.4.4.2.4 Risk-based decision-making

When the identified adverse incidents have been analysed and risks estimated (qualitatively or quantitatively, according to need), risk evaluation can be performed. In this phase judgements are made about the significance and acceptability of the risks, and finally decisions are made on how to deal with the risks. All the adverse incidents should be already analysed and stored into the risk database with documentation. If the analyses described above are performed in the Lifecon extent, there should be risks in all four main categories and their sub-categories. If quantitative analyses have been performed for the adverse incidents the risks can be summed by category. If only qualitative risk analysis has been performed, still the number of adverse incidents that have an impact on a certain Lifecon consequence category is easily obtained. With normal database commands the primary factors of these risks in a certain category can be easily listed, and consequently they can be dealt with.

If the risk is acceptable, it is enough that the decision-maker is aware of the risk attendant upon the decision, but the evaluated risk does not have to be reduced. The decision is then made according to Lifecon decision-making procedures. Whether the risk is, in that case, one of the factors influencing the decision is up to the end user. If the risk is estimated and evaluated quantitatively, it can be easily included in the decision tree or MADA as a criterion. In a decision tree the limit for risk criterion is decided by the end user and in MADA the impact of risk is taken into account by giving appropriate weight to the risk criterion.

If the risk is not acceptable, further considerations must be made. There are four options to choose from:

- lowering the probability of the adverse incident
- reducing the consequences of the adverse incident
- rejecting the risk and
- transferring the risk.

The best option is to lower the probability of the adverse incident. With the visual causes–consequences structure (created in step 2 of the risk procedure), it is easy to see which factors affect the top event, and consequently more effort can be put into the problematic factors. If quantitative risk analysis has been performed, the allocating of efforts is even easier, because sensitivity analysis automatically reveals the biggest contributors to the top event.

Another way of reducing risk is to reduce the consequences of the adverse incident. Sometimes it can be easier to accept the relatively high probability of an adverse incident and create safeguards against severe consequences than to overspend resources in trying to reduce the probability. For example, input errors – when inserting information manually into any system – are unavoidable, but the system can be created so that an input error does not affect the system. Floods cannot easily be prevented, but an old stone bridge in weak condition can be closed for the flood peaks to avoid casualties, for example.

Rejecting risk in this context means rejecting an option in which unacceptable risk is involved. Decisions are normally made between different alternatives; thus, rejecting one alternative because of too high risk can be very practical in the decision-making process.

Risk transfer is used a lot but it cannot be recommended if sustainable development is to be emphasised. If this means is chosen, the risk itself does not diminish at all, only the responsibility is transferred to another party. In practice risk transfer means taking out insurance against the risk.

Whatever the risk-based decision is, it must be well documented (who made the decision, what were the circumstances, etc.). This means the quality of the decision can be followed up and improvements and updatings made to the FTA and ETA for future needs.

2.6.4.5   Using the risk assessment and control procedure in practice

In theory, all problems that involve uncertainty are best solved with a risk analysis approach, but risk analysis forms only a part of the decision-making and optimisation procedure. The reason for this is quite clear: many time-dependent phenomena (e.g. corrosion or carbonation) are studied and modelled accurately, and those models are or are being widely approved. However, in concrete facility management there are a lot of moments where suspicion arises but no models are available. In these situations a risk analysis approach is the best. For example, in the following hypothetical decision-making situations risk analysis can offer help:

- Bridge is always congested but in very good shape. Suspicion: Is it safe for the users or should it be widened or replaced with a broader one?
- Old bridge seems to be in good condition, but in the same subsoil area settlements of abutments have been reported. Suspicion: Is there a danger of settlement with this bridge also?
- A certain MR&R method works perfectly in one country and is used a lot there, but has not been used in another. Suspicion: Is the method applicable in this other country also, or should the facility owner keep using the traditional method?
- Long dark underpass in a suburb is always full of graffiti, otherwise the condition is good. Suspicion: Do the imago and worth of the area suffer and do people have uneasy feeling because of the old underpass, and consequently should the underpass be modified?
- Old building needs rehabilitation urgently, but should the façade be replaced with the same method and materials as were used when first built? Suspicion: The old building is very expensive for people, and strong modifications can raise resistance.

Decision-making and optimisation are sometimes performed on two hierarchical levels – the network level and the object level. Network is the whole stock of facilities, e.g. all the bridges owned by community or road administration, all the tunnels, or all the lighthouses. An object is logically one of these facilities: a certain bridge, a certain tunnel, a certain lighthouse. As can be seen in the examples above, risk-based decision-making is best applied on object level, because only then all the local factors can be taken into account. When the identified adverse incidents are analysed using the presented risk procedure, there will also be found causes that can and should be treated on a network level (e.g. low quality of inspection, bad safety policy, difficult data storing system). However, sensitivity analyses always show that object level factors contribute more to the probability of adverse incidents than network level causes.

Given that risk is a very case-sensitive variable and depends a lot on the facility owner's company strategy and preferences, a lot of responsibility is left for the end user in implementing the risk procedure. Only general guidelines can be given about how the risk procedure can be introduced, and used in the long run (maintained and updated). The prerequisite for successful risk management is that it will be taken seriously as any part of management, and that there are enough resources reserved for risk management. The corporate strategy and preferences concerning the Lifecon risk consequence categories (human, cultural, ecological and economical) and the risk acceptance levels in those categories should be clarified at the outset. Also, it must be decided before the analyses how the results will be used: is category-wise risk going to be the only one criterion in MADA among the other criteria, or is it going to be used separately like veto, what is the importance (or weight) of risk in decision-making, and so on.

The risk assessment and control procedure should be introduced first only on object level, taking some well-studied object as a pilot project. Then the first two steps of the risk procedure should be performed as extensively as possible. In finding out causes and consequences, innovation and imagination should not be restricted but rules of FTA and ETA should be followed. When the qualitative analyses are ready, the contributors and consequences of the identified adverse incidents are ready to be quantified. This next step (quantification) is the giant one and the reason why a well-studied object was chosen as a pilot project. Normally some estimates can be found, but for most of the contributors even a guess can be difficult. However, this very moment of helplessness is a positive improvement to current practice: end users are forced to see these weak (or blank) points in their maintenance and management policy! Consequently, they should revise the analyses, cut off the most improbable (or almost impossible) scenario branches, and after trimming the logical trees allocate their efforts to the problematic contributors and scenarios.

Finally, end users make the decision and compare it with the decisions made in reality. If there is a lot of difference in the decisions, the reasons should be studied. Finding reasons can be difficult, because in real life the decisions may not be documented, or even if the company has a written strategy, it need not necessarily be followed very accurately in practice. An important point to remember is that in the management of concrete infrastructures, reasonable and optimised decisions have been made for years without taking a risk approach, so big differences normally indicate that the new system needs adjustments.

At the outset of establishing the risk assessment and control procedure, the depth of the analyses will certainly be at a rather general level, because the system cannot be introduced, installed, established and verified in a day. But when the fault and event trees exist and more information flows in, these analyses will become more detailed and consequently the results more accurate. However, at best the process of establishing and adjusting the new system will take years and needs a lot of patience and commitment.

2.6.4.6 Qualitative or quantitative risk analysis

In many cases of management troubles, the qualitative evaluation of risks is enough for decision-making. Qualitative evaluation shows the weakest links and areas to facility owners, and counsels them to put more effort into solving problems on those areas. The routines (modelling, inspections, MR&R actions, etc.) are well established, and if some deviations from the plans occur, they are probably the result of human activity (negligence, carelessness, etc.). Unfortunately, in the management (and especially with MR&R actions) of concrete facilities human labour is needed in all phases and, unlike with machines and processes, human behaviour in different situations is very difficult to predict.

If there is no real need or possibility to get exact numbers for risks, then the heavy-scale quantitative risk analysis should not be carried out. Qualitative analysis and comparative estimates are more reliable and readable than absolute values, especially when single numbers are used instead of distributions in quantification. It should always be remembered that running a rigorous, quantified risk analysis is extremely expensive at present (mostly due to lack of consistent source data) and in normal cases out of the question in maintenance policies.

So far there have not been demands for quantitative risk analyses from the authorities in the maintenance sector, but the trend is in favour of more accurately calculated and explained decisions, and in the future, quantitative risk analyses can be some kind of routine. For example, in the offshore oil industry there are already regulations about quantitative risk analysis. However, the quantified analyses in the oil industry are not applied as extensively as is the goal of Lifecon (economical, ecological, human and cultural aspects), but have concentrated more on the human safety and environmental issues.

Applying qualitative risk analyses in the maintenance policy will be an improvement to the present day practice. The risk procedure proposed above does not require any miracles or higher wisdom from facility owners when used in a qualitative way. In addition, by applying FTA- and ETA-based qualitative risk analysis to the management policy, the facility owner can now prepare for the future, because this qualitative phase always precedes the quantitative analysis.

The qualitative risk analysis versus quantitative risk analysis is a topic for endless discussion. Of course decisions are easier to explain and justify if they are based on numeric facts. Unfortunately, in the construction sector these numeric facts have not been easy to find. Methodologies exist, but without appropriate numeric input they do not give consistent numeric results. On the other hand, the qualitative versions of risk methods can be applied with good results, but they do not help decision-makers who play only with numeric values.

2.6.4.7 Appendix to Section 2.6.4

In this example of the quantitative use of Lifecon risk procedure is presented. The adverse incident to be analysed is 'falling from a road bridge'. The bridge has a concrete deck and steel railings. The steps to be taken in this quantification phase are:

1. estimation of probabilities of events in fault tree
2. estimation of probabilities of events in event tree
3. estimation of numeric values for consequences in event tree
4. calculation of the probability of top event in fault tree
5. calculation of the expected costs in case of falling
6. combining of fault tree and event tree calculations to obtain the value for risk.

In this example the risk is calculated in both deterministic way and probabilistic way in order to show the difference between the two approaches. The deterministic calculations are performed using normal spreadsheet software (Excel), but for the probabilistic calculations simulation is used. In this example the simulations are performed using the software called @RISK, which is an add-in programme to Excel.

2.6.4.7.1 ESTIMATION OF PROBABILITIES OF EVENTS IN FAULT TREE

The fault tree used in this quantification phase is the one presented in Figure 2.33. The probabilities must be estimated for the diamond, circle and house symbols. Rectangle values are then calculated according to the principle presented in Figure 2.30. For easier follow-up of calculations, the fault tree of Figure 2.33 is re-drawn and re-coded in Figure 2.40.

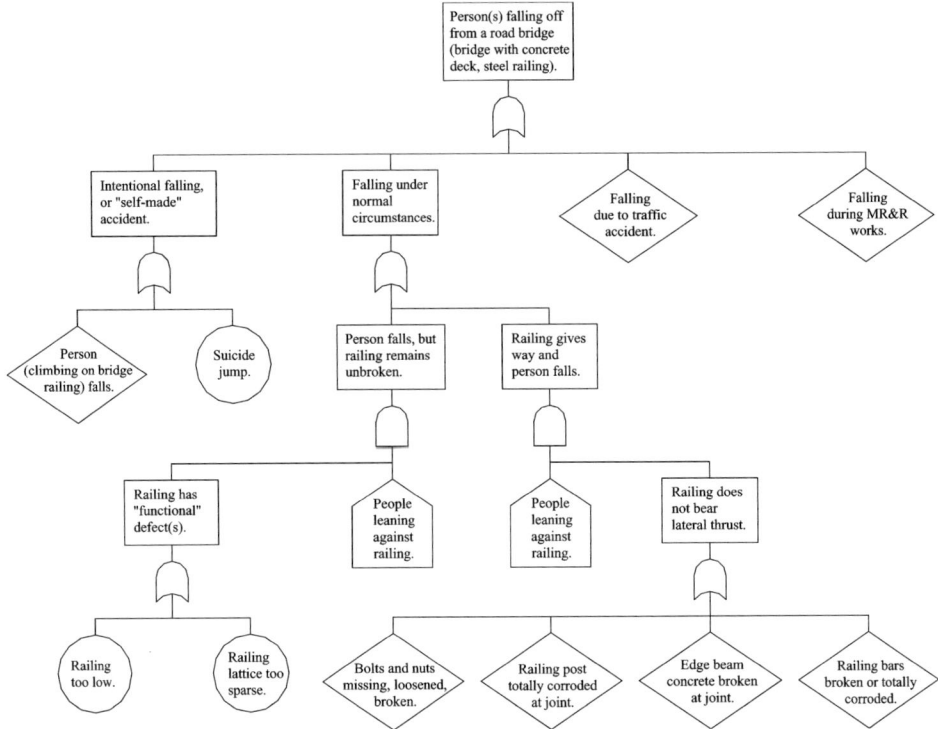

Figure 2.40 Redrawn and recoded fault tree of Figure 2.33

Note
Abbreviations for the fault tree events. The P's stand for probabilities of the corresponding events. The logic of numbering order is from left to right and from basic events – diamond, circle, house – through intermediate events to top.

The probabilities for diamond, circle and house events, presented in Table 2.8, are illustrative only. In true case they could be very different, like the fault tree itself, depending on the case.

2.6.4.7.2   ESTIMATION OF PROBABILITIES OF EVENTS IN EVENT TREE

The event tree used in this quantification phase is the one presented in Figure 2.39 in which four possible final outcomes from falling are presented. Those outcomes are:

- person dies
- person remains permanently disabled
- person injures badly, but recovers
- person gets only scratches or no harm at all.

Figure 2.39 is re-drawn for easy follow-up in Figure 2.41. The three subsequent events are numbered E_1, E_2, E_3 and their outcome options $E_{11}...E_{32}$. The outcome probabilities for these subsequent events are presented in Table 2.9. The probabilities are illustrative only.

Table 2.8 Illustrative frequency values for fault tree example

Frequency values for $P_1 \ldots P_{11}$ (basic events, basic failures or assumed basic failures)		
Event	Deterministic value	Probabilistic value (distribution with parameters, standard format)
P_1	0	0
P_2	0	0
P_3	0.08	Extvalue (0.07, 0.018, truncate (0, 0.2))
P_4	0.01	Extvalue (0.007, 0.005, truncate (0, 0.2))
P_5	1	1
P_6	0.008	Extvalue (0.006, 0.0035, truncate (0, 0.1))
P_7	0.007	Extvalue (0.005, 0.0033, truncate (0, 0.15))
P_8	0.002	Uniform (0, 0.004)
P_9	0.004	Uniform (0, 0.008)
P_{10}	0.001	Triang (0, 0.001, 0.002)
P_{11}	0.001	Triang (0, 0.001, 0.002)

2.6.4.7.3   ESTIMATION OF NUMERIC VALUES FOR CONSEQUENCES IN EVENT TREE

The outcomes shown in Figure 2.41 are:

- C_1 – person dies
- C_2 – person remains permanently disabled
- C_3 – person injures badly, but recovers
- C_4 – person gets only scratches or no harm at all.

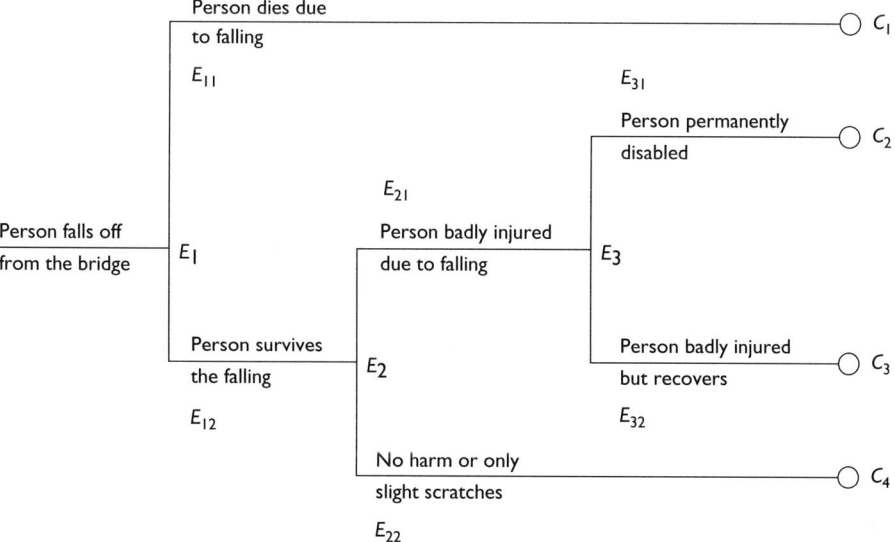

Figure 2.41 Event tree example, with events and outcomes numbered.

Table 2.9 Outcome probabilities for subsequent and final events of event tree

Outcome probabilities of subsequent events, in case the falling has happened.			
Event	Path	Deterministic value	Probabilistic value (distribution with parameters, standard format)
E_1	E_{11}	0.1	Uniform $(0,1)$; if $> 0.9 \Rightarrow$ path E_{11} is chosen, otherwise E_{12}
	E_{12}	$1 - E_{11} = 0.9$	
E_2	E_{21}	0.9	Uniform $(0,1)$; if $> 0.9 \Rightarrow$ path E_{22} is chosen, otherwise E_{21}
	E_{22}	$1 - E_{21} = 0.1$	
E_3	E_{31}	0.2	Uniform $(0,1)$; if $> 0.8 \Rightarrow$ path E_{31} is chosen, otherwise E_{32}
	E_{32}	$1 - E_{31} = 0.8$	

Table 2.10 Illustrative values for consequences

Consequence values (€) for $c_1 \ldots c_4$		
Event	Deterministic value	Probabilistic value (distribution with parameters, standard format)
C_1	1 000 000	Normal (1 000 000, 200 000, truncate (0, 5 000 000))
C_2	100 000	Normal (100 000, 10 000, truncate (0, 1 000 000))
C_3	10 000	Triang (0, 10 000, 20 000)
C_4	1000	Triang (0, 1000, 2000)

In Table 2.10, the consequences are numerically estimated and, once again, the numbers are more or less arbitrary and illustrative only.

2.6.4.7.4   CALCULATION OF THE PROBABILITY OF TOP EVENT IN FAULT TREE

With the help of Figures 2.33 and 2.40, the probabilities of the intermediate events and finally top event can be easily derived and calculated. The equations for solving the 'rectangle' probabilities are:

$$P_{12} = 1 - [(1 - P_3)(1 - P_4)]$$

$$P_{13} = 1 - [(1 - P_6)(1 - P_7)(1 - P_8)(1 - P_9)]$$

$$P_{14} = P_{12}\, P_5$$

$$P_{15} = P_5\, P_{13}$$

$$P_{16} = 1 - [(1 - P_1)(1 - P_2)]$$

$$P_{17} = 1 - [(1 - P_{14})(1 - P_{15})]$$

$$P_{\text{TOP}} = 1 - [(1 - P_{16})(1 - P_{17})(1 - P_{10})(1 - P_{11})].$$

Now the event frequencies can be calculated by inserting the starting values (either deterministic or distributions) from Table 2.8 into the equations. With deterministic starting values the frequencies for intermediate events and top event are obtained immediately. The result is:

$$P_{12} = 0.0892$$

$$P_{13} = 0.020846$$

$$P_{14} = 0.0892$$

$$P_{15} = 0.020846$$

$$P_{16} = 0$$

$$P_{17} = 0.108187$$

$$P_{\text{TOP}} = 0.10997.$$

When distributions and simulation is used, the result is a distribution also. With this example, 10 000 simulations were used. The resulting distribution for top event is presented in Figure 2.42. As can be seen in the figure, the mean of the distribution is close to the deterministic value, which is logical. However, with the distribution result the confidence interval can be easily seen. Also the sensitivity analysis is performed automatically, as can be seen in Figure 2.43. The two biggest contributors to the top event are the factors P_3 and P_4, which can be identified in Figure 2.40.

In the following figures (which are screenshots of the @RISK program) there appear letter–number codes preceded by slash on the titles and sides of the screenshots, for

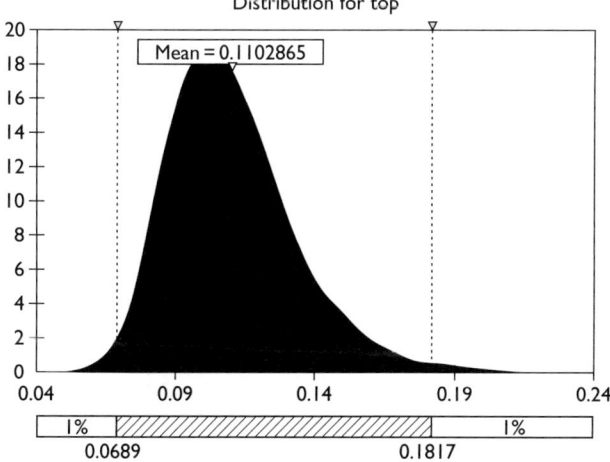

Figure 2.42 Simulated distribution for top event frequency.

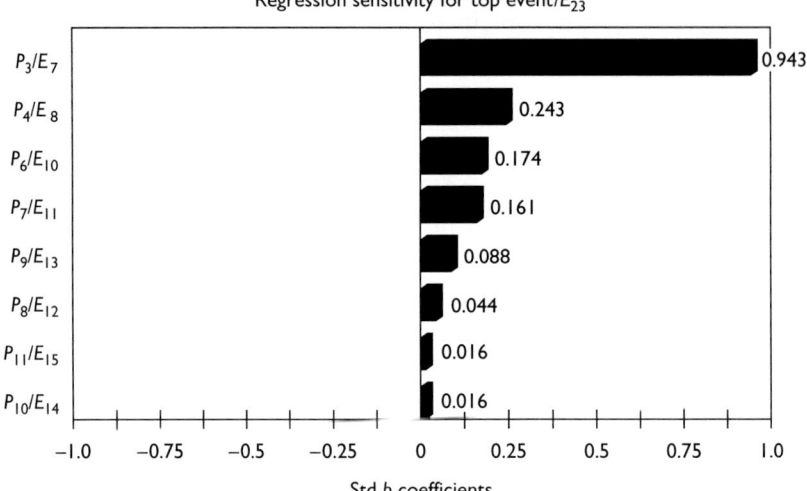

Figure 2.43 Sensitivity analysis for top event.

example in Figure 2.40 there appears a code/E_{23} in the title. These codes refer to the Excel worksheet cells used in this quantification example. This Excel worksheet is presented in Figure 2.48.

2.6.4.7.5 CALCULATION OF THE EXPECTED COSTS IN CASE OF FALLING

The different possible consequences of falling in safety category are presented in Figure 2.41, and in Tables 2.9 and 2.10 the numeric values have been estimated. Now the expected costs in case of falling can be easily calculated:

$$C_{falling} = E_{11}C_1 + E_{12}[E_{22}C_4 + E_{21}(E_{31}C_2 + E_{32}C_3)]$$

With deterministic values the result is obtained immediately by inserting the starting values into the equation above. The result is:

$$C_{falling} = \text{\euro}\,122\,770$$

When probabilistic values and simulation are used, the result differs a lot from the deterministic solution. This is due to difference in logics: while in deterministic model mean (or expected) values are used, simulation randomly chooses different scenarios, and repeats the procedure as long as is wanted. As there is discontinuity in consequence values of different outcomes, more than one peak is expected in the resulting distribution. The distribution of consequence costs after 10 000 simulations is presented in Figure 2.44, and in Figure 2.45 the result of sensitivity analysis is shown. The event E_1 has the biggest contribution to the consequences.

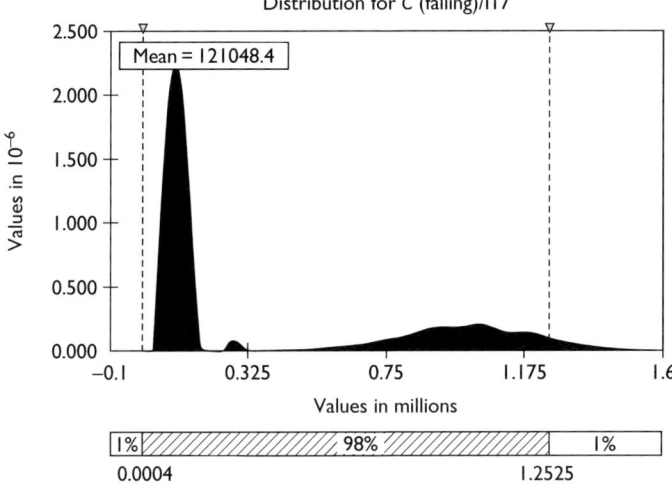

Figure 2.44 Distribution for consequence costs in case of falling.

Notes
1 YES means the criterion is desired to be minimised.
2 This criterion could be partly quantitative, partly qualitative (biodiversity for instance).

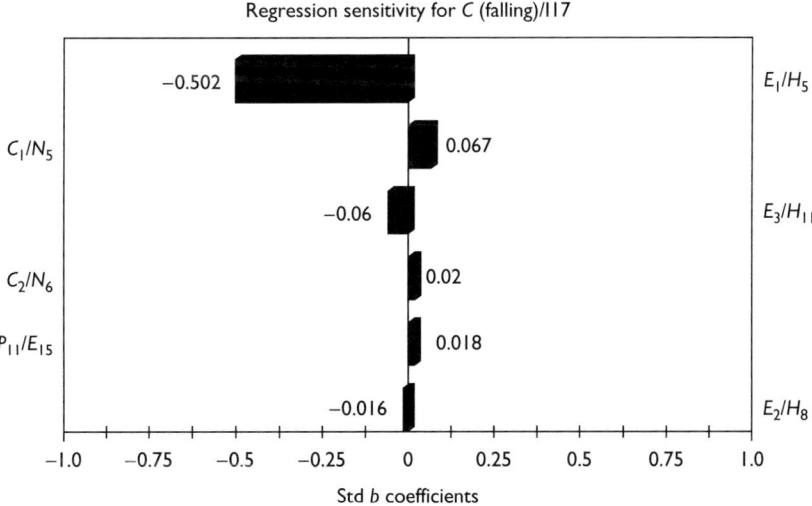

Figure 2.45 Sensitivity analysis for consequence costs in case of falling.

2.6.4.7.6   COMBINING OF FAULT TREE AND EVENT TREE CALCULATIONS TO OBTAIN THE VALUE FOR RISK

When both the probability of the top event and the expected costs have been calculated, the value of risk is obtained simply as a product of those two factors. In case of deterministic values the result is:

$$\text{Risk} = P_{\text{TOP}} \times C_{\text{falling}} = (0.10997/\text{year}) \times €122\,770 = €13\,500.98/\text{year}$$

When using simulation for calculation, the risk is (like in the deterministic case) the product of the top event probability and consequence cost. But instead of calculating value for the expected costs of falling, the consequence value is simulated on each simulation round and the resulting consequence value is multiplied with the top event value of the same simulation round. The resulting risk value is stored, and after certain number of simulations a distribution for risk can be drawn. The distribution in Figure 2.46 is obtained after 10 000 simulations. The sensitivity analysis is presented in Figure 2.47.

As can be seen, the simulation results reveal the advantage of using distributions. While the deterministic calculation gives only one value for risk (a mean), simulation presents in addition the whole range of possible values for risk. By changing the confidence interval limits the chance for high consequences can be examined. Most of the simulation programs perform also the sensitivity analysis automatically.

The example above was for illustrative purpose only, to show the quantified use of fault tree and event tree. However, the presented quantification possibility does not help an end user who does not have numerical source data to be inserted in the fault and event tree or expertise to perform. In those cases the analyses are performed only qualitatively.

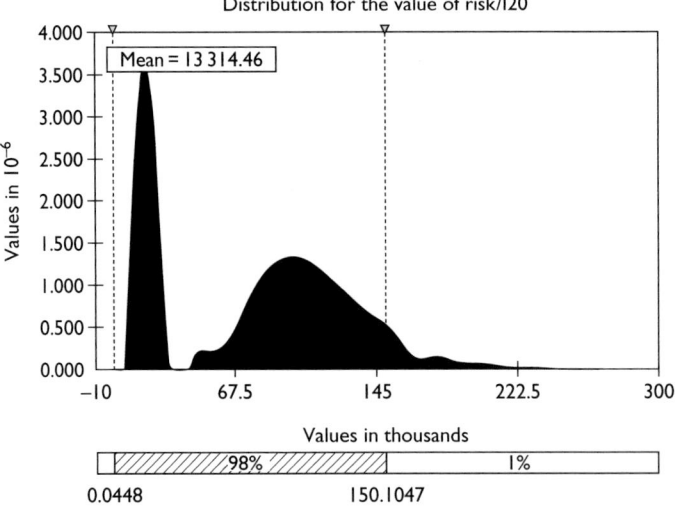

Figure 2.46 Distribution for the risk.

Regression sensitivity for the value of risk/I20

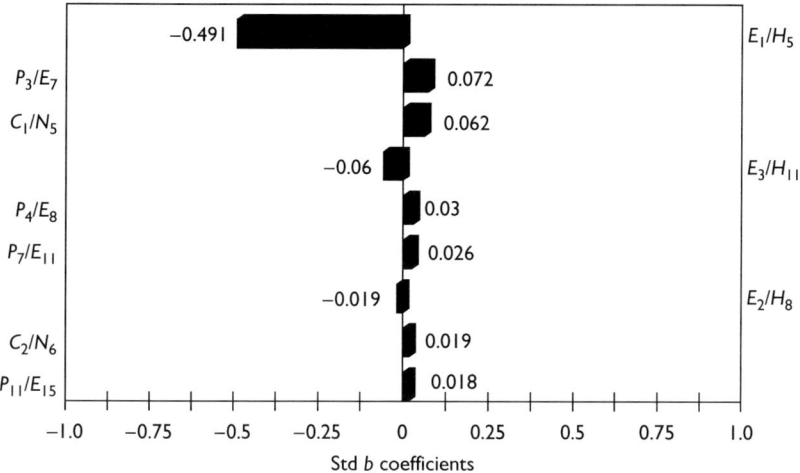

Figure 2.47 Sensitivity analysis for the risk.

	A	B	C	D	E	F	G	H	I	J	K	L	M	N
1	**Deterministic risk**			**Simulated risk** (Input distributions in light shaded cells, output distributions in dark shaded cells)										
2														
3	FTA deterministic			FTA simulation			ETA simulation		limit value	(Table 2.9)			ETA consequences	
4	(Table 2.8)			(Table 2.8)						E_{11}	E_{12}		(Table 2.9)	
5	P_1	0		P_1	0		E_1	0.5	0.9	0	1		C_1	1 000 000
6	P_2	0		P_2	0								C_2	100 000
7	P_3	0.08		P_3	0.08026					E_{21}	E_{22}		C_3	10 000
8	P_4	0.01		P_4	0.010061		E_2	0.5	0.9	1	0		C_4	1 000
9	P_5	1		P_5	1									
10	P_6	0.008		P_6	0.008042					E_{31}	E_{32}			
11	P_7	0.007		P_7	0.006974		E_3	0.5	0.8	0	1			
12	P_8	0.002		P_8	0.002									
13	P_9	0.004		P_9	0.004									
14	P_{10}	0.001		P_{10}	0.001				Consequence costs (simulated)					
15	P_{11}	0.001		P_{11}	0.001				in case falling has happened:					
16									C(falling):					
17	P_{12}	0.0892		P_{12}	0.089514				€10 000					
18	P_{13}	0.020846		P_{13}	0.020862									
19	P_{14}	0.0892		P_{14}	0.089514				The value of risk (simulated):					
20	P_{15}	0.020846		P_{15}	0.020862				€1 102.9082/year					
21	P_{16}	0		P_{16}	0									
22	P_{17}	0.108187		P_{17}	0.108509									
23	P_{TOP}	0.10997		P_{TOP}	0.110291									
24														
25	ETA deterministic		(Tables 2.9 and 2.10)				Expected cost (deterministic)							
26		probability		cons.value			in case of falling:							
27	E_{11}	0.1	C_1	1 000 000			€1 22 770							
28	E_{12}	0.9	C_2	100 000										
29	E_{21}	0.9	C_3	10 000			The value of risk (deterministic):							
30	E_{22}	0.1	C_4	1 000			€1 3 50 l/year							
31	E_{31}	0.2												
32	E_{32}	0.8												
33														

Figure 2.48 The Excel worksheet used for the quantification procedure.

Note

The comparison of deterministic and simulated risk cannot be made by comparing directly the numbers in cells F27 and I17 for the consequence costs, or the cells F30 and I20 for the value of risk. The comparable simulation results are presented in Figures 2.44 and 2.46.

2.7   Procedures in optimisation and decision-making

2.7.1   Optimisation and decision-making with Multiple Attribute Decision Aid (MADA)

2.7.1.1   Principles

This methodology is able to rank the alternatives in the order of preference (preference is measured by means of human requirements, lifetime economy, lifetime ecology and cultural criteria). In order to help the user, we have elaborated a framework identifying and explaining the six steps in Figure 2.49 [11].

2.7.1.2   Hierarchy

The hierarchy of buildings is divided into the following levels: network (stock of objects), object (building, bridge, dam, etc.), module, component, subcomponent, detail and material.

   The first step of MADA procedure consists in identifying the level of decision and the phase in the decision process.

   The decision-maker could decide at different phases of maintenance planning:

- Network level (among all the objects of the stock): Which one(s) is(are) identified as having priority for intervention?
- Object level (a building, bridge, etc.): Which part(s) of the object is(are) identified as having priority (e.g. during condition assessment)?
- Module, component, subcomponent, detail and material levels: What are the best solutions to keep or upgrade the level of requirements in performance?

2.7.1.3   Definition of an alternative

An alternative is defined according to the objectives. It could be:

- an entity amongst a set of objects (Bridge 1, Bridge 2, . . . Bridge i, . . . Bridge n),
- an action amongst a set of maintenance and repair solutions (M&R solutions).

As an example, once the need of intervention on an object is identified (by means of the condition assessment of the stock of objects), various actions (strategies for object management) are possible:

- no action
- maintenance solutions
- repair solutions
- restoration solutions
- rehabilitation solutions
- modernisation solutions
- demolition and new construction.

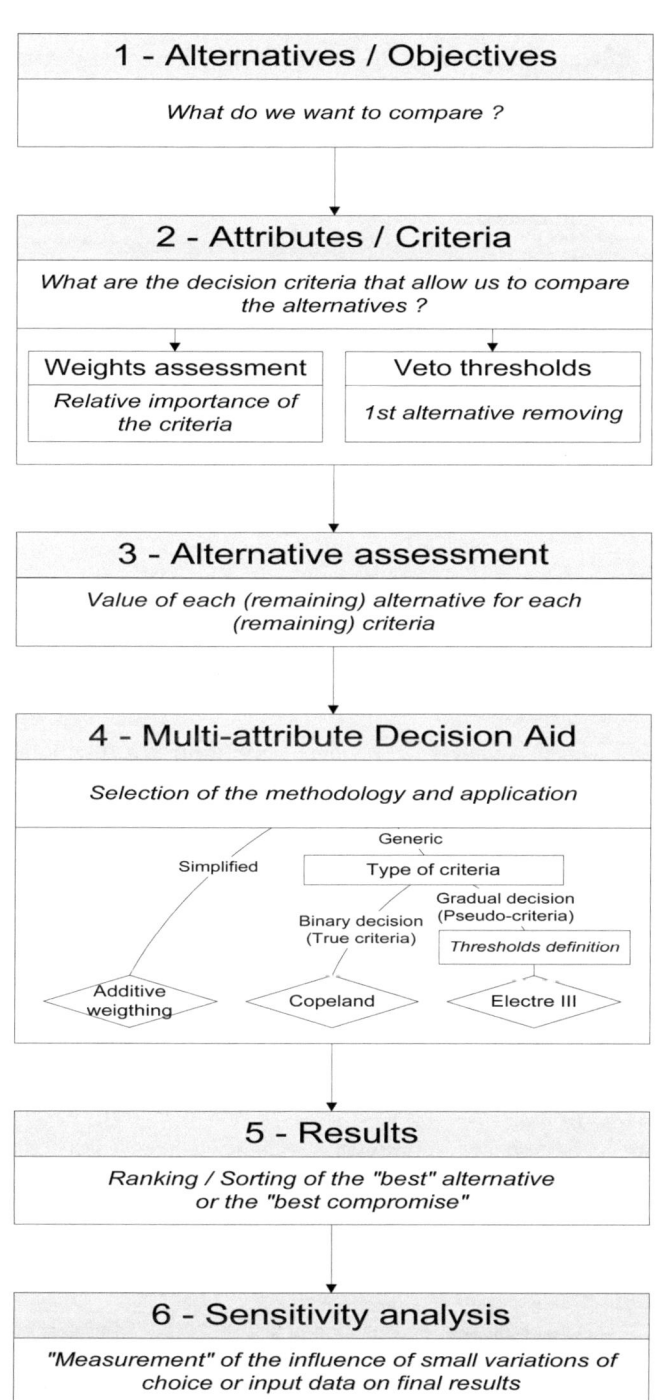

1 - Alternatives / Objectives

What do we want to compare ?

2 - Attributes / Criteria

What are the decision criteria that allow us to compare the alternatives ?

Weights assessment

Relative importance of the criteria

Veto thresholds

1st alternative removing

3 - Alternative assessment

Value of each (remaining) alternative for each (remaining) criteria

4 - Multi-attribute Decision Aid

Selection of the methodology and application

Generic

Simplified

Type of criteria

Gradual decision
(Pseudo-criteria)

Binary decision
(True criteria)

Thresholds definition

Additive
weigthing

Copeland

Electre III

5 - Results

Ranking / Sorting of the "best" alternative or the "best compromise"

6 - Sensitivity analysis

"Measurement" of the influence of small variations of choice or input data on final results

Figure 2.49 MADA flow-chart.

2.7.1.4   Generic and techno-economic attributes and criteria

What are the decision criteria that allow us to compare the alternatives? Once the alternatives are defined, we have to identify the various parameters (human, economical, ecological, cultural) characterising alternatives and allowing the comparison, as well as the importance of these parameters (by means of a weight).

An indicator (measure of the criteria) corresponds to each criterion. For example, in order to give some user guidelines, we propose:

- generic requirements and criteria
- techno-economic requirements and criteria.

The generic and techno-economic requirements can be classified as presented in this section and in Table 2.11. Generic requirements and criteria are refined into techno-economic requirements, indicators and criteria for fulfilling the primary criteria, as shown in Table 2.12. We can also establish a link between the general requirements and this techno-economic level, as shown in Table 2.13.

2.7.1.5   Assessment of the criteria

A method of assessment and a scale are associated with each criterion.

- Some are quantitative criteria and require a unit (Investment cost in M€, Future costs in M€/year).
- Some are qualitative criteria (5-level or 10-level scale) and are thus expressed with a textual description (For instance, a 5-level scale could be 'very good', 'good', 'medium', 'bad' and 'very bad').

We also have to define the characteristic called 'Minimisation (Y/N)'. Indeed, according to the criteria, less could be worse as far as quality is concerned (Minimisation = Yes), or less could be the best as for cost (Minimisation = No).

Table 2.11 Generic requirements of LMS [1]

Criteria	Attributes			
	A Human conditions	B Economy	C Ecology (Economy of nature)	D Culture
1	Functionality and usability	Investment economy	Raw materials resources economy	Building traditions
2	Safety	Building costs	Energy resources economy	Life style
3	Health	Life cycle costs	Pollution of air	Business culture
4	Comfort		Pollution of soil	Aesthetics
5			Pollution of water	Architectural styles and trends
6			Waste economy	Image
7			Loss of biodiversity	

Table 2.12 Relation between general requirements and techno-economic level

Criteria	Attributes				
	A Lifetime usability	B Lifetime economy	C Lifetime performance	D Lifetime environmental impact	E Recovery
1	Functioning of spaces	Investment economy	Static and dynamic safety and reliability in use	Non-energetic resources economy	Recycling of waste in manufacture of materials, components and modules
2	Functional connections between spaces	Construction cost	Service life	Energetic resources economy	Ability for selective dismantling
3	Health and internal air quality	Operation cost	Hygro-thermal performance	Production of pollutants into air	'Reusability' of components and modules
4	Accessibility	Maintenance cost	Safe quality of internal air	Production of pollutants into water	'Recycling-ability' of dismantling materials
5	Experience	Repair costs	Safe quality of drinking water	Production of pollutants into soil	Hazardous wastes
6	Flexibility in use	Restoration costs	Acoustic performance		
7	Maintainability	Rehabilitation costs	Changeability of structures and building services		
8	Refurbishability	Renewal costs	Operability		

Based on the assessment of the relative importance of each criterion over the others, the AHP methodology clearly takes into account expert opinion. This method was programmed in an Excel sheet to be used easily.

We assess the value of each alternative for each criterion. Software was developed to simplify the calculation. The problem is now totally defined and we can process the information. Each step is automated – from the value of each alternative for each criterion, as well as the weights, the software ranks to the alternatives.

2.7.1.5.1 SIMPLIFIED METHODOLOGY: ADDITIVE WEIGHTING

When using the software, a normalisation procedure is needed. The same results are usually attained but we recommend the use of the fourth method, which is the most powerful normalisation method in some very specific cases.

Table 2.13 Relation between generic and techno-economic requirements

Category	Requirement	Functionality and usability	Safety	Health	Comfort	Investment economy	Building costs	Life cycle costs	Raw materials resources economy	Energy resources economy	Pollution of air	Pollution of soil	Pollution of water	Waste economy	Loss of biodiversity	Building traditions	Life style	Business culture	Aesthetics	Architectural styles and trends	Image
A - Lifetime Usability	Functioning of spaces	X	X		X												X			X	
	Functional connections between spaces	X	X		X												X			X	
	Health and internal air quality	X	X	X	X						X	X	X				X				
	Accessibility	X	X	X	X														X	X	
	Experienceness																				
	Flexibility in use	X						X	X	X				X	X		X				
	Maintainability	X	X	X	X			X	X	X	X	X	X	X	X						
	Refurbishment-ability	X	X	X	X			X	X	X	X	X	X	X	X					X	
B - Lifetime Economy	Investment economy					X															
	Construction cost						X														
	Operation cost							X													
	Maintenance cost							X													
	Repair costs							X													
	Restoration costs							X													
	Rehabilitation costs							X													
	Renewal costs							X													
C - Lifetime Performance	Static and dynamic safety and reliability in use		X																		
	Service life							X	X	X	X	X	X	X	X			X			
	Hygro-thermal performance		X	X	X						X							X			
	Safe quality of internal air			X	X						X										
	Safe quality of drinking water			X									X								
	Acoustical performance			X	X																
	Changeability of structures and building services	X						X	X												
	Operability	X	X	X	X																
D - Lifetime Environmental impact	Non Energetic resources economy							X	X						X	X	X	X			
	Energetic resources economy							X			X	X			X	X	X	X			
	Production of pollutants into air		X	X				X			X	X	X		X	X					
	Production of pollutants into water			X				X				X	X		X	X					
	Production of pollutants into soil			X				X				X	X		X	X					
E - Recovery	Recycling of wastes of materials, components and modules							X	X	X		X		X		X					X
	Ability for Selective dismantling							X	X					X		X				X	
	"Reuse-ability" of components and modules							X	X	X	X	X	X	X		X				X	X
	"Recycling-ability" of dismantling materials							X	X	X	X	X	X	X							X
	Hazardous wastes			X	X			X			X	X	X	X							

(Column groups: A-HUMAN — Functionality and usability, Safety, Health, Comfort; B-ECONOMY — Investment economy, Building costs, Life cycle costs; C-ENVIRONMENT — Raw materials resources economy, Energy resources economy, Pollution of air, Pollution of soil, Pollution of water, Waste economy, Loss of biodiversity; D-CULTURE — Building traditions, Life style, Business culture, Aesthetics, Architectural styles and trends, Image)

2.7.1.5.2 COPELAND

COPELAND needs no intervention of the user.

2.7.1.5.3 ELECTRE III

Thresholds For ELECTRE III, we need the definition of three thresholds: strict preference threshold, indifference threshold and veto threshold [19, 20]. Let us take a simple example to illustrate the meaning of each threshold (Table 2.14)

(a) Indifference threshold means that we do not have preference between two entities for a given criterion, if the difference is lower than this threshold. For instance, if Future Costs (Object 1) = 5 and Future Costs (Object 2) = 10 then Object 1 is equivalent to Object 2 for the criterion 'Future Costs' $(10 - 5 < 10)$.
(b) Strict preference threshold means that we prefer one object to another for a given criterion if the difference between assessments is above the threshold. For instance, if Safety (Object 1) = 4 and Safety (Object 2) = 1 then Object 1 is preferred to Object 2 for the criterion 'Safety' $(4 - 1 > 2)$.
(c) Veto threshold means that we definitely prefer an alternative if the difference between assessments is above the threshold for at least one criterion. For instance, if Investment costs (Object 1) = 190 and Investment costs (Object 2) = 100 then Object 1 is definitely preferred to Object 2.

Calculations
Two different calculations have to be done successively: downward distillation and upward distillation.
   Each calculation gives a ranking. The final ranking is the mean of the two rankings:

- If the rankings are equivalent for the two calculations then the final ranking could be considered as the real ranking (the incomparableness indicator given in sheet 'Electre III' is equal to 0).
- If the rankings are totally different for the two calculations, then there is a doubt (the incomparableness indicator is far from 0). These alternatives are considered as disruptive elements. They have to be studied in more detail (checking the assessment for each criterion, doing pair-wise comparisons).

Table 2.14 Threshold illustration

	Safety	Health	Investment Costs	Future costs	Environmental impacts	Aesthetic
Strict preference threshold	2	4	50	15	4	8
Indifference threshold	1	2	20	10	2	4
Veto threshold	5	6	80	20	6	12

2.7.1.6   Ranking of alternatives

The MADA methodology leads to the ranking of alternatives by order of preference [14–18]. It is used to select the best alternative or possibly the best compromise and to select the actions to be applied to a stock of entities given a restricted budget (refer to the handbook).

2.7.1.7   Sensitivity analysis

The subjectivity of weights assessment as well as the uncertainty of the assessment of some criteria could lead to a great variation in the results. Therefore we have to measure the influence of variations on the ranking of alternatives, which shows the stability of the chosen MADA methodology. A suited method for this sensitivity analysis is the Monte-Carlo methodology.

In order to 'measure' the influence of decision on the results (during subjective steps), we will need to look at:

- the influence of the weights and
- the influence on alternatives assessments (range of value instead of deterministic values).

Note: The second aspect is partially taken into account in ELECTRE methods through thresholds.

With the first MADA analysis, from criteria, weights and alternatives assessments for these criteria, we obtain a ranking of alternatives.

Sensitivity analysis with Monte-Carlo simulation consists therefore of four steps (Figure 2.7):

1. Random assessment of the weights or alternatives assessments simulating small variations (e.g. $\pm 5\%, \pm 10\% \ldots$).
2. Application of the MADA methodology.
3. Ranking of alternatives.
4. Statistical analysis of the various rankings.

A simulated weight/alternative assessment is obtained by multiplying the initial weight/alternative assessment (given by the user) by a multiplicative factor (variation) modelling small variations (see Figure 2.8).

For instance, an initial weight $W = 30$, subjected to small variations in the interval $(-10\%, +10\%)$, will vary in the range $(30 \times 0.9 - 30 \times 1.1)$, i.e. $(27-33)$.

These small variations can be calculated by means of a bounded Gaussian distribution defined with:

$$\begin{cases} \text{Mean: } \mu = 1 \\ \text{Standard deviation: } \sigma = \dfrac{\text{variation}}{3} \end{cases}$$

It is then bounded in lower values and upper values respectively by $(1 - \text{variation})$ and $(1 + \text{variation})$.

The bounds and standard deviation are chosen this way to include 99.7% of the values (99.7% of a Gaussian distribution is included between $(\mu - 3\sigma)$ and $(\mu + 3\sigma)$).

After *n* simulations, we study the various ranking of alternatives and analyse the variations. If some alternatives are classified differently in the function of the simulations and with similar probabilities, then we could consider that the ranking is sensitive to the input parameters.

For instance, if an alternative is ranked 1st for 60% of the simulations and 2nd for 40% of the simulations then the results will be considered as sensitive to the input parameters.

Example

Let us assume that we have to manage a stock of objects (seven objects, to simplify the study). The objective is to 'measure' the importance of each object in order to select those that require an MR&R action. Each object is characterised by means of six criteria. Figure 2.50 gives the characteristics of each criterion.

1. *Weights*

The next step is the identification of the relative importance of the criteria. We suggest the establishment of an a priori ranking of the criteria. For instance:

> Aesthetic > Safety > Investment costs > Future costs > Health > Environmental impacts

Then AHP methodology is easier. The user just has to fill in the yellow cells with values representing the pair-wise preferences (according to the values given in Figure 2.51).

	Quantit./Qualitat.	Range	Unit	Minimisation[1]
Safety	Qualitative	[0–5]	–	No
Health	Qualitative	[0–5]	–	No
Investment costs	Quantitative	[0–200]	M€	Yes
Future costs	Quantitative	[0–50]	M€/year	Yes
Environmental impacts	Quantitative[2]	[0–10]	–	Yes
Aesthetic	Qualitative	[0–20]	–	No

Figure 2.50 Criteria definition.

		1 Safety	2 Health	3 Investment costs	4 Future costs	5 Environmental impacts	6 Aesthetic						
1	Safety	1	5	2	3	7	1	1	1	1	1	1	1
2	Health	1/5	1	1/4	1/3	1	1/7	1	1	1	1	1	1
3	Investment costs	1/2	4	1	3	6	1/3	1	1	1	1	1	1
4	Future costs	1/3	3	1/3	1	5	1/2	1	1	1	1	1	1
5	Environmental impacts	1/7	1	1/6	1/5	1	1/6	1	1	1	1	1	1
6	Aesthetic	1	7	3	2	6	1	1	1	1	1	1	1
								1	1	1	1	1	1
									1	1	1	1	1
										1	1	1	1
											1	1	1
												1	1
													1

Consistency ratio = 0.04108

Figure 2.51 AHP methodology for weight definition.
Reminder: The consistency ratio has to be lower than 10% (0.1) in order to have consistent comparisons.

The resulting weights are as shown in Figure 2.52 (column Q on the right of the table in the AHP sheet):

Criteria	Safety	Health	Investment costs	Future costs	Environmental impacts	Aesthetic
Weights	30	5	19	12	4	31

Figure 2.52 Weights definition.

2. *Alternative assessment*
For each object, we assess its values, for each criterion (Figure 2.53).

	Safety	Health	Investment costs	Future costs	Environmental impacts	Aesthetic
Weight	30	5	19	12	4	31
Minimization			Yes	Yes		
Alternative 1	4	0	100	20	4	16
Alternative 2	3	0	80	30	1	14
Alternative 3	2	0	85	30	5	6
Alternative 4	4	1	130	25	3	10
Alternative 5	1	0	30	35	6	2
Alternative 6	1	0	90	30	2	12
Alternative 7	2	0	88	20	7	8

Figure 2.53 Alternative assessment.

3. *MADA*
The problem is now totally defined and we can process the information. Each step is automated: from the value of each alternative for each criterion, as well as the weights, the software ranks the alternatives.

> *Launch the MADA macro in the Excel software*
> *(Tools menu → Macro → Macros → MADA)*

The user interface shown hereafter allows data input in the software (tick 'Yes' if the criteria has to be minimised) (Figure 2.54).

All the information concerning the criteria are stored in the sheet 'DATA' (Figure 2.55).

Note: When starting a new MADA study, the user has to tick the 'Erase historic' box. The results stored in the 'HISTO' sheet will be deleted.

This table is not screened automatically. The user has to select the sheet titled 'DATA'. Then click on the SELECT METHOD button to select one of the three methods:

- ADDITIVE WEIGHTING in the weighting method category,
- ELECTRE III in the outclassing method category,
- COPELAND in the ordinal method.

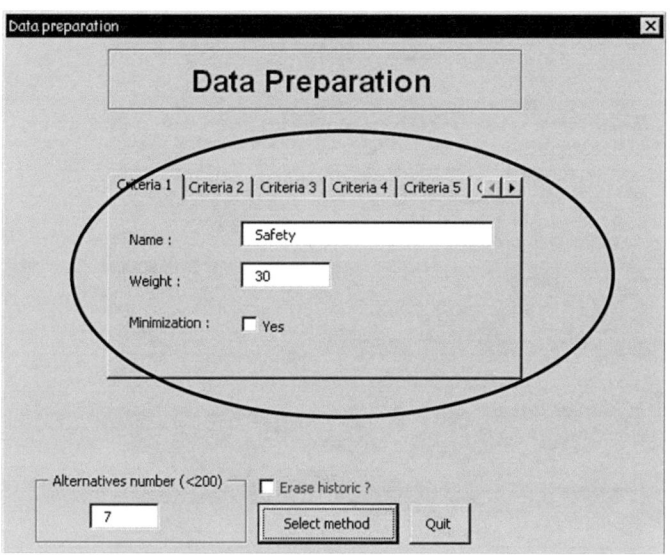

Figure 2.54 Main MADA software interface.

	A	B	C	D	E	F	G	H
1		Safety	Health	Investment costs	Future costs	Environmental impacts	Aesthetic	
2	Weight	30	5	19	12	4	31	
3	Minimization			Yes	Yes			
4	Alternative1	4	0	100	20	4	16	
5	Alternative2	3	0	80	30	1	14	
6	Alternative3	2	0	85	30	5	6	
7	Alternative4	4	1	130	25	3	10	
8	Alternative5	1	0	30	35	6	2	
9	Alternative6	1	0	90	30	2	12	
10	Alternative7	2	0	88	20	7	8	
11	Strict preference threshold	2	2	50	10	2	4	
12	Indifference threshold	1	2	20	15	4	8	
13	Veto threshold	5	2	80	20	6	12	
14								
15								
16								
17								
18								
19								
20								
21								
22								
23								
24								
25								
26								
27								
28								
29								

"DATA sheet" hinge

Figure 2.55 Data sheet.

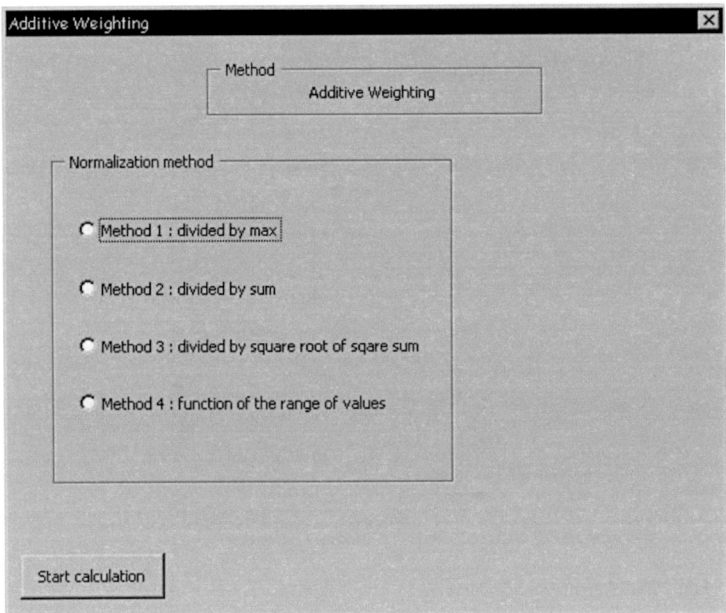

Figure 2.56 Additive weighting – Choice of the normalisation procedure.

4. *Additive weighting*

The following interface is opened. As seen previously, a normalisation procedure is needed (Figure 2.56). This usually gives the same results but we recommend the use of the fourth one, which is the most powerful.

The results are given in the following interface (Figure 2.57).

5. *COPELAND*

The calculation is done and the ranking is stored in the HISTO sheet.

6. *ELECTRE III*

For ELECTRE III method, we have seen that various threshold have to be defined. The user interface shown in Figure 2.58 allows the threshold input.

This discrimination threshold is a default value that could be left as default.

7. *Calculation*

- Click on the START CALCULATION button.
- Click on button Downward distillation.
- Click again on the START CALCULATION button and then on Ascending distillation.

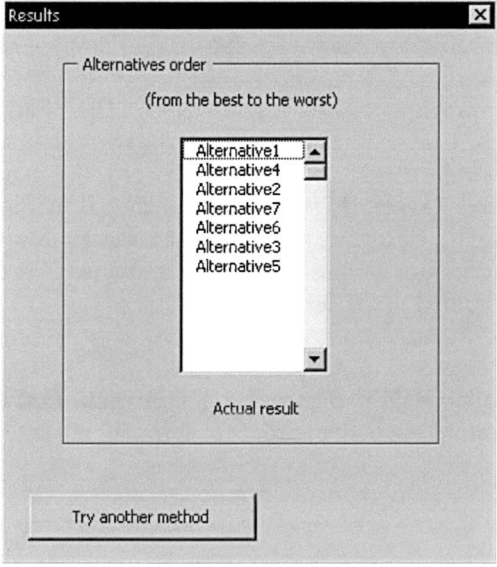

Figure 2.57 Additive weighting – Results.

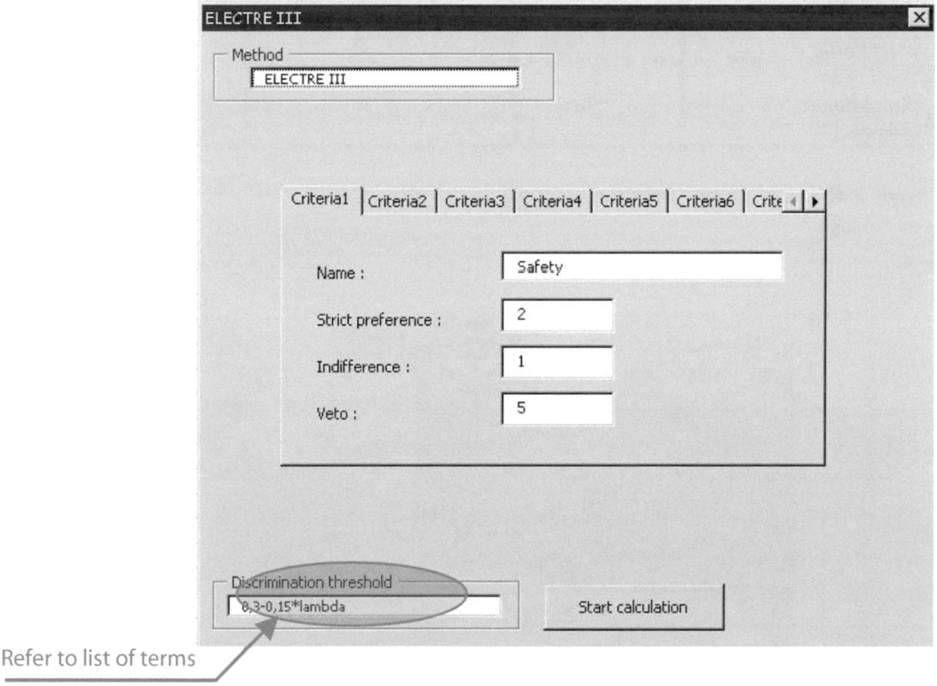

Refer to list of terms

Figure 2.58 ELECTRE III – Threshold definition.

8. *Results*

Two types of results are proposed. The first is a table (Figure 2.59) that gives the ranking for the descending and ascending row, as well as the median ranking and the incomparableness index.

The second is a plotting of the previous results (Figure 2.60). The alternatives located in the left upper corner are the best ones (ranked one for the two calculations). The more the dots go away (to the right lower corner of the graph), the worse the alternative. The alternative numbered 7 is clearly far from the diagonal (white line). Outside the light-shaded zone, there is a difference of two ranks between downward (medium ranking) and upward distillations (worst ranking). Alternative 7 is then a disruptive element, and should be studied more deeply.

9. *Results*

All the results are stored in a sheet titled HISTO (Figure 2.61). It gives the ranking for the three methods (and makes the distinction between the four different normalisation methods).

Alternatives	Row			Incomparableness
	Descending	Ascending	Median	
Alternative1	1	1	1	0
Alternative2	2	1	1.5	-0.5
Alternative5	3	2	2.5	-0.5
Alternative4	4	4	4	0
Alternative7	6	3	4.5	-1.5
Alternative6	5	5	5	0
Alternative3	6	6	6	0

Figure 2.59 ELECTRE III – Results.

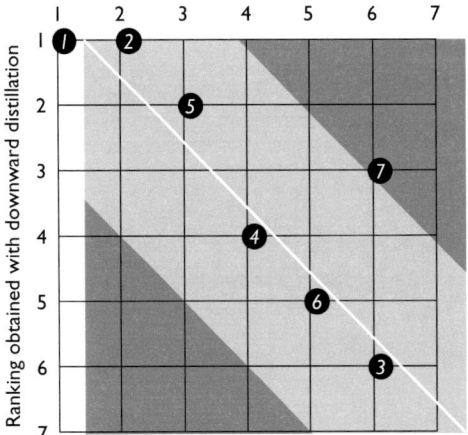

Figure 2.60 ELECTRE III – Plotting.

Additive weighting arrangement problems function of the range of utilities	ELECTRE III ranking problems not needed (outclassing method)	Copeland arrangement problems not needed
Alternative 1	Alternative 1 (1, 1)	Alternative 1
Alternative 4	Alternative 2 (2, 1)	Alternative 4
Alternative 2	Alternative 5 (3, 2)	Alternative 2
Alternative 7	Alternative 4 (4, 4)	Alternative 7 ex
Alternative 6	Alternative 7 (6, 3)	Alternative 3
Alternative 3	Alternative 6 (5, 5)	Alternative 5
Alternative 5	Alternative 3 (6, 6)	Alternative 6 ex

Figure 2.61 HISTO sheet.

Notes

1 For the Copeland method, the results indicate 'placed equal' alternatives (Alternatives 2 and 7, as well as 5 and 6 have to be considered as equivalent).

2 The results in this example differ according to the method used.

3 Searching for the commonalities between methods, we can draw a graph (Figure 2.62) showing the average ranking obtained with the three methods for each alternative (cross), as well as the minimum and maximum rankings (vertical line).

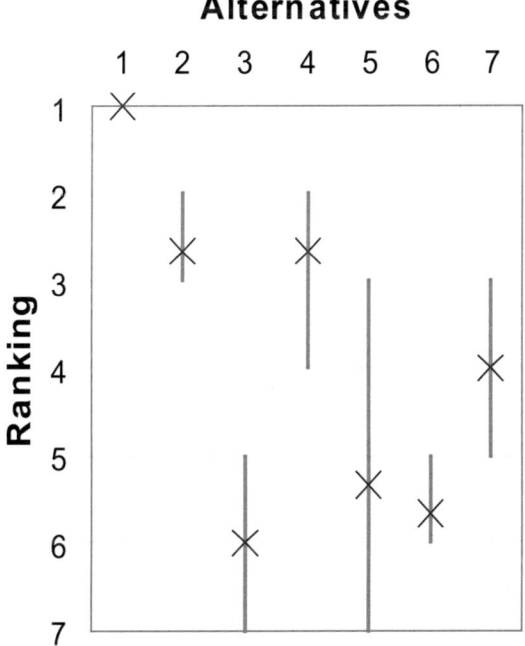

Figure 2.62 Comparison of the results.

We identify that:

- alternative 1 is clearly the best,
- alternatives 2 and 4 are acceptable, and seem to be joint second in rank (but we could not decide if one is better than the other),
- alternative 7 is a 'medium' alternative,
- alternatives 3 and 6 are bad alternatives.

As for alternative 5, it is difficult to obtain information from the MADA procedure and therefore it could be a medium or bad alternative.

10. *Sensitivity analysis*
This part was not totally developed in the Lifecon project. We will illustrate the sensitivity analysis with a basic example and give the code used to process this sensitivity analysis.

For instance, we have performed a sensitivity analysis with additive weighting methodology, first looking at the influence of a small variation of the weights, and second a small variation on the assessments.

11. *Variation of the weights*
Using the MATLAB code with variations of 20 and 50% on the weights, we obtain the results shown in Figures 2.63 and 2.64.

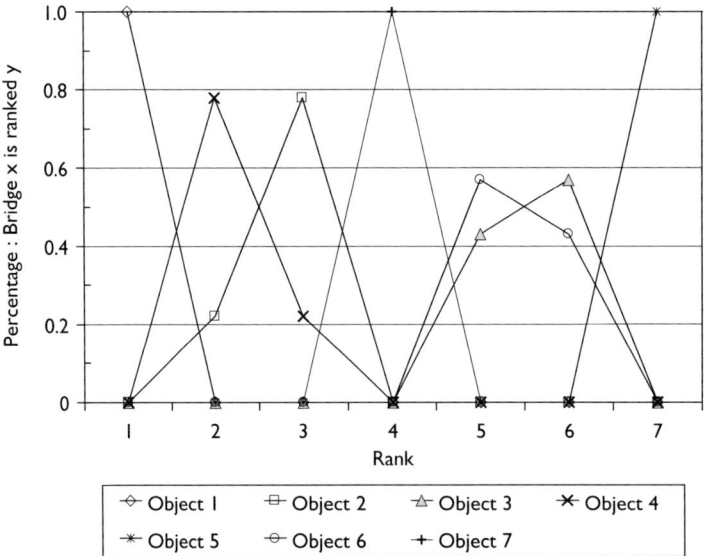

Figure 2.63 Additive weighting – Weights ±20%.

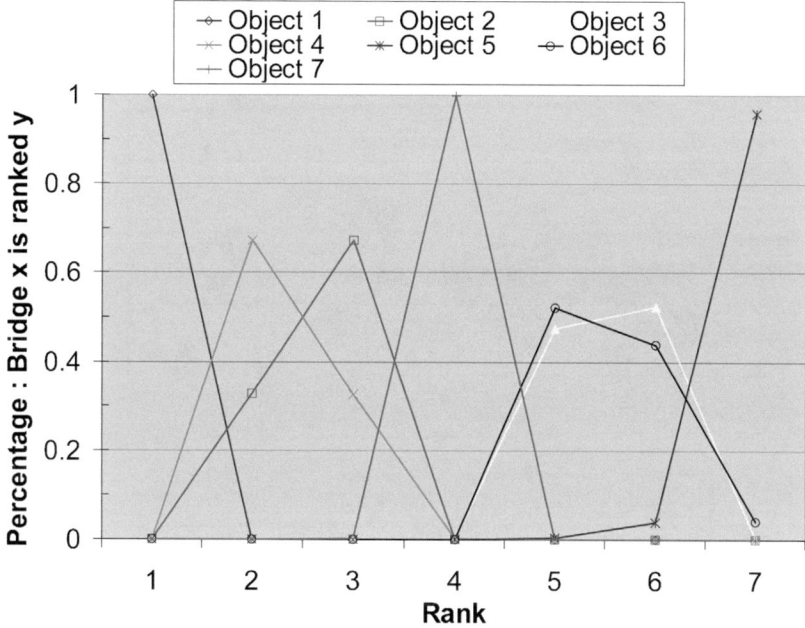

Figure 2.64 Additive weighting – Weights ±50%.

For variations of 20 or 50%, we notice that object 1 is always ranked number 1, object 7 is always ranked 4, object 5 is always ranked 7 (95% of the simulations for a 50% variation). This shows that the ranking is very reliable. Concerning the objects 2 and 4 on the one hand, and objects 3 and 6 on the other hand, we cannot be sure of the ranking for a 50% variation.

	Rank 2	Rank 3			Rank 6	Rank 7
Object 2	33%	67%		Object 3	48%	52%
Object 4	67%	33%		Object 6	52%	44%

Note
Object 6 is ranked 7 in 44% of the simulation.

The final ranking is then:

Rank	1	2	3	4	5	6	7
Object	1	4 or maybe 2	2 or maybe 4	7	3 or 6	3 or 6	5

12. *Variation of the assessments*
The same reasoning can be used for variations of assessments using the MATLAB code.

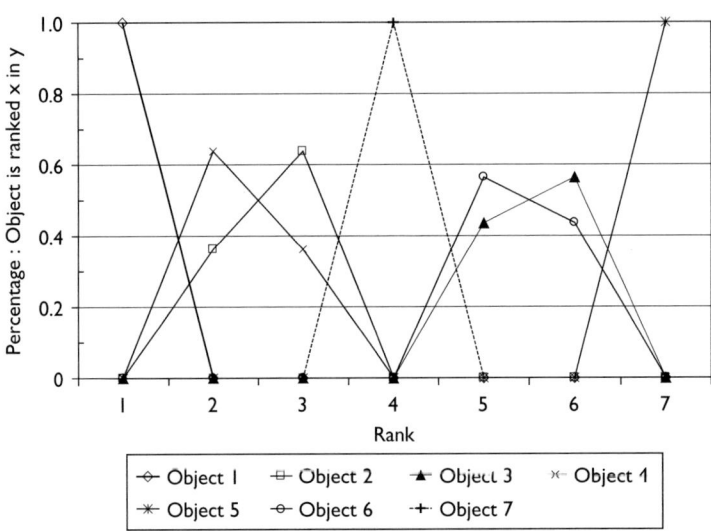

Figure 2.65 Additive weighting – Assessments ±10%.

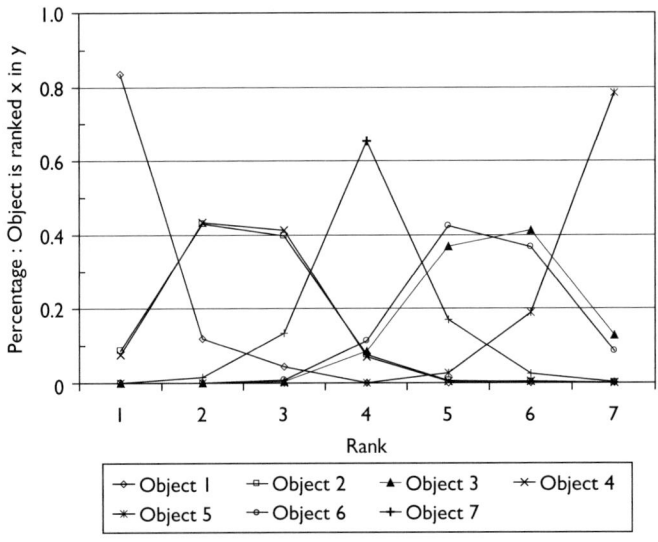

Figure 2.66 Additive weighting – Assessments ±30%.

In the case of variations of 10%, the results are rather stable (Figures 2.65 and 2.66). We just have uncertainty on the ranking of objects 4 and 2 on the one hand (Indecision 63%/37% for being ranked 2 or 3), and objects 6 and 3 on the other hand (Indecision 58%/42% for being ranked 5 or 6).

In case of variations of 30% (Figure 2.67), the results are:

Rank	1	2	3	4	5	6	7
Object	1	2 or 4	4 or 2	7	6 or 3	3 or 6	5

Figure 2.67 Final ranking.

The aim of sensitivity analysis is to control the influence of the user's choices (concerning weights and assessments) on the final results. Such a procedure gives more confidence in the results.

Acknowledgements

This Section 2.7.1 is based on the corresponding report of the contributor and Mr Jerome Lair from CSTB, who I would like to thank for his kind co-operation, especially with regard to the literature study and the computer application.

2.7.1.8 Appendix 1 to Section 2.7.1

MATLAB code for sensitivity analysis
Additive weighting method – Influence of the variation of the weights

```
Clear all;
% ---------------------------------------------------------------
% FIRST MODULE: PROBLEM DEFINITION
% ---------------------------------------------------------------
% Criteria weights
w=[30,5,19,12,4,31];
% Minimisation (1 → minimisation/0 → maximisation)
m=[0,0,1,1,0,0];
% Matrix with assessments
assessment = [4    0    100    20    4    16;3    0    80    30    1    14;
2    0    85    30    5    6;4    1    130    25    3    10;1    0    30
35    6    2;1    0    90    30    2    12;2    0    88    20    7    8];
% e = error (10% → e=0.1)
e = 0.5;
% n = number of iterations
n = 500;
% Number of alternatives (anb) and criteria (cnb)
anb=size(assessment,1);
cnb=size(assessment,2);
% ---------------------------------------------------------------
% SECOND MODULE: RANDOM WEIGHTS
% ---------------------------------------------------------------
% Creation of a matrix with all simulated weights
```

```
for i=1:n
for j=1:cnb
w_sim(i,j)=w(j)*(1+(2*rand-1)*e);
end
end
% ---------------------------------------------------------------
% THIRD MODULE: AHP METHOD
% ---------------------------------------------------------------
% Normalisation of weights
for i=1:n
for j=1:cnb
w_sim_norm(i,j)=w_sim(i,j)/sum(w_sim(i,:));
end
end
% Normalisation and minimisation of alternatives
for i=1:cnb
minimum(i)=min(assessment(:,i));
maximum(i)=max(assessment(:,i));
end
for i=1:anb
for j=1:cnb
assess_norm(i,j)=(assessment(i,j)-minimum(j))/(maximum(j)
-minimum(j));
end
end
for i=1:cnb
if m(i)==1
for j=1:anb
assess_norm(j,i)=1 - assess_norm(j,i);
end
end
end
% Agregation sum(weights*assessment)
for i=1:n
for j=1:anb
temp=0;
for k=1:cnb
temp=temp+w_sim_norm(i,k)*assess_norm(j,k);
end
res(i,j)=temp;
end
end
% Increasing order (rank)
for i=1:n
vect=res(i,:);
for j=1:floor((anb+1)/2)
mini=min(vect);
```

```
maxi=max(vect);
order(i,anb+1-j)=mini;
order(i,j)=maxi;
v=1;
test1=0; test2=0;
for k=1:anb+2-2*j
if vect(k)==mini | vect(k)==maxi
test1=1;
else
tem(v)=vect(k);
v=v+1;
test2=1;
end
end
if test1==1
vect=mini;
end
if test2==1
clear vect;
vect=tem;
clear tem;
end
end
end
end
% Assessment of rank
for i=1:n
for j=1:anb
for k=1:anb
if res(i,j)==order(i,k)
rank(i,j)=k;
end
end
end
end
```

2.7.1.9 Appendix 2 to Section 2.7.1

MATLAB code for the sensitivity analysis
Additive weighting method – Influence of the variation of the assessments
Clear all;

```
% -------------------------------------------------------------
% FIRST MODULE: PROBLEM DEFINITION
% -------------------------------------------------------------
% Criteria weights
w=[30,5,19,12,4,31];
```

```
% Minimisation (1→minimisation 0→maximisation)
m=[0,0,1,1,0,0];
% Matrix with assessments
assessment = [4     0   100  20   4  16; 3  0   80  30     1  14;
2         0    85   30 5    6; 4  1   130    25  3   10; 1  0  30
35        6     2; 1  0    90   30   2  12; 2  0   88  20     7   8];

% e = error (10%→e=0.1)
e = 0.1;
% n = number of iterations
n = 500;
% Number of alternatives (anb) and criteria (cnb)
anb=size(assessment,1);
cnb=size(assessment,2);
% -----------------------------------------------------------------
% SECOND MODULE: WEIGHTS
% -----------------------------------------------------------------
% Normalisation of weights
for i=1:cnb
w_norm(i)=w(i)/sum(w(:));
end
% -----------------------------------------------------------------
% THIRD MODULE: AHP METHOD
% -----------------------------------------------------------------
for nb=1:n
% Random assessment of alternatives
for i=1:anb
for j=1:cnb
assessment_sim(i,j)=assessment(i,j)*(1+(2*rand-1)*e);
end
end
% Normalisation and minimisation of alternatives
for i=1:cnb
minimum(i)=min(assessment_sim(:,i));
maximum(i)=max(assessment_sim(:,i));
end
for i=1:anb
for j=1:cnb
assessment_sim_norm(i,j)=(assessment_sim(i,j)−minimum(j))
/(maximum(j)−minimum(j));
end
end
for i=1:cnb
if m(i)==1
for j=1:anb
assessment_sim_norm(j,i)= 1 −assessment_sim_norm(j,i);
end
end
```

```
end
% Agregation sum(weights*assessment)
for j=1:anb
temp=0;
for k=1:cnb
temp=temp+w_norm(k)*assessment_sim_norm(j,k);
end
res(nb,j)=temp;
end
% Increasing order (rank)
vect=res(nb,:);
for j=1:floor((anb+1)/2)
mini=min(vect);
maxi=max(vect);
order(nb,anb+1-j)=mini;
order(nb,j)=maxi;
order;
v=1;
test1=0; test2=0;
for k=1:anb+2-2*j
if vect(k)==mini | vect(k)==maxi
test1=1;
else
tem(v)=vect(k);
v=v+1;
test2=1;
end
end
if test1==1
vect=mini;
end
if test2==1
clear vect;
vect=tem;
clear tem;
end
end
% Assessment of rank
for j=1:anb
for k=1:anb
if res(nb,j)==order(nb,k)
rank(nb,j)=k;
end
end
end
end
```

2.7.2 Procedure of QFD

2.7.2.1 Alternative applications

The QFD method basically means only handling the requirements and properties, analysing their interrelations and correlations as well as their weights and finally optimising the LCQ (Life Cycle Quality) properties and selecting from alternative solutions of asset management strategies or MR&R plans, designs, methods and products [33–36]. Hence, QFD can be applied in many variables, depending on the characteristic aims and contents of each application.

In Lifecon LMS, QFD can be used for the following purposes [4]:

- identifying functional requirements of owner, user and society;
- interpreting and aggregating functional requirements into primary performance properties;
- interpreting the performance properties into technical specifications of the actual object;
- optimising the performance properties and technical specifications in comparison to requirements;
- alternating the selection between different design and repair alternatives;
- alternating the selection between different products.

The QFD can be used on all levels of the Lifecon LMS system [4]:

- Network level: prioritising the requirements of users, owners and society, strategic optimisation and decision-making between alternative MR&R strategies;
- Object level: ranking of priorities between objects, optimising and decision between MR&R alternatives, technologies, methods and products;
- Module, component and detail/materials levels: refined optimising and decision between MR&R alternatives, technologies, methods and products.

2.7.2.2 Phases of the QFD procedure

The QFD procedure usually has three main phases, as presented in the application examples in the Appendices at the end of this section and in Deliverable D5.1 [4]:

1. Selection of the primary requirements and their weight factors from a set of numerous detailed requirements with the aid of QFD matrix.
2. Moving the primary requirements and weight factors into second QFD matrix for selection between the alternatives of plans, designs, methods or products.
3. Sensitivity analysis with simulation of variances of primary requirements and properties [D5.1].

The following detailed procedure can be applied in Lifecon LMS when using QFD for analysis of functional requirements against owner's and user's needs, technical specifications against functional requirements, and design alternatives or products against technical specifications:

1. Identify and list factors for 'Requirements' and 'Properties'.
2. Aggregate and select the requirements into primary requirements.
3. Evaluate and list priorities or weighting factors of 'Primary Requirements'.
4. Evaluate correlation between 'Requirements' and 'Properties'.
5. Calculate the factor: correlation multiplied by weight for each 'Property'.
6. Normalise the factor 'correlation multiplied by weight' of each 'Property' to use as a priority factor or weighting factor of each 'Property' in the subsequent steps.

2.7.2.3 Requirements and properties

2.7.2.3.1 GENERIC REQUIREMENTS AND PROPERTIES

Lifecon LMS is aiming to fulfil the requirements of sustainable development, which are defined in very general terms within society. Alternative systemisation to that used for other methods is required for the application of QFD in the practice of Lifecon LMS. Therefore, this systematisation of requirements and properties is presented in more details for QFD [4].

In QFD the following categories of compatible requirements and properties can be used:

1. generic requirements
2. generic performance properties as attributes of lifetime quality
3. application-specific performance properties as attributes of lifetime quality
4. aggregated or primary application-specific performance properties as attributes of lifetime quality
5. technical specifications of products [Lifecon D5.1].

Generic requirements are always applied as basic requirements independent of the application. These can be described and modelled with the aid of generic performance properties.

When moving into a specific application of QFD in practice, the generic performance properties will be interpreted into generic attributes, and further into application-specific attributes of the actual object.

Examples of this procedure were presented in Lifecon deliverable D5.1, and explain the practical application of the hierarchy of the requirements and properties in detail.

2.7.2.3.2 PERFORMANCE REQUIREMENTS AND PROPERTIES

The generic requirements of lifetime quality cannot be used directly in planning and design procedures of Lifecon LMS, because it is not possible to model and describe the solutions of planning, design, methods or products of the asset management and MR&R planning. Therefore the requirements and properties must be defined separately in each application. As a link between generic lifetime quality requirements and specific calculations, the lists presented in Tables 2.15, 2.16 and 2.17 [37] can be

Table 2.15 Specified categories of Generic Requirements [38–42]

A *Performance*
  A1 Conformity
     A1.1 Core processes
     A1.2 Supporting processes
     A1.3 Corporate image
     A1.4 Accessibility
  A2 Location
     A2.1 Site characteristics
     A2.2 Transportation
     A2.3 Services
     A2.4 Loadings to immediate surroundings
  A3 Indoor conditions
     A3.1 Indoor climate
     A3.2 Acoustics
     A3.3 Illumination
  A4 Service life and deterioration risks
     A4.1 Service life
     A4.2 Deterioration risks
  A5 Adaptability
     A5.1 Adaptability in design and use
     A5.2 Space systems and pathways
  A6 Safety
     A6.1 Structural safety
     A6.2 Fire safety
     A6.3 Safety in use
     A6.4 Intrusion safety
     A6.5 Natural catastrophes
  A7 Comfort

B *Cost and environmental properties*
  B1 Life cycle costs
     B1.1 Investment costs
     B1.2 Service costs
     B1.3 Maintenance costs
     B1.4 Disposal and value
  B2 Land use
  B3 Environmental burdens during operation
     B3.1 Consumption and loads, building
     B3.2 Consumption and loads, users
  B4 Embodied environmental impacts
     B4.1 Non-renewable natural materials
     B4.2 Total energy
     B4.3 Greenhouse gases
     B4.4 Photochemical oxidants
     B4.5 Other production-related environmental loads
     B4.6 Recycling

C Requirements of the process
  C1 Design and construction process
     C1.1 Design process
     C1.2 Site operations
  C2 Operations
     C2.1 Usability
     C2.2 Maintainability.

Table 2.16 Performance indicators of *civil infrastructures* in Lifecon LMS [D2.1, D2.3 Part I] [4]

Primary Requirements				
A Lifetime Usability	B Lifetime Economy	C Lifetime Performance	D Lifetime Ecology	E Culture
1 Functioning in use	Investment economy	Static and dynamic serviceability in use	Non-energetic resources economy	Compatibility with local building traditions
2 Flexibility in use	Construction cost	Service life	Energetic resources economy	compatibility with local natural and built environment
3 Health in construction	Operation cost	Hygro-thermal performance	Production of pollutants into air	aesthetic acceptability
4 Health in use and maintenance	Maintenance cost	Acoustic performance	Production of pollutants into water	acceptability in image requirements of the built environment
5 Comfort in use	Repair costs	Operability	Production of pollutants into soil	
6 Maintainability	Restoration costs	Changeability of structures	Reusability	
7 Safety in construction	Rehabilitation costs		Recycling-ability	
8 Safety in use	Renewal costs	Operability	Loss of biodiversity	

Table 2.17 Performance indicators of buildings in LIFECON LMS [D2.1, D2.3 Part I] [4]

Primary Requirements				
A Lifetime Usability	B Lifetime Economy	C Lifetime Performance	D Lifetime Ecology	E Culture
1 Functioning of spaces	Investment economy	Static and dynamic safety and reliability in use	Raw materials economy	building traditions
2 Functional connections between spaces	Construction cost	Service life	Energetic resources economy	life style
3 Health and internal air quality	Operation cost	Hygro-thermal performance	Production of pollutants into air	business working culture
4 Accessibility	Maintenance cost	Safety & quality of internal air	Production of pollutants into water	aesthetics

Table 2.17 (Continued)

Primary Requirements				
A Lifetime Usability	B Lifetime Economy	C Lifetime Performance	D Lifetime Ecology	E Culture
5 Experience	Repair costs	Safety & quality of drinking water	Production of pollutants into soil	architectural styles and trends
6 Flexibility in use	Rehabilitation costs	Acoustic performance	'Reuse-ability' of components and modules	image
7 Maintainability	Renewal costs	Changeability of structures and building services	Recycling of waste in manufacture and repair works	
8 Refurbishment ability	Demolition, recovery, recycling and disposal costs	Operability	Loss of biodiversity	

used. The requirements, which are presented in the column titles, must be interpreted for calculations with performance indicators. These performance indicators, which are presented in the cells, are variables, which can be expressed in quantitative (numerical) values, and thus can be used in numerical calculations. This is possible, when dealing with so-called laboratory problems [Lifecon D 2.1]. When dealing with so-called real world problems these indicators have to be expressed qualitatively [Lifecon D2.1].

2.7.2.4 Aggregation of life cycle performance requirements and properties

2.7.2.4.1 AGGREGATION METHODS

Because of the complexity of the building system, the decisions between design alternatives of the building, as well as between its technical system, module and product alternatives must be simplified limiting the number of parameters used during final decisions. With this aim in mind the aggregation of a number of design parameters will be outlined.

As described earlier, the final objective of LMS system is the optimised life cycle quality which consists of four dominant groups of parameters:

1. lifetime human requirements
2. lifetime economy
3. lifetime cultural aspects
4. lifetime ecology.

The optimisation and decision-making in lifetime management often includes quite numerous variables both on the level of generic requirements and on the level of technical and economic criteria. This can lead to very complex optimisation and decision-making procedures. Therefore, these parameters of generic and techno-economic levels are aggregated into primary parameters in the choice between repair alternatives and products (Table 2.18). They are referred to in this section as 'Primary Technical Properties'.

2.7.2.4.2 AGGREGATION PROCEDURES

An important phase of the optimisation or decision-making procedure is the aggregation of a large number of specific performance properties into the LCQ (Life Cycle Quality) properties. The aggregation procedure includes the following stages:

- listing the parameters to be aggregated
- defining the values and weights of these parameters
- summing the values times weights of the parameters into aggregated values.

Table 2.18 Methods used in aggregating the life cycle quality (LCQ) properties from technical life cycle properties [43–46]

Life Cycle Quality property	Aggregation method	Criterion
1. Life Cycle Functionality	Quality Function Deployment (QFD)	Functional efficiency Normative minimum requirements and classifications
2. Life Cycle Monetary Economy LCME	Life Cycle Costing	Economic efficiency (Normative minimum requirements and classifications)
3. Life Cycle Natural Economy (Ecology) LCNE	EPA Science Advisory Board study. Harvard University Study	Eco-efficiency Normative minimum requirements and classifications
4. Life Cycle Human Conditions LCHC	Analysis of Total Volatile Organic Compound (VOC) Emissions. Evaluation of fungi risk. Evaluation of risk of radioactive radiation from materials and from earth. Evaluation of ventilating air quality. Evaluation of health risks of water quality.	Quality classifications of indoor air quality and other indoor air conditions. Quality classifications of acoustic performance. Normative minimum criteria and classifications of safety, health and comfort.
5. Overall Life Cycle Quality	Multi-Attribute Decision-making	Life Cycle Quality (LCQ)

Examples of these aggregation schemes of LCQ-Parameters are presented in following categories:

- basic human conditions properties (Figure 2.68)
- functional properties (Figure 2.69)
- life cycle costs (LCC) (Figure 2.70)
- ecological properties (LCE) (Figure 2.71).

The weighting in aggregation of ecological properties is made on the following levels:

1. global level
2. regional level
3. local level.

Typical global properties, which always have high weight, are consumption of energy and air pollution, which include the factors of global climatic change. Typical regional properties are, for example, consumption of raw materials and water. In some areas and places these properties are extremely important, but in some other areas they, or some of them, hardly have any meaning.

The aggregation method of ecological properties, which is presented in Figure 2.71, can be applied regionally and locally with slightly different weightings [1, 43, 45, 46]. The weights presented are mainly done from the perspective of the USA. One example of regional as well as local application and simplified methods is the Nordic method, which is designed for Northern European conditions. Because most natural raw materials and water resources are not critical, this method includes only the factors which are related to the consumption of non-renewable energy and pollution of air, soil and water.

The weighting between safety, health and comfort can be made individually. Usually the weights of safety and health are very high, while the weight of comfort can vary

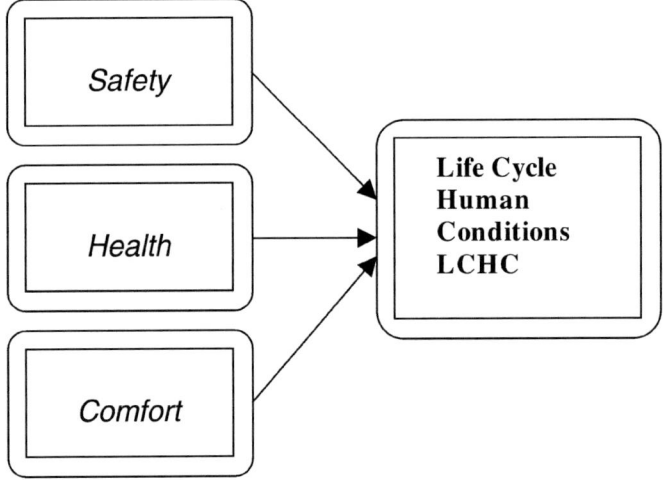

Figure 2.68 An example of the aggregation of basic human conditions properties.

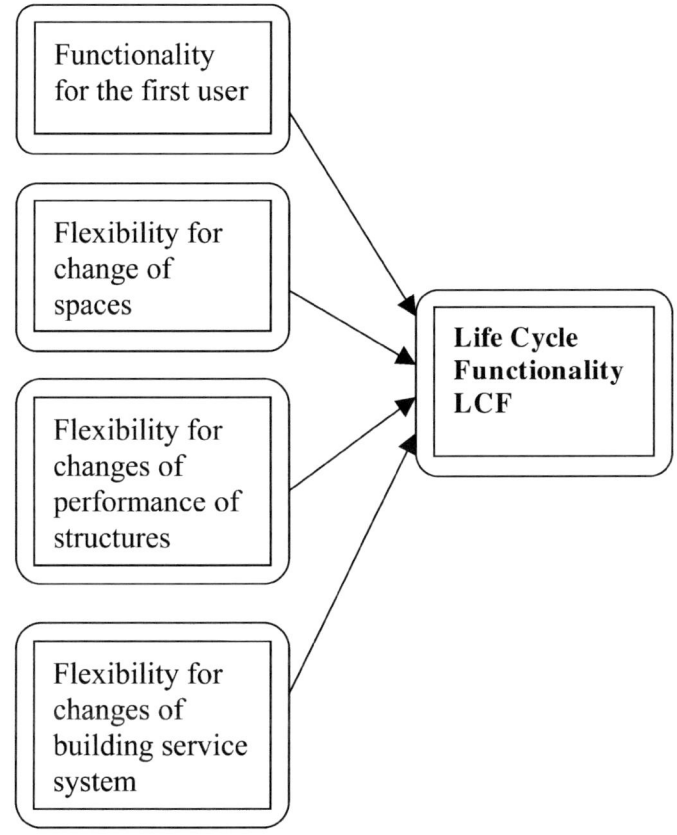

Figure 2.69 An example of the aggregation of functional performance properties.

according to a larger range. In any case, safety and health must fulfil the minimum regulatory requirements, which usually are quite strong.

As an example, we can take the weights in Northern Europe (Scandinavian countries). There is a weighting widely used where the factors of climatic change and air pollution, CO_{2eqv}, SO_{2eqv} and $ethene_{eqv}$, are taken into account (eqv. means equivalence).

The general aggregated environmental Life Cycle Natural Economy (LCNE) value, which is described above, can be used in calculating the normalised eco-efficiency property ECOEFF. The property ECOEFF can be calculated as a ratio of LCNE of a reference object (product, design solution, building concept, production method, etc.) to the LCNE of the actual object, using the equation [46]

$$ECOEFF = LNCE_{ref}/LNCE_{actual}$$

where

ECOEFF   is the normalised ecological efficiency property
$LNCE_{ref}$ Life Cycle Natural Economy parameter LCNE of the reference object
$LNCE_{actual}$ Life Cycle Natural Economy parameter LCNE of the actual object.

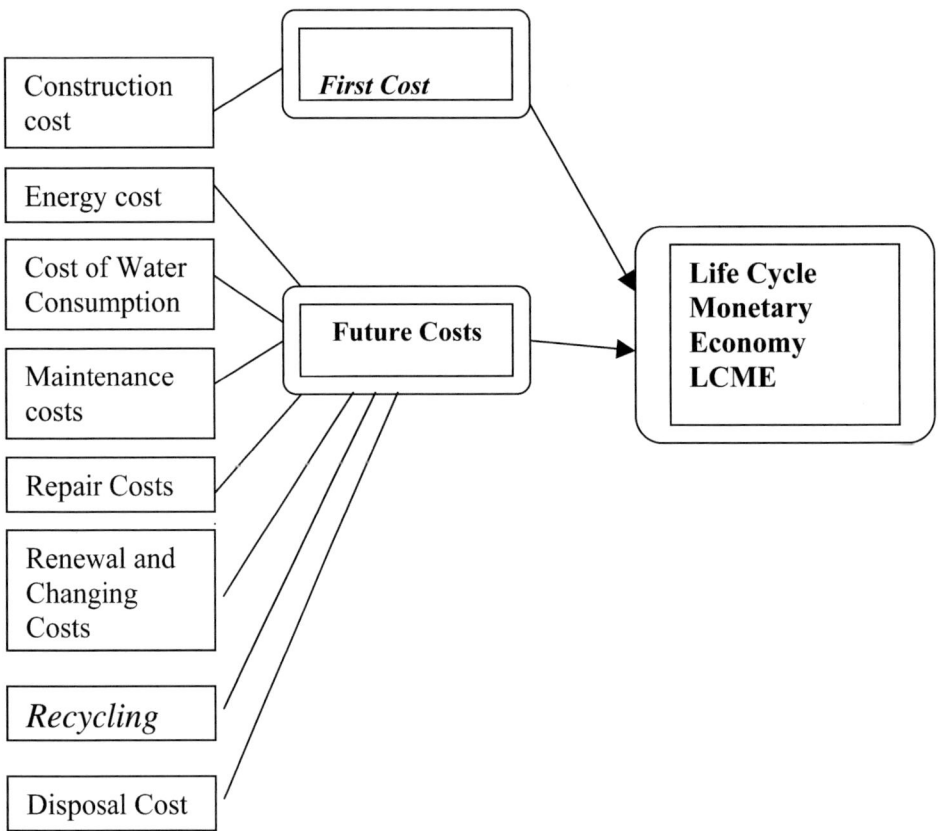

Figure 2.70 An example of aggregation of life cycle costs (LCC).

2.7.2.5   Selection of the primary requirements and properties

The selection of primary requirements and properties of each alternative under selection or optimisation can be based on some of the following, previously described methods:

1. direct strategic decisions of the user organisation between the generic Lifecon requirements;
2. analysis of weights of the multiple requirements and properties using the QFD matrix. The selection is made by ranking the requirements and properties directly into the order of their weights, which have been gained as a result of this QFD analysis;
3. handling all or some of the aggregated requirements and properties as primary requirements and properties.

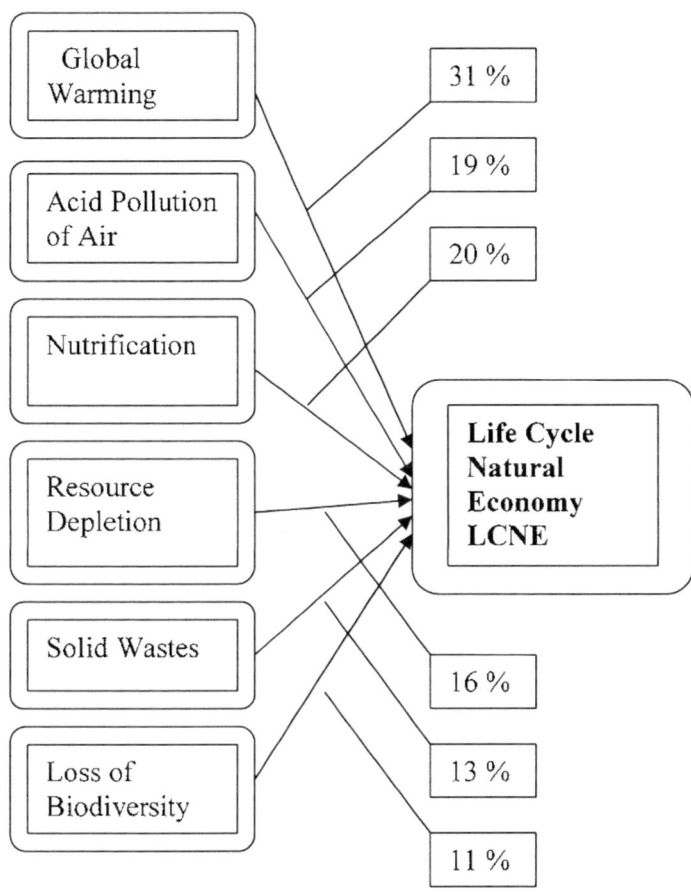

Figure 2.71 An example of aggregation of ecological properties, and the weighting coefficients in percentages (LCE).

2.7.2.6 IT support for QFD method

2.7.2.6.1 DIRECT EXCEL APPLICATIONS

Every user can in principle apply the Excel program directly to program the individual application of QFD matrix calculations.

2.7.2.6.2 COMMERCIAL PROGRAMS

In continuous use it is more practical to apply a commercial QFD program, which includes a user interface and Excel calculation procedures. Examples include one of the oldest commercial programs; 'QFD/Capture', which is a product of ITI (International Techno Group Incorporated); 'QFDwork' of Total Quality Software and 'QFD Designer' of the Qualisoft firm.

Extensive and updated information on QFD can be found on the website (http://www.qfdi.org/) of the QFD Institute, USA.

2.8   Content and structure of a generic and open facility management system

2.8.1   Systematics of life cycle management

The LMS system includes the following elements [4]:

- Terms and definitions of lifetime management
- Summary of general principles of lifetime engineering
- Procedure from generic requirements of sustainability (human requirements, lifetime economy, lifetime ecology and cultural acceptance) into lifetime management
- Integrated management of lifetime quality with reliability and the limit state approach, including generalised limit state methodologies and methods for:

  - the management of mechanical (static, dynamic and fatigue) safety and serviceability
  - condition management of assets with modelling of performance, service life and degradation
  - usability and functionality management under varying use and requirements with obsolescence analysis.

- Generic theory of systems as an application into management system for structural system, LMS structure and LMS process.

These elements link European and global normative regulations into the reliability approach of Lifecon LMS at different phases, especially in safety, serviceability and usability checks of condition assessment and MR&R planning.

Different parts of this system were applied for use in several modules of Lifecon LMS; especially in generic handbook [Lifecon D1.1], condition assessment protocol [Lifecon D3.1], statistical durability models [Lifecon D2.1 and D3.2] and MR&R planning [Lifecon D5.1] [4, 47, 48].

Mechanical industry as well as building and civil engineering sectors are aiming at the goal towards social, economic, ecological and cultural sustainable development. A technical approach for this objective is Lifetime Engineering (also called 'Life Cycle Engineering'). This can be defined as follows:

- *Integration*, which means that all requirement classes (human social, economic, ecological and cultural) are included in the MR&R planning, design and execution processes
- *Predictivity*, which means that the functional and performance quality of the facilities will be predicted for a planning and design period of the facility with integrated performance analysis, including:

  - predictive performance and service life modelling
  - modular product systematics

- methods of system technology, reliability theory and mathematical modelling
- residual service life prediction of structures
- quantitative classification of degradation loads

- *Openness*, which means:

  - freedom to apply the generic LMS into specific applications, using selected modules of the LMS for each application, and
  - freedom to select between methods given in Lifecon reports or outside these. The openness is valid for both the LMS description and the IT application.

The objective of this deliverable is to provide an integrated, systematic and uniform reliability-based methodology for modelling, analysing and optimising the lifetime quality in the Lifecon LMS under the constraints of normative reliability requirements. This reliability approach works as a link between life cycle management and generic sustainability requirements and European and international normative requirements, as shown in Figure 2.72 which shows the flow of reliability approach between generic requirements of sustainable building, European and global norms and standards, Lifecon D2.1 and reliability approaches of other modules of Lifecon LMS [5, 47, 48].

The generalised reliability-based methodology guarantees conformity of Lifecon LMS with existing normative requirements together with an efficient integrated optimisation of lifetime quality, which is based on generic requirements of sustainable building.

This deliverable is linked to several parts of Lifecon LMS, mainly those which are presented in Lifecon Deliverables: D1.1: 'Generic technical handbook for a predictive life cycle management system of concrete structures (Lifecon LMS)', D3.1: 'Prototype

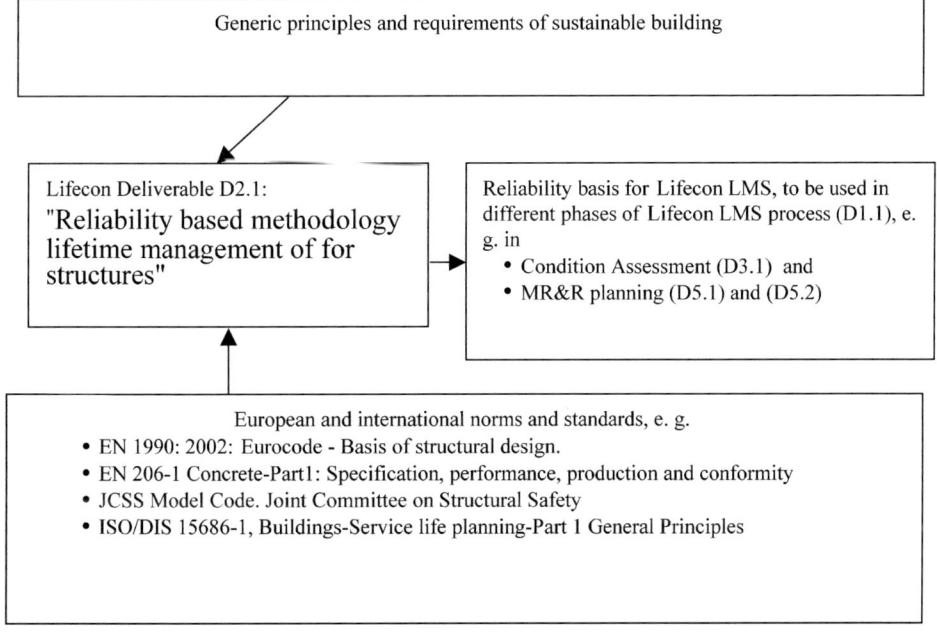

Figure 2.72 Generic principles and requirements of sustainable building.

of condition assessment protocol' and D5.1: 'Qualitative and quantitative description and classification of RAMS (Reliability, Availability, Maintainability, Safety) characteristics for different categories of repair materials and systems'. The main links of this deliverable are presented in Figure 2.72 [4].

2.8.2 System structure

Lifecon LMS is an outcome of an *open and generic European model of an integrated and predictive Life cycle Maintenance and management planning System (LMS)* that will facilitate the change of the facility maintenance and management *for sustainability*, and from a reactive approach into a predictive approach. LMS is working for sustainability on life cycle principles, and includes the following (integrated) requirements of sustainable building: human requirements, lifetime economy, lifetime ecology and cultural values. The content and use of these requirements will be explained in further detail later.

Lifecon LMS includes a generic system, methodology and methods for management of all kinds of assets. Only the durability management, condition assessment protocol and service life models, are focused on concrete structures. Hence, Lifecon LMS can be applied to all kinds of assets by replacing the condition assessment protocol and service life models with other descriptions and models.

The system makes it possible to organise and implement all the activities related to maintaining, repairing, rehabilitating and replacing assets in an optimised way, taking into account all generic requirements of sustainable building: life cycle human requirements (usability, performance, health, safety and comfort), life cycle economy, life cycle ecology, and cultural acceptance.

Lifecon LMS is an open system, which means:

- openness for applications in different environmental and cultural conditions of Europe
- openness for applications for different types of assets: bridges, tunnels, harbours, buildings, etc.
- openness for applications into networks (set of objects under management) of very different numbers of objects (bridge, harbour, tunnel, building, etc.): from several thousands of objects into individual object
- openness for different weightings of generic requirements, technical criteria and properties.

Open systems always have a modular structure; consisting of modules and components. In Lifecon the modular principle has several meanings:

- Real modular structure of objects: structural system, structural modules, components, details and materials (see Terms and Definitions). These are described and applied in Lifecon Deliverable D3.1.
- Modular structure of the Lifecon LMS structure, consisting of thematic modules, and model and method components.
- Modular structure of Lifecon LMS process. This is described and applied in Lifecon Deliverable D1.1.

Lifecon LMS has a modular structure, consisting of following thematic modules:

- System and Process Description: 'Generic Handbook' [Lifecon D1.1]
- IT tools [Lifecon D1.2, D1.3 and D1.4]
- Reliability-based methodology [Lifecon D2.1]
- Methods for optimisation and decision-making [Lifecon D2.3]
- Condition assessment protocol [Lifecon D3.1]
- Degradation models [Lifecon D3.2, D2.1 and D2.2]
- Planning of MR&R projects [Lifecon D5.1, D5.2 and D5.3].

These modules of Lifecon LMS system support the following activities in the Lifecon management system and process modules:

1. assistance in inspection and condition assessment of structures
2. determination of the network level condition statistics of a building stock
3. assessment of MR&R needs
4. life cycle analysis and optimisation for determination of optimal MR&R methods and life cycle action profiles (LCAPs) for structures
5. definition of the optimal timing for MR&R actions
6. evaluation of MR&R costs
7. combination of MR&R actions into projects
8. sorting and prioritising of projects
9. allocating funds for MR&R activity
10. performing budget check
11. preparation of annual project and resource plans
12. updating degradation and cost models using inspection and feedback data.

As can be seen in Figure 2.73, some modules include alternative methods and models. This property is aimed at helping the users to select best-suited methods of models for each specific application.

2.8.3 Modular product systematic

The modular product systematic is aimed for compatible object description for different kinds of objects, like bridges, harbours, airports, tunnels, buildings, etc. This is applied in several parts of the LMS: different levels and optimisation and decision-making procedures of an LMS process [Lifecon D1.1], condition assessment [Lifecon D3.1] and MR&R planning [Lifecon D5.1] [4].

In a modular systematic the modulation involves division of the whole asset into sub-entities, which to a significant extent are compatible and independent. The compatibility makes it possible to use interchangeable products that can be joined together according to connection rules to form a functional whole of the object [1, 48].

Typical modules of a building are:

- bearing frame
- envelope
- roofing system
- partition walls and
- building service systems.

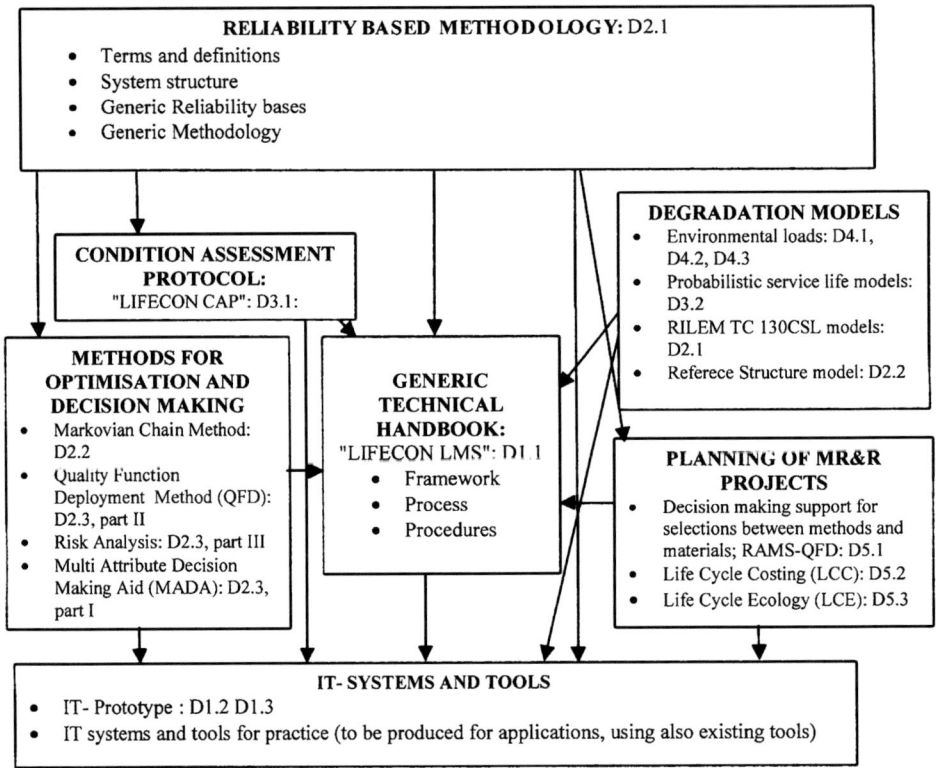

Figure 2.73 Thematic modules of the Lifecon LMS and their main interaction.

Typical modules of a bridge are:

- foundations (including pilings)
- supporting vertical structures
- bearing horizontal structures
- deck
- waterproofing of the deck
- pavement
- edge beams and
- railings.

The modular product systematic is firmly connected to the performance systematic of the object. As an example, the main performance requirements of floors of buildings can be classified in the following way:

1. Mechanical requirements, including

    - static and dynamic load bearing capacity,
    - serviceability behaviour: deflection limits, cracking limits and damping of vibrations.

2.   Physical requirements, including

- tightness of insulating parts (against water, vapour, etc.)
- thermal insulation between cold and warm spaces
- fire resistance and fire insulation
- acoustic insulation.

3.   Flexible compatibility with connecting structures and installation partition services: piping, wiring, heating and ventilating installations
4.   Other requirements:

- buildability
- maintainability
- changeability during the use
- easy demolition
- reuse, recycling and waste.

In the case of bridges the modulation, specification of major performance properties and design service life cost estimation can be done applying the schemes presented in Table 2.19.

Table 2.19 Specification of performance properties for the alternative structural solutions on a module level using a bridge as an example

Structural assembly (Module)	Central performance properties in specifications
1. Substructures foundations, retaining walls	Bearing capacity, target service life, estimated repair intervals, estimated maintenance costs, limits and targets of environmental impact profiles.
2. Superstructures Bearing structural system: • vertical • horizontal	Bearing capacity, target service life, estimated repair intervals, estimated maintenance costs, limits and targets of environmental impact profiles.
3. Deck over layers waterproofing concrete topping pavement	Target values of moisture insulation, target service life, estimated repair intervals, estimated maintenance costs, limits and targets of environmental impact profiles, estimated intervals of the renewal.
4. Installations • railings • lights, etc.	Target service life, estimated repair intervals, estimated maintenance costs, limits and targets of environmental impact profiles, estimated intervals of the renewal.

References

[1] Sarja, A. 2002. *Integrated life cycle design of structures*. London: Spon Press, 142pp. ISBN 0-415-25235-0.

[2] Sarja, A. 2004. *Reliability based methodology for lifetime management of structures*. EU GROWTH project no G1RD-CT-2000-00378: Lifecon, Life Cycle Management of Concrete Infrastuctures, Deliverable D2.1. http://lifecon.vtt.fi/.

[3] Sarja, A., *et al.* 2005. *Generic description of lifetime engineering of buildings, civil and industrial infrastructures*. Deliverable D3.1, Thematic Network Lifetime, EU GROWTH, CONTRACT N°: G1RT-CT-2002-05082. http://lifetime.vtt.fi/index.htm.

[4] Life Cycle Management of Concrete Infrastuctures for improved sustainability LIFE-CON. Co-ordinator: Professor Asko Sarja, Technical Research Centre of Finland (VTT). http://lifecon.vtt.fi/.

[5] EN 1990:2002. Eurocode – Basis of structural design. CEN: European Committee for Standardisation. Ref. No. EN 1990:2002 E, 87pp.

[6] Sarja, A. 2004. Generalised lifetime limit state design of structures. *Proceedings of the 2nd International Conference, Lifetime-oriented design concepts, ICDLOC*, Ruhr-Universität Bochum, Germany, pp. 51–60. ISBN 3-00-013257-0.

[7] ISO/DIS 15686-1, *Buildings-service life planning, Part 1, General principles*. Draft 1998

[8] Sarja, A., and Vesikari, E. eds. 1996. *Durability design of concrete structures*. RILEM Report of TC 130-CSL. RILEM Report Series 14. E&FN Spon, Chapman & Hall, 165pp.

[9] Sarja, A. 2005. *Applications of European standards into durability design guidelines*. International Conference "Concrete and Reinforced Concrete – Development Trends", Moscow, September 5–9, along with the 59th RILEM Week, in English: pp. 233–245; in Russian: pp. 218–232.

[10] EN 206-1. *Concrete, Part 1, Specification, performance, production and conformity*. CEN European Committee for Standardisation, December 2000. Ref. No EN 206-1:2000 E. 72pp.

[11] Lair, J., Sarja, A., and Rissanen, T. *Methods for optimisation and decision-making in lifetime management of structures*, EU GROWTH project no G1RD-CT-2000-00378: Lifecon, Life Cycle Management of Concrete Infrastuctures, Lifecon Deliverable D2.3. http://lifecon.vtt.fi/.

[12] Miller, John B., Miller, Iain H. B., and Sarkkinen, M. *Qualitative and quantitative description and classification of RAMS (Reliability, Availability, Maintainability, Safety) characteristics for different categories of repair materials and systems*. EU GROWTH project no G1RD-CT-2000-00378: Lifecon, Life Cycle Management of Concrete Infrastuctures, Lifecon Deliverable D2.3. http://www.vtt.fi/rte/strat/projects/lifecon/.

[13] Triantaphyllou, E., *et al.* 1997. Determining the most important criteria in maintenance decision-making. *Quality in Maintenance Engineering*, 3 (1): 16–28.

[14] ASTM. 1995. Standard practice for applying the analytic hierarchy process to multiattribute decision analysis of investments related to buildings and building systems. ASTM designation E 1765–95.

[15] ASTM Standard: E 1765–98. Standard practice for applying analytical hierarchy process (AHP) to multiattribute decision analysis of investments related to buildings and building systems.

[16] Norris, G. A., and Marshall, H. E. 1995. *Multiattribute decision analysis method for evaluating buildings and building systems*. NISIR 5663, National Institute of Technology, Gaithersburg, MD.

[17] Roozenburg, N., and Eekels, J. 1990. EVAD, Evaluation and decision in design. (Bewerten und Entscheiden beim Konstruiren). Schriftenreihe WDK 17, Edition HEURISTA: Zürich.

[18] Roy, B. 1985. *Méthodologie Multicritère d'Aide à la Décision*, vol. 1, *Collection Gestion*, Editions Economica.

[19] Pomerol, J.-C., and Barba-Romero, S. 1993. *Choix multicritère dans l'entreprise*, Hermes.

[20] Maystre, L. Y., Pictet, J., and Simos, J. 1994. Méthodes multicritères ELECTRE. *Collection Gérer l'environnement*, vol. 8, Presses polytechniques et universitaires, Romandes.

[21] Thoft-Christensen, P., Faber, M. H., Darbre, G., and Høj, N. P. 2001. *Risk and reliability in civil engineering – Short course*, Lecture Notes. Zürich.

[22] James, M., ed. 1996. *Risk management in civil, mechanical and structural engineering*, Conference Proceedings, London: Thomas Telford.

[23] *Safety, Risk and Reliability – Trends in Engineering*, Conference Report, Malta, March 2001.

[24] Rissanen, T. 2001. *Probabilistic traffic load model applied especially to composite girder bridges*, Master's Thesis, Espoo (in Finnish).

[25] Lair, J., and Sarja, A. 2003. *Multi-attribute decision aid methodologies*, Lifecon Deliverable, D2.3. http://lifecon.vtt.fi/.

[26] Modarres, M. 1993. *What every engineer should know about reliability and risk analysis*, New York: Marcel Dekker, Inc.

[27] Goldberg, B. E., Everhart, K., Stevens, R., Babbitt III, N., Clemens, P., and Stout, L. 1994. *System engineering "Toolbox" for design-oriented engineers*, NASA Reference Publication 1358.

[28] Vose, D. 1996. *Quantitative risk analysis: A guide to Monte Carlo simulation modelling*, Chichester: John Wiley & Sons, 328pp.

[29] Wang, J. X., and Roush, M. L. 2000. *What every engineer should know about risk engineering and management*, New York: Marcel Dekker, Inc.

[30] Faber, M. H. 2001. *Risk and Safety in Civil Engineering*, Lecture Notes, Swiss Federal Institute of Technology.

[31] *Risk Assessment and Risk Communication in Civil Engineering*, CIB report, Publication 259, March 2001.

[32] Söderqvist, M.-K., and Vesikari, E. 2003. *Generic technical handbook for a predictive life cycle management system of concrete structures (LMS)*, Lifecon Deliverable, D1.1. http://lifecon.vtt.fi/.

[33] Zairi, M., and Youssef, M. A. 1995. Quality Function Deployment: A main pillar for successful total quality management and product development. International Journal of Quality & Reliability Management 12 (6): 9–23.

[34] Akao, Y. 1990. Quality Function Deployment (QFD), Integrating customer requirements into product design. Cambridge, MA, USA: Productivity Press, 369pp.

[35] Quality function deployment, awareness manual. 1989. American Supplier Institute, Inc. Dearborn, Michican, pp. 106.

[36] Lakka, A., Laurikka, P., and Vainio, M. 1995. *Quality Function Deployment, QFD in construction*. Technical Research Centre of Finland VTT, Research Notes 1685. Espoo, Finland, pp. 54 + Appendix (in Finnish).

[37] Nieminen, J., and Huovila, P. 2000. Quality Function Deployment (QFD) in design process decision-making. In *Integrated life cycle design of materials and structures ILCDES*, ed. A. Sarja. Proceedings of the RILEM/CIB/ISO International Symposium. RILEM Proceedings PRO 14. RIL-Association of Finnish Civil Engineers. Helsinki, pp. 51–56.

[38] ISO 6240-1980, Performance standards in building – Contents and presentation.

[39] ISO 6241-1984, Performance standards in building – Principles for their preparation and factors to be considered.

[40] ISO 6242-Building performance – Expression of functional requirements of users – Thermal comfort, air purity, acoustical comfort, visual comfort and energy saving in heating.

[41] ISO 7162-1992, Performance standards in building – Contents and format of standards for evaluation of performance.

[42] ISO 9699-1994, Performance standards in building – Checklist for briefing – Contents of brief for building design.

[43] Lippiatt, B. 1998. *Building for Environmental and Economic Sustainability (BEES)*. Building and Fire Research Laboratory, National Institute of Standards and Technology, Gathersburg, USA. Manuscript for CIB/RILEM Symposium: "Materials and Technologies for Sustainable Construction", Royal Institute of Technology, Centre of Built Environment, Gävle, June, 8p.

[44] Lippiat, Barbara C. 1998. *BEES 1.0. Building for Environmental and Economic Sustainability. Technical Manual and User Guide*. NISTIR 6144. NIST U. S. Department of Commerce, Technology Administration. National Institute of Standards and Technology. Office of Applied Economics, Building and Fire Research Labooratory, Gathersburg, Maryland, April, 84p.

[45] Vicki, N.-B., *et al.* 1992. *International comparisons of environmental hazards: Development and evaluation of a method for linking environmental data with the strategic debate management priorities for risk management*, Center for Science & International Affairs, John F. Kennedy School of Government, Harvard University, October.

[46] Lindfors, L. G., *et al.* 1995. *Nordic guidelines on life cycle assessment*, Nord, Nordic Council of Ministers. Århus, 222pp.

[47] Sarja, A. 2003. Lifetime performance modelling of structures with limit state principles. *Proceedings of 2nd International Symposium ILCDES2003, Lifetime engineering of buildings and civil infrastructures*, Kuopio, Finland, December 1–3, pp. 59–65. Association of Finnish Civil Engineers, Helsinki.

[48] Sarja, A. 2002. *Reliability based life cycle design and maintenance planning*. Workshop on Reliability Based Code Calibration, Swiss Federal Institute of Technology, ETH Zürich, Switzerland, March 21–22. http://www.jcss.ethz.c.

Chapter 3

Durability control methodology

3.1 Classification of degradation models and methodologies

Asko Sarja

3.1.1 Criteria for selection of the degradation model

In the structural design as well as in the MR&R planning procedure the structural designer must select from existing degradation models, frequently presented in literature on the subject.

The main criteria in selecting the degradation model for each specific use are:

- dictating components of environmental loads onto structures (e.g. corrosion due to carbonation, corrosive pollutants or corrosive agents like chlorides, frost, temperature changes, moisture changes, etc.)
- availability of statistical data of variables of each model
- availability of data or testing method for the coefficients of each model
- accuracy of the model when using the available data in relation to the required accuracy level
- costs of IT tools and the work in calculations.

Some of these criteria can be evaluated roughly beforehand, but often some comparative test calculations are needed. A designer must determine which degradation factors and variables are decisive for service life. Preliminary evaluations of rates of degradation for different factors may be necessary.

3.1.2 Types of degradation models

Three types of degradation models are described in detail in this book, including some examples of application. These models are:

1. Statistical degradation models
2. RILEM TC130 CSL models
3. Reference structure model.

The characteristic properties of these models are as follows:

- Statistical degradation models are based on physical and chemical laws of thermodynamics, and thus have a strong theoretical base. They include parameters which have to be determined with specific laboratory or field tests. Therefore, some equipment and personnel requirements exist for the users. The application of the statistical 'Duracrete' method raises the need for a statistically sufficient number of tests. The statistical reliability method can be directly applied with these models.
- RILEM TC 130 CLS models are based on parameters which are available from the mix design of concrete. The asset of these models is the availability of the values from the documentation of the concrete mix design and of the structural design.
- Reference structure model is based on statistical treatment of the degradation process and condition of real reference structures which are in similar condition and own similar durability properties with the actual objects. This method is suited in the case of a large network of objects, for example bridges. It can be combined with the Markovian Chain method in the classification and statistical control of the condition of structures.

In an open LMS, each user can select the best-suited models for their use [1, 2]. There certainly exist many other suitable models, and new models are under development. They can be used in LMS after careful validation of the suitability and reliability. Special attention has to be paid to the compatibility of the entire chain of the procedure of reliability calculations.

3.1.2.1   Statistical degradation models

Statistical degradation models include the mathematical modelling of corrosion induction due to carbonation and chloride ingress, corrosion propagation, frost (internal damage and surface scaling) and alkali-aggregate reaction [3]. The models are presented on a semi-probabilistic and a full-probabilistic level. Semi-probabilistic models only include parameters obtainable throughout structure investigations, without making use of default material and environmental data. Full-probabilistic models are applicable for service life design purposes and for existing objects, including the effect of environmental parameters. For each full-probabilistic model, a parameter study was performed in order to classify environmental data.

The application of the models for real structures is outlined. The objects of the case studies have been assessed in order to obtain input data for calculations on residual service life. Each degradation mechanism will be treated separately thereby demonstrating:

- possible methods to assess concrete structures
- the sources for necessary input data
- approach used in durability design
- application of models for existing structures
- the precision of the applied models
- necessary assumptions due to lack of available data
- possible methods to update data gained from investigations throughout condition assessment

- default values for input data
- output of the calculations.

The use of full-probabilistic models for the calibration of the Markov Chain approach is described.

3.1.2.2 RILEM TC 130 CSL models

These degradation models include a set of selected calculation models consisting of parameters which are known from mix design and other material properties and ordinary tests [4, 5]. This enables an interactive optimisation between durability for required target service life (design life) and mix design of newly structured and repair concrete. Also the checking of durability of existing structures is easy if the mix design is known. Therefore, these models are usually easy to apply also in cases when no advanced laboratories and equipment are available. The following degradation processes are included in the RILEM TC 130 CSL models [4]:

- corrosion due to chloride penetration
- corrosion due to carbonation
- mechanical abrasion
- salt weathering
- surface deterioration
- frost attack.

   Degradation affects the concretet and/or the steel. Usually degradation takes place on the surface zone of concrete or steel, gradually destroying the material. The main structural effects of degradation in concrete and steel are the following:

- loss of concrete leading to reduced cross-sectional area of the concrete
- corrosion of reinforcement leading to reduced cross-sectional area of steel bars; corrosion may occur

  - at cracks
  - at all steel surfaces, assuming that the corrosion products are able to leach out through the pores of the concrete (general corrosion in wet conditions)

- Splitting and spalling of the concrete cover due to general corrosion of reinforcement, leading to a reduced cross-sectional area of the concrete, to a reduced bond between concrete and reinforcement, and to visual unfitness.

3.1.2.3 Reference structure models

Reference structure degradation prediction is aimed for use in cases when the network of objects (e.g. bridges) is so large in number that a sample of them can be selected for follow-up testing, and these experiences can be used to describe the degradation process of the entire population [6]. The reference structure models are of two types: (1) surface damage models and (2) crack damage models. Degradation factors such as frost damage, corrosion of reinforcement, carbonation and chloride penetration may have combined effects that may be of great importance to the service life of a structure. These combined effects are usually ignored by the traditional prediction methods of service life. However, in computer simulation they can be considered without great theoretical problems. The progress of the depth of carbonation or the depth of critical chloride content is promoted by both the frost-salt scaling of a concrete surface and the internal frost damage of concrete. The internal frost damage is evaluated using the theory of critical degree of saturation. The internal damage is evaluated as the reduction of the dynamic E-modulus of concrete.

The condition state (or damage index) of a structure is evaluated using the scale 0, 1, 2, 3, 4. This scale is also used throughout the bridge management system.

The degradation models for both surface damage and crack damage have been programmed on Excel worksheets. The surface damage models describe normal degradation processes on the surfaces of reinforced concrete structures combining the effects of frost-salt attack, internal frost damage attack, carbonation, chloride ingress and corrosion of reinforcement. The crack damage models emulate the processes of depassivation and corrosion at a crack of a concrete structure.

All management systems that include a prediction module, such as Lifecon LMS, need reliable environmental load data. In Lifecon deliverable D4.2 the relevant systematic requirements for quantitative classification of environmental loading onto structures, as well as sources of environmental exposure data are given [7]. Lifecon D4.2, Chapter 6, contains instructions and guidelines for how to characterise the environmental loads on concrete structures on object and network levels.

However, these guidelines have to be validated (and possibly adjusted) before they can finally be used in the LMS. In this the results from the practical validation are summarised. The EN 206–1 system and the standard prEN 13013 have been tested on the chosen objects and compared with detailed environmental characterisation of the same objects using the available data and methods for environmental characterisation. Such studies were undertaken in five countries (Norway, Sweden, Germany, Finland and the United Kingdom) to develop the necessary national annexes for a proper implementation of EN206–1 across Europe [8].

Strategies and methodologies for developing the quantitative environmental classification system for concrete are given. Those are, first, comparative case studies using the new European standard – 'EN 206–1 Concrete' and detailed environmental characterisation of the same objects, and, second, a more theoretical classification based upon parametric sensitivity analysis of the complex Duracrete damage functions under various set of conditions. In this way the determining factors are singled out and classified. Such a systematic classification is needed to enable sound prediction of service lives and maintenance intervals both on object and on network level. This in turn is a necessary prerequisite for change of current reactive practice into a proactive life cycle-based maintenance management.

References

[1] Sarja, A. 2004. Reliability based methodology for lifetime management of structures. EU GROWTH project no GIRD-CT-2000-00378: Lifecon, Life cycle management of concrete infrastructures, Deliverable D2.1. http://lifecon.vtt.fi/.

[2] Sarja, A. 2004. Generalised lifetime limit state design of structures. *Proceedings of the 2nd international conference, Lifetime-oriented design concepts, ICDLOC*, pp. 51–60. Bochum, Germany: Ruhr-Universität. ISBN 3-00-013257-0.

[3] Lay, S. and P. Schießl. 2003. *Service life models*. Life cycle management of concrete infrastructures for improved sustainability LIFECON.EU GROWTH Program, GIRD-CT-2000-00378. Deliverable D3.2. Technical University Munich. http://lifecon.vtt.fi/.

[4] Sarja, A. and E. Vesikari eds. 1996. Durability design of concrete structures. *RILEM Report of TC 130-CSL. RILEM Report Series 14*. London: E&FN Spon, Chapman & Hall, 165pp.

[5] Sarja, A. 2000. Development towards practical instructions of life cycle design in Finland. *RILEM Proceedings PRO 14, Proceedings of the RILEM/CIB/ISO international symposium: Integrated life cycle design of materials and structures, ILCDES*, pp. 1–5.

[6] Vesikari, E. 2004. *Reference structure model for prediction of degradation*. Life cycle management of concrete infrastructures for improved sustainability LIFECON.EU GROWTH Program, GIRD-CT-2000-00378. Deliverable D2.2, Part 2. Technical Research Centre of Finland. http://www.vtt.fi/rte/strat/projects/lifecon/.

[7] Haagenrud, S. E. and G. Krigsvoll. 2004. *Instructions for quantitative classification of environmental degradation loads onto structures*. LIFECON.EU GROWTH Program, GIRD-CT-2000-00378. Deliverable D4.2. Norwegian Building Research Institute NBI. http://www.vtt.fi/rte/strat/projects/lifecon/.

[8] EN 206-1, Concrete – Part 1: Specification, performance, production and conformity. CEN European Committee for Standardisation, December 2000. Ref. no EN 206-1:2000E, 72pp.

3.2 Statistical models and methodology for durability

Peter Schießl and Sascha Lay

3.2.1 Deterioration mechanisms affecting concrete and reinforced concrete structures

Concrete structures are exposed to environmental conditions which may lead to deterioration of the materials (concrete and steel), which is illustrated in Figure 3.2.1.

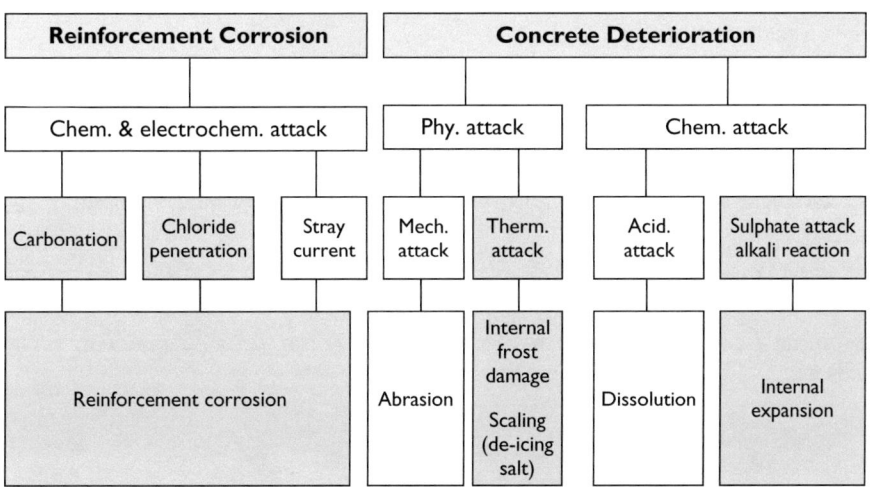

Figure 3.2.1 Overview of basic species and mechanisms leading to deterioration.

The mechanisms highlighted in Figure 3.2.1 will be treated separately with respect to semi-probabilistic and full-probabilistic models as well as possible synergistic effects between them [1].

3.2.2 Semi-probabilistic model of carbonation of concrete

3.2.2.1 Mathematical model

Most of today's models of the carbonation of concrete are based on Fick's first law of diffusion. The amount of CO_2 which penetrates the concrete due to the CO_2 gradient between the outer air of the environment and the content in the concrete can be balanced:

$$dm = -D \cdot A \cdot \frac{c_1 - c_2}{x} \cdot dt \qquad (3.2.1)$$

dm mass increment of CO_2 transported by diffusion during the time interval dt ($kg\,CO_2$)
D CO_2 diffusion coefficient of carbonated concrete ($m^2\,s^{-1}$)
A surface area considered (m^2)
c_1 CO_2 concentration of the environment ($kg\,CO_2\,m^{-3}$)
c_2 CO_2 concentration at the carbonation front in the concrete ($kg\,CO_2\,m^{-3}$)
dt time interval (s)
x depth of carbonated concrete (m).

At the carbonation front, CO_2 reacts with alkalis of the pore water solution to form various types of carbonate phases, which can be balanced:

$$dm = a \cdot A \cdot dx \qquad (3.2.2)$$

dm mass of CO_2 required for the complete carbonation of the depth increment dx ($kg\,CO_2$)
a CO_2 binding capacity of non-carbonated concrete ($kg\,CO_2\,m^{-3}$)
A surface area considered (m^2)
dx depth increment (m).

The balances of the diffusion and the reaction process can be combined:

$$x \cdot dx = -\frac{D}{a}(c_1 - c_2) \cdot dt \qquad (3.2.3)$$

Assuming D, a and $(c_1 - c_2)$ to be neither time- nor depth-dependent, integration leads to:

$$x^2 = \frac{2 \cdot D}{a}(c_1 - c_2) \cdot t^* \qquad (3.2.4)$$

where t^* is the exposure time.

Combining the single concentrations c_1 and c_2 into the concentration gradient ΔC_S and solving for the penetration depth gives:

$$x_c(t) = \sqrt{\frac{2 \cdot D \cdot \Delta C_S}{a}} \cdot \sqrt{t^*} \qquad (3.2.5)$$

Combining the material parameters D and a with the environmental parameter ΔC_S and expressing the exposure time t^* as the difference in the age t and the moment once the surface is exposed to CO_2 finally lead to a simple square root of time approach including only the carbonation rate K, which can be determined by structure investigations without further knowledge of the environmental conditions or material properties.

$$x_c(t) = K \cdot \sqrt{(t - t_{exp})} \qquad (3.2.6)$$

where
K the carbonation rate $[\text{mm}\,(\sqrt{s})^{-1}]$,
t the age of concrete at time of inspection (s),
t_{exp} the time until surface is exposed to CO_2 (s).

Usually the time until exposure t_{exp} can be set to zero as it is negligibly short compared to the service life. But for coated surfaces t_{exp} equals the time at which coating failed. From structure investigations with a low sample scope (e.g. three readings) the mean value of the carbonation depth x_c at the respective concrete surface can be measured. With the knowledge of the structure age and the exposure time, the equation can be solved for K.

For concrete surfaces in a dry atmosphere, sheltered from rain (e.g. laboratory climate of $T = 20\,°C$, $RH = 65\%$) the carbonation process is in fairly good agreement with the 'square root of time' law. The variables D, ΔC_S and a can be considered to be more or less independent of time and site. Precipitation at concrete surfaces will hinder the carbonation process for a certain period. Bearing this in mind, the 'square root of time' law is a simple model, which is always on the safe side.

3.2.2.2 Spatial deviation of carbonation depth

When performing calculations on the service life of concrete components, this is only possible if data were collected from surfaces produced with the same concrete quality and which are exposed to the same environmental conditions. Deviations of the carbonation depth will, under these circumstances, mainly be influenced by:

- the inhomogeneous character of the concrete
- measurement accuracy of personnel
- inspection technique.

The variation coefficient of the inverse carbonation resistance $R_{ACC,0}^{-1}$ (ACC – accelerated carbonation conditions) can be determined as an input to full-probabilistic design models, which will be dealt with later [14]. It can be demonstrated that the coefficient of variation decreases with increasing carbonation depth. This can be explained by the relatively constant inaccuracy of the carbonation depth measurement. The larger the penetration depth, the lesser is the influence of measurement accuracy. Therefore, the coefficient of variation must be expressed as a function of the carbonation resistance itself:

$$\text{CoV}_{R_{\text{ACC},0}^{-1}} = a \cdot R_{\text{ACC},0}^{-1} = a \cdot \left(\left(\frac{X_c}{\tau} \right)^2 \cdot 10^{11} \right)^b \qquad (3.2.7)$$

where
CoV	coefficient of variation (%)
a	regression parameters $[10^{11b}\,\text{m}^5\,(\text{s}\,\text{kg}\,CO_2)^{-1}]$; here $a = 68.9$
b	regression parameter (dimensionless); here $b = -0.22$
$R_{\text{ACC},0}^{-1}$	mean value of inverse carbonation resistance $[10^{-11}\,\text{m}^5\,(\text{s}\,\text{kg}\,CO_2)^{-1}]$
X_C	carbonation depth measured in accelerated carbonation test (m)
τ	time constant $[(\text{s}\,\text{kg}\,CO_2\,\text{m}^{-3})^{0.5}]$; here 420 for a CO_2 concentration of 2% by volume and a test duration of 28 days.

The carbonation depth X_C of laboratory specimens is measured after accelerated carbonation tests, thus taking into account the inhomogeneity of the concrete. As the carbonation depth is the only variable subjected to deviations in the test (test conditions and duration can be considered as constant), the relationship stated above can be adapted to the penetration depth itself:

$$\text{CoV}_{R_{\text{ACC},0}^{-1}} \approx \text{CoV}_{X_c} \qquad (3.2.8)$$

The regression analysis was based on the data obtained at a single laboratory (level of repeatability). The deviations of these measurements are below those which would be obtained at the construction site by various personnel (level of reproducibility). It was assumed that the level of reproducibility can be set to the 90% quantile of the level of repeatability. Under these conditions the regression parameters yield the values given above.

3.2.2.3 Calculation procedure

In Table 3.2.1 a summary of the necessary input data is given. The calculation procedure will be demonstrated by an example.

The following information is given:

- Concrete cover d_{cover}, which is a resistance parameter, was measured with a mean value of $\mu_{\text{cover}} = 30\,\text{mm}$ and a standard deviation of $\sigma_R = 8\,\text{mm}$. For the sake of simplicity the distribution is assumed to be normal.
- Carbonation depth X_C, which is considered as the stress variable, was measured at three spots at an age of 30a. The mean value was determined to be $\mu_S = 12\,\text{mm}$. The standard deviation should not be based only on these three measurements. Instead, the variation should be determined following the approach of the previous chapter.

Table 3.2.1 Summary of the necessary data input for the semi-probabilistic model of carbonation ingress

Parameter	Unit	Format	Source
Carbonation depth, X_C	mm	ND (μ, ρ)	Measurements according to Lifecon D3.1 [1] Appendix
Age of concrete, t	s	D	Structure documentation
Concrete cover, d_{cover}	mm	ND (μ, σ)	Measurements according to Lifecon D3.1 [1]

Notes
D – discrete.
ND – normal distribution.

With an average CO_2 content in the air of around $6 \times 10^{-4}\, kgCO_2\, m^{-3}$, the time constant τ in Equation (3.2.7) yields:

$$\tau = \sqrt{2 \cdot \Delta C_S \cdot t} = \sqrt{2 \cdot 6 \cdot 10^{-4} \cdot 30 \cdot 365 \cdot 24 \cdot 3600} = 1065 \qquad (3.2.9)$$

which leads to a coefficient of variation for the carbonation depth of:

$$\mathrm{CoV}_{X_c} = 68.9 \left(\left(\frac{0.012}{1065} \right)^2 \cdot 10^{11} \right)^{-0.22} = 39\% \qquad (3.2.10)$$

The mean value of the carbonation rate K can be determined from:

$$x_c(t) = K \cdot \sqrt{(t - t_{\exp})} = K \cdot \sqrt{30 - 0} = 12\, \text{mm} \qquad (3.2.11)$$

leading to $\mu_K = 2.191\, \text{mm}\, (\sqrt{a})^{-1}$. The standard deviation then yields $\sigma_K = 0.39\, \mu_K$. The target service life is assumed to be 100 years, with a minimum reliability index of $\beta_0 = 1.8$. The flow of the calculation to prove whether the requirements will be fulfilled the sequence of various steps in the calculation to check the fulfilment of the requirement is as follows:

1. Calculation of the stress variable S (carbonation depth at age $T = 100a$):

$$\mu_{X_c}(t) = 2.191 \cdot \sqrt{(100 - 0)} = 21.9\, \text{mm}, \quad \sigma_{X_c}(t) = 21.9 \cdot 39\% = 8.5\, \text{mm}$$
$$(3.2.12)$$

2. Transformation to standard space:

$$u_1 = \frac{R - 30}{8}, \quad u_2 = \frac{S - 21.9}{8.5} \qquad (3.2.13)$$

$$L(U_1 - U_2) = R - S = 8U_1 - 8.5U_2 + 8.1 = A \cdot U_1 + B \cdot U_2 + C = 0 \qquad (3.2.14)$$

3. Calculation of the distance of the limit state line from the origin using the so-called Hesse normal format (comparison of the coefficients A, B, C):

$$\beta = \frac{C}{\sqrt{A^2 + B^2}} = 0.69 < 1.8 \qquad (3.2.15)$$

This means that after 100 years the reliability index will drop to a level below the target reliability.

If the residual service life is sought, the reliability has to be set to $\beta = \beta_0 = 1.8$. The stress S equals the time-dependant carbonation depth X_C:

$$U_1 = \frac{R - 30}{8}, \quad U_2 = \frac{S - 2.191 \cdot \sqrt{t}}{2.191 \cdot \sqrt{t} \cdot 0.39} \qquad (3.2.16)$$

Comparison of the coefficients (A, B, C) as above leads to:

$$\beta = \frac{(30 - 2.191) \cdot \sqrt{t}}{\sqrt{8^2 + (0.85 \cdot \sqrt{t})^2}} = 1.8 \qquad (3.2.17)$$

With a little algebra (or the Microsoft Excel solver option), this results in a service life of $t = 35$ years.

3.2.3 Full-probabilistic model of carbonation of concrete

3.2.3.1 Mathematical model

The carbonation model combines two mechanisms – diffusion and binding of carbon dioxide – influenced by, for example, relative humidity, drying and wetting of the concrete, and inhomogeneity [8, 9, 10].

The model is based on Fick's first law of diffusion and includes [14]:

- carbonation resistance as a measurable property of concrete and
- factors for environment, execution and test method.

The carbonation depth X_C (in m) is given by

$$X_C = \sqrt{2 \cdot k_{RH} \cdot k_c \cdot (k_t \cdot R_{ACC,0}^{-1} + \varepsilon_t) \cdot \Delta C_S} \cdot \sqrt{t} \cdot \left(\frac{t_0}{t}\right)^w \qquad (3.2.18)$$

index 0	material parameter, prepared, cured and tested under defined reference conditions;
k_{RH}	influence of the real moisture history at concrete surface on D_{eff}
k_c	influence of the execution on D_{eff} (e.g. curing)
k_t	test method factor
ε_t	error term
$R_{ACC,0}^{-1}$	inverse effective carbonation resistance of concrete, determined in accelerated carbonation conditions (ACC) $[m^5 (s\,kg\,CO_2)^{-1}]$.

$R_{NAC,0}^{-1}$ inverse effective carbonation resistance of dry concrete, determined with specimens under defined preparation and storage conditions under natural carbonation conditions (NAC) $[m^5 (s \, kg \, CO_2)^{-1}]$ is given by:

$$R_{NAC,0}^{-1} = k_t \cdot R_{ACC,0}^{-1} + \varepsilon_t \qquad (3.2.19)$$

ΔC_S gradient of the CO_2 concentration $(kg \, CO_2 \, m^{-3})$
t time in service (s)
t_0 reference period (s)
w weather exponent, taking into account the microclimatic conditions of the concrete surface considered (e.g. $w = f(\text{ToW}, p_{splash})$, where ToW is time of wetness and p_{splash} is the probability of splashed surface due to rain event).

The carbonation resistance comprises the effective diffusion coefficient $D_{eff,0}$ of dry carbonated concrete $(m^{-2} \, s^{-1})$ and the CO_2-binding capacity a of the concrete $(kg \, CO_2 \, m^{-3})$:

$$R_{NAC} = \frac{a}{D_{eff,0}} \qquad (3.2.20)$$

where

$$a = 0.75 \cdot C \cdot c \cdot \text{DH} \cdot \frac{M_{CO_2}}{M_{CaO}} \qquad (3.2.21)$$

C   CaO content in cement (M%/cement)
c cement content $(kg \, m^{-3})$
DH   degree of hydration
M   molar masses of respective substance $(kg \, mol^{-1})$.

The inverse carbonation resistance $R_{ACC,0}^{-1}$ of concrete prepared, cured and tested in an accelerated test under defined laboratory conditions, is transformed to carbonation resistance $R_{NAC,0}^{-1}$ under natural carbonation conditions (NAC). The resistance $R_{NAC,0}^{-1}$ will then be multiplied by a factor k_c, accounting for concreting and curing procedures on the construction site, deviating from the reference laboratory conditions. As the carbonation resistance of concrete sampled from existing structures already includes the effect of curing, the following arrangement is proposed:

$$X_C = \sqrt{2 \cdot k_{RH} \cdot k_c \cdot (k_t \cdot R_{ACC,0}^{-1} + \varepsilon_t) \cdot \Delta C_S} \cdot \sqrt{t} \cdot \left(\frac{t_0}{t}\right)^w \qquad (3.2.22)$$

leading to

$$X_C = \sqrt{2 \cdot k_{RH} \cdot (k_t \cdot k_c \cdot R_{ACC,0}^{-1} + \varepsilon_t) \cdot \Delta C_S} \cdot \sqrt{t} \cdot \left(\frac{t_0}{t}\right)^w \qquad (3.2.23)$$

where

$$R_{ACC}^{-1} = k_c \cdot R_{ACC,0}^{-1} \qquad (3.2.24)$$

$R^{-1}_{\text{ACC},0}$ inverse carbonation resistance for concrete with a given curing, determined by the ACC test

and

$$k_c \cdot R^{-1}_{\text{NAC},0} = (k_t \cdot k_c \cdot R^{-1}_{\text{ACC},0}) + \varepsilon_t \qquad (3.2.25)$$

$R^{-1}_{\text{NAC},0}$ is the inverse carbonation resistance for concrete with defined preparation and curing conditions, determined by the NAC test.

The degree of water saturation controls the penetration rate of CO_2 into the concrete. The degree of water saturation is a function of the concrete composition, the relative humidity (RH) and the time of wetness (ToW). The degree of water saturation of concrete may be considered as constant in a certain depth x of concrete.

Since carbonation is restricted to the outer portion of the concrete cover, the approximation of using meteorological data on the quantity of rain and the relative humidity of the surrounding air seems reasonable.

For the carbonation of concrete, the following limit state can be considered [1]:

$$p_f = p(X_c \geq d_{\text{cover}}) \leq \Phi(-\beta) = p_{\text{target}} \qquad (3.2.26)$$

p probability
p_f failure probability
X_c carbonation depth (m)
d_{cover} cover depth (m)
Φ probability distribution function of standard normal distribution
β reliability index
p_{target} target failure probability, which has to be defined by the principal or a national standard.

Explicitly, Equation (3.2.26) expresses the probability that the carbonation front X_C reaches or exceeds the cover depth d_{cover}, and which should be smaller than a predefined value. The parameters to be statistically defined are given in Table 3.2.2.

Table 3.2.2 Input parameters for the carbonation model [1]

Parameter family	Parameter
Material	$R^{-1}_{\text{ACC},0}$
Environmental	k_{RH}, w, ΔC_S
Execution/curing	k_c
Test	k_t, ε_t
Geometry	d_{cover}

3.2.3.2 Inverse carbonation resistance

The effective carbonation resistance of the concrete can either be measured by an ACC, as is done for the durability design of new structures, or by measuring the carbonation depth of an existing structure. In the later case, the parameters may then be grouped according to:

$$X_c = \sqrt{2 \cdot \Delta C_S \cdot R_{Carb}^{-1}} \cdot \sqrt{t} \cdot \left(\frac{t_0}{t}\right)^w \tag{3.2.27}$$

with

$$R_{Carb}^{-1} = k_{RH} \cdot R_{NAC}^{-1} = k_{RH} \cdot (k_c \cdot k_t \cdot R_{ACC,0}^{-1}) + \varepsilon_t \tag{3.2.28}$$

R_{Carb} effective carbonation resistance of concrete on site $[\mathrm{m^5\,(s\,kg\,CO_2)^{-1}}]$

The effective carbonation resistance parameter R_{Carb} is dependent on the composition, placement and curing of the concrete as well as on the relative humidity. The accelerated carbonation test has been chosen as a compliance test for new structures [12]. The carbonation resistance $R_{ACC,0}$ is the result of this test, which can be regarded as the potential resistance (Figure 3.2.2).

A detailed description of the test procedure is presented in Lifecon D3.1, 'Condition assessment protocol'. For existing structures the following investigation options are proposed:

(a) If concrete composition is known, $R_{ACC,0}$ may be chosen from a database or existing literature data.
(b) Direct measurement of the carbonation depth in the existing structure (Lifecon D3.1, Appendix).
(c) Accelerated carbonation testing with specimens (e.g. cores) from the structure (Lifecon D3.1, Appendix).

Application of option (a), as is performed for new structures, determines the carbonation resistance under defined curing and environmental conditions ($R_{ACC,0}$) and requires knowledge of the concrete composition [most importantly the binder type and water–binder (w/b) ratio].

Option (b) is a simple and inexpensive test that can be applied to any concrete structure. The effective carbonation resistance R_{Carb} of the concrete includes the influence of concrete quality ($R_{ACC,0}$), relative humidity (k_{RH}) and curing (k_c). Structures with high resistance and/or low exposure aggressiveness (e.g. high humidity and/or high value of time of wetness) may show carbonation depth readings close to zero, because carbonation is not the decisive deterioration mechanism. However, it must be kept in mind that in young concrete structures carbonation may not yet have taken place to an extent which may already be measured, but may do so within the service life. In this special case, this option should not be applied.

Applying option (c) will determine $R_{ACC,0}^{-1}$ with inherent influences of the curing and production procedure (k_c). This approach should be followed especially for young

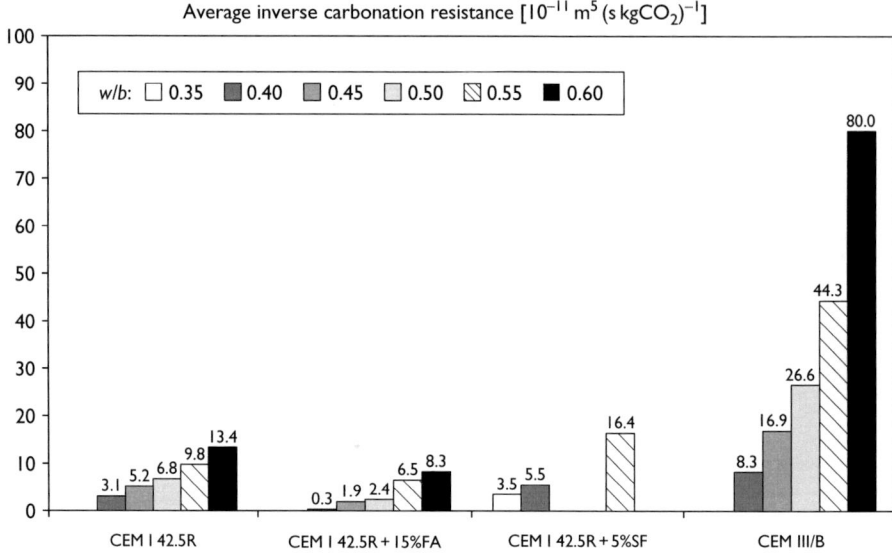

Figure 3.2.2 Default data for inverse carbonation resistance $R_{ACC,0}^{-1}$ $[10^{-11}\,m^5\,(s\,kg\,CO_2)^{-1}]$ determined in the accelerated carbonation (ACC) test [14].

structures, where carbonation has not yet taken place to a sufficient degree to be measured precisely.

In general, it can be stated that as the amount of information increases, the quality of the prediction model is increased, as these models can be used in Bayesian model updating.

3.2.3.3 Weather exponent, w

The exponent w is called a 'weather condition factor' [14] because a rain event will lead to saturation of the concrete surface, which will, at least temporarily, prevent further carbonation since the pores are widely filled with water (Figure 3.2.3).

The weather function $W(w, t)$ considers the derivation of the carbonation process of unsheltered structures using the 'square root of time' law (Figure 3.2.4):

$$W = \left(\frac{t_0}{t}\right)^w \tag{3.2.29}$$

where

t time (s)
t_0 reference time (s), which is the age when ACC test is performed (28d)
w weather exponent.

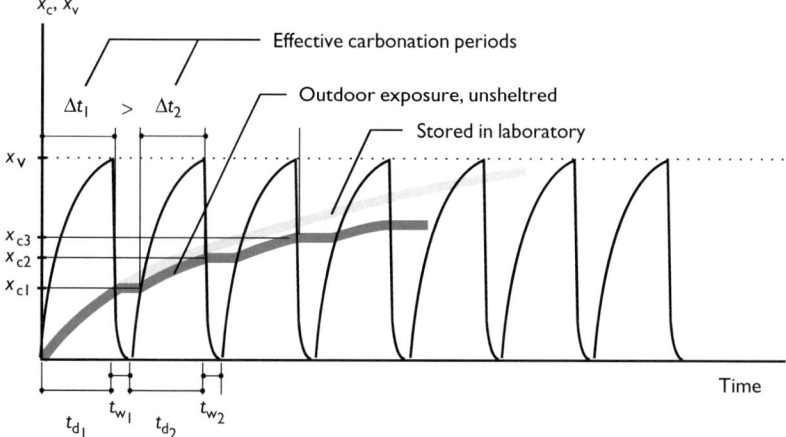

Figure 3.2.3 Progress of carbonation of specimens stored under different exposure conditions (laboratory, outdoor-unsheltered).

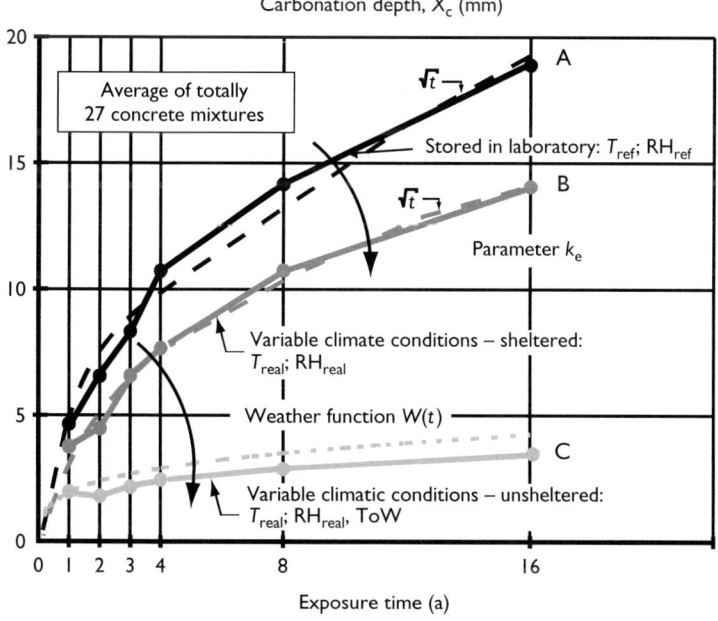

Figure 3.2.4 Progress of carbonation in different exposure environments.

The course of carbonation depends strongly on the frequency and distribution of the wetting periods:

$$w = a_{\mathrm{w}} \cdot \mathrm{ToW}^{b_{\mathrm{w}}} \tag{3.2.30}$$

where

ToW   time of wetness
a_{w} regression parameter, D ($\mu = 0.50$)
b_{w} regression parameter, ND ($\mu = 0.446, \sigma = 0.163$).

To quantify the time of wetness, a criterion is required to register a rain event (duration, intensity) as such. The moisture content prior to the rain event and the moisture saturation content of the concrete are decisive for the zone being affected by a given amount of precipitation, which leads to a high degree of model complexity. For simplicity, all days with amounts of rain above values of $h_{\mathrm{rain}} = 2.5\,\mathrm{mm\,d}^{-1}$ were chosen in Equation (3.2.31):

$$\mathrm{ToW} = \frac{\text{Number of days with amount of rain} \geq 2.5\,\mathrm{mm\,d}^{-1}}{365} \tag{3.2.31}$$

Data on carbonation depths of unsheltered structures have to be related to meteorological data in order to determine values for the regression parameters a_{w} and b_{w} in Equation 3.2.28. In this procedure two boundary conditions have to be met:

Boundary condition 1: $\mathrm{ToW} = 0 \rightarrow w = 0$

For the case of a sheltered structure, the time of wetness will be $\mathrm{ToW} = 0$. The carbonation will proceed according to the 'square root of time' law, resulting in a weather exponent of $w = 0$.

Boundary condition 2: $\mathrm{ToW} = 1.0 \rightarrow w = 0.50$

For a continuous rain load, the time of wetness is $\mathrm{ToW} = 1$; Progress of the carbonation mechanism is not to be expected, leading to the boundary of $w = 0.50$ (thereby cancelling the time variable); see Equation (3.2.27).

For vertical components the probability of being splashed by driving rain has to be taken into account, yielding the boundary conditions:

$$w = \frac{(p_{\mathrm{splash}} \cdot \mathrm{ToW})^{b_{\mathrm{w}}}}{2} \tag{3.2.32}$$

w weather exponent
ToW   time of wetness
b_{w} regression parameter, ND ($\mu = 0.446$; $\sigma = 0.163$)
p_{splash} probability of a splash event in the case of a decisive rain event, dependent on the orientation of the structure.

$$p_{\mathrm{splash,i}} = \frac{\sum d(w_i \cap r)}{\sum d(r)} \tag{3.2.33}$$

$\sum d(w_i \cap r)$ sum of days during 1 year with wind in the direction considered, i, while on the same day a decisive rain event (precipitation above a level of $h_{rain} \geq 2.5$ mm) is taking place

$\sum d(r)$ sum of days during 1 year with decisive rain events.

3.2.3.4   Relative humidity factor, k_{RH}

The relative humidity factor may be quantified by comparing the data obtained in the considered climate with the data for a reference climate. The parameter k_{RH} describes the effect of the average level of humidity, with $k_{RH} = 1$ for a reference climate, usually set to $T = +20\,°C$, RH $= 65\%$:

$$k_{RH} = \frac{X_{c,obs}}{X_{c,lab}} = \left(\frac{1 - RH^f}{1 - RH^f_{ref}} \right)^g \qquad (3.2.34)$$

where

$X_{c,obs}$ and $X_{c,lab}$ are observations of the carbonation depth made in the field and in the laboratory, respectively

f exponent in the range of 1–10
g exponent in the range of 2–5.

Data regarding RH must be collected from the nearest weather station and analysed statistically [e.g. using a Weibull (W) distribution]. Figure 3.2.5 shows the functional relationship of the relative humidity and the parameter k_{RH}. Figure 3.2.6 illustrates the relative and cumulative frequencies of RH in the city of Aachen, Germany, for the year 1996.

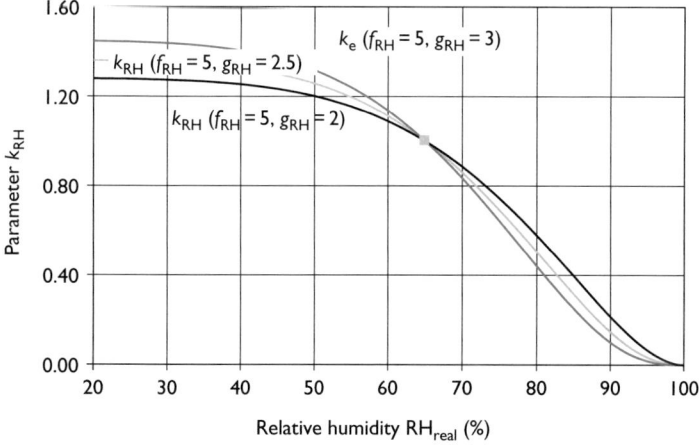

Figure 3.2.5 Functional relationship of the relative humidity and the parameter k_{RH}.

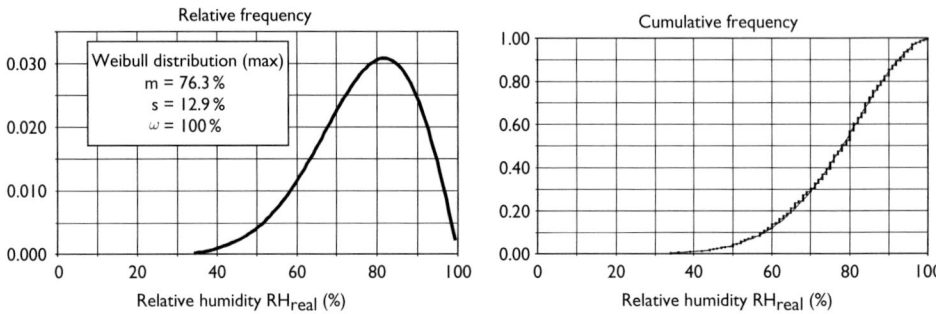

Figure 3.2.6 Relative frequency and cumulative frequency of the relative humidity in the city of Aachen (Germany) for the year of 1996 [14].

3.2.3.5   Surface concentration of carbon dioxide, ΔC_S

The CO_2 concentration of the atmosphere is influenced by two major factors:

- combustion of fossil fuels
- global reduction of vegetation.

Currently, the average CO_2 content of the air varies between 350 and 380 ppm, which equals 0.00057–0.00062 kg CO_2 m^{-3}. Data indicate a constant standard deviation of 10 ppm. Nevertheless, on a microenvironmental scale these values may change considerably due to natural dips (processes reducing CO_2 content, such as absorption from sea, photosynthesis) or low air exchange rates (e.g. in tunnels).

Extrapolating the current trends, an estimate of the increase rate in CO_2 concentration of the atmosphere $C_{S,Atm}$ is in the range of, for example, 1.5 ppm a^{-1}(1.63×10^{-5} kg CO_2 m^{-3} a^{-1}) [15]:

$$\Delta C_S = C_{S,Atm} + \Delta C_{S,Em} \tag{3.2.35}$$

where

$C_{S,Atm}$ CO_2 concentration of atmosphere (kg CO_2 m^{-3})
$\Delta C_{S,Em}$ local additions due to special emissions (e.g. in tunnels) must be measured (kg CO_2 m^{-3}).

with

$$\mu(\Delta C_{S,Atm}) = 5.7 \times 10^{-4} + t \times 2.44 \times 10^{-6} \quad [\text{kg } CO_2 \text{ m}^{-3}] \tag{3.2.36}$$

$\sigma(\Delta C_{S,Atm})$ 1.63×10^{-5}
t time (a).

3.2.3.6 Curing parameter, k_c

The influence of curing on the diffusion properties of concrete depends on the chosen material, as well as on the environment and the curing conditions. In order to analyse data on carbonation depth, sorting criteria have to be introduced, such as:

- time of inspection
- environmental classification (e.g. according to EN 206:XC1–4)
- material classification (binder type and water/cement ratio, w/c).

The data are related to a reference execution procedure, which in Equation (3.2.38) was chosen as 7-day moist curing, for which $k_{c,ref} = 1$ (without differentiation of the moist curing method) [11]:

$$\frac{X_{c,t}}{X_{c,ref}} = \sqrt{\frac{k_{c,t}}{k_{c,ref}}} = \sqrt{k_{c,t}} \qquad (3.2.37)$$

Based on Bayesian linear regression, the relationship of curing time with the curing factor k_c is given by:

$$k_c = a_c \cdot t_c^{b_c} \qquad (3.2.38)$$

where

$\quad k_c$ curing factor
$\quad t_c$ duration of curing (d)
$\quad a_c$ regression parameter, $a_c = 7^{bc}(1\,d^{-1})$
$\quad b_c$ regression parameter, b_c: ND ($\mu = -0.567$; $\sigma = 0.024$) [14].

3.2.3.7 Test method factors k_t and ε_t

The factors k_t and ε_t (slope and y-intercept in Figure 3.2.7) relate the data obtained in tests with accelerated carbonation (ACC, increased CO_2 concentration) to conditions of natural carbonation (NAC).

The regression analysis shows that inverse carbonation resistances $R_{NAC,0}^{-1}$ that are determined under natural carbonation conditions will be larger by an average factor of $k_t = 1.25$ (slope). This may be explained by the fact that in an accelerated test, due to the reduction in test duration resulting from increasing the CO_2 concentration, the drying front has not yet penetrated as deep as it does under natural conditions (though the test is performed under the same climatic conditions: 20°C, 65% RH). This will slightly retard the carbonation process under ACC conditions. This theoretically implies a value of $R_{NAC,0}^{-1} = 0$. As concrete may not possess infinite resistance, this leads to a so-called error term $\varepsilon_t > 0$ (y-intercept in Figure 3.2.7).

3.2.3.8 Concrete cover, d_c

The concrete cover is chosen during the design phase but will always be subjected to variation due to the precision of work. Values of the concrete cover are restricted to the positive domain (negative values are not possible). Additionally, the domain

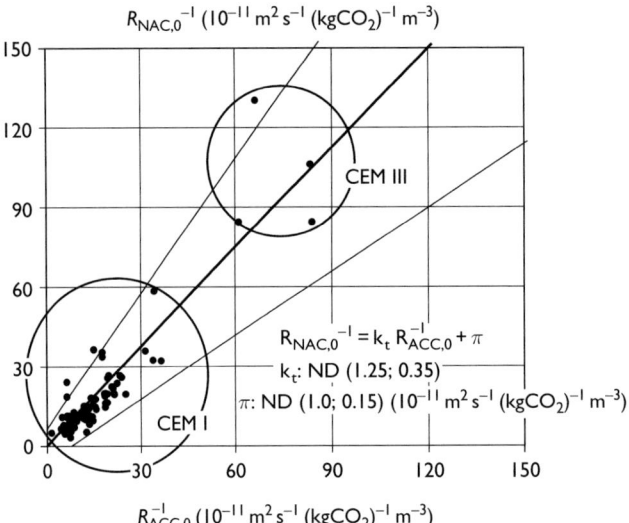

$R_{NAC,0}^{-1}\ (10^{-11}\ m^2\,s^{-1}\ (kgCO_2)^{-1}\ m^{-3})$

CEM III

$R_{NAC,0}^{-1} = k_t\ R_{ACC,0}^{-1} + \pi$

k_t: ND (1.25; 0.35)

π: ND (1.0; 0.15) $(10^{-11}\ m^2\,s^{-1}\ (kgCO_2)^{-1}\ m^{-3})$

CEM I

$R_{ACC,0}^{-1}\ (10^{-11}\ m^2\,s^{-1}\ (kgCO_2)^{-1}\ m^{-3})$

Figure 3.2.7 Relationship of the inverse carbonation resistances, obtained under natural conditions (NAC) and in an accelerated test (ACC) [14].

should have a maximum boundary as the cover cannot exceed the geometric boundaries of the component. These conditions lead to the application of a beta-distribution for the concrete cover. It is not possible to consider situating reinforcing steel directly next to the formwork with this type of distribution. These cross-errors need to be avoided by quality management. Three levels for the statistical quantification of the cover depth are proposed for durability design purposes:

1. Without requirements for workmanship:
    Beta: $\mu = d_{cover,nom}$; $\sigma = 10.0\,mm$; $a = 0\,mm$; $b = 5 \cdot c_{nom} \leq d_{element}$
2. Regular requirements for workmanship:
    Beta: $\mu = d_{cover,nom}$; $\sigma = 8.0\,mm$; $a = 0\,mm$; $b = 5 \cdot c_{nom} \leq d_{element}$
3. Special requirements for the workmanship:
    Beta: $\mu = d_{cover,nom}$; $\sigma = 6.0\,mm$; $a = 0\,mm$; $b = 5 \cdot c_{nom} \leq d_{element}$

where μ is the mean value and σ is the standard deviation. For existing structures, μ and σ are derived from measurements of the cover depth (Lifecon D3.1, Appendix) [1].

3.2.4 Shortcomings of the model

The model neglected the following aspects:

*   Different moisture properties of the carbonated and non-carbonated cover zone are not taken into account.
*   Effect of the concrete temperature. Due to higher temperatures, mainly the chemical reaction of CO_2 with the hydroxides of the pore solution is accelerated. This

chemical reaction takes place much faster than the ingress of CO_2. Temperature effects are thus of minor importance.

- A change in carbonation resistance with increasing age due to hydration. Age-dependent data is not available on this topic. So far only default data on concrete tested at an age of 28 days is provided for various binder types.
- Interaction of carbonation with other degradation mechanisms, such as frost.

3.2.5 Case studies of carbonation-induced corrosion

3.2.5.1 Television tower – an old structure

3.2.5.1.1 INFORMATION PROVIDED

The object considered is a television tower with an age of 32 years at the time of investigation. The tower is of an age when damage due to carbonation usually starts to occur. This implies that a carbonation depth can actually be measured on-site.

This section is meant to demonstrate what type of information usually can or will be provided by the principle. Here the owner provided:

- extracts of the service specifications
- data on the geometry of the tower and
- general notes on the concrete composition.

3.2.5.1.2 DESCRIPTION OF THE OBJECT

The tower was opened to the public on 22 February, 1968, in the area of Munich. The shank is a reinforced concrete tube with a diameter of 16.50 m at ground level (GL). Up to a height of 145 m the tower is designed as a cubic parabola. Higher up, the shape is conical. The thickness varies from bottom to top between 2.0 and 0.65 m (Figure 3.2.8).

The concrete of the shank was placed in sections with a width of 2 m using a climbing formwork. Three of these sections at various heights were chosen as investigation areas.

3.2.5.1.3 EXPOSURE ZONES

A major impact on the carbonation process and the subsequently induced corrosion is given by the moisture state of the component. In areas with a high degree of water saturation, lower carbonation depths will be found for the same concrete quality. Due to the fact that a tower is exposed to different loads of driving rain, distinct exposure zones should be identified. Thus, a durability calculation for an unsheltered vertical concrete surface, as is the case here, has to be performed for each orientation separately, since the concrete resistance and the environmental loading vary (Lifecon D3.1) [1]. The following exposure zones were investigated. The surface of the tower was divided into four zones according to orientation: (1) east, (2) north, (3) west and (4) south.

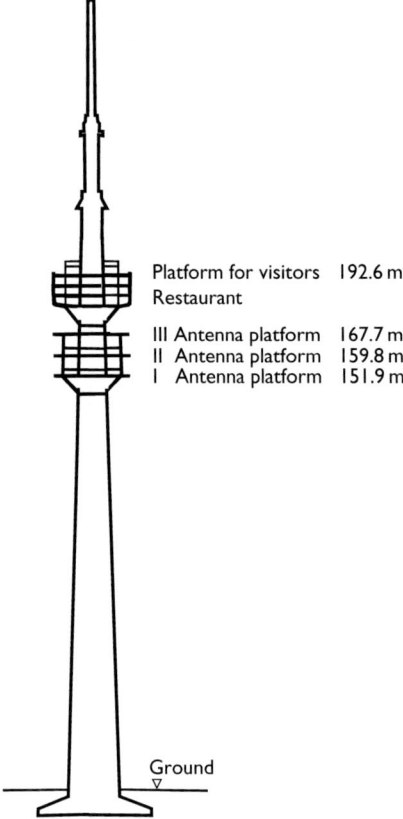

Figure 3.2.8 Scheme of television tower in Munich-City.

3.2.5.1.4 OVERVIEW OF INVESTIGATIONS PERFORMED

The tower has been investigated as follows [16, 17]:

- visually up to a height of 30 m (not treated here)
- measurements of the concrete cover and of the carbonation depths were performed in three levels from GL: 1.0 m, 8.0 m and 28.0 m.

3.2.5.1.5 MEASUREMENT OF CONCRETE COVER, d_{cover}

The concrete cover was measured with commercially available equipment, following the procedure explained in Lifecon D3.1 (Table 3.2.3) [1]. In each investigation section a field of width $(b) = 2$ m/height$(h) = 1$ m was investigated for each orientation. The planned nominal concrete cover was supposed to yield $c_{nom} = 50$ mm. The entrance square is situated at the east side of the tower. Measurement of the concrete cover could therefore not be performed at a height up to $h = 1$ m above GL.

Table 3.2.3 Compilation of concrete cover data

Exposure zone	Level	Height above GL	Minimum/maximum values		Result of data analysis		
		h(m)	$d_{cover,min}$ (mm)	$d_{cover,max}$ (mm)	$d_{cover,mean}$ (mm)	Standard deviation (mm)	Distribution type
1. East	1	1	Entrance	–	–	–	–
	2	8	37	69	50.2	8.0	Normal
	3	28	43	67	54.9	5.7	Normal
2. North	1	1	49	99	72.8	17.8	Normal
	2	8	35	99	69.3	17.2	Normal
	3	28	20	79	55.2	8.7	Normal
3. West	1	1	41	79	57.1	8.0	Normal
	2	8	50	99	62.8	11.1	Normal
	3	28	29	99	48.2	10.2	Normal
4. South	1	1	41	99	71.8	17.3	Normal
	2	8	14	50	38.4	10.1	Normal
	3	28	45	65	54.9	4.3	Normal

3.2.5.1.6 MEASUREMENT OF THE CARBONATION DEPTH, X_C

The carbonation depth was determined by drilling two holes close to each other. The holes were cleaned with a high-pressure air blower. The web between the two holes was broken with a hammer and chisel and cleaned with compressed air. The surfaces were sprayed with a solution of phenolphthalein. The distance from the concrete surface to the depth of the characteristic colour was recorded. Readings of the carbonation depth were taken for all of the investigation areas, four times each, at the north, west, south and east sites and are shown in Table 3.2.4, except at a height $h = 1\,m$ above GL on the north, south and east sides (entrance area).

3.2.5.1.7 SEMI-PROBABILISTIC RELIABILITY CALCULATION

With the information from the structure investigations, the current reliability level can be checked according to the procedure presented. The calculation will be performed for the exposure zone 4 ('south side'), because the analysis revealed that the relatively low concrete cover of the southern parts of the tower caused the lowest reliability index β over time t.

Transforming the stress variable (carbonation depth) and the resistance (concrete cover) into standard space gives:

$$U_1 = \frac{R - 38.4}{10.1}, \quad U_2 = \frac{S - 11.5}{1.2} \tag{3.2.39}$$

$$L(U_1, U_2) = R - S = 10.1 U_1 - 1.2 U_2 + 26.9 = A \cdot U_1 + B \cdot U_2 + C = 0 \tag{3.2.40}$$

Table 3.2.4 Carbonation depth X_c of the television tower

Site		X1 (mm)	X2 (mm)	X3 (mm)	X4 (mm)	Mean (mm)	Standard deviation for area(mm)	Mean for area(mm)
Orientation	H a. GL (m)							
North	1	n.d.				–	1.3	3.0
	8	4	2	1	2	2.5		
	28	5	4	3	3	3.8		
West	1	2	3	2	n.d.	2.3	1.0	2.8
	8	5	2	2	3	3.0		
	28	4	2	3	3	3.3		
South	1	n.d.				–	1.2	4.1
	8	6	4	5	2	4.3		
	28	3	4	5	2	4.0		
East	1	n.d. (entrance area)				–	4.4	10.6
	8	8	13	12	12	11.5		
	28	12	4	6	18	10.0		

Note
n.d. – not determined.

Calculation of the distance of the limit state line from origin using the so-called Hesse normal format (comparison of coefficients A, B, C) gives:

$$\beta = \frac{C}{\sqrt{A^2 + B^2}} = 2.65 > 1.8 \tag{3.2.41}$$

Currently, the target reliability of $\beta_0 = 1.8$ is fulfilled and a further in-depth investigation would not be required. However, for demonstration purposes a full-probabilistic calculation is demonstrated in the next step.

3.2.5.1.8 INVERSE CARBONATION RESISTANCE OF CONCRETE, $R_{\text{ACC},0}^{-1}$

From the service specifications arose:

- the concrete quality is B450 (obsolete terminology), with a compressive strength of $45\,\text{N}\,\text{mm}^{-2}$;
- the cement is a Z375 (obsolete terminology), with a compressive strength of $37.5\,\text{N}\,\text{mm}^{-2}$;
- the w/c ratio had to be chosen to be as low as possible, but requirements were not quantified (in the case study $w/c = 0.50$ was assumed).

According to investigations performed on different concrete compositions [14], the carbonation resistance $R_{\text{ACC},0}^{-1}$ of the concrete may be evaluated according to the following reference mixture:
CEMI, $c = 320\,\text{kg}\,\text{m}^{-3}$, $w/c = 0.50$

giving:

- distribution type: normal distribution (ND)
- mean value: $\mu = 6.8 \times 10^{-11}\,[\mathrm{m}^5\,(\mathrm{s\,kg\,CO_2})^{-1}]$
- standard deviation (SD): $\sigma = 0.45, \mu = 3.06 \times 10^{-11}\,[\mathrm{m}^5\,(\mathrm{s\,kg\,CO_2})^{-1}]$

3.2.5.1.9 RELATIVE HUMIDITY FACTOR, k_{RH}

The parameter k_{RH} describes the effect of the 'average' level of humidity. For reference climate $(T = +20\,^{\circ}\mathrm{C}, \mathrm{RH} = 65\%)$, the relative humidity factor $k_{\mathrm{RH}} = 1$.

The position of the nearest weather station is Munich city, only existing since July 1997. Therefore, data from the Munich–Nymphenburg weather station for the period from 1991 to 1996 was added. As an example, the 'distribution plot' of the relative humidity for the year 1999 measured in Munich city is given in Figure 3.2.9.

As can be seen in the figure, the relative humidity varies in the range from $\mathrm{RH}_{\mathrm{min}} = 34\%$ to $\mathrm{RH}_{\mathrm{max}} = 100\%$. According to data for the past 10 years, the following input data was calculated with the software package STRUREL [19, 20], which is used as an add-in tool:

Distribution type: Beta
Mean value: $\mu = 74\ (\%)$
Standard deviation: $\sigma = 14\ (\%)$
Boundaries: $34 \leq \mathrm{RH} \leq 100\ (\%)$

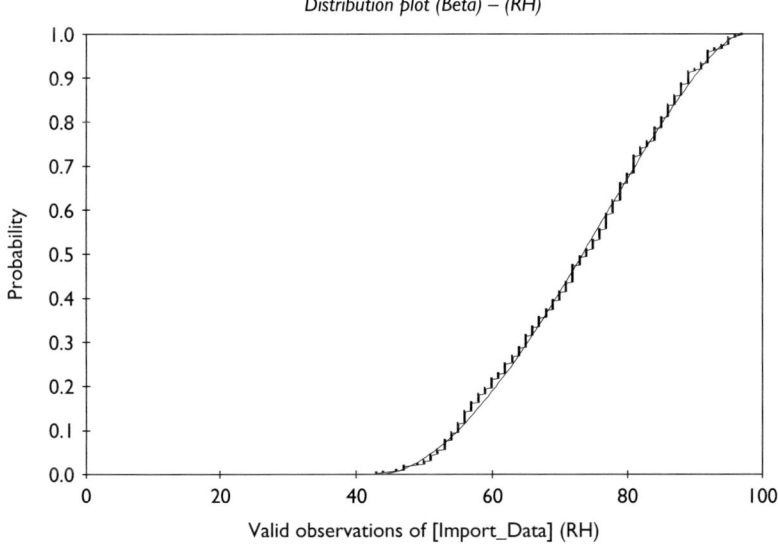

Figure 3.2.9 Probability distribution function of daily relative humidity RH (%) for the year 1991 measured at Munich city.

Table 3.2.5 Input data for the determination of the weather exponent *w*

Exposure zone	Position	ToW	p_i	b_w
1	North	27.3	0.021	ND
2	West	27.3	0.375	
3	South	27.3	0.037	$\mu = 0.446$
4	East	27.3	0.014	$\sigma = 0.163$

3.2.5.1.10   WEATHER FUNCTION, W

The necessary input data in the weather function for each of the four exposure zones considered are given in Table 3.2.5. The time of wetness ToW has been calculated from the data of the weather stations in Munich city and Munich–Nymphenburg. The probability of driving rain in a considered direction p_i was determined by the mean distribution of the wind direction p_{wind} during a rain event (for sufficient duration; here 10 years) (Lifecon D4.3) [21].

3.2.5.1.11   CURING FACTOR, k_c

For the current structure, data on the curing duration was not available and was thus assumed to be $t_c = 1$ day. The assumption can be regarded as being on the safe side, but reasonable for a climbing formwork.

$$k_c = a_c \cdot t_c^{b_c} = 3.01 \cdot 1^{b_c} \tag{3.2.42}$$

where

a_c regression parameters, set to $a_c = 3.01$
b_c regression parameters, normally distributed: $\mu = -0.567$; $\sigma = 0.024$ [14].

3.2.5.1.12   CARBON DIOXIDE CONCENTRATION OF THE ENVIRONMENT, ΔC_S

In the calculations for the television tower, the CO_2 concentration is assumed to be constant. On the safe side, the following value for the year 2100 has been calculated:

$$\mu(\Delta C_{S,Atm}) = 5.7 \cdot 10^{-4} + t \cdot 2.44 \cdot 10^{-6} = 8.2 \cdot 10^{-4} \quad (Kg\,CO_2\,m^{-3}) \tag{3.2.43}$$

$$\sigma(\Delta C_{S,Atm}) = 1.63 \cdot 10^{-5}$$

3.2.5.1.13   CONCRETE COVER, d_{cover}

As input for the prediction of service life, a statistical description of the concrete cover is essential. The measured data has been evaluated with the software [20]. The data fit well with to a normal distribution (ND), although distribution types with lower and upper limits (e.g. beta distributions) are physically more reasonable (Figure 3.2.10).

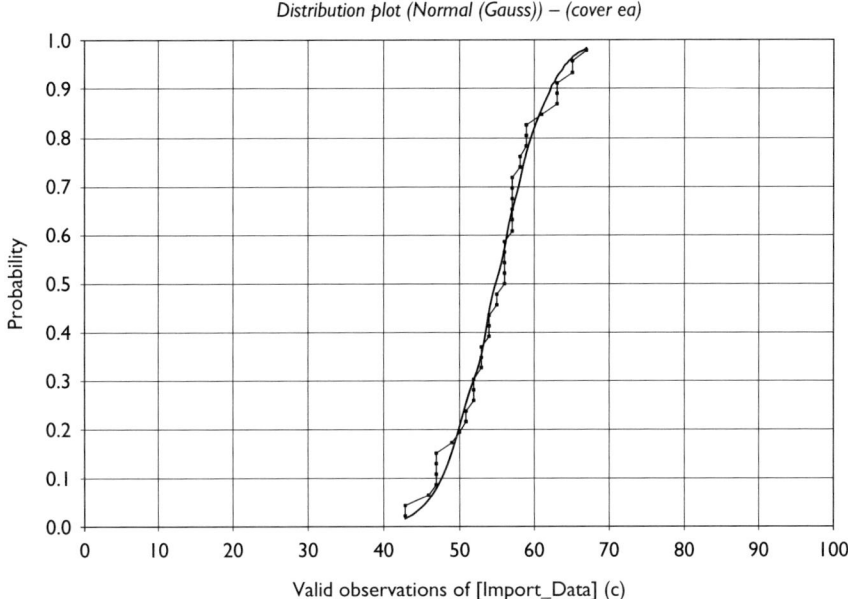

Figure 3.2.10 Statistical analysis of the actual concrete cover of the investigation level 3 (28 m above GL) for the east side (exposure zone 1).

3.2.5.1.14 CALCULATION WITHOUT DATA UPDATE

3.2.5.1.14.1 Development of the carbonation depth

Calculation of the time-dependent development of the carbonation depth, which may be performed with a simple pocket calculator or an Excel spreadsheet using the mean values of the quantified data, indicates that the calculated carbonation depths exceed the values actually measured (Figure 3.2.11 and Table 3.2.6).

One explanation for the deviations of calculated values from measured values is the fact that some of the input parameters were assumptions (e.g. w/c ratio, curing time t_c) since no reliable information was available.

In the axis of the common wind direction, west to east, the results of the prediction fit fairly well with the actual measurements (Figure 3.2.12). However, the deviations in the north–south axis are probably due to the geometry of the structure. This may be explained by the parameter p_i in the weather function. The probability p_i of driving rain hitting a surface facing direction i describes the average distribution of the wind direction during rain events. The data is collected separately for each orientation. For a round geometry, as is the case for the television tower investigated here, the water hitting the structure will run along the surface. This effect will especially influence the surfaces on the north and south sides (of uncommon wind direction). Water which mainly hits the west side, causing low carbonation depths (Table 3.2.6) will reach the northern and southern surfaces as well, thereby reducing the carbonation rate in these zones.

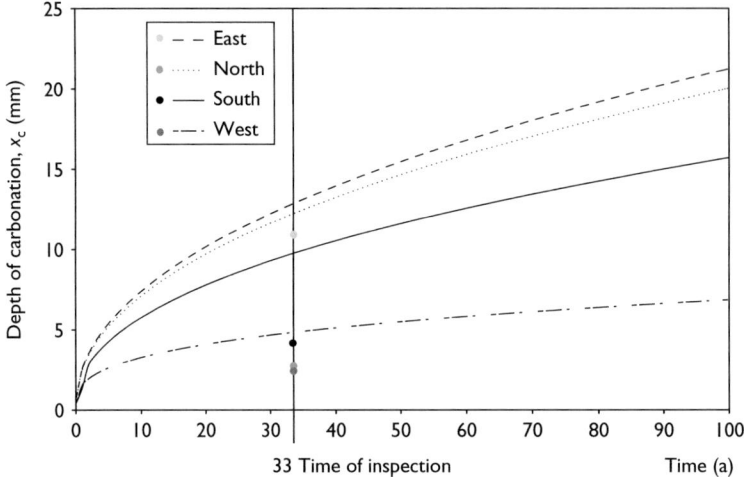

Figure 3.2.11 Calculated time-dependent development of the carbonation depth and actually measured values at an age of $t = 33a$.

Table 3.2.6 Comparison of calculated mean values and actually measured carbonation depths after $t = 33a$ in the four exposure zones

Carbonation depth at age $t = 33a$ (mm)	Exposure zone 1: east	Exposure zone 2: north	Exposure zone 3: west	Exposure zone 4: south
Calculated mean	12.8	12.0	4.8	9.7
Measured value (see Table 3.2.4)	10.6	3.0	2.8	4.1

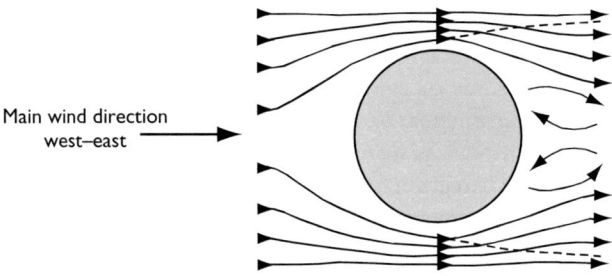

Main wind direction west–east

Figure 3.2.12 Scheme of wind streams around a round structure.

3.2.5.1.15   BOUNDARY CONDITIONS FOR CALCULATIONS WITH BAYESIAN UPDATE

By introducing inspection data as a boundary, the precision of calculations of the remaining service life will be increased. The implementation of inspection data (carbonation depth) is called Bayesian updating and can be performed, for example, with the program SYSREL of the STRUREL software package [20]. In addition to the limit state function, the formulation of boundary conditions can be inserted; this accounts for the actual investigations. For the present case study, the carbonation data from inspections at age $t_{\text{inspection}} = 33a$ are included.

The boundary condition to be considered (equality constraint) is:

$$X_{C,\text{inspection}}(t = 33a) = \sqrt{2 \cdot k_{\text{RH}}(k_t \cdot k_c \cdot R_{\text{ACC},0}^{-1} + \varepsilon_t) \cdot \Delta C_S} \cdot \sqrt{t = 33a} \cdot \left(\frac{t_0}{t = 33a} \right)^w$$

$$(3.2.44)$$

3.2.5.1.16 EXAMPLE OF CALCULATION WITH BAYESIAN UPDATE – EXPOSURE ZONE 4, 'SOUTH'

The calculation including a data update will be demonstrated for exposure zone 4, the south side, because the analysis revealed that the relatively low concrete cover of the southern parts of the tower caused the lowest reliability index β over time t. The quantified input parameters are given in Table 3.2.33.

The results of the calculation are the reliability index β and the failure probability p_f over time t in the time domain of t_{target} (which here has been set to 100 years).

For this decisive level in each exposure zone, the time-dependent reliability index β and failure probability p_f (%) have been calculated as displayed for the south side (Figures 3.2.13 and 3.2.14).

3.2.5.1.17 EVALUATION OF CALCULATION RESULTS

The calculations demonstrated that, due to the relatively low concrete cover, the south side of the tower has to be considered as the most critical area. The results were verified by the symptoms discovered during visual examinations, revealing corrosion

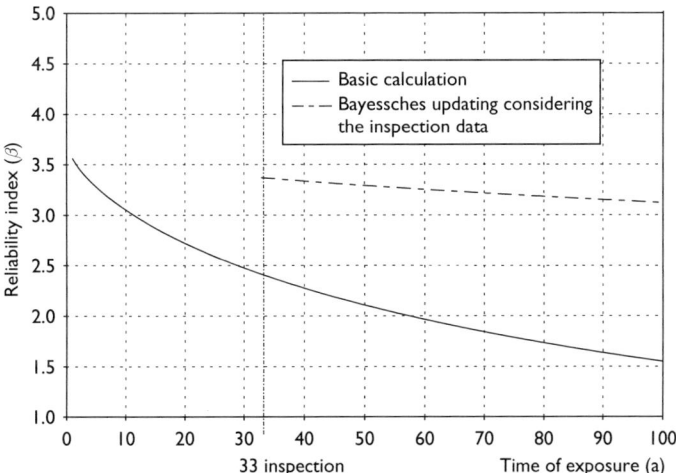

Figure 3.2.13 Time dependent reliability index β of exposure zone 4 'South'.

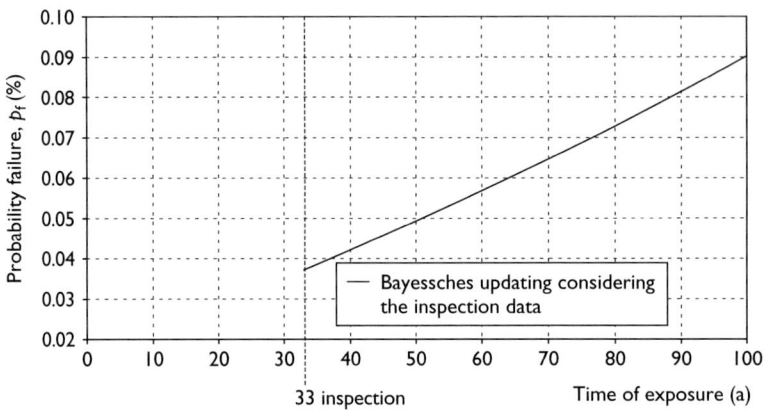

Figure 3.2.14 Time failure probability p_f (%) of exposure zone 4 'South'.

stains in this exposure zone. At age $t_{target} = 100$ years, approximately 0.09% of the reinforcement will be depassivated.

The calculation with the simple semi-probabilistic approach indicated a reliability index of around $\beta = 2.6$ at the time of inspection, which is slightly higher than the result obtained with the a priori model (without knowledge of inspection data), whereas the model which was improved by Bayesian updating showed a significantly higher reliability level at the time of inspection. The reason for this effect is that the a priori model includes a high level of uncertainty in the various input parameters. The semi-probabilistic calculation includes only two stochastic variables, which were determined by measurements. However, the least amount of uncertainty is inherent in the updated full-probabilistic model, which thus gives the highest reliability level.

The calculated results can be compared to requirements stated in codes of practice (EuroCode 1) [3] or by the owner, as presented in Table 3.2.7 and Table 3.2.8.

3.2.5.2   A young bridge

3.2.5.2.1   INFORMATION PROVIDED

The treated object is a bridge, which is less than 1 year old. Due to insufficient cover depth, as was detected during acceptance inspection, the necessity of calculating

Table 3.2.7 Planned service life according to EuroCode 1 (Part 1, Table 2.1) [3]

Class	Planned service life (a)	Examples
1	1–5	Structures with limited use
2	25	Exchangeable components, e.g. bearings
3	50	Buildings and other common structures
4	100	Monumental structures, bridges and other engineering structures

Table 3.2.8 Value indications for the reliability index β according to EuroCode I (Part I, Appendix A, Table A2)

Limit state	β (at end of planned service life)	β (after I year)
ULS	3.8	4.7
Fatigue	1.5–3.8[1]	—
Serviceability	1.5	3.0

Notes
ULS – ultimate limit state.
[1] Dependent on the degree of inspection and repair possibilities.

the remaining service life arose. The following data was accessible prior to the investigations:

1. Design documents

   - shuttering plan
   - reinforcement plans

2. As-built specifications

   - concrete catalogue
   - extract of the construction diary
   - information on quality management

3. Inspection reports

   - list of insufficiencies.

3.2.5.2.2 DESCRIPTION OF THE OBJECT

The bridge is a crossing of two municipal roads situated in a township close to Munich. The construction process was finished at the beginning of the fourth quarter of the year 2000. The bridge is designed as a frame with a total length amounting to $l_{tot} = 30\,m$ and a span of $l_s = 18\,m$ (Figures 3.2.15 and 3.2.16).

The frame design was a concrete B35 (compressive strength $35\,N\,mm^{-2}$ determined according to the German standard DIN 1045). The water/cement ratio was set to 0.50. According to the German additional technical contract requirements (ZTV-K 1996), the nominal value of the concrete cover was set to $c_{nom} = 45\,mm$, with a minimum cover depth of $c_{min} = 40\,mm$.

During the first assigned bridge investigation it was recognised that the minimum cover depth had partly fallen short. Areas with low concrete cover were cantilevers bottom areas of the abutments.

The background of such investigations is that a contractor may not have to carry out additional repairs at his own expense if it can be proved that, due to the concrete quality, local undershooting of the minimum cover depth will be compensated and therefore that the costs of maintenance and repair will not exceed a generally accepted amount.

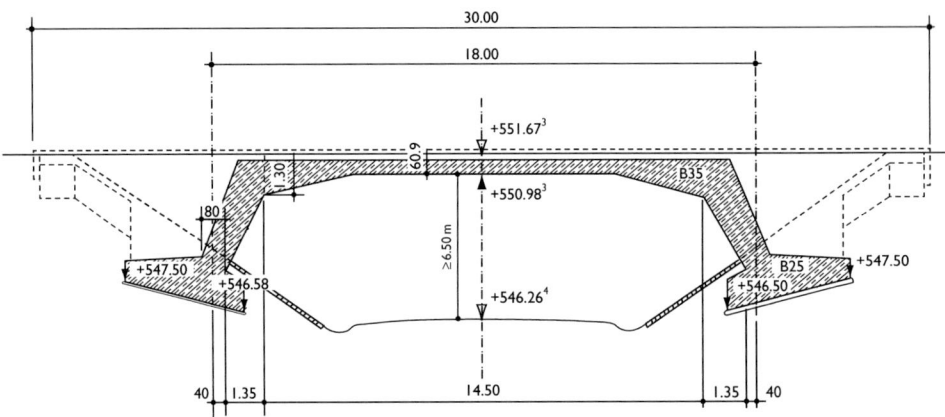

Figure 3.2.15 Longitudinal section of the bridge.

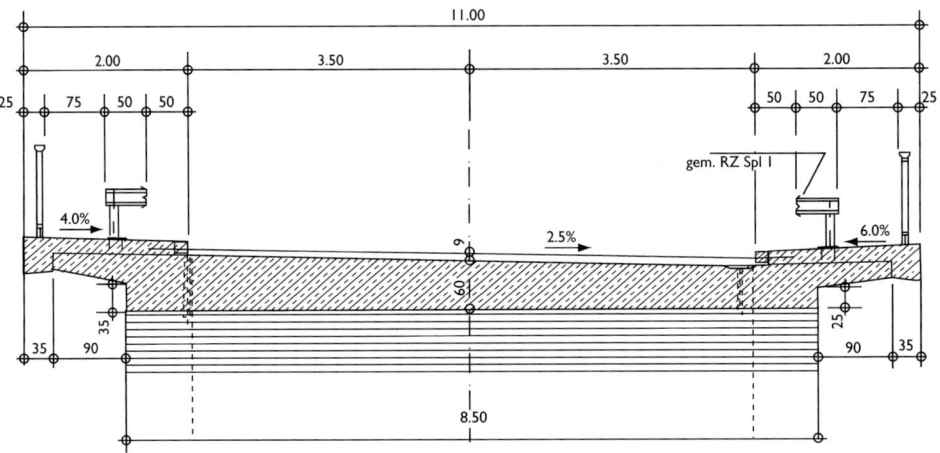

Figure 3.2.16 Cross section of the bridge.

3.2.5.2.3 EXPOSURE ZONES

In the present case study a division into exposure zones took place (Figure 3.2.17) [1]. The following surface areas were investigated:

- abutment at the west and east sides
- bottom of the cantilevers at north and south sides
- cantilevers north and south.

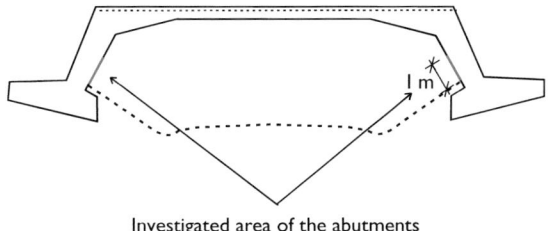

Investigated area of the abutments

Figure 3.2.17 Example for the division of the bridge into components and exposure zones.

3.2.5.2.4   OVERVIEW OF INVESTIGATIONS PERFORMED

The following investigations were performed [16]:

1. Core drilling

   - bridge deck
   - both abutments

2. Accelerated carbonation tests on cores
3. Measurement of the concrete cover

   - sides of the cantilevers
   - bottom of the cantilevers
   - both abutments.

3.2.5.2.5   CORE DRILLING AND ACCELERATED CARBONATION TESTING

In the first case study, the Munich television tower, the inverse carbonation resistance $R_{\mathrm{ACC},0}^{-1}$ was extracted from the literature (database) and was updated with measured data on the carbonation penetration front obtained throughout structure investigations. This approach is only applicable for structures for which a response to the environmental loading, i.e. an exposure to CO_2, can be measured in the form of a penetration depth. For the present object under observation, a penetration depth was not detectable, due to the short exposure duration. Therefore, accelerated carbonation testing had to be performed in order to determine the real carbonation resistance.

In every exposure zone a core with diameter $\Phi = 50\,\mathrm{mm}$ was sampled with a length of $l = 50\,\mathrm{mm}$. The cores were stored for 21 days under constant climatic conditions ($T = 20\,^{\circ}\mathrm{C}, \mathrm{RH} = 65\%$). The cores were kept for 28 days in a carbonation chamber with a CO_2 concentration of $\Delta C_s = 2.0\%$ by volume, thereby accelerating the carbonation process. After 28 days the cores were split and the carbonation depth was determined using a phenolphthalein indicator. The carbonation depth was measured at four positions for every core with a precision of $0.5\,\mathrm{mm}$ (Table 3.2.9).

Table 3.2.9 Mean values of the penetration depth measured in accelerated carbonation test

Location of withdrawal	Carbonation depth X_c (mm)				
	Core 1	Core 2	Core 3	Core 4	Mean
Abutment east	5.0	6.5	6.0	2.0	
Southern cantilever of superstructure	7.0	5.0	5.0	5.5	5.5
Northern cantilever of superstructure	4.5	7.0	7.0	5.0	

3.2.5.2.6 MEASUREMENT OF THE CONCRETE COVER, d_{cover}

The average concrete cover of all investigated areas was above a value of $d_{cover} = 45$ mm. Low cover depths were measured for the western abutments and the southern cantilevers: $d_{cover} = 42$ mm and $d_{cover} = 36$ mm respectively.

3.2.5.2.7 INVERSE CARBONATION RESISTANCE, $R_{ACC,0}^{-1}$

In order to determine the actual carbonation resistance of the built-in concrete quality, ACC were performed. The inverse carbonation resistance $R_{ACC,0}^{-1}$ is calculated by inserting the mean of the penetration depth into the equation:

$$R_{ACC,0}^{-1} = \left(\frac{X_C}{T}\right)^2 \tag{3.2.45}$$

X_C mean of penetration depth determined with ACC test
T constant $= 419.45 \, [((s \, kg \, CO_2) \, m^{-3})^{0.5}]$.

This yields $R_{ACC,0}^{-1}$ is normally distributed with mean $\mu = 6.8 \times 10^{-11} \, [m^5 \, (s \, kg \, CO_2)^{-1}]$ and standard deviation $\sigma = 0.45$, $\mu = 3.06 \times 10^{-11} \, [m^5 \, (s \, kg \, CO_2)^{-1}]$.

3.2.5.2.8 ENVIRONMENTAL PARAMETER: RELATIVE HUMIDITY FACTOR, k_{RH}

For the calculations of k_{RH} the same meteorological data as for the Munich television tower case study were used.

3.2.5.2.9 CURING FACTOR, k_c

In the ACC test, specimens were tested in which the curing conditions were inherent. Therefore, a curing factor could be set constant to $k_c = 1$.

3.2.5.2.10 CARBON DIOXIDE CONCENTRATION, ΔC_s

For ΔC_s the same assumptions were made as for the Munich television tower case study.

3.2.5.2.11 CALCULATIONS AND RESULTS

In contrast to the case study of the television tower, calculation of the remaining service life including a Bayesian update cannot be performed, because data on the carbonation penetration depth is not obtainable at an early age of the structure. The only updated information considered is the real concrete resistance of the structure, tested in an ACC test. Inserting all the input data into the limit state function, the result of the computations with the software package STRUREL [20] is the reliability index β or the failure probability p_f over time.

3.2.6 Semi-probabilistic model of chloride ingress

3.2.6.1 Mathematical background

3.2.6.1.1 FICK'S FIRST LAW

In water, dissolved chloride follows Brownian motion in a stochastic manner. In a concentration gradient, chloride ions aim for equilibrium and thus move from high levels of concentration to lower concentrations [22]. The diffusion of chloride in concrete can be expressed by means of ionic flow J, which is by definition positive, if the chlorides move in a positive x-direction. If the concentration decreases in the x-direction the flow is negative:

$$J = -D \cdot \frac{\partial c}{\partial x} \qquad (3.2.46)$$

J ionic flow $(kg\,m^{-2}\,s^{-1})$
D diffusion coefficient $(m^{-2}\,s^{-1})$
c chloride concentration in solution $(kg\,m^{-3})$
x distance (m).

3.2.6.1.2 FICK'S SECOND LAW

The diffusion process is fully described if the particle density is given as a function of space and time t. Starting with Fick's first law and considering the law of mass conservation, Fick's second law is obtained:

$$\frac{\partial c}{\partial t} = D \cdot \frac{\partial^2 c}{\partial x^2} \qquad (3.2.47)$$

The application of Fick's laws of diffusion to describe the transport processes in concrete is based on the simplified assumption that concrete is homogeneous, isotropic and inert [23]. Furthermore, the movement of negatively charged ions induces the movement of positively charged ions. The well-known fact that sodium and chloride move at different velocities is neglected when applying Fick's law [24].

While chlorides are diffusing in the pore system they are chemically and physically bound:

$$\frac{\partial c}{\partial t} = -\frac{\partial J}{\partial x} - \frac{\partial c_b}{\partial t} \qquad (3.2.48)$$

where c_b is the proportion of bound chlorides within the concrete.

If Equation (3.2.46) is inserted into Equation (3.2.48) and the diffusion coefficient is considered as time-dependent, the transport is described as follows:

$$\frac{\partial c}{\partial t} + \frac{\partial c_b}{\partial t} = \frac{\partial D(x, t)}{\partial x} \cdot \frac{\partial c}{\partial x} + D(x, t) \cdot \frac{\partial^2 c}{\partial x^2} \qquad (3.2.49)$$

Chloride binding may, for instance, be described by the so-called Freundlich isotherm. If the isotherm is related to the pore volume of concrete, one obtains [25]:

$$C_b = \frac{(1 + W_n^0) \cdot \alpha}{\left(\frac{1}{f_c} + W_n^0\right) \cdot \alpha} \cdot \frac{f_a}{V_p} \cdot c^B = A \cdot c^B \qquad (3.2.50)$$

where

W_n^0 the proportion of bound water
α degree of hydration
V_p pore volume ($m^{-3} kg^{-1}$ dry concrete)
f_c binder content (kg binder/kg concrete).

This results in:

$$\frac{\partial c_b}{\partial t} = A \cdot B \cdot c^{B-1} \cdot \frac{\partial c}{\partial t} \qquad (3.2.51)$$

If Equation (3.2.51) is inserted in Equation (3.2.49), the general relationship for the time-dependent non-steady-state diffusion of chlorides, considering chloride binding, is obtained:

$$\frac{\partial c}{\partial t} = \frac{1}{1 + A \cdot B \cdot c^{B-1}} \left(\frac{\partial D(x, t)}{\partial x} \cdot \frac{\partial c}{\partial x} + D(x, t) \cdot \frac{\partial^2 c}{\partial x^2} \right) \qquad (3.2.52)$$

An analytical solution is hard to derive, and calls for the application of numerical approaches.

3.2.6.1.3 TYPES OF DIFFUSION COEFFICIENT

In the current literature there is tremendous disagreement over the terminology of diffusion coefficients. There are real (D), effective (D_{eff}), apparent or achieved (D_{app} or D_a), steady-state (D_{ss}), non-steady-state (D_{nss}) and potential (D_p) diffusion coefficients, among others. In order to prevent any bewilderment of the reader, here only the following terms are used:

1. Effective diffusion coefficient, D_{eff}
   This is a steady-state diffusion coefficient including the effect of different transport velocities of anions and cations.
   The effective diffusion coefficient is the result of diffusion cell tests. In these a thin concrete specimen separates a chloride-free and a chloride-containing solution. The chloride concentration of the solution that is at first chloride-free is measured against time. Once a steady ingress rate is reached, D_{eff} is calculated according to Fick's first law. The value obtained must then be related to the porosity of the concrete.
2. Apparent diffusion coefficient, D_{app}
   This is a non-stationary diffusion coefficient including the effects of:

   - different transport velocities of anions and cations
   - hydration of cement paste
   - concentration-dependent and hence time-dependent chloride binding.

   This parameter is derived by evaluating chloride profiles, which can be obtained in immersion tests or structure investigations by applying a solution of Fick's second law of diffusion.

3.2.6.2   Choice of model

Due to the extreme complexity of the evolution and interactions of the mechanisms contributing to the ingress of chlorides into concrete, all current models are based on simplifications and assumptions and hence are more or less empirical. There are two constraints:

- The time constraint: depth $x > 0$ and time $t = 0$ – it follows that the concentration c is equal to the initial concentration: $c = c_i$.
- The geometrical constraint: for $x = 0$ and $t > 0$ – the concentration c is equal to that of the surrounding environment: $c = c_0$.

with these two constraints, the most common approach, known as the 'error function solution' of Fick's second law, can be derived:

$$\frac{c_0 - c}{c_0 - c_i} = \text{erf} \cdot \frac{x}{\sqrt{4(t - t_{exp}) \cdot D_{app}}} \tag{3.2.53}$$

where
$\quad c_o \quad$ chloride concentration of surrounding solution $(g\,l^{-1})$
$\quad c \quad$ free-chloride concentration of pore solution in depth x $(g\,l^{-1})$
$\quad i \quad$ 'initial' concentration in pore solution $(g\,l^{-1})$
$\quad \text{erf} \quad$ error function
$\quad x \quad$ distance from concrete surface (m)

t age of concrete (s)

t_{exp} time until concrete is exposed to chloride environment (s)

D_{app} apparent diffusion coefficient ($m^2 \, s^{-1}$).

This model is based on numerous simplifications:

- It is only valid for a constant apparent diffusion coefficient D_{app} and a constant surface concentration C_S.
- The approach is only valid for semi-infinite conditions, i.e. the object considered is sufficiently large, so that a constant concentration will be reached in the inner concrete zone.
- The approach assumes in its pure mathematical form that ingress is a stationary process, thereby neglecting the binding of chlorides. As the binding of chlorides is a non-linear process the free chloride concentration c may not be replaced by the total chloride content, which is convenient to measure [26]:

$$\frac{C_{ts} - C_t}{C_{ts} - C_{ti}} = \frac{\left(c_0 + \alpha \cdot c_0^\beta\right) - \left(c + \alpha \cdot c^\beta\right)}{\left(c_0 + \alpha \cdot c_0^\beta\right) - \left(c_i + \alpha \cdot c_i^\beta\right)} \neq \frac{c_0 - c}{c_0 - c_i} \qquad (3.2.54)$$

where

t total

s surface

i initial

c_o chloride concentration of surrounding solution

c free chloride concentration in pore solution in depth x

α, β regression parameters of the Freundlich binding isotherm.

As throughout chloride analysis usually only the total soluble chloride content is measured, the application of the error function solution is physically not correct. However, from an engineering point of view it is a good approximation and is convenient to use.

Typically the error function solution is written in the following way:

$$C(x, t) = C_i + (C_S - C_i) \cdot \operatorname{erf}\left(1 - \frac{x}{2\sqrt{(t - t_{exp}) \cdot D_{app}}}\right) \qquad (3.2.55)$$

where

$C(x, t)$ chloride concentration at depth x at age t in e.g. ($M\% \, c^{-1}$)

C_i initial chloride background level in e.g. ($M\% \, c^{-1}$)

C_S surface chloride content ($M\% \, c^{-1}$)

D_{app} time-dependent apparent diffusion coefficient ($m^2 \, s^{-1}$)

t_{exp} time until first exposure to chlorides (s)

t concrete age (s)

erf   error function.

The error function erf is defined by:

$$\mathrm{erf}\, x = \Phi\left(\sqrt{2x}\right) = \frac{2}{\sqrt{\pi}} \int\limits_{0}^{x} e^{-t^2}\, \mathrm{d}t$$

$$= \frac{2}{\sqrt{\pi}}\left(x - \frac{x^3}{1!3} + \frac{x^5}{2!5} + \cdots + \frac{(-1)^n x^{2n+1}}{n!(2n+1)}\right),$$

$$n = 0, 1, 2, \ldots \qquad\qquad (3.2.56)$$

The chloride ingress in concrete results in a chloride penetration profile, which is the distribution of the chloride content (e.g. related to the weight of concrete specimen or weight of binder) versus depth measured from the concrete surface. As a sample with infinitely small thickness cannot be taken, the surface concentration of the concrete is determined by extrapolation of the profile to depth $x = 0$. The surface concentration C_S is therefore not the real concentration at the time, but rather a regression parameter.

In practice, the surface chloride concentration C_S changes with time under certain environmental conditions, but usually reaches a constant value during a period which is relatively short compared to the intended service life of the structure (Figure 3.2.18). This commonly leads to the approach of assuming C_S to be constant. Nevertheless, there are cases in which C_S should be regarded as time dependent. In the case of low chloride loads of the environment (e.g. in the spray zone of road environments

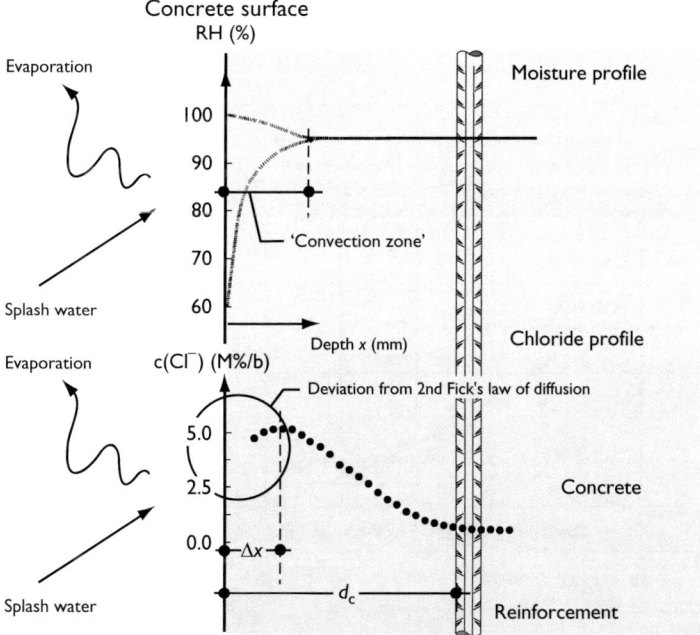

Figure 3.2.18 Distribution of the moisture content and the chloride concentration, as commonly observed in the splash-zone [14].

with application of de-icing salt) the surface level may take many years to reach its maximum.

Especially for the case of an intermittent impact of chlorides on concrete structures, practical observations have shown that Fick's second law of diffusion may not be applied without restraints. Close to the concrete surface of a component situated in a spray environment, the concrete is exposed to a continuous change of wetting and subsequent evaporation. This zone is usually referred to as the 'convection zone'. Due to the changes of the moisture content in the convection zone, the following effects may take place:

- so-called piggyback transport caused by capillary penetration of solutions trailing chlorides (convection);
- wash-off at times when the surface is subjected to chloride-free water; for instance, structures in the splash zone of the road environment during periods without salt application;
- change of the chloride-binding capacity induced by carbonation and leaching.

In order still to be able to describe the penetration of chlorides under intermittent loading using Fick's second law of diffusion, the data for the convection zone, which deviates from diffusion behaviour, may be neglected in the fitting process. The calculation starts with the substituted surface concentration $C_{s,\Delta x}$ at depth Δx.

This effect may be due to variation in chloride loading (mainly washout with chloride-free water) or carbonation of the surface zone. This type of profile may also be fitted using Fick's second law, by neglecting the data points near the surface (Figure 3.2.19).

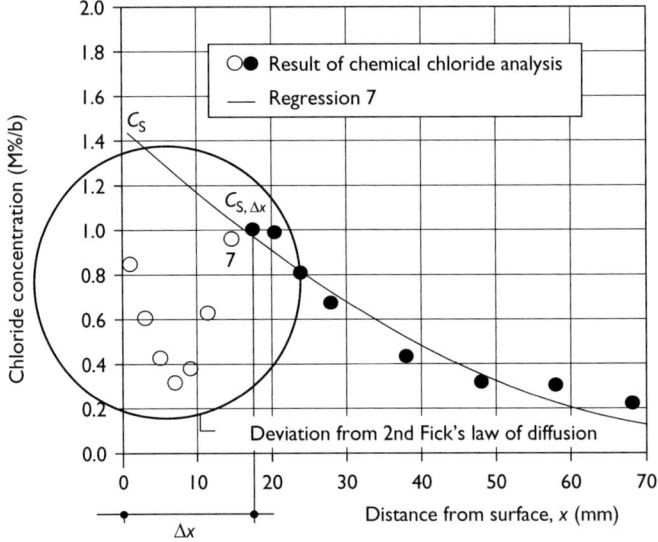

Figure 3.2.19 Detailed chloride profile in the splash zone of a bridge [27], which was fitted by neglecting outer data points.

The output parameters are then the chloride concentration $C_{\Delta x}$, the depth of the so-called convection zone Δx and the apparent diffusion coefficient D_{app}. A simple engineering model which incorporates the effect of capillary suction is thus [14]:

$$C(x, t) = C_i + (C_S - C_i) \cdot \mathrm{erf} \left(1 - \frac{x - \Delta x}{2\sqrt{(t - t_{exp}) \cdot D_{app}}} \right) \qquad (3.2.57)$$

The chloride concentrations determined in the chemical analysis are inserted, with the corresponding drilling depth, in a data sheet (e.g. Microsoft Excel) (Figure 3.2.20). These chloride profiles are fitted to Equation (3.2.57) using an optimisation program, such as the solver option of Microsoft Excel.

As can be seen in Figure 3.2.20, the chloride profile converges with increasing depth towards the initial chloride concentration C_i. The initial chloride content is the sum of the chloride contents in the particular components of the concrete. This value must be chosen visually from the given data points. For this reason it is important to collect sufficient data points in the tail of a profile, but at least two for the estimation of C_i. If chloride profiles have not been determined to a sufficient depth, this value must be assumed. For concrete produced with Portland cement a statistical analysis of approximately 640 chloride profiles in Germany results in a mean value of around 0.1 (wt%/cement) or approximately 0.015 (wt%/concrete).

The depth of the convection zone Δx is also visually determined from the chloride profile considered. The choice of this parameter may have a significant impact upon the result of the regression.

The exposure duration $(t - t_{exp})$ is the difference between the total age of the concrete and the moment of first contact with chlorides, which may usually be set equal to the concrete age t.

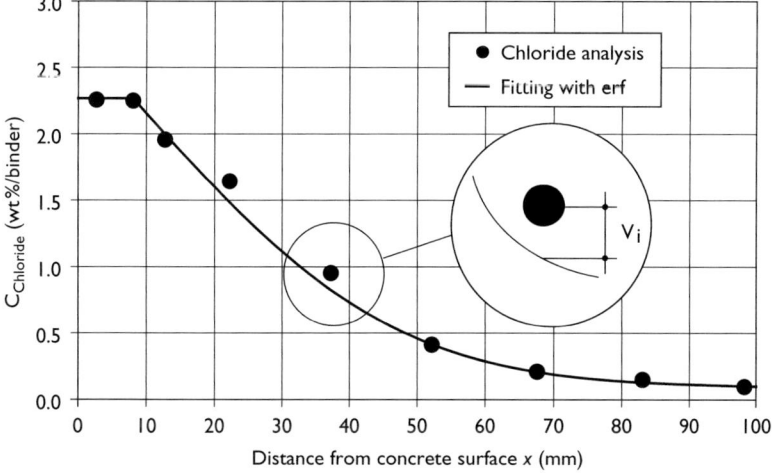

Figure 3.2.20 Typical chloride distribution in concrete and result of curve fitting obtained by minimizing the sum of the squared deviations v_i between (3.2.55) and the measured data.

Table 3.2.10 Example of spreadsheet to fit chloride profiles (data from case study on the Hofham Bridge – a pillar exposed to de-icing salts)

Sample interval		Depth X (mm)	$C(x)_{measured}$ (wt%/cement)	$C(x)_{calc}^{*}$ (wt%/cement)	$V_i^2(wt\%/cement)^2$
from (mm)	to (mm)				
1	2	3	4	5	$6 = ((4-5)/4)^2$
0	5.6	2.8	2.250	2.250	0.00000
5.6	10.8	8.2	2.243	2.338	0.00181
10.8	15.0	12.9	1.950	2.061	0.00323
15.0	29.8	22.4	1.635	1.528	0.00426
29.8	45.0	37.4	0.945	0.847	0.01065
45.0	59.6	52.3	0.405	0.424	0.00220
59.6	75.8	67.7	0.203	0.210	0.00143
75.8	90.6	83.2	0.143	0.130	0.00831
90.6	106.2	98.4	0.090	0.100	0.01235
				Sum (minimise!)	0.0443

Set input parameters			Regression variables (result)	
t (a)	C_i (wt%/cement)	Δx (mm)	$D_{app}(10^{-12}\,m^{-2}\,s^{-1})$	C_S (wt%/cement)
38	0.10	8.0	0.385	2.350

Note
* According to Equation (3.2.57)

If the exposure time t, the initial concentration C_i and the depth of the convection zone Δx are fixed, the remaining variables, C_S and D_{app}, can be calculated, as shown in the example in Table 3.2.10.

If these regression variables are known, the future chloride profiles may be predicted by extrapolation (Figure 3.2.21):

Today numerous models and modifications of these exist, and may be categorised as follows [28, 29]:

(a) Empirical models
   Chloride profiles are predicted by means of analytical or numerical solutions of Fick's second law of diffusion:

   1. Error function methods with a constant diffusion coefficient D and surface concentration C_S [30]

      • $D(t)$ and constant C_S [31, 32]

   2. Analytical solutions of Fick's second law besides the error function solution

      • $D_a(t)$ and $C_S(t)$ [33]

   3. Numerical solutions of Fick's second law

      • $D(t)$ and $C_S(t)$ [34]
      • $D(t)$ and $C_S(t)$ plus binding isotherms [35].

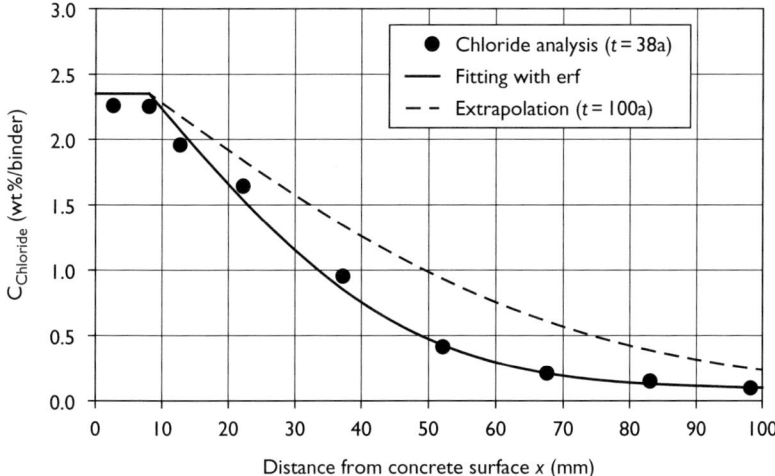

Figure 3.2.21 Chloride profile at inspection time of 38a, and predicted profile at $t = 100a$.

(b) Physical methods

The chloride transport and the binding of chloride is described by separate sub-models:

1. Based on Fick's first law of diffusion

   - Binding isotherms [25]
   - convection [36]

2. Based on the Nernst–Planck equation

   - Ion equilibrium [37].

On a the semi-probabilistic level within the Lifecon concept, an approach with constant values for D_{app} and C_S seems most appropriate. This is because a decrease in the diffusion coefficient with time is neglected to be on the safe side and the model is the most simple for application.

However, even the simple error function solution is not convenient when it comes to manual calculations. For this purpose, Equation (3.2.57) may be simplified by exchanging the error function (erf) approach for a parabolic function:

$$C(x, t) = C_i + (C_{\Delta x} - C_i) \cdot \left(1 - \frac{x - \Delta x}{2\sqrt{3(t - t_{exp}) \cdot D_{app}(t)}}\right)^2 \sqrt{b^2 - 4ac} \qquad \text{for } x \le x_i$$

(3.2.58)

which is of course only valid in the domain of constantly decreasing values of the chloride concentration, i.e. until depth x_i, where $C(x, t) = C_i$:

$$x_i = 2 \cdot \sqrt{3 \cdot (t - t_{exp}) \cdot D_{app}} + \Delta x$$

(3.2.59)

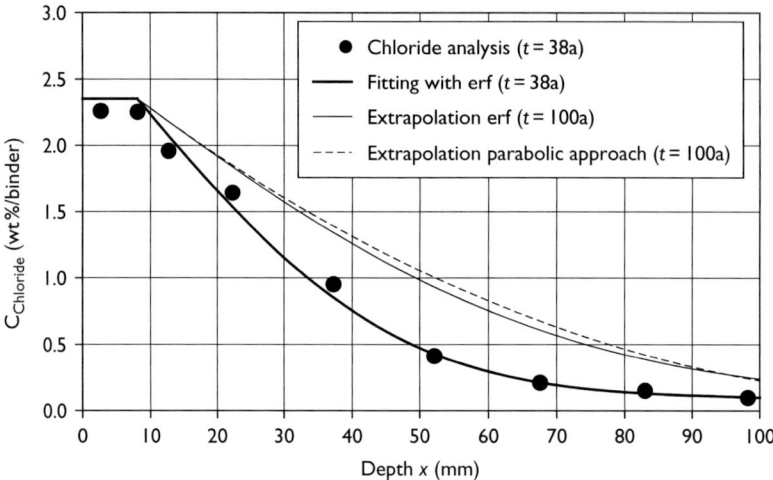

Figure 3.2.22 Chloride distribution at inspection time and predicted distributions using either the 'error function solution' (Equation (3.2.55)) or the parabolic relation given in Equation (3.2.56) for identical values of the aparent diffusion coefficient D_{app} and the surface concentration C_S.

Beyond the depth x_i, the chloride concentration $C(x, t)$ is equal to the initial concentration C_i. Figure 3.2.22 clearly demonstrates that even an extrapolation over large periods using the parabolic approach according to Equation (3.2.58) results in a nearly identical chloride profile.

The target is not to predict the future development of the chloride distribution in concrete, but rather to determine the probability of chloride-induced depassivation at any given time. Therefore the chloride concentration $C(x, t)$ in Equation (3.2.58) must be set equal to the critical chloride concentration x_{crit} and solved for penetration depth, thereby resulting in the penetration depth of the critical chloride concentration:

$$x_{crit} = 2\sqrt{3(t - t_{exp}) \cdot D_{app}(t)} \cdot \left(1 - \sqrt{\frac{C_{crit} - C_i}{C_S - C_i}}\right) + \Delta x \qquad (3.2.60)$$

which is compared to the concrete cover d_C.

3.2.6.3 Scatter for the depth of the critical chloride concentration, x_{crit}

From Equation (3.2.60) it follows that the penetration depth of the critical chloride concentration x_{crit} is a random variable which itself is a function of various random variables. For manual calculations of the probability of depassivation, the most simple way is to compare the random penetration depth x_{crit} (stress variable, S) with the random concrete cover d_C (resistance variable, R) following the basic safety concept (Figure 3.2.23). The mean value of the time-dependent penetration depth follows directly from Equation (3.2.60), when inserting the regression parameters obtained by fitting chloride profiles (C_i, C_S, D_{app}, Δx) and the critical chloride concentration C_{crit}.

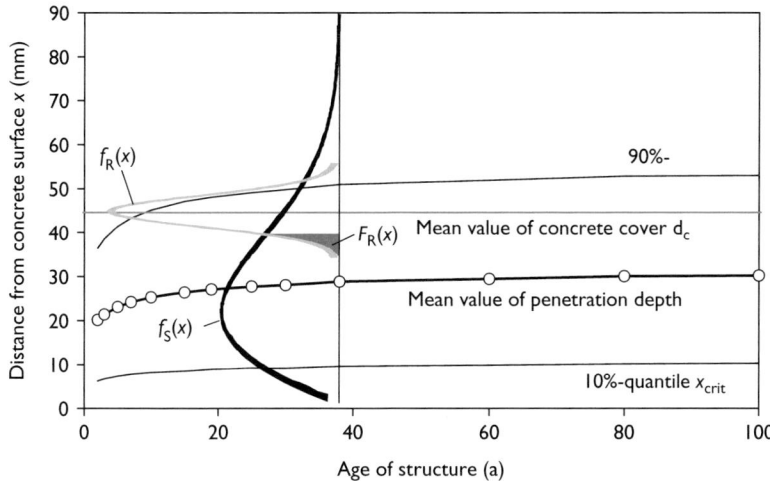

Figure 3.2.23 Comparison of constant concrete cover d_C (resistance R) and time dependent penetration depth of critical chloride concentration x_{crit} (stress S) for a case study object in the road environment.

The C_{crit} depends on the concrete composition and the environmental exposure conditions, as will be outlined in more detail when treating the full-probabilistic model approach. On the safe side, a very simple but convenient method is to assume a mean value of $C_{crit} = 0.48$ (weight%/cement) regardless of the exposure environment.

However, the standard deviation of x_{crit} cannot be determined directly without the application of professional software tools. The problem of modelling chloride-induced corrosion carries numerous and significant uncertainties. To overcome this problem, the statistical distribution of x_{crit} has been studied for case study objects using a full-probabilistic model, as detailed later.

The coefficient of variation CoV (ratio of standard deviation σ to mean μ) of the penetration depth x_{crit} must cover all uncertainties, including:

- spatial scatter of all variables in Equation (3.2.60)
- deviations of chloride analysis from the real values
- model uncertainties.

The calculations indicated that:

- CoV of the penetration depth is nearly constant over time
- CoV = 50% is a good assumption, on the safe side.

3.2.6.4 Calculation procedure

The calculation procedure is comparable to the example already demonstrated for carbonation-induced depassivation and will likewise be exemplified. Table 3.2.11 provides the summary of necessary input data.

Table 3.2.11 Summary of necessary input data for the semi-probabilistic model of chloride ingress

Parameter	Unit	Format	Source
Apparent diffusion coefficient $D_{app}(t = 38a)$	$(10 - 12\,m^{-2}\,s^{-1})$	ND (μ, σ)	Fitted from chloride profile as outlined in Chapter 13.2.6.21
Surface concentration C_S	(weight cement)	ND (μ, σ)	Fitted from chloride profile as outlined in Chapter 13.2.6.21
Initial chloride concentration C_i	(weight cement)	ND (μ, σ)	Assumed for fitting as outlined in Chapter 13.2.6.21
Critical chloride concentration	(weight cement)	ND (μ, σ)	Assumed as outlined in Chapter 13.2.6.31
Concrete cover d_{cover}	(mm)	ND (μ, σ)	Measurements according to (Lifecon D3.1)

The following information shall be available:

- A chloride profile was taken at age $t = 38a$ and fitted as shown in Table 3.2.11.
- The critical chloride concentration is assumed to be on average $C_{crit} = 0.48$ (weight of cement).
- Concrete cover d_{cover}, which is a resistance parameter, was measured with a mean value of $\mu_{cover} = 70\,mm$ and a standard deviation of $\sigma_{cover} = 8\,mm$. For the sake of simplicity the distribution is assumed to be normal.

From this information the mean value of the time-dependent penetration depth of the critical chloride concentration can be estimated according to:

$$\mu(x_{crit}) = 2\sqrt{3(t \cdot 365 \cdot 24 \cdot 3600) \cdot 0.385 \cdot 10^{-12}} \cdot \left(1 - \sqrt{\frac{0.48 - 0.1}{2.350 - 0.1}}\right) \cdot 10^3 + 8$$

$$= 7.1\sqrt{t} + 8 \tag{3.2.61}$$

where
t is the age of structure (a)

The standard deviation is thus:

$$\sigma(x_{crit}) = 50\% \cdot \mu(x_{crit}) = 3.6\sqrt{t} + 4.0 \tag{3.2.62}$$

The minimum reliability index is set to $\beta_{min} = 1.8$. The sequence of various calculation to prove whether the requirements are currently fulfilled is as follows:

1. Calculation of the stress variable S (penetration depth at age $t = 38a$):

$$\mu(x_{crit}) = 7.1 \cdot \sqrt{38} + 8 = 52 \quad \sigma(x_{crit}) = 3.6\sqrt{38} + 4.0 = 26 \tag{3.2.63}$$

2. Transformation of resistance and stress variables into standard space:

$$U_1 = \frac{R - 70}{8} \quad U_2 = \frac{S - 52}{26} \tag{3.2.64}$$

3. Limit state equation in standard space:

$$L(U_1, U_2) = R - S = 8U_1 - 26U_2 + 18 = A \cdot U_1 + B \cdot U_2 + C = 0 \qquad (3.2.65)$$

4. Calculation of the distance of the limit state line from origin using the so-called Hesse normal format (comparison of coefficients A, B, C):

$$\beta = \frac{C}{\sqrt{A^2 + B^2}} = 0.66 < 1.8 \qquad (3.2.66)$$

This means that at the current age the reliability requirements are not fulfilled and more thorough investigations are necessary.

3.2.7 Full-probabilistic model of chloride ingress

3.2.7.1 Mathematical model

The basic concept of the model is identical to that for the semi-probabilistic approach outlined in the previous section. Depassivation of the reinforcement will start when the critical corrosion-inducing chloride content C_{crit} will be exceeded in the depth of the concrete cover d_{cover}, which is expressed by a limit state function:

$$p_{depassivation} = p\{C_{crit} - C(x = d_{cover}, t) < 0\} < p_{set} \qquad (3.2.67)$$

where $C(x, t)$ is calculated according to Equation (3.2.57). The major difference is that all input parameters:

- are considered as random variables
- follow the performance concept, i.e. can be measured
- have been statistically quantified for durability design purposes and must thus not be determined throughout structure investigations.

The apparent diffusion coefficient is inserted according to the following sub-model:

$$D_{app}(t) = \frac{\int_{t_0}^{t} \left(D_{RCM,0}(t) \cdot k_T \cdot k_{RH} \cdot \left(\frac{t_0}{t}\right)^{n_1}\right) dt^*}{(t - t_0)} \qquad (3.2.68)$$

where

$D_{app}(t)$ apparent diffusion coefficient of concrete at the time of the inspection $(m^{-2}\,s^{-1})$

$D_{RCM,0}$ chloride migration coefficient of water-saturated concrete prepared and stored under predefined conditions, determined at the reference time t_0

n_1 exponent regarding the time-dependence of D_{app} due to the environmental exposure

k_T temperature parameter [14]

k_w factor to account for degree of water saturation w of concrete

t age of concrete (s)

t_0 reference time (s)

Note that D_{app} is introduced into the error function solution as the average over a regarded time period. This topic will be emphasised when treating the age exponent n.

A clear classification of the input parameters is not possible in a straightforward manner (as is the case for the carbonation model), because some of these are influenced by the concrete composition as well as by the environmental exposure conditions (Table 3.2.12).

3.2.7.1.1 MATERIAL INPUT PARAMETER, $D_{RCM,0}(t)$

The chloride diffusion coefficient is highly dependent on the concrete composition. The experimental determination of chloride diffusion coefficients of concrete by conventional methods, as there are diffusion-cell tests or immersion tests, is very time-consuming. In immersion tests concrete samples are kept submerged in a chloride-containing solution to measure chloride ingress after certain testing periods by withdrawal and chemical analysis of dust samples, thereby obtaining chloride ingress profiles. An effective diffusion coefficient can be derived by making use of curve fitting methods based on Fick's second law of diffusion on the chloride profiles obtained.

For the design of new structures, the rapid chloride migration (RCM) test (Figure 3.2.24) has been chosen because of the following reasons [12]:

- For equal concrete compositions the chloride migration coefficient $D_{RCM,0}$ shows a strong statistical correlation with the effective diffusion coefficient D_{eff} (determined by time-consuming immersion tests) (Figure 3.2.25)
- A short test duration is achievable with the RCM test
- The RCM test is a robust and precise method.

Details on the RCM test method and further investigation methods concerning the effective diffusion coefficient are provided in Lifecon D3.1, Appendix [1]. For

Table 3.2.12 Summary of input parameters for the chloride ingress model

Type of parameter	Parameter
Material	$D_{RCM,0}(t)$
Material/environmental	$k_{RH}, k_T, n, C_{S,\Delta x}, \Delta x, C_{crit}$
Test	k_t
Geometry	d_{cover}

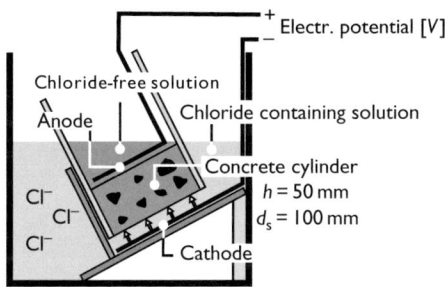

Figure 3.2.24 Experimental design of the Rapid Chloride Migration test (RCM).

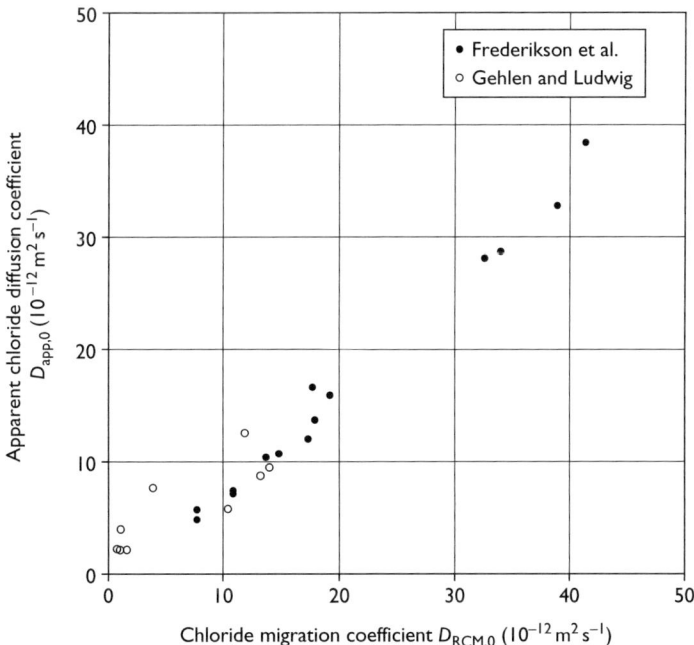

Figure 3.2.25 Correlation of the apparent diffusions coefficient obtained by immersion tests and the chloride migration coefficient $D_{RCM,0}$.

existing structures the apparent diffusion coefficient $D_{app}(t)$ can be obtained one of the following ways:

- using literature data with respect to D_{RCM}, if the concrete composition is known;
- withdrawal of dust samples and curve fitting of the chloride ingress profiles obtained, if exposure time was sufficiently large (old structures); and
- performing the RCM test with specimens (e.g. cores) from the structure, which is necessary for young structures but should be performed for every object.

In essence the content of cement, superplasticiser and air voids as well as the aggregate type can be neglected, whereas the binder type and the w/b ratio are of decisive relevance [5, 38].

3.2.7.1.2   EFFECT OF WATER–BINDER RATIO

Following the results (Figure 3.2.26) from exposure tests in sea water [39] an exponential approach seemed to be most suitable to describe the effect of the w/b ratio: This leads to the derivation of:

$$k_{w/b} = \frac{D_{RCM,w/b=i}}{D_{RCM,w/b=0.45}} = \left(\exp\left(a_{w/b} \cdot (w/b - 0.45)\right)\right) \cdot \varepsilon_{w/b} \qquad (3.2.69)$$

where

$k_{w/b}$ factor accounting for a change in water–binder ratio with respect to the chosen reference value $w/b = 0.45$

$a_{w/b}$ regression parameter

$\varepsilon_{w/b}$ error term.

[mean value $= 1$, coefficient of variation $\text{CoV}_{w/b}$ according to Table 3.2.13]. The regression parameter $a_{w/b}$ expresses the sensitivity with respect to change in w/b ratio. The model uncertainty of the function compared to the actually measured values is

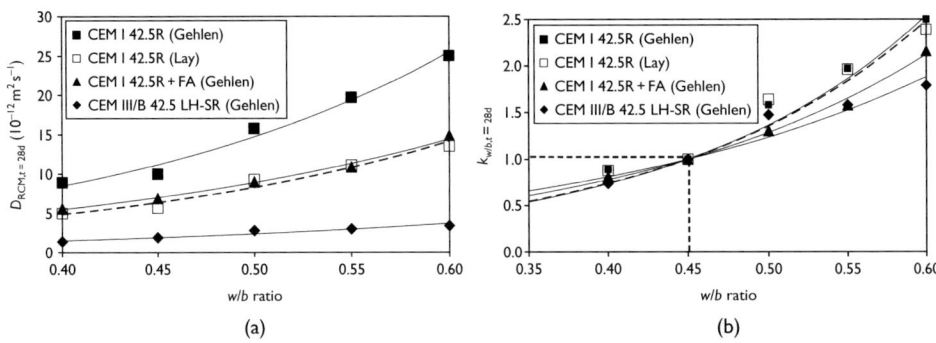

Figure 3.2.26 (a) $D_{RCM,t=28d}$ versus w/b ratio for different binder types; data of Gehlen and Lay [5, 14]. (b) Binder specific factor $k_{w/b, \ t=28d}$ to account for changes in w/b ratio calculated by relating original data given in (a) to the value measured at $w/b = 0.45$.

Table 3.2.13 Regression parameters describing the influence of the w/b ratio upon the rapid migration coefficient D_{RCM} of concrete for various binder types [40]

CEM	$a_{w/b}$	$CoV_{w/b}(\%)$
I*	6.0	11
I+FA(18%, k = 0.5)*	5.0	11
II/A-LL 42.5 R†	4.2	8
II/B-S 32.5 R†	3.8	11
II/B-T 32.5 R†	4.4	10
CEM III/B*	4.2	15

Notes
* Valid for $0.40 \leq w/b \leq 0.60$.
† Valid for $0.40 \leq w/b \leq 0.50$.

incorporated by means of the error term $\varepsilon_{w/b}$. This (Equation (3.2.69)) describes the dependence of D_{RCM} as a random sub-function.

The degree of hydration and therefore the degree of hydration at infinite concrete age theoretically increase for larger values of the w/b ratio [41]. However, in the practically relevant domain the effect of the w/b ratio can be regarded as independent of the concrete age, which leads to the conclusion that Equation (3.2.69) can be applied regardless of the considered moment in time. This relationship is exemplified in Figure 3.2.27, as the curves for various w/b ratios can be shifted parallel along the ordinate.

3.2.7.1.3 EFFECT OF BINDER TYPE

The cements used for the studies were CEM I (Portland cement) of strength class 32.5, 42.5, 52.5 and blended cements of the type CEM I+FA (15, 20 or 40% fly ash), CEM II/A-L 42.5R (15% limestone), CEM II/B-S 32.5R and 42.5R (31% slag), CEM II/B-T 32.5R (23% oil shale), CEM III/A 32.5 (51% slag) and CEM III/B 32.5 (72% slag).

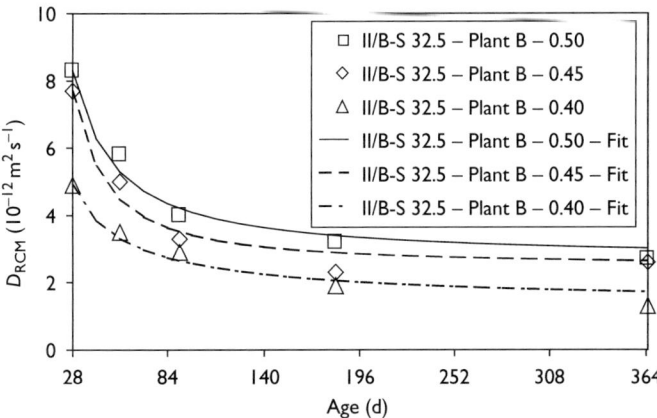

Figure 3.2.27 Chloride migration coefficients of concrete at different ages for three w/b-ratios produced with 360 kg/m³ CEM II/B-S 32.5 (plant B) and corresponding fitted functions (Fit).

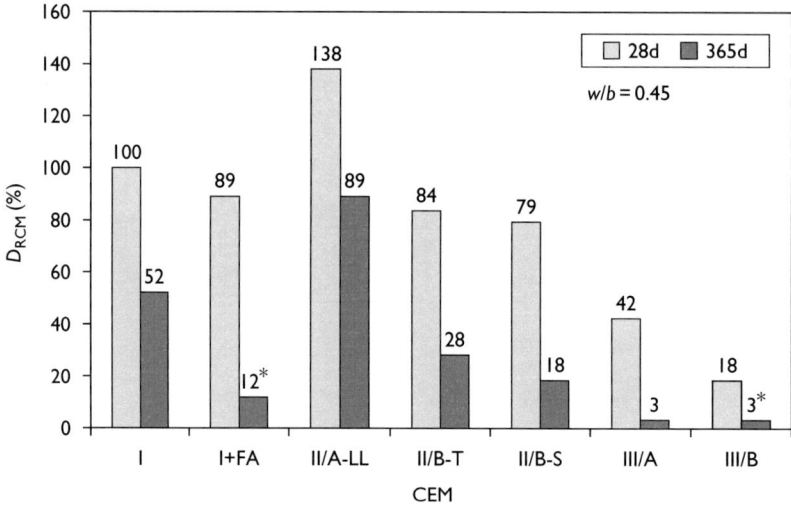

Figure 3.2.28 Mean values for D_{RCM} at an age of 28 and 365 days for various binder types (all concretes with $w/b = 0.45$) given as % related to CEM/(Portland cement) at an age of 28 days. *Data at an age of 182 days was used if measurements were not possible at 365 days (colorimetric reaction failed) [5].

The investigations revealed that a classification is most reasonable according to the type of binder, whereas data with different cement strength class and cement factory should be grouped, as was done to obtain the relationships given in Figure 3.2.28.

3.2.7.1.4   DERIVATION OF A MODEL FOR CHLORIDE MIGRATION COEFFICIENT

As a starting point for a regression analysis concerning the time-dependent decrease of D_{RCM}, the commonly used reference age of $t_0 = 28$ days was chosen. In the following, the migration coefficient at this age will be referred to as $D_{RCM,t=28d}$. The hydration of concrete proceeds in theory for a very long time. However, when compared to the common service life of concrete structures, the effect of hydration upon D_{RCM} becomes insignificant rather early (see e.g. Figure. 3.2.28, Table 3.2.14). To solve this problem a finite value, which is referred to as $D_{RCM,t=\infty}$, may be introduced, leading to the following expression:

$$D_{RCM}(t) = \left(D_{RCM,t=28d} - D_{RCM,t=\infty}\right) \cdot \left(\frac{t_0}{t}\right)^{n_2} + D_{RCM,t=\infty} \qquad (3.2.70)$$

where

$D_{RCM,t=28d}$ binder-specific migration coefficient at reference age $t_0 (m^2 s^{-1})$
$D_{RCM,t=\infty}$ binder-specific migrations coefficient after the end of hydration $(m^2 s^{-1})$
t concrete age
t_0 reference age, here 28 days (s)
n_2 age exponent due to hydration.

Table 3.2.14 Binder-specific chloride migration coefficients $D_{RCM,t=28d}$ and $D_{RCM,t=\infty}$ $(w/b = 0.45)$

CEM	I		I+FA		II/A-LL		II/B-T		II/B-S		III/A		III/B	
Additions (M %)	0		15		15*		23		31		51		71	
Number of samples	6		4		2		1		2		1		1	
Age (d)	28	∞	28	∞*	28	∞	28	∞	28	∞	28	∞	28	∞*
Mean $m\,(10^{-12}\,m^{-2}\,s^{-1})$	9.2	4.8	8.2	1.1	12.7	8.2	7.7	2.6	7.3	1.7	3.9	0.3	1.7	0.3
CoV (%)	52	73	37	32	–	–	–	–	–	–	–	–	–	–

Notes
CEM – cement type CEM II/B-S 32.5.
* Data at age 182 days was used if measurements were not possible at 365 days (colorimetric reaction failed) [5].

As, currently, data is only available at an age of 365 days the migration coefficient at this moment in time is set equal to $D_{RCM,t=\infty}$, which is good approximation on the safe side. The velocity decrease from $D_{RCM,t=28d}$ to $D_{RCM,t=\infty}$ due to hydration of the cement paste is expressed with the age exponent n_2.

If the values for $D_{RCM,t=28d}$ and $D_{RCM,t=\infty}$ are known, the age exponent n_2 can be calculated by fitting the measured results according to Equation (3.2.70). The results indicated that a classification is only reasonable with respect to the binder type (Table 3.2.15).

3.2.7.1.5 APPLICATION OF THE MODEL

To apply the given relationships in a full-probabilistic model for the prediction of the residual service life, all variables must be quantified in terms of statistical distribution type, mean value and coefficient of variation (or standard deviation) and where necessary boundaries of the distribution. Due to the low amount of available data at present, an analysis of the best suitable distribution function is not possible. However, as negative values for D_{RCM} are physically impossible a log-normal distribution seems reasonable for application.

Binder-specific mean values for D_{RCM} with $w/b = 0.45$ are provided. The number of samples is, however, too low to calculate the coefficient of variation for all binder types. Meanwhile, it is proposed to use values determined for CEM I irrespective of the binder type considered, which means 50% at an age of 28 days and 70% at infinite age.

Table 3.2.15 Binder-specific age exponent n_2

CEM	I	I+FA	II/A-LL	II/B-T	II/B-S	III/A	III/B
Number of samples	6	4	4	3	4	3	3
Mean m	0.964	1.244	1.303	0.860	1.041	1.159	1.624
CoV (%)	22	21	63	9	25	20	27

If for durability design purposes the actual concrete quality is tested or cores are sampled from an existing structure, the actual concrete quality can be tested, and will be outlined later. In this case, the coefficient of variation of the test method obtained from repeated tests of identical concretes under identical conditions should be used, which is only around 20% [14].

3.2.7.1.6 ENVIRONMENT- AND MATERIAL-DEPENDENT INPUT PARAMETERS

There are six input parameters which depend on the type of environmental exposure and on the concrete composition, which have to be quantified statistically: the depth of the convection zone Δx in which the chloride profile deviates from the behaviour predicted by the second law of diffusion, temperature parameter k_T, the age factor n, influence of partial water saturation k_w, chloride concentration $C_{\Delta x}$ and the critical chloride content c_{crit}.

The statistical quantification of environmental parameters based on data obtained throughout structure investigations may be performed as follows [11]:

- providing parameters as mean and standard deviation of the corresponding distribution type;
- presenting parameters as a function of time, applying Bayesian regression analysis on data of the same classification; and
- providing parameters as a function of other parameters (material or environmental) by multiple regression analysis.

The exposure environment may be classified as:

(a) Marine environment

- submerged zone
- tidal zone
- splash zone
- atmospheric zone.

(b) Road environment

- splash zone (exposed/not exposed to rain)
- spray zone (exposed/not exposed to rain).

The most important aspects of classification with respect to the concrete composition are the binder type and the water–binder ratio.

3.2.7.1.7 DEPTH OF THE CONVECTION ZONE, Δx

The convection zone Δx is the outer portion of the concrete cover where significant deviation from pure diffusion takes place. This zone can be identified by means of chloride profiles, as maximum chloride concentration is observable in the inner zone of the concrete cover. The main reason for this visible effect is the washout of chlorides at times without chloride loading or with low chloride loading, which will only be possible if the concrete surface is subjected to chloride-free water.

More importantly, it must be reckoned that, in periods with chloride loading, chlorides will penetrate very fast by means of capillary suction into the depth of the convection zone. Beyond this depth, diffusion is the decisive transport mechanism. In marine structures this is the case in the atmospheric zone. For structures in the road environment, chlorides are washed out close to the road edge by splash and spray water of the traffic or rain if the structure is not sheltered.

The existence of a convection zone is therefore dependent on the position of the surface considered with respect to the source of chloride-free water. Gehlen [14] analysed 127 chloride profiles in the marine environment without distinguishing the distance to the chloride source. Since negative values for Δx are not possible and convection is limited to a finite suction depth, a beta distribution (B) seemed most appropriate for analysis:

$$B(\mu = 8.9; \sigma = 63\% \ \mu; a = 0; b = 50)$$

If a surface is situated in the road environment within the domain of 3 m above the road level and 2 m from the road edge, a convection zone will most likely to exist. The statistical analysis resulted in:

$$a \leq 3 \, \text{m and } h \leq 2 \, \text{m:} \, \Delta x = B(\mu = 9.03; \sigma = 100\% \, \mu; a = 0; b = 60) \, (\text{mm})$$

$$a > 3 \, \text{m or } h > 2 \, \text{m:} \, \Delta x = 0 \, (\text{mm})$$

For structures which are not sheltered from rain little data is available. However, these showed the same order of magnitude for Δx [5]. Hence, the same statistical distributions as given above should be used for these.

Systematic data with respect to the influence of concrete composition on the depth of the convection zone is scarce (Figure 3.2.29). From the figure it can be concluded

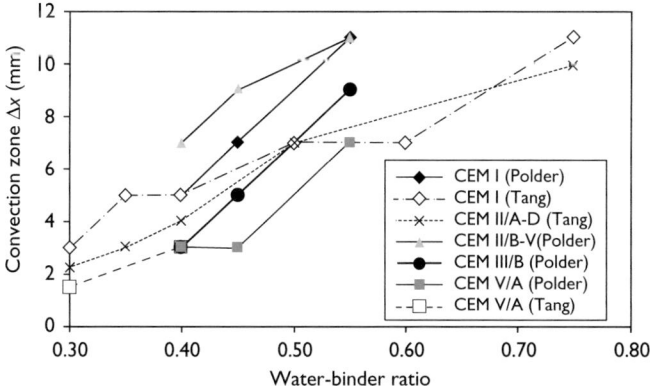

Figure 3.2.29 Effect of binder type and water–binder ratio on depth of convection zone Δx. Chloride profiles from exposure tests (field station Träslövsläge, Sweden) in atmospheric and tidal marine zone and wet–dry cycles in laboratory tests (26 cycles; 1 day in 3% NaCl; 6 days dry at 20°C/50% RH; each data point reflects the average of six chloride profiles).

that an increase of 1/10 in w/b ratio will lead to an increase of the convection zone of around 2–4 mm. Even at water–binder ratios as low as $w/b = 0.30$, a convection zone of 2–3 mm must be expected. The use of additions such as silica fume, fly ash or blast furnace slag (CEM III/A or B) will reduce the depth of the convection zone by up to around 60% compared to ordinary Portland cement (CEM I). A surprising feature of the results of Polder is the fact that low slag contents did not show an improvement but rather the opposite. This should not be overestimated, having in mind the large scatter when evaluating the convection zone.

3.2.7.1.8 TEMPERATURE PARAMETER k_T

The influence of the air temperature upon the diffusion coefficient can be taken into account by using the Arrhenius equation [14]:

$$k_T = \exp\left(b_T\left(\frac{1}{T_{ref}} - \frac{1}{T}\right)\right) \tag{3.2.71}$$

where
k_T temperature parameter
b_T regression parameter (normal distribution: $\mu = 4800$; $\sigma = 700$) (K)
T_{ref} reference temperature (K)
T temperature of the environment (microclimate) (K).

The parameter k_T accounts for the dependence of the apparent diffusion coefficient on the temperature of the surrounding environment (Figure 3.2.30).

3.2.7.1.9 THE AGE FACTOR n

As can be seen in Equation (3.2.68) the apparent diffusion coefficient D_{app} is a time-dependent variable. It has been shown that D_{app} may decrease considerably with

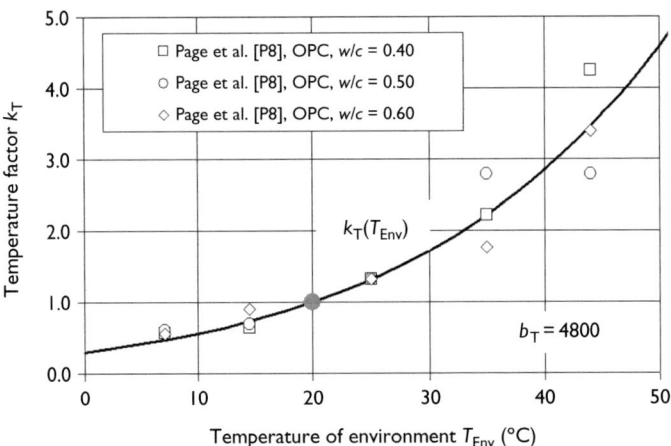

Figure 3.2.30 Function k_T accounts for the dependence of the apparent diffusion coefficient on the temperature of the surrounding environment (air, water).

increasing age of the concrete. There are several general effects which may contribute to a decrease in the diffusion coefficient with increasing time, e.g.:

- the development of a denser pore structure due to the hydration process subsequent to the reference time t_0, which can be separately accounted for;
- swelling of the cement paste and deposition effects in the pore structure [42];
- a pore-blocking effect due to chloride ingress [43];
- concentration dependence of the binding capacity inherent in the effective diffusion coefficient [53]; and
- oversimplification of the model (assumption of constant $C_{\Delta x}$), which will be outlined more in detail.

As an output the apparent diffusion coefficient D_{app} is obtained, which can be plotted against time [14, 45], as shown in Figure 3.2.31. Apparently the fitted diffusion coefficients seem to decrease over various orders of magnitude during exposure to a chloride-containing environment. This relation can be expressed as follows [46]:

$$D_{app,2} = D_{app,1} \cdot \left(\frac{t_1}{t_2}\right)^n \tag{3.2.72}$$

$D_{app,i}$ apparent diffusion coefficient at exposure time $t_i(\mathrm{m^{-2}\,s^{-1}})$
t_i exposure time (s)
n age exponent

If the apparent diffusion coefficient is known at two times, the age exponent n can be calculated:

$$n = \frac{\ln\left(\frac{D_{app,2}}{D_{app,1}}\right)}{\ln\left(\frac{t_1}{t_2}\right)} \tag{3.2.73}$$

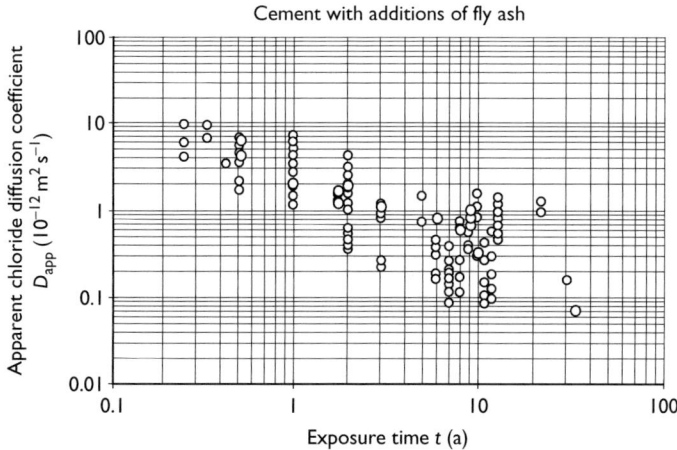

Figure 3.2.31 Apparent diffusion coefficient plotted versus exposure time.

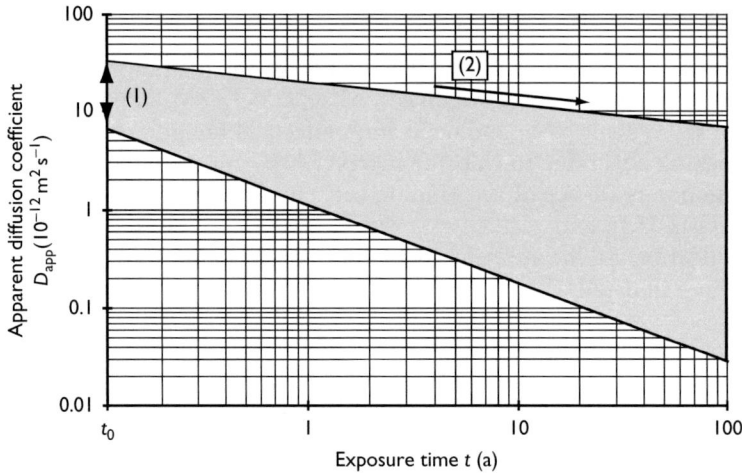

Figure 3.2.32 Regression analysis of time-dependent apparent diffusion coefficient. With (1) being the starting value obtained from chloride migration tests at $t_0 = 28$ d and (2) being the slope of the regression line on a double-logarithmic scale, which is equal to the age exponent n.

If more than two results are given, as is the case in Figure 3.2.31, a regression analysis can be performed. The starting point of such a regression analysis is the migration coefficient obtained at the reference age, which was chosen to be $t_0 = 28$ days, as shown in Figure 3.2.32 [14].

According to the approach given in Equation (3.2.72), the age exponent is only valid in the domain of $0 \leq n \leq 1$. For the case of $n = 1$, time t is cancelled out from Equation (3.2.57), which means that the chloride concentration $C(x,t)$ remains constant. This is referred to as 'blocking' of the chloride ingress [47]. For a value of $n > 1$ the concentration at any depth x would decrease in the next time interval. This implies that chlorides would flow out of the concrete, which is not consistent with the boundary condition for which the error function solution was derived, i.e. that $C_{\Delta x}$ remains constant over time. For $n < 0$ the diffusion coefficient would increase with time, which is not the case either. Therefore a beta distribution with the abovementioned boundaries is used for the statistical analysis of the age exponent.

The approach described above neglects the fact that the apparent diffusion coefficient obtained by fitting chloride profiles is the average diffusion coefficient in the time interval considered (Figure 3.2.33). This value should therefore be plotted against the time when this average value was achieved, which is unknown, as the age exponent n remains unknown at the time of plotting the value. The exact procedure for determining the age exponent is thus iterative [48]. However, plotting D_{app} against the time of inspection always results in lower values for the age exponent, which is a safe approach.

However, the time-dependent decrease in chloride diffusion coefficient not only may be material dependent (type of cement and additions) but has to be evaluated considering the actual exposure conditions as well (e.g. chloride-loading intensity because of the concentration dependence of D_{app}). Therefore, it is advisable to

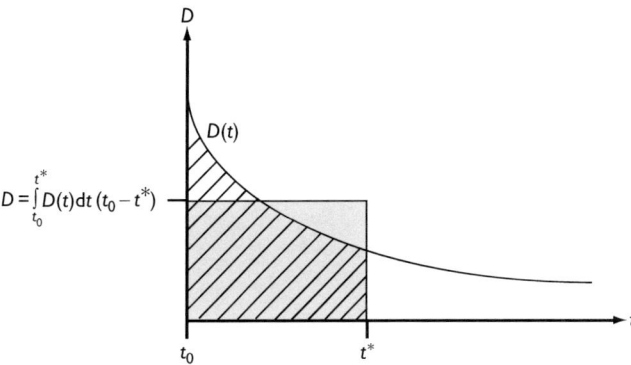

$$D = \int_{t_0}^{t^*} D(t)\,dt\,(t_0 - t^*)$$

Figure 3.2.33 Relationship of time-dependent diffusion coefficient with the resulting average.

Table 3.2.16 Result of the statistical quantification of age exponent *n* for exposure in the submerged, splash and tidal marine exposure zone

Binder type	Age exponent n
CEM I (OPC)	Beta($\mu = 0.30$; $\sigma = 0.12$; $a = 0$; $b = 1$)
CEM I + FA*	Beta($\mu = 0.60$; $\sigma = 0.15$; $a = 0$; $b = 1$)
CEM III/B (GBFS)	Beta($\mu = 0.45$; $\sigma = 0.20$; $a = 0$; $b = 1$)

Note
* Fly ash content FA $\geq$ 20%.

determine the age factor *n* for separate concrete compositions and exposure conditions (Table 3.2.16) [14].

3.2.7.1.10 INFLUENCE OF PARTIAL WATER SATURATION – PARAMETER, k_W

By introducing a convection zone Δx, the deviations from the diffusion transport mechanisms in the surface-near layer have been accounted for. In the layers beyond Δx a constant humidity may be assumed. Nevertheless, saturation of the concrete, as provided by the curing and storing conditions throughout the RCM test, may not be assumed in the inner layers of the structures. The moisture conditions of the concrete have to be accounted for. There are very few systematic investigations regarding the influence of partial water saturation of concrete upon the apparent diffusion coefficient D_{app}, as these are quiet difficult to perform (Figure 3.2.34).

The ratio of the rapid migration coefficient D_{RCM} under water saturation to the apparent diffusion coefficient D_{app} for a particular relative humidity is referred to the factor k_w [49]:

$$k_w = \frac{D_{RCM}}{D_{app}(\text{RH})} \tag{3.2.74}$$

which is course restricted to the domain of $0 \leq k_{RH} \leq 1$.

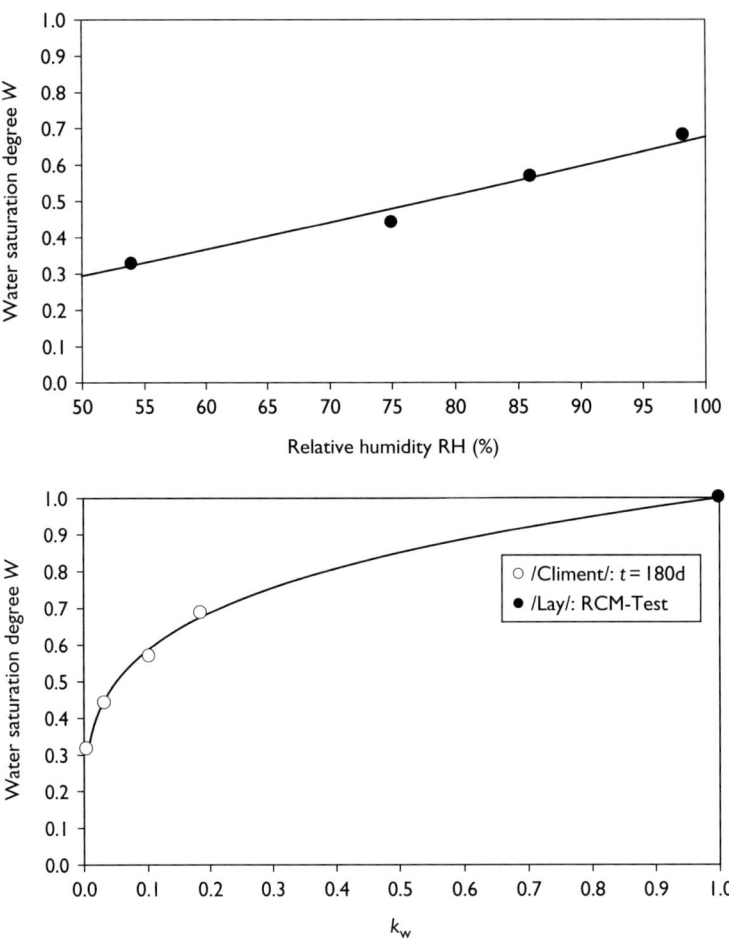

Figure 3.2.34 Influence of the relative humidity upon the degree of water saturation of concrete (left) and the reduction factor for the apparent diffusion coefficient as a function of the water saturation degree (right) [49].

The data in Figure 3.2.34 was used to derive the following deterministic relationship:

$$k_w = w^{a_{RH}} \qquad (3.2.75)$$

where
  k_{RH} a factor to take into account the effect of partial water saturation of concrete upon the apparent diffusion coefficient
  c regression parameter, here $a_{RH} = 4.297$.

However, the degree of water saturation is not only dependent on the relative humidity of the air but depends even more on the amount and frequency of liquid water reaching the concrete surface, which is the medium through which chlorides are always transported onto the concrete surface. The water saturation degree w, as an input to Equation (3.2.75), must thus be determined as a function of the relative humidity and the time of wetness ToW, dependent on the concrete composition (binder type, w/b ratio).

Simulations to obtain the water saturation degree are possible by using commercial software, e.g. WUFI. It should be realised, however, that these work well for materials such as sandstone which do not react with the penetrating substance (water). This is not so for concrete, which swells and seals up. Calculations with such programs will thus usually overestimate water ingress into concrete.

As there is currently not enough data existing for the concrete-specific effect of the water saturation degree w upon the apparent diffusion coefficient D_{app} and the water saturation degree is difficult to predict, it does not yet seem appropriate to introduce Equation (3.2.75) into the full-probabilistic concept for service life calculations. The simple approach of applying a constant value of k_{RH} for concrete which is splashed or sprayed with chloride-containing water seems to be an adequate solution, having in mind that for existing structures the actual apparent diffusion coefficient can be measured to update the model, as will be shown in later sections. This constant value can be determined by means of regression analysis, as will be shown in the following chapter.

3.2.7.1.11 SEPARATION OF AGE EXPONENT n AND FACTOR, k_w

The concept of the age exponent n includes the effects of cement paste hydration and the marine environment, effects which lead to a decrease in the apparent diffusion coefficient with time. The decreasing effect due to hydration can be neglected after a certain point in time, depending on the binder type. However, environmental effects still seem to cause a decrease in D_{app} after the contribution of the hydration process has come to an end. Lay took these relationships into account and developed the following relationship:

$$D_{app}(t) = k_T \cdot k_w \cdot \int_{t_0}^{t} D_{RCM}(t) \cdot \left(\frac{t_0}{t}\right)^n dt \cdot \frac{1}{(t - t_0)} \qquad (3.2.76)$$

$D_{app}(t)$ apparent diffusion coefficient of concrete at the time of inspection $(m^{-2}\,s^{-1})$

$D_{RCM}(t)$ binder-specific, time-dependent chloride migration coefficient of water-saturated concrete prepared and stored under predefined conditions

k_w environmental parameter accounting for the effect of moisture on D_{app}

n age exponent describing the time-dependent decrease in D_{app} due to environmental exposure

k_T temperature parameter

t age of concrete (s)

t_0 reference time (s).

In modelling the chloride ingress in the road environment, i.e. under the conditions for which concrete is most of the time in an unsaturated state (i.e. having a water

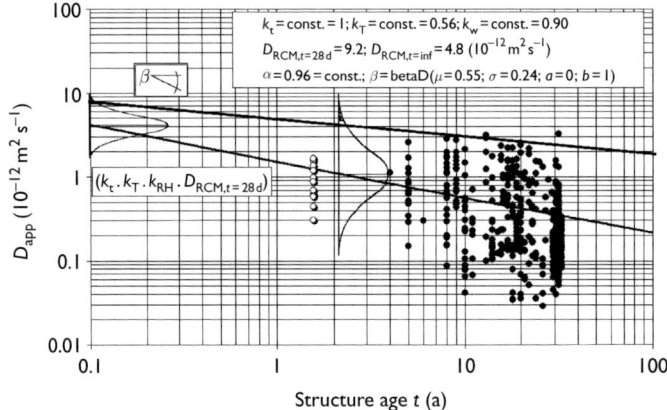

Figure 3.2.35 Quantification of relative humidity factor k_{RH} and age exponent n. White dots refer to exposure tests (CEM I only) in Sweden, black dots from structures in Germany.

saturation degree $w < 1$), the factor k_w was introduced. The apparent diffusion coefficient D_{app} obtained from collected chloride profiles was plotted against the age of the structure (Figure 3.2.35). A regression analysis of the data was performed applying Equation (3.2.76), which includes numerous variables.

The average effect of the temperature k_T can be accounted for using the average daily air temperature in Germany. The rapid time-dependent chloride migration coefficient is modelled assuming as an average composition concrete produced with CEM I and $w/b = 0.50$. Hence, k_{RH} and the age exponent n remained as the only variables which had to be fitted to the data. To simplify the analysis of the age exponent n, for which Equation (3.2.76) must be solved by iteration, the factor k_w is considered a deterministic (D) variable (single number).

Negative values for n are hardly possible, because for these values the diffusion coefficient would increase with time. For $n \geq 1$ chlorides will not be transported at all – the concrete would be blocked. The age exponent is hence restricted to the domain of $0 \leq n \leq 1$, which can be accounted for by a beta distribution.

Investigations indicated the following relationships for the average water saturation degree in the depth interval $x = 0$–50 mm and the age exponent n:

- At a depth of around 10–40 mm the water content reaches a constant level.
- With increasing height above road level the degree of water saturation increases within a narrow range.
- The age exponent n behaves inversely to the degree of water saturation.
- In the bottom parts of sheltered vertical surfaces the age exponent may even drop to the minimum value of $n = 0$.
- The intensity of chloride loading seems not to influence D_{app}. Chloride migration tests with chloride-containing specimens, for which a special test procedure was developed, showed the same results.

Based on these observations, it appears that the exponent n expresses the frequency of dry periods. During times at which the concrete dries out considerably, D_{app} is almost reduced to zero. With increasing distance from the road, the ratio of dry to wet periods is enlarged, leading to an increase in n. The age exponent n should therefore be a site-dependent parameter. Due to the large scatter of data collected from road administrations, no dependence of the age exponent n with respect to the distance from the road could be established. Nevertheless, as most of the chloride profiles used for statistical analysis were sampled close to the road (average height above street level $h = 85\,cm$; average horizontal distance $a = 170\,cm$), which is the decisive domain for design, the age exponent n can be applied independent of the distance to the road.

3.2.7.1.12 CHLORIDE CONCENTRATION, $C_{\Delta x}$

The chloride concentration $C_{\Delta x}$ at depth Δx depends on the concrete composition and on the environmental conditions. The decisive material parameters are the binder type and w/b ratio, which define the chloride-binding capacity as well as the pore volume.

The most important environmental variable influencing the chloride concentration of the concrete is the chloride concentration of the water to which a concrete surface is subjected. However, the resulting concentration $C_{\Delta x}$ in the concrete also depends on the rate of loading with chloride-containing water, which is a function of various environmental influences, which will be discussed in the following sections.

3.2.7.1.12.1 Chloride concentration of the environment, C_{env}

In the case of existing structures, the chloride concentration of the concrete $C_{\Delta x}$ can be directly determined from the evaluation of chloride profiles. For 'young' structures or for durability design purposes $C_{\Delta x}$ must be estimated by models. The first step may be to determine the chloride concentration of the water reaching the concrete surface. For off-shore structures the chloride load can be considered to be equal to the chloride content of the sea water:

$$C_{water} = C_{sea} \tag{3.2.77}$$

where
 C_{water} average content of the chloride source $(g\,l^{-1})$
 C_{sea} chloride content of sea water $(g\,l^{-1})$.

The chloride load of structures in a road environment is controlled by the de-icing salt applications:

$$C_{water} = C_{road} = \frac{n \cdot M_{Chloride}}{h_S} \tag{3.2.78}$$

where

n number of de-icing salt application incidents

C_{road} average concentration of the chloride-containing water on a street $(g\,l^{-1})$

$M_{Chloride}$ average mass of chlorides spread during each application of de-icing salts $(g\,m^{-2})$

h_s average precipitation (rain, snow) during salt application period $(l\,m^{-2})$.

The approach above neglects the facts that salt may be removed from the street by wind and driving cars, and salt-containing water and snow slush will be drained. These reduce the salt concentration C_{water}. For durability design purposes the above-mentioned approach is on the safe side. The average mass of chlorides spread $M_{chloride}$ depends on climatic conditions road type (degree of importance, velocity) and road maintenance policy, and must thus be based on data provided by the nearest road administration office. Data should be collected for the longest period available, as the scatter of annual data is extremely high. If no data is available from nearby stations, empirical models can be applied which link meteorological data and salt spread data, e.g. [51]:

$$M_{Chloride} = 1000 \cdot C_{Salt} \cdot (-9.56 + 0.52 \cdot SF + 0.38 \cdot SL + 0.14 \cdot FD - 0.20 \cdot ID)/w \; (g\,m^{-2})$$

(3.2.79)

where

Index FR   federal roads

$M_{chloride}$ average mass of chlorides spread during a single application event on federal roads $(t\,km^{-1})$

C_{salt} chloride concentration of de-icing salt spread (%)

SF   number of days per month with snow fall above 0.1 mm

SL   number of days per month with a layer of snow above 1.0 cm

FD   number of days per month with average daily temperature below 0°C

ID   number of days per month with ice on road surface

w average spread width (can be set equal to average street width) (m).

The correlation between salt spread data for federal roads (FR) and highways (HW) was found to be:

$$M_{Chloride,HW} = 1.51 + 0.235 \cdot M_{Chloride,FR} \qquad (g\,m^{-2})$$

(3.2.80)

The chloride concentration of the water and slush may also be measured directly. From Figure 3.2.36, large variation of the chloride concentration of snow slush during the winter becomes evident.

3.2.7.1.13 CHLORIDE SATURATION CONCENTRATION OF CONCRETE, $C_{S,0}$

For structures which are submerged or in constant contact with chloride-containing water, the chloride concentration $C\Delta_x$ can directly be set equal to the chloride

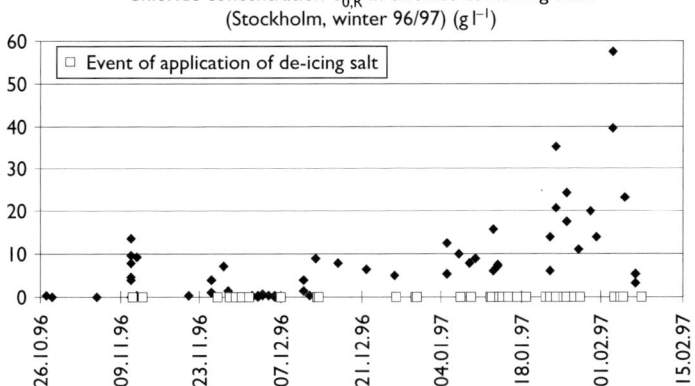

Figure 3.2.36 Chloride concentration in snow-slush and melted water (Stockholm, Winter 96/97) [52].

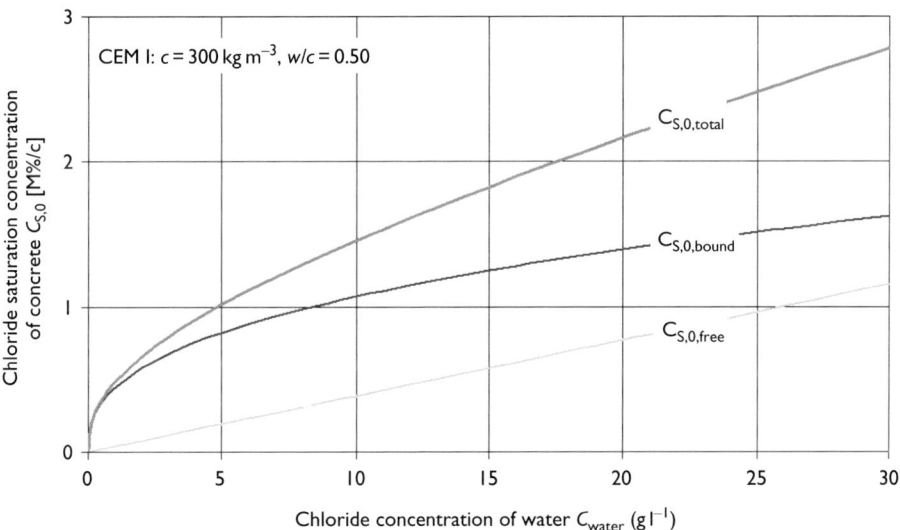

Figure 3.2.37 Chloride saturation concentration $C_{S,0}$ (free, bound and total) in dependence of the chloride concentration in the surrounding water C_{water}. Data of Tang [53].

saturation concentration $C_{S,0}$ (Figure 3.2.37), which can be estimated if the following information is given:

- chloride concentration of surrounding water C_{water}
- chloride adsorption isotherm, which can be taken from literature data [53] or determined in the laboratory, and
- concrete composition.

Table 3.2.17 Parameters of chloride-binding isotherm [53]

Binder type	f_b	β
100% CEM I (OPC)	3.57	0.38
30% slag + 70% CEM I	3.82	0.37
50% slag + 50% CEM I	5.87	0.29
30% fly ash + 70% CEM I	5.73	0.29

Tang describes the ratio of free to bound chlorides as a so-called Freundlich isotherm, in relation to binder type and porosity of the concrete (Table 3.2.17):

$$C_b = f_b \cdot \frac{W_{Gel}}{1000 \cdot \varepsilon} \cdot c^\beta \qquad (3.2.81)$$

C_b amount of bound chlorides in cement paste $(g\,l^{-1})$
f_b, β regression parameters
W_{Gel} cement paste content $(kg\,m^{-3})$
ε concrete porosity $(m^3_{pores}\,m^{-3}_{concrete})$
c concentration of solution (water) $(g\,l^{-1})$.

However, concrete surfaces which are not in constant contact with chloride-containing water, as is the case for structures in the splash zone of the sea or structures in the road environment, will show lower concentrations, which depend on a number of influences.

In the marine environment the main influences are:

- concrete composition
- distance of the considered surface to the sea (vertical and horizontal)
- orientation with respect to wind direction.

As negative values may not occur, a log-normal distribution seems to be the most appropriate to describe the chloride concentration $C_{\Delta x}$. The so-called windward/leeward effect causes chlorides to be deposited at the leeward side. During rain events, deposited chlorides are washed away at the windward side whereas the rain will not reach the leeward side, which emphasises this effect (Figure 3.2.38). The values given in Table 3.2.18 must thus be corrected by means of a so-called turbulence factor, which depends highly upon the geometry of the structure considered.

In the road environment the chloride concentration $C_{\Delta x}$ is a function of:

- concrete composition
- amount of de-icing salt spread
- drainage behaviour of the street (texture of street surface, pot holes, etc.)
- distance to the road
- traffic density
- climate
- orientation.

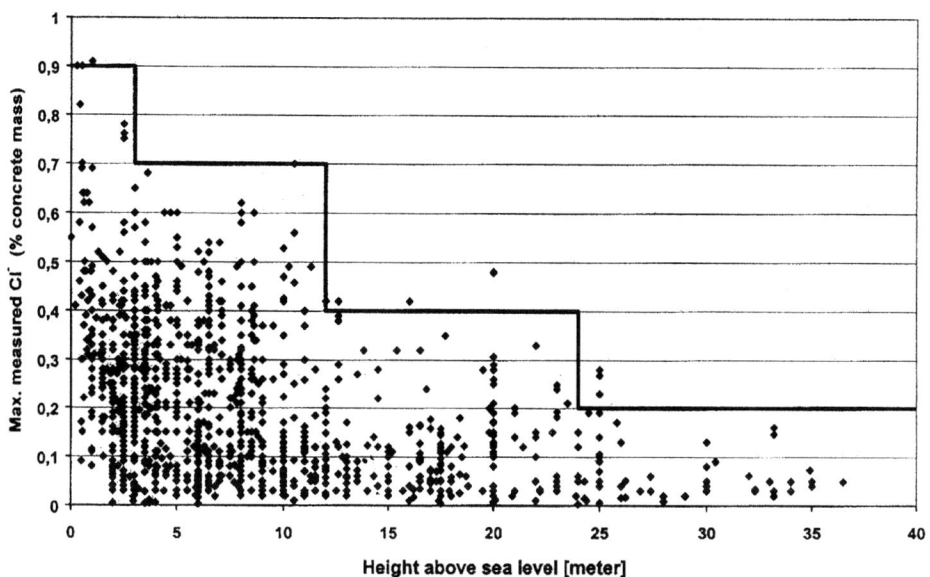

Figure 3.2.38 Maximum recorded chloride concentration $C_{\Delta x}$ at different heights above sea level of Gimsøystraumen bridge and 35 other coastal bridges in Norway (for details see Lifecon D4.2). Data includes both windward and seeward effects. The data represents 850 chloride profiles. Four environmental zones are distinguished.

Table 3.2.18 Chloride concentration $C_{\Delta x}$ in relation to height above sea level

Zone Height above sea level (m)	Mean value (wt%/concrete)	Standard deviation (wt%/concrete)
0–3	0.51	0.23
3–12	0.36	0.24
12–24	0.22	0.19
> 24	0.17	0.10

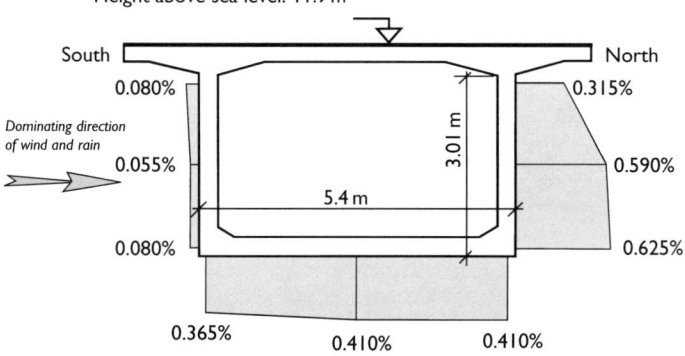

Figure 3.2.39 Influence of orientation on $C_{\Delta x}$ (Gimsøystraumen bridge, Norway).

Possibilities of predicting a long-term average for the chloride concentration of water and snow slush on the roads during de-icing salt applications were treated above. However, transformation from the concentration of the street water and slush to that at the concrete surface seems hard to be develop, bearing in mind all the abovementioned influences. The most promising solution to this problem seems be to statistically analyse the chloride concentration at depth Δx of the concrete directly.

Lay collected 640 chloride profiles from German road administrations (160 structures), which were analysed with respect to the most relevant influences. Although the particular concrete composition of each investigated structure was usually unknown it can be assumed that CEM I (OPC), with $w/b = 0.40-0.50$, was usually used. Site-dependent information, such as traffic volume and velocity and the regime of de-icing salt application, was not available and was not incorporated into the analysis. The chloride profiles were thus only analysed with respect to exposure time and distance to the road. The analysis revealed for the concentration $C_{\Delta x}$:

- a linear decrease with increasing height (h) above street level
- a logarithmic decrease with increasing horizontal distance (a) from the street edge
- no trend with respect to exposure time.

These relationships were used to derive a function in the form $C_{\Delta x} = f(h, a)$ (Figure 3.2.40):

$$C_{\Delta x, \text{mean}}(a, h) = 0.465 - 0.051 \cdot \ln(a + 1) - (0.00065 \cdot (a + 1)^{-0.187}) \cdot h \qquad (3.2.82)$$

where
$C_{\Delta x, \text{mean}}(a, h)$ mean chloride concentration at depth Δx (weight%/concrete)
a horizontal distance to the roadside (cm)
h height above road level (cm)

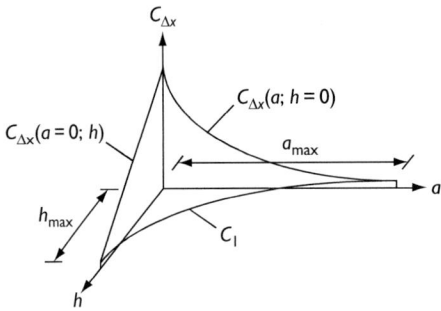

Figure 3.2.40 Chloride concentration $C_{\Delta x}$ as a function of height above street level (h) and horizontal distance to the street edge (a).

A model uncertainty ε_c was calculated, which is the ratio of each measured value $C_{\Delta x,\text{measured}}$ (weight%/concrete) to the calculated mean value according to Equation 3.2.82:

$$\varepsilon_C = \frac{C_{\Delta x,\text{measured}}}{C_{\Delta x,\text{mean}}(a, h)} \tag{3.2.83}$$

The statistical analysis showed that a log-normal distribution (L) gives the best description the model uncertainty ε_c. Due to the large scatter of the analysed data, the coefficient of variation (CoV $= 100\,\sigma/\mu$) is quite high, with CoV $= 75\%$. The statistical distribution of the concentration at depth Δx is thus defined as follows: L(mean $\mu = C_{\Delta x,\text{mean}}$; standard deviation $\sigma = \mu \times 0.75$) in weight%/concrete.

By extrapolation of Equation 3.2.82 it can be seen that elevated chloride concentrations above the initial background level are to be expected up to a height of $h_{\text{max}} = 7$ m and as far as a horizontal distance of around $a_{\text{max}} = 70$ m. However, Equation (3.2.82) should only be used within the limits of the collected data ($0 \le h \le 5.5$ m; $0 \le a \le 12$ m).

The mean value of the presented site-dependent function for the concentration $C_{\Delta x}$ is consequently only valid in Germany and for concrete produced with CEM I and water–binder ratios of around 0.40–0.50. If the salt application regime differs considerably from the German average or a different binder type is considered, the values may be adapted by using binding isotherms, which express the relationship between chloride supply C_{free} and the resulting total content C_{total} (Figure 3.2.41). The standard deviation of the inherent spatial scatter should meanwhile still be based on the model uncertainty obtained for CEM I (see Equation 3.2.83).

Example
If a concrete produced with 360 kg OPC and $w/b = 0.50$ shows a surface concentration of around 1.5 weight%/concrete in a certain environment, the use of cement with 50% slag (CEM III/A) would lead to a value of around 1.66 weight%/concrete according to the binding isotherms given above. A quite similar result is to be expected when changing the w/b ratio from 0.40 to 0.50 while all other parameters remain the same (Figure 3.2.42).

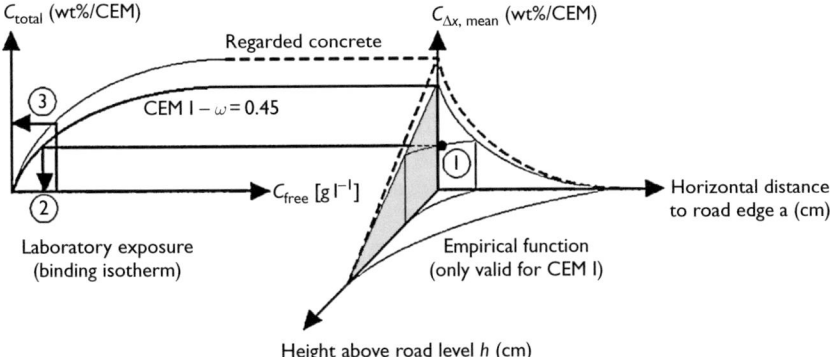

Figure 3.2.41 Procedure to determine concentration $C_{\Delta x,\text{mean}}$ for any concrete composition to be expected in the road environment.

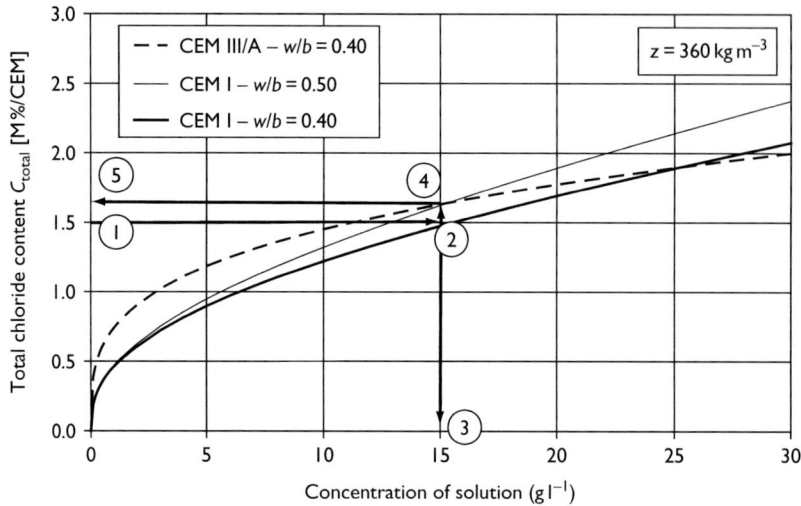

Figure 3.2.42 Using binding isotherms to estimate chloride surface concentration.

3.2.7.1.14 CRITICAL CHLORIDE CONTENT, C_{crit}

Chloride concentration at the depth of the reinforcement is compared with published information on threshold concentrations for the initiation of corrosion. If chloride concentration exceeds the threshold, it may be assumed that the steel has been depassivated, and that corrosion is possible.

The exact value for the chloride threshold required to initiate corrosion is unclear and is best judged as a risk of corrosion. A number of factors affect the chloride threshold level, including the chloride/hydroxyl ratio, chloride-binding capability of the cement, use of replacement materials, admixtures, carbonation of the concrete and the condition of the reinforcing steel. The critical level is generally taken as falling in the range of 0.2–0.4% chloride by weight of cement but values as high as 1% have been found in structures where corrosion has not been initiated, while corrosion has been evident in structures with values as low as 0.1% [61].

The British Standard BS 8110 sets limits of chloride concentration of between 0.1 and 0.4% by mass of cement, depending on the type of structural member, curing regime and cement type. These limits are intended for new constructions. BS 7361 states that no satisfactory threshold value for chloride concentration has been established, and notes that some authorities impose a limit of 0.3% as allowable before remedial action is necessary. BS 7361 also states that the presence of any chloride should always be assumed to place the structure at risk.

Chloride concentration thresholds are generally quoted by proportion of cement, but it is samples of concrete which are weighed and analysed in tests. This means either that assumptions must be made about the cement content of the mix or that analysis of the hardened concrete should be undertaken to determine cement content. Chemical analysis of the cement content of hardened concrete is notoriously tricky, particularly if a sample of the aggregate is unavailable.

Where chlorides are transported from the outside environment but have yet to reach the reinforcement, knowledge of the variation of chloride concentration with depth will enable an estimate of the diffusion coefficient and hence of the time to depassivation to be made. Computer programs are available to assist in this process.

No information on the critical chloride content for rebar corrosion in high-alumina cement concrete is known. Sulphate-resisting cement has a low binding capability, and critical (acid-soluble) chloride content can be expected to be significantly lower than for Portland cements. Similarly, blended cements containing PFA or silica fume also tend to have lower binding capabilities. The lesser binding capacity is more than offset by a reduction in permeability, however.

The critical chloride content is here considered as the content leading to depassivation of the steel surface, irrespective of whether corrosion damage becomes visible or not. The critical chloride content mainly is controlled by (Figure 3.2.43) [54]:

- the moisture content of the concrete
- the quality of concrete cover.

As a statistical parameter, the critical chloride content (following the first definition above) may be given [55] beta distribution ($m = 0.48$; $s = 0.15$; $a = 0.20$; $b = 2.0$).

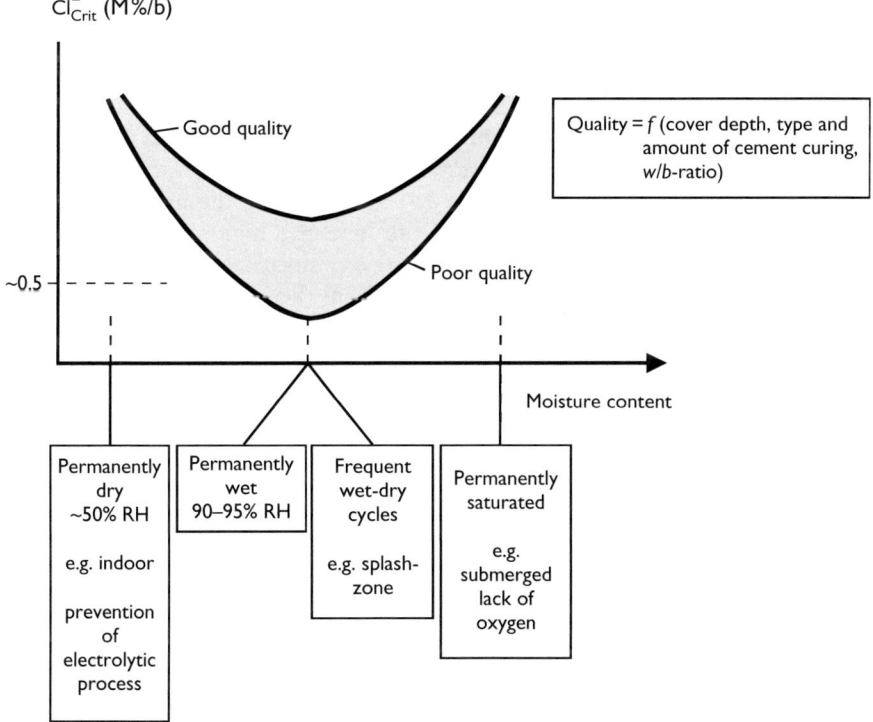

Figure 3.2.43 Qualitative relationship of the critical chloride content, the environmental conditions and quality of concrete cover [54].

3.2.7.1.15   SHORTCOMINGS OF THE MODEL

- There is a lack of knowledge about the dependence of the apparent diffusion coefficient D_{app} on the humidity conditions of the environment, i.e. the degree of saturation of the concrete.
- The effect of carbonation on chloride-binding capacity is only indirectly treated by introducing the parameter Δx, being the depth at which the data deviates from diffusion behaviour.
- The interaction of chloride ingress with other degradation mechanisms, such as frost, is not included.
- The effect of concrete on the effective diffusion coefficient is properly modelled. The temperature of concrete is assumed to be equal to environmental temperature. This is a sufficiently correct assumption for surfaces, which are sheltered from direct radiation. The effect of radiation must still be incorporated [13].

3.2.8   Case study of chloride-induced corrosion

3.2.8.1   Description of the object – Hofham Brücke

The model was applied to a bridge column, which was not part of the data analysis and can thus be regarded as independent for the purpose of verification. The bridge column is situated 1.8 m from the edge of a federal road in Bavaria (near Landshut) with a speed limit of $100\,km\,h^{-1}$. The concrete was composed of $320\,kg\,m^{-3}$ CEM I with $w/b = 0.48$. The structure was built in June 1963, and inspected in June 1982 and June 2001. The target of the durability design is either the mean value of the concrete cover d_C or the apparent diffusion coefficient D_{app}. In the case study we assume that the structure was planned with a mean value of $\mu = 45\,mm$ for the concrete cover [14]. For regular workmanship a standard deviation of $\sigma = 8.0\,mm$ can be assumed today for the concrete cover. Regarding the height above the road level, the bottom, $h = 0\,cm$, is decisive, as here the highest concentration $C_{\Delta x}$ is to be expected (see Equation 3.2.82). However, since in 1982 and 2001 chloride profiles were only taken at a height of $h = 60\,cm$, this is used as a reference site for model verification. As an input to the temperature function (Equation (3.2.71)), the daily average air temperature measured by the German weather service (DWD) at the nearest weather station (Landshut) was analysed for the last 10 years [21]. The summary of input parameters is given in Table 3.2.19

3.2.8.1.1   RELIABILITY ANALYSIS

In first step the service life of the column under case study is calculated using the information available without any further structure investigations. In a second step the additional information from investigations is used to improve the precision of the model. Regarding the height of the road level, the bottom area is the decisive domain for service life calculations. As in the years 1982 and 2001, chloride profiles were taken at a height of $h = 60\,cm$; this domain is taken as a reference for checking the model precision. If only the mean values given in Table 3.2.19 are inserted in Equation (3.2.57) (semi-probabilistic approach), the average expected chloride profile can be predicted.

Table 3.2.19 Summary of input parameters for the durability design

Parameter	Units	Distribution[‡]	Mean (μ)	Standard deviation (σ)
C_{ini}	wt%/cement	L	0.108	0.057
$C_{\Delta x}$	wt%/cement	L	1.39	1.04
C_{crit}	wt%/cement	B	0.48	0.15
		$0.2 \leq C_{crit} \leq 2.0$		
$\Delta x > 2^{*}$	mm	B	9.03	9.36
		$0 \leq \Delta x \leq 60$		
d_{C}	mm	B	45.0	8.0
		$0 \leq d_{C} \leq 700$		
DRCM (28d)	$10^{-12}\,m^{2}\,s^{-1}$	L	9.28	4.83
DRCM $(t = \infty)$	$10^{-12}\,m^{2}\,s^{-1}$	L	4.88	3.56
k_{ω}		L	1.20	0.13
n_{2}		L	0.964	0.212
$T_{air} > 3^{\dagger}$	(K)	B $260 \leq T \leq 302$	283	8
b_{T}	(K)	N	4800	700
k_{RH}		D	0.90	—
n_{1}		B $0 \leq \beta \leq 1$	0.55	0.24

Notes
* Statistical analysis of 640 chloride profiles of German road administrations.
† Statistical analysis of the daily average air temperature.
‡ Distribution types: N (normal); B (beta); L (log-normal); D (deterministic).

The comparison of chloride profiles (mean values until a certain moment in time) for the years 1982 and 2001 (Figure 3.2.44) revealed that the depth-dependent chloride loading of the concrete was overestimated by the a priori model (without inspection data), especially in the domain of the depth $10 \leq x \leq 70$ mm.

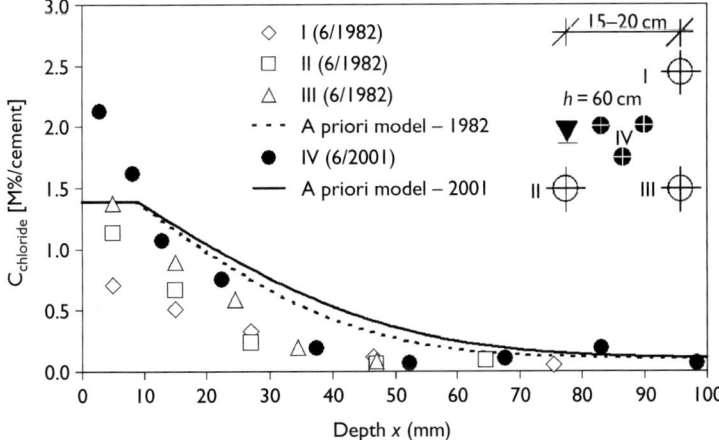

Figure 3.2.44 Measured and predicted chloride profiles (average values) in the year 1982 and 2001 at nearby sampling sites.

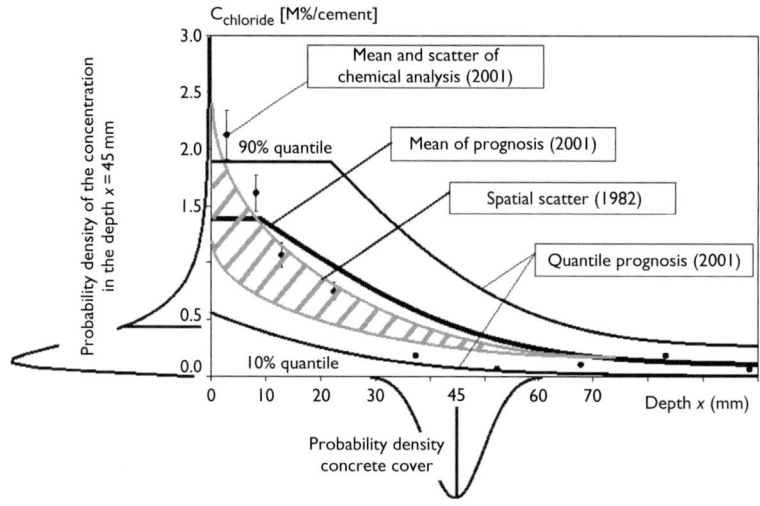

Figure 3.2.45 Scatter in the prognosis of chloride profiles.

When comparing predicted and measured values, the considerable spatial scatter must be taken into account. Moreover, deviations from the real values due to the chemical analysis should be considered (Figure 3.2.45). Round robin tests [56, 57] showed that, for concentrations in the domain of the critical chloride threshold, average deviations of around 25% will occur (for the most common technique of titration). With decreasing chloride concentration the deviations of the analysis will increase even more.

The probability density of the chloride concentration and the resulting quantile values can be calculated for every depth x according to the full-probabilistic model. Having the various sources of scatter in mind when modelling chloride-induced depassivation of the reinforcement, it becomes obvious that only probabilistic methods are appropriate to tackle the problem. This is done by comparing the critical chloride depth x_{crit} and the concrete cover.

The probability of depassivation $p_{\text{depassivation}}$ is calculated according to Equation (3.2.67). However, today it is common practice to consider the reliability index β instead of $p_{\text{depassivation}}$. As chloride profiles were determined in the years 1982 and 2001 (age 19 and 38 years), the apparent diffusion coefficient D_{app}, the chloride concentration $C_{\Delta x}$ and the depth of the convection zone Δx were known at these moments in time.

Measurements of the concrete cover resulted in a somewhat lower mean value, as expected, of $\mu = 44\,\text{mm}$ and a relatively high standard deviation of $\sigma = 12\,\text{mm}$.

Furthermore, the chloride migration coefficient D_{RCM} of the actually built concrete can be measured. For young concrete structures, which have not yet been subjected to a considerable chloride load (e.g. before first application of de-icing salt in the road environment), this is possible using the regular method according to Tang. For those concretes already loaded with chlorides, as was the case for the considered structure in the year 2001, a new test method was developed [58].

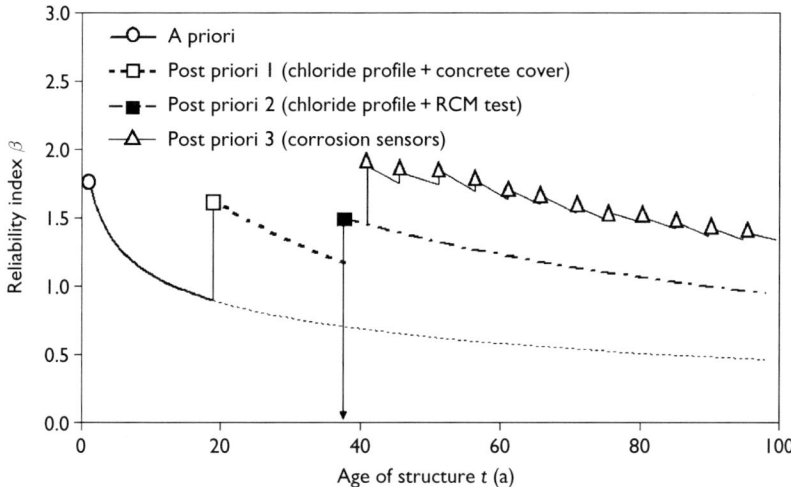

Figure 3.2.46 Bayesian update with data of structure investigation (until current age) and monitoring with corrosion sensors (assumption for future).

Using all of the additional investigation data improves the prediction of the velocity of chloride penetration considerably (Figure 3.2.46).

Compared to the a priori model, the post-priori model results in a higher reliability index versus time. The reason for this effect compared to the assumptions for design is a lower mean value and a lower standard deviation of the convection zone Δx and of the apparent diffusion coefficient D_{app}. This positive influence overwhelms the effect of a higher scatter of the concrete cover.

The level of reliability of today's requirements, which was determined with the post-priori model 2 at $t = 38$ a, lies in the domain of today's requirements of $\beta_{min} = 1.5$. As the structure was expected to fulfil these requirements throughout the entire service life, maintenance measures need to be taken if the expected service life is not achieved.

The uncertainty with respect to the chloride penetration velocity was reduced by inspections at the ages of $t = 19$ and 38 years. Besides the penetration velocity of chlorides, also the variable critical chloride concentration in concrete is subject to uncertainty with respect to the mean value and standard deviation. This uncertainty cannot be reduced by collecting chloride profiles, but can be readily reduced by integrating data from corrosion sensors (monitoring data). In the present case study, such sensors were not installed. Still, for demonstration purposes it is assumed that sensors were installed during the second inspection. The sensor elements, which are installed at different depths, give a corrosion current signal once the critical corrosion-inducing chloride content is reached at a certain depth. This information can be monitored continuously over time and be inserted for future predictions [60].

3.2.8.2 Conclusions for practice

The prognosis of chloride profiles revealed that the model developed for durability design (a priori model) gives on average a sufficiently good estimate of the penetration velocity of chlorides. When considering the variability of each input parameter, the reliability can be estimated using full-probabilistic methods. A verification of the time-dependent penetration depth (mean and standard deviation) by means of structure investigations will lead unavoidably to an increase in the reliability index, because the uncertainty in the model is reduced.

Improvement of an a priori model will, however, not always result in an increase in reliability. This is the case if assumptions in the a priori model with respect to the structure resistance (concrete cover) are overestimated and/or the stress variable (critical penetration depth) is underestimated.

Moreover, the case study demonstrates the necessity for inspection data, which are obtained at different times with different inspection methods (Figure 3.2.47). For newly built structures, the necessary concrete cover and material resistance are determined according to the environmental exposure. Immediately after completion of the structure the concrete cover and the chloride migration coefficient should be tested within the quality control system. The results of these investigations should be documented in a 'birth certificate' and used for another calculation of the residual service life.

When evaluating chloride profiles, chloride contamination versus the penetration depth is obtained and, thus, information regarding the penetration velocity of the chlorides. Correction sensors provide this information; such sensors also provide information about the critical chloride content under the particular environmental conditions.

If the probability of depassivation of the reinforcement is high, mapping of the electrochemical potential may be performed. The result of these measurements supplies site-dependent information regarding the corrosion activity of the reinforcement over the entire surface area of a component, which means that the information is

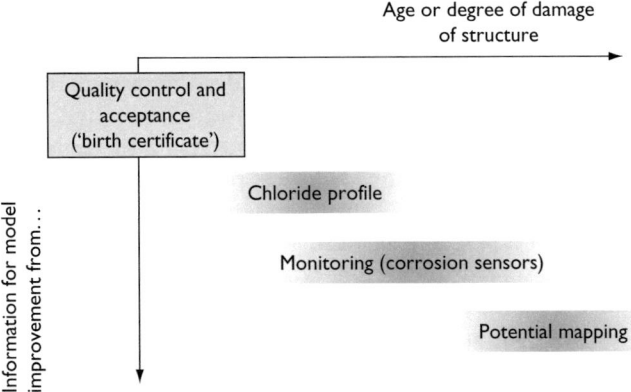

Figure 3.2.47 Moment in time and method to obtain additional information for a model update using Bayesian methods.

obtained at a certain depth. The combination of corrosion sensors and chloride profiles with data on potential mapping gives a good picture of the condition of the component.

3.2.9 Models of propagation of corrosion

3.2.9.1 Principles

The propagation period t_{prop} is the time during which the rebar is being actively corroded. The main parameter which has to be determined is the corrosion rate. Neglecting the propagation period in service life calculations is a common approach for the durability design of new structures. This procedure is sufficient when the indications are that the corrosion of steel will propagate at such a rapid rate that the propagation period is relatively short compared to the initiation phase, or when design considerations will not tolerate any form of rebar corrosion.

Since most existing structures suffer from rebar corrosion to a certain degree and decisions on the necessary measures have to be taken, the corrosion phase must be included in calculations of the residual service life.

The corrosion process can be modelled by an equivalent circuit of a galvanic element with a serial connection of resistances representing the anodic and cathodic charge transfer resistances R_{ct} of the steel surface, the electrical conductive resistance of the reinforcement R_{steel} and the concrete resistance $R_{concrete}$.

The corrosion current I_{corr}, which is proportional to the corroded mass of steel, depends on the potential difference ΔU between the anodic and cathodic areas, as well as on the resistances in the equivalent circuit. The corrosion rate V_{corr} can be expressed in terms of the current density $i_{corr}(\mu A\,cm^{-2})$ (corrosion current I_{corr} related to the steel surface area) or in terms of the rate of loss in bar section $(mm\,a^{-1})$. The relationship between electrical current and loss of steel section is as follows:

- mass of a single iron ion

$$m_{Fe} = \frac{M_{Fe}}{N_L} = \frac{55.847}{6.0234 \cdot 10^{23} \cdot 1000} = 9.272 \cdot 10^{-26} \qquad (3.2.84)$$

M_{Fe} mass of relative atom of iron $(g\,mol^{-1})$
N_L Loschmidt constant (mol^{-1})

- electrical current set free in the reaction

$$|L_{Fe^{++}}| = |2e^-| = 2 \cdot \frac{F}{N_L} = 2 \cdot \frac{96493}{6.0234 \cdot 10^{23}} = 3.204 \cdot 10^{-19} (C = As) \quad (3.2.85)$$

F Faraday constant $(C\,val^{-1} = As\,val^{-1})$

- corroded mass of steel for 1 amp per year

$$m_{corr} = \frac{m_{Fe} \cdot 3.154 \cdot 10^7 \, s \, a^{-1}}{|L_{Fe^{++}}|} = \frac{9.272 \cdot 10^{-26} \cdot 3.154 \cdot 10^7}{3.204 \cdot 10^{-19}} = 9.127 (kg \, Aa^{-1})$$

(3.2.86)

- corrosion rate related to surface area

$$V_{corr} = \frac{m_{corr}}{\rho_{Fe}} = \frac{9.127}{7.85 \cdot 10^{-6}} = 1.16 \cdot 10^6 \left[\frac{mm^3}{Aa} = \frac{mm \, a^{-1}}{A \, mm^{-2}} \right] = 11.6 \left[\frac{\mu m \, a^{-1}}{\mu A \, cm^2} \right]$$

(3.2.87)

ρ_{Fe} density of steel $(kg \, mm^{-3})$.

For a known corrosion current density i_{corr} $(\mu A \, cm^{-2})$, the annual penetration rate is thus:

$$V_{corr} = i_{corr} \cdot 11.6 \, (\mu m \, a^{-1})$$

(3.2.88)

i_{corr} corrosion rate density $(\mu A \, cm^{-2})$

V_{corr} may be constant along the propagation period or vary following different events suffered by the structure (environmental changes). There are numerous parameters related to concrete quality and the environment that influence V_{corr}. The mutual influences may be additive, synergistic or opposite. Four main parameters are [13]:

1. resistivity of concrete
2. galvanic effects
3. chloride content of concrete
4. humidity/temperature in concrete.

The resistivity of the concrete is the major factor affecting the corrosion of depassivated steel, being in turn influenced by the mix composition and the moisture content of the concrete.

The damage function $P(t)$ represents the loss of rebar diameter at time t:

$$P(t) = \int_{t_i}^{t} V_{corr}(\tau) d\tau$$

$P(t)$ corrosion depth (mm)
t_i initiation period (a)
V_{corr} corrosion rate $(mm \, a^{-1})$.

Due to the presence of chlorides or partial depassivation induced by carbonation, the loss of rebar diameter may vary locally because of the occurrence of pitting corrosion. The pitting factor is the ratio of the maximum pit depth to the average loss of rebar diameter.

Considering a constant corrosion rate, the progressive loss of rebar diameter is given by [8]:

$$P(t) = V_{corr} \cdot \alpha \cdot t_{prop} = V_{corr} \cdot \alpha \cdot (t - t_{ini}) \tag{3.2.89}$$

α pitting factor accounting for a non-uniform corrosion of the rebar (–)
t_{prop} duration of propagation phase (s)
t age of structure (s)
t_{ini} time until initiation of corrosion (s).

The initiation time t_{ini} is inserted in Equation (3.2.89) by solving the models for the onset of corrosion due to carbonation or chloride ingress for t. This is not possible for the model on chloride ingress in an analytical way for the proposed model structure; see Equations (3.2.57) and (3.2.76). There are two solutions to overcome this problem:

1. Change the structure of the model on chloride ingress by neglecting the fact that hydration only proceeds for a certain time and that the apparent diffusion coefficient is the average diffusion coefficient in the period considered.
2. Do not solve these equations for time, but determine t_{ini} in a different way.

The second option is proposed here. The user has to calculate and plot the probability of depassivation against time. This data can be fit to the probability distribution function of time until depassivation. For the distribution type, a Weibull (minimum Type III) seems to fit rather well, as shown in Figure 3.2.48. The results of fitting are the parameters of the distribution (here τ, w and k), which characterise the distribution for the time of depassivation unambiguously.

Statistical values of the pitting factor α and the ratio of the maximum penetration depth P_{max} to the mean penetration depth P_{mean} are presented in Table 3.2.20. As negative values for the pitting factor cannot occur, a log-normal distribution is suited for use.

The main task of modelling the damage function $P(t)$ concerns the establishment of the value of V_{corr}. For durability design applications or for young structures (corrosion has not yet started) it is possible to [8]:

(a) assume values in relation to exposure classes
(b) use empirical expressions
(c) perform direct measurements of specimens exposed under accelerated conditions simulating the considered environment.

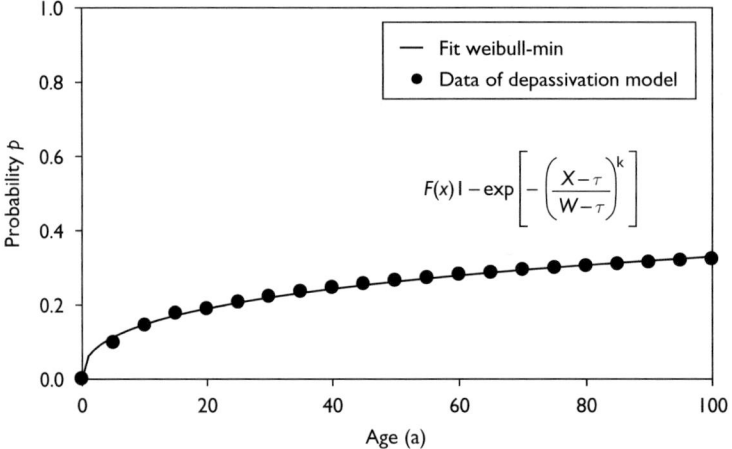

Figure 3.2.48 Probability distribution plot for the time until depassivation.

Table 3.2.20 Pitting factor α, according to DuraCrete [11]

Environment	Distribution	Mean	Standard deviation
Carbonation	Deterministic	2.00	—
Chloride	Log-normal	9.28	4.04

For existing objects, data obtained from alternatives (a) to (c) can be updated by:

(d) data from direct measurement on-site.

For the alternative (d) various possibilities exist, which are benchmarked and described in the Appendix of Lifecon D3.1 [1].

3.2.9.2   Corrosion rate in relation to exposure classes

The corrosion rate may be given as:

$$V_{corr} = V_{corr,a} \cdot \text{ToW} \tag{3.2.90}$$

ToW   wetness period, given as the fraction of the year
$V_{corr,a}$ mean corrosion rate when corrosion is active (mm year^{-1}).

Based on experience, values for ToW and $V_{corr,a}$ are given for classification in Tables 3.2.21 and 3.2.22 [11].

Table 3.2.21 Distribution of corrosion rate V_{corr} and weather exponent w for carbonation-induced corrosion independent of concrete type and quality

Exposure class	V_{corr} (mm a^{-1})			w		
	Mean	Standard deviation	Distribution type	Mean	Standard deviation	Distribution type
Sheltered	0.002	0.003	Weibull	0.50	0.12	Normal
Unsheltered	0.005	0.007		0.75	0.20	

Table 3.2.22 Distribution of corrosion rate V_{corr} and weather exponent w for chloride-induced corrosion independent of concrete type and quality for different exposure classes

Exposure class	$V_{corr,a}$ (mm a^{-1})			w		
	Mean	Standard deviation	Distribution type	Mean	Standard deviation	Distribution type
Wet–rarely dry	0.004	0.006	Weibull	1.00	0.25	Normal
Cyclic wet–dry	0.030	0.040		0.75	0.20	
Airborne sea water	0.030	0.040		0.50	0.12	
Submerged	Not expected, unless bad concrete quality or low cover					
Tidal zone	0.070	0.070	Weibull	1.00	0.25	Normal

3.2.9.3 Direct measurement of the corrosion rate

Applying Faraday's law, the corrosion rate can be defined from direct measurement [61]:

$$V_{corr} = 11.6 \cdot k_t \cdot i_{corr} \tag{3.2.91}$$

k_t testing factor
i_{corr} mean value of the corrosion current (mA cm^{-2}).

The corrosion current can be directly measured in specimens or structures by means of:

- the linear polarisation technique
- electrochemical impedance (EIS) or
- intersection method of the polarisation curve.

These methods measure the polarisation resistance R_P, which is used for the calculation of the corrosion current i_{corr}:

$$i_{corr} = B/R_p \tag{3.2.92}$$

B constant varying between 13 and 52; usually 26 is applied (mV).

Table 3.2.23 Values for I_{corr} and k_t

Case	Parameter	Unit	Distribution	Mean	Standard deviation
Laboratory specimen	I_{corr}	$\mu A\,cm^{-2}$	Log-normal	—	0.86
	k_t	—	Log-normal	1.0	0.80
On site	I_{corr}	$\mu A\,cm^{-2}$	Weibull	—	1.38
	k_t	—	Log-normal	1.0	0.80

The most frequently used method is the linear polarisation technique, which is the only one applicable for the measurement on real structures. Two possibilities of the test are feasible:

1. laboratory testing with simulated climates for new structures; and
2. measurement *in situ* of existing structures at several locations for a period of 1 year at least, in order to estimate the scatter with variation of the climatic conditions, because values of I_{corr} may change by several orders of magnitude due to changes in the moisture content and temperature of the concrete.

The values of I_{corr} change considerably also because corrosion is a dynamic process. In consequence, single measurements must be corrected for deviations due to these influences, as presented in Table 3.2.23.

3.2.9.4 Empirical modelling of the corrosion rate

In order to model the corrosion rate, the following relationship can be used:

$$i_{corr} = \frac{k_0}{\rho(t)} \cdot F_{Cl} \cdot F_{Galv} \cdot F_{O_2} \qquad (3.2.93)$$

i_{corr} corrosion rate $(\mu A\,cm^{-2})$
k_0 constant regression parameter $(\mu m\,\Omega m\,a^{-1})$
$\rho(t)$ actual resistivity of concrete at time t (Ωm)
F_{Cl} accounting for the influence of the chloride content
F_{Galv} influence of galvanic effects
F_{O_2} availability of oxygen.

The model for corrosion rate contains two main environment- and material-dependent parameters, k_0 and $\rho(t)$. A reasonable approach to modelling the material and environmental effects on concrete resistivity $\rho(t)$ is to introduce a potential resistivity ρ_0 for a reference concrete and defined environmental conditions (20 °C, 100% RH).

3.2.9.4.1   REGRESSION PARAMETER, k_0

The regression parameter k_0 can be obtained by performing a regression analysis of the corrosion current i_{corr} versus the resistivity ρ. Nilsson and Gehlen propose a deterministic value of $k_0 = 882$ for the time being. The scatter of the corrosion rate model is included in all other parameters. Should this parameter be quantified more precisely in the future, it would be a simple task to incorporate it as a stochastic parameter.

3.2.9.4.2 CONCRETE RESISTIVITY: MODEL APPROACH

This material parameter can be obtained with little effort and is already quantified for numerous concretes [12]. The difference in environmental conditions is covered by a number of environmental parameters. Since the potential resistivity is measured at an age of 28 days, an age factor n, analogous to the chloride ingress model, is introduced:

$$\rho(t) = \rho_0(t) \cdot k_c \cdot k_t \cdot k_{R,T} \cdot k_{R,RH} \cdot k_{R,Cl} \tag{3.2.94}$$

ρ_0 time-dependent potential concrete resistivity for a reference concrete and defined environment $(T = +20\,°C$; immersed in chloride-free water)
k_c curing factor
k_t test method factor
$k_{R,T}$ temperature factor
$k_{R,RH}$ relative humidity factor
$k_{R,Cl}$ chloride factor.

In a similar way as for the onset of corrosion, a limit state can be considered:

$$p_f = p\{P(t) > P_{crit}\} \le \Phi(-\beta) \tag{3.2.95}$$

$P(t)$ loss of rebar section [Equation (3.2.89)]
P_{crit} critical loss of rebar section, depending on the structural geometry and the chosen limit state.

An overview of parameters for the equations presented above is presented in Table 3.2.24.

Table 3.2.24 Overview of parameters in corrosion propagation model

Type of parameter	Parameter
Material/environment	$k_0, F_{Cl}, F_{O_2}, F_{Galv}$
Material	ρ_0, n
Execution	k_c
Test method	k_t
Environment	$k_{R,T}, k_{R,RH}, k_{R,Cl}$

3.2.9.4.3 MATERIAL-DEPENDENT RESISTIVITY, ρ_0

Applicable test procedures are given in the Appendix of Lifecon D3.1 [1]. The binder type has been identified to be the dominating material factor influencing the resistivity. Measurements [11] were continuously performed with the multi-ring method during the first 91 days of hydration (Figure 3.2.49), and fitted according to the potential approach [Equation (3.2.96)] to obtain the statistical input parameters given in Table 3.2.25.

$$\rho_0(t) = \rho_{0,t=28d} \left(\frac{t}{t_0}\right)^n \tag{3.2.96}$$

$\rho_{0,t=28d}$ specific resistance at age $t_0 = 28d$ $(k\Omega\,cm^{-1})$
n influence of ageing of concrete on resistivity.

Long-term observations indicate that the time-dependent increase in electrolytic resistivity of concrete stored in water will end with the termination of the hydration period, which is not taken into account with the potential approach according to Equation (3.2.96) (Figure 3.2.50).

The time until the hydration process starts does not contribute any more to the increase in electrical resistivity of the concrete and depends on the type of binder and the w/b ratio. Only the binder type and w/b ratio are significant [5], whereas cement strength class, cement content, air void content, aggregate type and superplasticiser

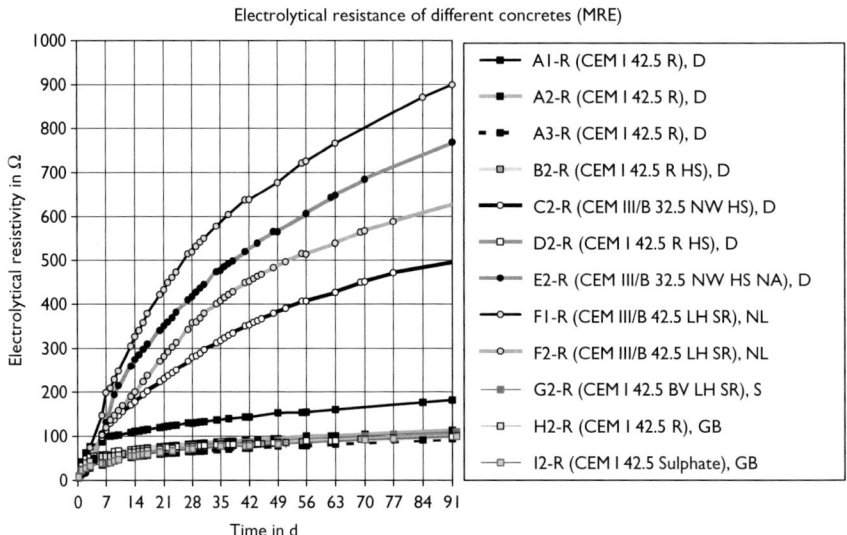

Figure 3.2.49 Electrolytic resistivity versus time (early hydration phase) for different concrete mixes submerged in chloride free water at $T = 20\,°C$.

Table 3.2.25 Statistical results for the specific electrical resistivity $\rho_{0,\text{TEM}}$ and the aging exponent n according to results shown in Figure 3.2.50

Binder	w/b	$\rho^{0,\text{TEM}*}(k\,cm^{-1})$			n		
		Distribution type	Mean	Standard deviation	Distribution type	Mean	Standard deviation
OPC (CEM I)	0.5	ND	7.7	1.2	ND	0.23	0.04
GGBS (CEM III)			35.2	15		0.54	0.12
CEM I + FA (20% replacement)			9.2	3.0		0.62	0.13

Note
* Resistivity measured with the two-electrode method (TEM)

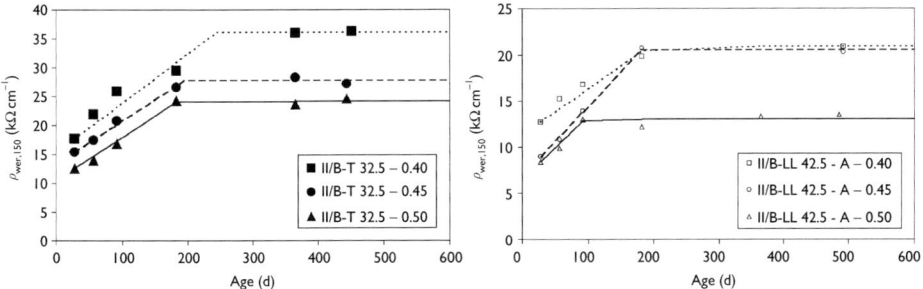

Figure 3.2.50 Electrical resistivity (Wenner-Method) versus time of concrete produced with CEM II/B-T 32,5R (oil slag cement) CEM II/A-LL (limestone cement) and with variation of the w/c ratio.

content have been showed to be negligible with respect to the electrical resistance of concrete. The following simplified linear model is proposed:

$$t < t_{\text{hyd}}: \rho_0(t) = \rho_{0,t=28d} + \frac{(\rho_{0,t-\infty} - \rho_{0,t=28d})}{(t_{\text{hyd}} - 28)} \cdot (t - 28) \tag{3.2.97}$$

$$t \geq t_{\text{hyd}}: \rho_0(t) = \rho_{0,t=\infty}$$

Index 0 indicates reference condition (water saturation)
$\rho_{0,t=28}$ specific resistance at an age of $t_0 = 28$ days $(k\Omega\,cm^{-1})$
$\rho_{0,t=\infty}$ specific resistance at infinite age $(k\Omega\,cm^{-1})$
t_0 reference age, here 28 days (s)
t concrete age (s)
t_{hyd} age until an increase in resistivity due to hydration can be measured (s).

To calculate with this approach, the average value of the concrete resistance during the time period considered has to be calculated:

$$\rho_{\text{AVG}}(t) = \frac{\frac{(\rho_{0,t=\infty} - \rho_{0,t=28d})}{2} \cdot (t_{\text{hyd}} - 28) + (t - t_{\text{hyd}}) \cdot \rho_{0,t=\infty}}{(t - 28)} \tag{3.2.98}$$

Table 3.2.26 Amount of investigated mixes n, mean value (kΩ cm⁻¹) and coefficient of variation (%) of $\rho_{0,Ref,t=28d}$ for w/b = 0.45 and various binder types

CEM	I	II/A-LL	II/B-T	II/B-S	III/A	III/B	I+FA
Additions (weight %)	0	15	23	31	51	71	15
n	6	2	1	2	1	1	4
$m(\rho_{0,Ref,t=28d})$	13.4	11.2	15.4	15.0	30.1	62.5	15.2
$CoV(\rho_{0,Ref,t=28d})$	16	–	–	–	–	–	13

Table 3.2.27 Amount of investigated mixes n, mean value (kΩ cm⁻¹) and coefficient of variation (%) of $\rho_{0,Ref,t=\infty}$ for w/b = 0.45 and various binder types

CEM	I	II/A-LL	II/B-T	II/B-S*	III/A*	III/B*	I+FA
Additions (weight %)	0	15	23	31	51	71	15
n	6	2	1	2	1	1	4
$m(\rho_{0,Ref,t=\infty})$	22.7	18.7	27.7	55.5	166.9	201.3	90.3
$CoV(\rho_{0,Ref,t=\infty})$	35	–	–	–	–	–	18

Note
* Value at the end of investigations (2 years). Hydration process is not yet finished (approach on safe side).

Table 3.2.28 Amount of investigated mixes n, mean value of m (d) and coefficient of variation (%) for the end of hydration for various binder types

CEM	I	II/A-LL	II/B-T	II/B-S	III/A	III/B	I+FA
Additions (weight %)	0	15	23	31	51	71	15
n	6	4	3	4	3	3	4
$m(t_{hyd})$	176	185	207	–*	–*	–*	–*
$CoV(t_{hyd})$	55	40	15	–*	–*	–*	–*

Note
* Not yet quantified as hydration still proceeds after 2 years of measurement.

The relationship of the electrical resistance of concrete and the w/b ratio can be expressed by means of an exponential approach, which corrects the results given in reference [58] for a change in w/b ratio:

$$\rho_0 = \rho_{WER,w/b=0.45} \cdot \exp^{a_{w/b,t=28d} \cdot ((w/b)-0.45)} \quad 0.40 \leq w/b \leq 0.60 \quad (3.2.99)$$

$\rho_{0,w/b=0.45}$ binder-specific electrical resistivity (Wenner method) for w/b = 0.45
$a_{w/b}$ regression exponent (CEM I, −2.9; CEM I+FA, −6.4; CEM III, −2.2).

3.2.9.4.4   CURING FACTOR k_c

The curing factor is not available at present.

3.2.9.4.5 TEST METHOD FACTOR k_t

It is necessary to provide information concerning the method used to test for electrolytic resistivity ρ_0. Relationships between the following test methods have been determined [12]:

- two-electrode method (TEM)
- multi-ring electrode (MRE)
- Wenner method (WER).

The relationship factors can be determined by testing identical materials with all of the methods given above. All quantifications should always be related to the Wenner method, as in this way capacitive charge effects are reduced. The following relationships are given [11]:

$$\rho_{TEM} = A \cdot \rho_{WER} + \varepsilon_A \tag{3.2.100}$$

$$\rho_{MRE} = B \cdot \rho_{WER} + \varepsilon_B \tag{3.2.101}$$

ρ specific electrolytic resistivity measured by TEM, WER or MRE $(k\,cm^{-1})$
A regression parameter; $A = 0.68$
B regression parameter; $B = 0.721$
ε_A offset; ND $(0; 3.4)$ $(k\,cm^{-1})$
ε_A offset; ND $(0; 2.8)$ $(k\,cm^{-1})$.

3.2.9.4.6 HUMIDITY FACTOR $k_{R,RH}$

For sheltered concrete structures, the humidity factor, which serves to correct the electrolytic resistivity ρ_0 (determined for saturated concrete) of the actual relative humidity on site, can be derived by comparing the resistivity of the concrete at different relative humidities with the reference humidity (100%). A reasonable assumption is to estimate relative humidity from the annual average humidity of the environment close to the structure. Due to lack of data, a quantification has so far only been performed for two binder types: an OPC (CEM I) and a blast-furnace slag cement (BFSC) (CEM III). Water–cement ratio varied from 0.45 to 0.65. The statistical quantification of the factor $k_{R,RH}$ is performed for different relative humidity as given in Table 3.2.29.

Table 3.2.29 Relative humidity factor $k_{R,RH}$ to correct the resistivity ρ_0 given in the form: distribution type (mean value), coefficient of variation (%), limit value

RH	CEM I	CEM III
50	sLN (17; 145; 3.2)	sLN (15; 120; 4.9)
65	sLN (6; 100; 2.41)	sLN (7; 70; 3.57)
80	sLN (3.2; 48; 1.33)	sLN (3.8; 27; 2.36)
95	LN (1.08; 13)	sLN (1.2; 10; 0.24)
100	D (1)	D (1)

sLN   shifted log-normal distribution (moments: ξ, δ, τ)
LN   log-normal distribution
D   deterministic.

From Figure 3.2.51 it can be seen that the mean value of the factor $k_{R,RH}$ increases exponentially and the coefficient of variation increases linearly with decreasing relative humidity.

Fitting of the data was performed [11] according to the following equations:

$$\mu(k_{R,RH}) = \left(\frac{100}{RH}\right)^{a} \tag{3.2.102}$$

$\mu(k_{R,RH})$ mean value of relative humidity factor
RH   relative humidity (%)
a regression parameter.

$$CoV(k_{R,RH}) = b \cdot (100 - RH) \tag{3.2.103}$$

$CoV(k_{R,RH})$ coefficient of variation of relative humidity factor (%)
b regression parameter.

For the distribution type, a log-normal distribution is proposed. The results indicate that distinction between binder types is not necessary for the factor $k_{R,RH}$ so that

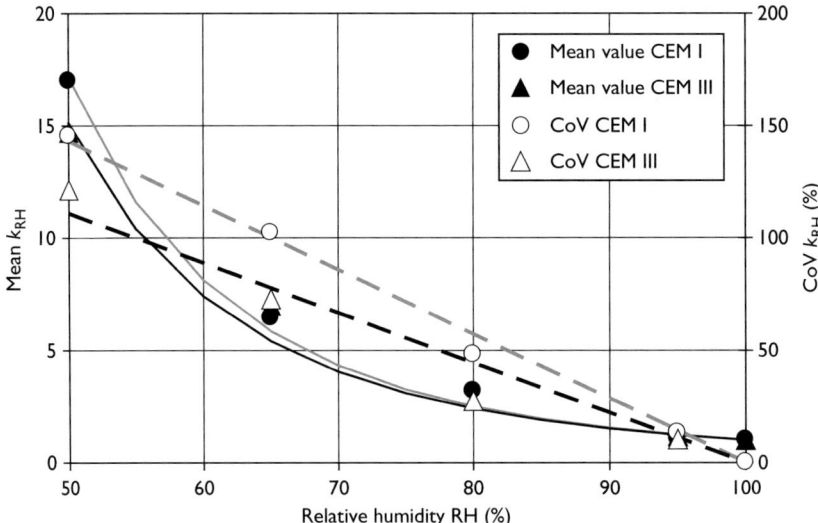

Figure 3.2.51 Mean value and coefficient of variation (CoV) of $k_{R,RH}$ as a function of the relative air humidity for concrete (w/c ranging from 0.45 to 0.65) produced with OPC and BFSC.

the above given regression parameters can be applied regardless of the binder type $(a = 4.0; b = 2.6)$.

From Equation (3.2.100) the following consequences can be derived:

- $RH = 0 \quad \Rightarrow \quad k_{R,RH} = \infty \quad \Rightarrow \rho = \infty \quad$ (dry concrete)
- $RH = 100 \quad \Rightarrow \quad k_{R,RH} = 1 \quad \Rightarrow \rho = \rho_0 \quad$ (saturated concrete).

For unsheltered conditions, the time of wetness ToW, i.e. the duration of rain and condensation at the concrete surface, must be considered. A possibility to account for wet periods in the model is to introduce ToW as follows:

$$\rho(t) = \rho_0(t) \cdot k \cdot (k_{R,RH})^d \qquad (3.2.104)$$

k product of all other influences apart from relative humidity

$$k = k_c \cdot k_t \cdot k_{R,T} \cdot k_{R,Cl} \qquad (3.2.105)$$

The exponent d expresses the time of dry periods. When formulating an expression for the duration of periods in which the concrete is wet, it must be realised that concrete is saturated rapidly once wetted but needs considerably more time to dry out afterwards. An example of measurement results of concrete resistance during cyclic wetting and drying is given in Figure 3.2.52.

The effect of a slow drying process may be accounted for by the following expression:

$$d = 1 - \text{ToW}^b \qquad (3.2.106)$$

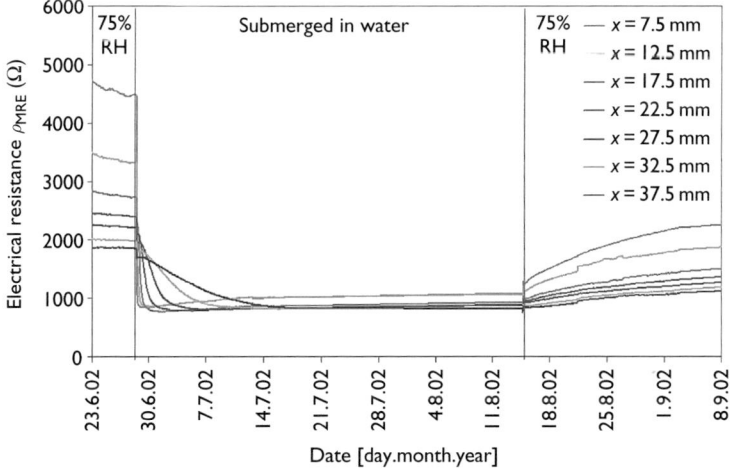

Figure 3.2.52 Measurement of concrete resistance at different depths x during cyclic wetting and drying.

d   dry period exponent
ToW   time of wetness
b   retardation parameter.

The time until concrete dries out to the same level as was the case before wetting depends on the concrete quality (binder type, *w/c* ratio). As sufficient data for the estimation of the retardation effect is lacking it is proposed to:

- set the exponent $d = 0$ if concrete is not sheltered from rain or is wetted by further sources; and
- set $d = 1$ if concrete is not wetted.

3.2.9.4.7   TEMPERATURE FACTOR, k_T

An increase in temperature will decrease the electrolytic resistivity of concrete. It is a common approach to use the Arrhenius equation to model the temperature effect upon the electrolytic resistivity:

$$k_{R,T} = \exp\left[b_{R,T} \cdot \left(\frac{1}{T} - \frac{1}{T_0} \right) \right]$$ (3.2.107)

$k_{R,T}$ factor to account for the change in concrete resistance with temperature
$b_{R,T}$ regression parameter (K)
T temperature of environment (K)
T_0 reference temperature, here 20 °C (293 K) (K).

To obtain the regression parameter $b_{R,T}$, data on the resistivity at different temperatures (with a reference temperature of usually $T_0 = +20$°C) has to be compared. A literature study by Raupach [59, 62] resulted in a range of values between $2130 \leq b_{R,T} \leq 5500$ (K). Here it is proposed to use the following quantification:

Log-normal N($\mu = 3815$; $\sigma = \mu \cdot 0.15 = 560$) (K)

The following relation can be used [11]:

$$k_{R,T} = \frac{1}{1 + K \cdot (T - 20)}$$ (3.2.108)

K temperature dependence of conductivity (°C^{-1}) (Table 3.2.30)
T temperature (°C)

However, a quantification dependent on different temperature regions is not suitable for a parameter study. The relationship given in Equation (3.2.109) is thus recommended.

Table 3.2.30 Parameter $K(°C^{-1})$ for Equation (3.2.108)

Environment	Distribution	Mean (μ)	Standard deviation (σ)
$T > +20°C$	Normal	0.073	0.015
$T < +20°C$	Normal	0.025	0.001

3.2.9.4.8 CHLORIDE FACTOR, $k_{R,Cl}$

The effect that an increasing chloride content has in reducing the electrolytic resistivity can be modelled as a factor $k_{R,Cl}$, a function of the chloride content c_{Cl} [11]:

$$K_{R,Cl} = 1 - \frac{1 - a_{Cl}}{2} \cdot c_{Cl} \qquad (3.2.109)$$

c_{Cl} chloride content (wt%/cem)
a_{Cl} regression parameter.

The chloride factor $k_{R,Cl}$ and the parameter a_{Cl} for different chloride contents a_{Cl} are presented in Table 3.2.31.

3.2.9.4.9 CHLORIDE CONCENTRATION FACTOR, F_{Cl}

The parameter F_{Cl} describes the effect of the chloride content on the corrosion rate, apart from the effect upon resistivity. The corrosion rate will increase with increasing chloride content for the same resistivity values. A simple approach may be a linear function, starting from $F_{Cl} = 1$ at a chloride content equal to the chloride threshold level C_{crit} with a slope of k [11]:

$$F_{Cl} = 1 + k(C - C_{crit}) \text{ with } C > C_{crit} \qquad (3.2.110)$$

k regression parameter, here shifted log-normal distribution (minimum $= 1.09$; $\mu = 2.63$; $\sigma - 3.51$)
C chloride concentration
C_{crit} critical chloride content.

3.2.9.4.10 THE GALVANIC FACTOR, F_{GALV}

There is so far no information available on the galvanic factor, which accounts for the possible effect of corrosion products in producing lower values of the corrosion rate,

Table 3.2.31 Chloride factor $k_{R,Cl}$ and parameter a_{Cl} used in Equation (3.2.109)

Environment	Parameter	Values
$c_{Cl} \geq 2\%$	$k_{R,Cl}$	ND (0.72; 0.11)
$0\% < c_{Cl} < 2\%$	a	ND (0.72; 0.11)

Note
ND = normal distribution.

because the oxides already generated may impede further access of aggressive media, thereby reducing the corrosion rate. Neglecting this factor, i.e. setting $F_{Galv} = 1$, is thus on the safe side.

3.2.9.4.11   THE OXYGEN FACTOR, F_{O_2}

The influence of the availability of oxygen may be taken into account as a factor. Data is not yet available. Nevertheless, except for submerged concrete, the oxygen supply is not usually limited, thus leading to $F_{O_2} = 1$ in most cases.

3.2.9.4.12   ASPECTS NOT TREATED BY THE CHOSEN MODEL

* Influence of cracks is not taken into account.
* Quantification of parameters is partly missing.

3.2.10   Parameter study for environmental classification

3.2.10.1   General procedure

The aim of the parameter study is to define the boundaries of classes for environmental data Haagenrud [64]. Once these boundaries are defined, environmental data can be mapped. The benefit for the user is a reduction of work during the process of collecting data as input to degradation models. For the mapping of environmental data, the general procedure is as follows:

1. Collect daily data for the last 10 years from weather stations to calculate mean values.
2. Classify each weather station.
3. Draw isolines of the class boundaries by interpolating classes of weather stations.
4. For the areas within isolines, the data of all weather stations must be combined to calculate the mean value and standard deviation using an appropriate distribution type.

The result is a map with classified areas. For each area, distribution type, mean value and boundary limits are stated.

For the meaningful classification of environmental parameters, a study of these can be performed. The boundaries of the five classes (0–4) need to be determined by dividing the maximum of the considered damage measure (e.g. corrosion penetration depth) into five equal sections. In class 4, the severest extent of damage is to be expected. In class 0 no damage at all will occur within the time scale.

The entire process, usually consisting of initiation and propagation periods, has to be modelled. This is necessary because some parameters may have a reverse influence on the duration of either phase. For instance, with increasing relative humidity the time until initiation will increase, because the carbonation process is retarded. Nevertheless, increasing relative humidity leads to an increase in electrical resistivity and thus to an increase in the corrosion rate during the propagation phase. Hence, a pessimum of relative humidity must exist for which the service lifetime of a reinforced concrete element tends to a minimum.

Table 3.2.32 Environmental parameters included in service life models for carbonation-induced corrosion

Initiation	Propagation
$k_{RH}(RH)$; $w(ToW)$; ΔC_S	$k_{R,T}(T)$; $k_{R,RH}(RH)$, ToW

3.2.10.2 Classification of carbonation-induced corrosion

The parameters given in Table 3.2.32 are presented for two qualities of concrete, both produced with CEM I cement:

- good: low w/c ratio (0.45), high concrete cover (50 mm), 7 days curing
- bad: high w/c ratio (0.60), low concrete cover (15 mm), 1 day curing.

The average time until depassivation of 'good' concrete is far beyond the scale of commonly requested service lives. The decisive parameter proves to be the concrete cover. Thus, the classification has only to be performed for 'bad' concrete. The result of the calculations for the corrosion due to carbonation is the corrosion penetration depth against the considered parameter (RH, ToW, T, ΔC_S) for the age of 30 and 100 years.

3.2.10.3 Input data for parameter study

For the study of one environmental parameter all other environmental data must be kept constant at an average level (European or regional), as can be seen from Tables 3.2.33 and 3.2.34. The time of wetness (number of days with rain above 2.5 mm) is set to 0 as surfaces exposed to rain usually will not suffer from carbonation-induced corrosion, unless the concrete cover is very low.

Table 3.2.33 Input data for parameter study on carbonation-induced corrosion (initiation phase)

No.	Input	Sub-input	Unit	Distribution type*	Mean Good	Mean Bad	Standard deviation† Good	Standard deviation† Bad	Limits Minimum	Limits Maximum
1	d_{cover}		mm	Beta	15		10		0	100
2	k_{RH}	RH	%	Weibull (maximum)	75 $(40–100)^{\ddagger}$		15		–	100
		RH_{ref}	%	Constant	65		–	–		
		g_{RH}	–	Constant	2.5		–	–		
		f_{RH}	–	Constant	5.0		–	–		
3	k_c	a_c	–	Constant	3.01		–	–		
		b_c	–	Normal	−0.567		0.024	–		
		t_c	–	Normal	1		–	–		
4	k_t		–	Normal	1.25		0.35	–		
5	ε_t		$m^5 (s\,kgCO_2)^{-1}$ $mm^5 (a\,kgCO_2)^{-1}$	Normal	1.0×10^{-11} (315.5)		0.15×10^{-11} (48)	–		

Table 3.2.33 (Continued)

No.	Input	Sub-input	Unit	Distribution type*	Mean		Standard deviation†		Limits	
					Good	Bad	Good	Bad	Minimum	Maximum
6	ΔC_S		$kgCO_2\ m^{-3} \times 10^{-4}$	Normal	$8.2^{\ddagger}$ (6.2–20)#		1.0		–	
7	T		a	Constant	30 or 100		–		–	
8	W	ToW	–	Constant	$0^{\ddagger}$ (0–100)#		–		–	
		b_w	–	Normal	0.446		0.163		–	
		p_{splash}	–	Constant	1		–		–	
		t_0	a	Constant	0.0767 (equals 28d)		–		–	
		a_w	–	Constant	0.5		–		–	
9	$R_{ACC,o}^{-1}$		$m^5\ (s\,kgCO_2)^{-1} \times 10^{-11}$ $mm^5\,(a\,kgCO_2)^{-1}$	Normal	3.1 (982)	13.4 (4243)	1.67 (149)	5.21 (465)	–	

Notes
* CEM I; $w/b = 0.40$.
‡ CEM I; $w/b = 0.60$.
‡ Value chosen as constant for the study of other environmental parameters.
# Range of parameter study.

Table 3.2.34 Input data for parameter study on carbonation-induced corrosion process (propagation phase)

No.	Input	Sub-input	Unit	Distribution type	Mean		Standard deviation		Moments	
					Good	Bad	Good	Bad	Minimum	Maximum
1	α		–	Constant	2		–		–	
2	k_0		$\mu m\,\Omega m\,a^{-1}$	Log-normal	882		–		–	
3	F_{Cl}		–	Shifted log-normal	2.63		3.51		1.09	–
4	F_{Galv}		–	Constant	1.0		–		–	
5	F_{O_2}		–	Constant	1.0		–		–	
6	$\rho(t)$	RH	%	Weibull maximum	75* (40–100)		15		–	100
		k_{RH}	–	Constant	3.16		2.05		–	
		ToW	–	Constant	0* 0–1		(0–1)		–	
		T	K	Beta	283 (273–293)		8		253	313
		b_T	K	Normal	3815		560		–	
		k_c	–	Constant	1.0		–		–	
		k_t	–	Constant	1.0		–		–	
		$k_{R,Cl}$	–	Constant	1.0		–		–	
		n	–	Normal	0.23		0.04		–	
		t_{hydr}	a	Constant	1		–		–	
		t_0	a	Constant	0.0767		–		–	
		ρ_0	$\Omega\ m$	Normal	115	41	14.4	6	–	

Note
* Value chosen as constant for the study of other environmental parameters.

3.2.10.4 Classification of relative humidity RH

A rise in relative humidity slows down the carbonation process but accelerates the subsequent corrosion once initiated. The corrosion penetration depth was used as a measure. The study of the effect revealed the following:

- A maximum of corrosion penetration depth can be observed. On the left side of the maximum corrosion is the limiting process. On the right, carbonation is the dominating factor.
- The relative humidity for which a maximum of corrosion penetration depth is observed depends on the ratio of initiation time to propagation time and hence on the service life considered.
- A change in corrosion rate due to a change in temperature or concrete porosity (not due to a change in relative humidity) alters the maximum value of corrosion penetration depth, but does not shift the pessimum of the relative humidity.
- As the service life considered increases the pessimum of relative humidity is shifted towards larger values (80% after 30a; 88% after 100a), as the corrosion process is given more weight in the study.

Figure 3.2.53 illustrates the corrosion penetration depth of reinforcement against relative humidity at ages of 30 and 100 years, as well as variation of temperature (T) and time of wetness (ToW).

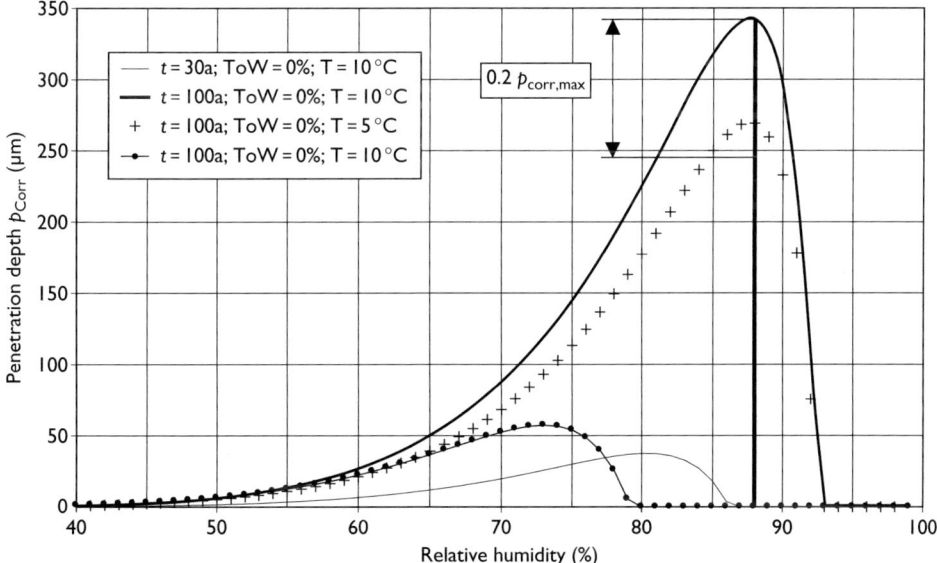

Figure 3.2.53 Corrosion penetration depth of reinforcement versus relative humidity at age of 30 and 100 years and variation of temperature (T) and time of wetness (Tow).

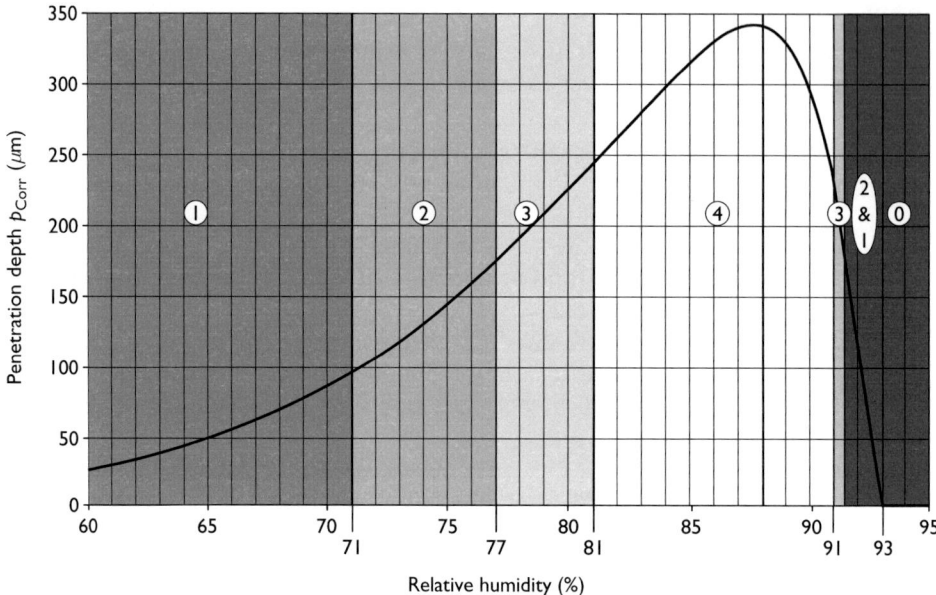

Figure 3.2.54 European classification of average annual relative humidity regarding carbonation induced corrosion, regarding 100 years of service life (Tow) = 0; $T = 10°C$.

For the classification, a period of 100 years was considered. In Europe, the yearly average relative humidity, calculated from hourly values, lies between 60 (Madrid, Spain) and 88% (Fichtelgebirge, Germany). In this study, the range of 60–92% (largest value for which depassivation will still occur within 100 years) was classified. The boundaries of five classes (0–4) were determined by dividing the maximum penetration depth into five equal sections as shown in Figure 3.2.54. In class 4, the severest corrosion damage and in class 0 the lowest extent of corrosion damage are to be expected. A difference of one class in relative humidity class equals 20% of the maximum of penetration depth at 88% relative humidity. The classification of yearly average relative humidity less than 60% and greater than 92% is shown in Table 3.2.35.

The same procedure can be adopted on a national level.

Table 3.2.35 Classification of average annual relative humidity regarding carbonation-induced corrosion

Class	(0)	1	2	3	4
Domain (%)	<60* or >92*	60–71	>71–77	> 77–81 or >90–91.5	>81–90

Note
* Beyond European range.

3.2.10.5 Classification of time of wetness, ToW

Like relative humidity (RH), the time of wetness (ToW) is a parameter with an inverse influence on the initiation (carbonation) phase and the propagation of corrosion. An increase in ToW retards the carbonation process severely, but reduces the resistivity of the concrete, causing an increase in corrosion rate. A study of the effect revealed the following:

- There is no pessimum of the time of wetness for which a maximum in penetration depth can be observed. The largest penetration depth will be obtained for ToW = 0 because carbonation is the dominating process.
- The dependence of penetration depth upon ToW is very sensitive to changes in the relative humidity (Figure 3.2.55).
- For an average relative humidity of 75% on a European scale, corrosion will only occur within a period of $t = 100$ years in a horizontal structure up to a value of ToW = 6.5% (Figure 3.2.56, Table 3.2.36).
- For vertical surfaces the probability of a surfacing being splashed by driving rain must be taken into account.

3.2.10.6 Classification of carbon dioxide concentration, ΔC_s

Although the carbon dioxide concentration is a decisive environmental input parameter, a classification does not seem reasonable as it does not change considerably on

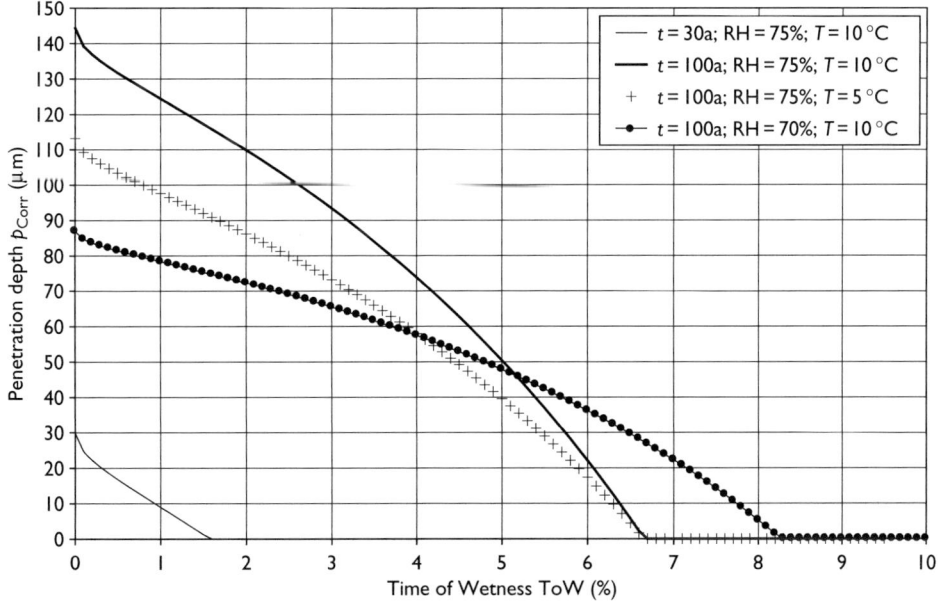

Figure 3.2.55 Corrosion penetration depth of reinforcement versus Time of Wetness (ToW) at age of 30 and 100 years and variation of temperature (T) and relative humidity (RH).

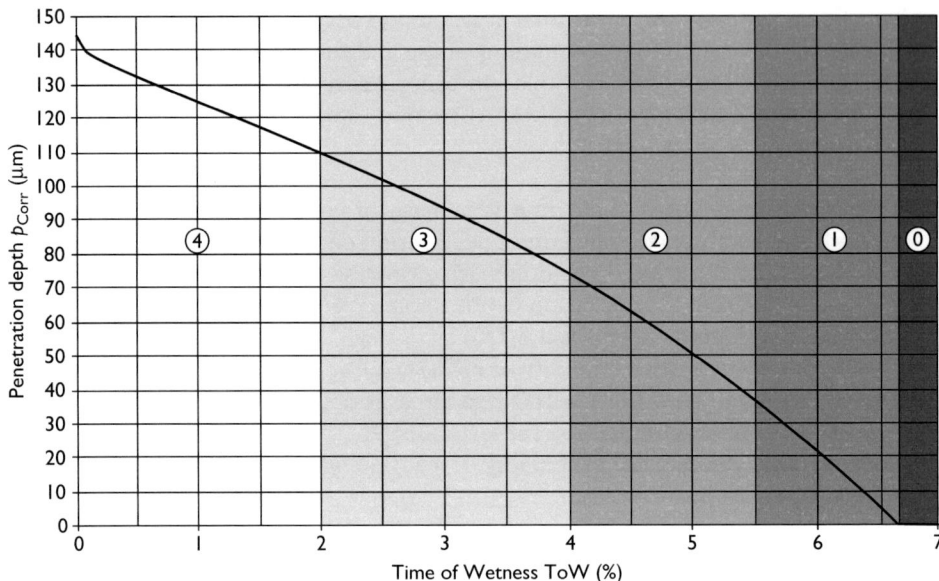

Figure 3.2.56 European classification of average annual time of wetness regarding carbonation induced corrosion.

Table 3.2.36 Classification of average annual time of wetness (ToW) with respect to carbonation-induced corrosion

Class	0	1	2	3	4
Domain (%)	>6.6	5.5–6.6	4.0–5.5	2.0–4.0	0–2.0

a macroenvironmental scale. Instead, it can be considered to be quite constant in the range of 350–380 ppm over Europe. These values may be exceeded in areas with high emissions of combustion gases, especially within such buildings as tunnels. Mapping of the CO_2 concentration is therefore not reasonable. More important for this aspect are models on the meso and micro levels, i.e. models for the determination of the CO_2 content in the vicinity of a building and the considered concrete surface.

3.2.10.7 Classification of temperature, T

The effect of changing temperatures is only included in the model for the progress of corrosion, whereas the carbonation process is considered to be independent of the temperature. A rise in temperature causes a decrease in concrete resistivity, which is modelled by the exponential approach of Arrhenius, and thus an increase in corrosion rate. As can be seen from the parameter study, the effect of other parameters in the models on the classification of the temperature is very low, since the curve of penetration depth versus temperature is only shifted more or less in parallel without changing the gradient of the curve to any great extent. For the classification of

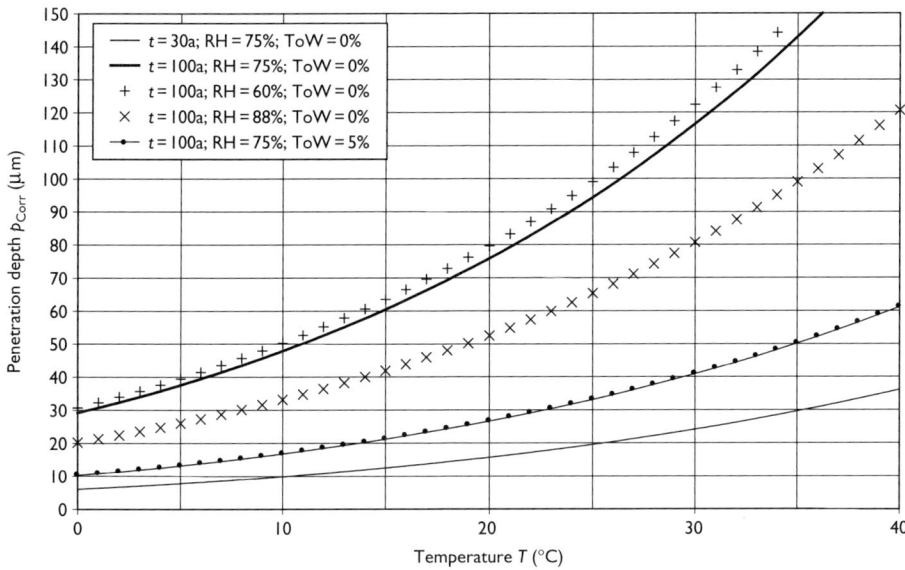

Figure 3.2.57 Corrosion penetration depth of reinforcement versus temperature (*T*) at age of 30 and 100 years under variation of time of wetness (ToW) and relative humidity (RH).

temperature, relative humidity was set to 75%, ToW was set to 0% and the service life considered is 100a (Figure 3.2.57). For classification on a European scale, the range of annual temperature of $T = 0–20°C$ was considered relevant (Figure 3.2.58). The class 0 is not defined, as corrosion will always occur if other environmental parameters are set to default values.

3.2.10.8 Classification of chloride-induced corrosion

3.2.10.8.1 PROCEDURE

The study of parameters given in Table 3.2.38 was performed distinguishing two qualities of concrete:

- Good: blast-furnace slag cement (BFSC)t; $w/c = 0.40$; concrete cover 60 mm
- Bad: OPC, $w/c = 0.60$; concrete cover 45 mm.

Table 3.2.37 Classification of average annual temperature with respect to carbonation-induced corrosion

Class	1	2	3	4
Domain (°C)	0–6.4	6.4–11.6	11.6–16.0	16.0–20.0

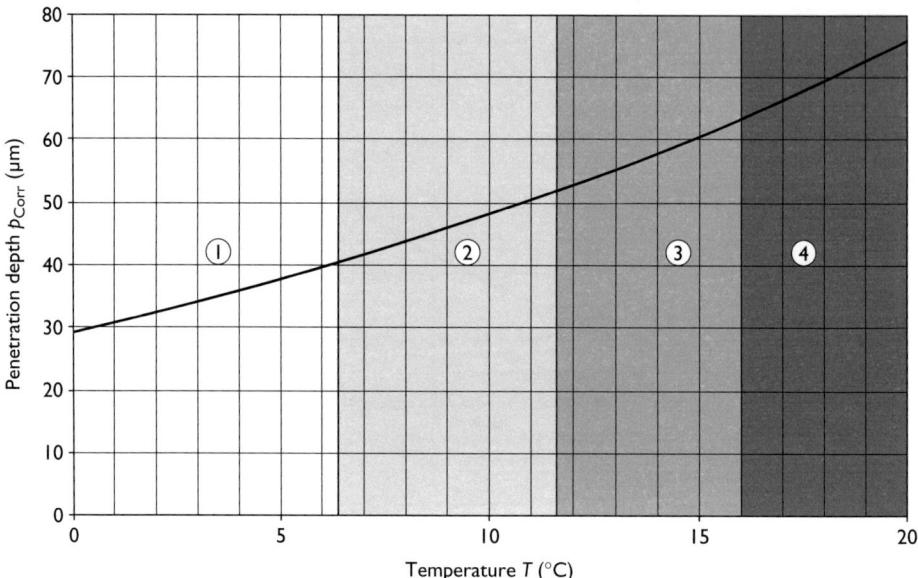

Figure 3.2.58 European classification of average annual temperature regarding carbonation-induced corrosion.

Table 3.2.38 Environmental parameters included in service life models for chloride-induced corrosion

Initiation	Propagation
$k_T(T)$; $C_{S,\Delta x}$; ΔX	$k_{R,T}(T)$; $k_{R,RH}(RH)$, ToW

In the parameter study on carbonation-induced corrosion, a simple spreadsheet with mean values of the model parameters was programmed, including the initiation phase (carbonation of concrete) and the propagation phase (corrosion of reinforcement). The study may also be conducted using a software which enables full-probabilistic parameter studies, as will be demonstrated in the following section. To do so, a limit state criterion must be set.

Here the loss of $Q_{max} = 10\%$ of the cross-sectional area of a bar with $\emptyset = 12\,mm$ was chosen (for which the concrete cover will in most cases already have been cracked for a long time). The choice of the limit state criterion does influence the result of the classification. If a larger value for the section loss is chosen, more weight is given to the propagation phase. A lower value for the limit state criterion gives more weight to the initiation phase. As 10% cross-sectional loss will usually have considerable impact on the residual strength, a value beyond this limit is not considered as reasonable in a management process, as immediate action will most likely have to be taken. With a value of 10% we thus investigate the whole range of the corrosion process managed by a life cycle management system.

3.2.10.8.2 INPUT DATA FOR PARAMETER STUDY

De-icing salt was chosen as the chloride source; it is dependent on the site of the structure. The typical case for which a durability design may be applied is chosen, i.e. a structure close to the street edge (horizontal distance $a = 1.0$ m; height above street level $h = 0.0$ m). The input data for both the initiation and the propagation phase are given in Tables 3.2.39 and 3.2.40, respectively.

3.2.10.9 Classification of temperature, T

The effect of the temperature upon the chloride-induced corrosion is twofold:

- An increase in temperature increases the apparent diffusion coefficient in an exponential manner, leading to a more ingress of chlorides.
- The resistivity of concrete is reduced, which causes a more rapid corrosion process.

Table 3.2.39 Input data for parameter study on chloride-induced corrosion (initiation phase)

No.	Input	Sub-input	Unit	Distribution type	Mean Good*	Mean Bad[†]	Standard deviation Good	Standard deviation Bad	Limits Minimum	Limits Maximum
1	$D_{RCM}(t)$	$D_{RCM,t=28d}$	10^{-12} m^{-2}s^{-1}	Log-normal	1.7	9.2	0.88	4.78	–	–
		$D_{RCM,t=\infty}$	10^{-12} m^{-2}s^{-1}		0.3	4.8	0.22	3.50	–	–
		n_2	–		1.624	0.964	0.438	0.212	–	–
		$k_{w/b}$	–	Log-normal	0.81	2.46	0.12	0.27	–	–
2	$C_{\Delta x}$		Wt %/cem	Log-normal	1.53	1.60	1.15	1.21	–	–
3	k_{RH}		–	Discrete	0.9	0.9	–	–	–	–
4	k_T	b_T	K	Normal	4800	4800	700	700	–	–
		T	K		283[‡] (273–293)[††]		8		253	313
5	N_1		–	Beta	0.55	0.55	0.24	0.24	0	1
6	ΔX		mm	Beta	9.03	9.03	9.03	9.03	0	60
7	C_{Crit}		Wt %/cem	Beta	0.48	0.48	0.15	0.15	0.2	2.0
8	C_{ini}		Weight %/cem	Log-normal	0.108	0.108	0.057	0.057	–	–
9	d_{cover}		mm	Beta	60	60	8	10	0	1000

Notes
* Good quality: CEM III; $w/b = 0.40$.
[†] Bad quality: CEM I; $w/b = 0.60$.
[‡] Value chosen as constant for the study of other environmental parameters.
[††] Range of parameter study.

Table 3.2.40 Input data for parameter study on chloride-induced corrosion process (propagation phase)

No.	Input	Sub-input	Unit	Distribution type	Mean Good*	Mean Bad†	Standard deviation Good	Standard deviation Bad	Moments Minimum	Moments Maximum
1	α		—	Log-normal	9.28		4.04			
2	k_0		μm Ω m a^{-1}	Log-normal	882		—			
3	F_{Cl}		—	Standard-log-normal	2.63		3.51		1.09	—
4	F_{Galv}		—	Constant	1.0		—			
5	F_{O2}		—	Constant	1.0		—			
6	$\rho(t)$	$k_{r,RH}$	—	Discrete	I		—			
		k_T T	K	Beta	283‡ (273–293)#		8		253	313
		b_T	K	Log-normal	3815		560			
		k_c	—	Constant	1.0		—			
		k_t	—	Constant	1.0		—			
		$k_{R,Cl}$	—	Normal	0.72		0.11			
		ρ_0 t_{hydr}	a	Constant	I		—			
		t_0	a	Constant	0.0767		—			
		$\rho_{0,Ref,t=28d}$	Ωm	Log-normal	625	134	100	21		
		$\rho_{0,Ref,t=\infty}$	Ωm	Log-normal	2010	227	704	80		
		t_{hyd}	a	Log-normal	1.5$^{⊕}$	0.5	0.8$^{⊕}$	0.3		
		$k_{w/b}$	—	Discrete	1.12	0.65	—			
	ΔQ_{max}		—	Discrete	0.1		—		—	—
	t		a	Discrete	100		—		—	—

Notes
* Good quality: CEM III; $w/b = 0.40$.
† Bad quality: CEM I; $w/b = 0.60$.
‡ Value chosen as constant for the study of other environmental parameters
# Range of parameter study.
⊕ Value from measurements after around 1.5 years (hydration not yet completed; there approach on safe side).

The severest corrosion problems are, thus, to be expected in warm climates. The time until depassivation t_{ini} cannot be solved directly in the proposed model of chloride ingress. A solution to this problem for the desired parameter study is to:

1. calculate and plot the probability of depassivation $p_{depassivation}(t = 100a, T(K))$ in relation to the temperature T (K)
2. determine the functional relationship for $p_{depassivation}(T)$
3. multiply the probability of depassivation for $p_{depassivation}(T)$ with the model for section loss at $t = 100a$:

$$P(t = 100a, T) = V_{corr} \cdot \alpha \cdot t_{prop} = V_{corr} \cdot \alpha \cdot (100 - t_{ini})$$

$$= V_{corr} \cdot \alpha \cdot 100 \cdot p_{depassivation}(t = 100, T) \qquad (3.2.111)$$

P	penetration depth (mm)
α	pitting factor
t_{prop}	duration of propagation phase (s)
t	age of structure (s)
t_{ini}	time until initiation of corrosion (s)
$p_{depassivation}(t = 100a, T)$	probability of depassivation at $t = 100a$ as a function of temperature.

As expected for bad quality concrete with a low concrete cover, high probabilities of depassivation were calculated in the temperature domain considered, which means that the upper tail of the cumulative probability (probability distribution function) is looked at. For good quality concrete with high concrete cover the opposite is the case, where we look at the bottom tail of the cumulative probability (Figures 3.2.59 and 3.2.60) .

The example above shows how sensitive such parameter analysis is with respect to the chosen boundary conditions, especially concrete composition and concrete cover. It is proposed to apply the average of the boundaries for classes as given for the two extremes of Figure 3.2.60 see also Table 3.2.41.

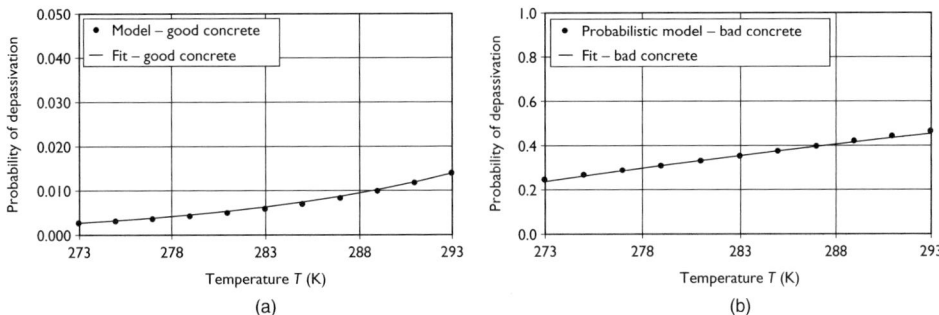

Figure 3.2.59 Cumulative probability of depassivation versus temperature of environment for good quality (a) and bad quality (b) concrete after t = 100 years.

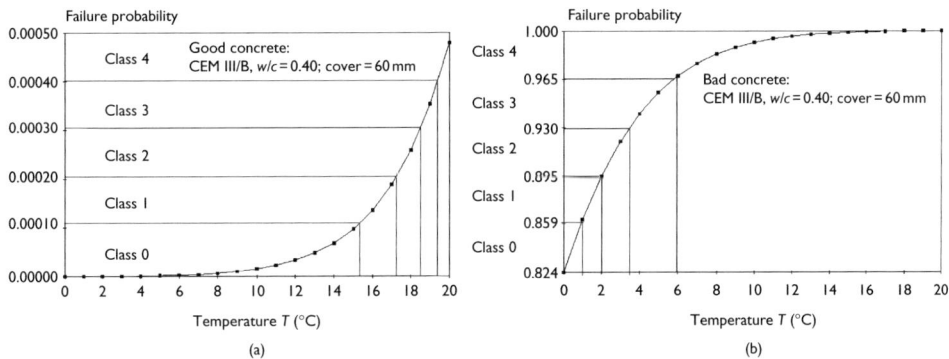

Figure 3.2.60 Cumulative probability for exceeding 10% section loss versus temperature of environment for good quality (a) and bad quality (b) concrete after t = 100 years and resulting classes.

Table 3.2.41 Classification of average annual temperature with respect to chloride-induced corrosion

Class	0	1	2	3	4
Domain (°C)	0–8	8–9.5	9.5–11	11–13	>13

References

[1] LIFECON. 2003. Condition Assessment Protocol. Deliverable D3.1, WP 3, Project G1RD-CT-2000-00378. http://lifecon.vtt.fi/.

[2] Schneider, J. 1996. *Sicherheit und Zuverlässigkeit im Bauwesen, Grundwissen für Ingenieure*, ETH Zürich. ISBN 3-519-15040-9.

[3] EuroCode 1, DIN V ENV. 1991. *Grundlagen der Tragwerksplanung und Einwirkung auf Tragwerke*, Teil 1–4.

[4] Faber, M. H. 2003. Risk and safety in civil, surveying and environmental engineering, Lecture notes, Swiss Federal Institute of Technology, Zürich.

[5] Lay, S. 2003. Service life design of reinforced concrete structures exposed to de-icing salts – A case study. In *Proceedings of 3rd international IABMAS workshop on Life-cycle cost analysis and design of civil infrastructure systems and fib WP 5.3–1, TG 5.6 The Joint Committee on Structural Safety Workshop on Probabilistic Modelling of Deterioration Process in Concrete Structures*. Lausanne.

[6] Rackwitz, R. 1999. Zuverlässigkeitsbetrachtungen bei Verlust der Dauerhaftigkeit von Bauteilen und Bauwerken. In *Kurzberichte aus der Bauforschung* 40(4): 297–301, Stuttgart IRB (1998).

[7] LIFECON. 2003. Generic technical handbook for a predictive life cycle management system of concrete structures (LMS), Deliverable D1.1, Working Party 1, Project G1RD-CT-2000-00378. http://www.vtt.fi/rte/strat/projects/lifecon/.

[8] Brite EuRam III. 1998 (December). Project BE95–1347 'DuraCrete' – Report R4-5, *Modeling of Degradation*.

[9] Schießl, P. 1997. New Approach to durability design: An example for carbonation induced corrsion, Lausanne. In *Bulletin d'Information* 238.

[10] Wierig, H.-J. 1984. Longtime Studies on the Carbonation of Concrete under normal Outdoor Exposure. In *Proceedings of the RILEM seminar on the durability of concrete structures under normal outdoor exposures*, Hannover, 26–29 March, 239–249.

[11] Brite EuRam III. 2000 (January). Project BE95-1347 'DuraCrete' – Report R9, *Statistical quantification of the variables in the limit state functions*.

[12] Brite EuRam III. 1999 (March). Project BE95-1347 'DuraCrete' – Report R8, *Compliance testing for probabilistic design purposes*.

[13] Brite EuRam III 1999 (March). Project BE95-1347 'DuraCrete' – Report R3 *Models for environmental actions on concrete structures*.

[14] Gehlen, C. 2000. Probabilistische Lebensdauerberechnung von Stahlbetonbauwerken – Zuverlässigkeitsbetrachtungen zur wirksamen Vermeidung von Bewehrungskorrosion, Dissertation an der RWTH-Aachen D82 (Diss. RWTH+Aachen), 7.

[15] Umweltbundesamt. 1997. *Daten zur Umwelt – Der Zustand der Umwelt in Deutschland*, 6th edition. Berlin: Erich Schmidt.

[16] IBPS: Ingenieurbüro Professor Schießl, Munich.

[17] Gehlen, C., and C. Sodeikat. 2002. Maintenance planning of reinforced concrete structures: Redesign in a probabilistic environment, inspection, update and derived decision-making. In *Durability of building materials and components, Proceedings of the 9th international conference*, Brisbane, Australia, 17–20 March.

[18] Profometer by Proceque, Switzerland, distributed by Form+Test in Germany.

[19] Gehlen, C. 2000. Probabilistische Lebensdauerbemessung von Stahlbetonbauweken – Zuverlässigkeitsbetrachtungen zur Wirksamen Vermeidung von Bewehrungskorrosion. *Schriftenreihe des DAfStb* 510.

[20] STRUREL: Software for structural reliability analysis, RCP GmbH, Munich, Germany.

[21] LIFECON. 2003. 'GIS based national exposure modules and national reports on quantitative environmental degradation loads for chosen objects and locations', Deliverable D4.3, Working Party 4, Project G1RD-CT-2000-00378. http://lifecon.vtt.fi/.

[22] Bergmann-Schaefer. *Lehrbuch der Experimental-Physik*, vol. 1, pp. 416–437.

[23] Dhir, R. K. *et al*. 1998. Prediction of chloride content profile and concentration – Time dependent diffusion coefficients for concrete, In *Magazine of Concrete Research*, No. 1, pp. 37–48.

[24] Volkwein, A. 1993. Untersuchungen über das Eindringen von Wasser und Chlorid in Beton. In *Beton- und Stahlbetonbau* 88 (8): 223–226.

[25] Tang, L., and L. O. Nilsson. A numerical method for prediction of chloride penetration into concrete structures. In *The modelling of microstructures and its potential for studying transport properties and durability*, The Netherlands: Kluwer Academic Publishers, 539–552.

[26] Tang, L., and L. O. Nilsson. 1993. Chloride binding capacity and binding isotherms of OPC pastes and mortars. In *Cement and Concrete Research* 23 (2): 247–253.

[27] Andersen, A. 1996. Investigations of chloride penetration into bridge columns exposed to de-Icing salts, HETEK Report No. 82, Danish Road Directorate.

[28] Nilsson, L. O. 2000. A test of prediction models for chloride ingress and corrosion initiation. In *Proceedings Nordic mini seminar on prediction models for chloride ingress and corrosion initiation*.

[29] Nilsson, L. O. 2000. On the uncertainty of service-life models for reinforced marine concrete structures. In *RILEM international workshop on life prediction and aging management of concrete structures*, 16–17 October, Cannes, France.

[30] Collepardi, M., A. Marcialis, and R. Turriziani. The Kinetics of chloride ions penetration in concrete. *Il Cemento* 67: 157–164.

[31] Takewaka, K., and S. Matsumoto. *Quality and cover thickness of concrete based on the estimation of chloride penetration in marine environments*. Detroit, USA: American Concrete Institute, ACI SP109-17, 381–400.

[32] Helland, S. 2001. The 'Selmer' chloride ingress model applied for the case of the Nordic Mini Seminar. In *Proceedings of Nordic mini seminar*, Göteborg.

[33] Mejlbro, L., and E. Poulsen. 2000. On a model of chloride ingress into concrete exposed to de-icing salt containing chloride. In *Proceedings of 2nd international workshop – Testing and modelling the chloride ingress into concrete, Paris*, 11–12 September, 337–354.

[34] LIFE-365. 2001. The Life-365 model. In *Proceedings of Nordic mini seminar – Prediction models for chloride ingress and corrosion initiation in concrete structures*, 22–23 May, Göteborg, Sweden.

[35] Snyder, K. A. 2001. Validation and modification of the 4SIGHT computer program, National Institute of Standards and Technology (NIST), NISTIR 6747, May.

[36] HETEK. 1997. Theoretical background: A system for estimation of chloride ingress into concrete, Report No. 83.

[37] Johannesson, B. 1998. Modelling of transport process involved in service life prediction of concrete – Important principles, Report TVBM-3083, Thesis, Lund Institute of Technology, Division of Building Materials.

[38] Lay, S., S. Zeller, and P. Schießl. 2003. Time dependent chloride migration coefficient of concrete as input to a probabilistic service life model. In *Proceedings of International workshop on management of durability in the building process*, Milano.

[39] Frederiksen, J. M., and M. Geiker. 2000. On an empirical model for estimation of chloride ingress into concrete. In *Proceedings of 2nd international RILEM workshop on testing and modelling the chloride ingress into concrete*, Pairs.

[40] DIN EN 197-1. 2000. Composition, specifications and conformity for common cements, Berlin.

[41] Krogbeumker, G. 1971. Beitrag zur Beurteilung des Zementsteingefüges in Abhängigkeit von der Mahlfeinheit, dem Wasserzementwert und der Hydratationstemperatur, Dissertation an RWTH Aachen.

[42] Schießl, P., and U. Wiens. 1995. Rapid determination of chloride diffusivity in concrete with blending agents. In *RILEM international workshop on chloride penetration into concrete*. St.Rémy-lès-Chevreuse, France.

[43] Volkwein, A. 1993. Untersuchungen über das Eindringen von Wasser und Chlorid in Beton. In *Beton- und Stahlbetonbau* 88 (8): 223–226.

[44] Zhang, T., and O. E. Gjorv. 1995. Diffusion behaviour of chloride ions in concrete. In *Chloride penetration into concrete, Proceedings of the International RILEM workshop*, 53–63.

[45] Bamforth, P. B. 1995. A new approach to the analysis of time-dependent changes in chloride profiles to determine effective diffusion coefficients for use in modelling of chloride ingress. In *Chloride penetration into concrete, Proceedings of the international RILEM workshop*, St.Rémy-lès-Chevreuse, France, 195–205.

[46] HETEK. 1996. Chloride penetration into concrete – State-of-the-art, Report No. 53, Danish Road Directorate.

[47] EuroLightCon. 1999. Chloride penetration into concrete with lightweight aggregates – European Union-Brite EuRam III. *Economic design and construction with light weight aggregate concrete*, Document BE96-3942/R3.

[48] Stanish, K., and M. Thomas. 2003. The use of bulk diffusion tests to establish time-dependent concrete chloride diffusion coefficients. In *Cement and Concrete Research* 33: 55–62.

[49] Climent, M. A., *et al.* 2000. Transport of chlorides through non saturated concrete after an initial limited chloride supply. *Proceedings of 2nd RILEM workshop: Testing and modelling the chloride ingress into concrete*, Paris.

[50] Nilsson, L. O., *et al.* 2000. *Chloride ingress data from field exposure in a Swedish road environment*. Göteborg: Chalmers University of Technology, Publication P-00: 5.

[51] Breitenstein, J. 1995. Entwicklung einer Kenngröße der Winterlichkeit zur Bewertung des Tausalzverbrauchs. *Berichte der Bundesanstalt für Straßenwesen, Verkehrstechnik Heft V 18 (3)*. ISBN 3-89429-603-8.

[52] Paulsson-Tralla, J. 1999. Service life of repaired bridge decks. *Stockholm, KTH Trita-BKN Bulletin 50*. ISBN-99-3024703-3.

[53] Tang, L. 1996. Chloride transport in concrete – Measurement and prediction. Thesis, Gothenburg: Chalmers University of Technology.

[54] Comité Euro-International du Béton (CEB), Schießl P. 1989. Durable concrete structures CEB design guide, 2nd edition. In *Bulletin d'information no. 182*.

[55] Breit, W. 1997. Untersuchungen zum kritischen korrosionsauslösenden Chloridgehalt für Stahl in Beton. In *Schriftenreihe Aachener Beiträge zur Bauforschung, Institut für Bauforschung der RWTH-Aachen*, Thesis, no. 8.

[56] Nustad, G. E. 1994. Production and use of standardised chloride bearing dusts for the calibration of equipment and procedures for chloride analysis. In *Proceedings of corrosion and corrosion protection of steel in concrete international conference* 24–28 July, University of Sheffield, ed. R. N. Swamy, 515–526. Sheffield: Sheffield Academic Press.

[57] Dorner, H. 1986. Ringanalyse zur quantitativen und halbquantitativen Bestimmung des Chloridgehaltes von Beton. *Schlussbericht zum Forschungsauftrag des Bundesministers für Verkehr*, FA-No. 15.134.R 84 H, Bonn.

[58] Lay, S., S. Liebl, H. Hilbig, and P. Schießl. 2005. New method to measure the rapid chloride migration coefficient of chloride containing concrete. *Cement and Concrete (akzeptierter Beitrag)*.

[59] Raupach, M. 2002. Smart structure: Development of sensors to monitor the corrosion risk for the reinforcement of concrete bridges. In *Proceedings of the first international conference on bridge maintenance, safety and management, IABMAS*, July.

[60] Gehlen, C., and G. Pabsch. 2003. Updating through measurement and incorporation of sensitive variables. In *DARTS-durable and reliable tunnel structures*. European Commission, Growths 2000, Contract G1RD-CT-2000-00467, Project GrD1-25633.

[61] Andrade, C., C. Alonso, and A. Arteaga. 1997. Models for predicting corrosion rates. Brussels: Brite-EuRam. Project No. BE95-1347.

[62] Raupach, M. 1992. Zur chloridinduzierten Makroelementkorrosion von Stahl in Beton. In *Beuth Verlag, DAfStB Heft* 433, Berlin.

[63] Cairns, J. 1998. Assessment of effects of reinforcement corrosion on residual strength of deteriorating concrete structures. In *Proceedings of the first international conference on behaviour of damaged structures*, Rio de Janerio, May. Rio de Janerio: Federal University of Fluminense, Niteroi.

[64] Haagenrud, S. E., and Krigsvoll, G. 2004. Instructions for quantitative classification of environmental degradation loads onto structures. LIFECON.EU GROWTH Program, G1RD-CT-2000-00378. Deliverable D4.2. Norwegian Building Research Institute NBI, 2004. http://lifecon.vtt.fi/.

3.3 Deterministic methodology and models of serviceability and service life design

Asko Sarja

3.3.1 Lifetime safety factor method for durability

In practice, it is reasonable to apply the lifetime safety factor method in the design procedure for durability, which was for the first time presented in the report of RILEM TC 130-CSL [1, 2]. The lifetime safety factor method is analogous with the static limit state design. The durability design using the lifetime safety factor method is related to controlling the risk of falling below the target service, while static limit state design is related to controlling the reliability of the structure against failure under external mechanical loading.

The durability design with lifetime safety factor method is always combined with static or dynamic design, and aims to control the serviceability and service life of a new or existing structure, while static and dynamic design controls the loading capacity. Durability limit states, design life and reliability principles are treated as described in Section 2.4.

3.3.2 Calculation of the lifetime safety factor

The calculation of lifetime safety factor is based on the reliability modelling presented in Section 2.4 and in the methodology of RILEM TC 130-CSL [1].

Performance behaviour can always be translated into degradation behaviour. By definition, degradation is a decrease in performance. The transformation is performed by the following substitutions:

$$R(0) - R(t) = D(t) \qquad (3.3.1)$$

$$R(0) - S = D_{max}$$

Or

$$R(0) - R_{min} = D_{max}$$

Let us consider that the degradation function is of the following form:

$$D(t) = a \cdot t^n \qquad (3.3.2)$$

where

- $D(t)$ is the mean value of degradation
- a the constant coefficient
- t time
- n the degradation mode coefficient.

The exponent n may in principle vary between $-\infty$ and $+\infty$. The values of n are defined as follows:

- for acelerating degradation process: $n > 1$
- for declerating degradation process: $n < 1$
- for linear degradation process: $n = 1$

The coefficient a is fixed when the mean service life is known:

$$a = D_{max}/\mu_{tL} \qquad (3.3.3)$$

Degradation is assumed to be normally distributed around the mean. It is also assumed that the standard deviation of D is proportional to the mean degradation, the coefficient of variation being constant, V_D. Figure 3.3.1 shows the degradation as a function of time.

The safety index β of standard normal distribution can be expressed as a function of mean values of R and S, and standard deviation of the difference $R_0 - S_0$, as follows:

$$\beta = \frac{\mu_R - \mu_S}{\sqrt{V_R^2 + V_S^2}} \qquad (3.3.4a)$$

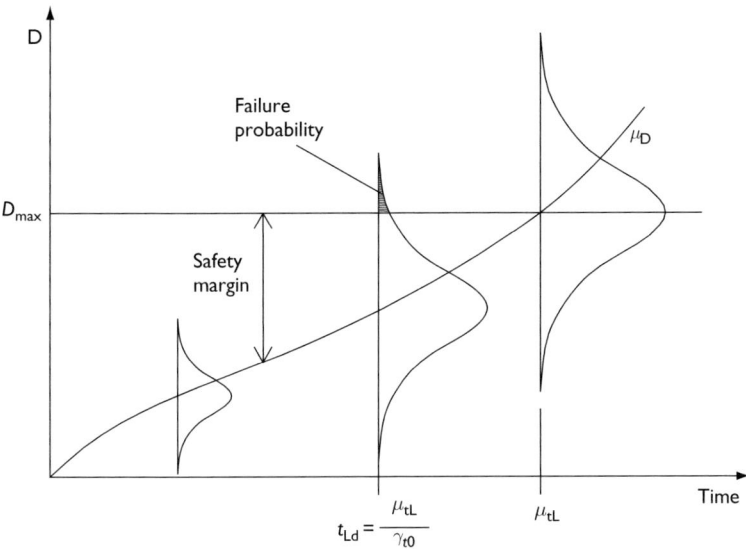

Figure 3.3.1 The meaning of lifetime safety factor in a degradation process.

In the degradation models we apply the statistical bases only for the capacity, because the environmental load is defined only as classified magnitudes. Applying into the degradation, and assuming S to be constant we get an estimate

$$\beta = \frac{D_{max} - D_g}{V_D D_g} = \frac{\left(\frac{D_{max}}{D_g} - 1\right)}{V_D} \tag{3.3.4b}$$

where

D_{max} is the maximum allowable degradation
D_g the mean degradation at t_g
V_D the coefficient of variation of degradation.

From Figure 3.3.1 and from Equation (3.3.4b) we get:

$$\frac{D_{max}}{D_g} = \frac{\left(\gamma_{t_0} t_g\right)^n}{\left(t_g\right)^n} = \gamma_t^{\,n} \tag{3.3.5}$$

By assigning this to Equation (2.8b), we obtain the central lifetime safety factor and mean value of the design life:

$$\gamma_{t_0} = (\beta \cdot V_D + 1)^{1/n} \tag{3.3.6}$$

$$t_{Ld} = \mu_{tL}/\gamma_{t0}$$

where

t_{Ld} the design life
μ_{tL} the mean value of the service life
β the safety index
V_D the coefficient of variation of the degradation.

The lifetime safety factor depends on statistical safety index β (respective to the maximum allowable failure probability at t_g), the coefficient of variation of D ($= V_D$) and the exponent n. Thus the lifetime safety factor is not directly dependent on design life (target service life) t_g itself.

If the degradation process is accelerating, $n > 1$. In the case of decelerating degradation $n < 1$. In the case of linear degradation process $n = 1$. The selection of the value of n can be done when knowing the degradation model. Often the degradation process in the degradation models is assumed to be linear. In these cases, or always when no exact information on the degradation process is known, the value $n = 1$ can be used.

The mean design life can be transformed into characteristic design life with the equation:

$$t_k = t_0(1 - kV_t) \tag{3.3.7}$$

$$t_{Ld} = t_{Lk}/\gamma_{tk} = \mu_{tL}/\gamma_{t_0}$$

where

t_{Lk} the characteristic service life
μ_{tL} the mean value of the service life
t_{Ld} the design life
k a statistical factor depending on the statistical reliability level expressed as a fractile of the cases under the characteristic value (usually the fractile is 5%)
V_t the coefficient of variation of the service life (if not known, an estimate V_D in the range 0.15–0.30 can usually be used).

The characteristic lifetime safety factor γ_{tk} can be calculated with the equation

$$t_{Ld} = t_{Lk}/\gamma_{tk} = \mu_{tL}/\gamma_{t_0} \tag{3.3.8}$$

$$\gamma_{tk} = \gamma_{t_0} \cdot t_{Lk}/\mu_{tL} = (\beta V_D + 1)^{1/n*}(1 - 1.645V_{tLd})$$

where

β the safety index,
V_D the coefficient of variation of the degradation (usually 0.2–0.4), and
V_{tLD} the coefficient of variation of the design life (usually 0.15–0.30).

In Equation (3.3.6) we obtain that $\sigma(t_d) = \sigma(V_D)$. This means that

$$V_{tLD} = V_D/\gamma_{t_0} = V_D/(\beta \cdot V_D + 1)^{1/n} \tag{3.3.9}$$

Assuming again that $n = 1$, we get the values of central and characteristic lifetime safety factors, which are presented in Table 3.3.1. Examples of central and

Table 3.3.1 Central and characteristic safety factors as function of reliability index and degradation coefficient of variation

Degradation C. o. V_{VD}	Service C. o. $V V_{td}$	Life Central Safety Fac γ_{t_o}	Character Safety Fac γ_{tk}
0.20	0.10	2.04	1.71
0.25	0.11	2.30	1.89
0.30	0.12	2.56	2.07
0.35	0.12	2.82	2.24
0.40	0.13	3.08	2.42
0.50	0.14	3.60	2.78
0.60	0.15	4.12	3.13
0.20	0.10	1.94	1.61
0.25	0.11	2.18	1.76
0.30	0.12	2.41	1.92
0.35	0.13	2.65	2.07
0.40	0.14	2.88	2.22
0.50	0.15	3.35	2.53
0.60	0.16	3.82	2.83
0.20	0.11	1.86	1.53
0.25	0.12	2.08	1.66
0.30	0.13	2.29	1.80
0.35	0.14	2.51	1.93
0.40	0.15	2.72	2.06
0.50	0.16	3.15	2.33
0.60	0.17	3.58	2.59
0.20	0.11	1.84	1.51
0.25	0.12	2.05	1.64
0.30	0.13	2.26	1.77
0.35	0.14	2.47	1.89
0.40	0.15	2.68	2.02
0.50	0.16	3.10	2.28
0.60	0.17	3.52	2.53
0.20	0.11	1.76	1.43
0.25	0.13	1.95	1.54
0 30	0.14	2.14	1.65
0.35	0.15	2.33	1.75
0.40	0.16	2.52	1.86
0.50	0.17	2.90	2.08
0.60	0.18	3.28	2.29
0.20	0.12	1.66	1.33
0.25	0.14	1.83	1.41
0.30	0.15	1.99	1.50
0.35	0.16	2.16	1.58
0.40	0.17	2.32	1.66
0.50	0.19	2.65	1.83
0.60	0.20	2.98	1.99
0.20	0.13	1.58	1.25
0.25	0.14	1.73	1.31
0.30	0.16	1.87	1.38
0.35	0.17	2.02	1.44
0.40	0.19	2.16	1.50
0.50	0.20	2.45	1.63
0.60	0.22	2.74	1.75

Table 3.3.1 (Continued)

Degradation C. o. V_{VD}	Service C. o. V V_{td}	Life Central Safety Fac γ_{t_o}	Character Safety Fac γ_{tk}
0.20	0.15	1.30	1.00
0.25	0.18	1.38	1.00
0.30	0.21	1.45	1.00
0.35	0.23	1.53	1.00
0.40	0.25	1.60	1.00
0.50	0.29	1.75	1.00
0.60	0.32	1.90	1.00

characteristic safety factors for different limit states and reliability classes in the cases $V_D = 0.3$ and $V_D = 0.4$ are presented in Table 3.3.2. In practice, it is recommended to use the characteristic values of the parameters, because they are used also in the static and dynamic calculations.

3.3.3 The procedure from environmental loadings into limit states

The environmental loadings are described as exposure classes, following the classification of European Standard EN 206–1 [4].

A summary of actual degradation factors, processes and performance limit states for design as well as for maintenance and repair planning for durability is presented in Table 3.3.3 [1].

A designer must determine which degradation factors are decisive for service life. Preliminary evaluations of rates of degradation for different factors may be necessary. The models presented in the report may be applied in these evaluations.

The following degradation factors are dealt with [1]:

1. corrosion due to chloride penetration
2. corrosion due to carbonation
3. mechanical abrasion
4. salt weathering
5. surface deterioration
6. frost attack.

Additionally there exist some internal degradation processes, such as alkaline-aggregate reaction, but they are not treated here as they can be solved by a proper selection of raw materials and an appropriate design of concrete mix.

Degradation factors affect either the concrete or the steel or both. Usually degradation takes place on the surface zone of concrete or steel, gradually destroying the material. The main structural effects of degradation in concrete and steel are the following:

1. loss of concrete leading to reduced cross-sectional area of the concrete;
2. corrosion of reinforcement leading to reduced cross-sectional area of steel bars.

Table 3.3.2 Central and characteristic safety factors in the cases $V_D = 0.3$ and $V_D = 0.4$. An application of EN 1990:2002 [3]

Reliability Class/Consequence Class	Safety index β		Lifetime safety factor							
			Central safety factor (γ_0)				Characteristic safety factor (γ_{bk})			
			1 year reference period		50 years reference periods		1 year reference period		50 years reference periods	
	1 year reference period	50 years reference periods	$V_D = 0.3$	$V_D = 0.4$	$V_D = 0.3$	$V_D = 0.4$	$V_D = 0.3$	$V_D = 0.4$	$V_D = 0.3$	$V_D = 0.4$
Ultimate limit stress										
RC3/CC3*	5.2	4.3	2.56	3.08	2.29	2.72	2.07	2.42	1.80	2.06
RC2/CC2†	4.7	3.8	2.41	2.88	2.14	2.52	1.92	2.22	1.65	1.86
RC1/CC1‡	4.2	3.3	2.26	2.68	1.99	2.32	1.77	2.02	1.50	1.66
Serviceability limit states										
RC3/CC3	No general recommendations. Will be evaluated in each case separately									
RC2/CC2	2.9	1.5	1.87	2.16	1.45	1.60	1.38	1.50	—	—
RC1/CC1	1.5	1.5	1.45	1.60	1.45	1.60	—	—	—	—

Notes
* High consequence for loss of human life, or economic, social or environmental consequences very great.
† Medium consequence for loss of human life, or economic, social or environmental consequences considerable.
‡ Low consequence for loss of human life, or economic, social or environmental consequences small or negligible.

Table 3.3.3 Typical durability-related performance limit states of concrete structures [5–9]

Degradation factor	Process	Degradation	Limit states Serviceability	Ultimate
Mechanical				
Static loading	Stress, strain, deformation	Deflection, cracking, failure	Deflection, cracking	Failure
Cyclic or pulsating loading	Fatigue, deformation	Reduced strength, cracking, deflection, failure	Deflection, cracking	Fatigue, failure
Impact loading	Peak loading, repeated impact, mass forces	Increase of load vibration, deflection, cracking, failure	Deflection, cracking, vibration	Failure
Physical				
Temperature changes	Expansion and contraction	Shortening, lengthening, cracking at restricted deformation	Surface cracking, surface scaling	
Relative humidity (RH) changes	Shrinkage, swelling	Volume changes, shortening and lengthening, surface cracking, surface scaling, structural cracking in case of restricted deformation	Surface cracking, surface scaling, structural cracking	
Freezing–melting cycles	Ice formation, ice pressure, swelling and shrinking	Cracking, disintegration of concrete	Surface cracking, surface scaling, strength weakening	Decrease of ultimate capacity
Combined de-icing – freezing–melting cycles	Heat transfer, salt-induced swelling and internal pressure	Cracking of concrete, scaling of concrete	Surface cracking, surface scaling	
Floating ice	Abrasion	Cracking, scaling	Surface cracking, surface scaling, surface abrasion	
Traffic	Abrasion	Rutting, wearing, tearing	Surface abrasion	
Running water	Erosion	Surface damage	Surface abrasion, surface scaling	
Turbulent water	Cavitation	Caves	Surface scaling, weakening of concrete	Decrease of ultimate capacity

Chemical

Soft water	Leaching	Disintegration of concrete	Surface abrasion, surface cracking, surface scaling	
Acids	Leaching, Neutralisation of concrete	Disintegration of concrete, depassivation of steel	Surface abrasion, surface cracking, surface scaling, steel corrosion	Decrease of ultimate capacity
Carbon dioxide	Carbonation of concrete	Depassivation of steel	Steel corrosion	Decrease of ultimate capacity
Sulphur dioxide	Sulphate reactions, formation of acids	Disintegration of concrete	Surface cracking, surface scaling, weakening of concrete	Decrease of ultimate capacity
Nitrogen dioxide	Formation of acids	Disintegration of concrete	Surface cracking, surface scaling, weakening of concrete	Decrease of ultimate capacity
Chlorides	Penetration, destruction of passive film of steel	Depassivation of steel, stress corrosion of steel	Steel corrosion, secondary effects: surface cracking, surface scaling	Decrease of ultimate capacity
Oxygen+water	Corrosion of depassivated steel	Loss of cross-sectional area of reinforcing steel, internal pressure in concrete due to expansion of steel, weakening of the steel surface	Surface cracking, surface scaling, changes of aesthetic colour of surface	Decrease of ultimate capacity due to loss of cross-sectional area of steel and loss of bond between reinforcing steel and concrete
Sulphates	Crystal pressure	Disintegration of concrete	Cracking, scaling, weakening of concrete	Decrease of ultimate capacity
Silicate aggregate, alkalis	Silicate reaction	Expansion, disintegration	Cracking, scaling, weakening of concrete	Decrease of ultimate capacity
Carbonate aggregate	Carbonate reaction	Expansion, disintegration	Cracking, scaling, weakening of concrete	Decrease of ultimate capacity

Biological

Micro-organisms	Acid production	Disintegration of concrete, depassivation of steel	Surface abrasion, surface cracking, surface scaling, steel corrosion	Decrease of ultimate capacity

Table 3.3.3 (Continued)

Degradation factor	Process	Degradation	Limit states Serviceability	Ultimate
Plants	Penetration of roots into concrete	Internal pressure, growing micro-organisms	Surface cracking, surface scaling	
Animals	Mechanical surface loading	Abrasion	Surface abrasion, surface scaling	
People	Painting of surfaces, impact and abrasion loading of surfaces	Penetration of colours into pores, abrasion	Aesthetic change of surface, scaling of surface	

Corrosion may occur at cracks at all steel surfaces, assuming that the corrosion products are able to leach out through the pores of the concrete (general corrosion in wet conditions). Splitting and spalling of the concrete cover due to general corrosion of reinforcement lead to:

- a reduced cross-sectional area of the concrete
- a reduced bond between concrete and reinforcement
- visual unfitness.

3.3.4 Application of factor method into environmental loads

The classifications described above do not always directly show the impact of the environmental loads in quantity. The method of ISO standard ISO/DIS 15686–1, which is called the 'factor method', can be applied in calculating the service life (design life) in specific conditions [10]. The factor method includes the following factors:

A: quality of components
B: design level
C: work execution level
D: indoor environment
E: outdoor environment
F: in-use conditions
G: maintenance level.

Estimated service life (ESLC) is calculated with the equation: ESLC = RSLC × factorA × factorB × factorC × factorD × factorE × factorF × factorG

where
RSLC   the reference service life.

For the purpose of reliability-based durability design this is applied in the form:

$$t^*_{Ld} = D \cdot x \cdot E \cdot x \cdot t_{Ld} \tag{3.3.10}$$

where
t^*_{Ld} the modified design life
D the indoor environmental load intensity factor, and
E the outdoor environmental load intensity factor.

The reference service life is a documented period, in years, that the component or assembly can be expected to last in a reference case under certain service conditions. It may be based on:

- service life calculation models, which are described above
- databased on experiments, theoretical calculations or combinations of these; provided by a manufacturer, a test house or an assessment regime; building codes may also give typical service life of components.

The modifying factors – the indoor environmental load intensity factor D and the outdoor environmental load intensity factor E – are in some cases included in service life models. This is the case in most of the Lifecon/Probabilistic service life models (Lifecon Deliverable D3.2) and Lifecon/RILEM TC130 CSL (Lifecon Deliverable D2.1) models [6].

The factor D is a deviation from assumed indoor conditions. Often, especially in buildings, the indoor environmental load is very small, and need not be calculated. However, in some buildings – factories, for example – the environmental load can be extremely high, because of acids or other chemical emissions resulting from the chemical processes.

Factor E means usually the environmental load of local level, but also the load of structural level, for example the direction of the surface (horizontal/vertical/inclined), the compass point (often south is more loading), salt-spray zone, etc. Factor E can be used also for combination of environmental loads (e.g. combination of wetting and freezing).

Usually the values of the factors are either 1 or vary between 0.8 and 2. In extreme conditions the values can be even higher.

3.3.5 Degradation models

The damages are determined using the design life, t_{Ld}. Selected calculation models are presented in the appendix of the TC 130-CSL report [1].

A designer must determine which degradation factors are decisive for service life. Preliminary evaluations of rates of degradation for different factors may be necessary. In the Lifecon system the following three groups of degradation models are presented in separate reports:

1. Probabilistic service life models (Lifecon Deliverable D3.2)
2. RILEM TC 130 CLS models (Lifecon Deliverable D2.1) [1]
3. Reference structure method (Lifecon Deliverable D2.2).

Characteristic properties of these models are as follows:

- Probabilistic service life models are based on physical and chemical laws of thermodynamics, and thus have a strong theoretical base. They include parameters which have to be determined with specific laboratory or field tests. Therefore, some equipment and personnel requirements exist for the users. The application of the probabilistic service life models method raises the need for a statistically sufficient number of tests. Statistical reliability method can be directly applied with these models.
- RILEM TC 130 CLS models are based on parameters which are available from the mix design of concrete. The asset of these models is the availability of the values from the documentation of the concrete mix design and of the structural design.
- Reference structure method is based on statistical treatment of the degradation process and condition of real reference structures, which are in similar conditions and own similar durability properties with the actual objects. This method is suited in the case of a large network of objects, for example bridges. It can be combined with Markovian Chain method in the classification and statistical control of the condition of structures.

Because of the openness principle of Lifecon LMS, each user can select the best-suited models for their use. There are certainly a lot of other suitable models, and new models are under development. They can be used in Lifecon LMS after careful validation of their suitability and reliability. Special attention has to be paid to the compatibility of the entire chain of the procedure of reliability calculations.

The main criteria in selecting the degradation model for each specific use are, for example:

- availability of statistical data of variables of each model
- availability of data or testing method for the coefficients of each model
- accuracy of the model when using the available data in relation to the required accuracy level
- costs of IT tools and the work in calculations.

Some of these criteria can be evaluated roughly beforehand, but often some comparative test calculations are needed.

3.3.6 Calculation procedure and phases

General phases of the service life and durability are as follows [1, 11]:

1. specification of the target service life and design service life;
2. analysis of environmental effects;
3. identification of durability factors and degradation mechanisms;
4. selection of a durability calculation model for each degradation mechanism;
5. calculation of durability parameters using available calculation models;

6. possible updating of the calculations of the ordinary mechanical design; and
7. transfer of the durability parameters into the final design.

The phases are presented as a schedule in Figure 3.3.2. The content of the phases of durability design procedure is as follows:

Phase 1: Specification of the design life   The design life is defined corresponding to the requirements given in common regulations, codes and standards in addition to possible special requirements of the client. Typical classes of design life are 10, 25, 50, 75, 100, etc. years. The safety classification of durability design is presented in Table 3.3.2.

   The calculated design life is compared with the required design life (also called target service life) using the formula:

$$t_{Ld} = \mu_{t_L}/\gamma_{t0} > \text{design life (required service life)} \qquad (3.3.11a)$$

$$t_{Ld} = t_{Lk}/\gamma_{tk} \geq \text{required service life}$$

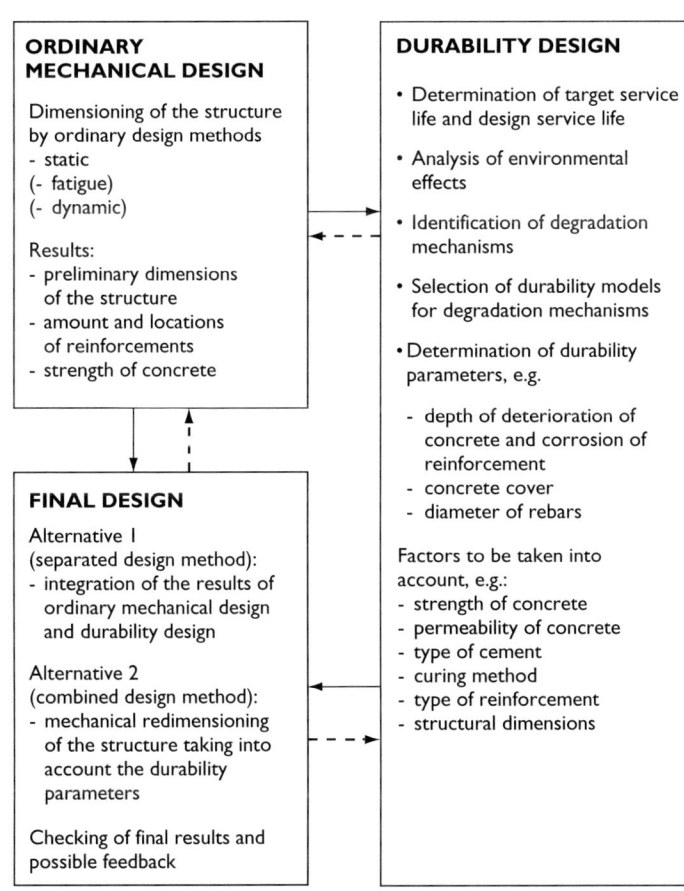

Figure 3.3.2 Flow chart of the durability design procedure [1].

Applying the environmental load intensity factors [7] of Equation (3.3.8) we get the final result:

$$t_{Ld}^* = D \times E \times t_{Ld} \qquad (3.3.11b)$$

where

t_{Ld} the design life
μ_{t_L} the calculated or experimental mean value of the service life
t_{Lk} the calculated or experimental characteristic value of the service life (5% fractile)
γ_{t_0} the central lifetime safety factor
γ_{t_k} the characteristic lifetime safety factor.

t_{Ld}^* the modified design life,
D the indoor environmental load intensity factor, and
E the outdoor environmental load intensity factor.

In some cases the environmental intensity factors are included in service life models. This is the case in most of the Lifecon/Probabilistic service life models (Lifecon Deliverable D3.2) and Lifecon/RILEM TC130-CSL models (Lifecon Deliverable D2.1) [6].

Phase 2: Analysis of environmental loads The analysis of environmental effects includes identification of the climatic conditions such as temperature and moisture variations, rain, condensation of moisture, freezing, solar radiation and air pollution, and the identification of geological conditions such as the location of ground water, possible contact with sea water, contamination of the soil by aggressive agents like sulphates and chlorides. Man-made actions such as salting of roads, abrasion by traffic, etc., must also be identified.

Phase 3: Identification of degradation factors and degradation mechanisms Based on the environmental effect analysis, the designer identifies the degradation factors to which the structure will most likely be subjected. Some kind of degradation process is usually assumed to take place in both the concrete and the reinforcement.

Phase 4: Selection of durability models for each degradation mechanism A designer must determine which degradation factors are decisive for service life. The models presented in the report may be applied in these evaluations. In *concrete structures exposed to normal outdoor conditions*, the effects of degradation mechanisms can be classified into the following structural deterioration mechanisms:

1. corrosion of reinforcement at cracks, causing a reduction in the cross-sectional area of steel bars;
2. surface deterioration or frost attack, causing a reduction in the cross-sectional area of concrete.

Phase 5: Calculation of durability parameters through calculation models Damage is determined using the design life, t_{Ld}. Selected calculation models are presented in the appendix of the TC 130-CSL report [1].

Phase 6: Possible updating of calculations in ordinary mechanical design Some durability parameters may influence the mechanical design. An increase in concrete dimensions increases the dead load, thus increasing the load effects on both the horizontal and vertical structures.

Phase 7: Transfer of durability parameters to the final design The parameters of the durability design are listed and transferred to the final design phase for use in the final dimensioning of the structure.

Phase 8: Final design The mechanical design and the durability design are separated. The ordinary structural design (phase 1) produces the mechanical safety and serviceability parameters whereas the durability design (phase 2) produces the durability parameters. Both of these groups of parameters are then combined in the final design of the structure.

3.3.7 Examples

3.3.7.1 Example 1: Service Life Design of a Beam

3.3.7.1.1 SETTING UP THE DESIGN PROBLEM

The beam presented in Figure 3.3.3 is presented as an example on durability design calculations [1]. This presentation is modified, corresponding to the modifications, which have been done later [6, 9, 11].

The beam is to be designed for the following loads:

$$M_g = 10 + 0.1d \, \text{kNm} \quad (d \text{ in mm})$$

$$M_p = 50 \, \text{kNm}$$

The cross section of the beam is assumed to be rectangular with the width of $b(\approx 300 \, \text{mm})$ and efficient height d. At the lower edge of the beam are three steel

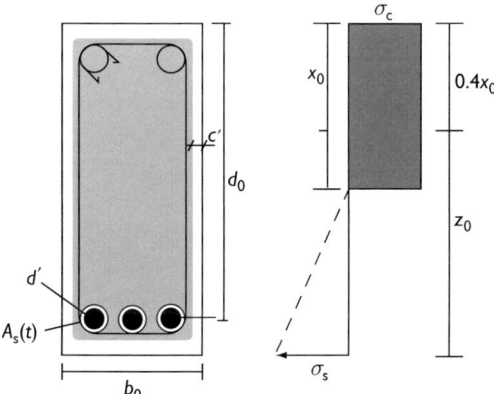

Figure 3.3.3 Design calculations for the beam example.

bars. The yield strength of steel is 400 MPa. The characteristic compressive strength is 40 MPa, the air content is 2% (not air-entrained), and the binding agent is Portland cement.

The beam is supposed to be maintenance free so that the corrosion of steel bars in the assumed cracks or the degradation of the concrete cover will not prevent the use of the column during its service life. The cross section of hoops (stirrups) must not be completely corroded at cracks. The concrete cover must be at least 20 mm after the service life and the cover must not be spalled off because of general corrosion.

3.3.7.1.2 ORDINARY MECHANICAL DESIGN

The ordinary mechanical design of the beam is performed using traditional design principles:

$$R_d \geq S_d \tag{3.3.12}$$

$$S_d = \gamma_g \cdot M_g + \gamma_p \cdot M_p \tag{3.3.13}$$

$$R_{ds} = A_s \, z f_y / \gamma_s \text{ (the stress of steel is decisive)} \tag{3.3.14}$$

$$R_{dc} = b \, x \, z \, f_c / (2 \, \gamma_c) \text{ (the stress of concrete is decisive)} \tag{3.3.15}$$

$$x = d\mu n \left(-1 + \sqrt{1 + (2/\mu n)}\right) \tag{3.3.16}$$

$$z = d - 0.4x \tag{3.3.17}$$

$$n = E_s / E_c$$

$$\mu = \frac{A_s}{b \cdot d} = \frac{N_s \cdot \pi \cdot D^2 / 4}{b \cdot d} \tag{3.3.18}$$

A_s is the cross-sectional area of steel bars:

$$A_s = 3 \cdot \mu \cdot D^2 / 4 \tag{3.3.19}$$

Taking $D = 15$ mm we get:

$$A_s = 530 \, \text{mm}^2$$

By setting R_{ds} equal to S_d we get:

$$d = 2083 \, \text{mm}$$

However, increasing the diameter of the steel bars quickly reduces the efficient height. By replacing $D = 20$ mm we get:

$$A_s = 942 \, \text{mm}^2$$
$$d = 543 \, \text{mm}$$

3.3.7.1.3 DURABILITY DESIGN

The design life (target service life) is 50 years. The central lifetime safety factor is assumed to be 3.3. Thus, the mean service life, t_0, is 165 years.

We apply the degradation model of corrosion degradation as presented in [1]. All sides of the beam are assumed to be exposed to frost action. The environmental factor for frost attack, c_{env}, is 40 and the anticipated curing time is 3 days. The curing factor is:

$$c_{cur} = \frac{1}{0.85 + 0.17\log_{10}(3)} = 1.074 \tag{3.3.20}$$

As concrete is made of Portland cement we conclude:

$$c_{age} = 1$$

Inserting these values into Formula 17 of Appendix 2 in [1] we get:

$$c' = 0.117\,t \tag{3.3.21}$$

At the same time corrosion is occurring in steel bars at cracks. The rate of corrosion is evaluated as $0.03\,\text{mm year}^{-1}$:

$$d' = 0.03\,t \tag{3.3.22}$$

The durability design parameters are as follows (depending on the design service life): Separated design method ($t_d = 50$ years, $t_0 = 165$ years): The depth of deterioration

$$c' = 0.117 \cdot 165 = 19.3\,\text{mm}$$

The required concrete cover is

$$C_{min} = 20 + 19.3\,\text{mm} = 39.3\,\text{mm}$$

We choose $C = 40\,\text{mm}$.
The depth of corrosion at cracks

$$d' = 0.03 \cdot 165 = 5.0\,\text{mm}$$

The diameter of hoops must be at least

$$D_{hmin} = 2 \cdot 5.0\,\text{mm} = 10.0\,\text{mm}$$

We choose $D_h = 10\,\text{mm}$.
The corrosion cracking limit state time of the concrete cover is then checked. The following values of parameters are inserted into the formula:

$C = 40\,\text{mm}$ (separated design method) or $35\,\text{mm}$ (combined design method)
$C_h = C - D_h = C - 10\,\text{mm}$
$f_{ck} = 40\,\text{MPa}$
$c_{env} = 1$
$c_{air} = 1$
$D_h = 10\,\text{mm}$
$r = 12\,\mu\text{m}$

By the separated method, we get $t_0 = 165$ years, which equals to the design life (50 years). So the concrete cover of 40 mm is adequate, and the concrete cover (35 mm) is increased to 40 mm. Then the design life is 50 years, which fulfils the requirement.

3.3.7.1.4 FINAL DESIGN

Separated design method:
The width of the beam at the beginning of service life is twice the deterioration depth of concrete added to the width obtained in the ordinary design:

$$b_0 = b + 2b' = 300 + 2 \cdot 19.3 = 339\,\text{mm}$$

The effective height of the beam is increased by the depth of deterioration:

$$d_0 = d + b' = 543 + 19.3\,\text{mm} = 562\,\text{mm}$$

The minimum diameter of the steel bars is:

$$D_{0\,\text{min}} = 20 + 2 \cdot 5.0\,\text{mm} = 30.0\,\text{mm}$$

We choose $D_0 = 30\,\text{mm}$.

3.3.7.2 Example 2: Service Life Design of a Bridge Slab

A spread sheet application for service life design is presented in Figure 3.3.4 [2].

3.3.8 Reliability requirements of existing structures

3.3.8.1 Design life

In MR&R planning the design life periods of EN 1990:2002 [3] can be basically applied. Often, the service life of an old structure has to be prolonged with a repair. This leads to a new term: 'residual design life'. The residual design life can be decided case by case, but it is usually the same or shorter than the design life of new structures. The residual design life can be optimised using MADA procedure. Proposed values of design life for MR&R planning are presented in Table 2.1.

Dimensioning of a slab   Point   2. 7   **Support T1**   Longitudinal direction
Preliminary ordinary design   **Durability design**

	Design values	Characteristic values			
γ_g	1.2	1	γ_t		2.5
γ_p	1.8	1	t_g		100
γ_c	1.35	1	t_d		250
γ_s	1.1	1	Concrete cover		
M_g	0.2220	0.1850			0.020
M_p	0.5400	0.3000	K_e		3
K	35	35	P		40
f_c	18.1	24.5	r_c		0.000075
f_{ct}	1.59	2.14	$S_c(t_d)$		0.019
f_y	455	500	C_{oreq}		0.039
b	1.000	1.000	$C_{ochosen}$		0.040
h	0.800	0.800	$\phi_{beneath}$		0.025
C_{min}	0.035	0.035	C_{tot}		0.065
ϕ	0.025	0.025	Steels in the direction of bending		
$\phi_{beneath}$	0.025	0.025	ϕ_{vaad}		0.020
C	0.060	0.060	r_s		0.000004
Number of rows	1	1	$S_s(t_d)$		0.001
d	0.728	0.728	ϕ_{oreq}		0.022
x	0.189	0.189	$\phi_{ochosen}$		0.025
z	0.652	0.652	Check of cracking of the cover		
M_{dim}	0.7620	0.4850	$C_{arb}=1, C_l=2$		1
A_s	0.002571	0.002571	C_e		1
k/k	0.191	0.191	C_b		1
Check for the capacity			C_a		0.9
S	0.7620	0.4850	a		2.027
R_s	0.7620	0.8382	C		0.065
R_c	1.7856	2.4105	ϕ_o		0.025
R	0.7620	0.8382	t_o		1028
$R-S$	0.0000	0.3532	t_l		52
			t_L		1080

Check for the crack width   **Final mechanical design**
Serviceability limit state

			Design values	Characteristic values
			$t=t_d$	$t=0$
M_g+M_q	0.4850	γ_g	1.2	1
M_t+M_s	0.0120	γ_p	1.8	1
M	0.4970	γ_c	1.35	1
C	0.065	γ_s	1.1	1
k_w	0.085	M_g	0.2220	0.1850
ϕ	0.025	M_p	0.5400	0.3000
A_s	0.003118	K	35	35
A_{ce}	0.265000	f_c	18.1	24.5
ρ_r	0.011765533	f_y	455	500
s	0.408	b	1.000	1.000
Strain of the steel		Number of rows	1	1
E_c	29580	d	0.709	0.728
E_s	200000	x	0.184	0.189
n	6.761	z	0.635	0.652
μ	0.00429	ϕ	0.023	0.025
x	0.155	M_{dim}	0.7620	0.4850
z	0.676	A_s	0.002639	0.003118
σ_s	235.9	k/k	0.157	0.157
ε_s	0.001179	A_s/A_{s0}	0.846	O.K.
Reduction coefficient		Check for the capacity		
W_{ce}	0.1067	S	0.7620	0.4850
M_r	0.2283	R_s	0.7620	1.0165
k	0.9007	R_c	1.6947	2.4105
Width of the crack		R	0.7620	1.0165
w_k	0.000434	$R-S$	0.0000	0.5315
Allowable crack width				
w_{ks}	0.000300			

Figure 3.3.4 Service life design of a bridge slab.

3.3.8.2   Reliability requirements for service life

The reliability requirements for service life are different from the requirements for structural safety in mechanical limit states. Therefore, it is recommended to use for mechanical safety indexes of EN 1990:2002 [3], and for corresponding reliability classes the safety indexes of service life in durability limit states and in obsolescence limit states as presented in Table 3.3.2.

It is important to notice that in each case of durability and obsolescence limit states, the safety of mechnical (static, dynamic and fatigue) limit states also has to be checked separately.

Because durability works in interaction with structural mechanical safety, the recommended reliability indexes of durability service life are close to the level of requirements for mechanical safety. The obsolescence does not usually have direct interaction to the structural mechanical safety, why the safety index recommendations are lower. The mechanical safety requirements of Table 2.1 always have to be checked separately in cases when obsolescence is caused by insufficient mechanical safety level in comparison to increased loading requirements or increased safety level requirements.

The required lifetime safety coefficients of durability limit states and obsolescence limit states can be found from Table 3.3.2 using the safety indexes of Table 3.3.4. For the safety index 1.5 the lifetime safety factor is 1. This means that the characteristic service is directly applied as design life.

Table 3.3.4 Recommended minimum values for reliability index β in ultimate limit states and in serviceability limit states of durability and obsolescence

Reliability Class of structures	Minimum values for reliability index β			
	Durability limit states		Obsolescence limit states	
	Ultimate limit states	Serviceability limit states	Ultimate limit states	Serviceability limit states
RC3/CC3: High consequence for loss of human life, or economic, social or environmental consequences very great	4.7	3.3	3.3	1.5
RC2/CC2: Medium consequence for loss of human life, or economic, social or environmental consequences considerable	4.3	1.5	1.5	1.5
RC1/CC1: Low consequence for loss of human life, or economic, social or environmental consequences small or negligible	3.3	1.5	1.5	1.5

References

[1] Sarja, A., and E. Vesikari, eds. 1996. Durability design of concrete structures. *RILEM Report of TC 130-CSL. RILEM Report Series 14.* London: E&FN Spon, Chapman & Hall, 165pp.

[2] Sarja, A. 1999. Environmental design methods in materials and structural engineering. *RILEM Journal: Materials and Structures 32* (December): 699–707.

[3] EN 1990:2002: Eurocode – Basis of structural design. CEN: European Committee for Standardisation. Ref. No. EN 1990:2002 E, 87pp.

[4] EN 206-1. *Concrete – Part 1: Specification, performance, production and conformity.* CEN European Committee for Standardisation, December 2000. Ref. No. EN 206-1:2000 E, 72pp.

[5] Sarja, A. 2004. *Reliability based methodology for lifetime management of structures.* EU GROWTH Project no G1RD-CT-2000-00378: Lifecon, Life cycle management of concrete. http://lifecon.vtt.fi/.

[6] Sarja, A. 2004. Generalised lifetime limit state design of structures, *Proceedings of the 2nd international conference, Lifetime-oriented design concepts, ICDLOC,* pp. 51–60. Bochum, Germany: Ruhr-Universität, ISBN 3-00-013257-0.

[7] Sarja, A. 2003. Integrated life cycle design of structures. System-based vision for strategic and creative design. *Proceedings of the 2nd international structural engineering and construction conference: ISEC02,* Rome, September. The Netherlands: A. A. Balkema Publishers, pp. 1697–1702.

[8] Sarja, A. 2003. Lifetime performance modelling of structures with limit state principles. *Proceedings of 2nd international symposium ILCDES2003, Lifetime engineering of buildings and civil infrastructures,* Kuopio, Finland, December 1–3, pp. 59–65. Helsinki: Association of Finnish Civil Engineers.

[9] Sarja, A. 2005. Applications of European standards into durability design guidelines. *International Conference "Concrete and reinforced concrete – Development trends",* Moscow, September 5–9, along with the 59th RILEM Week, in English: pp. 233–245; in Russian: pp. 218–232.

[10] ISO/DIS 15686-1. 1998. Buildings – Service life planning – Part 1: General Principles. Draft.

[11] Sarja, A. 2002. *Integrated life cycle design of structures.* London: Spon Press, 142pp. ISBN 0-415-25235-0.

[12] Lay, S., and P. Schießl. 2003. *Probabilistic service life models for reinforced concrete structures.* EU GROWTH project no G1RD-CT-2000-00378: Lifecon, Life cycle management of concrete infrastructures, Deliverable D3.2. http://lifecon.vtt.fi/.

[13] Vesikari, E. 2003. Statistical condition management and financial optimisation in lifetime management of structures. EU GROWTH project no G1RD-CT-2000-00378: Lifecon, Life cycle management of concrete infrastructures, Deliverable D2.2. http://lifecon.vtt.fi/.

3.4 Reference structure models and methodology for service life prediction

Erkki Vesikari

3.4.1 General scheme

The rate of degradation is usually highly dependent on the environmental conditions. Both macro- (resulting from the meteorological conditions of the building site) and

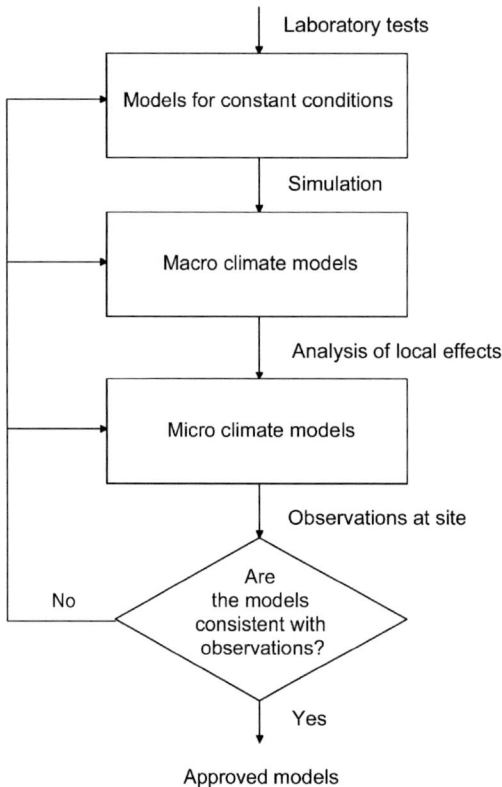

Figure 3.4.1 General scheme for the development of degradation models.

microclimate models (resulting from the performance and condition of the drainage system of the building, condition of joints, protective effects of neighbouring structures, etc.) are of significance.

A general approach for developing degradation models for specific structural parts is given in Figure 3.4.1. In principle, this approach can be applied to most degradation types of various structural parts [1]. As a result, approved degradation models can be developed.

3.4.2 Models for constant conditions (laboratory models)

In the first phase, 'laboratory models' that can be exposed to constant environmental conditions are developed, based on theoretical studies and laboratory tests. For some degradation types, such as chloride penetration, laws of diffusion can be applied. Under laboratory tests, the basic data of the effects of various material parameters on the rate of degradation can be studied; the effects of environmental parameters, such as moisture content and temperature, can also be studied. Many degradation types can also be easily studied under constant conditions. As a result of these tests, degradation curves as a function of time can be developed. Examples of such degradation types for concrete structures are carbonation, penetration of chlorides and corrosion of reinforcement.

More complicated degradation types are those which require fluctuating temperature cycles to occur. The models of frost attack can usually be based on the number of freeze–thaw cycles in a laboratory test, relating these results with the number of effective freeze–thaw cycles under real environmental conditions. The internal frost damage can be based on critical freezing events during which the critical degree of water saturation is exceeded. The number of critical freezing events can be determined using computer simulation.

3.4.3 Models for natural weather conditions (macroclimate models)

In the second phase 'macroclimate models' for natural weather conditions are developed. Under natural weather conditions, both temperature and moisture content of the model vary from time to time. So the laboratory models that are created for constant environmental conditions may not be applicable for natural weather conditions. However, they can serve as initial data in the development process which results in macroclimate models of degradation. By dividing the time into short discrete steps and assuming that constant conditions prevail during each time step, it is possible to determine the incremental degradation corresponding to the time steps. Then by integrating the incremental degradations, the total degradation in regional atmospheric conditions can be determined. In practice, this process is usually connected to computer simulation by which it is possible to determine the momentary temperature and moisture contents of the material in a model structure which is exposed to normal weather (reproduced by weather models – Figure 3.4.2).

The computer simulation method can be used as the 'missing link' between 'laboratory models' and 'macro climate models'. During simulation, temperatures, moisture contents and degradation processes of structures are emulated as they occur under constant conditions. The weather data are obtained from meteorological stations, they can be based on either a long-term average or observations of a single year of a local site. They consist of data on temperature and relative humidity of the air as well as the velocity and direction of wind, amount of rain and intensity of solar radiation. Both daily and seasonal variations are considered in the weather models. The models of rain and solar radiation are statistical models which yield the correct amount of rain and solar radiation but the exact moments of rain and direct solar radiation are not given in a deterministic way [2–4].

The model structure is burdened with virtual weather. The outdoor weather conditions form the boundary conditions of a thermal and moisture mechanical problem to solve the temperature and moisture distributions in a structural cross section. The incremental time may be one hour or shorter. The incremental degradation during each incremental time step is evaluated and added to the sum of degradation. The total calculation may cover some months, some years or even some hundreds of years from the lifetime of the structure.

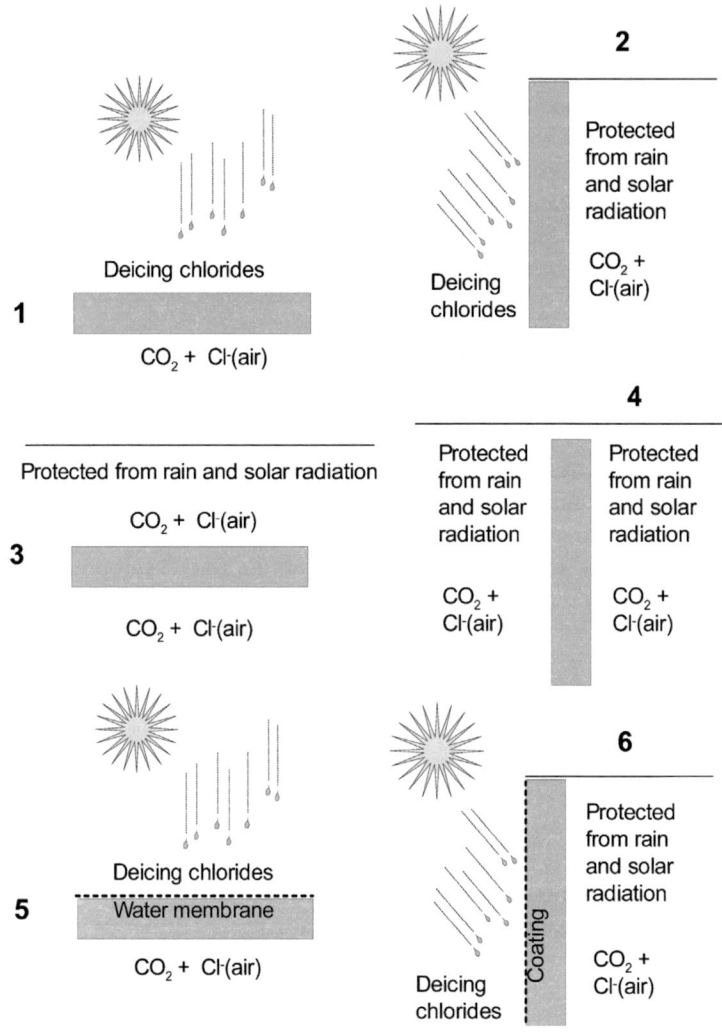

Figure 3.4.2 Different model structures and exposure types for computer simulation: 1 and 2 exposed to rain, solar radiation and de-icing chlorides, 3 and 4 sheltered from rain and solar radiation by other structures, 5 and 6 protected from rain and de-icing chlorides by water membrane or coating.

3.4.4 Models for specific structural parts (microclimate models)

Different structural parts may be exposed to different environmental loads, although the whole building or structure is under the same macro climatic conditions. As the amount of structural parts may be great, it is not usually practical to create separate degradation models for each structural part. A simplified but still reasonably sensitive method of modelling should be used instead. One possibility is to use index values to account for different environmental loads on which the rate of degradation is assumed to depend; indexes such as 'moisture index' and 'chloride index' by which the degree of moisture load and the degree of chloride load are evaluated. Applying index values results in interpolation between 'basic extreme conditions' representing extreme moisture and chloride index values.

The moisture and chloride burdens are, of course, dependent on the type of environment that is specific for the type of the structure. The environemental burdens differ for bridges, tunnels, canals, dams, pole basements, quays, lighthouses, etc. Each structure requires a careful examination of the environmental loads.

For example, in the case of bridge structures it is essential to know what are the obstacles that stand in its way: a gulf, river, road or railway line. Bridges crossing a gulf are exposed to a greater chloride burden than those crossing a river or lake. The waterway in a neighbourhood of a bridge affects also on the moisture burden of the bridge. Another factor affecting both chloride and moisture indexes is the amount of de-icing salt spread on the overpassing and the underpassing roads. Not only the total amount of de-icing salt is of concern, but also the location of the specific structural part in relation to the overpassing and the underpassing roads. Both the indexes are highly dependent on the distance of the structural part from these roads as also on the sheltering effects of other structural parts. The final chloride index of a structural part is obtained as the sum of the chloride burdens from the overpassing and underpassing roads, both being determined as the product of the degree of chloride exposure of the structural part and the relative de-icing burden of the road.

First, the degradation rates for the maximal and minimal burdens of moisture and chlorides (chloride index 0 and 1 and at moisture index 0 and 1) are determined for the bridge. Then the actual rate of degradation for a specific structural part is determined by interpolation using the specific moisture and chloride indexes of the structural part.

3.4.5 Mathematical formulation and calibration of degradation models

The mathematical formulation of degradation models depends, of course, on the type of degradation. However, a simple exponential formula, as presented in Equation (3.4.1), is proved to be applicable in many cases. This formulation is not a necessity but can usually be held as a starting point for modelling of many degradation types [1, 5]:

$$f = a \cdot t^n \tag{3.4.1}$$

where
   f the degradation
   t the time from the manufacture of the structure
   a the coefficient
   n the exponent of time.

Both the coefficient a and the exponent n may be a function of several material, structural and environmental parameters.

$$a = a(p_1, p_2 \ldots p_m) \tag{3.4.2}$$

$$n = n(p_1, p_2 \ldots p_m) \tag{3.4.3}$$

where $p_1, p_2 \ldots p_m$ are material, structural and environmental parameters effecting on the rate of degradation, and m is the total number of these parameters.

Depending on the value of n, the rate degradation can be determined. If:

$n > 1$,   the rate of degradation is increasing with time, i.e. accelerating;
$n = 1$,   the rate of degradation is directly proportional with time, i.e. steady; and
$n < 1$,   the rate of degradation is decreasing with time, i.e. decelerating.

The degradation functions can be normalised by dividing the function n by the maximum allowable degradation, that is the limit state of degradation. Considering Equation (3.4.1) stands for a normalised model, f starts from 0 and attains to 1 at the limit state.

A rating scale is a set of sequential whole numbers that express the observed or evaluated degradation state of a structure. The ratings usually correspond to 'degrees of damage' which are used in the visual inspection and condition assessment of structures (e.g. $0, 1, \ldots N_{max}$). Considering 0 is the starting state of degradation (no degradation) and N is the state corresponding to limit state, then a scaled degradation function which is called Degree-of-Damage (DoD) function can be easily derived from a normalised degradation function as stated in Equation (3.4.4a):

$$\mathrm{DoD} = N \cdot a \cdot t^{n_0}$$

or (3.4.4a)

$$\mathrm{DoD} = c \cdot t^{n_0}$$

where
DoD   the degree-of-degradation function for depassivation
N   the state corresponding to the limit state in a rating system that starts from 0
c the coefficient for a scaled degradation function ($c = N \cdot a$), t the time, year
n the exponent of time.

The calibration of a scaled degradation model for a specific component with field observations can easily be performed. Assuming that during an inspection that was performed at the time of t^* from the start of the life of the component, the observed degree of degradation was DoD^* the coefficient of depassivation can be calibrated using Equation (3.4.6) (assuming that the exponent of time stays the same).

$$c^* = \frac{\mathrm{DoD}^*}{(t^*)^n} \tag{3.4.4b}$$

where

c^* is the calibrated coefficient c.

The calibrated degradation curve is obtained by inserting c^* in Equation (3.4.4b).

3.4.6 Degradation models for structural damage

The load-bearing capacity of a structure can be evaluated with the help of the residual capacity factor (RCF) given in Equation (3.4.5) [1]. This factor expresses the load bearing capacity of a damaged structure in relation to its original undamaged load bearing capacity.

$$RCF = \frac{R'(t)}{R} \tag{3.4.5}$$

where

RCF the residual capacity factor
R the original capacity of a structure as undamaged
$R'(t)$ the load-bearing capacity of the structure as damaged.

The RCF of a damaged structure is evaluated by inserting the evaluated reduced cross-sectional area and reduced strength of materials (as a result of degradation) to the normal design equations of a structure, and by dividing the so-obtained load-bearing capacity by the original load-bearing capacity determined from the same design equations. RCF is usually a set of relationships covering, for example flexural capacity, shear capacity and anchorage/bond.

It is also possible to insert time-related models for material strengths and dimensions to the design formulae thus creating time-dependent models for the RCF. In general, the RCF can be expressed as shown in Equation (3.4.6)

$$RCF(t) = \frac{C'(p'_1, p'_2, p'_3 \cdots p'_m)}{C(p_1, p_2, p_3 \cdots p_m)} \tag{3.4.6}$$

where

C the original capacity of the component
C' the capacity of the component at time t from the start of service life
$p_1 \ldots p_m$ the original values of parameters (structural dimensions or strengths) of capacity
$p'_1 \ldots p'_m$ the values of parameters of capacity $p_1 \ldots p_m$ at time t from the start of the service life.

When calculating RCF, one can often observe that some of the parameters in Equation (3.4.8) are not time dependent and can thus be reduced off. The rest of the parameters that are not constant are evaluated by degradation models as applied to the material and possible losses of strength in a cross section of the component.

The ultimate limit state for the RCF can be derived from a safety analysis. The residual marginal factor (RMF) can be defined as in Equation (3.4.7):

$$\text{RMF} = \frac{R' - S'}{R - S'} \tag{3.4.7}$$

where
RMF is the residual marginal factor
S the original loading requirement and
S' the loading requirement at the time of treatment.

The minimum value for the RCF can be determined as derived from Equation (3.4.7):

$$\text{RCF}_{\min} = \text{RMF}_{\min}\left(1 - \frac{S}{R}\right) + \frac{S'}{R} \tag{3.4.8}$$

The minimum value of the residual marginal factor, $\text{RMF}_{\min}$, is evaluated based on the safety requirement in each case. Usually, no greater reduction than 10% of the original safety index (β) is allowed. Thus, the $\text{RMF}_{\min}$, Equation (3.4.8) is about 0.9. The quantity R/S is the total safety factor and is determined with characteristic values of the stress S and the resistance R for the structural part. The principle of the determination of the structural service life based on RCF is presented in Figure 3.4.3. By dividing Equation (3.4.10) by Equation (3.4.8), a model function for structural degradation which is closely consistent with Equation (3.4.1) is obtained. So the problem of structural degradation can be solved using the same methods as those used with problems of other degradation.

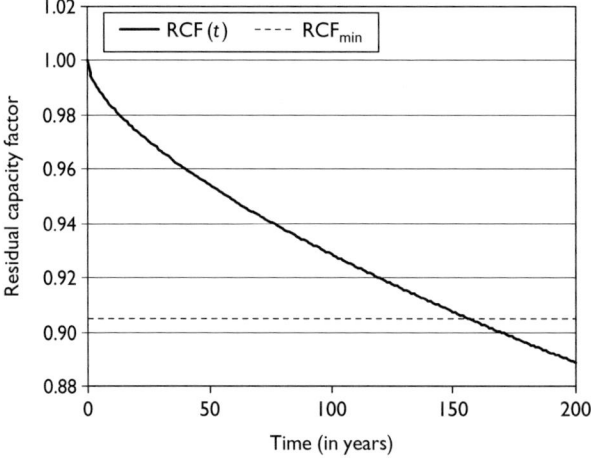

Figure 3.4.3 RCF method for determining the residual service life of a structure [2].

3.4.7 Calibration of models by field tests

3.4.7.1 Improving the validity of degradation models

Having inspections performed with measured degradation and parameter data, the general degradation models can be calibrated. The aim of this calibration is to improve the validity of the degradation models and thus also the validity of service life predictions and life cycle cost-analysis results. The process of updating degradation models should be an integrated part of the LMS.

An example of the calibration calculations is given for concrete bridge structures. For other structures, it can be applied with due modifications. For concrete bridge structures, the following data are assumed to be available for updating the degradation models:

- bridge (identification number, name, code, etc.)
- component (code)
- year of construction of the component
- age of component at the time of inspection
- depth of carbonation
- depth of critical chloride content
- degree of surface deterioration of concrete
- degree of corrosion of reinforcement (general corrosion)

- width of cracks
- length of cracks
- degree of corrosion at cracks
- concrete cover
- strength of concrete
- air content of concrete
- moisture index
- chloride index of the road
- chloride index of the crossing road
- method of protection
- condition class of the protection
- previous year of (re)protection.

The uncalibrated degradation models are used to determine the expected depth of carbonation, depth of critical chloride content, surface scaling and other degradation. Then a statistical analysis, such as a T-test, can be performed to study the statistical significance of the difference between the measured and model values of degradation. If a statistically significant difference is found, the analysis can be continued as presented in the following Example.

Example Definition of degradation model [5].

$$\mathrm{DoD} = N_0 \cdot a_0 \cdot t^{\mathrm{n}}$$

$$N_0 = 3 \text{ (limit state)}$$

Parameters
  K the nominal strength of concrete (MPa)
  C the thickness of concrete cover (mm)
  A the air content of concrete (%).

$$a_0 = k_{\mathrm{GIS}} \cdot k_{\mathrm{H/V}} \cdot k_{\mathrm{Cem}} \cdot k_{\mathrm{Cov}} \cdot k_{\mathrm{Frost}} \left(-0.01 + 3.88 \cdot 10^{-6} \cdot K^3 \right)$$

(*Continued*)

Regional coefficient	k_{GIS}
Coastal Finland	1
Middle Finland	0.78
Northern Finland	0.71

Coefficient of surface direction	$k_{H/V}$
Horizontal	1
Vertical	0.93

$$k_{Cov} = \frac{35}{Min(70; C)}$$

$$k_{Frost} = 0.8 \times \left(10 \cdot \frac{\left(\frac{28.8}{21.5+K}\right)^{1.25}}{A^{0.5}} - 1 \right)$$

$$n_0 = 0.976 - 1.3 \cdot 10^{-5} \cdot (55 - Min(50; K))^3$$

As an example, the data in Table 3.4.1 are composed on the basis of special inspections for bridge structures in the most severe conditions (both moisture and chloride indexes 1). There were 20 structural parts with different life span and the degree of degradation was evaluated for each of them. The values of nominal compressive strength, air content and concrete cover were measured and the values for the coefficients k_{GIS}, $k_{H/V}$ and k_{Cem} could be defined based on the geographical situation, direction of the surface and the type of cement of the structure. The task was to update the coefficients of the cover and frost, that is the constant coefficients in the equation of these coefficients which originally were 35 and 0.8 respectively, and the exponents in the equations of a_0 and n_0 which originally were 3.

In Table 3.4.1, under the head calculated values, the values for k_{cov}, k_{frost}, a_0 and n_0 are calculated according to the equations in Example 1. However, the searched coefficients and exponents in the equations of these quantities are placed outside the table. Originally, these values at the top of the table were 35, 0.8, 3 and 3. In the last two columns, the model values of DoD and the squared differences $(DoD_{model} - DoD_{observed})^2$ are determined. The sum of the squared differences is also determined.

The problem is solved by using linear programming (e.g. Microsoft Solver for Excel). The problem is defined simply by

1. minimising the sum of squared differences; and
2. changing the values of searched coefficients and exponents.

The values of the searched coefficients and exponents presented in Table 3.4.1 were determined in this way. These values should now be applied in the updated degradation equation instead of the original ones.

Table 3.4.1 Data table and a statistical analysis for improving the degradation models (example)

		Coeff 34.946	Coeff 0.5237	Exp 2.79013	Exp 3.0454

Case	Observed values									Calculated values					
	t*	DoD_{real}	K MPa	Air-%	Cover	k_{GIS}	$k_{H/V}$	k_{Cem}	k_{Cov}	k_{Frost}	a_0	n_0	DoD_{model}	$Diff^2$
1	27	0.2	25	2.6	33	1.00	1.00	—	1.06	1.27	0.0268	0.554	0.17	0.005
2	24	0.7	42	5.0	31	0.78	0.93	—	1.15	0.34	0.0357	0.947	0.74	0.000
3	11	0.0	22	1.9	13	0.71	1.00	—	2.73	1.74	0.0402	0.434	0.12	0.013
4	26	0.0	31	4.7	40	0.78	0.93	—	0.88	0.62	0.0179	0.765	0.22	0.048
5	14	3.1	42	2.2	6	0.78	0.93	0.83	5.91	0.78	0.3357	0.944	3.94	0.750
6	12	0.6	17	2.0	30	0.78	0.93	—	1.18	2.07	0.0011	0.138	0.00	0.419
7	10	0.0	26	3.1	31	0.78	0.93	—	1.11	1.07	0.221	0.620	0.09	0.009
8	11	2.6	42	1.2	13	0.71	0.93	—	2.63	1.23	0.2687	0.947	2.69	0.015
9	27	0.7	19	3.2	19	0.78	1.00	—	1.81	1.40	0.0075	0.246	0.02	0.530
10	30	0.0	20	3.3	31	0.93	1.00	—	1.12	1.34	0.0078	0.297	0.02	0.000
11	29	0.0	33	4.9	39	0.78	1.00	—	0.89	0.53	0.0220	0.826	0.36	0.128
12	21	0.0	30	2.6	12	0.78	0.93	—	2.82	1.05	0.0843	0.729	0.77	0.599
13	25	1.2	39	3.7	24	0.78	0.93	—	1.49	0.55	0.0588	0.918	1.14	0.002
14	17	0.7	15	2.6	38	0.93	1.00	—	0.93	1.90	−0.0039	0.003	0.00	0.542
15	13	0.0	25	3.0	38	0.78	0.93	—	0.92	1.15	0.0155	0.558	0.06	0.004
16	16	1.1	18	1.9	8	1.00	1.00	—	4.44	2.05	0.0161	0.181	0.03	1.147
17	25	2.9	32	2.7	5	0.78	0.93	—	6.64	0.94	0.2338	0.795	3.05	0.023
18	16	1.0	19	2.9	13	1.00	0.93	—	2.75	1.52	0.0137	0.238	0.03	0.954
19	14	0.2	17	1.2	16	0.78	0.93	—	2.17	2.70	0.0051	0.160	0.01	0.021
20	21	2.8	31	1.2	7	0.78	0.93	—	4.87	1.70	0.2740	0.765	2.85	0.004
												Σ		5.213

Although this example looks simple, the updation of degradation models is not always just a routine job. The original data on which the original degradation models were based should not be forgotten but they should be considered too. In general, it is recommended to leave the updation of degradation models to a specialist. The quality of degradation models is supposed to be improved slowly when the system is in use and when more and more special inspection data are available from structures.

3.4.8 Developing stochastic models from deterministic degradation models

The degradation models described in the previous sections are characterised as deterministic. The difference between deterministic and stochastic models is that not only the average value (or other characteristic value) of degradation is modelled but also the distribution.

The final purpose of the stochastic modelling is to find out the probability distribution of the service life (time when degradation attains its maximum with a certain probability). Thus the probability distribution of *degradation* may not as such be so interesting, but it can be used as a means for determination of the probability distribution of service life.

Another possibility is to evaluate the distribution of service life directly around the mean service life that is assumed to be known or determined by a model formula. Considering Equation (3.4.1) is a normalised model function for the average degradation (f starts from 0 and attains to 1 at the limit state), the mean value of service life can be approximated as:

$$\mu(t_L) = \left(\frac{1}{a}\right)^{\frac{1}{n}} \tag{3.4.9}$$

where
$\mu(t_L)$ is the mean value of service life.

The methods for evaluating the probability distribution of service life can be analytic or numeric calculation. Only various analytical calculation methods are discussed briefly.

Several analytic methods can be used for evaluating the distribution of service life. The main idea of these methods is to evaluate the scatter of service life based on the scatter of the parameter values in the degradation of service life models. The following methods are reviewed:

- partial derivation method;
- convolution integral method; and
- Monte Carlo simulation method.

3.4.8.1 Partial derivation method

If the service life can be estimated using a rather simple formula, which is easily derivable, the standard deviation of service life can be evaluated using the partial derivation method [4]. Let us assume that the formula of mean value service life is in a general form:

$$\mu(t_L) = t_L(p_1, p_2, \ldots, p_m) \tag{3.4.10}$$

where
- $\mu(t_L)$ is the mean value of service life
- p_i the design parameter (material property, structural measure or environmental parameter)
- m the number of parameters.

Then the standard distribution of service life can be determined from the equation:

$$\sigma(t_L) = \sqrt{\sum_{i=1}^{m} \left(\frac{\partial t_L}{\partial p_i} \cdot \sigma(p_i) \right)^2} \tag{3.4.11}$$

where
- $\sigma(t_L)$ the standard distribution of service life, year
- $\sigma(p_i)$ the standard deviation of parameter p_i
- $\sigma(t_L)\frac{\partial t_L}{\partial p_i}$ the partial derivative of service life with respect to parameter p_i.

The distribution of service life is usually assumed to be lognormal, Weibull or Gamma.

3.4.8.2 Convolution integral method

If the service life of a structure can be solved based on the load S and the resistance R, either one of which or both are time dependent, the distribution of service life can be determined using the convolution integral method. Then it is assumed that the probability of exceeding service life equals to the probability of the load being greater than the resistance. In Figure 3.4.4, the calculation principle is visualised. The distributions of the load and the resistance are at first far from each other. At this phase the probability that the load would be greater than the resistance is very small. However, by the time, the distributions come closer to each other and the overlapping area increases. The overlapping area illustrates the probability that the load is greater than the resistance at time t, i.e the probability of service life at t.

Mathematically, the probability of the load being greater than the resistance at time t can be determined using the convolution integral [6].

$$P(R < S) = F_R(s) \cdot f_S(s) \partial s \tag{3.4.12}$$

where
- $F_R(s)$ the cumulative probability function of resistance
- $f_S(s)$ the probability density function of load
- s the common parameter of load and resistance.

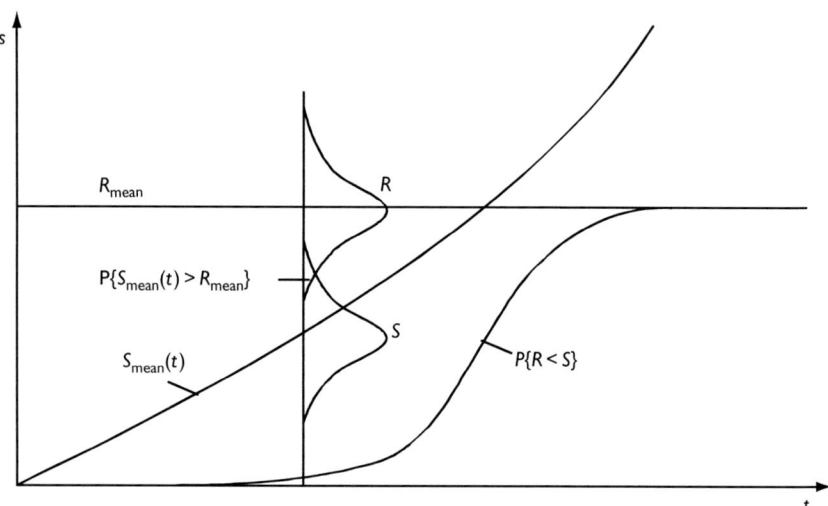

Figure 3.4.4 Principle of determination of service life.

To be able to determine the whole distribution of service life the convolution integral must be solved at several time values of t.

If the distributions of S and R are generic, the solution of the convolution integral may call for numeric integration. However, if these functions are of standard type, a ready solution for the convolution integral may be available.

3.4.8.3   Monte Carlo simulation method

A simple and still probably the most accurate way to determine the distribution of service life is the Monte Carlo simulation method. A prerequisite for using this method is that there is a simulation program, such as @RISK, available.

Monte Carlo simulation method is based on the use of random numbers in simulation. The term 'Monte Carlo' dates back to the Second World War and refers to the well-known roulette board – a kind of random number generator. In this context, the Monte Carlo simulation method is applied to the determination of the probability density function of service life based on the deterministic model function of service life. By random variation of the parameter values with all their possible combinations, an amount of possible service life values are obtained. The distribution of service life can then be defined based on these data points. In general, the analysis consists of the following phases:

1. definition of the deterministic model function of service life;
2. definition of uncertainty, that is the probability distributions of all parameters pertaining to the model function of service life;
3. analysing the model function using Monte Carlo simulation – service life is determined with, for example, 10 000 randomly selected value combinations of

variables and the probability function of service life is determined based on the results; and

4. decision-making phase – from the probability function the fulfilment of the service life requirement is checked as is the probability of service life smaller, equal to or greater than the demanded probability?

When the service life distribution is known, the application of reliability theory in service life design or life cycle planning is possible. However, if the probability functions $F_R(s)$ and $f_S(s)$ are of standard (Gaussian) type, an analytic solution for the convolution integral is possible.

Example Reference structure models for surface damage and corrosion at cracks

Depassivation

$$f_0 = a_0 \cdot t^{n_0} \tag{1}$$

where

f_0 the depth of carbonation (no chlorides) or critical chloride content in relation to the concrete cover

t the time from the manufacture of the structure

a_0 the coefficient, and

n_0 the exponent of time.

Active corrosion

$$f_1 = a_1 \cdot t^{n_1} \tag{2}$$

where

f_1 the depth of corrosion of reinforcing steel in relation to the critical depth of corrosion causing cracking of concrete cover,

t the time from depassivation,

a_1 the coefficient, and

n_1 the exponent of time.

Interpolation of coefficients a and exponents n

1. With respect to moisture burden:

$$a(m, 0) = a(0, 0) + m[a(1, 0) - a(0, 0)] \tag{3a}$$

$$a(m, 1) = a(0, 1) + m \cdot [a(1, 1) - a(0, 1)] \tag{3b}$$

$$n(m, 0) = n(0, 0) + m \cdot [n(1, 0) - n(0, 0)] \tag{3c}$$

$$n(m, 1) = n(0, 1) + m \cdot [n(1, 1) - n(0, 1)] \tag{3d}$$

(*Continued*)

where

    m the moisture index,

    $a(m, 0)$ the constant coefficient at moisture burden m and at chloride burden 0,

    $a(m, 1)$ the constant coefficient at moisture burden m and at chloride burden 1,

    $n(m, 0)$ the exponent of time at moisture burden m and chloride burden 0, and

    $n(m, 1)$ exponent of time at moisture burden m and chloride burden 1.

2.   With respect to the chloride burden:

$$a(m, c) = a(m, 0) + c \cdot [a(m, 1) - a(m, 0)] \tag{4a}$$

$$n(m, c) = n(m, 0) + c \cdot [0(m, 1) - n(m, 0)] \tag{4b}$$

where

    c is the chloride index,

    $a(m, c)$ the constant coefficient at moisture burden m and chloride burden c, and

    $n(m, c)$ the exponent of time at moisture burden m and chloride burden c.

Moisture index m

$$m = \text{Max(object-related default value;}$$

$$\text{component-related default value)} \tag{5}$$

Table 1 Degree of exposure with respect to moisture: Object-related default values for bridges

Obstacle to cross over	m
Road	0
Railroad	0
Sea	0.4
Other water	0.2

Table 2 Degree of exposure with respect to moisture: Component-related default values for bridges

Component	m
Side wall	0.7
Frontal wall	0.1
Wing wall	0.8
Bearing pad	0.1
Column at the edge	0.1
Column in the middle	1
Abutment at the edge	0.8
Abutment in the middle	0.5
Edge beam	0.8
Side of the deck	0.5
Deck without water membrane	1
Deck with water membrane	1
Deck, underside	0
Girder, beam	0
Girder, arch	0
Girder, box beam, underside	0
Secondary girder	0
Pylon	0.9

Chloride index c

The final chloride index of a structure is obtained as the sum of the chloride burdens from the overpassing and underpassing roads, each determined as the product of the degree of the chloride exposure of the structure and the level of the de-icing burden of the road.

$$c = (\text{degree of exposure of component with respect to road} \times \text{de-icing burden of the road}) + (\text{degree of exposure of component with respect to crossing road} \times \text{de-icing burden of crossing road}) \quad (6)$$

Table 3 Relative de-icing burden of the road or the crossing road depending on the winter maintenance class

Winter maintenance class	De-icing burden
Isk	1
Is	0.55
I	0.38
Ib	0.15
Lower	0

(Continued)

Table 4 Degree of exposure with respect to chlorides: Component-related default values for bridges

Component	Degree of exposure	
	For road	For crossing road
Side wall	0.2	0.4
Frontal wall	0	0.5
Wing wall	0.3	0.4
Bearing pad	0	0.4
Column at the edge	0	0.4
Column in the middle	I	0
Abutment at the edge	0.I	0.9
Abutment in the middle	0	0.6
Edge beam	0.I	0.9
Side of the deck	0	0.6
Deck without water membrane	I	0
Deck with water membrane	I	0
Deck, underside	0	0.2
Girder, beam	0	0.2
Girder, arch	0	0.2
Girder, box beam, underside	0	0.2
Secondary girder	0	0.2
Pylon	I	0

Coefficients for surface damage (a)
Chloride burden (0)/Moisture burden (0)
Parameters

K the nominal strength of concrete (MPa)
C the thickness of concrete cover (mm)

$$a_0 = k_{Cov} \cdot k_{Cem} \cdot \left(-0.012 + \frac{5.74}{K} \right) \tag{7}$$

$$k_{Cov} = \frac{10}{Min(70; C)} \tag{8}$$

Table 5 Coefficient of cement type for carbonation

Cement	k_{Cem}
CEM I A	1.00
CEM II/A-S	1.14
CEM II/B-S	1.25
CEM II/A-D	1.19
CEM II/A-LL	1.00
CEM II/A-M	1.00
CEM III/A	1.46
CEM III/B	1.56

$$n_0 = 0.409 + \frac{5.02}{K} \tag{9}$$

$$a_1 = k_{\text{GIS}} \cdot k_{\text{Cov}} \cdot \left(-0.01 + \frac{0.93}{K} \right) \tag{10}$$

Table 6 Regional coefficient for corrosion

Region	k_{GIS}
Coastal Finland	1
Middle Finland	0.86
Northern Finland	0.62

k_{Cov}: see Equation (8).

$$n_1 = 0.407 + 0.011 \cdot K \tag{11}$$

Coefficients for surface damage (b)
Chloride burden (0) / Moisture burden (1)
Parameters

- K the nominal strength of concrete (MPa)
- C the thickness of concrete cover (mm)
- A the air content of concrete (%).

$$a_0 = k_{\text{GIS}} \cdot k_{\text{H/V}} \cdot k_{\text{Cem}} \cdot k_{\text{Cov}} \cdot k_{\text{frost}} \cdot \left(-0.03 + \frac{2.84}{K} \right) \tag{12}$$

Table 7 Regional coefficient for carbonation

Region	k_{GIS}
Coastal Finland	1
Middle Finland	0.9
Northern Finland	0.76

Table 8 Coefficient of surface direction

Surface direction	$k_{\text{H/V}}$
Horizontal	1
Vertical	0.93

(*Continued*)

k_{Cem}: see Table 5.
k_{Cov}: see Equation (8).

$$k_{Frost} = \text{Max}\left(0.6; \frac{132}{27 + 1.48 \cdot A^{3.4}}\right) \cdot \text{Max}\left(0.5; \left(\frac{85.6 - 0.77 \cdot K}{21.5 + K}\right)^3\right) \quad (13)$$

$$n_0 = 0.71 - \frac{13.2}{K^{1.5}} \quad (14)$$

$$a_1 = k_{GIS} \cdot k_{Cov} \cdot k_{Frost} \cdot 0.075 \quad (15)$$

Table 9 Regional coefficient for corrosion

Region	k_{GIS}
Coastal Finland	1
Middle Finland	1
Northern Finland	0.76

k_{Cov}: see Equation (8).
k_{Frost}: see Equation (13).

$$n_1 = 1.18 \quad (16)$$

Coefficients for surface damage (c)
Chloride burden (1)/Moisture burden (0)
Parameters

K the nominal strength of concrete (MPa)
C the thickness of concrete cover (mm)
A the air content of concrete (%).

$$a_0 = 0.5 \cdot k_{Cov} \cdot k_{Cem} \cdot \left(-0.432 + \frac{27.7}{K}\right) \quad (17)$$

k_{Cov}: see Equation (8).

Table 10 Coefficient of cement type for chloride penetration

Cement	k_{Cem}
CEM I A	1.00
CEM II/A-S	0.83
CEM II/B-S	0.71
CEM II/A-D	0.23
CEM II/A-LL	1.00
CEM II/A-M	1.00
CEM III/A	0.43
CEM III/B	0.27

$$n_0 = 0.413 + \frac{85\,800}{K^{3.5}} \tag{18}$$

$$a_1 = k_{\text{GIS}} \cdot k_{\text{Cov}} \cdot \left(-0.15 + \frac{8.40}{K}\right) \tag{19}$$

Table 11 Regional coefficient for corrosion

Region	k_{GIS}
Coastal Finland	1
Middle Finland	0.84
Northern Finland	0.55

k_{Cov}: see Equation (8).

$$n_1 = 1.06 - \frac{65.9}{K^{1.5}} \tag{20}$$

Coefficients for surface damage (d)
Chloride burden (1)/Moisture burden (1)
Parameters

  K the nominal strength of concrete (MPa)
  C the thickness of concrete cover (mm)
  A the air content of concrete (%).

$$a_0 = k_{\text{GIS}} \cdot k_{\text{H/V}} \cdot k_{\text{Cem}} \cdot k_{\text{Cov}} \cdot k_{\text{Frost}} \cdot (-0.01 + 3.88 \cdot 10^{-6} \cdot K^3) \tag{21}$$

Table 12 Regional coefficient

Region	k_{GIS}
Coastal Finland	1
Middle Finland	0.78
Northern Finland	0.71

(*Continued*)

Table 13 Coefficient of surface direction

Surface direction	$k_{H/V}$
Horizontal	1
Vertical	0.93

$$k_{Cov} = \frac{35}{\text{Min}(70; C)} \tag{22}$$

$$k_{Frost} = 0.8 \cdot \left(10 \cdot \frac{\left(\frac{28.8}{21.5+K}\right)^{1.25}}{A^{0.5}} - 1 \right) \tag{23}$$

$$n_0 = 0.976 - 1.3 \cdot 10^{-5} \cdot (55 - \text{Min}(50; K))^3 \tag{24}$$

$$a_1 = k_{GIS} \cdot k_{Cov} \cdot k_{Frost} \cdot 0.3 \tag{25}$$

Table 14 Regional coefficient for corrosion

Region	k_{GIS}
Coastal Finland	1
Middle Finland	0.93
Northern Finland	0.71

k_{Frost}: see Equation (23).

$$n_1 = 1 \tag{26}$$

Coefficients for crack damage (a)
Chloride burden (0)/Moisture burden (0)
Parameters

K the nominal strength of concrete (MPa)
C the thickness of concrete cover (mm)
W the width of crack (mm)
D the diameter of the corroding steel bar at crack (mm).

$$a_0 = 4.75 \cdot k_{Cw}^{\frac{1}{2}} \cdot k_{Cem}^{\frac{1}{2}} \cdot k_{Cov}^{\frac{1}{2}} \cdot \left(-0.012 + \frac{5.74}{K} \right)^{\frac{1}{2}} \tag{27}$$

$$k_{Cw} = \text{Min}\left(1; \frac{W}{0.2} \right) \tag{28}$$

k_{Cov}: see Equation (8).
k_{Cem}: see Table 5.

$$n_0 = 0.5 \cdot \left(0.409 + \frac{5.02}{K} \right) \tag{29}$$

$$a_1 = 0.11 \cdot k_{Cw} \cdot k_D \cdot k_{GIS} \cdot k_{Cov} \cdot \left(-0.01 + \frac{0.93}{K} \right) \tag{30}$$

$$k_{Cw} = 1 + \frac{1}{1 + \frac{1}{W}} \tag{31}$$

$$k_D = \frac{D}{20} \tag{32}$$

Table 15 Regional coefficient for corrosion

Region	k_{GIS}
Coastal Finland	1
Middle Finland	0.86
Northern Finland	0.62

k_{Cov}: see Equation (8).

Coefficients for crack damage (b)
Chloride burden (0)/Moisture burden (1)
Parameters

K   the nominal strength of concrete (MPa)
C   the thickness of concrete cover (mm)
A   the air content of concrete (%)
W   the width of crack (mm)
D   the diameter of the corroding steel bar at crack (mm).

$$a_0 = 4.75 \cdot k_{Cw}^{\frac{1}{2}} \cdot k_{H/V}^{\frac{1}{2}} \cdot k_{GIS}^{\frac{1}{2}} \cdot k_{Cem}^{\frac{1}{2}} \cdot k_{Cov}^{\frac{1}{2}} \cdot k_{Frost}^{\frac{1}{2}} \cdot \left(-0.033 + \frac{2.85}{K} \right)^{\frac{1}{2}} \tag{33}$$

k_{Cw}: see Equation (28).

Table 16 Regional coefficient for carbonation

Region	k_{GIS}
Coastal Finland	1
Middle Finland	0.9
Northern Finland	0.76

(*Continued*)

Table 17 Coefficient of surface direction

Surface direction	$k_{H/V}$
Horizontal	1
Vertical	0.93

k_{Cem}: see Table 5.
k_{Cov}: see Equation (8).
k_{Frost}: see Equation (13).

$$n_0 = 0.5 \cdot \left(0.71 - \frac{13.24}{K^{1.5}} \right) \tag{34}$$

$$a_1 = 0.11 \cdot k_{Cw} \cdot k_D \cdot k_{GIS} \cdot k_{Cov} \cdot 0.075 \tag{35}$$

k_{Cw}: see Equation (31).
k_D: see Equation (32).
k_{Cov}: see Equation (8).

Table 18 Regional coefficient for corrosion

Region	k_{GIS}
Coastal Finland	1
Middle Finland	1
Northern Finland	0.76

$$n_1 = 1.18 \tag{36}$$

Coefficients for crack damage (c)
Chloride burden (1)/Moisture burden (0)
Parameters

  K   the nominal strength of concrete (MPa)
  C   the thickness of concrete cover (mm)
  A   the air content of concrete (%)
  W   the width of crack (mm)
  D   the diameter of the corroding steel bar at crack (mm).

$$a_0 = 10 \cdot 0.5^{\frac{1}{2}} \cdot k_{Cw}^{\frac{1}{2}} \cdot k_{Cem}^{\frac{1}{2}} \cdot k_{Cov}^{\frac{1}{2}} \cdot k_{Frost}^{\frac{1}{2}} \cdot \left(-0.43 + \frac{27.7}{K} \right)^{\frac{1}{2}} \tag{37}$$

k_{Cw}: see Equation (28).
k_{Cov}: see Equation (8).
k_{Cem}: see Table 10.

$$n_0 = 0.5 \cdot \left(0.413 + \frac{85\,800}{K^3}\right) \tag{38}$$

$$a_1 = 0.11 \cdot k_{Cw} \cdot k_D \cdot k_{GIS} \cdot k_{Cov} \cdot \left(-0.15 + \frac{8.40}{K}\right) \tag{39}$$

k_{Cw}: see Equation (31).
k_D: see Equation (32).

Table 19 Regional coefficient for corrosion

Region	k_{GIS}
Coastal Finland	1
Middle Finland	0.84
Northern Finland	0.55

k_{Cov}: See Equation (8).

$$n_1 = 1.06 - \frac{65.9}{K^{1.5}} \tag{40}$$

Coefficients for crack damage (d)
Chloride burden (1)/Moisture burden (1)
Parameters

K   the nominal strength of concrete (MPa)
C   the thickness of concrete cover (mm)
A   the air content of concrete (%)
W   the width of crack (mm)
D   the diameter of the corroding steel bar at crack (mm)

$$a_0 = 10 \cdot k_{Cw}^{\frac{1}{2}} \cdot k_{H/V}^{\frac{1}{2}} \cdot k_{GIS}^{\frac{1}{2}} \cdot k_{Cem}^{\frac{1}{2}} \cdot k_{Cov}^{\frac{1}{2}} \cdot k_{Frost}^{\frac{1}{2}} \cdot$$
$$\left(-0.01 + 3.88 \cdot 10^{-6} \cdot (55 - \mathrm{Min}(50; K))^3\right)^{\frac{1}{2}} \tag{41}$$

k_{Cw}: see Equation (28).
k_{Cov}: see Equation (22).
k_{Cem}: see Table 10.

(*Continued*)

Table 20 Regional coefficient for corrosion

Region	k_{GIS}
Coastal Finland	I
Middle Finland	0.78
Northern Finland	0.71

Table 21 Coefficient of surface direction

Surface direction	$k_{H/V}$
Horizontal	I
Vertical	0.93

k_{Frost}: see Equation (23).

$$n_0 = 0.5 \cdot \left(0.97 - 1.3 \cdot 10^{-5} \cdot (55 - \text{Min}(50; K))^3\right) \tag{42}$$

$$a_1 = k_{GIS} \cdot k_{Cw} \cdot k_D \cdot k_{Cov} \cdot k_{Frost} \cdot 0.3 \cdot K \tag{43}$$

k_{Cw}: see Equation (31).
k_D: see Equation (32).

Table 22 Regional coefficient for corrosion

Region	k_{GIS}
Coastal Finland	I
Middle Finland	0.93
Northern Finland	0.71

k_{Cov}: see Equation (22).
k_{Frost}: see Equation (23).

$$n_1 = 1.0 \tag{44}$$

Coefficients for frost attack only
The degree of damage is evaluated using equation:

$$f = a \cdot t^n \tag{45}$$

where

f the amount of damage in relation to the maximum allowable amount of damage,

t the time from the manufacture of the structure,

a the coefficient, and

n the exponent of time.

The interpolation method presented by Equations (3) and (4) is used.
For *Chloride burden (0)/Moisture burden (0)* and *Chloride burden (1)/Moisture burden (0)* the value of Equation (45) is 0.
Chloride burden (0)/Moisture burden (1)
Parameters

K the nominal strength of concrete (MPa)

A the air content of concrete (%).

$$a = 0.0069 \cdot k_{\text{GIS}} \cdot \text{Max}\left(0.6; \frac{132}{27+1.48 \cdot A^{3.4}}\right) \cdot$$

$$\text{Max}\left(0.5; \left(\frac{85.6 - 0.77 \cdot K}{21.5 + K}\right)^3\right) \tag{46}$$

Table 23 Regional coefficient for frost attack

Region	k_{GIS}
Coastal Finland	1
Middle Finland	0.95
Northern Finland	0.83

$n = 1$
Chloride burden (1)/Moisture burden (1)
Parameters

K the nominal strength of concrete (MPa)

A the air content of concrete (%).

$$a = 0.0174 \cdot k_{\text{GIS}}\left(10 \cdot \frac{\left(\frac{28.8}{21.5+K}\right)^{1.25}}{A^{0.5}} - 1\right) \tag{47}$$

k_{GIS}: see Table 23.

$$n = 1 \tag{48}$$

References

[1] Vesikari, E. 2004. *Statistical condition management and financial optimisation in lifetime management of structures. Part 1: Markov chain based LCC analysis. Part 2: Reference structure models for prediction of degradation*, Lifecon, Deliverable D2.2. http://lifecon.vtt.fi/.

[2] Vesikari, E. 1999. Computer simulation technique for prediction of service life in concrete structures, *Proceedings of the international conference life prediction and ageing management of concrete structures*, RILEM, July, Expertcentrum, Bratislava, 17–23.

[3] Vesikari, E. 1999. Prediction of service life of concrete structures with regard to frost attack by computer simulation, *Proceedings of the Nordic residential seminar on frost resistance of building materials*, August 31–September 1, Lund Institute of Technology, Division of Building Materials, Lund.

[4] Vesikari, E. 1998. *Prediction of service life of concrete structures by computer simulation*. Licentiate's thesis, Helsinki University of Technology. Faculty of Civil and Environmental Engineering (in Finnish).

[5] Söderqvist, M. K., and E. Vesikari. 2004. *Generic technical handbook for a predictive life cycle management system of concrete structures*, Lifecon, Deliverable D1.1. http://lifecon.vtt.fi/.

[6] Sarja, A., and E. Vesikari, eds. 1996. Durability design of concrete structures. *RILEM report of TC 130-CSL. RILEM report series 14*. London: Spon, Chapman & Hall.

3.5 Quantitative characterisation and classification of environmental degradation loads

Svein Haagenrud, Christer Sjöström, and Guri Krigsvoll

3.5.1 State of the art

In general, requirements for establishing and implementing quantitative classification systems for durability of materials and components are:

1. well-defined and relatively simple damage functions for the materials in question;
2. availability of environmental exposure data/loads, including methods and models for assessing their geographical distribution; and
3. user-friendly IT systems for storing, processing and modelling of the environmental loads onto the buildings and infrastructures, at the object and network levels.

Establishing proper dose–response and damage functions for families of common building materials has been the subject of extensive studies for more than a decade [1]. Although many models and functions are now available, the lack of knowledge and implementation of the damage function approach still constitutes a major barrier to progress in the durability and service life aspects within the building and construction community.

In most European countries, environmental data and models are available from meteorological offices and the environmental research community, and these data and the research efforts are directly applicable to life cycle management of the built environment. However, lack of knowledge of environmental exposure data among the players in the building sector is a serious barrier to further progress in service life prediction. Such gathering of environmental exposure data is necessary to provide a basis for quantitative classification systems when the service lives of building products have to be stated in quantitative terms [2]. Instructions/guidelines for characterising the environmental loads on buildings and infrastructures on the object and network levels, are provided in this section.

For *quantitative* classification of atmospheric environmental loads, there are two options. One, some systems aim to classify the *generic atmospheric aggressivity* on a global to local scale (ISO 15686 'Service Life Planning – Part 4' and [3]), without specific knowledge of damage functions, but based on overall experience of materials degradation in general. The other option, which is systematic, is material (family) specific and is based on a knowledge of damage functions, such as ISO 9223, that are specific for particular metals [4]. These systems are directly applicable to many of the building materials, but are not directly applicable to complicated damage functions of buildings and infrastructures. The quantitative classification of buildings and infrastructures is therefore very complicated and still lacking [5].

3.5.2 Environmental degradation factors

The degradation of buildings and infrastructures is influenced by a whole set of factors such as environmental degradation agents, type and quality of the materials and components, protective treatment, etc., as described in ISO 15686-1 [6, 7, 8].

The relationship between the environmental degradation agents and the observed effects is expressed as dose–response functions, which are not directly suitable for service life assessments. To transform the degradation into service life terms, performance requirements or limit states for allowable degradation before maintenance or complete renewal of material or component have to be decided. The dose–response function is then transformed into a damage function, which is also a performance over time function, and a service life assessment can be made.

In order to characterise and report the right type and form of the environmental degradation loads, they have to be first related to the degradation mechanism and dose–response functions for the specific materials in question. Further, a holistic approach modelling the physical processes controlling the corrosion needs to be considered across a wide range of physical scales, from macro through meso/regional to local, micro and lastly micron. These scales are defined in line with CIB, see Figure 3.5.1 [1, 2], and see Figure 3.5.2 [9].

Macro refers to gross meteorological conditions (polar, subtropical, etc.), *meso/regional* refers to regions with dimensions up to 100 km, *local* is in the immediate vicinity of an object (building, civil infrastructure, industrial infrastructure, etc.), and *micro* refers to the immediate proximity of a structural component, surface or detail.

Surface response then refers to largely physical responses of a surface, such as deposition and retention of pollutants or condensation and evaporation. *Micron* refers to interactions within the metal/oxide/electrolytic interfaces of buildings and infrastructures. In this approach, models on different dimensional scales are linked together so that the models on micron level are informed by models on the macro-, meso-, micro- and surface-response regimes (see Figure 3.5.2).

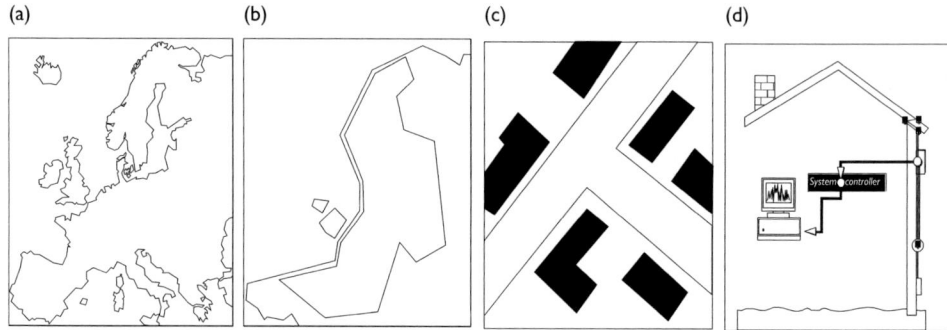

Figure 3.5.1 Exposure environment on different geographical scales [1]: (a) *macro* − Europe map; (b) *meso* − urban area; (c) *local* − road/building; and (d) *micro* − building.

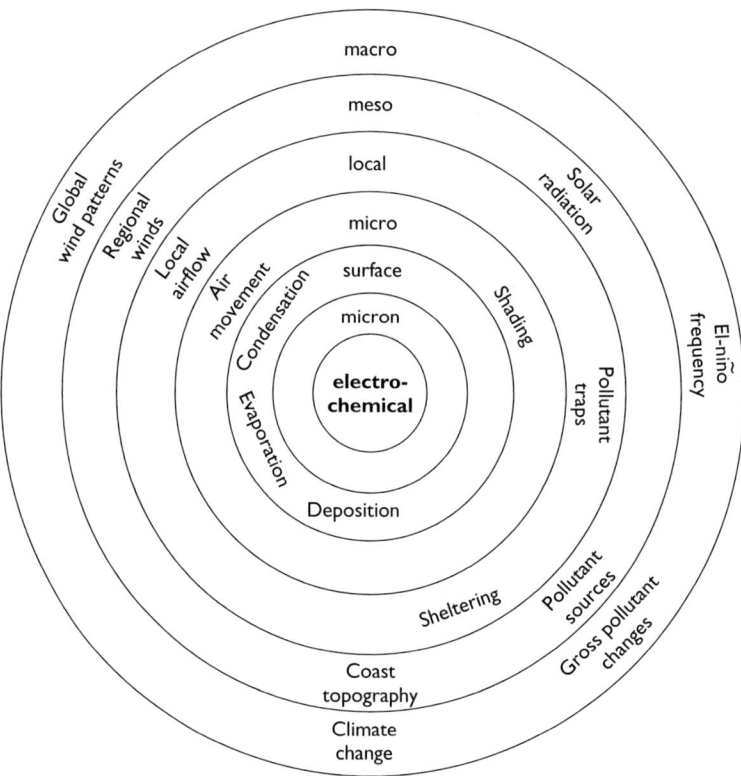

Figure 3.5.2 Framework for holistic model of corrosion, from Cole [9].

The microenvironmental conditions which are crucial to the degradation of materials can vary enormously throughout real construction. An example of such differences is shown in Figure 3.5.3, where the influence of wind and rain has caused a more than 10-fold difference in resulting chloride concentrations on the various parts of the

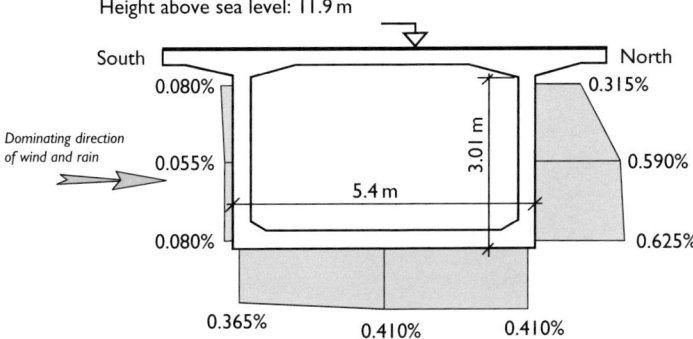

Figure 3.5.3 Gimsøystraumen bridge. Influence of microclimate on the environmental load. The chloride concentration is given as a percentage of the mass of the construction.

Gimsøystraumen bridge. To make precise predictions of the deterioration of buildings and infrastructures, a knowledge of the structure's response to the environmental actions should also be recorded.

The lack of knowledge and implementation of the damage function approach within the building and construction community constitutes a major barrier to progress in the durability and service life aspects. This approach is now widely used in, for instance, high-tech industries and in the medical, biological and agricultural communities.

Worldwide environmental research studies have made valuable contributions to establish such damage functions, which are a necessary basis for the cost–benefit analysis that must precede policy making in these studies. The materials studied comprise structural metals (carbon steel, weathering steel, zinc, aluminum, copper, etc.); stone (limestone and sandstone); paint coatings on coil, steel and wood; electric contact materials; and glass and polymer materials.

Recent research on concrete structures has demonstrated that rational reliability-based service life design methods are possible for a very complex system such as reinforced concrete [10]. This has been developed within the European research projects DuraCrete (Probabilistic performance-based durability design of concrete structures, 1996–1999), and disseminated through the network DuraNet (1999–2002) and further applied in the object specific project DARTS (Durable and Reliable Tunnel Structures, 2001–2004), as well as LIFECON (Life-Cycle Management of Concrete Infrastructures, 2001–2004) [11].

Table 3.5.1 shows the required environmental primary parameters/data as extracted from the DuraCrete models for concrete, and damage functions for some of the most important supplementary materials.

The derived parameters (e.g. time of wetness) and formats for the parameters have to be extracted from the complete functions. The characterisation should concentrate on the primary parameters, and the data should be given as time series, from which the mean values with standard deviations can be calculated.

Table 3.5.1 Relevant environmental primary data for degradation models linked to buildings and infrastructures

Deterioration mechanism	RH	Temp.	CO$_2$	Precipitation	Wind	Radiation	Chloride Conc.	Freeze–thaw cycles	[SO$_2$]	[O$_3$]
Reinforced concrete (DuraCrete models)										
• Carbonation-induced corrosion	X	(X)	X	X		X				
• Chloride-induced corrosion	X	X		X			X			
• Propagation of corrosion	X	X		X			X			
• Alkali-aggregate reaction	No model									
• Frost attack internal/scaling	(X)	X		X	(X)	(X)	(X)	X		
Supplementary materials (dose–response functions)										
• Galvanised steel/zinc coating	X	X		X			X		X	
• Coil-coated steel	X	X		X					X	
• Sealants/bitumen	No function									
• Polymers	No function									
• Aluminium				X			X		X	X

Note
(X) = contained as derived parameters.

3.5.3 Methods and data for assessments, modelling and mapping of degradation agents

3.5.3.1 Data from meteorological and environmental research networks

In principle, the characterisation of degradation agents has to be based on existing data. Such data and models are available from meteorological offices and environmental research community, and are directly applicable to the assessment of durability and service life of the built environment.

The measuring, testing and evaluation of air quality (pollutants) are gaining growing importance in developed countries as elements of a comprehensive clean air policy geared to sustainable development. All European countries have extensive meteorological and air pollution monitoring networks, many with GIS-based information and management systems, allowing for the necessary assessment, modelling and mapping of the relevant environmental degradation parameters on various scales down to the local/micro scale on the object level. Point measurements are very expensive, and for a broader assessment of air quality – needed for policy development and assessment, public information, etc. – the measured data need to be combined with modelling based on emissions inventories, to properly assess the exposure to and thus the effects of pollution on public health or on buildings.

The European Environment Agency (EEA) (www.eea.dk) was established in 1994, with the objective 'to provide to the European Community and its Member States objective, reliable, and comparable information at a European level enabling the Member States to take the requisite measures to protect the environment, to assess the results of such measures and to ensure that the public is properly informed about the State of the environment' [1]. Thus EEA provides annual summaries of the state of air pollution–monitoring situation in Europe, providing detailed information on networks, sites, compounds, reporting, etc. by country, covering more than 30 countries from which data are available.

On the *regional scale*, there is extensive monitoring in addition to the EMEP network, and about 750 sites are in operation totally in Europe. This monitoring is very extensive for sulphur and nitrogen compounds in air (gases and particles) and deposition, and also for ozone.

On the *local/urban scale*, monitoring is carried out at more than 5000 sites in Europe, operated by local, regional or national authorities. Most of these sites seem to be general urban background sites, while hot-spot sites (traffic, industry) are less well represented. The compounds mentioned in EU Directives (SO_2, particles, NO_2, ozone, lead) are extensively covered.

3.5.3.2 Mapping of environmental data and corrosion

Mapping of environmental parameters, and of areas and building stock at high risk of corrosion, is performed and co-ordinated for Europe within the UN ECE Working Group on Effects (WGE) under the Convention of Long Range Transport of Air Pollutants (CLRTAP), and specifically under its International Cooperative Programmes (ICPs), for Modelling and Mapping and for Materials respectively ([12, 13] see more on www.rivm.nl/cce).

Mapping in 50×50 km grids is possible for the whole of Europe, and for many regions a resolution of 1×1 km is possible. This will give information on regional differences in parameters between within each country. Regions may be geographical or 'topographical' (coast/inland, mountain/low-lying country, rural/urban).

It is obvious that the data and methodologies developed within these environmental research frameworks are directly applicable to asset management tasks, and great synergy could be achieved by linking them directly. The concepts have many similarities: establishing damage functions, measuring and co-ordinating all existing and relevant European environmental data in order to do proper mapping of degradation parameters, areas and stock at risk, etc. The network of its co-operating partners would be invaluable data providers for asset managers, and vice versa. Asset management, on the other hand, could provide important data on mapping of stock at risk, corrosion costs, etc., which would constitute important inputs to the work on improving the corrosion cost mapping for Europe.

3.5.3.3 Chloride exposure from sea salt and from de-icing salt

Chlorides are important degradation factors. Data on land-transported and wet-deposited chloride in Europe are available from the environmental research network, as shown by the example, Sweden, given later in this section. To map the near-sea (<1 km) influence, however, models for surf-produced chloride have to be implemented.

Cole has established the models for production, transport and deposition of marine salts (marine aerosols) [9], as well as the integration of these models into GIS-based computerised tools [14]. These models will allow calculation of the marine aerosol impact in coastal areas of Europe, i.e. from macro down to local level (Figure 3.5.4).

Some of the conclusions are:

- salt production by breaking surf is estimated to lead to peak aerosol concentrations up to 40 times higher than salt production by whitecaps on ocean waves;
- the concentration of surf-generated aerosol falls dramatically with distance from the coast so that under typical conditions little aerosol is transported more than 1 km;
- the concentration of ocean-generated aerosol falls off more gradually so that this aerosol may be transported >50 km under appropriate conditions;
- the transport of aerosol produced by both surf and ocean whitecaps is strongly affected by wind speed;
- the transport of surf-generated aerosol is dramatically affected by terrain roughness, whilst that of ocean-generated aerosol is more affected by variations in relative humidity and rainfall.

In a cooperative effort within the LIFECON project, Cole's models were used to characterise and classify the marine exposure to greater aerosols at the chosen objects in the Oslo fjord bay and along the coast.

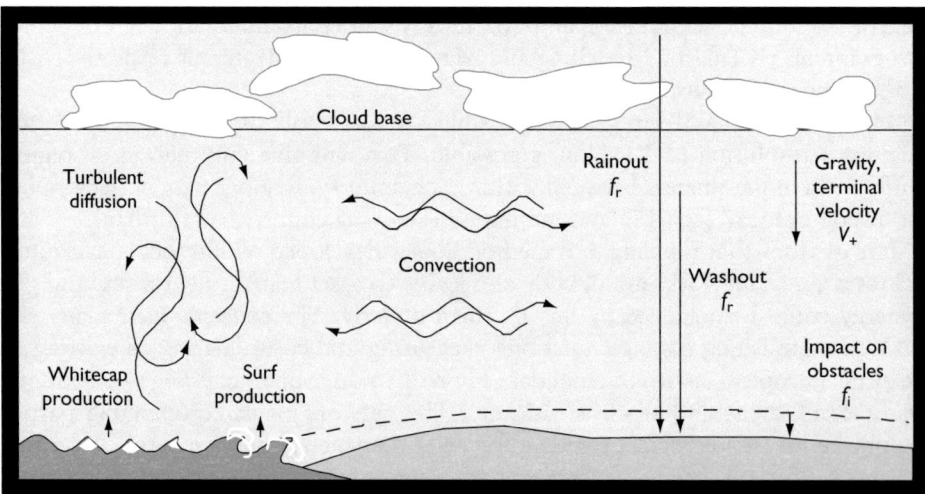

Figure 3.5.4 Chloride sea-salt aerosol mass transport model [9].

Where relevant, information concerning *chlorides from de-icing* salts must be obtained from the appropriate pollution authorities. The formula to be used is:

$$Cr = 1000(-9.56 + 0.52SF + 0.38SL + 0.14FD - 0.20ID)/w$$

where
  Cr   average amount of de-icing salt for each application incident $(g\,m^{-2})$
  SF   number of days with snow fall $>0.1\,mm$
  SL   number of days with a snow layer $>100\,mm$
  FD   number of days with an average daily temperature $>0\,°C$,
  ID   number of days with ice on the road surface
  w average spread width (street width) in metres.

Lay [15] collected 640 chloride profiles from German road administrations (160 structures) and analysed them with respect to the most relevant influences. The analysis revealed for the concentration of chloride, $C_{\Delta x}$:

• a linear decrease with increasing height (h) above street level;
• a logarithmic decrease with increasing horizontal distance (a) from the street edge; and
• no trend with respect to exposure time.

These relationships were combined to derive a function for $C_{\Delta x}$, mean, the mean chloride concentration at depth Δx (wt – %/construction mass):

$$C_{\Delta x,\,mean}(a, h) = [(0.465 - 0.051) \times \ln(a + 1)] - [0.00065 \times (a + 1)^{-0.187} \times h]$$

3.5.3.4   ICT systems

Today, surveillance and management of air quality can be facilitated and performed by means of total information systems. The air quality information system, AirQuis (www.airquis.com), developed by the Norwegian Institute for Air Research, represents the air pollution part of a modern air quality management system. The combination of online data collection, statistical evaluations and numerical modelling enables the user to obtain information and to carry out forecasting and future planning of air quality. The system can be used for monitoring, and to estimate environmental impacts from planned measures to reduce air pollution. Thus the AirQuis system also contains the module CorrCost, which is a Geographic Information System (GIS) platform for modelling material damage, and cost estimates based on environmental parameters. Dose–response functions and lifetime equations can be calculated and the resulting costs for maintenance, repair and re-placement estimated [16].

Cole has also used GIS-based systems to model and exhibit climatic and pollutant parameters for Australia and Southeast Asia [14, 17].

An ICT-based life cycle management system (LMS) was developed in the LIFECON project [18]. When the degradation factors have been assessed and modelled, the LMS Environmental Risk Factor (ERF) and Service Life Prediction (SLP) modules contain tools that also allow for modelling and mapping of the corresponding damage and service life functions.

3.5.3.5 Microenvironmental exposure

The available regional exposure data can, after appropriate adjustment, be used for characterisation of the local – environment and microenvironment of a building or construction at the object level.

The microclimate is heavily influenced by the macroclimate. The importance of various factors will, of course, vary for different types of construction objects, and where these objects are used in relation to the orientation of construction and their position on or within the construction.

The prediction of the hazard for a component within the construction envelop is more complex, and in general can only be estimated by some kind of transfer function which considers the relevant external parameters and construction design and material characteristics. In order to estimate the environmental conditions for construction surfaces and the internal material conditions, an approximation of mass (water, gases) and heat transfer is required.

The moisture content or water availability is important for the corrosion processes. Precipitation and relative or absolute humidity are measured at meteorological stations. Time of wetness (ToW) may be calculated from these meteorological data.

Different methods may be used to describe or express the quantity of water at a wall or construction. In addition to the methods using the measured data directly, there are standards such as prEN 13013–3, 'Calculation of driving rain index for vertical surfaces from hourly wind and rain data' [19]. The standard specifies a procedure for analysing hourly rainfall and wind data derived from meteorological observations so as to provide an estimate of the quantity of water likely to impact on a wall of any given orientation. It takes account of topography, local sheltering and the type of building and wall. It specifies the method of calculation of:

- the annual airfield index (I_A), which influences the moisture content of a masonry wall; and
- the spell index (I_S), which influences the likelihood of rain penetration through a masonry wall.

After calculating the I_S for a certain period of time, the next step is to estimate the location and exposure of a building compared to an airfield. This is done by estimating the values of four different parameters: the roughness coefficient (C_R), the topography coefficient (C_T), an obstruction factor (O), and a wall factor (W). And converting the airfield indices into wall spell indices (I_{WS}) by the formula:

$$I_{WS} = I_S \times C_R \times C_T \times O \times W$$

The prEN 13013–3 has categorised, described, and illustrated the C_R, C_T, O and W factors.

The effect of road traffic pollution on urban populations is expected to increase during the next few years. Traffic planners often require practical tools for studying the effects of this on the environment. Several air dispersion models exist and can be used for this purpose [1].

The AirQuis application contains the road model *RoadAir*, for quantitative descriptions of air pollution along road networks. RoadAir calculates total emissions, concentrations along each road segment and the air pollution exposure of the population and *buildings* along each road. Calculations can be carried out for road networks, defined by road and traffic data. The model was primarily developed for conditions in Scandinavia, but can easily be adapted to conditions in other parts of the world.

The use of Computational Fluid Dynamics (CFD) makes it possible to calculate the microclimate around the construction as well as at the surfaces. The wind pattern around the construction is determined by solving Navier–Stokes equations, and the distribution of snow, rain and gaseous components is found using the wind pattern and transport equations.

Cole [14] has used CFD models to map the transport of chloride aerosols in the environment. This methodology is now also extensively used by the Norwegian Building Research Institute (NBI), which used it to calculate the microenvironmental pattern around the two chosen objects for Norwegian Coast Directorate in the LIFECON project [20].

3.5.4 Systematic and options for classification of environmental degradation loads

Characterising and subsequently classifying the exposure environment in order to assess its aggressivity towards buildings and infrastructures has been attempted for about three decades, and some of the most relevant systems are now described.

This is the preferred systematic but it requires that the type and format of the ingoing environmental agents are defined, and that the function(s) are relatively simple for practical purposes. This is seldom the case, but an example is given below of how very complex functions, like the DuraCrete models for concrete corrosion, can be simplified for classification purposes.

3.5.4.1 EOTA – Annex A Building context [3]

As stated in the EOTA document 'Working Life of Building Products':

> The wide variation in European climatic conditions and in the user stresses imposed on structures depending upon type of structure and use intensity will make it necessary with many construction products to restrict their usage to defined situations in order that these achieve the predicted working life.

Examples of possible sub-divisions of climatic zones in Europe, of orientation of products/components in structures, of internal exposure environments in buildings, etc., are also provided by the document. The EOTA-proposed macroclimatic subdivision is used in the condition assessment protocol for environmental characterisation of selected objects (see Table 3.5.5).

3.5.4.2 ISO 15686 Service Life Planning – Part 4

The suggestions for classification are from ISO/WD 15686-4 Buildings – Service Life Data Sets – Part 4 – Service Life Prediction Data Requirements [21].

1. Global climatic classification
    A simplified classification of the climate with respect to two main factors, *rainfall/humidity* and *temperature*.

    1. *Rainfall/humidity* may be divided into four main classification:

        *   Dry – rainfall less than 400 mm per year or average yearly 9 am relative humidity <50%
        *   Sub-humid – rainfall is between 400 and 800 mm per year or average yearly 9 am relative humidity >50% and <70%
        *   Humid – rainfall is between 800 and 1300 mm or average yearly 9 am relative humidity is >70% and <80%
        *   Very humid – rainfall exceeds 1300 mm or average yearly 9 am relative humidity >80%.

    2. The *temperature* dimension has been divided into the following ranges:

        *   Cold – the average monthly minimum temperature is above $-5\,°C$ for more than 2 months of the year; alternatively, the average monthly maximum temperature for the hottest month is below $10\,°C$
        *   Temperate – the average monthly minimum temperature is above $-50\,°C$ for no more than 1 month of the year and the average monthly maximum temperature is below $35\,°C$ for no more than one month
        *   Hot – the average monthly temperature is below $35\,°C$ for more than 1 month of the year.

2. Global pollutant classification

    1. Determination of the level of pollution that building materials are subjected to require determination of a number of factors such as:

        *   airborne salinity
        *   airborne pollutant level (SO_2)
        *   rain acidity
        *   rain-ionic content
        *   relative frequency of salt deposition/rain events
        *   relative frequency of high pollutant/rain events
        *   average wind speed
        *   extreme wind speed.

    2. The pollutant classification is divided into two main areas, *industrial pollution* and *marine pollution* with the following definitions:

- severe marine (SM) – airborne salinity exceeds a daily average of $300\,mg\,m^{-2}\,day^{-1}$
- marine (M) – average daily airborne salinity is between 60 and $300\,mg\,m^{-2}\,day^{-1}$
- severe industrial (SI) – airborne SO_x level exceeds $200\,mg\,m^{-2}\,day^{-1}$
- industrial (I) – airborne SO_x level is between 60 and $200\,mg\,m^{-2}\,day^{-1}$
- severe industrial and marine (SI + M) – airborne salinity exceeds $300\,mg\,m^{-2}\,day^{-1}$ and SO_x level exceeds $200\,mg\,m^{-2}\,day^{-1}$
- light marine or industrial – any one of the following criteria must be met:

    - airborne salinity is between 15 and $60\,mg\,m^{-2}\,day^{-1}$
    - airborne SO_x level is between 10 and $80\,mg\,m^{-2}\,day^{-1}$ or
    - rain water pH <5.5

- benign (B) – All the following criteria must be met:

    - airborne salinity $<15\,mg\,m^{-2}\,day^{-1}$
    - airborne SO_x $<10\,mg\,m^{-2}\,day^{-1}$
    - rain water pH >5.5.

From those classes, a combined system can be established by combining the climate with its subclass versus the pollutant source. The environment can be defined by a three-figure number: the first number defines the pollutant sources, severe marine and severe industrial (SM + SI = 1) to bening (B = 9); the second defines the major climatic class (1 for dry to 4 for very humid); and the third defines the sub-class (1 for cold to 3 for hot).

3. Detailed classification of moisture from rainfall and relative humidity. Another, more detailed approach for classification of moisture is to use the annual rainfall and annual relative humidity.
4. Detailed pollutant classification of airborne salinity, frequency of significant salt deposition and frequency of rain.

3.5.4.3 ISO 9223–26 Classification of atmospheric corrosivity for metals

The standards ISO 9223 *Corrosion of metals and alloys* to ISO 9226 *Corrosivity of atmospheres* have been developed for the classification of atmospheric corrosivity of metals and alloys. The development was based on the Czech approach, which was used as early as 1975 to map atmospheric corrosivity in northern Bohemia. Based on a huge amount of experimental data for empirical dose–response functions, the standards use the approach of classifying both the degradation factors and the corrosion rates.

The ISO 9223 classification specifies the key factors in the atmospheric corrosion of metals and alloys, which are *time of wetness* (τ), *sulphur dioxide (P)* and *airborne salinity (S)*. Corrosivity categories are defined on the basis of these three factors and used for the classification of atmospheres for the metals (alloys and unalloyed steel, zinc and copper, and aluminium). TOW is described in five classes, SO_2 and chloride in four classes and corrosivity in five classes. The classification can be used directly to evaluate the corrosivity of atmospheres under known conditions of these environmental factors, and for technical and economical analyses of corrosion damage and choice of protection measures.

The ISO 9223 approach has, since the mid-1980, been used by many researchers to classify and map atmospheric corrosivity [1].

3.5.4.4 ISO DIS 12944–2 Paints and varnishes

The ISO 9223 approach for classification of atmospheric corrosivity has also, with appropriate amendments, been used to describe the degradation environment for non-metals, as in ISO DIS 12944–2:
Classification of Environments. This standard deals with the classification of the principal environments to which steel structures are exposed, and the corrosivity of these environments. It defines categories atmospheric of corrosivity, based on ISO 9223 and ISO 9226.

3.5.5 Quantitative classification of concrete

3.5.5.1 European standard – EN 206–1

The recently endorsed European standard 'EN 206–1 Concrete – Specification, performance, production, and conformity' [22] is a good basis for developing the quantitative system. EN 206–1 contains an agreed *qualitative* classification system as a synthesis of 'best available' knowledge, covering the relevant degradation mechanisms and exposures in atmospheres, fresh water, seawater and soil, indicating the decisive character of moisture and chloride. One way in which a quantitative classification system for environmental exposure can be developed is thus by extensive assessment of the degradation modes and the microenvironment on a sufficient sample of objects in practice. This implies also that national annexes describing the environmental classes in relation to geography have to be developed, as shown by the work by the Norwegian Building Standardisation Organisation [23].

The EN 206–1 should therefore be tested out on the chosen objects, and then eventually the qualitative classes should be replaced by more quantitative classes, by comparing the observed degradation on the objects with the characterisation of environmental parameters on the same objects.

3.5.5.2 Quantitative classification of environmental loads

3.5.5.2.1 GENERAL PROCEDURE

The statistical degradation models, which describe the complex processes occurring on the various levels (see the holistic model in Figure 3.5.2), should form the basis for classification. However, the DuraCrete models are too complex, and have to be simplified for classification purposes. Lay [15] did this by means of a parameter study, aiming to define class boundaries for environmental data. Once these classes are defined, environmental data can be mapped as described.

In the parametric study, the depth of corrosion penetration in the damage functions is taken as the performance characteristic for classification, together with relevant environmental parameters. A framework of five classes (0–4) was chosen, and the class boundaries are to be determined by dividing the maximum of the damage measure (e.g. corrosion penetration depth) into five equal sections. The severest extent of damage will be found in class 4, and in class 0, *no damage* will occur within the time scale.

The entire process, usually consisting of the initiation and propagation periods, has to be modelled. This is necessary because some parameters may have an inverse influence on the duration of either phase. For instance, with rising RH, the time until initiation will increase, because the carbonation process is retarded. Nevertheless, rising RH leads to an increase in electrical conductivity and thus to an increase in the corrosion rate during the propagation phase. Hence, a value of RH must exist for which the service lifetime of a reinforced concrete element tends to a minimum.

The whole process is illustrated by the classification of RH in carbonation-induced corrosion (XC-class).

A parametric study (given in Table 3.5.2) was performed to distinguish two qualities of concrete, both produced with CEM I cement:

1. good low water–cement (0.45), high concrete cover (50 mm), 7 days curing; and
2. bad high water–cement ratio (0.60), low concrete cover (15 mm), 1 day curing.

The study revealed, as expected, that the average time until depassivation of 'good' concrete is far beyond the scale of commonly requested service lives, and that the decisive parameter was the concrete cover. Thus, the classification was only performed for 'bad' concrete. The result of the calculations for the corrosion due to carbonation is the corrosion penetration depth with respect to the parameters (RH, ToW, T and ΔC_S) for periods of 30 and 100 years.

For the study of one environmental parameter, all other environmental data must be kept constant on an average level (European or national), as can be seen from Table 3.5.3a and b. The TOW (days with rain above 2.5 mm) was set to 0, as surfaces exposed to rain will not usually suffer from carbonation-induced corrosion, unless the concrete cover is very low.

Table 3.5.2 The environmental parameters included in the service life models for carbonation-induced corrosion (see Table 3.5.3a for definitions)

Carbonation-induced Corrosion	
Initiation	*Propagation*
$k_{RH}(RH)$, $w(ToW)$, ΔC_S	$k_{R,T}(T)$, $k_{R,RH}(RH)$, ToW

Table 3.5.3a Input data for parameter study on carbonation-induced corrosion (initiation phase)

No.	Input	Sub-input	Unit	D type	Mean Good	Mean Bad	StD Good	StD Bad	Limits Min	Limits Max
1	d_{cover}		mm	Beta	15		10		0	100
		RH	%	Weibull (max)	$75^1 (40-100)^2$		15		–	100
2	k_{RH}	RH_{ref}	%	Const.	65		–		–	
		g_{RH}	–	Const.	2.5		–		–	
		f_{RH}	–	Const.	5.0		–		–	
		a_c	–	Const.	3.01		–		–	
3	k_c	b_c	–	ND	-0.567		0.024		–	
		t_c	–	ND	1		–		–	
4	k_t		–	ND	1.25		0.35		–	
5	ε_t		$m^5 (s\,kgCO_2)$		$1.0 \cdot 10^{-11}$		$0.15 \cdot 10^{-11}$		–	
			$mm^5 (a\,kgCo_2)$	ND	(315.5)		(48)		–	
6	ΔC_s		$kgCO_2 m^{-3} \times 10^{-4}$	ND	$8.2^1 (6.2-20)^2$		1.0		–	
7	T		a	Const.	30 or 100		–		–	
		ToW	–	Const.	$0^1 (0-100)^2$		–		–	
		b_w	–	ND	0.446		0.163		–	
8	W	p_{splash}	–	Const.	1		–		–	
		t_0	a	Const.	0.0767 (equals 28d)		–		–	
		a_w	–	Const.	0.5		–		–	
9	$R^{-1}_{ACC,o}$		$m^5 (s\,kgCO_2) \cdot 10^{-11}$	ND	3.1 (982)	13.4 (4243)	1.67 (149)	5.21 (465)	–	
			$mm^5 (a\,kgCO_2)$							

Note
D type – distribution type. StD – standard deviation. 1 value chosen as constant for the study of other environmental parameters. 2 range of parameter study. 3 CEM I; $w/b = 0.40$. 4 CEM I; $w/b = 0.60$.

Table 3.5.3b Input data for parameter study on carbonation-induced corrosion process (propagation phase)

No.	Input	Sub-input	Unit	D type	Mean Good	Mean Bad	StD Good	StD Bad	Moments Min	Moments Max
1	α		–	Const.	2		–		–	
2	k_0		$\mu m\,\Omega m a^{-1}$	LogN	882		–		–	
3	F_{Cl}		–	sLN	2.63		3.51		1.09	–
4	F_{Galv}		–	Const.	1.0		–		–	
5	F_{O_2}		–	Const.	1.0		–		–	
6	$\rho(t)$	RH	%	Weibull max	751 (40–100)		15		–	100
		k_{RH}	–	Const.	3.16		2.05		–	
		ToW	–	Const.	0^1 0–1		(0–1)		–	
		T	K	Beta	283 (273–293)		8		253	313
		b_T	K	ND	3815		560		–	
		k_c	–	Const.	1.0		–		–	
		k_t	–	Const.	1.0		–		–	
		$k_{R,Cl}$	–	Const.	1.0		–		–	
		n	–	ND	0.23		0.04		–	
		t_{hydr}	a	Const.	1		–		–	
		t_0	a	Const.	0.0767		–		–	
		ρ_0	Ωm	ND	115	41	14.4	6	–	

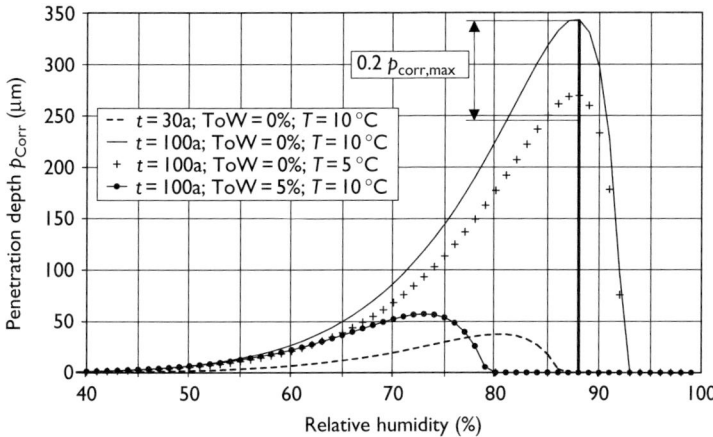

Figure 3.5.5 Corrosion penetration depth of reinforcement with respect to RH at ages 30 and 100 years (a) and variation of temperature (*T*) and time of wetness (ToW).

3.5.5.2.2   CLASSIFICATION OF RH

A rise in RH slows down the carbonation process but accelerates the subsequent corrosion once initiated. The depth of corrosion penetration was regarded as a measure. The study of the effect revealed the following (see Figure 3.5.5):

- A maximum depth of corrosion penetration can be observed. On the left side of the maximum, corrosion is the limiting process. On the right, carbonation is the limiting factor.
- The RH for which a maximum depth of corrosion penetration is observed depends on the ratio of initiation and propagation time, and hence on the intended service life.
- With increasing intended service life, the pessimum of RH is shifted towards larger values (80% after 30 years; 88% after 100 years), as the corrosion process is given more weight in the study.
- A change in corrosion rate due to a change in temperature or concrete porosity (not due to a change in RH) alters the maximum depth of corrosion penetration, but does not shift the pessimum of the RH.

For the purpose of classification, a period of 100 years was considered. In continental Europe, the yearly average RH, calculated from hourly values, lies between 60% (Spain, Madrid) and 88% (Germany, Fichtelgebirge). *In this study, the range of 60–92% (the largest value for which depassivation will still occur within 100 years) was classified.* The boundaries of five classes (0–4) were determined by dividing the maximum penetration depth into five equal sections as shown in Figure 3.5.6. The severest extent of damage will be found in class 4, and in class 0, the lowest. A difference of one class in RH equals 20% of the maximum of penetration depth at 88% RH.

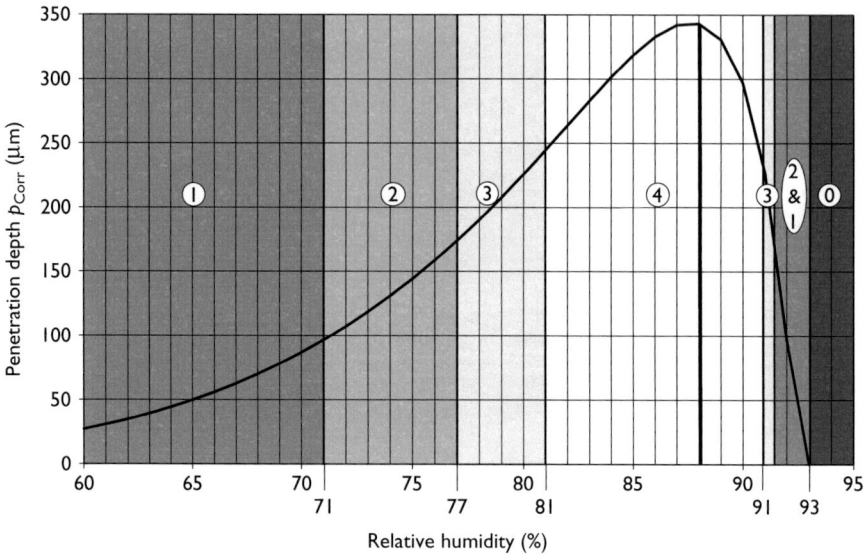

Figure 3.5.6 European classification of average annual RH for carbonation-induced corrosion, assuming 100 years of service life, $ToW = 0$; $T = 10\,^\circ C$.

3.5.5.2.3 CLASSIFICATION OF ENVIRONMENTAL PARAMETERS FOR CARBONATION- AND CHLORIDE-INDUCED CONCRETE CORROSION

Lay [6] conducted a parametric analysis and concluded with the proposed quantitative classification of important environmental degradation parameters for concrete in Europe, as shown in Table 3.5.4. Both of these types of corrosion can and should be modelled and mapped on the macro to local scale in Europe, thus serving as basis for validation.

Table 3.5.4 Proposed quantitative classification scheme of important degradation agents for carbonation- and chloride-induced corrosion of concrete (after Lay [15])

Agent/Class	0	1	2	3	4
Carbonation-induced corrosion					
Av.Ann.	<60[1]	60–71	>71–77	>77–81	>81–90
RH	or			or	
domain (%)	>93[1]			>90–91.5	
Av. Ann. ToW domain (%)	>6.6	>5.5–6.6	>4.0–5.5	>2.0–4.0	0–2.0
Av. Ann. temp domain (°C)	>0–6.4	>6.4–11.6	>11.6–16.0	>16.0–20.0	
Chloride-induced corrosion					
Temp (°C)		0–8	8–9.5	9.5–11	11–13	>13

3.5.6 Guidance for classification of exposure environment of structures

When characterising the exposure loading of an object, both the surroundings and the components of the object must be taken into account. The different components are exposed in different ways and to different extents, according to their orientation, sheltering, sun/shadow, distance from source for exposure, among other factors, and all of these have to be taken into account.

A stepwise procedure for characterising of the environmental parameters on the surface of the structure, giving a basis for subsequent classification, can be described as follows:

1. Choose object.
2. Divide the structure/construction into components with different expected categories of location (due to orientation, sheltering, etc.) Various systems should be used, depending on the expected impact.
3. For concrete structures, attain EN 206–1 exposure classes to the various components/parts of the construction.
4. Adjust for the effect of sheltering, etc., on driving rain and deposition on other agents to the structure by calculation of C_R, C_T, O and W from prEN 13013.
5. Find climatic information from nearby meteorological stations. Necessary information:

   (a) *Temperature* Preferably, if available, time series for a long period (>10 years). Main information is average temperature for summer and winter conditions, max/min and average monthly temperature.
   (b) *Moisture* Preferably, if available, time series for a long period (>10 years). Main information is average annual precipitation and monthly number of days with rain >0.1 mm and rain >2.5 mm. Monthly or seasonal RH.
   (c) *Wind* Preferably, if available, time series for a long period (>10 years). Main information is wind rose showing frequencies of wind speed and direction.

6. Check the correlation or relevance of obtained meteorological data for the object in question, by assessing:

   (a) From some (2–4) nearby stations – any significant difference in meteorological data?
   (b) Distance from meteorological station.
   (c) Height above sea level. Normally the average temperature decreases 0.6–0.7 °C per 100 m.
   (d) Sunny/shadowed areas (for instance in valleys). A difference of 0.5–1°C in air temperature may be expected.
   (e) Topography – differences in wind speed and direction.

7. Calculation of spell index, wall spell index and driving rain (Section 3.5.3.5). For concrete and the various parts of the structure classified under EN 206–1:
8. Characterisation of RH

9. Characterisation of moisture: total time with moisture is calculated as time with rain + condensation + high RH
10. Characterisation of temperature profiles on construction
11. Characterisation of chloride, using Cole's models for sea-salt or mapping authorities for land-transported sea-salt, ref example from Germany, or deicing salts (see Section 3.5.3.3).
12. Characterisation of pollutants such as SO_2, O_3, H^+ and CO_2. Contact national (and local) ICP Modelling and Mapping groups to obtain already mapped information. Find available environmental data from national or local authorities (Table 3.5.5).

3.5.7 Modelling and mapping of damage functions

Once the degradation factors have been assessed and modelled, the LMS ERf and SLP modules contain tools for modelling and mapping of the corresponding damage – and service life functions, or for that matter the dose–response functions as seen for Sweden [18].

Table 3.5.5 Protocol for environmental assessment of case studies

Component	Category of Location EOTA system			
	1 (horizontal or low slope surfaces (<20°)	2 (Steep slope (>20°)	3 (Vertical)	4 (Underside of horizontal and sloping surface)

Component	Exposure classes (EN 206-1)

Environmental parameter characteristic surroundings							
RH (%) summer	RH (%) winter	Days w/rain >2.5 mm	Precipitation (mm y^{-1})	Condensation	Av. air temp (summer)	Av. air temp (winter)	Main wind direction

Environmental parameter characteristic components					
Component	Surface T (summer)	Surface T (winter)	CR	CT	O	W

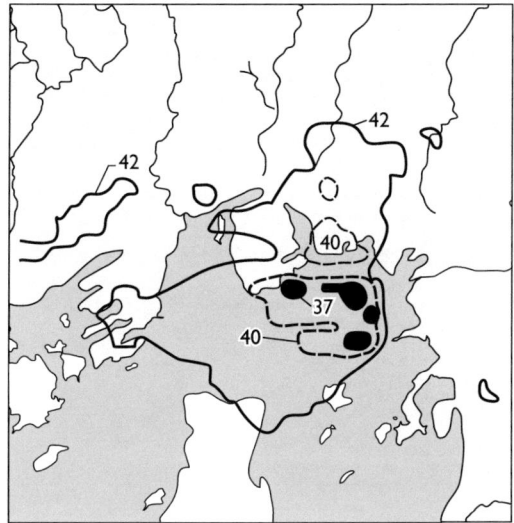

Figure 3.5.7 Maintenance intervals (lifetime) for Zinc-coated steel in years for Oslo in 1994 given in 500 × 500 m grids from the CosBen model [16].

Corrosion maps for some supplementary materials, such as weathering steel, zinc, aluminium, copper and bronze, already exit. An example is shown in Figure 3.5.7 which shows a map for the service life/maintenance intervals for Zinc-coated steel, in Oslo, based on 1994 exposure data.

Cole has also used GIS-based systems to model and exhibit climatic and pollutant parameters for Australia and South-east Asia [14, 17]. Under contract to the Australian Galvanizers Association, he has also mapped the corrosivity and service life for galvanised coatings in Australia and South-east Asia as shown in Figure 3.5.8 (http://www.dbce.csiro.au/biex/indgalv/).

3.5.8   Classification of environmental degradation loads: Examples

3.5.8.1   Concrete structures in the Oslofjord marine environment

3.5.8.1.1   INTRODUCTION

The exposure environment for selected pilot objects in the five participating countries was characterised in the LIFECON project [11]. The priority parameters to be characterised for deterioration of concrete are RH, temperature, precipitation/ToW ($d > 2.5$ mm), chloride (from sea-salt deposition, de-icing salt or precipitation), freeze–thaw cycles (number of days below $0°C$), and SO_2 as supplementary material.

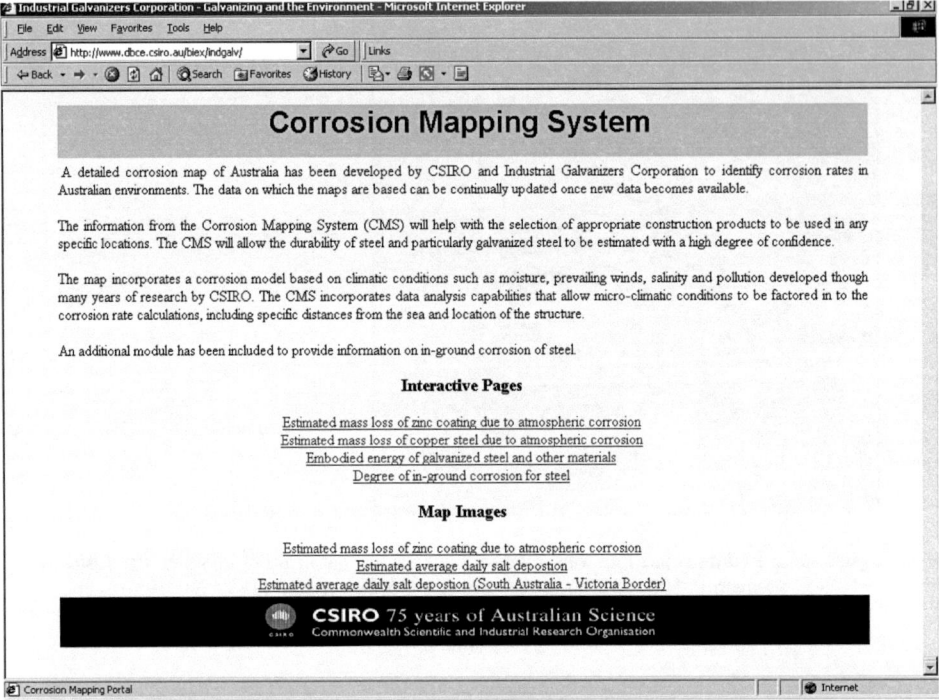

Figure 3.5.8 Web page showing the CSIRO Corrosion Mapping System.

Some of the objects studied were also subject to comparative assessment of conditions according to EN 206–1 [11]. Examples are Ormsund quay and Homborsund lighthouse in the marine environment of the Oslo fjord, Norway, (see Figure 3.5.9).

Ormsund quay located in Oslo Harbour has a total length of 272 m and consists of two parts, which were built in 1977 and 1986, see Figure 3.5.10. The structural system consists of 27 sections, each deck slab supported by 2 secondary beams one at the front and one at the back, and 2 main beams supported by 3–9 reinforced concrete pillars. The heights above mean seawater level are 2.2 m for the top of the the wharf deck, 1.7 m for the Bottom slab and 1.0 m for the bottom of the secondary beams; and the difference between high water autumnal equinox and average water level at this location is 0.3 m.

Three-dimensioned models of the Ormsund quay were constructed on the basis of 1:20 and 1:100 scale drawings and photographic documentation. Figure 3.5.10 shows the underside of the quay. The 3D models were used as input to CFD simulations.

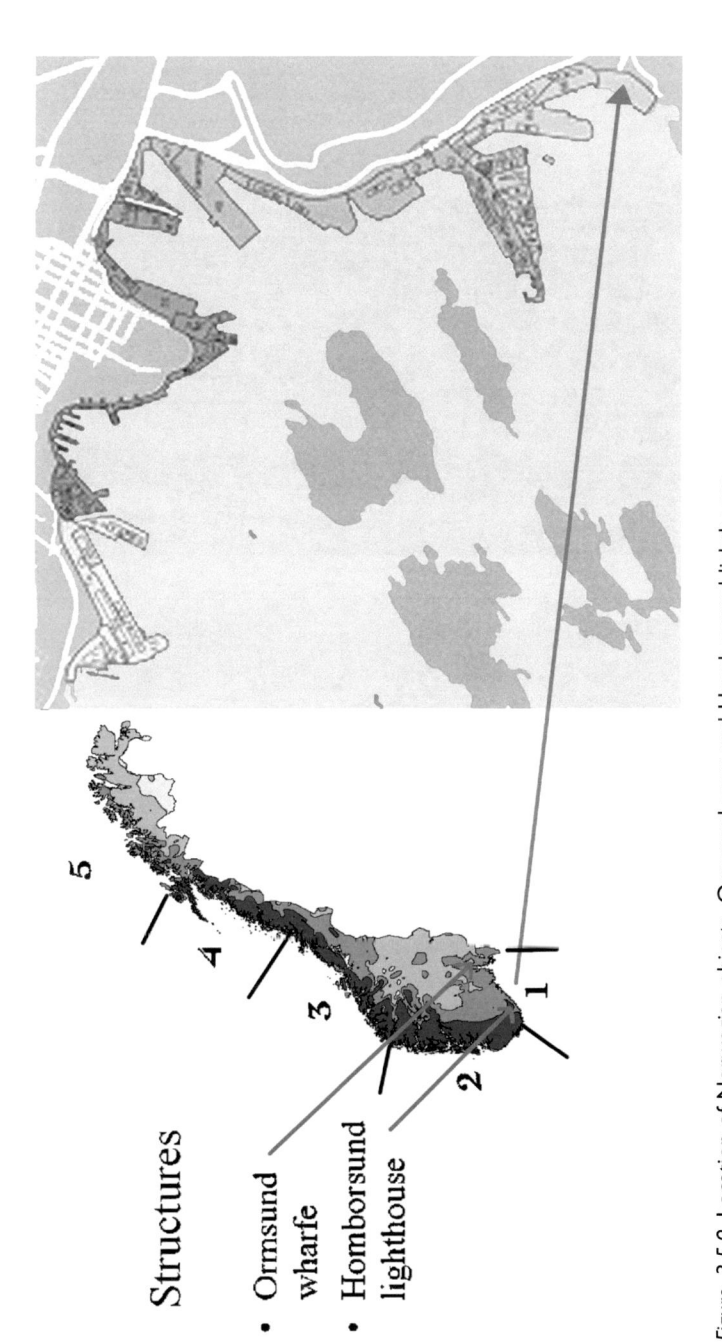

Structures

- Ormsund wharfe
- Homborsund lighthouse

Figure 3.5.9 Location of Norwegian objects, Ormsund quay and Homborsund lighthouse.

Figure 3.5.10 Ormsund quay–as seen from below at right.

3.5.8.1.2 REGIONAL CHARACTERISATION AND CLASSIFICATION

The important parameters, *average annual temperature* and *average yearly precipitation*, are shown for Norway in Figure 3.5.11. Mapping is performed based on data from Norwegian Meteorological Institute, and the class boundaries may be chosen

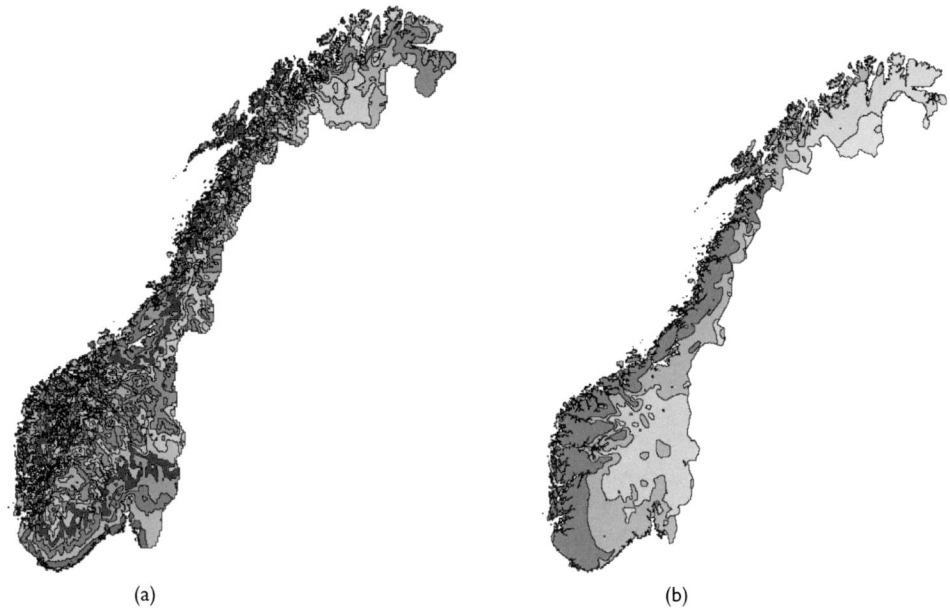

Figure 3.5.11 Regional classification of Norway: (a) – average yearly temperature – the temperatures are classified from −6 to 8 °C; (b) – average yearly precipitation is divided into four zones – Zone 1 <400 mm, zone 2 <800 mm, zone 3 <1300 mm and zone 4 >1300 mm.

from any classification system. The one shown here is for ISO 15686-4, but one for developed concrete classes could be used [15].

3.5.8.1.3   LOCAL CHARACTERISATION AND CLASSIFICATION OF CLIMATE

Based on data from nearby meteorological stations and pollutant monitoring stations and models, the Ormsund quay is classfied as follows:

- There are no months with average temperature below −5 °C or >35 °C, so according to ISO 15686 – Part 4, the climatic classification is *temperate*. A temperature correction due to difference in height above sea level will give a temperature about 0.5 °C higher at Ormsund than at Blindern.
- The average annual precipitation of 763 mm so according to ISO 15686 – Part 4, the climatic classification is *sub-humid*; however, the 1999 value gives a climatic classification of *humid*. The quay is in an area where information taken from a map may not be accurate enough for such classification.
- *Pollutant classification*:

  - airborne SO_x < 10 mg m^{-2} day^{-1}
  - rainwater pH >5.5

- According to the proposed classification in ISO 15686 – Part 4, the Ormsund quay is classified as 8–2–2.
- Blindern, with average winter temperature −2.7 °C, average summer temperature 15.9 °C and maximum temperature 28.4 °C, is in *zone B* in the EOTA system [3].

As for classification of the marine exposure environment, the models and tools of Cole have been applied for modelling and assessing the salinity in the Oslo-fjord relevant to the two objects of study, the Torungen lighthouse and Ormsund quay.

3.5.8.1.4 PREDICTED SALINITY IN OSLO FJORD: BASIC THEORY [24]

Salt aerosols may be generated from either the ocean or the surf, through bursting of bubbles generated by ocean whitecaps (breaking waves that generated white foam), particles torn from the crests of ocean whitecaps or breakers on the shore. To apply the formulations of Monahan *et al.* [25] and McKay *et al.* [26] to a coastline, it is necessary to know the exposure of the coast to the ocean, as well as the activity of the ocean adjacent to the coast. A GIS is used as a framework for modelling, as it allows integration of geographical information and climatic and other databases. The coast-line is divided into a series of points and at each point the level of whitecap activity in the adjacent ocean, the fetch in cardinal wind directions and wind conditions are defined. These parameters, in conjunction with surf and ocean aerosol production terms, are then used to determine marine salt production.

The next stage in the modelling process is to determine the marine aerosol transport to the location of interest. Cole *et al.* [24] carried out extensive CFD studies on aerosol transport and verified these by long-range studies of salt deposition. An example of the results of such studies is given in Figure 3.5.12.

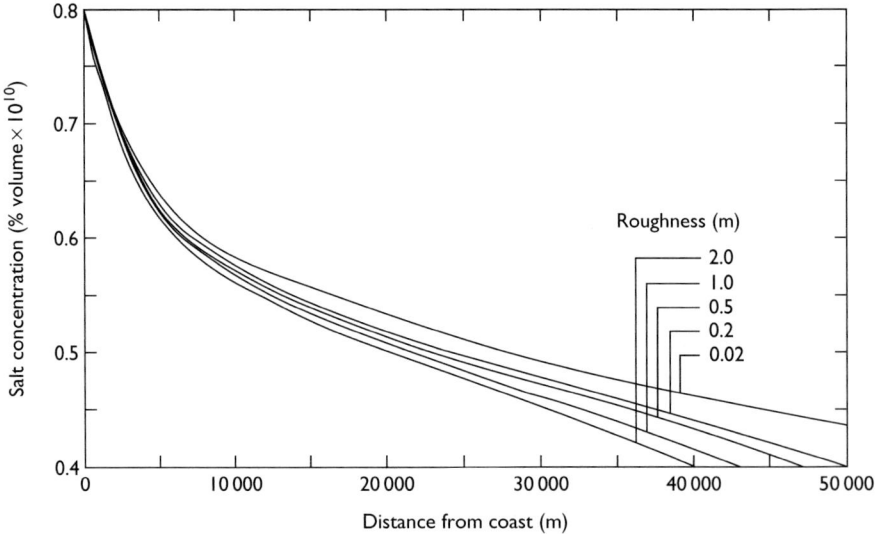

Figure 3.5.12 Effect of Roughness on Transport of Marine salts [24].

These studies show that salt transport can reasonably be modelled by an exponential decrease with distance, where the rate of decrease is influenced by terrain roughness, RH and rainfall. Separate functions are applied to salt produced by whitecaps on the ocean and breaking surf, with the former having a slow decrease and the latter a fast decrease with distance.

3.5.8.1.5 OCEAN STATE AND SURFACE ROUGHNESS FOR SOUTH NORWAY

Satellite images have been downloaded and correlated to give seasonal images of ocean activity Figure (3.5.13a). The wind cap activity up to 10 km from the coast is crucial for determining the level of salt transported to the coast.

Calculations based on the paths shown in Figure 3.5.13b, indicate that for the airstream travelling up the Fjord, an appropriate roughness is 0.02, while for that travelling across ground it would be 1.0.

As can be seen from Figure 3.5.13b, increasing roughness decreases the transport of ocean-produced aerosols. However, for roughness values of 0.2 and 1 m, the difference in transport is not dramatic. In fact, given a transport path of 50 km, RH of 70%, rainfall of 800 mm and wind speed of $7.5\,\mathrm{m\,s^{-1}}$, the ocean-produced airborne salinity is reduced to 55% of its coastal value when moving across smooth terrain (roughness 0.02 m). If the roughness is increased to 0.2 or 1 m, there is a further reduction to 93% or 84% of the smooth transport value respectively.

Figure 3.5.14 shows the wind climate for Oslo. Normally, different data are applied for each season, but in the absence of such seasonal data, the data assumed to be applicable to all seasons. Because of the position and topography, only the sector ranging from NW to SW is significant for the wind transportation of salt aerosols. In this sector, the most frequent wind episodes are from SW.

3.5.8.1.6 RESULTS

The results of the simulation for the points defined in Figure 3.5.13b are given in Table 3.5.6. It is apparent that there is a fall-off with distance from the open ocean, as expected from past studies. Further, the higher levels of whitecap activity in the vicinity of Torungen result in a higher salinity. Little seasonal variation is predicted near Oslo; however, as annual climate and wind data have been used, this is not an unexpected result. The values are in accordance with expectation from previous studies in Australia and South-east Asia [14].

Table 3.5.7 gives the estimated values and corresponding classes for these levels of the parameters.

3.5.8.1.7 MICROENVIRONMENTAL CHARACTERISATION AND CLASSIFICATION AT THE OBJECT LEVEL BY CFD SIMULATION

Numerical simulation of wind in complex terrain and around buildings has become a good supplement to traditional wind-tunnel simulations, and sometimes a substitute for them. Several different environmental loads acting on the Hombsund lighthouse and Ormsund quay have been numerically simulated using CFD [20].

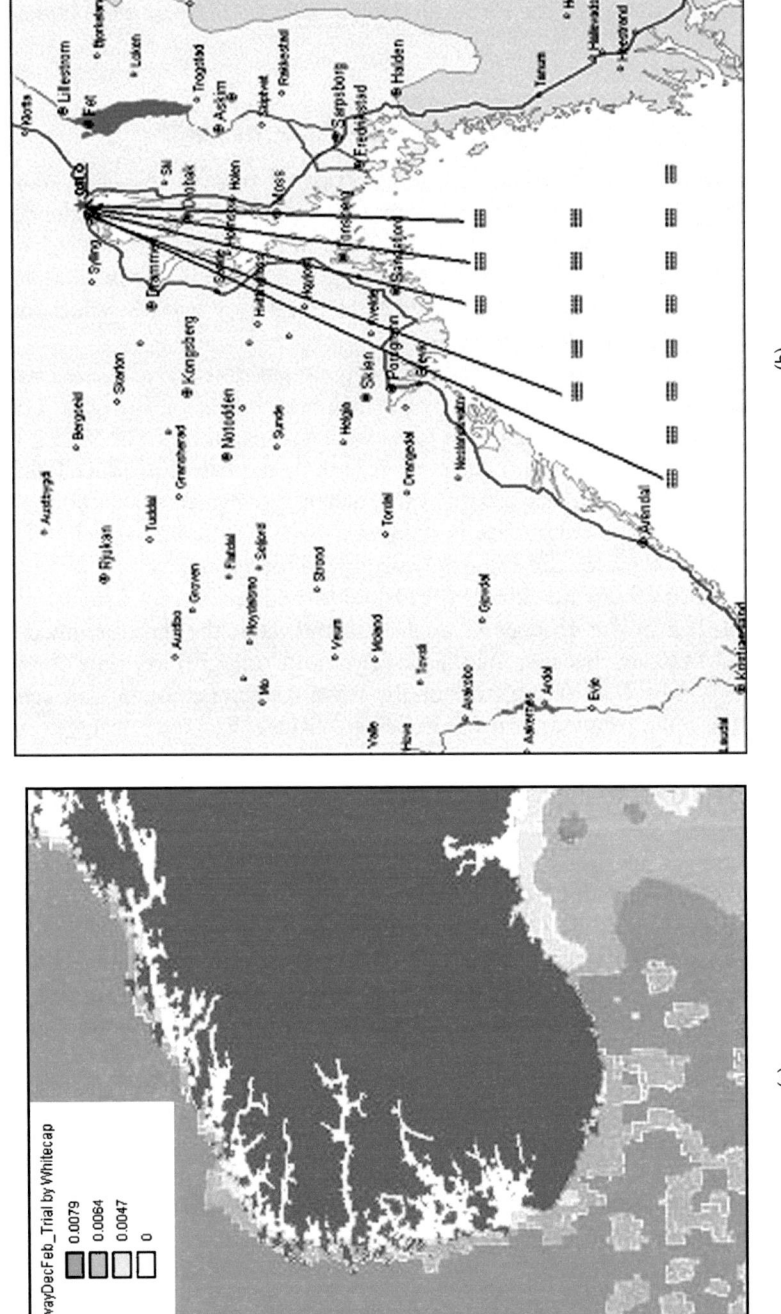

Figure 3.5.13 (a) – Ocean state for Norway (December to February), with percentage of whitecap coverage and (b) – paths used to calculate roughness.

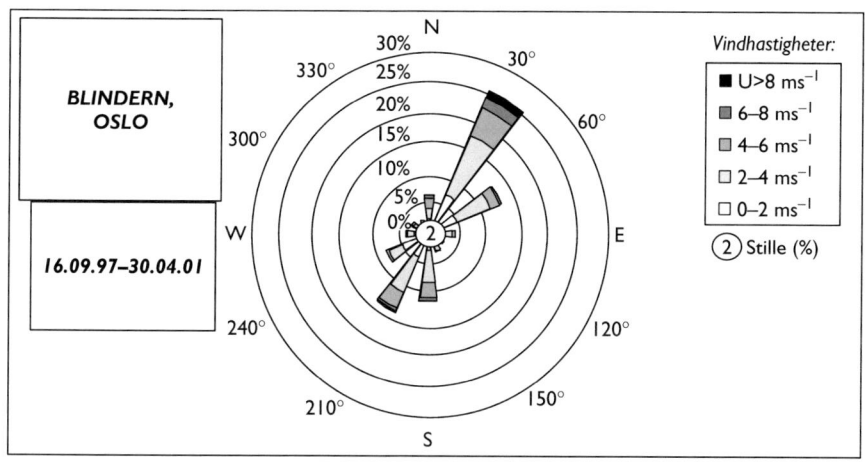

Figure 3.5.14 Wind climate at Blindern, Oslo.

Table 3.5.6 Salinity at various locations in Norway

Location	Distance from coast	Summer	Autumn	Winter	Spring	AnnualAverage
Loc1	0.2	5.1	5.2	5.1	5.0	5.1
Loc2	1.1	4.6	4.7	4.5	4.5	4.6
Loc3	1.5	4.4	4.6	4.4	4.4	4.4
Loc4	6.0	6.5	7.1	7.0	6.1	6.7
Loc5	3.8	7.1	8.2	7.9	6.4	7.4
Loc6	3.9	7.1	8.2	7.9	6.3	7.4
Loc7	5.0	7.6	9.5	8.5	6.6	8.1
Loc8	6.2	13.6	11.5	9.3	6.9	10.3
Loc9	1.9	10.9	15.4	12.5	7.9	11.7
Loc10	5.9	15.7	10.1	9.2	6.7	10.4
Loc11	0.9	16.6	22.9	20.3	12.6	18.1
Loc12	1.1	9.6	14.0	8.5	7.0	9.8
Loc13	1.2	11.6	17.6	10.6	8.0	12.0
Loc14	0.7	165.0	50.4	66.7	24.2	76.5

Table 3.5.7 Classification of airborne salinity, acidity and airborne pollutants (SO_x and CO_2)

Class	1	2	3	4	5
Salinity (mg m^{-2} day^{-1}) CO_2	<3	3–15	15–60	60–300	>300
pH	>5.5	4.5–5.5	3.5–4.5	3.0–3.5	<3.0
SO_x (mg m^{-2} day^{-1})	<10	10–35	35–80	80–200	>200

Three-dimensional models were constructed on the basis of 1:5000 maps and graphs, and used as input to CFD simulations.

The general-purpose finite-volume CFD code used solved the incompressible, time-averaged Navier–Stokes equations, using the RNG turbulence model to close the equations. The outlet boundary is continuative, meaning that the normal derivatives of all quantities are set to zero. On the surface boundaries, a law-of-the-wall velocity profile is assumed. The lateral and top boundaries have symmetry conditions.

The salt aerosol is simulated as a tracer gas in the airflow. The inlet boundary has aerosol concentration 1.0 for the lowest 0.5 m. The resulting aerosol concentration on the surface of the dock can, therefore, only be interpreted as a relative salt concentration.

For Ormsund quay, the wind pattern for the wind directions SW and WSW shows that the difference in angle of attack leads to different flow patterns around the structure. In both wind directions, the airflow becomes stagnant below the quay. The effect of this is that some air is deflected out of the space between the quay and the water surface. The streamline visualisation shows that the angle between the deflected airflow and the edge of the quay is largest in the WSW wind direction. The complete simulations are reported elsewhere (Haagenrud and Krigsvoll, 2004).

Figure 3.5.15 shows the distribution of sea-salt aerosol on the underside of the quay with wind directions SW and WSW. The distribution is influenced by the deflection of the air. With wind direction WSW, the concentration of salt increases near the edge of the quay where the deflected air is leaving the space between the quay and the water surface. In this area, more of the air approaching the quay is being pushed over the edge of the quay by the air leaving the underside of the quay. The same effect is visible for the SW wind direction. The difference is that with an SW wind, the effect is taking place further downwind, to the right in the picture. With the SW

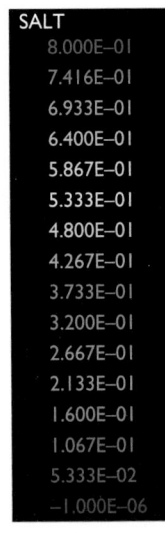

SALT
8.000E–01
7.416E–01
6.933E–01
6.400E–01
5.867E–01
5.333E–01
4.800E–01
4.267E–01
3.733E–01
3.200E–01
2.667E–01
2.133E–01
1.600E–01
1.067E–01
5.333E–02
–1.000E–06

Figure 3.5.15 Stimulated sea-salt concentration on the underside of Ormsund quay with an SW wind and WSW wind.

wind, the salt concentration is generally higher near the rear wall of the quay than at the front. Towards the downwind end of the quay, the salt concentration is higher near the front of the quay.

The study shows that CFD can be a useful tool for assessing environmental loads, and thus degradation risk zones, on quay structures. Such studies are a 'first approach', and simulations need to be validated with measurements. Point measurements performed at Ormsund quay indicate a good correlation with the CFD studies [11].

This could be combined with a condition assessment according to EN 206–1. In Figure 3.5.16 are the Ormsund quay and the estimated corrosion risk zones according to EN 206–1. Total validation could now be performed by modelling, measurements and extensive condition assessments.

3.5.8.1.8 THE WORK OF ICP MATERIALS

The ICP on Materials, including Historic and Cultural Monuments (ICP Materials), was set up in 1985. A Task Force organises the programme, with Sweden as lead country, and the Swedish Corrosion Institute serving as the main research centre (www.corr-institute.se). The research carried out within the programme and coordinated by the WGE are of the utmost relevance and importance for the improved life cycle management of Europe's buildings and infrastructures. Much of the results from projects like LIFECON-such as stock at risk mapping, corrosion costs, etc would constitute important input to the work on improving the corrosion cost mapping for Europe. Therefore some of the methodologies and recent results both of the ICP Materials and its exploitation within the WGE mapping activity is presented in more detail.

The main aim is to perform a quantitative evaluation of the effects of multi-pollutants such as sulphur and nitrogen compounds, ozone and particles as well as climate parameters on the atmospheric corrosion of important materials, including materials used in objects forming the cultural heritage. The primary objective is to collect information on corrosion and environmental data in order to evaluate

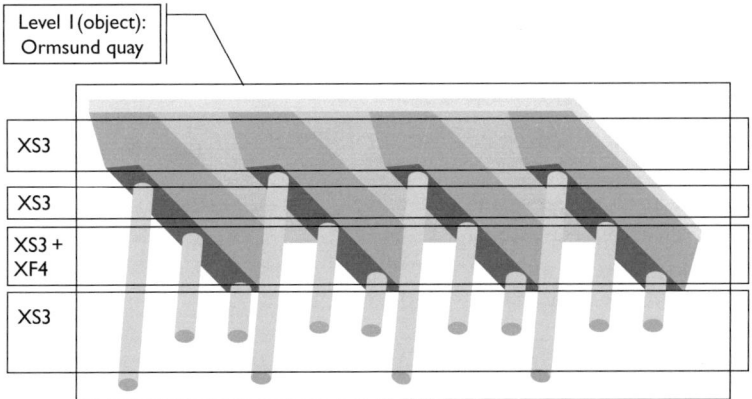

Figure 3.5.16 Estimated corrosion risk zones on the Ormsund quay according to EN 206-1.

dose–response functions and trend effects. This is achieved by exposing specimens of materials in a network of field test sites, by measuring gaseous pollutants, precipitation and climate parameters at or nearby each test site, and by evaluating the corrosion effects on the materials [13].

Two networks of test sites have been used for exposures. The original network, used from 1987 to 1997, consisted of 39 exposure sites in 12 European countries and in the United States and Canada. The present network consists of 30 exposure sites in 15 European countries and in Israel, the United States and Canada. The reduction in the number of testsites represented an improvement of the efficiency of the network; redundant sites from the original network were excluded, and only 21 of the original 39 sites were kept.

3.5.8.1.9 DOSE–RESPONSE FUNCTIONS

The dose–response functions constitute the main results of the programme, and are important in the development of systems for classifying the corrosivity of environments, for mapping of areas with increased risk of corrosion, and for calculation of cost of damage caused by deterioration of materials [27, 28]. A dose–response function links the dose of pollution, measured in ambient concentration and/or deposition, to the rate of material corrosion. For unsheltered positions, the damage to materials is usually discussed in terms of dry and wet deposition. *Wet deposition* includes transport by means of precipitation and *dry deposition* transport by any other process. One important task for the programme has been to estimate the relative contribution of dry and wet depositios to the degradations of materials. Therefore, and also because it makes sense from a mechanistic point of view, the dose–response functions for unsheltered materials are of the type

$$K = \mathrm{dry}(T, \mathrm{RH}, [\mathrm{SO}_2], [\mathrm{NO}_2], [\mathrm{O}_3], t) + \mathrm{wet}\,(\mathrm{Rain}\,[\mathrm{H}^+], t)$$

where
K	the corrosion attack
T	the temperature in °C
RH	the relative humidity in %
[]	the concentration in $\mu\mathrm{g\,m}^{-3}$ (SO_2, NO_2 and O_3)
t	the time in years
Rain	the amount of precipitation in mm
$[\mathrm{H}^+]$	the acidity of precipitation in $\mathrm{mg\,l}^{-1}$.

The corrosion attack can, depending on material, be quantified as either mass loss (ML, $\mathrm{g\,m}^{-2}$), surface recession (R, $\mu\mathrm{m}$), ASTM D 1150-55 1987 (ASTM, 1–10), depth of leached layer (LL, nm) or weight increase (WI, $\mu\mathrm{g\,cm}^{-2}$).

For example, the dose–response function for zinc is:

$$\mathrm{ML} = 1.4[\mathrm{SO}_2]^{0.22}\exp\{0.018\mathrm{RH} + f(T)\}t^{0.85} + 0.029\mathrm{Rain}[\mathrm{H}^+]t$$

An extensive statistical analysis not only confirmed the corrosive effect of SO_2 but also enabled the quantification of the effects of other environmental parameters for a wide range of materials. For most unsheltered materials, the effect of wet deposition (acid precipitation) has also been quantified, and forms the second most important contribution to the corrosion rate. For selected materials, the effect of ozone (O_3) and (NO_2) have been demonstrated.

The *trend exposure* consisted of repeated one-year exposures of steel and zinc. Of the environmental parameters, SO_2, NO_2 and H^+ exhibit decreasing trends of concentration with SO_2 having the strongest trend and NO_2 the weakest. This has resulted in decreasing corrosion rates for both carbon steel and zinc in unsheltered as well as sheltered positions. SO_2 is the largest single contributing factor to the decreasing corrosion trends; the effect of decreasing H^+ is much smaller. The decrease in corrosivity is generally larger than expected from the fall of SO_2 and H^+ concentrations. This cannot be directly related to a specific pollutant, and reflects the multi-pollutant character of the process of material degradation. This finding is the basis for establishing of the new four-year multi-pollutant programme.

3.5.8.1.10 MAPPING AIR POLLUTION, ACTUAL CORROSION RATES AND EXCEEDANCES IN SWEDEN: SELECTION OF GRID FOR CALCULATION AND PRESENTATION

The available data come from different sources, in different formats and in different coordinate systems. Before using the dose–response functions for calculating the corrosion attack, the data have to be transformed to a common set of geographical coordinates and expressed at the same resolution or scale. The SO_2 concentration data sets are available for Sweden on a so-called RT90 grid ($20\,km \times 20\,km$). Since the SO_2 concentration shows significant local variation, it is not correct to interpolate or transform these data to a finer grid. This grid resolution and projection has, therefore, been used as a basis for the presentation, and all other data have been interpolated or transformed to fit it (Figure 3.5.17).

3.5.8.1.11 ENVIRONMENTAL DATA

The climate data are shown in Figure 3.5.18. Climatic norms (1961–90) of *temperature* and *precipitation* on a $0.5° \times 0.5°$ resolution were obtained from the Intergovernmental Panel on Climate Change (IPCC); http://ipccddc.cru. uea.ac.uk/cru_data/datadownload/download_index.html. Further description of the data set is given by New *et al.* [29]. Relative humidity (RH) was calculated from IPCC temperature (T) and vapour pressure (p_w) data using the relations:

$$RH/100 = p_w/p_{w,sat}$$

where
$$p_{w,sat} = 6.108 \cdot \exp\{17.2661T/(237.3 + T)\}$$

Data of SO_2, NO_2 and O_3 were obtained from the Swedish Meteorological and Hydrological Institute (SMHI). http://www.smhi.se/ (Figure 3.5.19). The data were

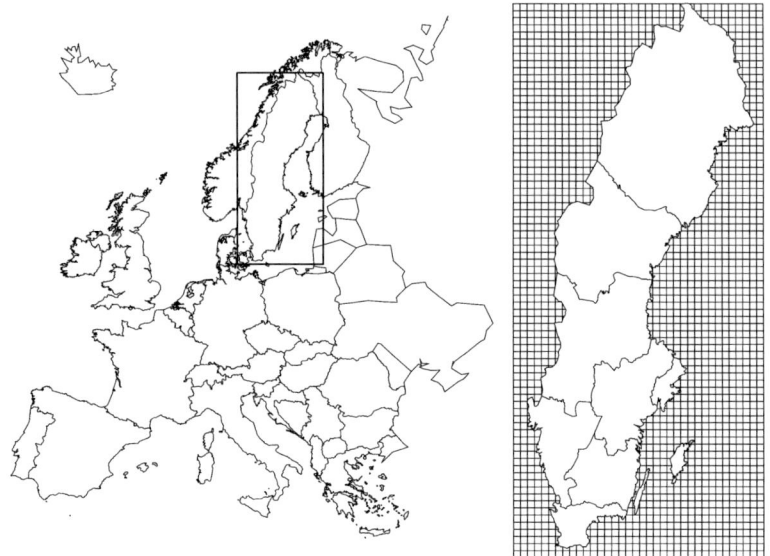

Figure 3.5.17 The 20 km × 20 km grid used for calculation and presentation (Projection: RT90 – 'Rikets nät') and its location relative to Europe and Sweden.

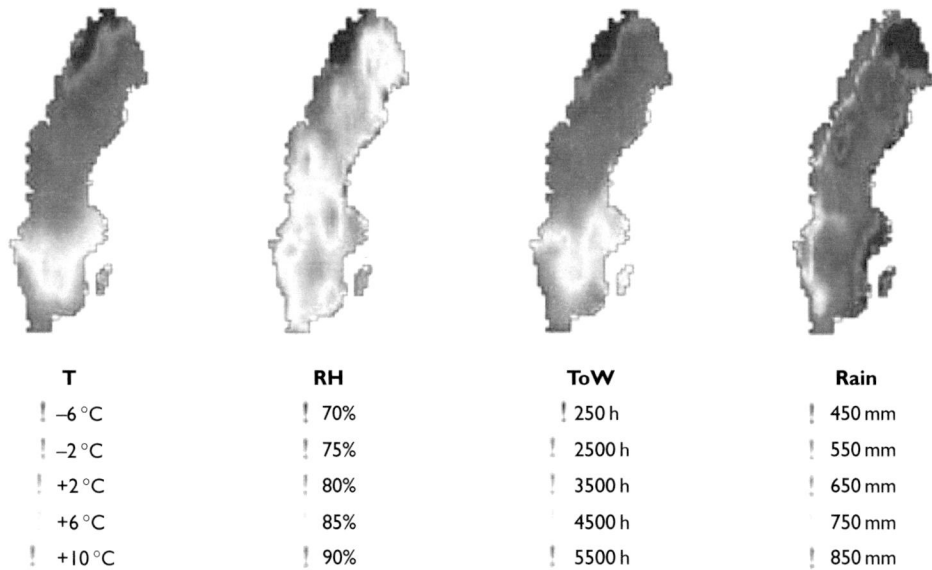

T	RH	ToW	Rain
! –6 °C	! 70%	! 250 h	! 450 mm
! –2 °C	! 75%	! 2500 h	! 550 mm
! +2 °C	! 80%	! 3500 h	! 650 mm
+6 °C	85%	4500 h	750 mm
! +10 °C	! 90%	! 5500 h	! 850 mm

Figure 3.5.18 Climatic norms (1961–90) of temperature (T), relative humidity (RH), time of wetness (ToW, calculated from T and RH) and amount of precipitation (Rain). Source: New et al. [29].

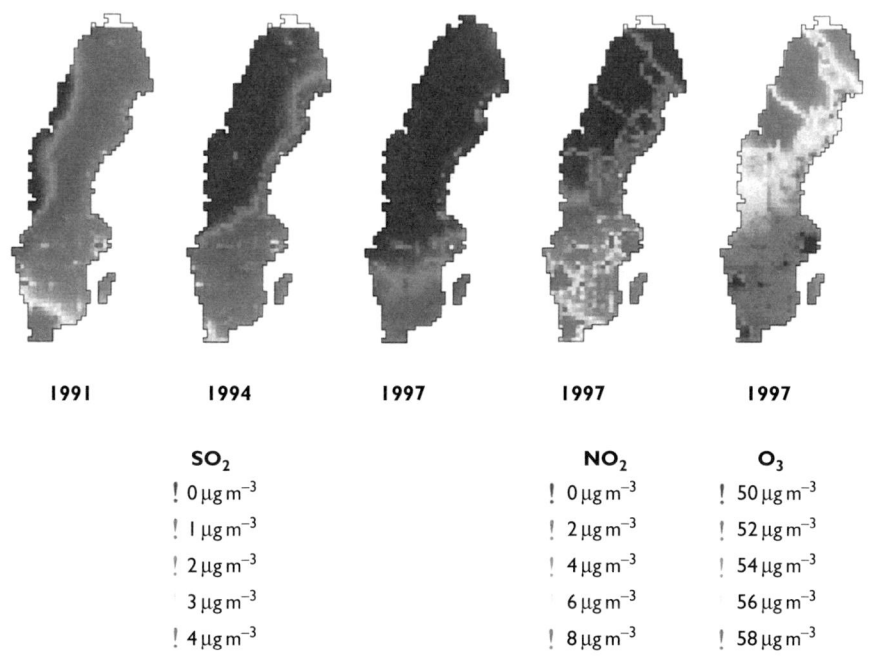

1991	1994	1997	1997	1997

SO$_2$			NO$_2$	O$_3$
! 0 µg m^{-3}			! 0 µg m^{-3}	! 50 µg m^{-3}
! 1 µg m^{-3}			! 2 µg m^{-3}	! 52 µg m^{-3}
! 2 µg m^{-3}			! 4 µg m^{-3}	! 54 µg m^{-3}
3 µg m^{-3}			6 µg m^{-3}	56 µg m^{-3}
! 4 µg m^{-3}			! 8 µg m^{-3}	! 58 µg m^{-3}

Figure 3.5.19 SO$_2$, NO$_2$ and O$_3$ concentration (1991–1997). Source: www.smhi.se (SHMI's MATCH-model). O$_3$ is calculated from NO$_2$ according to the Mapping Manual.

produced by SMHI's MATCH (Mesoscale Atmospheric Transport and Chemistry) model, which is used to study air pollution in Sweden. The area of the model covers all Sweden and has a resolution of 20×20 km. The SMHI also provides data on the total wet deposition of chloride (mg m^{-2}) at the same resolution. This quantity is identical to Rain[Cl$^-$], which is the quantity used in the dose–response functions for aluminium and bronze (mm mg l^{-1} = mg m^{-2}). The chloride map is shown along with the pH data.

Acidity of precipitation (pH) was interpolated from station data using ordinary kriging. Data on pH were compiled as annual averages during the period 1986–2000 on 89 stations, (Figure 3.5.20). Four pH data sets were combined: Swedish National network (32 sites), Finnish National network (14 sites), Norwegian national network (30 sites) and the EMEP network (13 sites in Germany, Denmark, Estonia, Lithuania, Latvia, Poland and Russia). Data on precipitation and acidity of precipitation in the EMEP network are available free of charge from the Chemical Coordinating Centre (CCC) at the Norwegian Institute for Air Research (NILU) (http://www.nilu.no//projects/ccc/download_data.html). Station data in the Swedish National network are available from the Swedish Environmental Research Institute Ltd. (IVL), http://www.ivl.se/miljo/db/. Norwegian National data were compiled from Aas *et al.* [30] and Finnish National data were obtained free of charge from the Finnish Meteorological Institute (FMI).

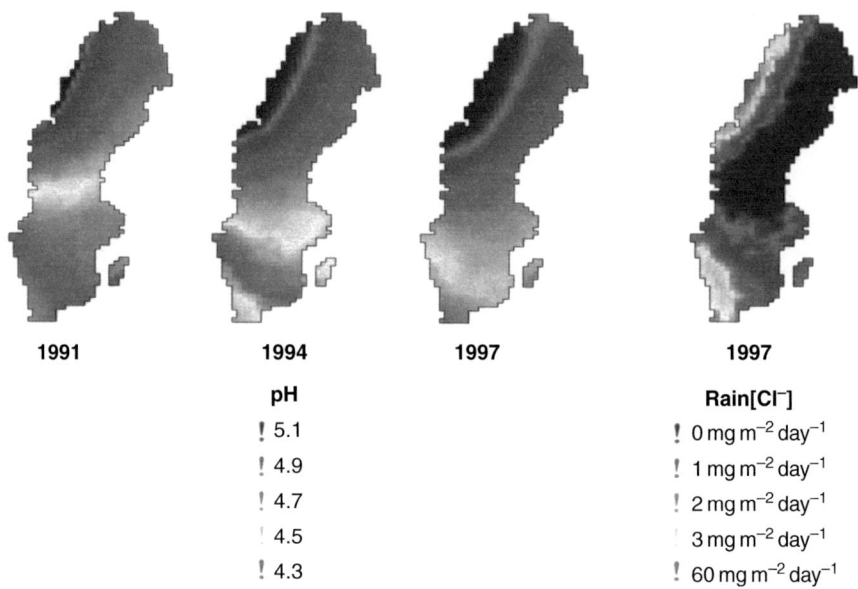

1991	1994	1997	1997

pH	Rain[Cl⁻]
5.1	$0\,mg\,m^{-2}\,day^{-1}$
4.9	$1\,mg\,m^{-2}\,day^{-1}$
4.7	$2\,mg\,m^{-2}\,day^{-1}$
4.5	$3\,mg\,m^{-2}\,day^{-1}$
4.3	$60\,mg\,m^{-2}\,day^{-1}$

Figure 3.5.20 Acidity (pH) of precipitation (1991–1997) and total wet chloride wet deposition (Rain[Cl⁻]), Source: www.smhi.se (SHMI's MATCH model).

3.5.8.1.12 CORROSION MAPPING

The visual appearance of all maps is similar and reflects the pattern of the SO_2 maps, with increasing values from north to south and the greatest risk on the west coast and in the Stockholm area, as is shown in Figure 3.5.21 for zinc weathering steel, copper, bronze and aluminium. What differs between the maps is the magnitudes of corrosion attack, where weathering steel shows the highest values ($27–177\,g\,m^{-2}$), followed by zinc ($1.2–7.0\,g\,m^{-2}$), copper ($1.0–6.4\,g\,m^{-2}$), bronze ($0.9–3.5\,g\,m^{-2}$) and aluminium ($0.06–0.22\,g\,m^{-2}$).

3.5.8.1.13 CONCLUSIONS

The development of a methodology for mapping areas with increased corrosion risk was successful and applied for a case study 'Mapping on a $20\,km \times 20\,km$ grid for Sweden'. Corrosion attack was calculated for zinc, weathering steel, copper, bronze and aluminium. The visual appearance of all maps of metal corrosion attack is similar with increasing values from north to south due to a combination of climatic factors and pollutants. The Stockholm area and the west coast of are Sweden worthy of further studies, as the corrosion levels, there are unacceptably high and the stock of materials at risk is also high.

At the $20\,km \times 20\,km$ resolution, it is possible to distinguish the elevated corrosion and pollution levels in the Stockholm area, which is a significant improvement over what can be resolved in the larger EMEP scale ($50\,km \times 50\,km$). However, to identify

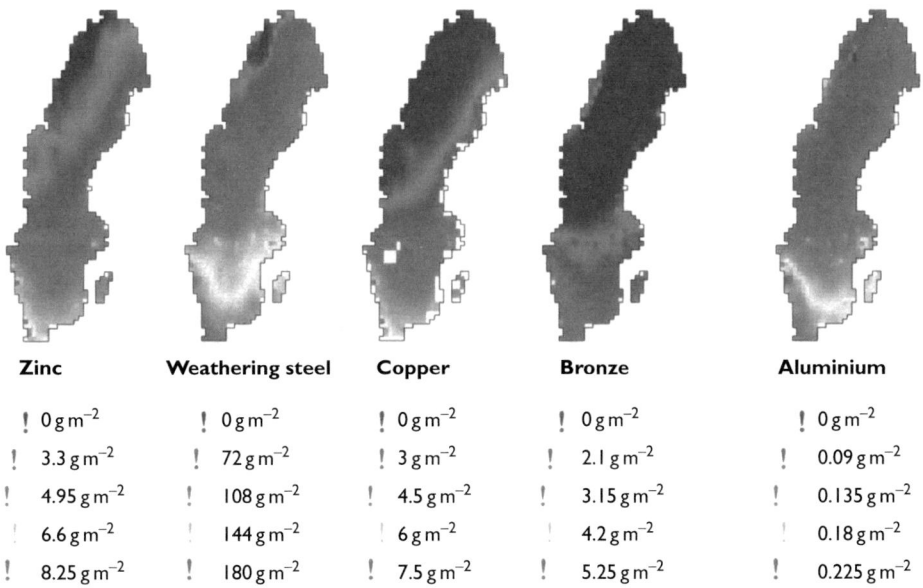

Zinc	Weathering steel	Copper	Bronze	Aluminium
! 0 g m^{-2}	! 0 g m^{-2}	! 0 g m^{-2}	! 0 g m^{-2}	! 0 g m^{-2}
! 3.3 g m^{-2}	! 72 g m^{-2}	! 3 g m^{-2}	! 2.1 g m^{-2}	! 0.09 g m^{-2}
! 4.95 g m^{-2}	! 108 g m^{-2}	! 4.5 g m^{-2}	! 3.15 g m^{-2}	! 0.135 g m^{-2}
6.6 g m^{-2}	144 g m^{-2}	6 g m^{-2}	4.2 g m^{-2}	0.18 g m^{-2}
! 8.25 g m^{-2}	! 180 g m^{-2}	! 7.5 g m^{-2}	! 5.25 g m^{-2}	! 0.225 g m^{-2}

Figure 3.5.21 Calculated corrosion of zinc, weathering steel, copper, bronze and aluminium after one year of exposure (1997).

smaller sources in the maps, such as smaller towns or suburbs of Stockholm, it is necessary to increase the resolution.

The main pollutant in the dose–response functions used for calculation of corrosion attack is the SO_2, concentration. With decreasing levels of SO_2, other parameters become relatively more important, and it is therefore important to develop dose–response functions to include other effects, such as dry deposition of chlorides, when mapping areas up to about 10 km from the west coast of Sweden.

References

[1] Jernberg, P., M. A. Lacasse, S. E. Haagenrud, and C. Sjöström. 2004. *Guide to bibliography to service life and durability research for building materials and components.* CIB W080/RILEM 175–SLM Publication 295, CIB, Rotterdam.

[2] Sjöström, C., and J. Lair. *Performance based building – some implications on construction, Materials and Components,* in Press.

[3] European Organisation for Technical approvals. 1999. *Assessments of working life of products.* PT3 Durability (TB97/24/9.3.1).

[4] International Organisation for Standardisation. 1992. *Corrosion of metals and alloys– Corrosivity of atmospheres – Classification.* Geneva (ISO 9223:1992).

[5] British Standards Institution. 1992. *Code of Practice for assessing exposure of walls to wind-driven rain.* (BS8104).

[6] International Organisation for Standardisation. 2000. *Service life planning of buildings and constructed assets. Part 1: General principles.* Geneva (ISO 15686-1).

[7] Sjöström, P., P. Jernberg, S. Caluwaerts, S. Kelly, S. Haagenrud, and J. L. Chevalier. 2002. Implementation of the European Construction Products Directive via the ISO 15686 standards, *9th conference of durability of building materials and components*, Brisbane.

[8] International Organisation for Standardisation. (2000) *Service life planning of buildings and constructed assets. – Part 1: – General principles*. Geneva (ISO 15686-2, 2001).

[9] Cole, I. S., D. A. Paterson, and W. D. Ganther. 2003. A holistic model for atmospheric corrosion. Part 1: Theoretical framework for production, transport and deposition of marine salts. *Corrosion Science* 38 (2).

[10] Markeset, G. 2001. Service life predictions of marine concrete structures – consequences of uncertainties in model parameters. *Proceedings, Nordic Mini Seminar: 'Prediction Models for Chloride Ingress and Corrosion Initiation in Concrete Structures'*, Sweden: Chalmers Institute of Technology Gothenburg, May 22–23, pp. 10.

[11] Haagenrud, S. E. and G. Krigsvoll. 2004. *Instructions for Quantitative Classification of environmental degradation loads onto structures*. LIFECON.EU GROWTH Program, G1RD-CT-2000-00378. Deliverable D4.2. Norwegian Building Research Institute NBI. http://lifecon.vtt.fi/.

[12] Mapping Manual. 2001. Manual on methodologies and criteria for mapping critical levels/loads and geographical areas where they are exceeded, UN ECE Convention on Long-range Transboundary Air Pollution, International Co-operative Programme on Modelling and Mapping, URL: www.icpmapping.com.

[13] Tidblad, J., V. Kucera, and A. A. Mikhailov. 1998. Statistical analysis of 4-year materials exposure and acceptable deterioration and pollution levels. Convention on long-range transboundary air pollution. Prepared by the main research centre, Swedish Corrosion Institute, Stockholm. (UN ECE ICP on effects on materials including historic and cultural monuments, Report No. 30).

[14] Cole, I. S., G. A. King, G. S. Trinidad, W. Y. Chan, and D. A. Paterson. 1999. An Australian-wide map of corrosivity: A GIS approach. IN *8th DBMC*, Canada.

[15] Lay, S., and P. Schießl. 2003. *Service life models*. Life cycle management of concrete infrastructures for improved sustainability. LIFECON.EU GROWTH Program, G1RD-CT-2000-00378. Deliverable D3.2. Technical University Munich. http://lifecon.vtt.fi/.

[16] Haagenrud, S. E., and J. F. Henriksen. 1996. Degradation of built environment – Review of cost assessment model and dose response functions. *7th international confernece on the durability of building materials and components*. Stockholm. Proceedings, ed. C. Sjöström. London: E & FN Spon, pp. 85–96.

[17] CIB report Publication 256. 2000. *GIS and the built environment*, eds S. E. Haagenrud, B. Rystedt and C. Sjöström, October, Rotterdam.

[18] Sarja, A. (co-ordinator). 2004. EU project G1RD-CT-2000-00378 Life cycle management of concrete infrastuctures for improved sustainability (LIFECON). Deliverable reports D1-D15. http://lifecon.vtt.fi/.

[19] European Committee for standardization-CEN 1997. *Hygrothermal performance of buildings – Climatic data. Part 3: Calculation of driving rain index for vertical surfaces from hourly wind and rain data* (prEN 13013-3).

[20] Thiis, T., J. Tidblad, and V. Kucera. 2003. *Mapping of acid deposition effects on materials*, Swedish Corrosion Institute, Stockholm: Sweden.

[21] ISO/WD 15686-4. 2002. *Building – Service life data sets. Part 4: Service life prediction data requirements*.

[22] European Committee for standardization. 2001. *EN 206 Concrete. Part 1: Specification, performance, production and conformity*.

[23] Fluge, F. 2001. Marine chlorides – A probabilistic approach to derive provision for EN 206-1, DURANET, Third Workshop, Tromsø 10th–12th, June.

[24] Cole, I. G. A., G. S. Trinidad, W. Y. Chan, and D. A. Paterson. 2003. Predicted salinity. In *The Oslo fjord, Internal report to NBI*. Spring.

[25] Monahan, E. C., D. E. Spiel, and K. L. Davidson. 1986. A model of marine aerosol generation via whitecaps and wave disruption. In *Oceanic whitecaps and their role in air–sea exchange processes*, eds E. C. Monahan, G. Macnioaill, pp. 167–174. Norwell, MA: D. Reidell Publishing Company.

[26] McKay, W. A., J. A. Garland, D. Livesley, C. M. Halliwell, and M. I. Walker. 1994. *Atmos. Environ.* 28 (3): 299–309.

[27] Kucera, V., J. F. Henriksen, D. Knotkova, and C. Sjöström. 1993. Model for calculations of corrosion cost caused by air pollution and its application in three cities. In *Progress in the understanding and prevention of corrosion, 10th European Corrosion Congress*, Barcelona, vol. 1, July, eds J. M. Costa and A. D. Mercer. London: Institute of material, pp. 24–32.

[28] Kucera, V., and H. D. Gregor. 2000. Mapping air pollution effects on materials including stock at risk. *Proceedings of a UN ECE Workshop*, SCI Bulletin 108E.

[29] New M., M. Hulme, and P. Jones. 1999. Representing twentieth-century space-time climate variability. Part I: Development of a 1961–90 mean monthly terrestrial climatology. *Journal of Climate* 12: 829–856.

[30] Aas W., K. Tørseth, S. Solberg, T. Berg, and S. Manø. 2000. Overvåkning av langtransportert forurenset luft og nedbør. Atmosfærisk tilførsel, 1999. Norwegian Institute for Air Research (NILU), O-8118/O-90077, ISBN 82-425-1176-4, Norway: Kjeller.

3.6   Structural consequences of degradation

Asko Sarja

3.6.1   Structural consequences

The direct effects of corrosion are loss of bar cross section, increase in bar diameter resulting from the volumetric expansion of the corrosion products and a change in the mechanical characteristics of the bar–concrete interface on the formation of corrosion products. Effects of corrosion on residual structural capacity are divided into those aspects which affect the reinforcement itself, those which affect the surrounding concrete, and those which affect interaction between the two [1, 2]. The consequences of each of these aspects and their interrelated effects on the load-carrying capacity of reinforced concrete structures are summarised in Figure 3.6.1 [3].

The most widely adopted measure of corrosion is the weight of metal lost after the completion of strength tests. Some investigators quantify corrosion simply as section loss, that is weight loss divided by original weight, expressed as a percentage. This is evidently an appropriate parameter to adopt where residual tensile strength of the bars is of principal concern. However, there is evidence that average corrosion penetration, that is the thickness of steel lost through corrosion assessed as an average over the original bar surface, offers a better measure of corrosion for assessment of residual bond strength (Figure 3.6.2).

Weight loss per unit area of bar surface provides an equivalent measure. Figure 3.6.2 compares the amount of corrosion to cracking required by bars of different diameter (but with constant cover). The results are quantified by weight loss on the left y-axis and by radius of steel lost to corrosion (corrosion penetration) on the right y-axis. Corrosion to cracking is almost constant when assessed by corrosion

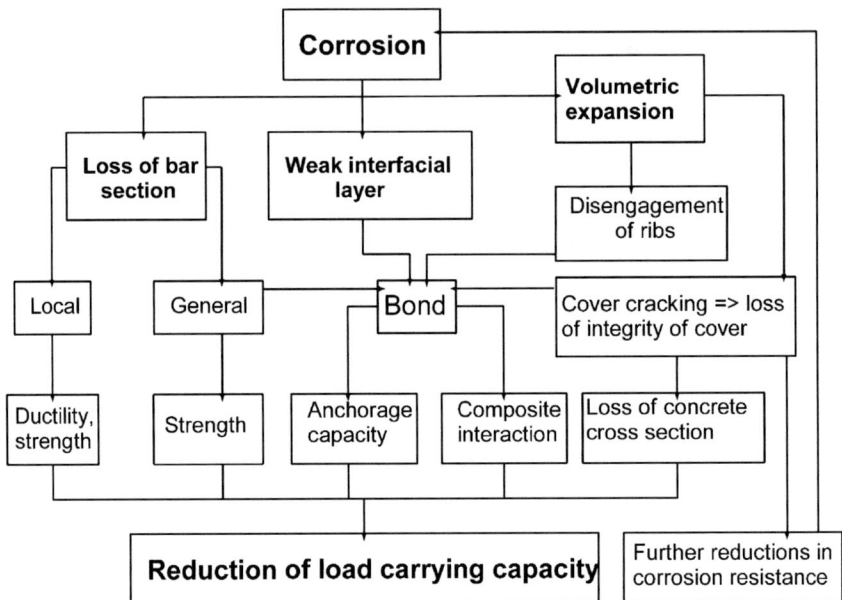

Figure 3.6.1 Effects of corrosion on residual strength [3].

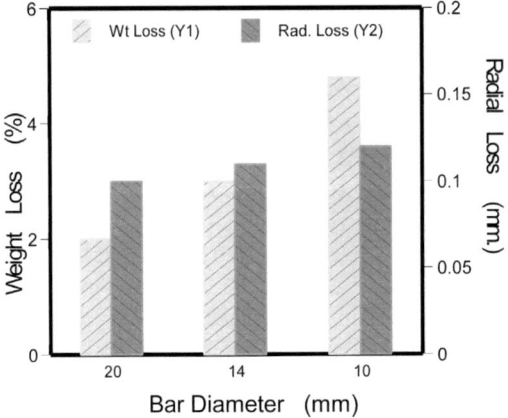

Figure 3.6.2 Amount of corrosion necessary for crack initiation [4].

penetration, but increases markedly with reducing bar diameter when assessed by section loss, suggesting that the first parameter may be a more useful measure. Bond strength will certainly be affected by crack width, and the relationship between loss of bar section and thickness of corrosion products depends on environment [5, 6]. An understanding of the mechanisms affecting bond strength of corroding bars and the establishment of an appropriate measure of damage is clearly a priority in reconciling conflicting test data and in the development of assessment guidelines [7].

3.6.2 Effect of corrosion on structural behaviour

3.6.2.1 Section loss

Dissolution of iron from steel reinforcement results in a loss of bar cross section, which may either be predominantly uniformly distributed over the length and circumference of the bar (general corrosion) or show concentration at localised sites (pitting corrosion) (Figure 3.6.3). The structural effects of these two forms of corrosion damage differ significantly, and will be examined in more detail.

General corrosion is caused by ingress of chlorides or carbonation of concrete. It is generally associated with the formation of 'brown rust' iron oxides, which occupy a greater volume than the parent metal, and expansion of the bar as it corrodes, which leads to cracking and eventually spalling of the concrete cover. The residual cross-sectional area A_{res} may be evaluated by:

$$A_{res} = A_0 - A_{corr} = \pi \cdot (d_b - 2p(t))^2/4 \tag{3.6.1}$$

A_0 original cross-sectional area (mm^2)
A_{corr} loss in cross-sectional area (mm^2)
d_b original bar diameter (mm)
$p(t)$ corrosion penetration depth (mm).

Local or *pitting corrosion* is invariably associated with chloride contamination and not with carbonation. In local corrosion, the area of the anode (where dissolution of metal occurs) may be relatively small. Once a pit has been initiated, the resulting electric field attracts negative (Cl^-) ions towards the pit. Hydrolysis of the corrosion product in the pit causes a decrease in pH. In the resulting saline and acidic conditions, corrosion may occur rapidly. The need to balance the release and consumption of electrons at anode and cathode means that current density, and hence rate of loss of metal at the anode, will be relatively high. Anodic and cathodic sites are separated by from tens of millimetres up to metres, and may develop on a single bar or between different layers of reinforcement.

Because the corrosion rate is rapid and the supply of oxygen restricted, the products of the corrosion reactions exhibit a lower degree of volumetric expansion as brown rust, and the tendency to split the concrete cover is consequently less. Extreme loss

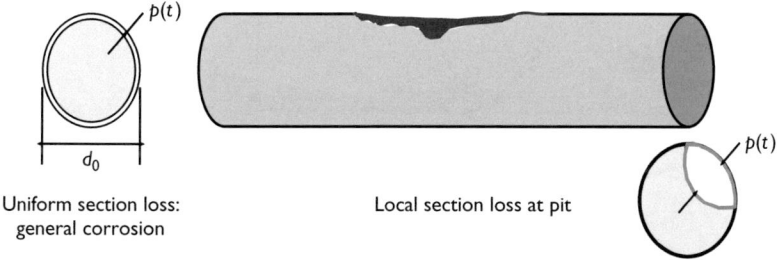

Uniform section loss:
general corrosion

Local section loss at pit

Figure 3.6.3 Section loss due to uniform and pitting corrosion [3].

of bar section may occur without external visual signs of cracking, although surface staining will usually be noticeable. However, local corrosion sites are readily detectable by the half cell method, where it appears as a strongly negative potential surrounded by a high potential gradient. Local corrosion can only be sustained where resistivity of the concrete is low. The residual cross-sectional area A_{res} from local corrosion may be evaluated with the help of Figure 3.6.4 if the pitting depth $P(t)$ is known.

$$A_{res} = \pi \cdot d_b^2/4 - A_1 - A_2 \quad [mm^2] \tag{3.6.2}$$

where

$$A_1 = \frac{\pi \cdot d_b^2 \theta}{4 \cdot 180} - a \cdot d_b \cdot \cos \theta \quad [mm^2] \tag{3.6.3}$$

$$A_2 = \frac{\pi \cdot P(t)^2 \cdot \phi}{180} - a \cdot p(t) \cdot \cos \phi \quad [mm^2] \tag{3.6.4}$$

$$\theta = 2 \cdot \arcsin(p(t)/d_b) \quad [°] \tag{3.6.5}$$

$$a = \frac{d_b}{2} \cdot \sin \theta \quad [mm] \tag{3.6.6}$$

$$\phi = \arcsin\left(\frac{a}{p(t)}\right) \quad [°] \tag{3.6.7}$$

Although it is convenient to describe the two forms of section loss separately here, both will occur together with chloride-induced corrosion, and loss of section will never be completely uniform. Note that the rate of corrosion penetration $dP(t)/dt$ will be different for general and for pitting corrosion.

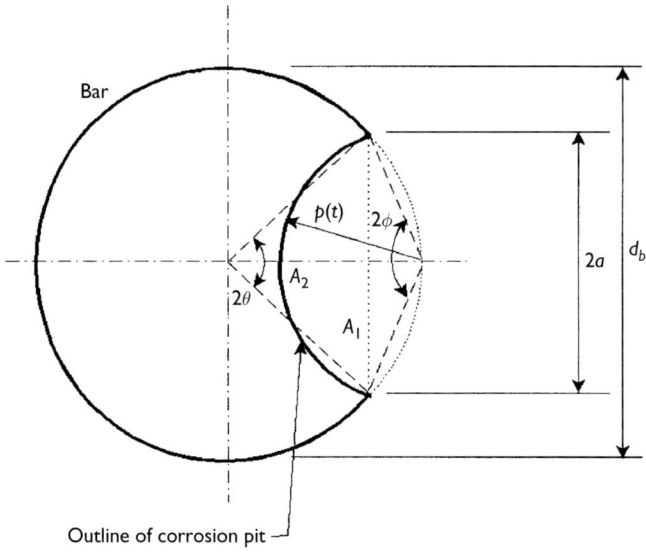

Figure 3.6.4 Calculation of residual area of an idealised pitted bar [3].

3.6.2.2 Strength and ductility of reinforcement

It is evident that loss of section will affect the strength of reinforcement and hence member strength. Perhaps less obviously, non-uniform corrosion may also affect ductility. Models for loss of strength and ductility are at present confined to empirical correlations with section loss, expressed as a percentage of original cross section:

$$f_y = (1.0 - \alpha_y \cdot A_{corr}) \cdot f_{y0} \qquad (3.6.8)$$

$$f_u = (1.0 - \alpha_u \cdot A_{corr}) \cdot f_{u0} \qquad (3.6.9)$$

$$\lambda = (1.0 - \alpha_1 \cdot A_{corr}) \cdot \lambda_0 \qquad (3.6.10)$$

f_y, f_u yield strength, ultimate tensile strength in corroded state (N mm^{-2})

λ strain corroded state

$f_{y0}, f_{u0}, \lambda_0$ yield strength, ultimate tensile strength and elongation of non-corroded bar

A_{corr} section loss

$\alpha_y, \alpha_u, \alpha_1$ regression coefficients.

To model the uncertainty in such variables the Model Code of the Joint Committee of Structural Safety (JCSS) [8] can be consulted (Tables 3.6.1 and 3.6.2). Almost all the investigators found in Table 3.6.2 report reductions in yield strength, ultimate tensile strength and elongation with corrosion. Maslehuddin *et al.* alone concluded that corrosion did not significantly affect mechanical characteristics. Given the relatively low weight loss of their test samples, any trend may have been obscured by variations in materials. These results were thus not regarded for the above given mean values and CoV.

All the investigators except Maslehuddin *et al.* report the reduction in bar strength to exceed the reduction in average cross-sectional area; in other words, corrosion effectively reduces the strength of the steel. The largest reduction in strength was reported by Morinaga, the least by Du. For example, let us assume a loss in section of 10%. In the case of Du this leads to a residual force of $F_y = 0.9 \cdot A_0 \cdot (1.0 - 0.005 \cdot 10) = 0.86 \cdot F_{y,0}$; in the case of Morinaga, $F_y = 0.9 \cdot A_0 \cdot (1.0 - 0.016 \cdot 10) = 0.76 \cdot F_{y,0}$. If only the loss of section is considered, the residual force would be $F_y = 0.9 \cdot F_{y,0}$. Consequently, Du's result indicates that the reduction in the force at which a bar

Table 3.6.1 Uncertainty in structural parameters according to JCSS [8]

Parameter	Unit	Distribution	CoV(%)
Yield stress f_y	(N mm^{-2})	Log-normal	7
Ultimate stress f_u	(N mm^{-2})	Log-normal	4
Elastic modulus of steel E_S	(N mm^{-2})	Log-normal	3
Ultimate strain of steel ε_u	—	Log-normal	6
Nominal steel cross-sectional area	(mm^2)	Log-normal	2

Table 3.6.2 Empirical coefficients for strength and ductility reduction of reinforcement assembled by Du [9]

Investigators	Cover	Exposure	A_{corr}	α_y	α_u	α_l
Maslehuddin et al. [10]	Bare	Service, marine	0–1%	0.0	0.0	0.0
Andrade et al. [11]	Bare	Accelerated, 1.0 mA cm^{-2}	0–11%	0.015	0.013	0.017
Lee et al. [12]	Concrete	Accelerated, 13.0 mA cm^{-2}	0–25%	0.012	NS	NS
Clark and Saifullah [13]	Concrete	Accelerated, 0.5 mA cm^{-2}	0–28%	0.013 and 0.012[a]	0.017 and 0.014[a]	NS
Morinaga [14]	Concrete	Service, chlorides	0–25%	0.016	0.026	0.060
Zhang et al. [15]	Concrete	Service, carbonation	0–67%	0.010	0.010	0.014
Du [9]	Bare	Accelerated, 0.5 – 2.0 mA cm^{-2}	0–25%	0.005	0.005	0.027
	Concrete	Accelerated, 1.0 mA cm^{-2}	0–18%	0.005	0.005	0.039
			Mean	0.011	0.013	0.029
			CoV (%)	41	63	77

Notes
NS – not stated.
[a] Clark and Saifullah tested both plain and ribbed bars. The two steels produced slightly different coefficients.

yields is around 40% greater than that would be estimated on the basis of the average loss of section. Morinaga's results even suggest the reduction would be 1.4 times greater than that estimated from average loss of cross section. Others have proposed intermediate values. The reduction in elongation is always greater than the reduction in strength. Du reported a tendency for greater reductions (i.e. larger values of α coefficients) to be measured with smaller-diameter bars.

The reductions are attributable to the non-uniform nature of corrosion attack (Du also carried out tests on machined bars to verify that the changes observed were not due to removal of a stronger outer layer of steel from the bar). Yielding of the bar first develops at the points of local attack, where the cross-sectional area is most highly reduced, while the remainder of the bar is still elastic. As load is further increased, strain hardening at the pitted section allows stress to increase above yield. Strains increase much more rapidly post-yield, and even if the length of the defect is not particularly large, the overall axial stiffness of the bar is altered and reduces the apparent yield strength of the steel f_y. Bar force can be increased until fracture occurs at the ultimate tensile strength of the reduced section. Fracture load will be controlled by the narrowest section, when strains throughout the remainder of the bar are still well below the fracture strain. Ductility and elongation at fracture are thus reduced. The reduction will be dependent on the ratio of local to mean corrosion penetration, and also on the mechanical characteristics of the steel, including the ratio of yield to ultimate tensile strength and strain at maximum stress.

Cross-study conclusions regarding the effect of other parameters must be highly tentative, due to the various differences in exposure/conditioning of test samples, test procedures, steel properties, bar diameters and concrete qualities between studies. For the present, models must remain semi-empirical, and their predictions must be treated with caution. The effect of the loss of reinforcement section on residual strength of statistically determinate structural elements may be estimated using conventional calculation procedures, but allowing for the reduction in cross-sectional area of the bar and in 'apparent' yield strength (ultimate tensile strength is not used in conventional strength calculations). It should also be verified that ductility is not reduced below the value assumed by the design standard in use.

Where design assumes development of plastic deformations at the ultimate limit state, that is, if based on yield line or redistribution, the reduction in residual capacity may be markedly greater as a reduction in ductility of a corroded bar affects the ability of a member to redistribute moments at a plastic hinge. It is essential to verify that bar ductility complies with the design standard in use. Should this not be the case, special measures which recognise the limitation on ductility will be required to verify strength.

It may also be necessary to reduce the strength of longitudinal compression on bars if links become ineffective either through loss of section to corrosion or because spalling of cover affects anchorage of links.

3.6.2.3 Longitudinal cracking and cover integrity

3.6.2.3.1 MODELLING CRACK INITIATION

The increase in the diameter of a bar as a result of the volumetric expansion of the corrosion-induced products generates tensile hoop strains in the concrete surrounding the bar. Once the strain capacity of the concrete is exhausted, cracking develops along the line of the bar. If corrosion is allowed to continue, spalling of the concrete cover eventually results, which would be a direct hazard to anyone in the immediate vicinity of an affected structure. Structural capacity of members will be impaired as a result of the loss of concrete cross section in the compression zone or in the web of a member. Member strength will also be impaired where a reduction in confinement to a bar affects bond resistance or the dowel action component of shear resistance.

The onset of longitudinal cracking along a corroding bar has been investigated experimentally and analytically. Table 3.6.3 compares section loss and corrosion penetration to the onset of cracking reported in various experimental studies.

The time until crack initiation, in Table 3.6.3, was calculated using the relationships between corrosion current density $i_{corr}(\mu A\,cm^{-2})$, penetration depth (mm) and test duration correcting for the fact that the corrosion current while testing was accelerated in by a factor $(i_{corr,test}/(5\mu A\,cm^{-2})$. The value of this factor varies between 2 and 2000, the average being 350. The results indicate the very short time until first cracking occurs, when compared to the initiation phase.

At impressed currents of up to $0.5\,\mu A\,cm^{-2}$, corresponding to corrosion rates of up to 50 times the maximum values observed in the field, it can be concluded with a fair degree of confidence that cracking first develops at corrosion penetrations (i.e. causing reductions in bar radius) of between 0.01 mm and 0.04 mm.

Table 3.6.3 Corrosion to cause cracking of concrete cover

Investigators	bar cross section Ø(mm)	Cover/bar ratio	Impressed current (mA cm^{-2})	Section loss(%)	Corrosion depth (mm)	Time until cracking[1] (a)
Al-Sulaimani et al. [4]	20	3.75	2	2	0.10	1.72
	14	5.36	2	3	0.11	1.90
	10	7.50	2	5	0.13	2.24
Cabrera and Ghodussi [5]	12	Large	3 V	1–2	0.03–0.06	–
Clark and Saifullah [13]	8	0.5	0.5	0.4	0.008	0.14
	8	1.0	0.5	0.6	0.012	0.21
	8	2.0	0.5	1.3	0.026	0.45
Andrade et al. [11]	16	1.25–1.88	0.1	0.4–0.5	0.015–0.02	0.30
		1.25	0.01	0.45	0.018	0.31
Clark and Saifullah [13]	8	1.0	2	0.3	0.006	0.10
	8	1.0	0.5	0.55	0.011	0.19
Rodriguez et al. [7]	16	2.0–4.0	0.003–0.10	0.4–1.0	0.015–0.04	0.47
Almusallam et al. [16]	12	5.0	10	4.0	0.12	2.07
			Mean	1.6	0.049	0.84
			CoV (%)	98	98	102

[1] assuming that a corrosion current density of $i_{corr} = 5\,\mu\text{A cm}^{-2}$ is a realistic average value.

It can also be surmised that cover cracking is delayed by increased cover and by increased concrete tensile strength, and is accelerated by more rapid impressed corrosion and where bars are cast near the top of the pour. (Note that beneficial effects of cover on time to cracking reported in these studies concern only the propagation phase of the deterioration process, and exclude the beneficial effect on the initiation phase.)

Andrade *et al.* [11] reported similar corrosion to first cracking, but found cover–bar diameter ratio to have a negligible effect. High corrosion rates probably produce less expansive corrosion products, and hence greater corrosion penetrations are required to split the cover. Concrete is weaker and more porous near the top of a pour, and sedimentation of fresh concrete may leave a void under top cast bars, thus allowing dispersion and space for corrosion products, which delays splitting. A number of researchers have attempted to use Finite Element Method (FEM) models to investigate longitudinal cracking. The models describe cracking of concrete sections containing corroding bars through 2-dimensional plan-strain analysis, aiming to describe the progression of cracking through the concrete cover and to determine the thickness of corrosion products or the internal pressure. The models idealised the expansive behaviour of corrosion products either as an internal pressure or as a prescribed displacement.

The FEM penetrations for cracking were generally appreciably lower than measured values, probably because they ignore the dispersion of corrosion products into the cement matrix. It is worth pointing out, however, that the significance of initiation of longitudinal cracking as a limit state criterion remains a matter of debate. Only very few engineering models exist currently which tackle the problem of crack initiation on a probabilistic basis. The most suitable seems to be the one of Gehlen and Banholzer [17], who established the following limit state equation:

$$p_{\text{crack initiation}} = p(\Delta r_{\text{r}} - \Delta r_{\text{s}} \leq 0) \tag{3.6.11}$$

p probability
Δr_{r} increase of steel radius necessary for initial cracking of concrete (resistance R) (μm)
Δr_{s} increase of steel radius due to the onsetting corrosion products (stress S) (μm).

Basically, the increase of the rebar radius is determined by the penetration depth of the reinforcement due to corrosion and a factor describing the expansion (K_{E}). Moreover, the parameter V_{p} takes account of the volume of pores in which corrosion products may expand without inducing hoop stresses. Possible influences of external radial loads are considered by the variable L:

$$\Delta r_{\text{s}} = L \cdot (V_{\text{corr}} \cdot t_{\text{prop}} - V_{\text{p}}) \cdot K_{\text{E}} \tag{3.6.12}$$

V_{corr} corrosion rate (μm a^{-1})
V_{p} volume of pores available for corrosion products to be accommodated without development of expansion stress (μm)
K_{E} expansion factor $R - 1$
R volume ratio of corrosion products and corroded steel
L load-dependent regression variable for extra radial load upon concrete
t_{prop} propagation period (difference of structure age and initiation period) (a).

The model for the resistance of the concrete against the formation of cracks is based on the elastic theory of material behaviour and complemented by variables to account for non-linear behaviour (K_{n}), relaxation (K_{R}) and reinforcement configuration (K_{S}). The quantification of these variables was performed by using FEM analysis:

$$\Delta r_{\text{r}} = \Delta r_{\text{r}}' \cdot K_{\text{n}} \cdot K_{\text{R}} \cdot K_{\text{S}} \tag{3.6.13}$$

$\Delta r_{\text{r}}'$ maximal increase of the radius according to the elastic theory bearable by the concrete without formation of cracks (μm)
K_{n} correction factor for non-linear material behaviour
K_{R} correction factor for relaxation ability of concrete
K_{S} factor to account for reinforcement detailing.

$$\Delta r_{\text{r}}' = \frac{f_{\text{t}}}{E_{\text{c}}} \cdot \left(0.60 \cdot \frac{(d_{\text{c}} + r_{\text{s}})^2}{r_{\text{s}}} + 0.40 \cdot r_{\text{s}} \right) \tag{3.6.14}$$

f_t tensile strength of concrete $(N\,mm^{-2})$
E_c E-Modulus of concrete $(N\,mm^{-2})$
d_C concrete cover (mm)
r_S radius of reinforcement (mm).

$$K_n = \frac{1}{\left(\frac{C_1}{r_s}+C_2\right)\cdot d_c + C_3\cdot r_s + C_4} \qquad (3.6.15)$$

C_1 regression parameter $= 3.71\cdot 10^{-1}$
C_2 regression parameter $= 5.83\cdot 10^{-3}(mm^{-1})$
C_3 regression parameter $= 2.27\cdot 10^{-4}(mm^{-1})$
C_4 regression parameter $= 1.48$.

$$K_s = C_s\cdot \left(\frac{s}{2\cdot r_s}-1\right) \qquad (3.6.16)$$

C_5 regression parameter $= 1.19\cdot 10^{-1}$
S spacing of reinforcement (mm).

The model is only valid to determine the time until a first microcrack of 0.005 mm is developed, which runs from the steel surface to the concrete surface.

The tensile strength of concrete f_t, elastic E-Modulus of concrete E_c and relaxation parameter depend upon the concrete composition. Mean values of these input variables can be derived from current standards or may be determined from measurements of specimens withdrawn from the actual structure. To model the uncertainty in such variables the Model Code of the Joint Committee of Structural Safety (JCSS) [8] can be consulted (Table 3.6.4).

In a sensitivity analysis, factors mainly responsible for the scatter for the time until crack initiation t_{crack} were found to be in order of significance: the expansion factor K_E in combination with the volume for expansion of corrosion products in

Table 3.6.4 Uncertainty in structural parameters according to JCSS [8]

Parameter	Unit	Distribution	CoV (%)
Concrete compression strength f_c	$(N\,mm^{-2})$	Log-normal	15
Concrete tensile strength f_t	$(N\,mm^{-2})$	Log-normal	30
E-Modulus of concrete E_c	$(N\,mm^{-2})$	Log-normal	0.10
Relaxation factor K_R	(—)		

concrete pore structure V_P, the E-Modulus of concrete E_c, the corrosion penetration depth V_{corr} and the concrete cover d_c. More precise information in regard to these variables, for example from inspections, will reduce the variability of t_{crack}. However, the largest degree of assumption and uncertainty is inherent in the expansion factor K_E and the volume V_P. In structure investigations, K_E may be quantified by chemical analysis of the corrosion products, but will in most cases not be conducted in practice. The volume V_P is still a matter for future quantification. In a case study of corrosion initiation due to carbonation Gehlen and Banholzer [17] assumed the mean value of $\mu(K_E) = 2 - 1 = 1$, see Equation (3.6.12). The standard deviation was set to $\sigma(K_E) = 0.50\,\mu$. As distribution type, a log-normal distribution was chosen. The volume V_P was already incorporated into K_E. The predicted time span between 0.125 and 2 years with an average of $\mu(t_{crack}) = 0.62a$ and coefficient of variation $CoV(t_{crack}) = 60\%$ until crack initiation seems quite realistic. For chloride-induced corrosion, no quantification of the expansion factor K_E is yet available. However, as corrosion products due to chlorides usually are less expansive because oxygen supply is reduced due to the wet concrete, values as used for carbonation seem to be on the safe side and the best choice available.

If inspection data regarding crack initiation are available, these can be used for model updating according to Bayesian methods. As input must be collected, the percentage of cracked concrete cover along the rebar length at inspection time is t_{insp}.

It is evident that the relationship between section loss and formation of longitudinal splitting cracks is dependent on a number of factors including the method employed to condition test specimens.

A corrosion penetration of 0.05 mm for general corrosion, which is the average value found in literature to induce corrosion, represents a reduction of less than 2.5% of cross-sectional area for an 8 mm diameter bar, and less for larger bars (for comparison, UK production tolerances on reinforcement are 6.5% for 8 mm to 10 mm diameter reinforcement and 4.5% for sizes 12 and above. Clearly, cracking will develop well before loss of bar section becomes significant. At least where general corrosion is concerned, it can safely be concluded that longitudinal cracking will precede significant loss of bar section.

3.6.2.3.2 MODELLING CRACK GROWTH

Rodriguez *et al.* [18] also derived relationships between crack width and corrosion penetration as corrosion progressed; Figure 3.6.5 shows a mean surface crack width of 1 mm to be associated with an attack penetration of around 0.2 mm (this would appear to correspond to a volume of corrosion products of 2.5 times the corrosion penetration).

A uniform metal loss of 0.2 mm around the circumference of 8, 10 and 25 mm diameter bars represents cross-sectional area losses of 9.8, 7.8 and 3.2% respectively. Cover–bar diameter ratio varied between 2 and 4 in these tests. It may also be inferred from the paper that concrete cover did not spall even where cracks reached 2 mm in width. Cabrera and Ghodussi [5] independently developed expressions for the relationship between crack width and corrosion, based on different parameters. Predictions of the two sets of expressions are broadly consistent despite significant

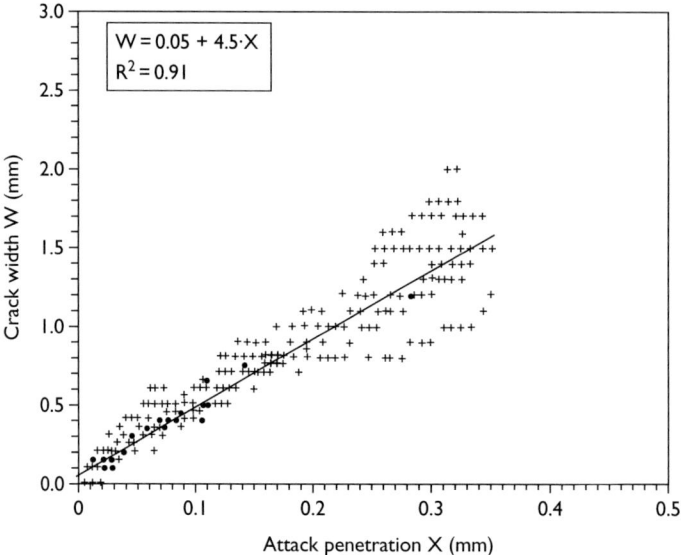

Figure 3.6.5 Relationship between surface crack width and corrosion [15].

differences in conditioning. Others report crack widths in excess of 0.6 mm without spalling.

Models for crack growth must be considered as a matter of debate. In this context, it should be kept in mind that – as was already concluded for the initiation time of cracking – the time until a crack width is reached, which may be regarded as critical until spalling occurs (0.3–1.0 mm), is rather short compared to initiation phase. Highly sophisticated modelling of crack growth versus time thus does not seem to be of importance either.

3.6.2.3.3   EFFECT OF LONGITUDINAL CRACKING ON RESIDUAL CAPACITY

Longitudinal cracking leads to loss of resistance of cover concrete. Bond resistance is also affected, and will be considered separately.

3.6.2.3.4   SHEAR

An estimation of ultimate residual shear force may be obtained by means of the standard method established in EuroCode 2 (EC2), with the following modifications:

(a) The effective depth should be reduced from d to $d - r_2$ if:
$V_{sd}/V_{cd} < 2$

   1. $s > 0.6d$ and $P(t) > 0.2$ mm
   or
   2. $\rho_2 > 0.5\%$ and $P(t) > 0.2$ mm

$V_{sd}/V_{cd} > 2$

1. $\rho_2 < 0.5\%$ and $P(t) > 0.2$ mm

or

2. $\rho_2 > 0.5\%$ and $P(t) > 0.1$ mm.

(b) The section width should be reduced from b to $b - 2r_w$, if:
$V_{sd}/V_{cd} > 2$

1. $s > 0.6$d and $P(t) > 0.4$ mm

or

2. $s < 0.6$d and $P(t) > 0.3$ mm.

Similar rules for concrete section loss are presented for columns. In addition, an additional load eccentricity should be added in order to determine the design moment. The magnitude of the additional eccentricity, which is equal to the cover thickness, should be taken in both directions (conservatively).

3.6.2.4 Bond

3.6.2.4.I MODELLING OF BOND LOSS

Bond is necessary to anchor reinforcement and to ensure composite interaction between reinforcement and concrete. Bond may conveniently be regarded as a shear stress over the surface of a bar, even if this represents a considerable simplification of the real behaviour of ribbed bars. Although the factors which influence bond strength are generally agreed, opinions differ markedly on the magnitude of their effects and the mechanisms through which they take effect. Friction is the major component of strength in plain round bars. With ribbed bars, bond depends principally on the bearing, or mechanical interlock, between ribs rolled on the surface of the bar and the surrounding concrete.

Bond action of ribbed bars generates bursting forces, which tend to split the surrounding concrete in the same manner as expansive products of corrosion. In many practical circumstances, bond failure load is limited by the resistance provided by concrete cover and confining reinforcement to these bursting forces.

Corrosion could be expected to affect bond strength in the following ways:

*   Increases in the diameter of a corroding bar at first increase radial stresses between bar and concrete and hence increase the frictional component of the bond. Further corrosion will lead to development of longitudinal cracking and a reduction in friction and in the resistance to the bursting forces generated by bond action of ribbed bars.
*   Corrosion products at the bar–concrete interface affect friction at the interface. Some suggest that a firmly adherent layer of rust may contribute to an enhancement in strength at early stages of corrosion. Cabrera and Ghodussi [6] suggest that, at more advanced stages, weak and friable material between bar and concrete will be at least partially responsible for reductions in strength. Cairns [19] and Du [9] have produced data that contradict this suggestion, however.

- Corrosion may reduce the height of the ribs of a deformed bar above the bar core. This is unlikely to be significant except at advanced stages of corrosion, however.
- Disengagement of ribs and concrete. The layer of corrosion products formed by oxidation of the steel may force the concrete away from the bar and reduce the effective bearing area of the ribs.

Research into the influence of corrosion on bond has used a wide variety of bond specimens and bar types, and it is therefore not surprising that the magnitudes of the bond strengths reported and the effect of corrosion on those bond strengths differ widely. Furthermore, there have been considerable variations in conditioning of specimens for corrosion studies, and this has been proven to influence residual bond strength. Despite wide variations in test specimens and in conditioning techniques, the general trends reported are the same in almost all studies, as illustrated in Figure 3.6.6.

The same pattern is observed in tests on plain and ribbed surface bars. Initially, bond strength is increased by small amounts of corrosion, but with further increases in corrosion, the bond starts to reduce. It appears, however, that bond strength does not reduce below the 'as new' value prior to development of externally visible longitudinal cover cracks. For purposes of assessment of residual strength of concrete structures suffering from general corrosion, the bond can be assumed to be sound in the absence of visible corrosion-induced cracking.

Once cracking develops, appreciable loss of bond strength may develop, particularly if no confining reinforcement is present. Residual strength decreases with increasing corrosion, and reductions of over 50% are reported. Table 3.6.5 provides a comparison of bond strength loss measured by various investigators at fairly advanced levels of corrosion.

The comparison has been estimated for a nominal corrosion penetration (thickness of material lost averaged over the bar surface, and equivalent to the reduction in bar radius) of 0.25 mm and represents conditions well beyond the corrosion levels

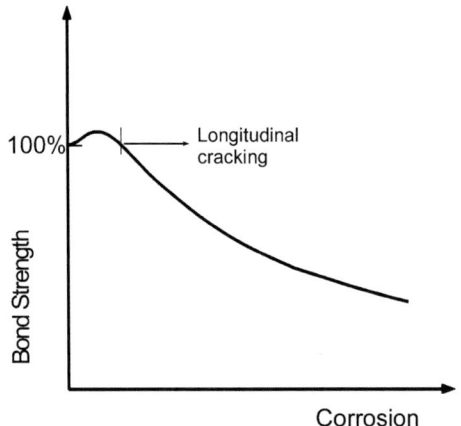

Figure 3.6.6 Variation in bond strength with corrosion (schematic).

Table 3.6.5 Summary of tests on bond strength of corroded reinforcement, all with a corrosion penetration of $P = 0.25$ mm (interpolated where necessary)

Investigators	Cover–bar ratio	Corrosion rate (mA cm^{-2})	Links	Crack width (mm)	Section loss (%)	Residual bond strength (%)
Al-Sulaimani et al. [4]	3.75	2	N	—	5	35
	5.36	2	N	—	7	15
	7.50	2	N	—	10	0
	Large	3 V	N	—	8	65
Cabrera and Ghodussi [5]	Large	3 V	Y	—	8	95
	Large	3 V	Y	—	8	110
Clark and Saifullah [13]	0.5–2	0.5	N	—	12.5	75
Clark and Saifullah [13]	1.0	0.04	N	—	12.5	60
	1.0	0.25	N	—	12.5	105
	1.0	4.0	N	—	12.5	50
Rodriguez et al. [7, 18]	1.5	0.1	N	1.6–2.2	5.4–9.1	20
	1.5–2.5	0.1	Y	1.0–2.0	5.4–9.1	65
	1.5	0.1	Y	1.1–1.4	14	75
Almusallam et al. [16]	5.0	10	N	0.4	8.0	15
Coronelli [20]	2.5	0.05	Y	—		110
Stanish [21]	2.0	0.21	N	—	10	60

required to crack the concrete cover. In Figure 3.6.5, for example, a corrosion penetration of 0.25 mm corresponds to a surface crack width of 0.8–1.5 mm. Some values given in Table 3.6.5 have been obtained by interpolation between original results in order to obtain a more consistent basis for comparison.

Residual bond strengths quoted by the various investigators could vary more widely, with values from 0% (complete loss of bond strength) to 110% (a modest gain strength) reported. The majority of results were obtained using specimens without confining reinforcement. Al-Sulaimani et al. report strength losses exceeding 70%, as do Rodriguez et al. Cabrera and Ghodussi report a reduction of only 35%, not dissimilar to that reported by Clark and Saifullah, although these authors have also pointed out that the results are markedly influenced by corrosion rate. Lesser reductions are reported in specimens provided with links. Rodriguez et al., for example, found strength losses to be reduced by half to a third where links were provided [22, 23].

Berra et al. [24] alone report a continuing increase in bond strength with increasing corrosion to 8% section loss, equivalent to 0.3 mm penetration, despite extensive longitudinal cracking. This is probably because stirrups, being at some distance outside the anchored bar, were particularly well arranged to maintain confinement in these specimens.

There appear to be inconsistencies between reductions reported in these studies, and apart from the presence of confining reinforcement, no clear pattern of influencing parameters emerges. However, in the majority of cases the rate of reduction in bond strength attributable to corrosion clearly exceeds the corresponding rate of reduction in tensile strength of the bar.

In what is probably the most extensive study undertaken to date, Rodriguez *et al.* [18] developed empirical expressions for residual bond strength of corroded bars. The experimental work was mainly based on tests carried out with cubic specimens reinforced with four bars, one in each corner, reproducing part of a beam subjected to constant shear force. These tests allowed bond strength values representative of the splitting failure modes considered in design Codes of Practice to be developed. They will, however, underestimate failure strengths where bars are well confined by high cover or large amounts of transverse reinforcement. Tests were conducted with and without stirrups, and used 16 and 10 mm diameter bars. Corrosion was accelerated by impressed current, current densities in the range of 3–100 μA cm^{-2}. Neither the concrete quality nor the cover size–bar diameter ratio influenced residual bond strength when cover was cracked by reinforcement corrosion.

A statistical analysis was carried out with these results and expressions obtained to predict the characteristic value of residual bond strength. If the ratio ρ_{tr} of the transverse reinforcement area at anchorage length (considering the reduction due to corrosion) to the area of the main bars is higher than 0.25 (minimum value established in EC2), the bond strength f_b (in N mm^{-2}) can be predicted as follows:

$$f_b = 4.75 - 4.64 p(t) \tag{3.6.17}$$

$p(t)$ attack penetration (bar radius reduction) (mm)

On the other hand, if ρ_{tr} is lower than 0.25, f_b can be estimated by:

$$f_b = 10.04 + (-6.62 + 1.98 \cdot (\rho_{tr}/0.25)) \cdot (1.14 + p(t)) \tag{3.6.18}$$

where $p(t)$ as defined above.

It is well known that external pressure, as occurs at support regions, for example, enhances the confinement of the bars. Consequently it improves bond strength, and its positive effect is considered in EC2 when calculating ultimate bond stresses in sound ribbed bars. In order to explore this bond enhancement in corroded bars, tests were carried out on beams designed to fail at anchorage of main tension bars. A similar expression to that of EC2 was obtained from the experimental results to predict the characteristic value of the residual bond strength f_b:

$$f_b = (4.75 - 4.64 \cdot p(t))/(1 - 0.08 f) \tag{3.6.19}$$

f external pressure at anchorage (N mm^{-2})

This expression can be used in the assessment of the bar anchorage at the support zone, independently of the amount of stirrups and their level of corrosion. This fact was verified in beam tests where bars anchored at support zones did not slip although heavy deterioration had been produced with significant stirrup corrosion and concrete cracking.

Equations (3.6.17) to (3.6.19) are applicable once longitudinal cover cracks develop, and are not applicable to small amounts of corrosion. Although the experimental values of the attack penetration $p(t)$ ranged from 0.04 to 0.5 mm, the authors believe extrapolation up to $p(t) = 1.0$ mm is reasonable (although bond strengths in (3.6.17) actually become negative at corrosion penetrations greater than 0.4 mm if no links are present). This proposal gives bond strength values for each attack penetration, taking account of the actual residual stirrup section at the anchorage length. The expressions can be used to demonstrate a more rapid reduction in bond strength where links are not provided, Figure 3.6.7.

According to the investigators, these expressions can also be applied to the evaluation of composite interaction at intermediate parts of the bar. In these cases, the estimation of the ρ_{tr} value has to be made by replacing the number of stirrups within the anchorage length by $200s$, where s is the stirrup spacing in mm.

Clark and Saifullah have also suggested semi-empirical expressions for the ratio of residual to original bond strength based on their tests on 8 mm diameter ribbed bars, although it was noted that residual bond strength was dependent on current intensity. Tests were conducted on beam end specimens similar to those used by Rodriguez *et al.* [18], but 8 mm diameter bars were used and links were not provided. Averaging the two sets of coefficients for current intensities within the range used by Rodriguez *et al.* leads to:

$$f_{corr}/f_{control} = 1.05 - 0.03 \cdot X \tag{3.6.20}$$

$f_{corr}/f_{control}$ ratio of bond strength of corroded specimen and non-corroded control specimen

X weight loss due to corrosion (%).

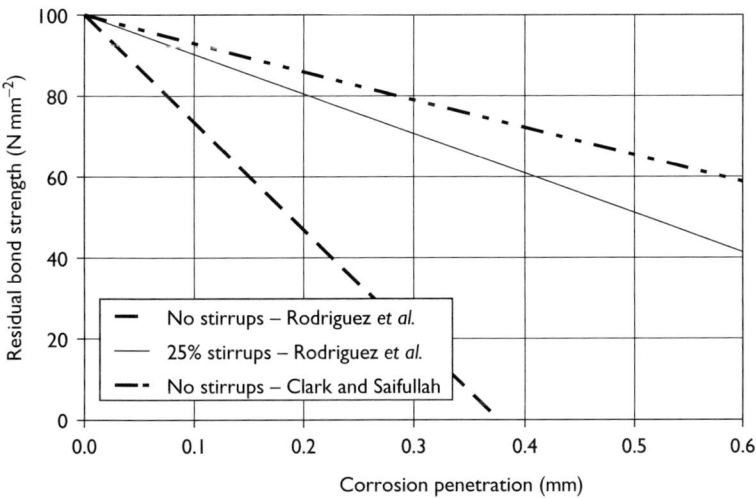

Figure 3.6.7 Effect of links on residual bond strength of 16 mm diameter bar.

It is evident that (3.6.17) predicts a much more rapid loss of bond in the absence of links than (3.6.20) does, see Figure 3.6.7. The difference in strength reductions reported in the two studies demonstrates the inherent difficulty in reaching reliable expressions for residual capacity when the mechanism controlling loss of strength is not fully understood.

3.6.2.4.2   EFFECT OF BOND LOSS ON RESIDUAL CAPACITY

A reduction in bond may affect element strength in two ways:

1. The stress that can be developed in reinforcement may be limited by a reduction in anchorage capacity at laps and points of bar curtailment (including end anchorages). This could influence flexural and shear strength of beams and slabs as well as axial strength of columns and walls.
2. Partial or complete loss of composite interaction between reinforcement and concrete over the affected length and beyond may occur due to marked reductions in bond stiffness. The plane strain assumptions normally made in section analysis/design would then no longer hold, and the pattern of strains in a member would be altered. Structural behaviour of a beam will tend to move away from purely flexural action towards a tied-arch form of action (provided that end anchorage is maintained).

A combination of (1) and (2), where the loss of plane section behaviour could increase the force to be anchored at points of bar curtailment, may also be a reason for the reduction in bond. A summary of tests on strength of beams and slabs with corroded reinforcement is given in Table 3.6.6. The largest reductions, reported by Kawamura *et al.* [24] and by Daly [25], arose from bond/anchorage failure. In general, however, strength loss measured on structural elements is less than might be feared given the relatively large values measured in bond tests at similar levels of deterioration in similarly reinforced specimens. The loss of element strength reported is greater than the average loss of cross section, probably attributable to a reduction in apparent yield strength as a result of uneven corrosion along the bars and to delamination of concrete cover. However, it is also apparent that many of the beam specimens tested would have been insensitive to loss of bond within the span, and end anchorage would have been enhanced by lateral pressure at simple supports. Many contained links which would have helped to maintain splitting resistance of cover concrete after longitudinal cracks developed, and structural elements without links might have shown greater susceptibility to bond loss. In the majority of laboratory tests, reinforcement was not lapped or curtailed within the span, and there appears a strong possibility that residual strength in real structures will be strongly affected by reinforcement details.

Findings thus tend to show that loss of anchorage capacity is more likely to be significant than loss of local bond. There is little evidence to suggest that conventional calculation procedures for flexural and shear strength are unsafe, provided allowance is made for reductions in cross-sectional area. Nonetheless, there is limited evidence that loss of local bond may influence behaviour of corrosion-damaged structures, particularly where plain round bars are used for main flexural reinforcement or in the absence of confining reinforcement to maintain residual bond strength, and tied-arch

Table 3.6.6 Summary of tests on strength of beams and slabs with corroded reinforcement

Investigators	Cross Section lost (%)	Thick-ness lost (mm)	Long. crack width (mm)	Links	Details of end anchorage	Local bond stress (N mm^{-2})	Residual strength [%]
Okada et al. [25]	—	—	0.02–0.15	Y	End hook	1.7	≥ 91
Tachibana et al. [26]	≤ 5	≤ 0.2	≤ 0.75	N	15 d_b straight	1.8	≥ 87
Al-Sulaimani et al. [4]	≤ 4	≤ 0.12	≤ 1.3	Y	12 d_b straight	5.3	≥ 90
Cabrera [5, 6] Series 1	≤ 9.2	≤ 0.28	≤ 0.6	Y	None	4.3	≥ 80
Series 2	≤ 7.8	≤ 0.24	≤ 0.6	Y	None	7.2	100
Rodriguez et al. [22, 23]	10–25	0.30– 0.53	—	Y	12.5/15 d_b straight	1.6 –1.9	75–50
	—	—	≤ 1.0	Y	End hook	2.1	Av. 99
Kawamura et al. [27]	—	—	≤ 1.0	N	End hook	2.1	Av. 94
	—	—	0.08–0.20	Y	Lapped joint	N/A	80–60
	—	—	0.04	N	Lapped joint	N/A	25
Daly [28]	≤ 17	≤ 0.5	—	Y	12 d_b straight	1.3–4.1	≥ 70
Almusallam et al. [16]	≤ 75	≤ 1.5	—	N	8.5 d_b straight	5.7	100–15

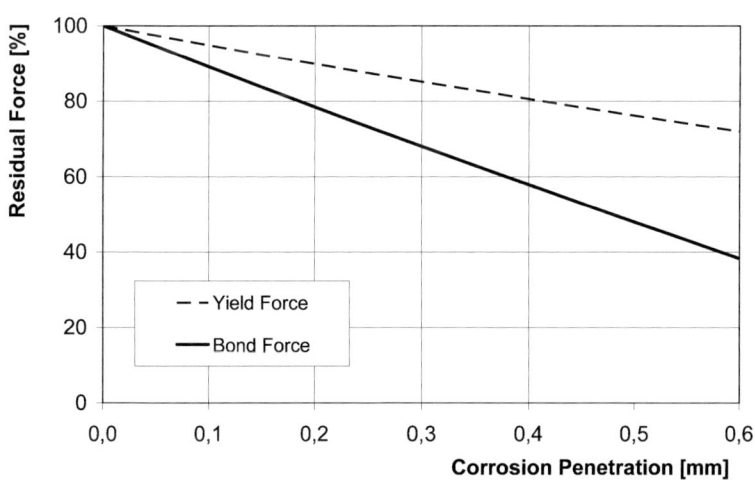

Figure 3.6.8 Comparison of residual bond and yield strengths of 16 mm bar.

action may come into play in such circumstances. Obviously tied-arch action can only develop where ends of bars are adequately anchored.

The difficulties inherent in the analysis of results of physical tests on deteriorated specimens should be noted, as the effect of a change in bond characteristics on member behaviour cannot be entirely separated from the effects of section loss of reinforcement and of spalling of concrete. No information is available on the effect of corrosion on anchorage capacity of hooks or bends. Figure 3.6.8 compares the reductions in bond force arising from corrosion with the reduction in bar yield strength and section loss. The reduction in bar strength has been calculated using a value $\alpha = 0.011$ in Equation (3.6.8). The reduction in bond force has been calculated using Equation (3.6.19) to account for a reduction in bond strength in combination with a reduced circumference of the rebar due to corrosion. Note that Equations (3.6.17) and (3.6.18) both indicate a more rapid loss in bond with corrosion. Equation (3.6.5) shows the same rate of reduction as (3.6.19). Still, Figure 3.6.8 shows the reduction in bond to be more rapid than the reduction in yield strength.

Acknowledgement

The author thanks Professor John Cairns from Heriot-Watt University for contributing a literature review on the structural consequences of reinforcement corrosion, which has provided the basis for this chapter.

References

[1] CONTECVET. 2001. A validated users manual for assessing the residual service life of concrete structures – Manual for assessing structures affected by ASR. EC Innovation Programme IN309021.

[2] Cairns, J., and Z. Zhao. 1993. Structural behaviour of concrete beams with reinforcement exposed. *Proceedings of the Institution of Civil Engineers: Structures and Buildings* 99: 141–54.

[3] Schießl and Lay 1998, Brite EuRam III Project BE95-1347 'Duracrete' – Report R4-5, *Modeling of degradation*, EU Growth Program database.

[4] Al-Sulaimani, G. J., M. Kaleemullah, I. A. Basunbul, and Rasheeduzzafar. 1990. Influence of corrosion and cracking on bond behaviour and strength of reinforced concrete members, *Proceedings American Concrete Institute* 87(2): 220–3.

[5] Cabrera, J. G., and P. Ghodussi. 1993. The effect of corrosion level on the crack characteristics of reinforced concrete. In *Deterioration and repair of reinforced concrete in the Arabian Gulf*, Vol. 2, ed. G. L. Macmillan, pp. 797–814.

[6] Cabrera G. J., and P. Ghodussi. 1992. Effect of reinforcement corrosion on the strength of steel concrete bond. *Proceedings of the international conference on bond in concrete – From research to practice*, Riga, Latvia, pp. 10.11–10.24.

[7] Rodriguez, J., L. M. Ortega, and J. Casal. 1994. Corrosion of reinforcing bars and service life of reinforced concrete structures: Corrosion and bond deterioration, *Proceedings of international conference on concrete across borders*, Odense, Denmark, Vol. II, 315–26.

[8] JCSS. 2000. *Probabilistic model code*. The Joint Committee on Structural Safety.

[9] Du, Y. 2001. Effect of reinforcement corrosion on structural concrete ductility, PhD Thesis, University of Birmingham, UK.

[10] Maslehuddin, M., I. M. Allam, G. J. Al-Sulaimani, A. L. Al-Mana, and S. N. Abduijauwad. 1990. Effect of rusting of reinforcing steel on its mechanical properties and bond with concrete, *ACI Materials Journal*, September–October, 496–502.

[11] Andrade, C., C. Alonso, D. Garcia, and J. Rodriguez. 1991. Remaining lifetime of reinforced concrete structures: Effect of corrosion in the mechanical properties of the steel, *Life prediction of corrodible structures*, NACE, UK: Cambridge, 12/1–12/11.

[12] Lee, H. S., F. Tomosawa, and T. Noguchi. 1996. Effect of rebar corrosion on the structural performance of singly reinforced beams. In *Durability of building materials and components*, vol. 7, ed. C. Sjostrom. London: Spon, 571–80.

[13] Clark, L. A., and M. Saifullah. 1994. Effect of corrosion rate on the bond strength of corroded reinforcement. In *Corrosion and corrosion protection of steel in concrete*, ed. R. N. Swamy. Sheffield: Sheffield Academic Press, 591–602.

[14] Morinaga, S. 1996. Remaining life of reinforced concrete structures after corrosion cracking, In *Durability of building materials and components*, vol. 7, ed. C. Sjostrom. London: Spon, 127–37.

[15] Zhang, P. S., M. Lu, and X. Y. Li. 1995. The mechanical behaviour of corroded bar, *Journal of Industrial Buildings* 25: 257, 41–4.

[16] Almusallam, A. A., A. S. Al-Gahtani, and A. R. Aziz. 1996. Rasheeduzzafar: Effect of reinforcement corrosion on bond strength, *Construction and Building Materials* 10: 123–129.

[17] Gehlen, C., and B. Banholzer. 2000. Time to initiate a corrosion-induced crack after depassivation of the reinforcement. In *Proceedings of the RILEM/CIB/ISO/symposium on integrated life cycle design of material and structures*, ILCDES, ed. A. Sarja, May 22–24, Helsinki, Finland, pp. 361–366.

[18] Rodriguez, J., L. M. Ortega, J. Casal and J. M. Diez. 1996. Assessing structural conditions of concrete structures with corroded reinforcement, *4th international congress on concrete in service mankind*, Dundee, UK.

[19] Cairns, J. 1995. Strength in shear of concrete beams with exposed reinforcement. *Proceedings of the Institution of Civil Engineers: Structures and Buildings* 110(2): 176–85.

[20] Coronelli, D. 1997. Bond of corroded bars in confined concrete: Test results and mechanical modeling, *Studi E Richerche* 18: 137–211.

[21] Stanish, K. 1997. *Corrosion effects on bond strength in reinforcement in concrete*, MSc Thesis, Department of Civil Engineering, University of Toronto.

[22] Rodriguez, J., L. M. Ortega, and J. Casal. 1995. Load carrying capacity of concrete structures with corroded reinforcement. *Structural faults and repair.*

[23] Rodriguez, J., L. M. Ortega, and J. Casal. 1996. Load bearing capacity of concrete columns with corroded reinforcement. In *Proceedings 4th international conference on corrosion of reinforcement in concrete structures*, eds C. L. Page, P. B. Bamforth and J. W. Figg. SCI, Cambridge, UK.

[24] Berra, M., A. Castellani, and D. Coronelli 1997. Bond in reinforced concrete and corrosion of bars. In *Proceedings, Conference structural faults and repair*, ed. M. Forde. Edinburgh: Engineering Technics Press, Vol. 2, 349–56.

[25] Okada, K., K. Kobayashi, and T. Miyagawa. 1988. Influence of longitudinal cracking due to reinforcement corrosion on characteristics on reinforced concrete members, *ACI Structural Journal* 85(2): 134–40.

[26] Tachibana, Y., K. I. Maeda, Y. Kajikawa, and M. Kawamura. 1990. Mechanical behaviour of RC beams damaged by corrosion of reinforcement. In *Corrosion of reinforcement in concrete*, eds C. L. Page, K. W. J. Treadaway and P. B. Bamforth. London: Elsevier.

[27] Kawamura, A., K. Maruyama, S. Yoshida, and T. Masuda. 1995. Residual capacity of concrete beams damaged by salt attack, concrete under severe conditions. In *Environment and loading*, eds K. Sakai, N. Banthia and O. E. Gjorv. London: Spon, vol. 2, 1449–57.

[28] Daly, A. F. 1995. Effects of accelerated corrosion on the shear behaviour of small scale beams, TRL Research Report PR/CE/97/95, Transport Research Lab, Crowthorne, UK.

3.7 Maintenance, repair and rehabilitation (MR&R) planning

John B. Miller

3.7.1 Concepts

Maintenance, repair and rehabilitation are related activities that merge and overlap without any clearly defined boundaries between them. *Maintenance* is usually perceived as being upkeep characterised by operations such as cleaning, touching-up, minor repairing, replacing worn materials and the like, which are expected during a structure's lifetime, and are regarded as being reasonable and inevitable. Maintenance becomes *repair* when the extent and type of work required tends to fall outside the routinely and reasonably expected. The necessity for repair arises when deterioration cannot, or can no longer, be prevented by routine maintenance. Examples of deterioration that lead to the necessity for repairs are those caused by agencies such as carbonation, chloride contamination, frost, moisture ingress, chemical attack, thermal movement and the like. When deterioration has advanced to a stage where repair work becomes global to the structure, or major parts of it, the concept of *rehabilitation* arises.

Maintenance, repair and rehabilitation are also relative concepts. The maintenance in one type of structure which may reasonably be expected during its lifetime may be viewed as catastrophic failure in others. For example, one would not be surprised by the damage caused to industrial floors subject to thermal shock in foundries, and the routine maintenance scheme would be expected to deal with it. Damage of similar severity in a parking facility, from whatever cause, would not be expected and could not be handled by normal routine maintenance. Likewise, in the agricultural facilities of some countries, rehabilitation is often not considered until the structure is near collapse. For civil and domestic structures, rehabilitation is normally required long before damage has advanced to such a degree.

When planning maintenance and repair of a structure, its maintainability and reparability have to be considered. Rehabilitation has no such corresponding concept. In rehabilitation, the aim is to bring the structure back to near its original condition, often with some improvement, and one is not to the same degree bound by the constraints of working with existing configurations, or even with the same materials. Nevertheless, maintainability and reparability are concepts that have to be considered in rehabilitation work also, since nothing is everlasting and maintenance and repair will be required at some point in time after rehabilitation.

In passing, it is mentioned that the term *restoration* is reserved by the author to apply to the specialist works required for the preservation and repair of listed, protected and historic structures that have little to do with the upkeep of ordinary commercial, civil and domestic structures. Sometimes, also, the term 'remediation' is used in connection with repair and maintenance work. This term is unfortunate since it applies to counteracting or removing destructive influences, rather than to the repair or maintenance work made necessary by these influences. Another much-bandied word is 'refurbishment'. This word properly refers to what may be regarded as 'brightening-up' or 'redecoration', rather than to the relatively intensive intrusions required in the repair and rehabilitation of concrete structures.

The terms *repair* and *rehabilitation* have dictionary definitions that are close to the concepts understood and applied to concrete structures by contractors, end users, authorities and consultants. However, the concept of maintenance has a wider compass and includes contingent actions, which are not connoted by dictionary definitions. Thus, in the Lifecon project [1], maintenance is defined to be the 'combination of all technical and associated administrative actions during the service life (required) to retain a repair and/or upgrade in a state in which it can perform its required functions'.

3.7.2 Necessary planning elements

There are many possible ways to make MR&R plans, but there are five overriding, basic elements that should be incorporated into any proper plan. These elements are:

- Survey and condition assessment
- Reliability, Availability, Maintainability and Safety (RAMS) of materials, techniques and personnel
- Life cycle cost (LCC)
- Life cycle ecology (LCE) and
- Combined quantification of results.

First, and necessary to get anywhere at all, is a proper survey than allows an accurate condition assessment and that is sufficiently predictive to allow sensible plans for future MR&R to be made. Second, the types of damage and deterioration suffered by a structure should be classified according to their RAMS characteristics, as should the RAMS characteristics of the possible repair materials and techniques which may be used to rectify that damage and deterioration. Third, the LCC of the MR&R actions necessary to keep a structure serviceable for its projected lifetime needs to be estimated. Fourth, the LCE of the MR&R actions throughout the lifetime of the structure needs to be estimated. Finally, a tool is needed for the combined quantification of all aspects of MR&R planning, including the more diffuse aspects such as, for example, health and happiness, culture, tradition, etc. in order that totalities may be compared.

To make MR&R plans encompassing these five elements is a complex task that is practically insurmountable without effective tools. The following is an attempt to give some basic guidance on how such tools can be structured, and to indicate how they can be used.

3.7.3 Survey and condition assessment

In order to manage a structure properly, it is necessary to collect and systematise as much information as possible about its history, design, construction, and present and probable future condition. Thus, the management of structures involves the registration, systematising and treatment of large amount of data in the form of drawings, notes, building descriptions, diaries, photographs, weather conditions, loads, environmental factors, concrete mix designs, anomalies, survey results, etc. MR&R planning,

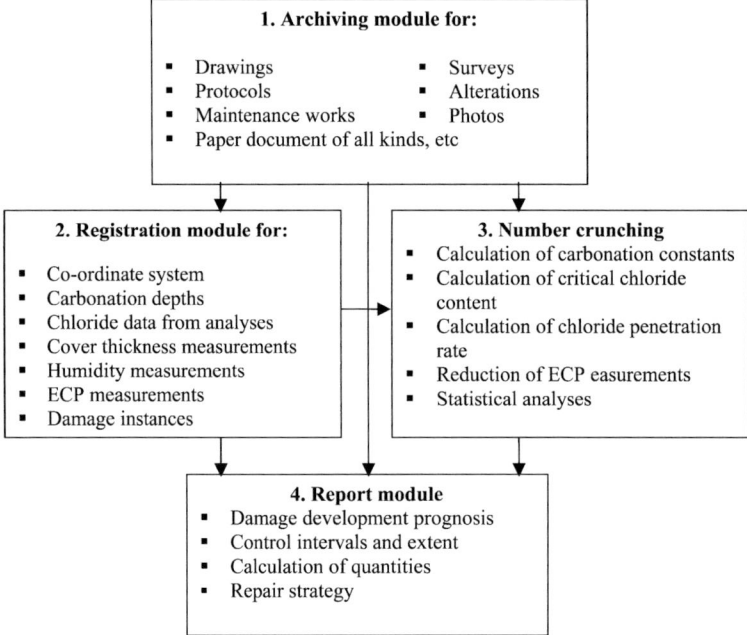

Figure 3.7.1 Model for the construction of a tool for MR&R information collection, treatment and administration.

therefore, involves the administration of documents of all kinds, and includes the registration of field and laboratory survey data, its mathematical treatment and, finally, the generation of reports.

To do this efficiently requires the use of a suitable IT tool, such as that used by the author [2], which consists of the four basic modules shown in Figure 3.7.1 and described below. The tool is in the form of a computer program that can be installed on an ordinary computer.

3.7.3.1   Module 1 – Archiving module

End users and their caretakers have a need to register, store and maintain much information pertaining to the properties and structures for which they are responsible. Very often, much of this information is lost due to lack of proper systemisation and safe deposition in an easily overviewed and easily accessible archiving system. The core of Module 1 is, therefore, a relatively simple program for the administration of documents. Data are entered principally by scanning paper documents (to the extent that digital material is not available) with a facility for the conversion of varying formats to common formats. Digital enhancement and clean-up features for photographs, faded texts and soiled and creased drawings are necessary, and should be provided by the installation, where necessary, of third-party photo and drawing enhancement software. The author's program integrates with such software, which thus becomes seamlessly available to the user.

The module makes it possible to keep all documentation safe, in multiple copies, and in easily accessible form, for example by storage on CD or DVD ROM disks, or on any other suitable medium.

3.7.3.2 Module 2 – Registration module

This module is used for the registration of field and laboratory data from surveys. The module allows the construction of tables for the entering of measurement data such as carbonation depths, cover thicknesses, chloride values, sample numbers, electrochemical potential (ECP) values, etc., all related to a coordinate system.

The module is relatively flexible in the way data are entered. For example, the coordinate system to which all data are referred can be mathematical in nature, or existing axes can be used, or even simple 'by-eye systems' can be used, such as 'Level 1,2,3,...' or 'right part, uppermost,..., lower part, downwards'. A table can consist of a single cell or several hundred cells. There is no requirement for entering data into all cells, and there is room for remarks and notes.

The numerical material is sample and measurement data from field and laboratory investigations. The values are always related to points on surfaces independently of the type of measurements which has taken place. For example, in a parking garage, there may be several different types of surface, such as the flooring, the soffits, the walls, and the surfaces of pillars and beams. It is, therefore, logical to use the same general table format for all types of measurements, since every number, or series of numbers, corresponds to a sampling location and sample type somewhere on a surface. An example of a table format is shown in Table 3.7.1, where the coordinates are the numbers of metres from the origin at 0.0 m.

Table 3.7.1 Example of a table for entering survey data into a computer

Origin located at the left front edge of every pillar surface	Notes	XXX Building Co-op Cover thickness measurements over stirrups in mm Measured on pillars in axes A2 to C2 in parking garage								
		Axis A2			Axis B2			Axis C2		
		Side 1	Side 2	Side 3	Side 1	Side 2	Side 3	Side 1	Side 2	Side 3
0.0		0.25	0.25	0.25	0.25	0.25	0.25	0.25	0.25	0.25
0.2		12	12	15	13	19	22	18	18	15
0.4		10	13	15	9	11	15	14	13	14
0.6		16	18	21	20	22	22	21	19	17
0.8		13	11	14	12	18	21	17	17	14
1.0		13	13	16	15	21	23	19	19	16
1.2		12	15	17	11	13	17	16	15	16
1.4		8	11	13	7	9	12	12	11	12
1.6		15	15	18	16	22	25	22	21	18
1.8		16	18	21	20	22	22	21	19	17
2.0		13	11	14	12	18	21	17	17	14
2.2		13	13	16	15	21	23	19	19	16
2.4		12	15	17	11	13	17	16	15	16

The table is adjustable to a convenient form and size. That part of the table upon which numerical calculations may be performed is marked in grey. For the purpose of editing and recording data and text, editing tools are available. Descriptions of sample locations can be entered as text with complete freedom. This means that the size and shape of the table can be adjusted in every case. The grey area of the table contains numbers upon which calculations are to be performed, and this area will vary in size from table to table. The numbers to be operated upon can be linked to parts of the descriptive text and to the distances from the origin.

This module allows fieldwork to be conducted in an efficient manner, and the efficiency gains will increase as more data-logging instruments that can be connected directly to the input interface become available.

3.7.3.3   Module 3 – Number crunching

This is a purely mathematical module. It picks the necessary data out of the tables in Module 2 and treats them in different ways. Examples of possible calculations are:

- Characteristic carbonation depth and statistical variations
- Statistical treatment of chloride data
- Comparison of carbonation depths with cover thicknesses
- Comparison of chloride data with cover thicknesses
- Treatment of ECP measurements (visualisation of areas subject to corrosion)
- Damage development prognosis and
- Calculation of damage frequency and rehabilitation needs related to desired damage free periods with reliability indications.

This module is perhaps the one that results in the greatest gain in efficiency, since data-entering into spreadsheets is eliminated. It also allows various ways of treating results by means of simple choices of pre-programmed calculation models.

3.7.3.3.1   NUMERICAL MATERIAL

Various types of measurements are listed below, together with their associated numerical ranges:

- Cover thickness measurements (positive numbers between 0 and 360 mm)
- Carbonation depth measurements (positive numbers between 0 and 100 mm)
- Chloride penetration measurements (positive numbers showing the penetration depth in mm combined with the chloride content in per cent at that depth)
- ECP measurements (numbers between ± 760 mV for Cu/CuSO4 half-cells)
- Crack widths (positive numbers between 0 and 50 mm)
- Delaminated areas (positive numbers between 0.1 and 100 m^2)

3.7.3.3.2   MARKING OF NUMBERS IN A TABLE

In many cases, it is practical to be able to mark values that are higher or lower than a given value. This can be done, for example, by changing the font in the table to bold in combination with shading or colour changes. This happens upon selecting one or more cells in one or more tables, whereupon a subprogramme marks the numbers which are higher or lower than the given number. In this connection, it is important that the routine is able to distinguish between negative and positive numbers, since ECP measurements can comprise both.

3.7.3.3.3 SELECTION OF TABLES CONTAINING VALUES HIGHER OR LOWER THAN A GIVEN VALUE

A labour-saving feature is the capability of the program to tell how many tables in a set of tables contain values lower or higher than a given value. For example, supposing there are 100 tables containing data from 12 different types of surfaces. The software makes it easy to find how many tables contain the exceeding values. The program can then be instructed as to which tables are to be evaluated simultaneously with respect to a given value and whether that value is to be exceeded or not.

3.7.3.3.4 CALCULATIONS

By means of pre-programmed subroutines, calculations can be performed on the recorded data. The pre-programmed number of available routines may, of course, be increased whenever desired. The author's present program allows several types of calculation, including commonly performed calculations relating to carbonation, chloride penetration and statistical treatment.

Rate of carbonation and standard deviation In practical work, the carbonation rate may be expressed by the equation:

$$x = x_0 + k_C \sqrt{t} \qquad (3.7.1)$$

where
  x the measured carbonation depth in mm
  x_0 the instantaneous carbonation depth from day one (usually, but not necessarily, zero)
  k_C a material constant for the particular concrete and location
  t the time in years

A typical calculation would be as follows:

A 35-year-old building has five surfaces, which all face in the same direction. Carbonation depths have been measured on each of the surfaces, and the measurements were 12, 9, 13, 10 and 21 mm. What is the general carbonation depth likely to be 25 years from now?

The standard deviation, σ_x, of these five measurements is 6 mm, and the average is 13 mm. Equation (3.7.1) can be used to find k_C:

$$k_C = \frac{x_{35}}{\sqrt{35}} = \frac{13}{5.916} = 2.20$$

In 25 years time, the building will be 60 years old ($t = 60$ years) thus

$$x_{60} = k_C \sqrt{60} = 2.2 \sqrt{60} = 17 \, \text{mm}$$

So 25 years from now, the average carbonation depth will be 17 mm with a variation of ±6 mm over 66% of the surfaces.

This type of information can be used to define the levels of repair which are necessary in order to preclude more MR&R actions within a given time frame with a desired level of certainty.

Chloride penetration The calculation of chloride ion penetration rate is a complicated business, since the penetration rate of these ions follows Fick's second law of diffusion:

$$\frac{dc}{dt} = D\frac{d^2c}{dx^2}$$

There are several mathematical models in existence for the calculation of future chloride distribution in concrete based on this law. None of them are suitable for practical work. In practical work, the author, therefore, chooses to use the same type of model as for carbonation, i.e. the penetration depth is a function of the square root of time:

$$D_{Cl} = k_{Cl} \sqrt{t} \tag{3.7.2}$$

where
t the time in years
k_{Cl} a material constant for the particular concrete and location
D_{Cl} the depth where the critical content has been found.

This gives a somewhat conservative estimate of the rate of penetration, but it is a practical tool for treating chloride penetration problems.

For example, supposing a chloride profile has been measured on a 25-year-old concrete parking deck which was originally free of chlorides. The cover thickness is 45 mm, and the profile shows that the critical chloride content has reached a depth of 30 mm. How long is it going to take for the critical content to reach the steel?

Applying Equation (3.7.2) gives:

$$30 = k_{Cl} \sqrt{25} \quad \therefore k_{Cl} = \frac{30}{5} = 6$$

By substitution in Equation (3.7.2):

$$45 = 6\sqrt{t_{45}}$$

$$t_{45} = \left(\frac{45}{6}\right)^2 = 56 \, \text{years}$$

The remaining time is thus $56 - 25 = 24$ years, i.e. it will take a further 24 years for the chloride content of the concrete at the steel to reach a critical level.

Statistical analysis The module allows effective, statistically based damage development prognoses to be made for given scenarios. For example, as indicated in section 'Rate of carbonation and standard deviation', it is possible to estimate the future regimen of carbonation damage or chloride damage in terms of when and where the carbonation front or critical chloride content will reach the steel, or to delineate the present necessary level of repair work for the structure to remain damage free for a given number of years, with a given probability.

3.7.3.4 Module 4 – Report module

The report module is used to generate report documents, such as the in-service inspection report, damage prognosis, inspection regimens, repair recommendations, design basis, basis for bills of quantities, etc. It makes use of whatever text editor the user prefers and is already installed on the computer in question, and is capable of fetching and editing files from Module 1, and arraying the results from Module 3.

3.7.4 RAMS (Reliability, Availability, Maintainability and Safety)

When planning repair or maintenance, the four characterising aspects of reliability, availability, maintainability and Safety have to be considered. In the Lifecon project [1], these terms have been defined as follows:

- *reliability* – the ability to reduce maintenance to a minimum during service life;
- *availability* – the ease of supply of methods, systems, materials or qualified personnel;
- *maintainability* – the ease with which the combination of all technical and associated administrative actions during the service life can retain a repair and/or upgrade in a state in which it can perform its required functions; and
- *safety* – the health and accident risks that can be directly connected to methods, systems, products and their end results.

Each of these characteristics is important in its own right, but their relative importances to a structure vary and have to be understood before relevant planning of maintenance, repair and rehabilitation can sensibly be undertaken. Thus, these four concepts will be discussed before proceeding to the task of quantifying their qualitative aspects.

Reliability If, in an ideal case, a structure is easily accessible with high maintainability, and the required materials are cheap, plentiful and easily applied by non-skilled personnel with near zero health and accident risks, then the reliability of the materials or methods used does not have to be very high; it can, in fact, be rather low. An example of such a case would be the painting of an outside wall at ground level with a simple latex paint because the old paint was worn or unsightly.

If, on the other hand, paint for protection against corrosion has to be renewed on aerial masts on top of a communications tower, the reliability of the paint and its method of application has to be very high because the maintainability is low, while the risk of health injuries, perhaps due to chemical substances in the paint, or to accident, could be relatively high. In this case, also, the availability of the material would be of importance, but not necessarily primarily so.

Availability   A contractor would tend to consider the ease with which materials, trained personnel, equipment, methods and/or systems can be had, and could sensibly measure their availability by the time they were needed for use on site. An end user, on the other hand, would tend to be more concerned with the number of qualified contractors available for the particular kind of work to be done. These quantities have a direct bearing on reliability and can also have a bearing on maintainability. Obviously, for example, if a particular product is difficult to obtain with a long lead time, then the reliability should be high compared to that of more easily acquired products. In cases where particular materials, trained personnel, equipment, methods and/or systems no longer exist or cannot be obtained, a structure requiring their use becomes difficult to maintain and its maintainability would therefore be low. This can happen when products or equipment go out of production for one reason or another, or when necessary licences are denied, unattainable or undesirable, or when skills have been lost.

Availability of materials, trained personnels, equipment, methods and/or systems has a decided impact on maintenance and repair planning. Apart from numerous instances of materials which have fallen out of production, or materials, the use of which has become restricted for health and/or environmental reasons, restrictions or difficulties imposed by patents or licences have to be considered, as do difficulties in obtaining planning permission. On occasion, materials and trained personnel can be difficult to find and lead times can be long. Sometimes, personnel capable of applying older materials or techniques, or the materials themselves, simply cannot be had. In addition, the future of the supply of maintenance and repair materials has to be considered before proper planning can proceed.

Maintainability   Circumstances causing structures to have low maintainability, besides the obvious and common one of difficult access, could be the availability issues discussed above, such as difficulties in obtaining permission to carry out the maintenance, or the materials used in the structure being inherently difficult, or perhaps even impossible, to repair.

Maintainability can also be low because of onerous but necessary health and environmental precautions, such as those required when radiation or toxic substances are present. Work in confined spaces, in high or very low temperatures, in foul atmospheres, in high winds, on unstable footing, in poor visibility or in heavy seas can all lead to low maintainability.

The question of access is often of overriding importance in maintenance planning. If access to concrete structures entails the building of disproportionately extensive scaffolding systems, or the closure of important traffic facilities or the removal of extensive installations, then these would be examples of structures of low maintainability.

Imagine, for example, a service tunnel to a subway station where the station also serves as an atomic shelter. Such a tunnel could be crammed with technical equipment such as warning, signalling, measuring and communication equipment, control panels and associated electronic and electrical equipment, air filters, decontaminators, pipes and cables, emergency generators with necessary fuel and exhaust systems, transformers and converters, battery banks, medical supplies and equipment, emergency rations, kitchen equipment, etc. To repair the surface of such a tunnel, say to stop water leaks, could entail the entire removal of all of the technical installations with the consequent closure of the station and perhaps even of some subway routes for a lengthy period of time. In such a case, maintainability would be near zero, and the demand for reliability would be extremely high while availability would tend to become a secondary aspect, providing that whatever was required was actually obtainable. This is a far cry from our ground level paint job, where the janitor can buy a can of paint at the nearest store and slap it on with a brush in fair weather, troubling no one.

Safety   Safety in (MR&R) planning has to do with the risks to which the workforce, the general public and the environment are exposed during the performance of the relevant works. Safety is a function of the structure under consideration. Safety aspects are very different in the cases of our ground level paint job and the painting of the aerial masts atop the communications tower.

It is thus obvious that the type of structure, its use, its location, its exposure, its artistry and the installations it contains or supports all have an influence on its maintainability, reparability or ease of rehabilitation, and consequently on the planning of its MR&R.

In addition to these four characterising aspects, there are others at least as important, but more difficult to quantify, which have to be considered. Chief among these are environmental impingements and effects on human well-being, health and social functioning.

Finally, no MR&R plan would be truly complete unless some consideration is given to the ultimate demise of the structure in question. At the end of a structure's serviceable life, it does not simply disappear. We have all seen unsightly, derelict buildings which have stood neglected and vandalised, often for decades, before a decision is made to remove them or convert them into useable structures. Letting structures slowly decay and contaminate the environment and human well-being is becoming less and less of an option, and the necessity to plan for the old age of structures is coming more and more to the fore. Although this aspect falls outside the realms of maintenance, repair and perhaps even rehabilitation, it should be borne in mind that the removal or conversion of structures has a cost, both financially and environmentally. For some types of structures, the cost can be very high. Cases in point are the removal of large North Sea concrete gravity platforms and of atomic reactors. In passing, it may be mentioned that the Second World War submarine harbour at Trondheim in Norway proved so difficult to remove after the war that the attempts were abandoned. Since then, the structures, in view of the great expense of removing them, have been converted into a combined activity centre and storage facility.

3.7.4.1 Reiterative decision-making

Existing standards and norms relating to MR&R planning describe condition assessment works, rehabilitation methods and suitable repair materials, but they do not offer a complete decision-making tool, nor do they include the environmental and human aspects found in today's legislation. At present, MR&R strategies are developed by the end users using some form of in-house decision-making technique, often using poor quality information from inadequate or inaccurate surveys. A decision-making tool which integrates lifetime considerations, environmental aspects, total costs and accurate and adequate surveys, all balanced against the actual available resources, is clearly needed. This need can be met by a decision-making process which can be tailored to meet the end user's needs, such as that illustrated in Figure 3.7.2 and described below. The flow diagram shows the reiterative process of finding the best MR&R strategy within the budgeted resources. Note that the diagram includes LCE and LCC considerations. Note too that the concept of 'total cost' is introduced and encompasses the monetary value placed on health and environmental factors. At present, these last two factors are not always taken into account by the end user, and are unlikely to be so until governing legislation is in place. The former can nevertheless be estimated, and for the latter, models for calculating environmental consequences in a quantitative fashion exist [3–8].

In Figure 3.7.2, box 1 contains the end user, normally the end user of the structure in question. An end user can be an individual, a company, an authority, a cooperative, a municipality, the military or even a government. For our purposes, the end user is whoever, or whatever, is footing the bill. It may be objected that the proper end users are those who actually use the structure; for example, motorists in the case of roads, passengers in the case of public transport, and the like. However, such a view is untenable in that these people are not normally in a position to make the necessary strategy decisions, and they have therefore somehow delegated their end-user-ship to representatives whose job it is to take care of their interests. How well the representation does is another matter.

As a starting point in the planning of the MR&R of a structure, the end user would normally commission a survey to be performed in order to gain sufficient information on its condition, as indicated in box 2. Obviously, such planning is only as good as the information upon which it is based, and it is thus of paramount importance that the survey provides sufficient relevant, good quality data, which unfortunately is not always the case. The intention contained in box 6 is that data must provide information on the degradation mechanisms which are at work, on the extent to which they have wreaked their damage, and, not least, on the rate of deterioration. In addition, the survey must take a note of the destructive effect of singularities which may have arisen at some time in the structure's past, whether from constructional errors or from singular agencies such as fire, explosion, collision, earthquake, flooding and the like.

Another ingredient in the making of plans is the end user's familiarity with the structure, indicated in box 3; its history, how it has been used, its environment, its operation and its value. The end user will also have decisive views on the desired lifetime of the structure with respect to its usefulness, as indicated in box 4. Much has been written on the various aspects of the lifetime of structures, but it is nevertheless

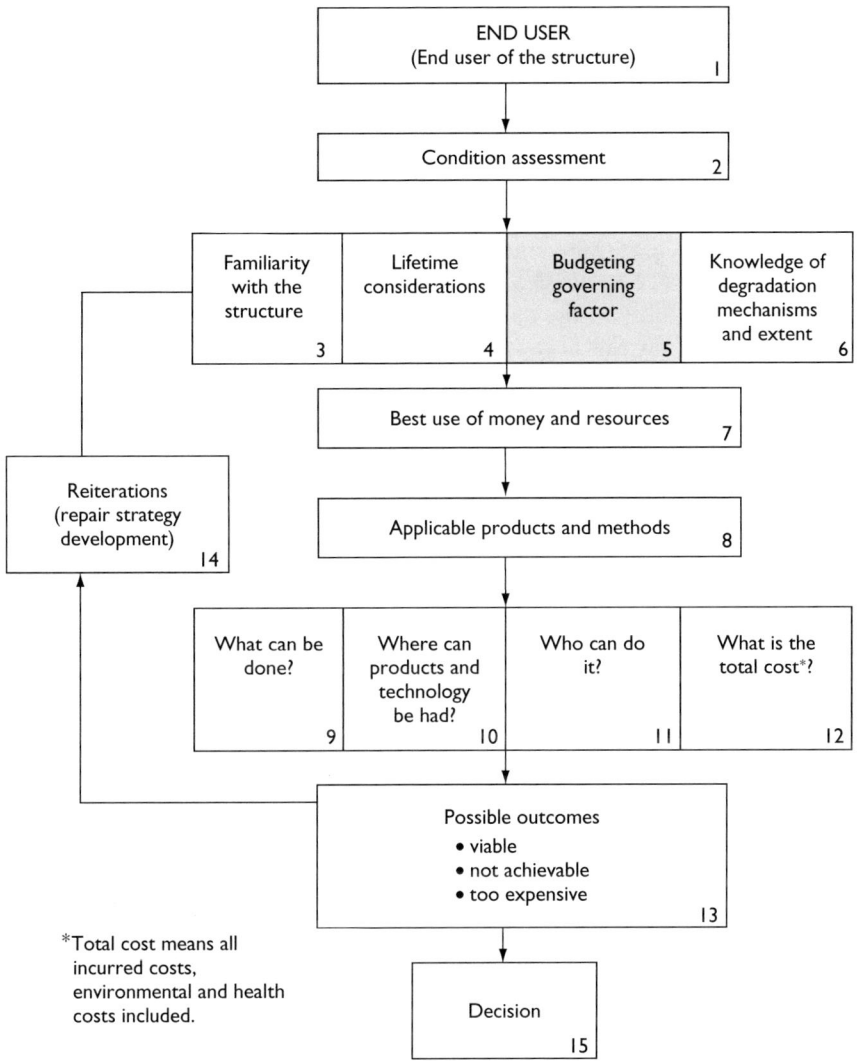

Figure 3.7.2 The re-iterative decision-making process.

the end user who has the dominant view of a structure's lifetime in terms of its cost-effectiveness. This brings us to the final, and overriding, ingredient in the making of plans, and that is money.

It is important to realise that there is always only a certain amount of money available for the upkeep and repair of structures, as emphasised by the shading of box 5. There is always a limit, which cannot be surpassed. Even, and perhaps some would say especially, governments have budgetary limits, and it is folly to try to exceed them, no matter how good the argument may be. When there is no more money, there is no more money. Financing is thus the absolute limiting factor on the extent and type of MR&R that may be undertaken. In this connection, it should be borne in mind that intrusions which require closures or limitations of the use of the structure often have a decided effect on the overall financial situation. Loss of income incurred during MR&R can be onerous to the extent that the project may not be viable.

Having collected the information represented in boxes 3–6, it is now possible for the end user to study the best use of the available money and resources, box 7, with respect to the maximum return from the investment. The prognostic results of the survey work of box 6 are here crucial to the evaluation of future damage development and residual lifetime. Surveys should thus be done in such a way that it is possible to evaluate the probability of the damage development to expect at various times.

Having decided on the best scenario, it is now time to examine the means by which the MR&R may be implemented. At this stage, possible methods, systems and materials, box 8, should be identified to single out those that are most suitable to the structure in hand, and the ambition to be achieved. This exercise can be subdivided into four more or less distinct aspects represented by boxes 9–12. The question of 'What can be done?' posed in box 9 has to be answered. In each case, there will be only a limited number of possible actions that can be undertaken from a technical point of view. These have to be identified before the next questions in boxes 10 and 11 of 'Where can products and technology be had?' and 'Who can do it?' can be answered. These questions have to do with the availability characteristic of RAMS and are important ones. Here it should be borne in mind that it is not usually sufficient to consider present day availability, but also to evaluate similar future needs. Care must also be taken, as far as possible, to avoid one way of repairing from excluding or exacerbating alternative methods or products that may be needed in the future. To cite examples, the use of coatings could easily preclude corrosion monitoring, or render more it difficult; the use of hydrophobing agents could make future electrochemical realkalisation or chloride extraction much more difficult to perform than they otherwise would have been; and the use of water glass–based sealers could impair the adhesion of coatings or the application of hydrophobation.

Last, but not least, the question in box 12 concerning the total cost should be answered as far as possible. At present, an end user will have a tendency to relate to the actual, financial cost of MR&R. However, there is now a trend towards the realisation that factors such as environmental impact and influences on health and human well-being are extremely important, and that indeed the financial costs of taking due care of these factors are part and parcel of the MR&R bill. This tendency is being driven by public opinion, by new legislation and by simple recognition by end users and others of the importance of these issues.

All of the foregoing considerations culminate in the three possible outcomes in box 13. Either one or more viable solutions will have been found, or the solutions thus far evaluated have been found to be not achievable, or to be too costly. In the latter two cases, the entire evaluation process represented by boxes 3–12 inclusive will have to be gone through again as indicated in box 14 until a viable solution or solutions have been found, and it becomes possible to reach the decision of box 15. This may require the re-examination of available resources, methods, materials and personnel, supplementary survey work, or a reduction in the end user's ambition.

The final outcome of the decision-making process will be the MR&R strategy adopted by the end user. This strategy could be any one of a number, ranging from doing nothing at all for financial or operational reasons and simply taking out the remaining lifetime as is, to complete rehabilitation, or perhaps even to closure or demolition of the entire structure. The strategy adopted will be dictated by cost-effectiveness within the confines of the total available financing.

3.7.4.2 Introduction to qualitative RAMS classification

In the decision-making process, it is useful to classify the data produced by surveys on deterioration and damage, and relate them to classifiable and quantifiable data available on materials, systems and methods with respect to the four RAMS characteristics of reliability, availability, maintainability and safety. This may be sensibly undertaken by constructing a classification system, the essentials of which are illustrated in Figure 3.7.3 and by the following description.

In the upper part of the figure, there are two parallel mainstreams. On the left, the flow diagram represents the collection and processing of survey data on deterioration and damage, while on the right, the diagram represents the collection and processing of LCE and LCC data on those products and techniques that can be used for dealing with the flaws found by the surveys.

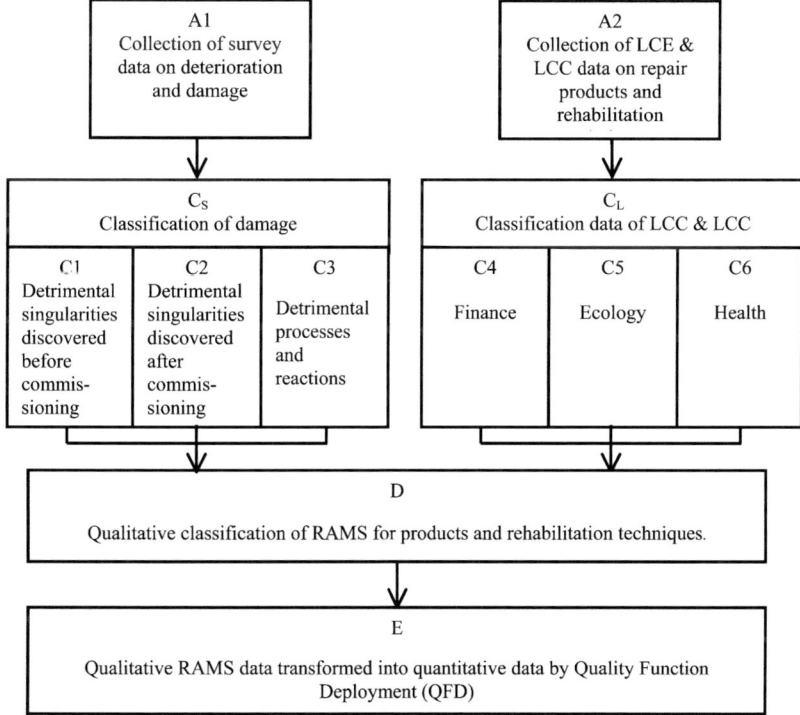

Figure 3.7.3 RAMS classification flow diagram.

Thus, box A1 represents the information on damage and deterioration collected during the surveying of the structure. This information is then classified in box C_S into the categories C1, C2 and C3. Likewise, box A2 represents collected information on the LCC and LCE of materials, techniques and systems that could be used to rectify the types of damage and deterioration classified in box C_S. The information collected is classified in box C_L into the three categories: C4 – financial cost, C5 – ecological cost and C6 – health cost.

Before commissioning a structure, a survey of some kind is normally carried out to discover flaws arising during construction or repair, since commissioning before flaws are remedied can result in severe future damage. Category C1 encompasses detrimental singularities discovered during pre-commissioning surveys. For our purposes, singularities are damage types that are caused by discrete events as opposed to continual processes such as, for example, carbonation. An example of such a singularity would be severe thermal or shrinkage cracking of concrete slabs in a parking facility, which could lead to both rapid chloride ingress and accelerated carbonation along the cracks – a condition that can become critical relatively quickly if unchecked.

Category C2 encompasses conditions caused by singularities occurring at any time after commissioning and which can cause reductions in lifetime or restrictions in functionality if not dealt with. An example of a singularity that could reduce lifetime would be fire damage resulting in accelerated carbonation or chloride ingress. An example of a singularity that could reduce functionality would be explosion or collision damage resulting in reduced bearing capacity.

Category C3 encompasses detrimental processes and reactions that tend to progress relatively slowly in a regular fashion, and which do not require immediate attention. An example is carbonation of concrete where action should be taken before reinforcement corrosion is initiated, or becomes severe. In this category, we find the processes that, in principle at least, are amenable to mathematical modelling, such as carbonation, chloride penetration, frost action and recurring differential thermal movement.

These three categories classify the types of damage found during the surveys of structures. As we shall see later, the results of the classification are used as input for deriving the qualitative RAMS characteristics, which are represented by box D.

Ideally, box A2 contains LCC and LCE data on the possible products and techniques that could be used to deal with the defects occurring in the structure in question. With some notable exceptions, data on the LCC and LCE aspects of products and techniques are both quantitatively and qualitatively sketchy at present. For example, there are few systematic studies of the lifetime of actual repairs made under, and exposed to, actual conditions. Much of the experience that exists is empirical and uncontrolled. Much work has been done by hundreds of manufacturers and researchers with regard to the contents of mortar bags, but very little scientific work has been done on the quality and performance of repairs done on site after the stuff comes out of the bags. Most manufacturers concentrate on making their products easy to apply, thus reducing the need for craftsmanship, which does not necessarily lead to better repair work. Most researchers work with materials in the laboratory under very controlled conditions that are not representative of actual site application conditions, or of actual site exposure conditions. Much of the information on the performance of repairs is fragmented and conflicting, depending, as it does, on any number of presently indefinable qualities such as handiwork, experience, personal preferences, trade secrets and the like. Added to this are variations in mixing procedures, weather, substrates, methods of application and the personal form of the workmen at the time of application. In parts of Europe and Scandinavia [9], attempts are under way to bring some of these factors under control by the introduction of training courses for the workforce which aim to certify them as repairers. No doubt these courses are called for, but the foundation of prior, scientifically controlled studies of performance and of what constitutes good practice seems to be missing.

With regard to LCE, the situation is almost the opposite, for in this area there are numerical models available [3, 4] that allow the LCE to be quantified, given sufficient input data. The status here is that the necessary data are not generally available, but when they are, the corresponding LCE can be easily estimated quantitatively.

Box D represents the qualitative classification of RAMS characteristics for those products and techniques that are technically suited to eliminating the defects in the categories C1, C2 and C3. As we have seen, systematic performance data on products and techniques are sketchy, and recourse has to be made to the empirical perception of the performance of previously executed MR&R works. Classifying on this basis clearly introduces an unfortunate, but presently unavoidable, element of subjectivity, since the classification needs must rest on personal preferences, loyalties and experience. Nevertheless, it is possible to construct a classification scheme which can serve as a basis for comparison, and which will become better as more well-founded data become available.

The classification into the broad categories C1, C2 and C3 is relatively easy, since each of the three categories tends to contain distinct types of defects that are not contained in the other two. Thus, defects occurring before commissioning, C1, all have causes attributable to some mishap or unfortunate circumstance during construction. Typical defects here would be leaking joints, displaced or missing reinforcement, thermal cracks, premature drying, freezing of fresh concrete, incorrect concrete quality or cast-in chloride or other contamination; in short, anything depreciative occurring during construction. This type of damage is rarely acute, though it is often treated as such because of the pressure to commission. It can, however, be serious and extensive, and, if not rectified, have major adverse effects on the lifetime of structures. To rectify C1 damage would typically encompass operations such as injection, strengthening, coating, replacing materials that are below specification, decontamination and electrochemical treatments.

C2 damage is typically caused by dramatic incidents such as explosion, fire, earthquake, collision, flooding and ground settlement or heave. This type of damage tends to be serious and acute. Combating techniques would include various kinds of strengthening, replacement, underpinning, coating and drainage, and the implementation of preventive measures.

C3 damage is that caused by time-dependent, comparatively regular processes such as carbonation, freeze and thaw, chloride penetration, sulphate attack, reinforcement corrosion and alkali-aggregate reaction expansion. In principle, the progression of this type of damage with time can be represented by precise mathematical functions and is quantitatively predictable. Combating techniques would typically be to install protection against, or the removal of, the destructive agency or its carrier. Commonly employed techniques here would be the replacement of contaminated or deteriorated concrete and reinforcement, the application of sealers, coatings and hydrophobing agents, electrochemical treatments such as chloride extraction, realkalisation, cathodic protection and electro-osmotic drying, and the diversion of contaminants.

Deficiencies in categories C1 and C2 ought not to in themselves call for an MR&R plan since they are caused by singularities and ought to be treated singularly in one-off operations. Unfortunately, particularly in the C1 category, they commonly go undiscovered until the effects of the damage become apparent some time after its occurrence. Thus, these types of defect often nevertheless result in the need for MR&R plans, since coincidental damage often occurs before the defects are rectified. An example would be penetration of seawater into a crack system in a new quay caused by yielding shuttering during casting. If the cracks are undiscovered and concrete is not replaced, a need for MR&R to deal with the resulting chloride-induced corrosion would at some point become apparent. Likewise, in a fire in which PVC in quantity is consumed, significant amounts of hydrochloric acid gas could be released, which could then penetrate concrete and condense directly on the surface of the embedded reinforcement steel, again calling for corrosion control and repair measures.

3.7.4.3 Qualitative RAMS classification system

Bearing in mind the subjectivity mentioned above arising from the lack of precise knowledge of the performance of repairs and materials, a RAMS classification system for repair methods can be constructed as indicated in Table 3.7.2.

In Table 3.7.2, the index column keeps track of which methods, materials or systems have been classified for future reference, while the second column contains short-form descriptions of them for easy recognition. The remaining columns contain qualitative evaluations of their RAMS characteristics. These are five in number, and range from excellent, through very good, good and medium, to poor.

In the present state of knowledge, the evaluations contained in these latter columns will be largely subjective and will vary according to the experience and preferences of the user. For example, in the author's experience, which in this specific connection dates from 1987, electrochemical realkalisation has proved to be a highly reliable method in that, to his knowledge, reinforcement in carbonated concrete has never started to corrode again when realkalisation has been properly applied to the concrete surface in any of the more than 2000 treated structures [10]. Thus, the reliability is evaluated as being 'excellent'. This is in sharp contrast to his experience with, for example, index no. 7, 'paints and coatings, category 1', with which, generally speaking, there is almost always trouble in one way or another, and the reliability of which is consequently evaluated as being 'poor'.

It will be apparent to the reader by now, that should a person of experience different from that of the author fill out such a table, the evaluations would be different depending on their own experience and preference. However, the point is that it is possible to construct a classification system which will usefully allow the evaluation of methods, systems and materials within the context of local and personal experience until such time as more scientifically founded or standardised

Table 3.7.2 Qualitative RAMS classification of repair methods

Index No.	Methods, systems and materials	Reliability	Availability	Maintainability	Safety
1	Realkalisation	Excellent	Good	Excellent	Excellent
2	Mechanical repair	Medium to poor	Excellent	Excellent	Very good
3	Cathodic protection, category 1: embedded rod anodes	Very good	Good	Poor to Good	Excellent
4	Cathodic protection, category 2: embedded mesh	Medium	Very good	Poor to Good	Excellent
5	Cathodic protection, category 3: embedded ribbon	Very good to Medium	Very good	Poor to Good	Excellent
6	Cathodic protection, category 4: conductive coating	Poor	Medium to Poor	Poor to Good	Very good
7	Paints and coatings, category 1: solvent and resin-free organic paints and coatings, CO_2 barriers	Poor	Excellent	Excellent	Excellent
8	Protective paints and coatings, category 2: solvent and resin-free organic paints and coatings	Medium	Good	Excellent	Excellent
9	Protective paints and coatings, category 3: cementitious paints and coatings containing polymer ad-mixtures and enhancers	Fair	Good	Excellent	Excellent
10	Aesthetic coatings: Cementitious paints and coatings	Very good	Good	Excellent	Excellent

data become available. It should also be apparent that such a classification system is open-ended in that entries may be added or deleted at any time, and that the number of evaluation grades may be extended as the reader sees fit. Similar considerations apply to the contents of the availability, maintainability and safety columns.

3.7.4.4 Quantifying RAMS characteristics

So far, the RAMS characteristics of MR&R planning have been dealt with in a qualitative fashion which, though it may serve the purposes of many end users, particularly those responsible for small, self-contained structures, leaves something to be desired in planning the MR&R of large or complex structures, or of numbers of structures, where quantitative measures are more important. Before qualitative RAMS aspects, such as those given in Table 3.7.2, can be transformed into quantitative figures, a revised set of the RAMS characteristics is needed. To do this, numbers have to be assigned to the four RAMS characteristics in such a way that each of them may encompass a ranking order describing the relative importance to the structure in hand of each of the elements that they contain. To be commensurate, the numbers assigned in this exercise must be devoid of specific units and be simply points awarded according to quality ranking. As before, for the present, the range of the scales and the number of points assigned within them are largely a matter of personal experience and preference. Nevertheless, the quantification is useful in that it exposes the relative merits of MR&R actions and makes it more difficult to suppress or distort quality differences. Each of the four characteristics is thus assigned numerical values as described in the following sections.

3.7.4.4.1   THE RELIABILITY CHARACTERISTIC FOR QUANTITATIVE CLASSIFICATION

Recalling the definition of reliability as 'the ability to reduce maintenance to a minimum during service life', it would be sensible to assign numbers in accordance with the elapsed time in years before a maintenance action is needed. This number of years is used to quantify 'reliability', but it needs to be stripped of its time dimension to become simply 'points'. To do this, the concept of reliability must first be evaluated for the structure in hand, and the question 'What is the minimum desired time before a maintenance action is needed?' must be answered. For the service tunnel in the subway station, this could easily be 100 years, whereas in the case of the janitor painting the outside ground level wall maybe even five years would be acceptable. Whatever this number of years is, its span is divided into five, where four is the highest rating and zero is the lowest. Thus, if 100 years were the desired span, then 0–20 years would be assigned to zero, 20–40 to one, 40–60 to two, 60–80 to three and 80–100 to four. Methods or materials that last for longer than the desired minimum time before an MR&R action is required would be assigned to the number five.

Let us assume that the minimum time desired before a maintenance action is needed is 25 years. The reliability column in Table 3.7.2 then becomes as in Table 3.7.3. Obviously, the end user may decide that 25 years is an unsuitable time, and thus expand or contract the scale to some other number of points.

3.7.4.4.2   THE AVAILABILITY CHARACTERISTIC FOR QUANTITATIVE CLASSIFICATION

The primary criterion chosen depends upon one's point of view. For example, if one is a contractor planning an MR&R action, then the concern would be with issues such as the normal lead time in weeks, from firm order until delivery, of goods, services or materials, or the number of available skilled personnel, or the amount of available

Table 3.7.3 Quantitative RAMS classification of the Reliability characteristic

Index No.	Methods, systems and materials	Reliability	Availability	Maintainability	Safety
1	Realkalisation	5	Good	Excellent	Excellent
2	Mechanical repair	1	Excellent	Excellent	Very good
3	Cathodic protection, category 1: embedded rod anodes	4	Good	Poor to Good	Excellent
4	Cathodic protection, category 2: embedded mesh	2	Very good	Poor to Good	Excellent
5	Cathodic protection, category 3: embedded ribbon	3	Very good	Poor to Good	Excellent
6	Cathodic protection, category 4: conductive coating	0	Medium to Poor	Poor to Good	Very good
7	Paints and coatings, category 1: solvent and resin-free organic paints and coatings, CO_2 barriers	0	Excellent	Excellent	Excellent
8	Protective paints and coatings, category 2: solvent and resin-free organic paints and coatings	2	Good	Excellent	Excellent
9	Protective paints and coatings, category 3: cementitious paints and coatings containing polymer admixtures and enhancers	1	Good	Excellent	Excellent
10	Aesthetic coatings: Cementitious paints and coatings	4	Good	Excellent	Excellent

equipment. An end user, on the other hand, would tend to be more pre occupied with the number of contractors qualified to do the work being planned. These two points of view can both be transformed in similar fashion into a number of points, which reflect the difficulty of availability.

In the following, we shall take the view of the end user and measure availability by the number of contractors who are qualified to do the necessary work. For the assignment of points to this availability, we shall normalise the number by referring it to a population of 1 000 000. Thus, our scale for availability, for a population of 1 000 000, becomes 0 for one to two contractors, 1 for three to four contractors, 2 for five to six contractors, 3 for seven to eight contractors, 4 for nine to ten contractors, and 5 for more than ten contractors. The availability column of Table 3.7.2 then becomes as in Table 3.7.4.

Rather than referring to a population of 1 000 000, it is of course possible to use other criteria, such as the total number of qualified national contractors, or the number of contractors available in a particular region, or the number of qualified contractors larger than a certain size, etc.

Table 3.7.4 Quantitative RAMS classification of the Availability characteristic

Index No.	Methods, systems and materials	Reliability	Availability	Maintainability	Safety
1	Realkalisation	5	3	Excellent	Excellent
2	Mechanical repair	1	5	Excellent	Very good
3	Cathodic protection, category 1: embedded ceramic rod anodes	4	3	Poor to Good	Excellent
4	Cathodic protection, category 2: embedded mesh	2	4	Poor to Good	Excellent
5	Cathodic protection, category 3: embedded ribbon	3	4	Poor to Good	Excellent
6	Cathodic protection, category 4: conductive coating	0	1	Poor to Good	Very good
7	Paints and coatings, category 1: solvent- and resin-free organic paints and coatings, CO_2 barriers	0	5	Excellent	Excellent
8	Protective paints and coatings, category 2: solvent and resin-free organic paints and coatings	2	3	Excellent	Excellent
9	Protective paints and coatings, category 3: cementitious paints and coatings containing polymer admixtures and enhancers	1	3	Excellent	Excellent
10	Aesthetic coatings: cementitious paints and coatings	4	3	Excellent	Excellent

3.7.4.4.3   THE MAINTAINABILITY CHARACTERISTIC FOR QUANTITATIVE CLASSIFICATION

The criterion chosen is that of how many of five special requirements need to be fulfilled in order to implement a given MR&R action. The five requirements are:

1. special procedures
2. highly specialised equipment
3. specially trained personnel
4. highly specialised materials
5. other special requirements.

The resulting scale ranges from 0 to 5, where one point is deducted from 5 for each special requirement that is necessary. Thus:

5 represents excellent (no special requirements)
4 represents very good (1 special requirement)
3 represents good (2 special requirements)
2 represents medium (3 special requirements)

1 represents fair (4 special requirements)
0 represents poor (5 special requirements).

The characteristic is thus defined on a scale of six ranging from 0 to 5, whereupon the maintainability column in Table 3.7.2 becomes as in Table 3.7.5, where examples of special requirements are indicated. Again, at present, the scale may be extended and the special requirements changed according to personal experience and preference.

Table 3.7.5 Quantitative RAMS classification of the Maintainability characteristic

Index No.	Methods, systems and materials	Reliability	Availability	Maintainability	Safety
1	Realkalisation	5	3	5	Excellent
2	Mechanical repair	1	5	5	Very good
3	Cathodic protection, category 1: embedded ceramic rod anodes	4	3	3 (Highly trained personnel) (Highly specialised equipment)	Excellent
4	Cathodic protection, category 2: embedded mesh	2	4	3 (Highly trained personnel) (Highly specialised equipment)	Excellent
5	Cathodic protection, category 3: embedded ribbon	3	4	3 (Highly trained personnel) (Highly specialised equipment)	Excellent
6	Cathodic protection, category 4: conductive coating	0	1	3 (Highly trained personnel) (Highly specialised equipment)	Very good
7	Paints and coatings, category 1: solvent and resin-free organic paints and coatings, CO_2 barriers	0	5	5	Excellent
8	Protective paints and coatings, category 2: solvent and resin-free organic paints and coatings	2	3	5	Excellent
9	Protective paints and coatings, category 3: cementitious paints and coatings containing polymer admixtures and enhancers	1	3	5	Excellent
10	Aesthetic coatings: cementitious paints and coatings	4	3	5	Excellent

3.7.4.4.4 THE SAFETY CHARACTERISTIC FOR QUANTITATIVE CLASSIFICATION

The criterion for the safety characteristic is defined in the same way as for the maintainability characteristic, that is, on a scale ranging from 0 to 5 according to how many of 5 requirements, this time related to safety, have to be fulfilled. As for maintainability, this gives the following scale:

5 represents excellent (no special requirements)
4 represents very good (1 special requirement)
3 represents good (2 special requirements)
2 represents medium (3 special requirements)
1 represents fair (4 special requirements)
0 represents poor (5 special requirements).

The safety column of Table 3.7.2 thus becomes as in Table 3.7.6, in which types of special requirements are indicated by way of example.

The RAMS characteristics are now represented by commensurable quantitative data, which can be manipulated by any method capable of performing total ranking exercises. One such possible method is the Quality Function Deployment (QFD) method.

Table 3.7.6 Quantitative RAMS classification of the Safety characteristic

Index no.	Methods, systems and materials	Reliability	Availability	Maintainability	Safety
I	Realkalisation	5	3	5	5
2	Mechanical repair	I	5	5	4 (Special procedures)
3	Cathodic protection, category 1: embedded ceramic rod anodes	4	3	3 (Highly trained personnel) (Highly specialised equipment)	5
4	Cathodic protection, category 2: embedded mesh	2	4	3 (Highly trained personnel) (Highly specialised equipment)	5
5	Cathodic protection, category 3: embedded ribbon	3	4	3 (Highly trained personnel) (Highly specialised equipment)	5
6	Cathodic protection, category 4: conductive coating	0	I	3 (Highly trained personnel) (Highly specialised equipment)	4 (Protective clothing)

7	Paints and coatings, category 1: solvent and resin-free organic paints and coatings, CO_2 barriers	0	5	5	5
8	Protective paints and coatings, category 2: solvent and resin-free organic paints and coatings	2	3	5	5
9	Protective paints and coatings, category 2: cementitious paints and coatings containing polymer admixtures and enhancers	1	3	5	5
10	Aesthetic coatings: cementitious paints and coatings	4	3	5	5

3.7.4.5 Quantitative and classified information on RAMS treated by the QFD method

As indicated by box E in Figure 3.7.3, QFD can be of help when transmuting requirements into specifications, which can be either performance or technical based. In this connection, QFD serves as an optimising and selective linking tool between alternative repair methods and products, and their performance properties in terms of quantified RAMS characteristics, LCC and LCE. QFD is an adjuvant decision-making tool that allows the basic requirements of human well-being, economy, culture and ecology to be taken into account when considering best choices between different repair methods, systems and materials [11]. The method is particularly useful in that it is potentially capable of handling more abstract factors such as the quality of human life, the importance of culture and tradition, environmental issues and the like. It is thus a tool that allows combination with aspects other than the purely technical and financial. In this respect, it can be made to lay importance on, for example, political and artistic aspects of MR&R, and it can be made to combine RAMS characteristics with LCC and LCE information.

3.7.5 Life cycle costs

Thus far in box D in Figure 3.7.3, the qualitative classifications of methods, systems and materials have been made on the basis of RAMS characteristics within the context of local and personal subjective experience and preference. End users are, however, also concerned with the financial cost-effectiveness of materials used in MR&R in order to gain an idea of their cost-efficiency. To address this question, an extension of RAMS classifications to encompass LCC is necessary, though again the exercise is hampered by the lack of reliable data.

Life cycle cost is the total discounted monetary cost of owning, operating, maintaining and disposing of a building, building system or infrastructure over a period of time. LCC analysis can be used to evaluate and compare different MR&R methods, systems and materials. To do this, calculations are made of the MR&R for the whole service life of a structure and the relevant costs are converted to their equivalent present value. The alternative which the analysis shows to have the lowest total present value would be the most cost-effective. LCC analysis is particularly well suited to determine whether the higher initial cost of an alternative is economically justified by reductions in future costs when compared with an alternative with lower initial but higher future costs.

When the applicable methods, systems and materials to be used for MR&R are chosen, the mathematics of LCC analysis is relatively straightforward. As is the case with most evaluation techniques in the state of present knowledge, the challenge is to make unbiased assumptions, which are crucial to the production of fair comparisons of alternative designs.

The greatest benefit is obtained from LCC analysis when it is used from the start of a new project. In the planning phase, designs may often be radically changed for the better without accompanying radical increases in costs. In LCC analysis for new-build, the initial cost part has a great effect on the result. However, here we are dealing with existing structures to which MR&R plans or LCC analysis have not been previously applied. Thus, our emphasis is on the costs of MR&R with the initial costs excluded, or playing only a minor role, because in many cases, the repayment period of the initial investment has already passed with the structure still being in use.

In the following, guidelines are presented to enable the end user to select the methods, systems and materials for MR&R on an LCC basis. However, it should be remembered that decisions should be made not only on a cost basis but also with due consideration of ecological and cultural aspects, and of human health and happiness. At present, such factors are unquantifiable in an LCC context, and needs must be evaluated by some other method such as the QFD method, as we shall see later.

3.7.5.1   Life cycle costing

In the literature, many different LCC methods are to be found [12, 13], but the basic objective is always the same, that is, to reduce the total cost of a product, system or asset. Neither the many equations for the calculations involved in different LCC analyses nor the extensive economic terminology of general LCC analysis are repeated here (for equations for the calculation of different LCC analysis and for the economic terminology of general LCC analysis, the reader is referred to references [13–15]). Here, we shall confine [14, 15] ourselves to the basics.

As mentioned earlier, LCC is the total discounted monetary cost of owning, operating, maintaining and disposing of a building, building system or infrastructure facility over a period of time. Keeping this in mind, the main terms in the LCC equation can be understood as comprising the following three variables:

1. the pertinent costs of end-user-ship;
2. the period of time over which these costs are incurred; and
3. the discount rate to be applied to future costs to equate them to present day costs.

In the LCC equation, the period of time over which end-user-ship and operational expenses are to be evaluated is called the 'study period'. This period varies, depending on the end-user's preferences. For governmental organisations, the study period may mean the whole lifetime of the facility from cradle to grave, while for a speculative investor 10 years may be a lengthy period.

The pertinent costs in the equation embrace all costs incurred during the study period, before and after the commissioning of the structure. Costs include, for example, those of inspections, condition surveys, yearly maintenance works, monitoring, cleaning, repairs, etc. In the case of new-build, the initial investment would also be taken into account.

The discount rate may be defined as 'the rate of interest reflecting the investor's time value of money' [16]. Basically, it is the interest rate that would make an investor indifferent as to whether a payment is received now, or a greater payment at some time in the future.

The advantage of LCC is apparent when a selection of alternatives can be compared. Different MR&R plans would result in interventions at different intervals; thus, costs arise at different times. To be able to compare these costs occurring at some time in the future requires the introduction of the term 'present value' (PV). This may be defined as 'the time-equivalent value of past, present or future cash flows as of the beginning of the base year' [16].

The basic LCC equation can now be written as follows:

$$C_{PV} = \sum_{i=0}^{t} \sum_{j=1}^{n_i} C_{j,i} \frac{1}{(1+r)^i}$$ (3.7.3)

where
  $C_{j,I}$ costs of the jth maintenance action in year i
  n_i number of maintenance actions in year i
  t number of years in the treated time frame (in the study period)
  C_{PV} sum of discounted (present value) costs from the study period
  r discount rate.

Discounted PV costs refer to maintenance costs discounted to the present day by the discount factor. As the discount factor diminishes with time, the PV costs of actions scheduled at a late date (from the beginning of the study period) are smaller than the PV costs of the same actions scheduled near to the start of the study time [15]. The residual value or the disposal costs of the structure should be included when comparing different alternatives (for further information on LCC analyses in general, and for standard practice procedures and examples of LCC calculations, the reader is referred arbitrarily to reference [13]).

3.7.5.2 LCC calculation procedure

ASTM E 917 presents a simple and logical procedure for calculating the LCC of structures [13]. The procedure consists of five steps:

1. Identifying objectives, alternatives and constraints
    The objective must be specified and clear boundaries set. Within those boundaries, alternatives that accomplish the objective must be defined. To be meaningful, LCC analyses for at least three different alternatives should be performed. It is very important to remember that each alternative should be capable of satisfying

the set requirements and that, if the decision is made on an LCC basis alone, then the alternative with the lowest LCC is the preferred choice.

Self-inflicted or official restraints may be introduced. Examples would be that only MR&R actions that can be performed by domestic contractors are to be considered to the exclusion of foreign contractors, or that only traditional materials are allowed as opposed to modern ones.

2. Establishing basic assumptions
   This includes choosing:

   - a consistent method of calculation (PV or annual value method)
   - a base time
   - study period (reflecting the investor's time horizon)
   - general inflation rate
   - discount rate
   - comprehensiveness of the LCC analysis.

3. Compiling cost data
   In its simplest form, this involves the timing of each cost as it is expected to occur during the study period. This is the most crucial part of making an LCC analysis, and requires considerable expertise and knowledge to be done adequately. Costs that do not have significant differences between alternatives may be omitted from the analysis where its purpose is solely comparative. The primary objective of LCC analysis is to rank alternatives on an LCC basis. The attainment of exact numbers is, at best, a secondary objective.

4. Computing the LCC for each alternative
   When the cost compilation for each alternative is ready, the LCCs of the alternatives can be made comparable by using Equation (3.7.3) and its derivatives.

5. Comparing LCCs of each alternative to determine the one with the minimum LCC
   If the decision is made on an LCC basis only, then the alternative with the lowest LCC will be the preferred choice for financial reasons.

If the decisions on MR&R strategy were based only on costs, the described five steps would be sufficient. However, if final decisions are to be made after also taking into consideration human, cultural and ecological effects and factors, the ranking order of the alternatives may be changed depending on the importance and weighting of those factors. Nevertheless, the LCC analyses of MR&R plans will have a great impact on the decision-making process.

3.7.5.3 Preliminary choice of MR&R methods for LCC analysis

ENV 1504-9:1997 [17] gives principles and methods related to the repair of defects in reinforced concrete that incorporate an understanding of the underlying chemical, electrochemical and physical processes that govern the various deterioration mechanisms. There are eleven principles, six of which are concerned primarily with defects caused by mechanical, chemical or physical actions on the concrete itself, and five of which are specifically concerned with corrosion of its reinforcement. The ENV 1504-9:1997 approach to the repair of defects in reinforced concrete is summarised in

Table 3.7.7. The tabulation of the principles and methods is not necessarily exhaustive nor definitive, but serves as an overview. It should be borne in mind that the selection of a maintenance or repair method will also be influenced by the condition of the structure when an action is carried out.

Preliminary selection of applicable MR&R methods can be made when the degradation mechanisms are known and the condition of the structure has been assessed. This information is obtained from the condition assessment of the structure [2, 18]. At least three different alternative repair methods should be studied [16].

Table 3.7.7 Principle and methods related to the repair of defects in reinforced concrete

Principle number	Definition	Methods based on the principle
1	Protection against ingress of adverse agents	Surface impregnation, surface coating, locally bandaged cracks, crack filling, crack to joint remakes, external panelling and membranes
2	Moisture control	Hydrophobation, surface coating, sheltering and overcladding, and electrochemical treatment
3	Concrete restoration	Manual mortaring, recasting with concrete, sprayed concrete or mortar, element replacement and encasement
4	Structural strengthening	Adding or replacing reinforcement, installing bonded rebars in holes, plate bonding, adding mortar or concrete, injecting cracks or voids and pre-stressing.
5	Physical resistance (to physical and mechanical attack)	Overlays or coatings and impregnation
6	Resistance to chemicals	Overlays or coatings and impregnation
7	Preserving or restoring passivity	Cover increase with mortar or concrete, replacing contaminated or carbonated concrete, electrochemical realkalisation and electrochemical chloride extraction
8	Increasing electrical resistivity	Moisture content decrease using surface treatment, coatings or sheltering
9	Cathodic control	Limiting oxygen availability by saturation or coatings.
10	Cathodic protection	Applying electrochemical potential using impressed current or sacrificial anodes
11	Control of anodic areas	Pigmented rebar coatings, rebar barrier coatings and applying corrosion inhibitors

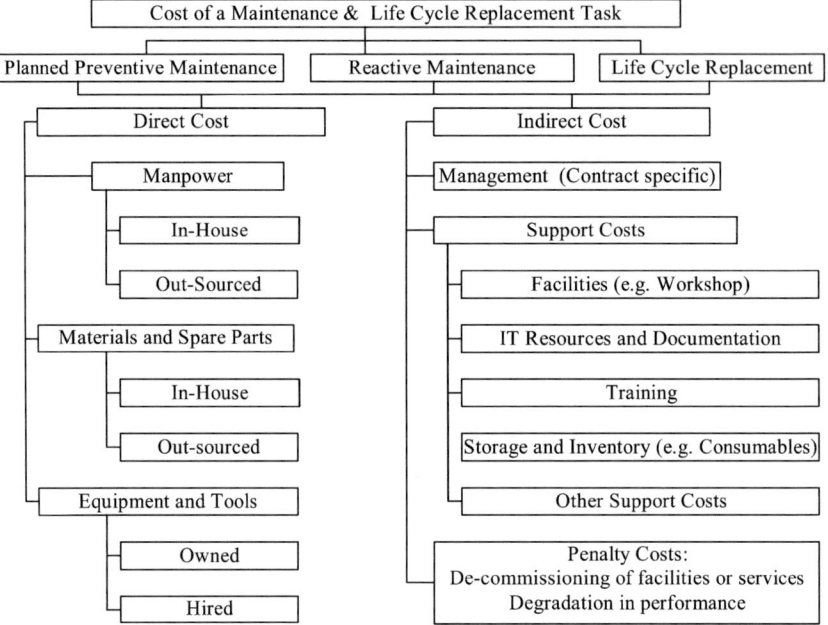

Figure 3.7.4 Cost breakdown structure.

3.7.5.4   Cost breakdown structure (CBS)

The real challenge of successful LCC analysis lies in making unbiased assumptions to produce fair comparisons of alternative designs or maintenance policies. As with any evaluation process, it is always easier to assess or evaluate smaller constellations. For this reason, it is best to build a CBS for various MR&R elements using consistent subtitles and units in order to easily compare the costs of the different elements [12, 14, 16, 19–22]. The factors that build up the total LCC sum of all MR&R actions should be categorised into a few subcategories (manpower, materials and spare parts, etc.) that are common to all MR&R methods. An example of a possible CBS [22] is presented in Figure 3.7.4.

Problems often arise when trying to find the information to be placed into the subcategories and the corresponding cost estimates. The variations can be great, depending, for example, on the location of the facility (centre of town, backcountry, etc.), the quality of the execution of MR&R works, the season of the year, the climate, etc. This is why no extensive, unequivocal catalogues or databases of different MR&R costs at European, or even national, level exist. However, with the help of consistent CBSs, the problem can be alleviated if not overcome.

First, the maintenance action or repair is divided into clear phases, which are:

- condition survey and analysis;
- design and planning of maintenance or repair actions;

- decommissioning, wholly or partially, if necessary;
- execution of maintenance or repair actions; and
- recommissioning.

Normally, the execution phase is the most costly one, though the cost of decommissioning can also be very high depending on the nature of the structure. The costs incurred in the execution phase accumulate from different elements, which are, for example:

- preparation;
- temporary support;
- repair materials (patching mortars and concrete, rebar coatings, bonding agents, coatings, etc.);
- repair systems and components (e.g. cathodic protection system components);
- application of the repair and/or system components;
- quality control;
- decommissioning costs (loss of production, interruption to operations, loss of income, etc.); and
- recommissioning costs.

For the execution phase, further divisions should be made. One such logical division is presented on the left side of Figure 3.7.4. All MR&R actions require manpower, materials and equipment. In the manpower category, the cost-bearing factors are, for example:

- management
- man hours
- personnel transportation
- subsistence
- special expertise.

In the materials category, total costs arise due to:

- materials
- transportation
- availability of materials
- handling
- storage.

Similar subdivisions can be made for the category of equipment and tools. Finally, the cost of quality assurance, control and documentation must not be forgotten.

The CBS can be made in matrix form, for example, using spreadsheets. Tables 3.7.8–3.7.13 are examples of LCC spreadsheets for compiling the cost estimates for different MR&R methods. These tables show the costs of the execution phase since those are predominant, the cost distinctions between different MR&R

Table 3.7.8 CBS summary matrix for the LCC analysis of MR&R of damage due to chloride penetration

Damage mechanism 1 Corrosion due to chloride penetration	Replacing contaminated concrete/ Applying mortar by hand	Replacing contaminated concrete/ Recasting with concrete	Replacing contaminated concrete/ Spraying concrete or mortar	Protection against ingress/ Surface inorganic coatings	Protection against ingress/ Surface organic coatings	Protection against ingress/ Filling cracks	Protection against ingress/ Surface impregnation	Structural strengthening/ Adding mortar or concrete, injecting	Cathodic protection	Electrochemical desalination	Other
Preparation (subtotal, €)											
Manpower (own work)											
Manpower (contract work)											
Material/spare parts (own)											
Material/spare parts (purchased)											
Equipment/tools (transport)											
Equipment/tools (machinery on site)											
Quality assurance/control											
Consequential costs											
Execution of the work (subtotal, €)											
Manpower (own work)											
Manpower (contract work)											
Material/spare parts (own)											
Material/spare parts (purchased)											
Equipment/tools (transport)											
Equipment/tools (machinery on site)											
Quality assurance/control											
Consequential costs											
Finishing (subtotal, €)											
Manpower (own work)											
Manpower (contract work)											
Material/spare parts (own)											
Material/spare parts (purchased)											
Equipment/tools (transport)											
Equipment/tools (machinery on site)											
Quality assurance/control											
Consequential costs											
Total Costs (€)											
Time to next intervention (years)											

Table 3.7.9 CBS summary matrix for the LCC analysis of MR&R of damage due to carbonation

Damage mechanism 2 Corrosion caused by carbonation	Replacing carbonated concrete/ Applying mortar by hand	Replacing carbonated concrete/ Recasting with concrete	Replacing carbonated concrete/ Spraying concrete or mortar	Increasing resistivity/ Surface inorganic coatings/CO_2-barriers	Increasing resistivity/ Surface organic coatings/CO_2-barriers	Preserving passivity by increasing cover with cementitious material	Electrochemical realkalisation	Cathodic protection	Other
Preparation, (subtotal, €)									
Manpower (own work)									
Manpower (contract work)									
Material/spare parts (own)									
Material/spare parts (purchased)									
Equipment/tools (transport)									
Equipment/tools (machinery on site)									
Quality assurance/control									
Consequential costs									
Execution of the work (subtotal, €)									
Manpower (own work)									
Manpower (contract work)									
Material/spare parts (own)									
Material/spare parts (purchased)									
Equipment/tools (transport)									
Equipment/tools (machinery on site)									
Quality assurance/control									
Consequential costs									
Finishing (subtotal, €)									
Manpower (own work)									
Manpower (contract work)									
Material/spare parts (own)									
Material/spare parts (purchased)									
Equipment/tools (transport)									
Equipment/tools (machinery on site)									
Quality assurance/control									
Consequential costs									
Total Costs (€)									
Time to next intervention (years)									

Table 3.7.10 CBS summary matrix for the LCC analysis of MR&R of damage due to frost action

Damage mechanism 3 Frost action	Concrete restoration/ Applying mortar by hand	Concrete restoration/ Recasting with concrete	Concrete restoration/ Spraying concrete or mortar	Protection against water ingress/ Surface inorganic coatings	Protection against water ingress/ Surface organic coatings	Protection against ingress/ Filling cracks	Protection against ingress/ Impregnation	Structural strengthening/ Adding mortar or concrete, injecting	Increasing resistance to physical attack/ Overlays or coatings	Other
Preparation (subtotal, €)										
Manpower (own work)										
Manpower (contract work)										
Material/spare parts (own)										
Material/spare parts (purchased)										
Equipment/tools (transport)										
Equipment/tools (machinery on site)										
Quality assurance/control										
Consequential costs										
Execution of the work (subtotal, €)										
Manpower (own work)										
Manpower (contract work)										
Material/spare parts (own)										
Material/spare parts (purchased)										
Equipment/tools (transport)										
Equipment/tools (machinery on site)										
Quality assurance/control										
Consequential costs										
Finishing (subtotal, €)										
Manpower (own work)										
Manpower (contract work)										
Material/spare parts (own)										
Material/spare parts (purchased)										
Equipment/tools (transport)										
Equipment/tools (machinery on site)										
Quality assurance/control										
Consequential costs										
Total Costs (€)										
Time to next intervention (years)										

Table 3.7.11 CBS summary matrix for the LCC analysis of MR&R of damage due to sulphate attack

Damage mechanism 4 Chemical attack: Sulphate	Concrete restoration/ Applying mortar by hand	Concrete restoration/ Recasting by concrete	Concrete restoration/ Spraying concrete or mortar	Protection against ingress/ Surface coating	Protection against ingress/ Filling cracks	Structural strengthening/ Adding mortar or concrete, injecting	Increasing surface resistance to chemical attack/Coating	Other
Preparation (subtotal, €)								
Manpower (own work)								
Manpower (contract work)								
Material/spare parts (own)								
Material/spare parts (purchased)								
Equipment/tools (transport)								
Equipment/tools (machinery on site)								
Quality assurance/control								
Consequential costs								
Execution of the work (subtotal, €)								
Manpower (own work)								
Manpower (contract work)								
Material/spare parts (own)								
Material/spare parts (purchased)								
Equipment/tools (transport)								
Equipment/tools (machinery on site)								
Quality assurance/control								
Consequential costs								
Finishing (subtotal, €)								
Manpower (own work)								
Manpower (contract work)								
Material/spare parts (own)								
Material/spare parts (purchased)								
Equipment/tools (transport)								
Equipment/tools (machinery on site)								
Quality assurance/control								
Consequential costs								
Total Costs (€)								
Time to next intervention (years)								

Table 3.7.12 CBS summary matrix for the LCC analysis of MR&R of damage due to elution: surface deterioration, mechanical abrasion, de calcification, etc.

Damage mechanism 5 Deterioration due to elution: Surface deterioration, Mechanical abrasion, De-calcification, etc.	Concrete restoration/ Applying mortar by hand	Concrete restoration/ Recasting with concrete	Concrete restoration/ Spraying concrete or mortar	Protection against ingress/ Surface coating	Protection against ingress/ Filling cracks	Protection against ingress/ impregnation	Structural strengthening/ Adding mortar or concrete, injecting	Increasing resistance to physical attack/ Overlays or coatings
Preparation (subtotal, €)								
Manpower (own work)								
Manpower (contract work)								
Material/spare parts (own)								
Material/spare parts (purchased)								
Equipment/tools (transport)								
Equipment/tools (machinery on site)								
Quality assurance/control								
Consequential costs								
Execution of the work (subtotal, €)								
Manpower (own work)								
Manpower (contract work)								
Material/spare parts (own)								
Material/spare parts (purchased)								
Equipment/tools (transport)								
Equipment/tools (machinery on site)								
Quality assurance/control								
Consequential costs								
Finishing (subtotal, €)								
Manpower (own work)								
Manpower (contract work)								
Material/spare parts (own)								
Material/spare parts (purchased)								
Equipment/tools (transport)								
Equipment/tools (machinery on site)								
Quality assurance/control								
Consequential costs								
Total Costs (€)								
Time to next intervention (years)								

Table 3.7.13 CBS summary matrix for the LCC analysis of MR&R of damage due to alkali-aggregate reaction

Damage mechanism 6 AAR (Alkali-aggregate reaction)	Concrete restoration/ Applying mortar by hand	Concrete restoration/ Recasting	Concrete restoration/ Spraying concrete or mortar	Structural strengthening/ Adding mortar or concrete	Structural strengthening/ Injecting or filling cracks	Other
Preparation (subtotal, €)						
Manpower (own work)						
Manpower (contract work)						
Material/spare parts (own)						
Material/spare parts (purchased)						
Equipment/tools (transport)						
Equipment/tools (machinery on site)						
Quality assurance/control						
Consequential costs						
Execution of the work (subtotal, €)						
Manpower (own work)						
Manpower (contract work)						
Material/spare parts (own)						
Material/spare parts (purchased)						
Equipment/tools (transport)						
Equipment/tools (machinery on site)						
Quality assurance/control						
Consequential costs						
Finishing (subtotal, €)						
Manpower (own work)						
Manpower (contract work)						
Material/spare parts (own)						
Material/spare parts (purchased)						
Equipment/tools (transport)						
Equipment/tools (machinery on site)						
Quality assurance/control						
Consequential costs						
Total Costs (€)						
Time to next intervention (years)						

methods resulting chiefly from this phase. In the tables, MR&R methods for the following six degradation mechanisms are addressed:

1. chloride-induced corrosion
2. carbonation
3. frost action
4. sulphate attack
5. elution
6. alkali-aggregate reaction.

The condition assessment [2, 18] of the structure gives the background information for the selection of repair methods. The CBS matrices presented are for illustration purposes and are open to modification; the number of lines and columns as well as their headings can be changed, according to the end user's needs and preferences.

Note that the tables are the means of comparison for one intervention at a certain time at a certain location. The number of headings has been limited for easier comparison of methods. The numbers required to populate the tables are generated by other sheets, not presented here, which encompass basic input variables and calculation routines. The basic input variables could be, for example, estimates of:

•   time needed for certain repair phases;
•   transportation distances or unit costs; and
•   manpower needed for certain repair phases.

The calculation routines would convert the estimates to monetary values by:

•   multiplying transport unit costs and the distances;
•   multiplying manpower unit costs and the time used; and
•   multiplying repair material unit costs and the amount of material used, etc.

The MR&R execution cost matrices are built up for one intervention at a certain time and at a certain location, and are divided into preparation, execution and finishing. The tables are illustrative of the LCC analysis output, the commensurate unit being the euro. The calculations themselves are performed elsewhere and are not referred to here. The number of cost elements can be increased in accordance with the desired accuracy of the analysis. Also, methods can be removed or new ones added, according to the end user's desires.

The number, and the possible combinations, of interventions during the study period, and estimates of discount rates, etc. are taken into account in other subsequent phases of the analysis. The whole LCC analysis procedure can easily be executed on the same spreadsheet; the difficulty lies in obtaining reliable information for the estimates, and subsequently keeping the table updated. Logically, the other MR&R phases (condition survey, planning of actions, commissioning, etc.) can also be broken down into matrix form with corresponding cost estimates.

When all of the elements in the CBSs have been assigned costs, and time and discount rate estimates, the calculation of the LCCs is performed using Equation (3.7.3), and consequently a comparison between the MR&R methods can be made. Here, it must be stressed that, when making LCC comparisons, the study period must be same for each alternative. The given LCC matrices are for one facility and one intervention at the present time. Cost estimates are usually based on current costs; that is, neglecting the effect of economic inflation on prices. The discount rate is correspondingly taken net of inflation.

3.7.5.5 Difficulties in obtaining information

When the CBS is prepared, it is easier to give cost estimates for the elements. However, as previously indicated, local micro and macro conditions (climate, general cost level, local workmanship, etc.) can have large impacts on unit costs, thus complicating the creation of universal cost catalogues.

The major difficulty is caused by the time factor. It is not easy to predict the future, but that is basically what has to be done when making LCC calculations in view of the lengths of the study periods, which in some cases can be more than 100 years. Uncertainties in prophesying discount rates, the general development of the prices of raw material, manpower and expertise, the rate of development of new materials or MR&R methods, the possible prohibitions of some methods and materials, etc. all result to compound the uncertainty in the LCC calculations.

The prediction of future degradation is complex and is also subject to considerable uncertainty. The producers of repair materials, for example, may assert that their materials have a 100-year durability, but the success of a repair is not a function of the durability of repair material as such. Local climate, the condition of substrates, local loads, the accuracy of condition surveys, quality of workmanship, etc. all have impacts on the success of the repair work. The quality of the executed repair work thus has a direct impact on the time that will pass before the next intervention is needed, which logically has an impact on the overall LCC. In the Lifecon deliverable D5.1 [23], some examples of times elapsed between interventions are given, but they are by no means absolute nor universal. The elapsed time between consecutive interventions differs markedly for the various maintenance or repair methods.

Predicting the length of the maintenance-free period is also problematic. For many MR&R methods and materials, such as different types of coatings, different CP anode systems, external facade insulation, etc. lengthy maintenance-free periods are often asserted, even though some of these methods and materials have not been in existence for comparable periods. Often, relevant track records do not exist, or they are sparse, or they are not properly documented. Valid real-life statistics are thus, to a great extent, unavailable. The durability of materials and methods asserted by the manufacturers of repair materials or the purveyors of methods and systems may sometimes be correct under ideal, laboratory-controlled conditions, but in realty conditions are never ideal. In passing, in fairness, it should be noted that for some materials and methods, durability sometimes proves to far surpass that asserted by their commercial protagonists.

3.7.5.6 Sources of information and presentation of cost estimates

There are numerous sources of information for making cost or time estimates, some of which are listed below:

- statistics and databases;
- experience from executed MR&R actions and works;
- laboratory tests and simulations (time to next MR&R action or repair);
- materials producers (cost of material, time to next MR&R action or repair . . .);
- contractors (cost of manpower, cost of equipment, time to next MR&R action or repair . . .);
- consultants (cost of condition survey and repair planning, time to next MR&R action or repair . . .);
- societal sources (interest rates, inflation rates, discount rates . . .); and
- authorities (changes in loads, traffic volumes, environmental loads, time to next MR&R action or repair . . .).

In making cost estimates and summing them by means of Equation (3.7.3), two different methods are available, one being deterministic and the other probabilistic. The difference is in the way of presenting estimates. In the deterministic method, the estimates are given with simple numbers, while in the probabilistic approach statistical distributions are used. When dealing with uncertainties, probabilities and simulation offer better results [22]. However, practical engineers are not yet sufficiently familiar with probabilistic methods, so they might find the traditional deterministic way of handling costs easier to adopt.

A combination of the two methods is also possible. For example, some costs may be known very exactly, such as the current price of repair materials, while other estimates may encompass large variations and could be better expressed by distributions. An example here would be the expected maintenance-free period between interventions.

3.7.5.7 Sensitivity analysis

Whatever method is applied, and indeed whenever uncertainty is involved in modelling, it is advisable to apply sensitivity analysis. Originally, sensitivity analysis was created to deal with uncertainties in input factors, but lately, the idea has been extended to incorporate model conceptual uncertainty; that is, uncertainty in model structures, assumptions and specifications [24]. There are many methods for performing sensitivity analysis, and they are not treated here, but the underlying principles in all of them are to determine:

- whether a model resembles the system or processes under study;
- those factors which contribute most to output variability and which require additional research to strengthen the knowledge base;
- the model parameters, or those parts of the model itself, which are insignificant, and which may be eliminated from the final model;
- whether there is some region in the space of input factors for which the model variation is a maximum;

- the optimal regions within the space of the factors for use in a subsequent calibration study; and
- whether factors or group of factors interact with each other, and if so, which ones.

The sensitivity analysis of the LCC analysis should reveal those cost factors, small variations in which changes the ranking order of the MR&R alternatives. On determining these factors, effort can be directed more precisely towards solving or alleviating the uncertainties (for help in finding the best method for sensitivity analysis for different LCC cases, the reader is arbitrarily referred to one of the great many publications on the subject [24]).

3.7.5.8 Case study

To clarify the LCC calculation, we shall apply it to a highway bridge in a simplified manner but nevertheless preserving the key issues. This particular bridge has a carbonation problem over an area of $1000\,m^2$, but with no cracking of concrete or visible problems with reinforcement corrosion as yet. The concrete cover over the reinforcement is stipulated as being about 25 mm; the carbonation depth at the time of observation is around 20 mm and progressing. The study period is 50 years forward from the present. The costs of different repair actions have been estimated and are shown in Table 3.7.14. Other MR&R actions, such as renewal of the bridge deck membrane, or of metallic components, are not taken into account in this example, nor are repairs necessitated by other agents such as chloride attack, frost, etc.

As Tables 3.7.8–3.7.13 show, there are a number of alternative techniques for the treatment of carbonation in concrete. Some of them are not particularly intrusive, nor expensive, ways of maintaining the structure. Some are more intrusive, but add significantly to the service life of the structure. The variations in the initial costs are noticeable, and the cost-efficiency of a particular repair action can be seen over longer periods of time. Different techniques may be compared by using LCC analysis. Their

Table 3.7.14 Cost comparison of alternative repair methods

Damage: Corrosion due to carbonation, area of $1000\,m^2$	1. Surface coating with organic CO_2 barrier	2. Shotcrete + coating	3. Cathodic protection with CarboCath	4. Electrochemical realkalisation
Unit cost ($€\,m^{-2}$)*	60	100	180	110
Total initial costs ($€$ per $1000\,m^2$)	60 000	100 000	180 000	110 000
Time to next intervention (years, estimated)	10–15	30–35	20–25	20–25

* Estimated, including material, work and equipment.
Note
Surface coating is applied in year zero to slow carbonation [25]. The other techniques allow carbonation to develop further until the carbonation front is closer to the reinforcement. Electrochemical realkalisation is performed in year five, and the more intrusive techniques such as shotcreting and cathodic protection are applied in year ten (see table 3.7.15).

reliability is dependent on the precision of the initial information, both factual and estimated. It is the function of the survey and condition assessment [2, 15, 18] to provide this information to the end user.

The methods we shall examine are as follows:

Repair method 1 – Surface coatings with semi-organic/organic CO₂ barriers

Here, we shall assume that the coatings have lifetimes of about 10–15 years and therefore need to be applied several times over the study period of 50 years. We realise that CO_2 barriers do not completely stop, but slow, the carbonation rate, so that more intrusive work within 25–30 years, such as mechanical repair, shotcreting, cathodic protection, etc. or a combination of them, is to be expected. This is, in a rough manner, accounted for in the calculation. For CO_2 barriers to be of benefit, they would have to be applied immediately.

Repair method 2 – Shotcreting

Shotcrete (that is, sprayed concrete or mortar) consists of cement, aggregate and water, sometimes with admixtures or fibres. It is applied through a nozzle by means of a pump or an airstream depending on whether the wet or dry method is used. The wet spray technique is used particularly in cases of structural strengthening or large scale concrete repair work. Dry spray is more suited to small scale repair work, which is the more usual in MR&R. Because of the high velocity of the emergent jet, a dense concrete layer is formed which is well bonded to the substrate and which effectively repassivates exposed steel. A 20 mm layer of shotcrete will protect the concrete from carbonation for many decades; in this calculation for 30 years. Some aesthetic repairs may be expected in 50 years.

Repair method 3 – CarboCath cathodic protection

The system consists of carbon fibre anode net with carbon and copper current distributors. The net is covered, and the surface is finished, with an inorganic coating. The system is asserted to protect reinforcement for many decades, the anode being said to have a lifetime of more than 50 years, and to need little maintenance. The surface of the concrete will need refurbishment (see Section 3.7.1) at certain intervals (15–20 years). The installation of the cathodic protection system could be deferred for some time, as there is no benefit from installation until the carbonation has reached the steel. Note that this is a case where there is little track experience, and thus the durability of the system is in reality unknown.

Repair method 4 – Electrochemical realkalisation

Protection against carbonation-induced corrosion is durably regained by re-establishing the lost alkalinity of the concrete simultaneously by the penetration of alkaline solution under the combined influences of electro-osmosis, ionic migration and diffusion, by cathodic reaction, and by strong re-passivation of the steel in the highly alkaline environment so created. The effect can be expected to last for well over 50 years by extrapolation of the present track record (20 years). In theory, some aspects of the treatment are everlasting. The protection can be further improved by inorganic coatings. It has been assumed that also, in this case, the structure will need some surface refurbishment (see Section 3.7.1) in 50 years' time, and this has been taken into account. Re alkalisation could be deferred for some time (albeit slightly less than for cathodic protection), as there is little cost-efficiency benefit from the treatment if it is applied before carbonation has reached the steel.

Usually, MR&R actions disturb the normal use of the structure in that traffic often has to be diverted or restricted, or special protective measures have to be installed. The estimated monetary value of these extra measures can be taken into account as de commissioning costs in the LCC calculation.

The results of the LCC calculation presented in Table 3.7.15 show that, in this case, the shotcrete method is the most cost-effective, with electrochemical realkalisation coming in second. However, it should be borne in mind that other circumstances can upset the LCC choice. For example, the realkalisation method would most likely be less intrusive, and this may weigh determiningly heavily in its favour, although the calculation quite clearly shows that on an LCC basis, the shotcrete method should be preferred.

Table 3.7.15 LCC of alternative repair methods with minimum time to the next intervention: discount rate of 5%, study period of 50 years, cost is €1000

Year	Discount factor	Repair method 1		Repair method 2		Repair method 3		Repair method 4	
		Real cost	Discounted cost	Real cost	Discounted cost	Real cost	Discounted cost	Real cost	Discounted cost
Year 0	1.000	60	60.00	0	0.00	0	0.00	0	0.00
Year 1	0.952	0	60.00	0	0.00	0	0.00	0	0.00
Year 2	0.907	0	60.00	0	0.00	0	0.00	0	0.00
Year 3	0.864	0	60.00	0	0.00	0	0.00	0	0.00
Year 4	0.823	0	60.00	0	0.00	0	0.00	0	0.00
Year 5	0.784	0	60.00	0	0.00	0	0.00	110**	86.19
Year 6	0.746	0	60.00	0	0.00	0	0.00	0	86.19
Year 7	0.711	0	60.00	0	0.00	0	0.00	0	86.19
Year 8	0.677	0	60.00	0	0.00	0	0.00	0	86.19
Year 9	0.645	0	60.00	0	0.00	0	0.00	0	86.19
Year 10	0.614	50*	90.70	100** 100**	61.39	180**	110.50	0	86.19
Year 11	0.585	0	90.70	0	61.39	0	110.50	0	86.19
Year 12	0.557	0	90.70	0	61.39	0	110.50	0	86.19
Year 13	0.530	0	90.70	0	61.39	0	110.50	0	86.19
Year 14	0.505	0	90.70	0	61.39	0	110.50	0	86.19
Year 15	0.481	0	90.70	0	61.39	0	110.50	0	86.19
Year 16	0.458	0	90.70	0	61.39	0	110.50	0	86.19
Year 17	0.436	0	90.70	0	61.39	0	110.50	0	86.19
Year 18	0.416	0	90.70	0	61.39	0	110.50	0	86.19
Year 19	0.396	0	90.70	0	61.39	0	110.50	0	86.19
Year 20	0.377	50*	109.54	0	61.39	0	110.50	50*	105.03
Year 21	0.359	0	109.54	0	61.39	0	110.50	0	105.03
Year 22	0.342	0	109.54	0	61.39	0	110.50	0	105.03
Year 23	0.326	0	109.54	0	61.39	0	110.50	0	105.03
Year 24	0.310	0	109.54	0	61.39	0	110.50	0	105.03
Year 25	0.295	0	109.54	50*	76.16	50*	125.27	0	105.03
Year 26	0.281	0	109.54	0	76.16	0	125.27	0	105.03
Year 27	0.268	0	109.54	0	76.16	0	125.27	0	105.03
Year 28	0.255	0	109.54	0	76.16	0	125.27	0	105.03
Year 29	0.243	0	109.54	0	76.16	0	125.27	0	105.03
Year 30	0.231	120**	137.31	0	76.16	0	125.27	0	105.03

Table 3.7.15 (Continued)

Year	Discount factor	Repair method 1		Repair method 2		Repair method 3		Repair method 4	
		Real cost	Discounted cost	Real cost	Discounted cost	Real cost	Discounted cost	Real cost	Discounted cost
Year 31	0.220	0	137.31	0	76.16	0	125.27	0	105.03
Year 32	0.210	0	137.31	0	76.16	0	125.27	0	105.03
Year 33	0.200	0	137.31	0	76.16	0	125.27	0	105.03
Year 34	0.190	0	137.31	0	76.16	0	125.27	0	105.03
Year 35	0.181	0	137.31	0	76.16	0	125.27	50*	114.10
Year 36	0.173	0	137.31	0	76.16	0	125.27	0	114.10
Year 37	0.164	0	137.31	0	76.16	0	125.27	0	114.10
Year 38	0.157	0	137.31	0	76.16	0	125.27	0	114.10
Year 39	0.149	0	137.31	0	76.16	0	125.27	0	114.10
Year 40	0.142	0	137.31	100**	90.36	50*	132.37	0	114.10
Year 41	0.135	0	137.31	0	90.36	0	132.37	0	114.10
Year 42	0.129	0	137.31	0	90.36	0	132.37	0	114.10
Year 43	0.123	0	137.31	0	90.36	0	132.37	0	114.10
Year 44	0.117	0	137.31	0	90.36	0	132.37	0	114.10
Year 45	0.111	50*	142.87	0	90.36	0	132.37	0	114.10
Year 46	0.106	0	142.87	0	90.36	0	132.37	0	114.10
Year 47	0.101	0	142.87	0	90.36	0	132.37	0	114.10
Year 48	0.096	0	142.87	0	90.36	0	132.37	0	114.10
Year 49	0.092	0	142.87	0	90.36	0	132.37	0	114.10
Year 50	0.087	0	142.87	0	90.36	0	132.37	0	114.10
Remaining value of repair (50 years)		−25	−2.18	−33	−2.88	−17	−1.48	0	0
Discounted cost in € 1000			140.7		87.5		130.9		114.1

Notes
* light MR&R action (e.g. aesthetic repair, coating).
** heavy MR&R action.

Table 3.7.16 The effect of the discount rate on the discounted cost level (cost in € 1000)

Discount rate (%)	Repair method 1	Repair method 2	Repair method 3	Repair method 4
2	212	146	195	158
5	144	88	131	114
8	107	58	93	89

In Table 3.7.16, the effect of the discount rate on the discounted cost levels of the four repair methods is shown. In Table 3.7.17, the four methods are ranked in the order of increasing costs. The table shows that the ranking order is not disturbed by the discount rate, and thus it can be concluded that the choice of the repair method is not sensitive to this rate.

It should be appreciated that the above examples are purely illustrative and that the figures may be ranked differently depending on the labour and material cost, or on the location or the country in which the object is situated. The evaluation of the time to the next intervention is one of the most difficult tasks, and the method of calculation is very sensitive to the costs which accumulate during the first decades.

Table 3.7.17 Sensitivity analysis – Ranking order

Discount rate (%)	Repair method 1	Repair method 2	Repair method 3	Repair method 4
2	4	1	3	2
5	4	1	3	2
8	4	1	3	2

This may change the LCC ranking order of different repair techniques with respect to the most economical. When optimising the time of certain MR&R actions, a structure's condition and the rate of degradation should be taken into account.

3.7.6 Life cycle ecology

No plan for MR&R would be complete if it were to ignore environmental aspects. Awareness of the importance of these is constantly increasing, as is corresponding regulatory legislation. Considering the increasing numbers of concrete structures that are in need of MR&R, and considering the amount of resources involved in their execution, that legislation is justified. The expectation is that the MR&R of concrete structures will be subjected to ever stricter regulations with regard to environmental impacts, and that it will consequently become increasingly important to take environmental issues into consideration when planning and executing MR&R of concrete structures.

The MR&R which is not properly planned and controlled is likely to result in inferior durability with concomitant inferior cost-efficiency, both in financial terms and in environmental terms. A tool is therefore needed which is capable of quantifying the environmental aspects of MR&R with regard to the environmental loading imposed by the involved materials, methods and resources. Here, we shall attempt to delineate the framework of a method for quantifying the environmental burden imposed by materials and systems for the MR&R of concrete structures which will encompass materials, energy consumption, waste generation, emissions and toxicity.

ISO 14040:1997 [5] describes LCA as a technique for assessing the environmental aspects and potential impacts associated with a product, by:

- compiling an inventory of relevant inputs and outputs for a product system;
- evaluating the potential environmental impacts associated with these inputs and outputs; and
- interpreting the results of the inventory analysis and impacts of assessment phases in relation to the objective of the study.

3.7.6.1 Impact categories and category end points

In LCA calculations, the concepts of impact categories and category end points arise. Impact categories are:

- global warming
- ozone depletion

- acidification
- photo-oxidant creation
- eutrophication
- human and eco-toxicity.

Category end points may be simply understood as being the final depositories of pollutants, such as forests, lakes, animals, humans, etc. In Figure 3.7.5, an attempt is made to illustrate how the concept of category end points derives from the life cycle inventory (LCI) with respect to the impact category of acidification. A similar derivatory flow can be constructed for the other impact categories.

3.7.6.2 LCA and LCE

Life Cycle Assessment (LCA) should ideally include the assessment of environmental impacts caused by all human activities throughout the whole life cycle of a structure. This is, however, a very difficult process since the relationship between the external environment and the category end points can be very complex. Normally, the LCA will stop at the step before the category end point, showing only the impact categories, which is a fairly easy exercise, and then interpret the results from the various category indicators. For this reason, the author has chosen to describe the following method as LCE, rather than the more all-encompassing but presently more impractical LCA.

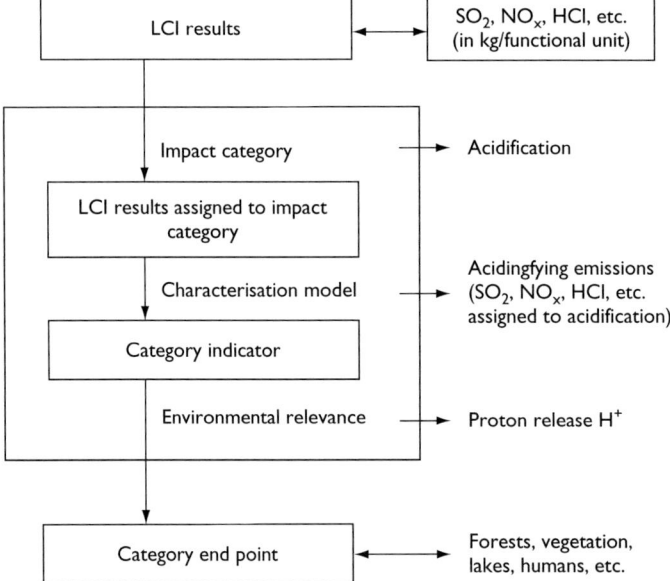

Figure 3.7.5 Concept of category end points [4].

3.7.6.3 LCE calculations

The methodological framework for the assessment of environmental impacts of the MR&R of concrete structures is shown in Figure 3.7.6, which is based on ISO standards 14040 to 14043 [5–8].

To calculate the LCE, the natural place to start is with a properly executed condition survey, which allows the MR&R methods and materials to be selected. [2, 17, 26–28]. In Figure 3.7.6, the selection boxes have been populated with four main possible MR&R methods, with several alternatives under each of them. The analysis is thus to be performed for each of the alternatives, which will result in a comparison from the environmental point of view.

Thereafter, it is necessary to choose the functional unit. This is used as the reference unit in a life cycle study [5]. All energy and materials consumption, and all emissions, arising during the MR&R action are related to this unit. The functional unit must be measurable and will depend on the purpose and scope of the analysis. The purpose of the LCE assessment should be clearly defined since the choice of the functional unit depends on it. For example, the functional unit for electrochemical treatment may be defined as the unit surface (m^2) protected for a specified time period, whereas the functional unit for a scaffolding system could be the unit surface accessed for a specified period of time. In another area, for example that of ventilation systems, the unit could be the unit volume of air processed in a specified time period.

Having decided upon the functional unit, it is the opportune time to set up the MR&R LCI. To do this, it is necessary to carry out the following:

- Quantify the amount of all raw materials [8], chemicals [29] and equipment [30] that are necessary to fulfil the MR&R action. This quantification gives the reference flow, to which all inputs and outputs are referred. It is closely connected to the functional unit.
- Provide environmental data [6] on consumed raw materials [8], chemicals [29] and equipment [30]. Specific data may be obtained from suppliers, and generic data may be found in databases. Sometimes data may be had from a LCI carried out by a supplier or manufacturer. All materials used should ideally be environmentally declared for their entire lifetime from cradle to grave. The environmental declaration should include the use of resources such as energy (renewable, non-renewable), materials (renewable, non-renewable), and water, and should quantify waste generation (recoverable, non-recoverable) and emissions to air and water.
- Quantify and classify waste generated by the MR&R action according to whether it is recyclable, recoverable, disposable or hazardous.

Conversions to impact categories are now performed as indicated in Figure 3.7.5 by assigning the LCI results to the relevant categories. All of the estimated effects should be potential effects. Conversion and characterisation should be carried out according to ISO 14042, using effect factors from IPCC in the Montreal protocol [29] as shown in Tables 3.7.18a and 3.7.18b. Emission of a specific gas may be assigned to more than one category. An example is the emission of NO_x, which is assigned to the categories of both eutrophication and acidification. The final result may be displayed as impact categories or weighted as an environmental index.

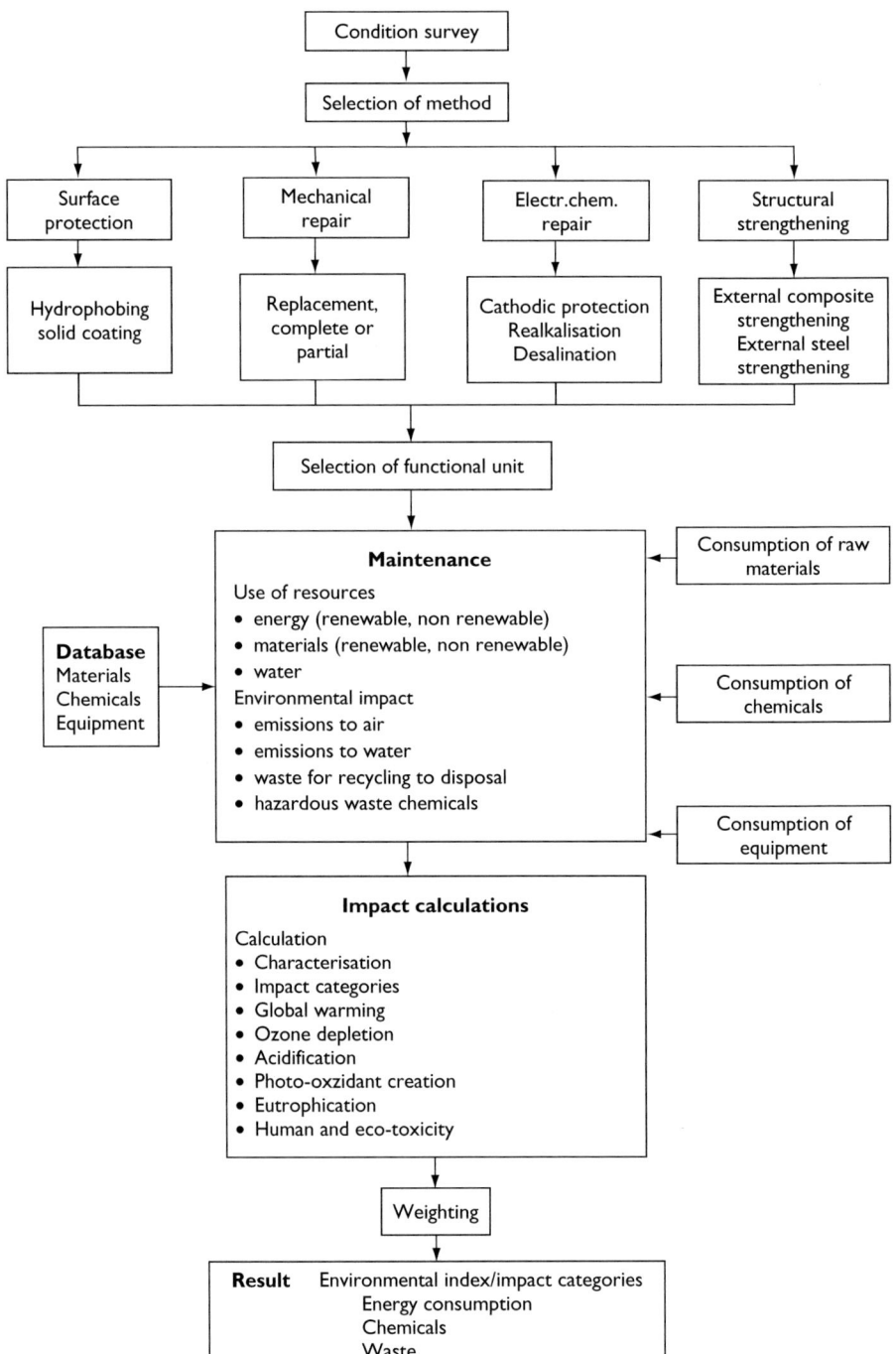

Figure 3.7.6 Flow diagram for the assessment of environmental impacts of MR&R.

Table 3.7.18a IPCC 1995 (Montreal protocol) effect factors for emissions to air

Global warming(100 years) $g\,CO_2$ eq. g^{-1}

C_2F_6	9 200	HCFC-124	430
C_3F_8	7 000	HCFC-141b	370
C_4F_{10}	7 000	HCFC142b	1 700
C_5F_{12}	7 500	HCFC-22	1 400
C_6F_{14}	7 400	HFC-125	2 800
$C-C_4F_8$	8 700	HFC-134	1 000
CF_4	6 500	HFC-134a	1 300
CFC-11	2 100	HFC-143	300
CFC-113	3 600	HFC-143a	3 800
CFC-114	7 000	HFC152a	140
CFC-115	7 000	HFC-227ea	2 900
CFC-12	7 100	HFC-23	11 700
CFC-13	13 000	HFC-236fa	6 300
CH_2Cl_2	9	HFC-245ca	560
CH_4	21	HFC-32	650
$CHCl_3$	4	HFC-41	150
CO_2	1	HFC-43-10mee	1 300
CO	2	N_2O	310
Halon 1211	4 900	SF_6	23 900
HCFC-123	50		

Table 3.7.18b IPCC 1995 (Montreal protocol) effect factors for emissions to air

Acidification $g\,SO_2$ eq. g^{-1}

HCl	0.88
HF	1.6
NH_3	1.88
NO	1.07
NO_2	0.7
NO_x (as NO_2)	0.7
HNO_3	0.51
H_3PO_4	0.98
H_2S	1.88
SO_2	1

Ozone depletion $g\,CFC$-11 eq. g^{-1}

CCl_4	1.1
CFC-11	1
CFC-113	0.8
CFC-114	1
CFC-115	0.6
CFC-12	1
CFC-13	1
CH_3CCl_3	0.1
$CHCl_3$	0.12
HALON-1201	1.4
HALON-1202	1.25
HALON-1211	3
HALON-1301	10
HCFC-123	0.006
HCFC-124	0.04
HCFC-141b	0.11
HCFC142b	0.065
HCFC-22	0.055
Other CFC	1

Eutrophication $g\,PO_4$ eq. g^{-1}

N_2O	0.13
NH_3	0.35
NO	0.2
NO_2	0.13
NO_x (as NO_2)	0.13

Discharge to water $g\,PO_4$ eq. g^{-1}

Ammoniacal N	0.33
BOD	0.11
COD	0.022
Nitrate	0.1
Orthophosphate	1
Total Nitrogen	0.42
Phosphorous	3.06

Table 3.7.18c IPCC 1995 (Montreal Protocol) effect factors for emissions to air

Formation of photo-oxidants g ethane eq g^{-1}

1,2,4-Trimethylbenzene	1.2	Heptane	0.529
2,2-Timethylpropane	0.398	Hexane	0.421
2-methylhexane	0.492	HFC-125	0.021
2-methylpentane	0.524	HFC-134	0.021
3-methylhexane	0.492	HFC-134a	0.021
3-methylpentane	0.431	HFC-143	0.021
4-methylpentan2one	0.178	HFC-143a	0.021
Acetone	0.326	HFC152a	0.021
Acetylene	0.168	HFC-227ea	0.021
Aromatic Hydrocarbons	0.761	HFC-23	0.021
Benzene	0.189	HFC-236fa	0.021
But2ene	0.992	HFC-245ca	0.021
Butan2one	0.326	HFC-32	0.021
Butane	0.41	HFC-41	0.021
Butanols	0.196	HFC-43-10mee	0.021
Butyl acetate	0.323	Isobutane	0.315
CCl$_4$	0.021	Isopentane	0.296
CFC-11	0.021	Methyl heptanes	0.469
CFC-113	0.021	m-ethyl toluene	0.794
CFC-114	0.021	m-xylene	0.993
CFC-115	0.021	NMVOC	0.416
CFC-12	0.021	Octane	0.493
CFC-13	0.021	Other CFC	0.021
CFC-50$_2$	0.021	Other HCFC	0.021
CH$_2$Cl$_2$	00.02	Other paraffins	0.761
CH$_3$CCl$_3$	1.021	Other unknown VOC	0.337
CH$_4$	0.007	Other VOC	0.337
CHCl$_3$	0.001	o-xylene	0.666
Ethane	0.082	PCB's	0.021
Ethanol	0.268	Pent2ene	0.93
Ethene	1	Pentane	0.408
Ethyl acetate	0.218	Pentane isomers	0.296
Ethyl benzene	0.593	p-ethyltoluene	0.725
Formaldehyde	0.421	Propan1ol	0.196
Glycols	0.196	Propan2ol	0.196
HALON-1201	0.021	Propane	0.42
HALON-1202	0.021	Propylene	1.03
HALON-1211	0.021	p-xylene	0.888
HALON-1301	0.021	Aliphatic hydrocarbons	0.398
HCFC-123	0.021	Tetrachloroethene	0.005
HCFC-124	0.021	Toluene	0.563
HCFC-141b	0.021	Trichloroethene	0.066
HCFC142b	0.021	White spirit	0.761
HCFC-22	0.021	Xylenes	0.888

Weighting is optional, and is the process of converting indicator results for different impact categories by using numerical factors based on value choices. It is an optional element in ISO 14042 [7]. Thus, factors from value choices may be based on political targets, for example the Kyoto protocol, or on other preferences. Interpretation of the results should be based on ISO 14043 [8] and should identify, qualify, evaluate and present the findings of significant issues.

As aforesaid, the assessment of environment impacts resulting from human activities throughout the life cycle of a structure can be both complex and difficult. However, a methodological framework for the assessment of LCE or LCA is slowly being established through a number of international standards and guidelines. LCE analysis would greatly benefit if, in addition, an international database were to be established, to which material suppliers, manufacturers, consultants, end users, governments, in short all concerned could contribute with relevant environmental information on materials, systems and methods. This would perhaps eventually necessitate the establishment of a controlling organisation concerned with the quality of the data supplied, but the creation of the data base itself is the primary concern. Controlling organsations, if deemed necessary, can always be created later.

For the present, although a number of assumptions have to be made, the application of LCE to actual cases allows useful conclusions to be drawn, as we shall see in the next section.

3.7.6.4 Case studies

In order to demonstrate how the methodological framework for the assessment of environmental impacts can be applied to various types of repair and maintenance systems for concrete structures, two examples of commonly used MR&R methods have been selected for analysis. One example is that of a patch repair using shotcrete where the damage has been caused by chloride-induced corrosion of embedded steel. The other example is of preventive maintenance using a hydrophobic surface treatment for protection of the concrete against ingress of moisture and moisture-borne chloride.

A major difficulty when dealing with commercial products is the obtaining of specific data. A case in point is that of the hydrophobic surface treatment for which the supplier divulged only comparative information and was unwilling to give specific data on energy inputs, raw materials inputs and other physical data. However, it is possible to get around this sort of unfortunate difficulty by using whatever data is supplied, in combination with other generally available information, to perform a reverse calculation, thus enabling the various impact categories to be determined as demonstrated in the following.

For both case studies, the following common assumptions for the calculation of environmental impacts were made:

- transport distance forth and back (60 km);
- materials and equipment transported by truck;
- fuel consumption (diesel – 0.2 kg per ton-km); and
- functional unit (1 m^2 of repaired or protected concrete surface with a durability of 10 years).

Quantification of the waste generated and of assessment of human and eco-toxicity from the processes were omitted due to lack of relevant data. The environmental profile of 1 kg of diesel [4] is shown in Table 3.7.19.

Table 3.7.19 Environmental profile of diesel

Use of energy (MJ kg^{-1})	Global warming (g CO$_2$ eq. kg^{-1})	Acidification (g SO$_2$ eq. kg^{-1})	Eutrophication (g PO$_4$ eq. kg^{-1})	Photo-oxidant formation (g Ethene eq. kg^{-1})
43.2	3150	2.8	50.0	10.0

3.7.6.4.1 PATCH REPAIR

The LCE analysis of the patch repair was based on the following assumptions:

- surface area repaired ($30\,m^2$);
- rebound of shotcrete (25%); and
- power on the construction site supplied by diesel engines.

It was further assumed that the concrete cover was removed to an average depth of 50 mm. The assumed average thickness of the shotcrete layer was 50 mm. The various stages of the process included:

- removal of concrete cover to an average depth of 50 mm by high pressure (1000 bar) hydro-jetting;
- cleaning of the reinforcing bars by sandblasting;
- protective coating of the reinforcement;
- application of the shotcrete layer; and
- curing measures applied to the shotcrete surface.

The composition of the shotcrete used is shown in Table 3.7.20, and the environmental profile of one cubic metre of this concrete is shown in Table 3.7.21.

Table 3.7.20 Shotcrete composition

Water/cement ratio	0.43
Cement	500 kg m^{-3}
Admixtures	1.0 kg m^{-3}
Aggregate	1500 kg m^{-3}

Table 3.7.21 Energy use and environmental impact per m^3 of shotcrete

Impact category				
Use of energy MJ m^{-3}	Global warming kg CO$_2$ eq. m^{-3}	Acidification kg SO$_2$ eq. m^{-3}	Eutrophication kg PO$_4$ eq. m^{-3}	Photo-oxidant formation Ethene kg eq. m^{-3}
2795	405	1.59	0.234	0.191

Table 3.7.22 Energy consumption and environmental impacts of patch repair using shotcrete

Stage	Impact category				
	Energy use (MJ m^{-2})	Global warming (kg CO$_2$ eq. m^{-2})	Acidification (g SO$_2$ eq. m^{-2})	Eutrophication (g PO$_4$ eq. m^{-2})	Photo-oxidant formation (g Ethene eq. m^{-2})
Hydro-jetting	677	84	75	1330	266
Cleaning of reinforcement	296	22	4	350	70
Protective coating on reinforcement	35	1.4	19	2.4	3
Application of shotcrete	59	4.4	19	70	14
Transportation	127	10	8	150	30
Sum	1194	122	125	1902	383

The total use of energy and the environmental impact of the patch repair technique are shown in Table 3.7.22.

3.7.6.4.2 HYDROPHOBIC SURFACE PROTECTION

Silane-based materials are often used for hydrophobic surface treatment, the protective mechanism of which is said to be the capillary suction of the material into concrete pores close to the surface, whereupon a very thin film of a silicon resin is formed by chemical reaction with substances in the cement. As a result, the concrete surface becomes hydrophobic, the treatment only slightly reducing the water vapour permeability. For the efficiency and durability of the hydrophobic treatment, the penetration depth and content of active substances in the concrete are decisive parameters. For a good treatment, a minimum penetration depth of 5 mm or more is desirable.

Concentrated or diluted silanes are commercially available in three forms – for example liquid, cream and gel – and are often applied to the concrete surface by spraying. For our purposes, we shall examine the use of a gel system. It should be pointed out that the example serves only for the illustration of LCE analysis and is not to be taken as necessarily advocating the use of hydrophobation as such. Nor shall we plumb the depths of silicon chemistry, remarking merely that silanes are silico-organic compounds where silicon is the central atom bound to four organic groups.

In our case, the gel was composed of iso-octyl triethoxysilane mixed with a mineral thickener, and the surface area treated was 150 m^2.

The various stages involved in our hydrophobic treatment of concrete include the following:

- In the process of surface preparation, all surfaces treated were cleaned with hot water (60–90°C) at a pressure of 160 bar.

- The hydrophobic agent was applied to a thickness of not less than 0.25 mm by means of a high-pressure spray gun. The amount of material reaching the actual surface was estimated to be about 45% of the total consumption; that is $500 \, g \, m^{-2}$. The amount released to the environment was thus about $600 \, g \, m^{-2}$ and contained iso-octyl triethoxysilane and ethanol, which are both volatile.
- Ethanol was emitted during the chemical reaction inside the concrete which forms silanols and ethanol. It was assumed that the emission of ethanol from 1000 g of iso-octyl triethoxysilane is 500 g.
- Long-term degradation processes include emissions of methane (CH_4) and carbon dioxide (CO_2). It is assumed that 500 g of iso-octyl triethoxysilane lead to the emission of $101 \, g \, CH_4$ and $50 \, g \, CO_2$.

The service life of a silane-based surface treatment may range from 7 to 20 years. It was assumed that the surface treatment would have a service life of 10 years, which is in accordance with Swedish regulations for highway bridges.

As can be seen from Table 3.7.23, it turns out that the major part of the environmental impact of the surface treatment arises from the production of the hydrophobic agent.

3.7.6.4.3 COMPARISON OF THE TWO CASES

On comparing the two cases, it can be seen from Table 3.7.24 that the ecological impact of the patch repair technique greatly exceeds that of the hydrophobic surface protection. The results demonstrate that the hydrophobic surface treatment can be repeated more than five times before the environmental impact of photo-oxidant formation approaches that of patch repair by shotcreting.

Although the assessment of impacts on the environment caused by human activities throughout the life cycle of a structure can be complex and, in the present state of our development, requires the making of a number of assumptions, the results of

Table 3.7.23 Energy consumption and environmental impacts of hydrophobic surface protection

Stage	Impact category				
	Use of energy $(MJ \, m^{-2})$	Global warming $(g \, CO_2 \, eq. \, m^{-2})$	Acidification $(g \, SO_2 \, eq. \, m^{-2})$	Eutrophication $(g \, PO_4 \, eq. \, m^{-2})$	Photo-oxidant formation $(g \, Ethene \, eq. \, m^{-2})$
Production of hydrophobic agent	47	295	0.5	6	2
Surface preparation	17	13	0.4	7	1
Transportation and surface treatment	12	80	0.1	2	66
Long-term degradation		2171			1
Sum	76	2559	1	15	70

Table 3.7.24 Comparison between patch repair and hydrophobic surface treatment

MR&R action	Impact category				
	Use of energy (MJ m^{-2})	Global warming (kg CO$_2$ eq. m^{-2})	Acidification (g SO$_2$ eq. m^{-2})	Eutrophication (g SO$_2$ eq. m^{-2})	Photo oxidant formation (g Ethene eq. m^{-2})
Hydrophobic treatment	76	2.6	1	15	70
Patch repair	1194	122	125	1902	383

our exercise clearly demonstrate that, from an ecological point of view, it appears to be a very good strategy to carry out preventive maintenance on a concrete structure before a stage is reached where patch repairs or other MR&R actions may be necessary. This is of course assuming that the hydrophobing agent works efficiently as intended.

3.7.7 Quality Function Deployment (QFD) method

We have previously examined how the RAMS characteristics of MR&R may be categorised qualitatively and quantitatively on the basis of technical quality and financial cost-effectiveness. However, the objective of integrated lifetime repair planning methodology is to optimise and guarantee the life cycle, taking due care and consideration of human well-being, economy, cultural compatibility and ecology, and moderate it with technical performance parameters. It is thus desirable to control and optimise human well being, the monetary economy and the ecology, while taking social aspects into account.

To combine these softer, perhaps more abstract though obviously important, aspects with the hard, technical world of RAMS characteristics, a further tool is needed. One such possible tool is represented by QFD, which is a method that was originally developed in Japan's shipbuilding industry in the 1970s in order to ensure the optimisation of the total quality of their products from the combined point of view of rational production and customer needs and desires. The method allows the integration of customer desires into the development of products, and is primarily used during the advance planning activity. It can also be used throughout product development from first concepts to final production, but it is most used as a way of defining the development programme itself.

Within the field of MR&R, the method has the potential to combine and reduce any requirements, which can be related to either performance or technical properties, to a common basis for comparison. In this connection, QFD could serve as an adjuvant optimising and selective tool to link alternative repair methods and products and their performance properties (RAMS) to how well they satisfy more abstract, softer requirements.

3.7.7.1   RAMS and Generic Functional Requirements

We have already seen how the quantification of the four RAMS aspects of methods or materials can be performed. The quantification results in the assignment of numbers according to how well a material or method performs with respect to its reliability, availability, maintainability and safety from the point of view of repairing a particular structure or group of structures. The problem to be addressed in the following is that of combining, in a systematic fashion, these aspects with those which are presented by the users of the structure or facility to be repaired, and by society at large. In the following, we shall refer to the former RAMS performance aspects as 'technical RAMS', and to the latter as 'Generic Functional Requirements'.

The Generic Functional Requirements, which figured in the Lifecon project [1], were:

- functionality and usability
- health
- safety
- comfort
- image
- investment economy
- building costs
- LCC
- building tradition
- life style
- business culture
- aesthetics
- architecture styles and trends
- raw materials resource economy
- energy resources economy
- environmental burdens economy
- loss of biodiversity
- waste economy.

Here it is perhaps appropriate to point out that the list of Generic Functional Requirements could, in principle, be of anything deemed appropriate. For example, a political party could perhaps set up a list containing items such as taxation, public relations, popularity, image, influence, etc. and use QFD to gain the maximum political benefit from a MR&R action, whereas an environmentalist could include only items directly related to ecology and use of resources and their ecological impingements in an attempt to minimise environmental impact.

For our immediate purposes, we shall use the following modified Lifecon list of Generic Functional Requirements:

- functionality (& usability)
- health
- safety
- comfort
- investment (viability)

- building costs
- LCC (Life cycle costs)
- life style
- building tradition
- aesthetics
- architecture
- image
- energy use
- pollution
- (Loss of) Biodiversity
- recycling
- LCE (Life Cycle Ecology).

A point, which must be made very clear before proceeding, is that when setting up a QFD to compare materials or methods, it is extremely important that *only those materials or methods which do the same or equivalent jobs are studied*. If this principle is not observed, then a QFD reduction will lead to some very strange conclusions. However, often MR&R actions comprise various tasks for diverse purposes where it makes no sense to compare them with each other. In these cases, QFD can still usefully be applied to ensure that none of the various methods or materials under consideration performs so poorly that it would significantly detract from the overall performance and so nullify the benefits of those that do perform well.

3.7.7.2 The Quality Table

In order to systematise relationships between Generic Functional Requirements and RAMS properties, some form of tabulation of the various correlations and their inter-dependencies is necessary. We shall use the form of tabulation presented as the Quality Table, Table 3.7.25.

In the upper part of the Quality Table, the Generic Functional Requirements are listed on the left-hand side. In the middle of this part of the table, there is a column entitled 'Relative importance of dependencies'. Here, each of the functional requirements is rated on a scale of 1–5 according to how important it is deemed to be for the result of the particular MR&R action being planned, 1 being of least importance and 5 being of the greatest. For example, if functionality is of prime importance, then the rating for this may be 5. If Life Style (or its preservation) is not particularly important, then it may be judged to merit no more than a 1. If one of the listed requirements cannot be seen to have any importance at all, or is found to be irrelevant, then it may be omitted (as we did with the Lifecon 'Business culture' in the lists above). The point of this exercise is to force the end user to think about what is important, and to differentiate these importances to some extent. It is a prerequisite that this column is populated before any of the other remaining columns in order to avoid false influences.

Table 3.7.25 Reasoning behind the assignment of dependencies

Generic Functional Requirement	D	Reason for dependency evaluation
		Reliability
Functionality		
Health		
Safety		
Comfort		
Investment		
Building Costs		
LCC		
Life style		
Building tradition		
Aesthetics		
Architechure		
Image		
Energy use		
Pollution		
Biodiversity		
Recycling		
LCE		

Generally speaking, in the opinion of the author, and at the risk of incurring the wrath of others of different propensities, the following considerations should apply.

• Human conditions and economy should be assigned stronger dependencies on the properties of RAMS than the requirements of culture and ecology.
• When looking at each requirement individually, safety and LCC should have the strongest dependencies on RAMS properties, and the requirements of image and lifestyle the weakest.
• Requirements of human conditions should have stronger dependencies than those of culture.
• When assigning individual weights, health and safety should have the highest weights, and life style and image the lowest.

Having thus evaluated the relative importances of the Generic Functional Requirements, the extent to which they interact with the four general RAMS characteristics is evaluated. Here, a scale of 1, 3, 6 and 9 is used, where 1 represents no dependency whatsoever, 3 represents a weak dependency, 6 indicates a moderately strong dependency and 9 represents an obviously strong dependency. This is perhaps the most difficult part of QFD reduction in connection with MR&R, since at first sight it may be difficult to see dependencies at all, far less rate them. For example, it may not be immediately obvious how health, safety and comfort depend on the RAMS safety, but a few moments of reflection should lead to the realisation that the safer the repair method is for the MR&R workforce, the safer it is likely to be for the general public. If a repair method does not require the workforce to use special equipment, such as protective clothing, or dust masks, or scaffolding, then it is unlikely that the public will be subjected to dangerous or nuisance substances, or falling objects. Even lifestyle could depend to some extent on the repair method, since an 'upmarket' lifestyle could be impaired by the use of certain repair methods. An example here could be the installation of scaffolding resulting in disfiguring bolt marks on an otherwise impeccable surface. A method which eliminates the need for scaffolding would be 'RAMS safer' and would therefore have less influence on lifestyle. It is thus important to evaluate the dependencies *for the particular structure or structures for which MR&R plans are to be made*, but not yet related to specific repair methods or materials.

In evaluating these dependencies, it is wise to use auxiliary tables, one for each of the four RAMS characteristics, where the reasons for the dependency evaluations are noted. This makes it easier to revert to the reasoning behind the evaluations at a later date, and thus to consider whether the first evaluations were sensible and whether changes are needed. Such tabulation is simple and can follow the scheme presented in Table 3.7.26, which is for dependencies on reliability. Similar tabulations are used for the other three RAMS dependencies.

The dependencies are then transferred to the 'D' columns of Table 3.7.26.

If we now multiply a RAMS dependency (D) by the relative importance of its corresponding Generic Functional Requirement (W), we create a weighted RAMS dependency; that is, a dependency which has been adjusted for the relative importance of the corresponding Generic Functional Requirement. In this way, the right-hand part of the table bearing the legend 'Weighted RAMS dependencies' is populated.

Having thus populated the upper part of the Quality Table, it remains to sum the weighted RAMS dependencies and enter the results into the bottom-most row of the upper part of the Quality Table. On the left of the lower part of the Quality Table, methods or materials which can be used for a particular MR&R purpose are listed. The immediately adjacent columns entitled 'Technical RAMS' are populated with the figures attained during the RAMS quantification carried out according to the principles set forth in Section 3.7.4.4. Thereafter, the right hand columns are filled in by multiplying each of the values in the left-hand columns successively by the weighted sums of the RAMS dependencies ($RAMS_D$). The resulting values are the Global Priority Orders and are placed in right-most bottom column of the Quality Table. These values are a measure of to what extent the various repair methods or materials fulfil the Generic Functional Requirements when adjusted for the relative importance of the latter and for their interaction with the general RAMS characteristics. The higher the value, the better the fulfilment.

Table 3.7.26 The quality table for combining generic functional requirements with RAMS

Combination of technical RAMS characteristics with Generic Functional Requirements			Dependence of RAMS characteristics on Generic Functional Requirements (D)				Relative importance of dependencies (W)	Weighted RAMS dependencies (W x D)			
			Reliability	Availibility	Maintainability	Safety		Reliability	Availibility	Maintainability	Safety
Generic Functional Requirements	Human Well-being	Functionality									
		Health									
		Safety									
		Comfort									
	Economy	Investment									
		Building Costs									
		LCC									
	Culture	Life style									
		Building tradition									
		Aesthetics									
		Architechure									
		Image									
	Ecology	Energy use									
		Pollution									
		Biodiversity									
		Recycling									
		LCE									
Weighted sums of RAMS dependencies ($RAMS_D$)											

		Technical $RAMS_T$					Weighted RAMS $RAMS_D$ x $RAMS_T$					
		R	A	M	S		R	A	M	S		
Materials or methods under evaluation						→						Global Priority Order GPO = $\Sigma[RAMS_D$ x $RAMS_T]$

Acknowledgement

The author is indebted to the contributors to the Lifecon project in general [1], and in particular to Professor Asko Sarja of VTT and Minna Sarkinnen of CT Laastit OY whose works have inspired the sections on QFD, to Minna Kesäläinen of Optiroc Oy Ab for a thorough base material for the section on LCC, and to Professor Odd E. Gjørv of the Norwegian University of Science and Technology, Sverre Fossdal of the Norwegian Building Research Institute, and Assistant Professor Vemund Årskog of Ålesund College whose Lifecon work has formed the basis of the section on LCE.

Acknowledgement must also be especially made to the author's son, Iain H. B. Miller of Millab Consult a.s. whose support, criticisms and suggestions greatly forwarded the writing of the entire section.

References

[1] The Lifecon project (Life cycle management of concrete infrastructures for improved sustainability), Project funded by the European community under the competitive and sustainable growth programme (1998–2002). http://lifecon.vtt.fi/.

[2] Abrams, N., J. B. Miller, and I. H. B. Miller. 2004. NICOLA 'Numerical Integrated Computer Organised Life Assessment': Computer programme for document archiving, data registration, mathematical treatment and report generation. Millab Consult a.s., Oslo, Norway.

[3] Årskog, V., and S. Fossdal. 1998–2002. Deliverable no D5.3: Methodology and data for calculation of Life Cycle Assessment (LCA) in repair planning. The Lifecon project (Life cycle management of concrete infrastructures for improved sustainability), Project funded by the European community under the competitive and sustainable growth programme). http://lifecon.vtt.fi/.

[4] Fossdal, S. 1995. Energy and environmental accounts for buildings, Project report 173-1995, Norwegian Building Research Institute, Oslo, Norway (in Norwegian).

[5] ISO 14040:1997. Environmental management, Life cycle assessment, Principles and framework. Published by national standardisation bodies.

[6] ISO 14041:1998. Environmental management, Life cycle assessment, Goal and scope definition and inventory analysis. Published by national standardisation bodies.

[7] ISO 14042:2000. Environmental management, Life cycle assessment, Life cycle impact assessment, Published by national standardisation bodies.

[8] ISO 14043:2000. Environmental management, Life cycle assessment, Life cycle interpretation. Published by national standardisation bodies.

[9] The Norwegian Society for Concrete Rehabilitation, The Norwegian Concrete Society, and Tekna. 2002–2004. Concrete rehabilitation – Control and leadership and concrete rehabilitation for site managers, foremen and craftsmen. Courses nos. R1 and R2, The Norwegian Society for Concrete Rehabilitation, Oslo, Norway.

[10] Fosroc International Ltd. 1987 – to date. Reference lists for the Norcure™ methods and Galvashield XP™ anodes. Marketing division, Fosroc International Ltd., Tamworth, England.

[11] Sarja, A. 2001. Lifecon Deliverable D2.3: Quality Function Deployment (QFD) Method. The Lifecon project (Life cycle management of concrete infrastructures for improved sustainability), Project funded by the European community under the competitive and sustainable growth programme. http://lifecon.vtt.fi/.

[12] Durairaj, S. K. et al. 2002. Evaluation of life cycle cost analysis methodologies, Corporate Environmental Strategy, 9(1).

[13] ASTM E 917-94. Standard practice for measuring life cycle costs of buildings and building Systems.

[14] ISO 15686. Buildings and constructed assets – Service life planning. Part 5: Whole life costing (Draft 2002-09-11).

[15] Søderquist, M.-K., and E. Vesikari. 2003. Generic technical handbook for a predictive life cycle management system of concrete structures (LMS). Lifecon Deliverable D1.1 The Lifecon project (Life Cycle Management of Concrete Infrastructures for Improved Sustainability), Project funded by the European Community under the Competitive and Sustainable Growth Programme. http://lifecon.vtt.fi/.

[16] State of Alaska – Department of Education & Early Development. 1999. Life cycle cost analysis handbook.

[17] ENV 1504–9:1997. Products and systems for the protection and repair of concrete structures – Definitions, requirements, quality control and evaluation of conformity. Part 9: General principles for the use of products and systems. Published by national standardisation bodies in all European countries.

[18] Lay, S., and P. Schießl. 2003. Condition assessment protocol. Lifecon Deliverable D3.1. The Lifecon project (Life cycle management of concrete infrastructures for improved sustainability), Project funded by the European Community under the Competitive and Sustainable Growth Programme. http://lifecon.vtt.fi/.

[19] Peitonen, M., and S. Pursio. 1999. Elinjaksokustannus (LCC)- ja - tuotto (LCP) laskenta – Laskentamallin kehittäminen. Report no. RIS B008, VTT Automaatio, Tampere (in Finnish).

[20] Barringer, P. H. 1998. Life cycle cost and good practices. NPRA Maintenance Conference May 19–22, 1998, San Antonio Convention Center, San Antonio, Texas.

[21] Barrett, P. J. 2001. Life cycle costing, Better practice guide. Australian National Audit Office.

[22] Kirkham, R. J., M. Alisa, A. Pimenta da Silvia, T. Grindley, and J. Brøndsted. 2002. A probabilistic whole life cycle performance model for buildings and civil infrastructure. Lifetime Cluster – EuroLifeForm Report.

[23] Miller, J. B., I. H. B. Miller, and M. Sarkkinen. Qualitative and quantitative descriptions and classifications of RAMS (Reliability, Availability, Maintainability, Safety), characteristics for different categories of repair materials and systems subjected to classified environmental exposures: Lifecon Deliverable D5.1. The Lifecon project (Life cycle management of concrete infrastructures for improved sustainability), Project funded by the European Community under the Competitive and Sustainable Growth Programme. http://lifecon.vtt.fi/.

[24] Saltelli, A., K. Chan, and E. M. Scott, eds. 2000. *Sensitivity analysis*. Chichester, England: John Wiley & Sons Ltd.

[25] Gerdes, A., and F. H. Wittmann. 2001. Decisive factors for the transport of Silicon-organic compounds into surfaces near zones of concrete, Hydrophobe III – 3rd International Conference on surface technology with water repellent agents. Aedificatio Publishers.

[26] Ho, D. W. S., S. L. Mak, and K. K. Sagoe-Crentsil. 2000. Clean concrete construction: An Australian perspective, *Proceedings, Concrete technology for a sustainable development in the 21st century*, ed. O. E. Gjørv and K. Sakai, pp. 236–245. London and New York: E & FN Spon.

[27] Gjørv, O. E., and K. Sakai. 2000. Concrete technology for a sustainable development in the 21st century, *Proceedings of an international workshop in Lofoten, Norway*, June, 1998, p. 386. London and New York: E & FN Spon.

[28] Horrigmoe, G. 2000. Future needs in concrete repair technology, *Proceedings, Concrete technology for a sustainable development in the 21st century*, ed. O. E. Gjørv and K. Sakai, pp. 332–340. London and New York: E & FN Spon.

[29] Centrum voor Milieukunde, Leiden, Holland. 1999. Environmental life cycle assessment of products.

[30] De Vries, H., and R. B. Polder. Hydrophobic treatment of concrete. *International Journal of Restoration* 2: 145–160.

3.8 RAMS and QFD in cases of MR&R planning

John B. Miller

In Section 3.7 of this book, the concepts of RAMS (Reliability, Availability, Maintainability and Safety) characteristics were explained, and it was shown how these characteristics could be classified first qualitatively and then quantitatively. As part of these exercises, it was also shown how LCC (Life Cycle Costs) could be calculated and how the LCE (Life Cycle Ecology) could be estimated. In addition, it was mentioned that the QFD (Quality Function Deployment) method could be used for the comparative evaluation of the totalities of MR&R plans, paying due attention not only to RAMS but also to less material aspects such as human well-being, culture, tradition, etc. In this section, we shall apply the concepts to actual cases in order to illustrate how the various principles can be embodied in practice.

We shall apply the concepts to two different structures which have been well documented by extensive surveys. First, we shall perform qualitative RAMS classifications. Thereafter, these will be transmuted into quantitative data by redefining the RAMS parameters. The quantitative data will then be reduced by the QFD method taking Generic Functional Requirements into account, such as economy, pollution, human well-being, etc.

3.8.1 The structures

Two different types of structures will be used in this study – a large public building complex and a subway station tunnel system. Each of these structures is different from the other in terms of accessibility, maintainability, types of damage and the facilities it contains and provides. Both of them are structures which are important to sectors of society at large.

3.8.1.1 Oslo Congress Centre Folkets Hus A/L, Oslo, Norway

The Oslo Congress Centre Folkets Hus A/L (hereafter referred as Oslo Congress Centre) is a large, 10-storey public building complex situated in the centre of Oslo. The complex, which serves both as a congress centre and office premises, was built in three separate stages (Figure 3.8.1–3.8.3), as shown in Table 3.8.1, together with the corresponding façades.

The pre-1956 Stage 1 is mainly of steel and masonry, except for the top three floors which are of reinforced concrete and breeze block. The steel and masonry parts of this building will not be used for the classification of damage categories since, in this book, the emphasis is on concrete structures. In passing, it is perhaps appropriate to remark that the same methods employed herein on concrete structures can, of course, be adapted to structures made of other materials, be they of wood, brick, stone or whatever.

Table 3.8.1 The three building stages (Oslo Congress Centre)

Stage	When built	Corresponding façades
1	Pre-1956 (with a roof storey added in the 1960s)	Parts of the façade along Henrik Ibsens gate and Møllergate
2	1956–1958	The entire façade along Torggate. Part of the façades along Youngs Market and Henriks Ibsens gate
3	1958–1962	The remainders of the façades along Youngs gate and Møllergata

Figure 3.8.1 Building stage 1 (Oslo Congress Centre).

During the summer of the year 2000, the author's firm, Millab Consult a.s., conducted an extensive survey of the structures, which included carbonation tests, chloride tests, reinforcement cover measurements, damage mapping, relative humidity measurements, and removal of loose, spalled concrete [1]. In passing, it may be mentioned that the concrete surfaces exposed to the atmosphere had never been repaired, painted or coated.

3.8.1.1.1 DAMAGE DESCRIPTION: STAGE 1

On the façades and roof edges of the upper three storeys, which are both partly recessed and partly constructed from reinforced concrete, there was spalled concrete due to reinforcement corrosion caused by carbonation. In addition, on some areas which were rendered, the render had delaminated from the concrete substrate. In one

Figure 3.8.2 Building stage 2 (Oslo Congress Centre).

part, on a wall which had been covered with cellular concrete blocks for insulation, the blocks had loosened.

The concrete cover over the reinforcement varied between 10 and 30 mm. Phenolphthalein tests showed that the concrete was carbonated to an average depth of 25 mm, though in some areas as much as 35 mm of carbonation was found. Since the affected concrete was about 35 years old at the time of the survey, the time needed for further carbonation was considerable. Nevertheless, it was necessary to bring the reinforcement corrosion under control since the carbonation depth had exceeded the cover depth in many places.

The concrete did not contain chlorides in concentrations above levels considered harmful to reinforcement surrounded in carbonated, but otherwise sound, concrete.

Figure 3.8.3 Buidling stage 3 (Oslo Congress Centre).

3.8.1.1.2 DAMAGE DESCRIPTION: STAGE 2

The façades consist of concrete columns, beams and the exposed end surfaces of decks. The concrete was frequently spalled due to reinforcement corrosion caused by carbonation. The concrete cover of the reinforcement varied between 0 and 100 mm. Phenolphthalein tests showed that the concrete was carbonated to an average depth of 25 mm, but in some areas as much as 45 mm of carbonation was found.

The most serious and frequently occurring type of damage was to be found on the upper third of the façade columns between each floor, where the reinforcement cover varied from 0 to 30 mm. The variation was caused by misplaced reinforcement during construction as indicated in Figure 3.8.4.

As for Stage 1, since the affected concrete was about 35 years old at the time of the survey, the time needed for further carbonation was considerable. Nevertheless, it was necessary here too to bring the reinforcement corrosion under control since the carbonation depth had exceeded the cover depth in many places.

Again, the concrete did not contain chloride concentrations above levels considered harmful to reinforcement surrounded by sound or carbonated concrete.

3.8.1.1.3 DAMAGE DESCRIPTION: STAGE 3

The façades consist of concrete columns, beams and the exposed end surfaces of decks. Spalled concrete was found at the time of the survey, but only in areas where the reinforcement cover was less than 15 mm, and the damage frequency was low. Misplaced reinforcement did not occur on this part of the building.

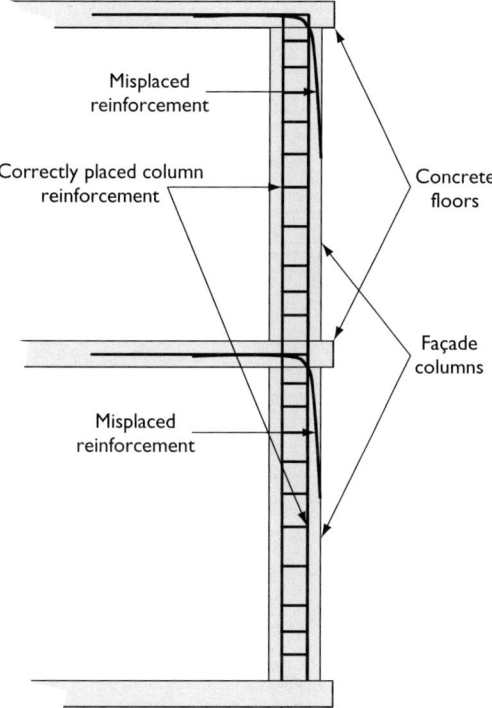

Figure 3.8.4 Sketch showing misplaced reinforcement in façade columns (Oslo Congress Centre).

The concrete cover of the reinforcement varied between 5 and 70 mm. Phenolphthalein tests showed that the concrete was carbonated to an average depth of 12 mm. In some areas, as much as 20 mm of carbonation was found. As for Stages 1 and 2, although the affected concrete was about 35 years old at the time of the survey and the time needed for further carbonation was therefore considerable, it was considered necessary to bring the reinforcement corrosion under control since the carbonation depth had again exceeded the cover depth in many places.

Once again, the concrete did not contain chloride concentrations above levels considered harmful to reinforcement surrounded by sound or carbonated concrete.

3.8.1.2 *Central Station (now Parliament Station), Oslo, Norway*

Central Station (now Parliament Station), hereafter referred as Central Station, with its connecting tunnels and facilities shown in Figure 3.8.5, forms a major part of A/S Oslo Sporveier's (Oslo Tramways) subway system below Oslo. The station is also Oslo's main atomic shelter, and in time of emergency is meant to house and protect people.

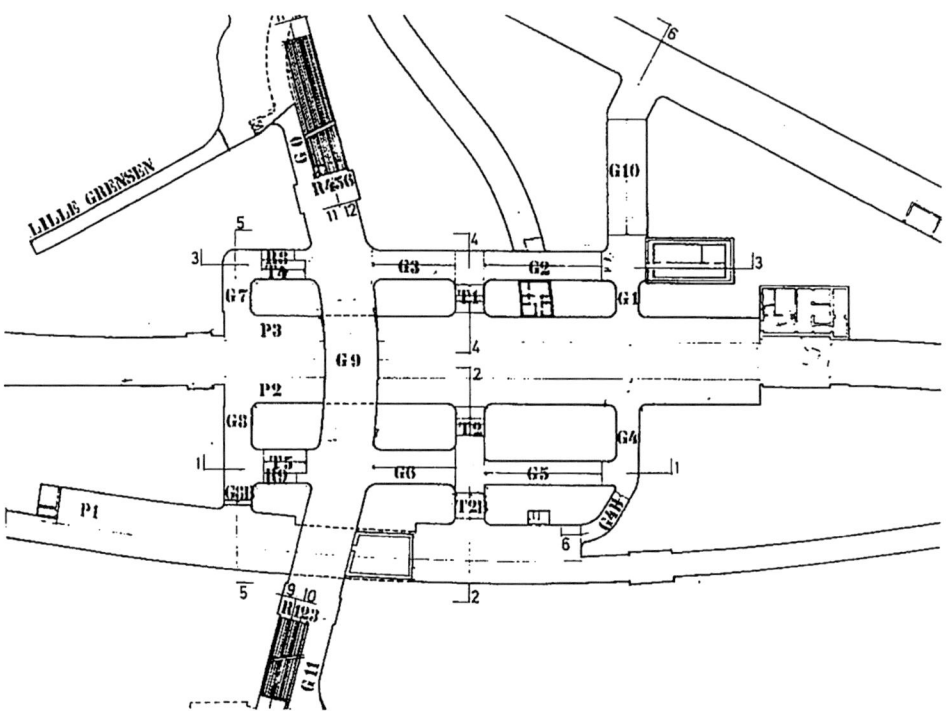

Figure 3.8.5 General plan of the subway tunnel complex (Central Station).

A prerequisite of the tunnels is that they should be watertight in order not to drain the neighbouring ground in the central parts of Oslo. Bedrock in the area is chiefly the so-called alum shale, which upon weathering caused by lowering the water table, can produce corrosive and aggressive sulphatic solutions as well as giving rise to ground heave.

Note: The matrix of alum shale contains finely divided sulphide minerals. These minerals are the main cause of chemical reactions occurring in alum shale. When the ground water table is lowered, oxygen is introduced to the alum shale matrix causing oxidisation of the sulphide minerals. The process releases iron sulphates, which can decompose ordinary Portland concrete, and sulphuric acid, which can deteriorate concrete and cause reinforcement corrosion. In addition to releasing harmful substances, alum shale swells upon weathering, creating swelling pressures up to 20 MPa, and has been known to cause heave of up to 300 mm depending on thickness.

In the station's area of influence, the loose deposits consist of soft marine clays, which, should leakages occur in the tunnels, would be subject to increases in effective stress, which in turn would result in the settlement of buildings. In addition, the tunnels were to be dry, out of consideration for the employees, the public, and the technical installations.

In 1978, one year after its opening, the station was already plagued by troublesome leakages practically in all areas and at all levels. These became gradually worse, and

Figure 3.8.6 Typical situation at the transition between station and train tunnel

as their compass increased, damage to technical equipment and installations became apparent. In 1980, a large piece of concrete fell from the vault arch in one of the train tunnels. This led to a survey which uncovered conditions so alarming that the station was closed in 1983 for extensive repairs (Figure 3.8.6) [2].

The principles of construction are illustrated in Figure 3.8.7. The tunnels were constructed by first casting a 20–30 cm thick layer of un-reinforced blind concrete directly against the raw blasted rock surface. The function of this concrete was to stabilise the rock surface during the construction period, and to form a base for the gluing of one to three layers of watertightening membrane. This membrane was of an approximate 2 mm thick, glass-fibre reinforced, asphaltic textile material.

The structural concrete, which was to have a thickness of between 35 and 100 cm, was then cast against the membrane on the walls and vaults using jackable, hinged formwork. This concrete is cross-reinforced on the atmospheric side of walls and vaults. The cement used was Norwegian produced, sulphate resistant cement.

3.8.1.2.1 DAMAGE DESCRIPTION: ERRORS MADE DURING CONSTRUCTION

3.8.1.2.11 Leakage
There is some uncertainty as to how the rebars were placed in the formwork prior to casting, but they appear to have been partly pushed in from the open end of the formwork, partly placed on the formwork before jacking it into place, and partly tied to metal framework bolted onto blind concrete and to rock. Whichever the case, it is obvious that the installation of the steel led to innumerable punctures in the membrane, and correspondingly to innumerable leakages (Figure 3.8.8). To make matters worse, the casting joints had not been tightened by waterstops or in any other way.

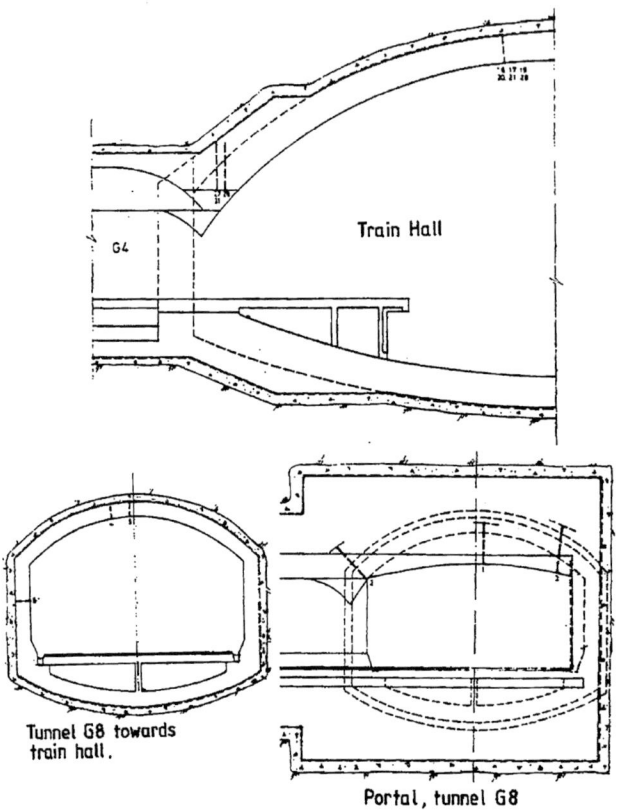

Train Hall

G4

Tunnel G8 towards
train hall.

Portal, tunnel G8

Figure 3.8.7 Cross-section of one of the passenger access tunnels showing principle of construction (Central Station).

Figure 3.8.8 Leakages seen from the passenger bridge in the station.

A general characteristic of the concrete is its high porosity and poor workability. This is due to stiff concrete consistency, and to the fact that all vibrating had to be done through the timber formwork, which absorbed almost all vibration energy. In many sections, the lack of vibration was almost total, resulting in the formation of cavities, honeycombs, pores, segregation, channels along the reinforcement, cracks, etc. In addition, yielding of the shuttering during casting operations resulted in extensive delaminations.

The most significant errors occurring during construction resulting in damage to the structure and its surroundings were the following:

- Water from the surrounding rock leaked through innumerable punctures in the membrane, thus lowering the water table and initiating alum shale oxidation.
- Direct consequences of the lowering of the water table were alum shale decay causing ground heave, and effective stress increase on clay soil causing subsidence, which gave numerous problems to cellars and foundations in the surrounding area.
- Water from the surrounding shale pressed against the membrane and tore it from its substrate where unsupported in cavities. The water then found its way between the concrete layers to the many leakage points along rebar channels, cracks and other paths in the structural concrete.

3.8.1.2.12 Carbonation
The concrete cover of the reinforcement varied between 0 and 120 mm. Phenolphthalein tests showed that the concrete was carbonated to an average depth of 30 mm. In some areas, as much as 100–150 mm of carbonation was found. Taken into account that the concrete was only 6–8 years old, that the depth of carbonation was 30 mm, and that the average cover thickness of the reinforcement was approximately 50 mm, the prediction could be made that all the concrete would be carbonated to the first layer of reinforcement in another 10–12 years approximately.

Due to the extreme carbonation depths that were found, a laboratory investigation was initiated to examine whether sulphate resistant cement carbonates faster than ordinary Portland cement. It was found that the carbonation rate in the Norwegian sulphate resistant cement used was three to four times faster than in the ordinary Portland cement then produced [2].

3.8.1.2.13 Reinforcement corrosion
Almost all the observed cases of corrosion damage on reinforcement steel could be ascribed to alum shale water in channels along the reinforcement.

3.8.1.2.14 Chemical attack
Secondarily oxidised alum shale water containing much sulphuric acid and seeping through construction joints and other leakage paths caused severe dissolution of the concrete in many places, even where sulphate resistant cement had been used. In addition, structural parts composed of ordinary concrete exposed to alum shale water containing sulphates and sulphuric acid suffered from swelling and disintegration caused by the formation of ettringite, jarosite, gypsum, thaulite and other sulphate compounds.

3.8.1.2.15   Degradation of technical installations

Electrical installations, ventilation and heating equipment, ticket offices, shops, plumbing fixtures, air filters, etc. were subjected to damage caused by leakage of corrosive alum shale water.

3.8.1.2.16   Employed rehabilitation techniques

Because of the complexity and scale of the problems involved in this structure, the rehabilitation techniques will be described before moving on to the task proper. In this connection, it will be useful to refer to Figure 3.8.9.

3.8.1.2.17   Concrete repair

As described earlier, there were more or less continuous gaps and cavities between the structural concrete and the membrane in the tunnels and vaults. As the gaps and cavities acted as water distribution reservoirs, and also resulted in considerably reduced bearing capacity due to lack of monolithic support behind the vaults, they had to be filled as completely as possible.

Gaps and cavities were filled by cement grout injection. A stable cement grout containing 20 parts per weight of cement and 1 part of silica fume, with a water/cement ratio of approximately 0.6 was used. The maximum allowed injection pressure was 1 MPa. The small gaps that formed between the injected grout and the cavity surfaces due to water separation in the grout were filled by secondary injections of a low viscosity silicate grout (Siprogel), which on setting gave a stable gel.

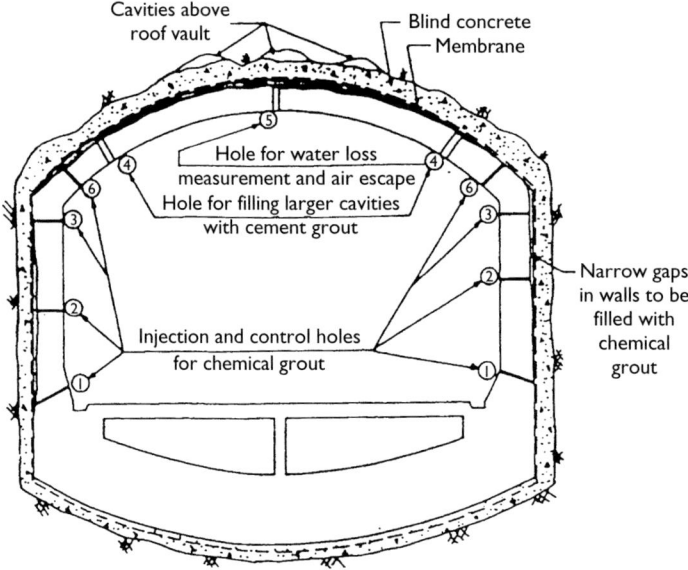

Figure 3.8.9 The main principles of rehabilitation (Central Station). (This shows a section through one of the pedestrian tunnels; the numbering indicates the sequence of injection; Levels lower than 1 are pressured watertight.)

3.8.1.2.18   Strengthening of vaults and leakage cut-off by injection of cement and chemical grouts

Some of the delaminated areas in the tunnel vault arches were deemed too weak to withstand the weights and pressure loads associated with cement grout injection. In such areas, delaminated concrete was chiselled off and the cavities repaired by wet spraying of concrete.

3.8.1.2.19   Sealing of surfaces, watertightening, and passive realkalisation

On coating the free surface of concrete subjected to an anterior water pressure with an impervious, adhering membrane, transverse transport of water will stop completely, and the concrete will gradually become saturated. The correct type of soventless polyurethane coating applied in the most commonly used thickness of 1.2 mm, will, for all practical purposes, be watertight, diffusion proof to atmospheric gasses including water vapour and to dissolved substances including sulphates and chlorides, and be able to bridge across cracks that open from 0 to about 1 mm.

The water and gas pressure that can be resisted by such a coating is limited by the tensile strength of the surface concrete. For lesser concrete qualities, that is up to C35, it is not difficult to obtain tensile strengths of up to 2–3 MPa. Compared to commonly encountered water pressures, which rarely exceed 20–30 m, even 1 MPa, corresponding to a hydraulic head of 100 m, should be a safe adhesion value in most cases.

Prior to applying the coating, the surfaces were prepared by high pressure jetting, sandblasting, scabbing or flame cleaning. After cleaning was completed, the surfaces were brushed and vacuumed. To control problems connected with water leakages remaining after injection works and to create optimal conditions for the application of the coating, it was necessary to either drain or inject the largest leakages to allow preparation of the working area, at least locally or in part. Application of the polyurethane coating was performed by brush, roller and spray gun, depending on the access to, and the size of, the area being treated.

The concrete coating also stops the carbonation process because the coating itself is diffusion proof and it allows complete saturation of the concrete. The extremely low diffusivity of the coating limits the access to oxygen and thereby contributes to slowing the corrosion process. More importantly, on sealing the concrete surface, the access to carbon dioxide is cut off and the concrete saturates with water. When the concrete is completely saturated, all water transport stops. Now what happens is that the excess alkaline agents in the greater, uncarbonated mass of concrete behind the reinforcement and the carbonated zone dissolves and diffuses into the latter and raises its pH to protective levels.

3.8.2   The RAMS classification process

Having examined the two structures, we are ready to apply the principles of RAMS characteristic classification, LCC, LCA and QFD as aids to the making of MR&R plans for them.

3.8.2.1   Oslo Congress Centre

The first step here, as in every case, is to make a preliminary systemisation of the damage found by a condition survey of the building as shown in Table 3.8.2. This is followed by the slightly more rigorous step of classifying the damage types as indicated under Box C_S in Figure 3.7.3 of Section 3.7.4.2 in Section 3.7. The result of this step is shown in Table 3.8.3. The third step is to qualitatively describe the suggested repair methods according to their RAMS characteristics. To recapitulate, the

Table 3.8.2 Preliminary systemisation of damage and deterioration (Oslo Congress Centre)

Component	Condition	Degradation mechanisms and/or causes	Consequences
Stage 1, 8th storey eves	Spalling concrete	Carbonation and reinforcement corrosion	Loose concrete – hazardous to car and pedestrian traffic
	Disintegration of render and concrete	Frost damage	
Stage 1, 9th storey walls	Spalling concrete and render	Carbonation and reinforcement corrosion	Loose concrete – hazardous to car and pedestrian traffic
Stage 1, 10th storey walls	Loose blocks of cellular concrete	Constructional error or carbonation & reinforcement corrosion	Loose concrete – hazardous to car and pedestrian traffic
Stage 2, Columns, 2nd to 10th storey	Concrete spalls weighing from 0.1 kg to 20 kg	Systematic low cover on the upper third of column, carbonation and reinforcement corrosion	Structural damage and loose concrete – hazardous to car and pedestrian traffic
Stage 2, 8th and 10th storey eves	Spalling concrete	Generally low concrete cover, carbonation and reinforcement corrosion	Long-term degradation and spalls will be hazardous to car and pedestrian traffic
Joint between Stages 1 and 2, 1st to 8th storey	Deterioration of cellulose based joint filler	Age, absorption of humidity into concrete and joint, frost damage	Concrete degradation
Stage 2, Deck end surfaces between 1st and 2nd storey	Concrete spalls weighing up to 1 kg	Generally low concrete cover, carbonation and reinforcement corrosion	Loose concrete – hazardous to pedestrians
Joint between Stages 2 and 3, 2nd storey	Spalling concrete at joint	Age, absorption of humidity into concrete and joint, frost damage	Concrete degradation and loose concrete – hazardous to pedestrians
Stage 3, Corner column, 2nd to 10th storey	Spalling concrete	Generally low concrete cover, carbonation and reinforcement corrosion	Concrete degradation and loose concrete – hazardous to pedestrians and cars

Table 3.8.3 RAMS classification of damage (Oslo Congress Centre)

Component	Condition	Degradation mechanisms and/or causes	Classification	Reason for classification	Rehabilitation techniques
Stage 1, 8th storey eves	Spalling concrete	Carbonation and reinforcement corrosion	C3	Detrimental process	Mechanical repair, wet or dry spray mortar, realkalisation, cathodic protection
	Disintegration of render and concrete	Frost damage		Detrimental process	
Stage 1, 9th storey walls	Spalling concrete and render	Carbonation and reinforcement corrosion	C3	Detrimental process	Mechanical repair, wet or dry spray mortar, cathodic protection, realkalisation
Stage 1, 10th storey walls	Loose blocks of cellular concrete		C2	Singularity	Re-fixing using helical ties
		Carbonation and reinforcement corrosion	C3	Detrimental process	Mechanical repair, wet or dry spray mortar, cathodic protection, realkalisation. Replace cellular concrete
Stage 2, Columns, 2nd to 10th storey	Concrete spalls weighing from 0.1 to 20 kg	Systematic low cover on the upper third of column.	C2	Singularity	Mechanical repair, wet or dry spray mortar, cathodic protection, realkalisation
		Carbonation and reinforcement corrosion	C3	Detrimental process	
Stage 2, 8th and 10th storey eves	Spalling concrete	Generally low concrete cover	C2	Singularity	Mechanical repair, cover increase, wet or dry spray mortar, cathodic protection, realkalisation
		Carbonation and reinforcement corrosion	C3	Detrimental process	
Joint between Stages 1 and 2, 1st to 8th storey	Deterioration of cellulose-based joint filler	Age, absorption of humidity into concrete and joint, frost damage	C3	Detrimental process	Mechanical repair, wet or dray spray mortar, replace joint filler, hydrophobation

Table 3.8.3 (Continued)

Component	Condition	Degradation mechanisms and/or causes	Classification	Reason for classification	Rehabilitation techniques
Stage 2, Transition beam between 1st and 2nd storey	Concrete spalls weighing up to 1 kg	Generally low concrete cover	C2	Singularity	Mechanical repair, wet or dry spray mortar, cathodic protection, realkalisation
		Carbonation and reinforcement corrosion	C3	Detrimental process	
Joint between Stages 2 and 3, 2nd storey	Spalling concrete at joint	Age, absorption of humidity into concrete and joint, frost damage	C3	Detrimental process	Mechanical repair, wet or dry spray mortar Replace joint filler Hydrophobation
Stage 3 Corner column, 2nd to 10th storey	Spalling concrete	Generally low concrete cover	C2	Singularity	Mechanical repair, wet or dry spray mortar, cathodic protection, realkalisation
		Carbonation and reinforcement corrosion	C3	Detrimental process	

definitions of the RAMS characteristics upon which these qualitative descriptions are based are:

- *Reliability*: The ability to reduce maintenance to a minimum during service life.
- *Availability*: The ease of supply of methods, systems, materials or qualified personnel and/or of the number of available qualified contractors.
- *Maintainability*: The ease with which the combination of all technical and associated administrative actions during the service life can retain a repair and/or upgrade in a state in which it can perform its required functions.
- *Safety*: The health and accident risks that can be directly connected to the methods, systems, products and their end results.

Keeping in mind the subjectivity described in Sections 3.7.4.2 and 3.7.4.3 arising from the lack of precise knowledge of the performance of repairs and materials, a RAMS classification for the repair methods suggested for the building can be compiled as shown in Table 3.8.4.

Table 3.8.4 RAMS classification of MR&R methods (Oslo Congress Centre)

Index no.	Methods, systems and materials	Reliability	Availability	Maintainability	Safety
1	Norcure realkalisation	Excellent	Fair	Good	Very Good
2	Mechanical repair	Medium to Poor	Excellent	Very good	Good
3	Cathodic protection, category 1: Embedded rod anodes	Very Good	Good	Poor to Good	Very Good
4	Cathodic protection, category 2: Embedded mesh anode	Good	Poor	Medium	Good
5	Cathodic protection, category 3: Embedded ribbon or wire	Good	Good	Good	Good
6	Cathodic protection, category 4: Conductive coating	Poor	Poor	Medium	Good
7	Paints and coatings, category 1: Organic paints and coatings, CO_2 barriers	Medium	Excellent	Medium	Good
8	Aesthetic organic paints and coatings	Medium	Excellent	Poor to Very Good	Very Good
9	Protective paints and coatings, category 2: Cementitious paints and coatings containing polymer admixtures and enhancers	Very good	Excellent	Poor to Medium	Very Good
10	Aesthetic coatings, cementitious paints and coatings	Excellent	Excellent	Excellent	Excellent

3.8.2.1.1   REASONING

Index number 1: Realkalisation

RELIABILITY   Excellent
Reason   To this day, about 1.5 million square meters of reinforced concrete have been realkalised with no reports of re-occurring reinforcement corrosion.

AVAILABILITY Fair
Reason   Four licensed contractors can offer realkalisation works in the Oslo region.

MAINTAINABILITY   Good
Reason   The achieved corrosion protection of the reinforcement steel is practically maintenance free. Especially no difficult measures needed for maintenance.

SAFETY   Very Good
Reason   The realkalisation process reduces the extent of heavy labour and potential hazards associated with mechanical repair works. In comparison to mechanical repair and some cathodic protection systems, the process offers reduced noise emissions, which during refurbishment improves human health conditions for the building users.

Index number 2: Mechanical repair

RELIABILITY   Medium to Poor
Reason   It is known that mechanical repair can give rise to incipient anodes and the creation of macro-cells which induce reinforcement corrosion. Its reliability depends on the extent of removal of carbonated and/or chloride contaminated concrete, which in many cases is insufficient, and on how well the electrolytic properties of the replacement material match those of the substrate. Re-occurring of corrosion is a frequent problem.

AVAILABILITY   Excellent
Reason   Numerous contractors can offer mechanical repair works in the Oslo region.

MAINTAINABILITY   Very Good
Reason   Re-occurring of corrosion is frequent, requiring maintenance. The maintenance itself is relatively easy.

SAFETY   Good
Reason   Mechanical repair work is labour intensive, and employs chiselling tools which are harmful to human health. Water jetting tools relieve the worker of harmful chiselling stresses, but increases risks of serious injury. The noise emissions from mechanical repair works are high.

Index number 3: Cathodic protection – embedded rod anodes

RELIABILITY   Very Good
Reason   If correctly designed, this cathodic protection system can offer reliable corrosion protection.

AVAILABILITY   Good
Reason   Several contractors can offer this type of cathodic protection works in the Oslo region.

MAINTAINABILITY	Poor to Good
Reason	The system is not maintenance-free, but has a trouble-free service life expectancy of 10–20 years. Troubleshooting is in principle easy to perform. Malfunction may require re-installation or repair of system elements.

SAFETY	Very Good
Reason	Cathodic protection reduces the extent of heavy labour and potential hazards associated with mechanical repair works. This type of cathodic protection (CP) requires drilling of numerous holes which gives rise to relatively high noise emissions.

Index number 4: Cathodic protection – embedded mesh anode

RELIABILITY	Good
Reason	If correctly designed, this cathodic protection system can offer reliable corrosion protection. The installation process is difficult, which can give rise to flaws and reduction in service life.

AVAILABILITY	Poor
Reason	Few contractors offer this type of cathodic protection works in the Oslo region.

MAINTAINABILITY	Medium
Reason	The system is not maintenance free, but has a trouble-free service life expectancy of 10 years or more. Trouble-shooting is difficult to perform. Malfunction may require removal and reinstallation of large parts of the system.

SAFETY	Good
Reason	The installation process does not involve high noise emissions, but does use hazardous equipment. Dust emissions can be high.

Index number 5: Cathodic protection – embedded ribbon or wire

RELIABILITY	Good
Reason	If correctly designed, this cathodic protection system gives reliable corrosion protection.

AVAILABILITY	Good
Reason	Some contractors can offer this type of cathodic protection in the Oslo region.

MAINTAINABILITY	Good
Reason	The system is not maintenance free, but should have a trouble-free service life of 10 years or more. Trouble-shooting is in principle easy to perform. Malfunction may require reinstallation or repair of system elements.

SAFETY   Good
Reason   Installation involves relatively high dust and noise emis-
                        sions, and may involve the use of hazardous equipment.

Index number 6: Cathodic protection – conductive coating

RELIABILITY   Poor
Reason   It is doubtful whether organic-based conductive coatings
                        can deliver sufficient current densities to passivate the rein-
                        forcement steel. In addition, the coating itself is subjected
                        to wear and tear from weather and to chemical incom-
                        patibility with the substrate, which can reduce its design
                        life.

AVAILABILITY   Poor
Reason   Only one contractor can offer this type of cathodic protec-
                        tion in the Oslo region.

MAINTAINABILITY   Medium
Reason   The system is not maintenance free and has a trouble-free
                        service life expectancy of less than 10 years. Failed coating
                        needs removal and renewal. Trouble-shooting is difficult
                        to perform.

SAFETY   Good
Reason   Installation does not involve high noise or dust emissions,
                        nor the use of hazardous equipment.

Index number 7: Protective paints and coatings – Category 1: Organic paints and coatings, CO_2 barriers

RELIABILITY   Medium
Reason   Achieved film thickness is difficult to control during appli-
                        cation due to the substrate's ability to absorb paint, which
                        can lead to large film thickness. Thus, laboratory tests per-
                        formed on even film thickness are often misleading. Large
                        film thickness can result in water vapour impermeability,
                        causing water to condense in the substrate material. This
                        may in turn lead to frost damage and the deterioration of
                        paint and substrate. This is especially the case on rendered
                        concrete surfaces. Some organic paints are incompatible
                        with concrete substrates due to their thermal expansion
                        coefficients differing from that of concrete, particularly in
                        winter.
AVAILABILITY   Excellent
Reason   Numerous suppliers and contractors can offer products and
                        their application in the Oslo region.

| MAINTAINABILITY | Medium |
| Reason | Failure is common. Removal is required before repainting. |

| SAFETY | Good |
| Reason | Some products contain solvents that can be harmful on inhalation. Some products contain harmful substances. Many products are considered safe. |

Index number 8: Aesthetic organic paints and coatings

| RELIABILITY | Medium |
| Reason | Achieved film thickness is difficult to control during application due to the substrate's ability to absorb paint, which leads to large film thickness. Thus, laboratory tests performed on even film thickness are often misleading. Large film thickness can cause water vapour impermeability, causing water to condense in the substrate material. This may in turn lead to frost damage and the deterioration of paint and substrate. This is especially the case on rendered concrete surfaces. Some organic paints are incompatible with concrete substrates due to their thermal expansion coefficients differing from that of concrete, particularly in winter. |

| AVAILABILITY | Excellent |
| Reason | Many suppliers and contractors can apply such products in the Oslo region. |

| MAINTAINABILITY | Poor to Very Good |
| Reason | Failure is common. Removal, which can be quite difficult, is required before re-painting. |

| SAFETY | Very Good |
| Reason | Some products contain solvents that can be harmful on inhalation. Some products contain harmful substances. Many products are considered safe. |

Index number 9: Protective paints and coatings Category 2: Polymer-modified cementitious paints and coatings

| RELIABILITY | Very Good |
| Reason | Cases of thermal and alkali incompatibility can occur. Formation of films can give reduced diffusivity to water vapour and give condensation and frost problems. |

| AVAILABILITY | Excellent |
| Reason | Many suppliers and contractors can apply such products in the Oslo region. |

MAINTAINABILITY	Poor to Medium
Reason	Removal, which can be difficult, is required before repainting.
SAFETY	Very Good
Reason	Cementitious polymer-modified products do not usually contain solvents or substances that are harmful on inhalation, although some primers do. Polymeric modifying substances are considered relatively safe. Only ordinary precautions for the handling of cementitious products are necessary.

Index number 10: Aesthetic coatings – Polymer-free cementitious paints and coatings

RELIABILITY	Excellent
Reason	Cementitious paints and coatings are applied on pre-wetted substrates, causing the substrates to absorb paint reasonably uniformly. This contributes to a controlled paint thickness, with predictable permeability and diffusivity parameters. Cementitious paints and coatings are compatible with concrete substrates, provided their performance is not based on additives such as polymeric modifiers.
AVAILABILITY	Excellent
Reason	One supplier and many contractors can offer such products in the Oslo region.
MAINTAINABILITY	Excellent
Reason	Flaking paint seldom occurs. Re-painting can be performed without removing old paint.
SAFETY	Excellent
Reason	Pure cementitious products do not contain solvents or substances that are harmful on inhalation. Only ordinary precautions for the handling of cementitious products are necessary.

3.8.2.2 Central Station

The RAMS classifications for the Central Station corresponding to those for the Oslo Congress Centre are shown in Tables 3.8.5, 3.8.6 and 3.8.7.

The methods referred to by indices 6, 7 and 8 in Table 3.8.7, were not options which viably existed at the time of the rehabilitation. For the sake of completeness of comparison, the evaluations in the table, and in the following, are made on the basis

Table 3.8.5 Preliminary systemisation of damage and deterioration (Central Station)

Component	Conditions	Degradation mechanisms and causes	Consequences
Watertight membrane between blind and structural concrete in tunnels and culverts	Perforated membrane allowed ground water leakage into the structural complex. Cavities between the blind concrete and the structural concrete caused the membrane to tear off due to the water pressure on the membrane	Lowering of the ground water level leading to oxidation of alum shale bedrock and release of sulphates and sulphuric acid.	• Swelling of alum shale bedrock and lowering of the water table caused structural damage to surrounding buildings by heave and subsidence. • Reinforcement corrosion • Sulphuric acid attack on sulphate resistant concrete • Sulphate attack on ordinary concrete • Electrical problems • Stoppage of drains • Closure of public toilets • Public discomfort • Reduced serviceability and final closure
Reinforced structural concrete in tunnels and vaults	Cavities, high porosity, honeycombing, cracks, delaminations and channel formation below and along the reinforcement steel.	Insufficient compaction and yielding formwork during casting	• Multiple water passages throughout the structure • Reinforcement corrosion
	Spalled concrete, some pieces weighing up to 300 kg	Rapid carbonation of sulphate resistant cement, high porosity concrete, high CO_2 levels and pressure fluctuations due to pumping action of trains.	• Reinforcement corrosion • Structural damage • Hazards to pedestrian and rail traffic • Reinforcement corrosion
	Disintegrated concrete	Sulphate and sulphuric acid attack	• Disintegration of sulphate resistant concrete at leakage points.
Non-structural concrete	Disintegration of concrete	Sulphate and sulphuric acid attack	• Damage to secondary concrete features

Table 3.8.6 RAMS classification of damage and deterioration (Central Station)

Component	Conditions	Degradation mechanisms and/or causes	Classification	Reason for classification	MR&R methods
Watertight membrane between blind and structural concrete in tunnels and culverts	Perforated membrane and ground water leakage.	Ground heave and lowering of the ground water level	C2	Singularity	Application of watertight membrane on concrete surface
	Cavities between blind and structural concrete caused the membrane to be torn off by water pressure	Oxidation of alum shale bedrock and release of sulphates and sulphuric acid	C3	Detrimental process	Application of watertight membrane on concrete surface
		Ground heave and lowering of the ground water level	C2	Singularity	Injection works
		Oxidation of alum shale bedrock and release of sulphates and sulphuric acid	C3	Detrimental process	Injection works Application of watertight membrane on concrete surface
Reinforced structural concrete in tunnels and vaults	Cavities, high porosity, honey-combing cracks, channels below and along reinforcement, delaminations	Insufficiently compacted concrete and yielding form-work during casting	C2	Singularities	Injection works Mechanical repair and wet spray mortaring
	Spalled concrete, some pieces weighing up to 300 kg	Rapid carbonation of sulphate resistant cement, high porosity concrete, high CO_2 levels and pressure fluctuations due to pumping action of trains.	C3	Detrimental process	Mechanical repair, wet spray mortaring, application of water- and gas-tight membrane on the concrete surface
	Disintegrated concrete	Sulphate and sulphuric acid attack	C3	Detrimental process	Mechanical repair and wet spray mortaring.
Non-structural concrete	Disintegrated concrete	Sulphate and sulphuric acid attack	C3	Detrimental process	Mechanical repair and wet spray mortaring

Table 3.8.7 RAMS classification of (Central Station) MR&R Methods

Index No.	Methods, systems and materials	Reliability	Availability	Maintainability	Safety
1	Mechanical repair	Very Good	Excellent	Poor to Good	Good
2	Wet concrete spray	Excellent	Medium	Poor to Good	Good
3	Cement grout injection	Excellent	Medium	Poor	Very Good
4	Silicate grout injection	Good	Medium	Poor	Good
5	Sealing by application of watertightening, diffusion-proof, polyurethane coating	Excellent	Poor	Poor to Good	Good
6	Electrochemical realkalisation	Very Good to Excellent	Fair	Poor to Good	Very Good
7	Cathodic protection – rods	Good	Good	Poor	Very Good
8	Cathodic protection – ribbons or wire	Good	Good	Poor	Good

of the modern methods as they would have applied to the station had they existed at that time as they do today.

3.8.2.2.1   REASONING

Index number 1: Mechanical repair

RELIABILITY
Reason

Very Good
In this case, mechanical repair is seen as a part of the rehabilitation technique and not as a repair measure in itself. Since the concrete saturates and thus becomes passively realkalised, the steel reinforcement is uniformly passivated. The risk of creating incipient anodes is therefore slight.

AVAILABILITY
Reason

Excellent
Numerous contractors can offer mechanical repair works in the Oslo region.

MAINTAINABILITY
Reason

Poor to Good
Repair is easy to perform in itself, but may involve puncturing of the watertightening membrane, which would need patching. In addition, repair is likely to involve removal of technical or architectural fixtures and may reduce the station's serviceability towards the public.

SAFETY Good

Reason Mechanical repair works is labour intensive and employs chiselling tools which are harmful to human health. Water jetting tools relieve the worker of harmful chiselling tools, but increases risks of serious injury. The noise emissions from mechanical repair works are high.

Index number 2: Wet-sprayed concrete

RELIABILITY Excellent

Reason Few problems are associated with wet-sprayed concrete, which is deemed a well-functioning replacement for the original concrete.

AVAILABILITY Medium

Reason Several contractors can offer wet spray concrete in the Oslo region.

MAINTAINABILITY Poor to Good

Reason Repair is easy to perform in itself, but is likely to involve removal of technical or architectural fixtures, and may reduce the station's serviceability towards the public.

SAFETY Good

Reason The wet-spraying process involves relatively laborious operations, which can be considered as detrimental to human health. Noise and dust emissions are acceptably low. The end result is normally very safe.

Index number 3: Cement grout injection

RELIABILITY Excellent

Reason Successful injections are stable and trouble free.

AVAILABILITY Medium

Reason Several contractors can perform cement grout injection in the Oslo region.

MAINTAINABILITY Poor

Reason The injection process is inherently difficult, and repairs even more so.

SAFETY Very Good

Reason Cement grout injection involves heavy labour, much the same type of operation as wet spraying, but little exposure to harmful chemicals. Noise emissions can be relatively high. The end result is normally very safe.

Index number 4: Silicate grout injection

RELIABILITY	Good
Reason	Successful injections are fairly stable and trouble free.
AVAILABILITY	Medium
Reason	Several contractors can offer silicate grout injection works in the Oslo region.
MAINTAINABILITY	Poor
Reason	The injection process is inherently difficult, and repairs even more so.
SAFETY	Good
Reason	Silicate grout injection does not involve heavy labour and causes only limited exposure to harmful chemicals. Noise emissions can be relatively high. The end result is normally very safe.

Index number 5: Polyurethane watertightening, diffusion-proof coating

RELIABILITY	Excellent
Reason	To this day, more than 20 years after the rehabilitation works were finished, no new water leakages, reinforcement corrosion, ground heave, or surface subsidence have been reported. This is as was expected on the basis of track records.
AVAILABILITY	Poor
Reason	Only one or two suppliers, contractors and consultants can offer appropriate products and associated services in the Oslo region.
MAINTAINABILITY	Poor to Good
Reason	Repair is easy to perform in itself, but is likely to involve removal of technical or architectural fixtures and may reduce the station's serviceability towards the public.
SAFETY	Good
Reason	Surface preparation involves the use of abrasive equipment driven by compressed air, which causes high dust and noise emissions. The coating work itself does not involve heavy labour, nor dust and noise emissions. Polyurethane work is regarded by some to be harmful. However, no cases, known to the author, of health problems have ever been reported in connection with the MDI type of polyurethane used in Norway, France, Denmark, and Switzerland for coating works. The end result is considered very safe.

Index 6: Electrochemical realkalisation

RELIABILITY	Very Good to Excellent
Reason	To this day, about 1.5 million square metres of reinforced concrete have been realkalised with no reports of re-occurring reinforcement corrosion. However, if electrochemical realkalisation had been employed instead of the diffusion-proof polyurethane membrane that ensured passive realkalisation, it is possible that the alkaline compounds introduced into the concrete by realkalisation could, in some places, have been leached out of the concrete by ground water.
AVAILABILITY	Fair
Reason	Four licensed contractors can offer realkalisation work in the Oslo region.
MAINTAINABILITY	Poor to Good
Reason	The achieved corrosion protection of the reinforcement steel is practically maintenance free. However, if maintenance were needed, it is likely that it would involve the removal of electrical installations and architectural fixtures.
SAFETY	Very Good
Reason	The realkalisation process reduces the extent of heavy labour and potential hazards associated with mechanical repair works. In comparison to mechanical repair and some cathodic protection systems, this process offers reduced noise emissions during MR&R, which do not affect the health conditions of the public and the workers.

Index 7: Cathodic protection – embedded rod anodes

RELIABILITY	Good
Reason	If correctly designed, this cathodic protection system can offer reliable corrosion protection. However, the typical service life of this CP system is 25–30 years. As the desired service life of the rehabilitation was at least 100 years, much of the installation would have to be replaced three to four times.
AVAILABILITY	Good
Reason	A number of contractors can offer this type of cathodic protection system in Oslo.
MAINTAINABILITY	Poor
Reason	Maintenance would involve the removal of electrical installations and architectural fixtures and can cause interruption to normal services.

SAFETY	Very Good
Reason	Cathodic protection reduces the extent of heavy labour and potential hazards associated with mechanical repair works. This type of CP requires drilling of numerous holes which gives rise to relatively high noise emissions.

Index 8: Cathodic protection – embedded ribbon or wire

RELIABILITY	Good
Reason	If correctly designed, this cathodic protection system can offer reliable corrosion protection. However, the typical service life of this CP system is 25–30 years. As the desired service life of the rehabilitation was at least 100 years, much of the installation would have to be replaced three to four times.

AVAILABILITY	Good
Reason	A number of contractors can offer this type of cathodic protection system in Oslo.

MAINTAINABILITY	Poor
Reason	Maintenance would involve the removal of electrical installations and architectural fixtures and can cause interruption to normal services.

SAFETY	Good
Reason	Installation involves relatively high dust and noise emissions and may involve the use of hazardous equipment.

3.8.3 Transforming the qualitative RAMS descriptions into quantitative characteristics

Before the qualitative RAMS descriptions can be transformed into quantitative figures, a revised set of the RAMS characteristics is needed. Looking at the contents of Tables 3.8.4 and 3.8.7 and recalling the point system described in Section 3.7.4.4, the basis of which is shown in condensed form in Table 3.8.8, certain numerical values can be assigned to the RAMS characteristics. This has been done for the Oslo Congress Centre in Table 3.8.9 and for the Central Station in Table 3.8.10.

This scheme allows the RAMS characteristics to be represented by quantitative data which can be manipulated by the Quality Function Deployment (QFD) method. Although the specific data used in the examples refers to two specific structures, exactly the same exercise can be performed for any object for which the necessary data are available. In this connection, the reader is reminded that the relativity of the RAMS characteristics will change with the type of structure according to its accessibility, maintainability and safety aspects. The reader is also reminded that the methods referred to by indices 6, 7 and 8 in Table 3.8.10 were not options which viably existed at the time of the rehabilitation of the Central Station; for the sake of

Table 3.8.8 Assigning numerical values to the RAMS characteristics of MR&R methods

Reliability	4 minus 1 point for every 1/5th of the desired time to the next MR&R action not attained 5 points if the time to the next MR&R action is longer than the desired time
Availability	1 point for every 2 qualified contractors, up to 10 in all, starting at 0 points for the first 2 5 points if there are more than 10 qualified contractors.

Maintainability	5 minus 1 point for each of the following requirements: • Special procedures • Highly specialised equipment • Specially trained personnel • Highly specialised materials • Other special requirements	The resulting scale ranges from 0–5 where one point is subtracted for each special requirement that is necessary. Thus: • 5 represents Excellent (no special requirements) • 4 represents Very Good • 3 represents Good • 2 represents Medium • 1 represents Poor (5 special requirements)
Safety	5 minus 1 point for each of the following requirements: • Special procedures • Highly specialised equipment • Specially trained personnel • Highly specialised materials • Other special requirements	The resulting scale ranges from 0–5 where one point is subtracted for each special requirement that is necessary. Thus: • 5 represents Excellent (no special requirements) • 4 represents Very Good • 3 represents Good • 2 represents Medium • 1 represents Poor (5 special requirements)

completeness of comparison, they are treated as though they did. Should the RAMS classification procedure highlight weaknesses in particular methods or materials, steps may be taken to reduce them. For example, the reliability of mechanical repair can be improved by more thorough procedures and by adjusting repair materials to match their mechanical and electrochemical properties to those of the substrate. It is then the RAMS characteristics of the re-adjusted methods or materials should be used.

3.8.4 The Quality Function Deployment (QFD) method for data with immaterial values

As explained in Sections 3.7.4 and 3.7.7, the QFD method [3] can be used as an adjuvant tool when converting requirements to specifications which can be either performance- or technical-oriented. In this connection, QFD serves as an optimising and selective linking tool between alternative repair methods and materials and their performance properties (RAMS). The method can be made to encompass immaterial

Table 3.8.9 Quantification of RAMS classification of MR&R methods (Oslo Congress Centre)

Index No.	Methods, systems and materials	Reliability	Availability	Maintainability	Safety
1	Realkalisation	5	1	3	4
2	Mechanical repair	1	5	4	3
3	Cathodic protection, category 1: Embedded rod anodes	4	3	3	4
4	Cathodic protection, category 2: Embedded mesh anode	3	0	2	3
5	Cathodic protection, category 3: Embedded ribbon or wire	3	3	3	3
6	Cathodic protection, category 4: Conductive coating	0	0	2	3
7	Paints and coatings, category 1: Organic paints and coatings, CO_2 barriers	2	5	2	5
8	Aesthetic organic paints and coatings	2	5	3	4
9	Protective paints and coatings, category 2: Cementitious paints and coatings containing polymer admixtures and enhancers	4	5	2	4
10	Aesthetic coatings, cementitious paints and coatings	5	5	5	5

Table 3.8.10 Quantification of RAMS classification of MR&R methods (Central Station)

Index No.	Methods, systems and materials	Reliability	Availability	Maintainability	Safety
1	Mechanical repair	4	5	2	3
2	Wet concrete spray	5	2	2	3
3	Cement grout injection	5	2	2	4
4	Silicate grout injection	3	2	2	4
5	Polyurethane sealer coating	5	0	2	3
6	Electrochemical realkalisation	4	1	3	4
7	Cathodic protection – rod anodes	0	2	1	4
8	Cathodic protection – ribbon anodes or wire	0	3	1	3

values such as the Generic Functional Requirements associated with environmental burdens, human conditions, safety, etc.

The objective of integrated lifetime repair planning methodology is to optimise and guarantee the life cycle with respect to technical performance parameters, moderated by the requirements of human well-being, economy, cultural compatibility and ecology. The QFD method makes it possible to control and optimise these requirements, the monetary economy and the ecology and taking social aspects also into account. Fundamental to the method are:

- Identifying the functional requirements of the owners, the users and society (Generic Functional Requirements)
- Interpreting and relating generic requirements first to Performance Properties (RAMS) and then to alternative repair methods and materials of structures
- Optimising the alternative repair methods or materials in relation to Performance Properties (RAMS)
- Selecting MR&R alternatives.

It should be clearly borne in mind that in each particular case the QFD method is applied for a particular purpose. For example, in the case of the Oslo Congress Centre, we are dealing mainly with simple carbonation damage and deterioration, and therefore QFD can be applied for the comparison of MR&R methods which are focused on this single task. Thus in this case, the QFD result will allow us to see better which of the examined methods best take care of the combined technical RAMS and the Generic Functional Requirements. However, in the case of the Central Station, the tasks to be undertaken are much more diverse and complex, and involve particular techniques of cement and chemical grouting, leakage control, carbonation control, coating and mechanical repair. Thus, in this case, the primary objective is not to compare the various methods with a view to choose the best, but rather to gain an overview of how well each of the chosen methods perform in relation to each other and to confirm that no inappropriate methods are used. Here, the QFD should reveal whether any of the applied MR&R methods are so poor compared to the others that they undermine the fulfilment of the Generic Functional Requirements.

To pursue the following with proper understanding, it is first necessary to have read Section 3.7.7.2, which gives explanations of some perhaps unfamiliar concepts, and which sets forth the principles underlying the construction of the Quality Tables (see Tables 3.8.20 and 3.8.21).

The QFD method entails first assigning numerical weights to the RAMS characteristics of materials and products involved in MR&R actions to be planned for specific structures or groups of structures, as we have already done in the previous section. Thereafter, the Generic Functional Requirements which are to be considered are listed, their relative importance estimated and used to adjust the dependencies of these requirements on RAMS characteristics in general. Finally, the so weighted values are used in conjunction with the specific, numerical, technical RAMS weighting values (RAM_T) are combined measures of how well the alternatives both perform technically, and satisfy the Generic Functional Requirements.

3.8.4.1 Determination weighting factors of RAMS characteristics

In both the examples, the Lifecon Generic Functional Requirements [4], presented in the modified list in Section 3.7.7.1, have been used in the QFD reductions. Evaluating the relative dependencies, D, of these requirements may at first sight seem to be a difficult task; dependencies between RAMS and Generic Functional Requirements may not be immediately apparent. However, considerable reflection will usually lead to the realisation that there are indeed dependencies. For example, it may seem that the availability of a product has no effect on energy use, health or recycling. However, if long lead times result in temporary or emergency repairs (stop-gap measures) which later have to be redone, then these will entail extra work, extra hazards, extra energy use and extra pollution. One way to avoid such repairs is to maintain a buffer store to ensure that products are always on hand, but this in itself requires extra resources. It is important, and later very helpful for reference purposes, to note the reasoning behind the dependency assignments. This has been done for the Oslo Congress Centre in Tables 3.8.11, 3.8.12, 3.8.13 and 3.8.14, and for the Central Station in Tables 3.8.15, 3.8.16, 3.8.17 and 3.8.18. Each set of tables has been filled in by a different person, and thus the wording and reasoning vary between the two cases, principally due to the differences in the type of structure.

The dependencies are then transferred to the 'D' columns of Tables 3.8.20 and 3.8.21 for the Oslo Congress Centre and the Central Station respectively.

If we now multiply a RAMS dependency (D) by the relative importance of its corresponding Generic Functional Requirement (W), we create a weighted RAMS dependency, i.e. a dependency which has been adjusted for the relative importance of the corresponding Generic Functional Requirement. In this way, the right-hand part of the table bearing the legend 'Weighted RAMS dependencies' is populated. Having thus populated the upper parts of the Quality Tables, it remains to sum the weighted RAMS dependencies and enter the results into the bottom-most rows of the upper parts of the two tables.

Table 3.8.11 Assignment of Reliability dependencies on generic RAMS with reasoning (Oslo Congress Centre)

Generic Functional Requirement	D	Reason for dependency evaluation: Reliability
Functionality	9	Functionality depends greatly on reliability
Health	3	Early failure lead to premature repair and more noise, dust, stress, etc.
Safety	9	Reliable façade gives safety to the public
Comfort	9	Reliable façade gives public comfort
Investment	9	Reliability has a decided influence on investments
Building costs	9	Reliability has a decided influence on costs
LCC	9	Increased reliability tends to lower LCC
Life style	9	Frequent repair detracts the urban life style
Building tradition	9	Influences the choice of the repair method
Aesthetics	9	Frequent failures are detrimental to the appearance
Architecture	9	The higher the building, the higher the required reliability of MR&R
Image	9	Frequent MR&R detracts the public's image of the building
Energy use	9	Energy consumption increases with repair frequency
Pollution	9	Pollution increases with repair frequency
Biodiversity	3	Reliable repairs give less disturbance to nesting birds
Recycling	6	Frequent repairs generate more waste
LCE	9	Frequent repairs detract LCE

Table 3.8.12 Assignment of Availability dependencies on generic RAMS with reasoning (Oslo Congress Centre)

Generic Functional Requirement	D	Reason for dependency evaluation: Availability
Functionality	9	Short lead times to give reduced disturbance
Health	3	Long lead times would perhaps give increased exposure to health hazards
Safety	9	Long lead times can increase hazards
Comfort	9	Short lead times increase the comfort of the building
Investment	9	Short lead times improve investment economy
Building costs	9	Short lead times reduce costs
LCC	3	Short lead times have only a slight influence on LCC
Lifestyle	9	Short lead times decrease time to reversion of urban lifestyle
Building tradition	9	Short lead times are the expected. Departures cause trouble
Aesthetics	6	Short lead times avoid stop-gap measures which could give less pleasing results
Architecture	3	Lead time has only a minor influence on the architecture
Image	9	Short lead times improve the public image (things happen faster)
Energy use	6	Long lead times tend to increase energy consumption
Pollution	3	Lead times have only a minor influence on pollution
Biodiversity	1	No dependency perceived
Recycling	1	No dependency perceived
LCE	3	Lead times have only a minor influence on LCE

Table 3.8.13 Assignment of Maintainability dependencies on generic RAMS with reasoning (Oslo Congress Centre)

Generic Functional Requirement	D	Reason for dependency evaluation: Maintainability
Functionality	9	Maintainability has a direct bearing on functionality
Health	9	Easy maintenance reduces health hazards
Safety	9	Easy maintenance reduces health hazards
Comfort	9	Easy maintenance increases public comfort
Investment	9	Maintainability has a direct influence on investment economy
Building costs	9	Maintainability influences costs
LCC	9	Maintainability has a direct bearing on LCC
Life style	6	Easy maintenance detracts less the urban life style
Building tradition	9	Traditional maintainability is expected. Departures could be problematic

Aesthetics	9	Easy maintenance gives less disturbance to the appearance
Architecture	9	Easy maintenance better preserves architectural detail
Image	6	Easy maintenance detracts less the public's image of the building
Energy use	6	Maintainability influences energy consumption
Pollution	3	Maintainability has a slight influence on pollution
Biodiversity	1	No dependency perceived
Recycling	3	Maintainability does not have a strong influence on recycling
LCE	6	Maintainability has an effect on LCE

Table 3.8.14 Assignment of Safety dependencies on generic RAMS with reasoning (Oslo Congress Centre)

Generic Functional Requirement	D	Reason for dependency evaluation: Safety
Functionality	9	A well-functioning façade is safe
Health	9	A safe façade does not impair health or cause injury
Safety	9	A safe façade does not present any hazard
Comfort	9	The public comfort is heightened by the safety of the façade
Investment	9	Safety influences the investment economy, but not necessarily positively
Building costs	9	Safety tends to increase costs
LCC	9	Safety influences LCC, but not necessarily positively
Lifestyle	9	Safety is perceived to Improve the urban life style
Building tradition	9	Traditionally, façades are expected to be safe
Aesthetics	1	No dependency perceived
Architecture	9	Architectural features must not present hazards
Image	9	The public's image of the building depends on its safety level
Energy use	9	Safety influences energy consumption, but not necessarily positively
Pollution	9	Safety influences pollution, but not necessarily positively
Biodiversity	1	No known effect
Recycling	9	Safety influences waste production, but not necessarily positively
LCE	6	Safety has an influence on LCE, but not necessarily positively

Table 3.8.15 Assignment of Reliability dependencies on generic RAMS with reasoning (Central Station)

Generic Functional Requirement	D	Reason for dependency evaluation: Reliability
Functionality	9	Functionality is highly dependant on reliable MR&R
Health	9	Public health and health-related need high reliability of MR&R
Safety	9	Public safety requires dry, reliable facilities
Comfort	9	Public comfort highly dependant on reliable MR&R
Investment	9	Reliability has a strong influence on investment economy
Building costs	9	Strongly related to reliability
LCC	9	Strongly related to reliability
Lifestyle	9	Urban lifestyle depends on reliable sub-ways
Building tradition	9	Sub-ways are traditionally expected to have high reliability
Aesthetics	9	Low reliability impairs the aesthetic experience
Architecture	3	Low reliability impairs architectural details when repairs are frequent
Image	9	Low reliability detracts greatly the public image of the station
Energy use	6	Energy consumption is influenced by reliability
Pollution	6	Pollution is adversely affected by dampness, bacterial and fungal growth, etc.
Biodiversity	1	No known relation
Recycling	6	Reliability directly affects production
LCE	6	Reliability influences the results of LCE analysis

Table 3.8.16 Assignment of Availability dependencies on generic RAMS with reasoning (Central Station)

Generic Functional Requirement	D	Reason for dependency evaluation: Availability
Functionality	9	Availability of repair materials is crucial to functionality – shortens MR&R time
Health	6	Short lead times relieves stress and extra labour
Safety	3	Short lead times give better repairs – decreases stop-gap measures
Comfort	9	Short lead times increases public comfort by decreasing MR&R time
Investment	3	Lead timing has only a weak influence on the investment economy
Building costs	3	Lead timing has a little influence on costs (stop-gap measures avoidance)
LCC	3	Little influence

Lifestyle	9	The shorter the lead time the better the MR&R which improves urban lifestyle
Building tradition	9	Short lead times are normally expected. Departures from this can be difficult
Aesthetics	3	Long lead times may affect choice of products
Architecture	1	No influence perceived
Image	3	Shortened MR&R and correct choice of products improves the station's image
Energy use	6	Energy use increases energy consumption (stop-gap measures, buffer storage, etc.)
Pollution	6	As for energy use
Biodiversity	1	No influence perceived
Recycling	6	Stop-gap measures due to long lead times increases waste production
LCE	6	Influenced by the lead time

Table 3.8.17 Assignment of Maintainability dependencies on generic RAMS with reasoning (Central Station)

Generic Functional Requirement	D	Reason for dependency evaluation: Maintainability
Functionality	9	Functionality depends directly on maintainability
Health	9	The less the maintenance, the less the health hazard
Safety	9	The less the maintenance, the greater the safety
Comfort	9	The less the maintenance, the greater the comfort
Investment	9	Maintainability strongly influences the investment economy
Building costs	9	Maintainability has a strong influence on costs
LCC	9	Maintainability has a strong Influence on LCC
Lifestyle	9	Easy maintenance, or low maintenance improves the urban lifestyle
Building tradition	9	Traditionally, sub-ways are expected to run without problems – low maintenance
Aesthetics	6	Maintainability affects the aesthetic experience (maintenance detracts)
Architecture	3	Maintainability is thought to have only a light influence on the appreciation
Image	9	Maintainability strongly affects the public's image of the station
Energy use	9	Maintainability strongly affects the use of energy
Pollution	6	Maintainability has a decided effect on pollution
Biodiversity	1	No influence perceived.
Recycling	6	Maintainability affects waste production
LCE	9	Maintainability strongly affects LCE

Table 3.8.18 Assignment of Safety dependencies on generic RAMS with reasoning (Central Station)

Generic Functional Requirement	D	Reason for dependency evaluation: Safety
Functionality	9	Depends strongly on safety
Health	9	Depends strongly on the station environment and the state of public facilities
Safety	9	The station's safety and the safety of the MR&R are interrelated
Comfort	9	The comfort of the station and the type and frequency of MR&R are related
Investment	9	The safety level affects the investment economy, but not necessarily positively
Building costs	6	The safety level affects costs, but not necessarily positively
LCC	6	The safety level affects the LCC, but not necessarily positively
Life style	9	Safety is important to the urban life style
Building tradition	9	All MR&R are expected to be safe before, during and after completion
Aesthetics	3	Safety affects the aesthetic experience to some extent
Architecture	3	Safety affects architectural detail to some extent
Image	9	Safety has a direct bearing on the public image of the station
Energy use	9	Safety strongly affects the energy consumption, but not necessarily positively
Pollution	9	Safety strongly affects pollution, but not necessarily positively
Biodiversity	1	No dependency perceived
Recycling	9	Safety strongly affects waste production, but not necessarily positively
LCE	9	Safety strongly affects LCE, but not necessarily positively

On the left of the lower parts of the Quality Tables, methods or materials which can be used for a particular MR&R purpose are listed. The immediately adjacent columns entitled 'Technical RAMS' are populated with the figures attained during the RAMS quantification carried out according to the principles set forth in Section 3.7.4.4. Thereafter, the right hand columns are filled in by multiplying each of the values in the left-hand columns successively by the weighted sums of the RAMS dependencies ($RAMS_D$). The resulting values are the Global Priority Orders and are placed in the right-most bottom columns of the Quality Tables. These values are a measure of to what extent the various repair methods or materials fulfil the Generic Functional Requirements when adjusted for the relative importance of the latter and for their interaction with the general RAMS characteristics – the higher the value, the better the fulfilment. When assigning values to correlations and weighting factors, it is necessary that the underlying ideas are as clear as possible. Definitions are repeated for convenience, though in condensed form, in Table 3.8.19.

Table 3.8.19 Clarification of concepts

Definitions of RAMS characteristics

Reliability	The elapsed time in years before a maintenance action is needed
Availability	The normal lead time in weeks, from firm order until delivery, of goods, services or materials, and/or number of skilled workers and/or amount of necessary equipment and/or the number of qualified contractors
Maintainability and Safety	The need for special procedures, highly specialised equipment, highly trained personnel or highly specialised materials

Definitions of some Generic Functional Requirements

Health	All kinds of effects on health during the manufacturing of materials, operation and use of an object, including its repair, demolition, reuse and recycling. Examples of typical health factors are poisonous or otherwise unhealthy emissions from materials, and allergic reactions to materials.
Safety	Safety relates to a structure as used, and to the safety of the workplace(s) during its construction, and to the manufacture, operation and use of any object, material or equipment required, and to their repair, demolition, reuse and recycling.
Investment economy	The lifetime economy of the investment, including income/cost/investment ratios relating to the investment's general profitability.

3.8.4.2 Evaluation of QFD results

The Global Priority Orders from Quality Tables 3.8.20 and 3.8.21 are reproduced below in descending order:

Oslo Congress Centre
Cathodic protection, rod anodes: 6711
Electrochemical realkalisation: 6422
Mechanical repair: 5910
Cathodic protection, ribbon anodes: 5685
Cathodic protection, mesh anodes: 4028
Cathodic protection, conductive coat: 2978
Cathodic protection, rod anodes: 3545

Central Station
Electrochemical realkalisation: 5919
Mechanical repair: 5758
Wet Spray Concrete: 5745
Cement grout: 5730
Polyurethane coating: 5065
Silicate grout: 4218
Cathodic protection, ribbon anodes: 3043

When evaluating the results of QFD reductions, the first point to note is that the reductions cannot be compared to each other unless they are performed on similar objects for similar purposes. Thus, it makes no sense to compare results for our two cases with each other because the structures are very different, the methods used are either different or have different consequences.

In the case of the Oslo Congress Centre, the QFD method has been applied to a large building, but in a building the deterioration, though perhaps extensive, was simple in character. The QFD reduction could therefore be applied to assess MR&R

Table 3.8.20 The Quality Table for combining Generic Functional Requirements with RAMS – Global Priority Order (Oslo Congress Centre)

Oslo Congress Centre
Combination of technical RAMS characteristics with Generic Functional Requirements

Generic Functional Requirements			Dependence of RAMS characteristics on Generic Functional Requirements (D)				Relative importance of dependencies (W)	Weighted RAMS dependencies (W x D)			
			Reliability	Availability	Maintainability	Safety		Reliability	Availability	Maintainability	Safety
	Human Well-being	Functionality	9	9	9	9	5	45	45	45	45
		Health	3	3	9	9	5	15	15	45	45
		Safety	9	9	9	9	5	45	45	45	45
		Comfort	9	9	9	9	4	36	36	36	36
	Economy	Investment	9	9	9	9	4	36	36	36	36
		Building costs	9	9	9	9	4	36	36	36	36
		LCC	9	3	9	9	4	36	12	36	36
	Culture	Life style	9	9	6	9	3	27	27	18	27
		Building tradition	9	9	9	9	3	27	27	27	27
		Aesthetics	9	6	9	1	4	36	24	36	4
		Architecture	9	2	9	9	4	36	12	36	36
		Image	9	9	6	9	3	36	36	18	36
	Ecology	Energy use	9	6	6	9	3	27	18	18	27
		Pollution	9	3	3	9	3	27	9	9	27
		Biodiversity	3	1	1	1	2	6	2	2	2
		Recycling	6	1	3	9	2	12	2	6	18
		LCE	9	3	6	6	4	36	12	24	24
Weighted sums of RAMS dependencies (RAMS$_D$)							519		394	475	07

Materials or methods under evaluation	Technical RAMS$_T$				Weighted RAMS = RAMS$_D$ x RAMS$_T$					Global Priority Order GPO = Σ (RAMS$_D$ x RAMS$_T$)
	R	A	M	S	R	A	M	S		
Mechanical repair	1	5	4	3	519	1970	1900	1521		5910
El. realkalisation	5	1	3	4	2595	394	1425	2028		6442
CP, rod anodes	4	3	3	4	2076	1182	1425	2028		6711
CP, mesh anodes	3	0	2	3	1557	0	50	1521		4028
CP, ribbon anodes	3	3	3	3	1557	1182	1425	1521		5685
CP, coating anodes	0	0	2	4	0	0	950	2028		2978

Table 3.8.21 The Quality Table for combining Generic Functional Requirements with RAMS (Central Station – Global Priority Order)

Central Station Combination of technical RAMS characteristics with Generic Functional Requirements			Dependence of RAMS characteristics on Generic Functional Requirements (D)				Relative importance of dependencies (W)	Weighted RAMS dependencies (W x D)			
			Reliability	Availability	Maintainability	Safety		Reliability	Availability	Maintainability	Safety
Generic Functional Requirements	Human Well-being	Functionality	9	9	9	9	5	45	45	45	45
		Health	9	6	9	9	5	45	30	45	45
		Safety	9	3	9	9	5	45	15	45	45
		Comfort	9	9	9	9	4	36	36	36	36
	Economy	Investment	9	3	9	9	4	36	12	36	36
		Building costs	9	3	9	6	4	36	12	36	24
		LCC	9	3	9	6	4	36	12	36	24
	Culture	Life style	9	9	9	9	4	36	36	36	36
		Building tradition	9	9	9	9	3	27	27	27	27
		Aesthetics	9	3	6	3	4	36	12	24	12
		Architecture	3	1	3	3	3	9	3	9	9
		Image	9	3	9	9	3	27	9	27	27
	Ecology	Energy use	6	6	9	9	4	24	24	36	36
		Pollution	6	6	6	9	4	24	24	24	36
		Biodiversity	1	1	1	1	1	1	1	1	1
		Recycling	6	6	6	9	3	18	18	18	27
		LCE	6	6	9	9	4	24	24	36	36
Weighted sums of RAMS dependencies (RAMS$_D$)								505	340	517	502

		Technical RAMS$_T$					Weighted RAMS = RAMS$_D$ x RAMS$_T$					
		R	A	M	S		R	A	M	S		
Materials or methods under evaluation	Mechanical repair	4	5	2	2	→	2020	1700	1034	1004		5758
	Wet spray concrete	5	2	2	3	→	2525	680	1034	1506		5745
	Cement grout	5	2	1	4	→	2525	680	517	2008		5730
	Silicate grout	3	2	1	3	→	1515	680	517	1506		4218
	Polyurethane coat	5	0	2	3	→	2525	0	1034	1506		5065
	El. realkalisation	4	1	3	4	→	2020	340	1551	2008	Global Priority Order GPO = Σ (RAMS$_D$ x RAMS$_T$)	5919
	CP, rod anodes	0	2	1	4	→	0	1020	517	2008		3545
	CP, ribbon anodes	0	3	1	3	→	0	1020	517	1506		3043
						→						
						→						
						→						
						→						
						→						
						→						
						→						

methods which more or less serve the same purpose, that is to repair and counteract damage caused by carbonation. Thus, in this case, the QFD reduction gives true, comparable, global measures of the six different ways of curing the carbonation problem. Of the six, it is seen in Table 3.8.20 and 3.8.21, that cathodic protection using drilled-in rod anodes gives the best overall technical result and fulfilment of the Generic Functional Requirements, closely followed by electrochemical realkalisation.

Then follows mechanical repair, with cathodic protection using ribbon anodes in cut grooves as the fourth best. Cathodic protection by means of sprayed-in mesh anodes is the second last, with conductive coating at the bottom of the list.

In the case of the Central Station, the QFD reduction is not used to directly compare methods, since each of the methods examined serves a different purpose or purposes. Here, the QFD reduction allows us to check whether any of the methods chosen are so poor that they undermine the GPO levels attained by the others. The listed GPOs clearly show that had the modern CP methods using rod or ribbon anodes been available at the time of the rehabilitation, they would likely not have been chosen on the basis of QFD reduction since the levels of their GPOs are seriously lower than those of the remaining methods. Using these types of CP would have seriously detracted the overall MR&R result.

Although the GPOs indicate the best overall compromises between technical performance and the Generic Functional Requirements, it may be that certain overriding considerations are more important than the best total compromise. If, for example, Availability is of minor importance compared to Reliability, Maintainability and Safety, then the lower part of the Quality Table could be adjusted by leaving the Availability columns blank. In that case, the Quality Tables 3.8.20 and 3.8.21 become as in Tables 3.8.22 and 3.8.23.

Ignoring Availability has an effect also on the priority order, which is no longer global, but simply a Priority Order ex-Availability (POA). The new orders are thus:

Oslo Congress Centre
Electrochemical realkalisation: 6048
Cathodic protection, rod anodes: 5529
Cathodic protection, ribbon anodes: 4503
Cathodic protection, mesh anodes: 4028
Mechanical repair: 3940
Cathodic protection, conductive coat: 2978
Cathodic protection, rod anodes: 3545

Central Station
Electrochemical realkalisation: 5579
Polyurethane coating: 5065
Wet Spray Concrete: 5065
Cement grout: 5050
Mechanical repair: 4058
Silicate grout: 3538
Cathodic protection, ribbon anodes: 3043

In the case of the Oslo Congress Centre, electrochemical realkalisation takes the first place ahead of cathodic protection using rod anodes when Availability is not an issue, with mechanical repair now coming in the fourth. In actual fact, the centre was repaired using mechanical repair combined with electrochemical realkalisation.

In the case of the Central Station, electrochemical realkalisation retains its first place and would have been seriously considered as a option had this method existed at that time. Polyurethane coating now comes in the second place together with wet spray concrete closely followed by cement grout, which were the three main rehabilitation methods actually used. Silicate grout takes the third last place and was possibly not the best choice of secondary filling material. The two cathodic protection methods, had they been viable at the time of the rehabilitation, would most likely not have been chosen in view of their low ratings.

Table 3.8.22 The Quality Table for combining Generic Functional Requirements with RAMS Priority Order ex-Availability (Oslo Congress Centre)

Oslo Congress Centre / Combination of technical RAMS characteristics with Generic Functional Requirements			Dependence of RAMS characteristics on Generic Functional Requirements (D)				Relative importance of dependencies (W)	Weighted RAMS dependencies (W x D)			
			Reliability	Availability	Maintainability	Safety		Reliability	Availability	Maintainability	Safety
Generic Functional Requirements	Human Well-being	Functionality	9	9	9	9	5	45	45	45	45
		Health	3	3	9	9	5	15	15	45	45
		Safety	9	9	9	9	5	45	45	45	45
		Comfort	9	9	9	9	4	36	36	36	36
	Economy	Investment	9	9	9	9	4	36	36	36	36
		Building costs	9	9	9	9	4	36	36	36	36
		LCC	9	3	9	9	4	36	12	36	36
	Culture	Life style	9	9	6	9	3	27	27	18	27
		Building tradition	9	9	9	9	3	27	27	27	27
		Aesthetics	9	6	9	1	4	36	24	36	4
		Architecture	9	2	9	9	4	36	12	36	36
		Image	9	9	6	9	3	36	36	18	36
	Ecology	Energy use	9	6	6	9	3	27	18	18	27
		Pollution	9	3	3	9	3	27	9	9	27
		Biodiversity	3	1	1	1	2	6	2	2	2
		Recycling	6	1	3	9	2	12	2	6	18
		LCE	9	3	6	6	4	36	12	24	24
		Weighted sums of RAMS dependencies (RAMS$_D$)						519	394	475	507

		Technical RAMS$_T$					Weighted RAMS = RAMS$_D$ x RAMS$_T$					Priority Order ex Availability POA = Σ (RAMS$_D$ x RAMS$_T$)
		R	A	M	S		R	A	M	S		
Materials or methods under evaluation	Mechanical repair	1		4	3	→	519		1900	1521		3940
	Electrochemical realkalisation	5		3	4	→	2595		1425	2028		6048
	CP, rod anodes	4		3	4	→	2076		1425	2028		5529
	CP, mesh anodes	3		2	3	→	1557		950	1521		4028
	CP, ribbon anodes	3		3	3	→	1557		1425	1521		4503
	CP, coating anodes	0		2	4	→	0		950	2028		2978
						→						
						→						
						→						
						→						
						→						
						→						
						→						
						→						

Similar exercises to the above can, of course, be undertaken by ignoring any one or more of the RAMS characteristics to create the corresponding Priority Orders. Thus we see that the QFD method can form the basis of a ranking system which allows not only totalities to be compared, but also comparisons on the basis of single characteristics.

Table 3.8.23 The Quality Table for combining Generic Functional Requirements with RAMS Priority Order ex-Availability (Central Station)

Central Station — Combination of technical RAMS characteristics with Generic Functional Requirements			Dependence of RAMS characteristics on Generic Functional Requirements (D)				Relative importance of dependencies (W)	Weighted RAMS dependencies (W x D)			
			Reliability	Availability	Maintainability	Safety		Reliability	Availability	Maintainability	Safety
Generic Functional Requirements	Human Well-being	Functionality	9	9	9	9	5	45	45	45	45
		Health	9	6	9	9	5	45	30	45	45
		Safety	9	3	9	9	5	45	15	45	45
		Comfort	9	9	9	9	4	36	36	36	36
	Economy	Investment	9	3	9	9	4	36	12	36	36
		Building costs	9	3	9	6	4	36	12	36	24
		LCC	9	3	9	6	4	36	12	36	24
	Culture	Life style	9	9	9	9	4	36	36	36	36
		Building tradition	9	9	9	9	3	27	27	27	27
		Aesthetics	9	3	6	3	4	36	12	24	12
		Architecture	3	1	3	3	3	9	3	9	9
		Image	9	3	9	9	3	27	9	27	27
	Ecology	Energy use	6	6	9	9	4	24	24	36	36
		Pollution	6	6	6	9	4	24	24	24	36
		Biodiversity	1	1	1	1	1	1	1	1	1
		Recycling	6	6	6	9	3	18	18	18	27
		LCE	6	6	9	9	4	24	24	36	36
Weighted sums of RAMS dependencies (RAMS$_D$)								505	340	517	502

		Technical RAMS$_T$					Weighted RAMS = RAMS$_D$ x RAMS$_T$					
		R	A	M	S		R	A	M	S		
Materials or methods under evaluation	Mechanical repair	4		2	2	→	2020		1034	1004	Priority Order ex Availability POA = Σ(RAMS$_D$ x RAMS$_T$)	4058
	Wet spray concrete	5		2	3	→	2525		1034	1506		5065
	Cement grout	5		1	4	→	2525		517	2008		5050
	Silicate grout	3		1	3	→	1515		517	1506		3538
	Polyurethane coat	5		2	3	→	2525		1034	1506		5065
	Electrochemical realkalisation	4		3	4	→	2020		1551	2008		5579
	CP, rod anodes	0		1	4	→	0		517	2008		3545
	CP, ribbon anodes	0		1	3	→	0		517	1506		3043
						→						
						→						
						→						
						→						
						→						
						→						
						→						

Further, by changing the list of Generic Functional Requirements or by changing their relative importance, the QFD method can be made to take cognisance of, and lay importance upon, aspects of any kind – political, environmental, artistic or otherwise.

Acknowledgement

The author wishes to thank his son, Iain H. B. Miller of Millab Consult a.s., whose support, criticisms and suggestions greatly accelerated the production of this section.

References

[1] Miller, I. H. B. 2000. Kongressenteret Folkets Hus a/l, Sikring av fasader, tilstandskontroll, prøvedata. Report no 1-11-00, Millab Consult a.s., Oslo, Norway (in Norwegian).
[2] Miller, J. B. 1987. Rehabilitation of Sentrum Stasjon (Parliament Station). Contract no. 20002, 1982–1987, Norsk Teknisk Byggekontroll a/s (Noteby), Oslo, Norway.
[3] Lifecon deliverable D2.3, Quality Function of Deployment Index QFD by Asko Sarja. http://lifecon.vtt.fi/.
[4] The Lifecon project (Life cycle management of concrete infrastructures for improved sustainability). 1998–2002. Project funded by the European community under the competitive and sustainable growth programme. http://lifecon.vtt.fi/.

Chapter 4

Condition assessment protocol

Sascha Lay

4.1 List of terms, definitions and symbols

AAR	Alkali-aggregate reaction
ASR	Alkali–silica reaction
CA	Condition assessment
CAP	Condition assessment protocol
CR	Condition rating – Inspection-based assignment of the condition to pre-defined service life stages which are expressed by discrete numbers on a given scale
Damage	An unfavourable change in the condition of a structure that may affect structural performance [1]
Extent of damage	To be expressed by physical units such as size (surface area, etc.), amount, depth, geometry (e.g. crack pattern)
Inspection	On-site examination to establish the present condition of the structure [1]
Investigation	Collection and evaluation of information through inspection, document research, load testing and other testing [1]
LMS	Life cycle management system
Model	Mathematical formulation which describes a certain measure as a function of various influences
MR&R	Maintenance, repair and rehabilitation
Symptom	Indicator of the condition of an item [2]
Target reliability level	Level of reliability required to ensure acceptable safety and serviceability

4.2 Framework of condition assessment

4.2.1 Principles

Condition assessment protocol (CAP) aims to incorporate models for the prediction of the future state in the current concept for condition assessment (CA) [3]. Default data for parameters in existing models can and should be updated by obtaining new information from inspections. To obtain suchlike data for continuous model building, the current phases and levels of inspections are reorganised and extended.

Structures are usually only inspected once, before 'in use' condition. However, predictions of models as used for durability design can be improved by collecting data immediately after the completion of the structure. This phase is here called 'acceptance' and is an addition to the current approach. The same accounts for inspections performed during and after repair works. Planning is essential prior to any inspection regardless of the level of sophistication (visual, general or structural assessment). The planning phase is thus described separately regardless of the inspection phase or level. Every subsequent level includes the actions taken in the previous one; for example, a visual inspection is always performed before starting the general inspection. The result of a CA of single objects must be transformed into a discrete rating system to be of use on a network level (set of objects). Condition rating (CR) is performed during each inspection phase irrespective of the inspection level and is thus also discussed in Section 4.6.

4.2.2 Phases and levels of inspections

A classification of the phases and levels of inspections is presented in Table 4.1 and is explained in detail in this section.

Acceptance inspections: Recently built structures can be checked with respect to the compliance of the target performance. These structures will, in general, not yet exhibit any deterioration due to environmental action. However, geometrical parameters (e.g. concrete cover) and material quality (e.g. chloride migration properties) should already be gathered as an input to service life models for future evaluations to the largest feasible extent. The mere measurement of the dominating parameters after construction is here referred to as 'birth certificate'. This concerns especially those components which will exhibit restricted possibilities for access once the structure is in use. This step may and in fact should be considered as part of the quality assurance concept. In the quality assurance of today data for condition predictions are only rarely collected. Usually mere visual inspections are conducted at this stage. Therefore an acceptance inspection, which serves to provide the basis for future condition predictions, is considered as an innovation compared to the current practice.

In-use inspections: They are performed during the regular service life of a structure. In-use inspections may be performed at various levels:

- *Visual inspections* are routine inspections and are performed at regular intervals. These are today's common practice and mainly serve for the identification of obvious defects without the application of further inspection tools.
- *General inspection* is also known as 'regular' or 'principal' inspections [4]. In addition to visual inspection, *destructive* and *non-destructive testings* are conducted. In this chapter, the current concept is extended by a statistical evaluation of the collected data and samples in combination with the application of service life models for planning of future inspections.
- *Structural assessment* is called for when there is concern over the structural safety as a result of degradation or any other reason as is common practice too.

Table 4.1 Summary of inspection types

Phases of structure	Level of inspection	Frequency	Purpose	Skills (Duration)	Possible actions
Acceptance	General	Once, before 'in-use' condition	− calibration of parameters in service life models (durability design models) − certification of performance.	Engineer (h to d)	If performance according to service life calculations is not met, additional protective measures are required.
In-use	Visual	1 to 3 years	detection of obvious defects; especially important to detect unexpected defects due to irregularities during fabrication, unreported accidents or misuse of a structure	Technician (h)	− Short-term repair on non-structural defects − Call for general inspection
	General	Flexible, according to development of condition state	− determination of condition state − calibration of parameters in service life models	Engineer (d)	− Set date for next inspection − Call for repair action − Call for structural assessment
	Structural assessment	On demand	assessment of structural safety (static calculations, FEM analysis)	Engineer (d to w)	− Call for strengthening and repair − Temporary load restriction
Repair	General	During repair action	− adaptation of MR&R extent − calibration of service life models − certification of performance	Engineer (d)	− Call for additional actions, if performance requirements are not fulfilled

Note
h − hours; d − days; w − weeks.

In-use inspections may affect the usability of a structure, e.g. traffic lanes may have to be closed. An assessment may be initiated under the following circumstances [5]:

• structural deterioration due to time-dependent actions (e.g. corrosion);
• change in use or extension of design working life; and
• reliability check (e.g. in case of earthquakes and increased traffic action).

Repair inspections: The purpose of repair inspections is comparable to that of the 'acceptance', but it has a larger scope. If a structure was foreseen to repair measures, these should be accompanied by inspections to (a) adapt the actually necessary repair extent, which was only based on localised inspections, (b) ensure that the demanded quality level is achieved and (c) collect data for the prediction of the future behaviour in the repaired condition (e.g. concrete cover or concrete quality may have been changed). Repair inspections are thus on the level of a general inspection.

Every investigation regardless of the level consists of three basic phases:

1. planning
2. inspection
3. processing of data.

'When is a higher inspection level appropriate?'

The main idea is to start with a low inspection effort (amount of samples, inspected surface portion) and with basic investigation methods which will be increased or become more sophisticated if intermediate results suggest so, according to the criteria in Table 4.2 [6].

The applied models include parameters which have been quantified in a statistical manner using data from either prior structure investigations or laboratory tests. It is evident that such models only reflect the as-built condition to a certain degree.

Table 4.2 Examples for criteria demanding for a more intensive inspection

Criterion	Example
Damage cause unknown	cracks in the superstructure
Susceptibility	Similar structures/components suffered from known but possibly hidden problems
Damage of unknown extent	chloride ingress, corrosion of re-bars
Damage of large extent	evaluation of secondary effects (significant reduction of safety margin index)
Damage which cannot be sufficiently investigated by conventional (visual) methods	cracks due to corrosion or ASR
Development of damage, which deviates strongly from expected values	chloride content or amount of cracks due to ASR increase very fast and deviate from prognosis
Damage development unknown	collection of necessary input data for service life models
Residual service life seems not sufficient	high variance of data → increase of sample amount models are too simple → use sophisticated models
High extent of statistical uncertainty	if the uncertainty due to the scatter of data can be reduced by additional sampling, such samples should be collected to an extent for which the additional inspection costs do not yet balance out the hereby-achieved benefits

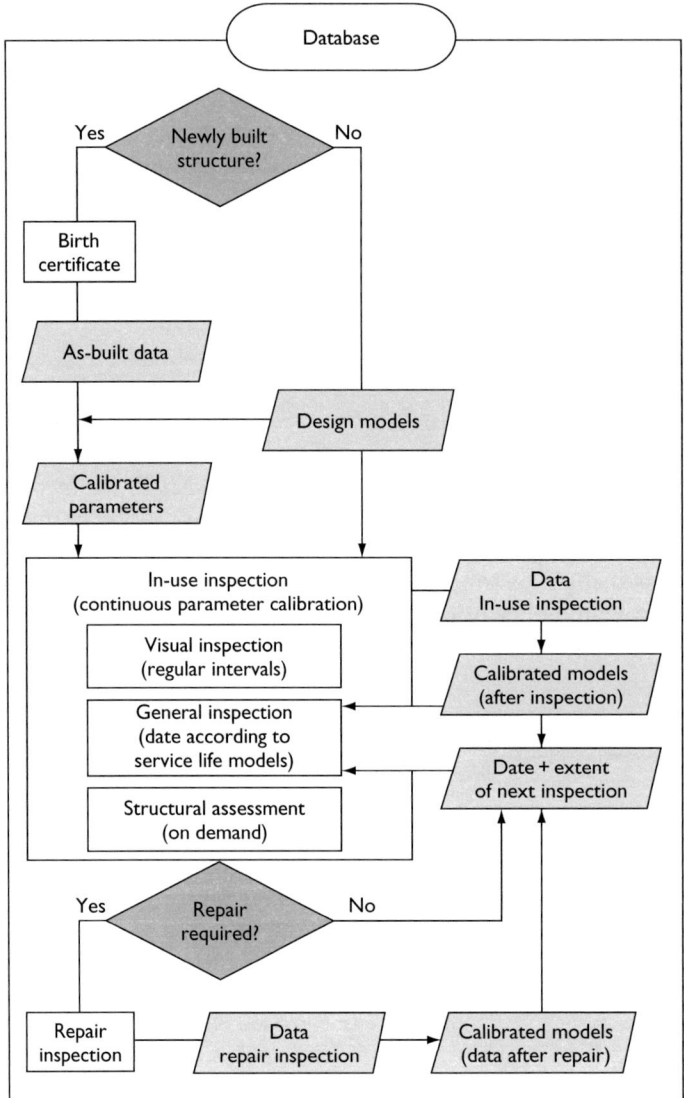

Figure 4.1 Basic framework for a condition assessment.

The first time that assumed input parameters can be compared to those measured in the as-built condition is directly after construction. A distinction must be made between parameters which depend on the environmental conditions and those which are independent of the exposure. Those parameters depending on the environmental conditions reflect the response of the material (concrete) to the environmental exposure (e.g. carbonation depth). As a consequence, such environment-dependent parameters cannot be measured directly after construction, as a certain exposure time is required to obtain a measurable response. However, those parameters which are independent of the environment and can thus usually be regarded as time-independent (e.g. concrete cover, if no abrasion occurs) can and should be measured after construction as soon as possible to check whether the demands are fulfilled. As most parameters are subjected to spatial scatter, sufficient measurements must be taken to express the parameter as a statistical distribution. The statistical distribution of the regarded parameter assumed before construction is thus updated by the measured results in the real world. To do so, reliability theory techniques are applied, which are here referred to as 'Bayesian Updating'. Most of today's structures in service were not designed using durability design models. Nevertheless, these durability design models can also be applied to structures already in service to obtain an estimation on their time-dependent degradation. The parameters in these models should be based on data representing the as-built condition.

Inspections performed during the service life will provide further data, which can be used for parameter updating. This is a continuous process. It must however be realised that visual inspections commonly provide information on the degradation state at a rather late stage of the degradation. Parameters which can be used in service life models are thus usually based on testing on site or in the laboratory with destructive or non-destructive techniques.

Models with calibrated parameters can be used for various purposes. First of all they provide the basis for the decisions on the best strategy for maintenance, repair and rehabilitation (MR&R) works. Moreover, they can be used to schedule the time and the extent of the next inspection in a more cost-effective fashion.

The whole process of CA is connected to a database. Available data on a single object or a group of objects may be recalled once collected and inserted. The basic flow of a CA is indicated in Figure 4.1.

4.3 Inspection planning

4.3.1 Aims and procedure

Planning of inspections makes use of all available information of a structure including judgements based on engineering understanding and the experiences gained from comparable structures. Planning aims to answer the following questions to inspect:

- What?
- How?
- Where?
- How often?

Although inspections are necessary to control the degradation of structures, they do cause significant direct costs. In consequence, inspections need to be planned carefully. A balance must be achieved between the reduction of service life costs due to the gathering of information and the hereby optimally scheduled damage-preventing maintenance on the one hand and the direct costs involved in the inspections on the other. This may imply more effort before starting the actual inspection, compared to today's practice, but will in the long run save costs for (a) unnecessary investigations and (b) repair at later stages due to lack of information at an earlier time, when simple maintenance would have solved the problem.

This chapter will primarily deal with the planning of inspections, assuming that we know 'what' we are looking for. Local experience of the performance of structures executed in comparable manner (type of cement, water–cement ratio, concrete cover, producer, etc.) are helpful indicators to come up with an idea of what kind of irregularities may be incorporated into the study.

However, a certain portion of inspections should always search for unexpected failure states and deterioration mechanisms, such as may be caused by irregularities during fabrication, unreported accidents or misuse of a structure.

Planning is considered to consist of a very brief preliminary site inspection, followed by an intensive desk-study in the office. The purpose is to provide all the information needed for the later on-site inspection. The chronological flow of necessary decisions, actions with corresponding tools and the evolving results is outlined in Figure 4.2.

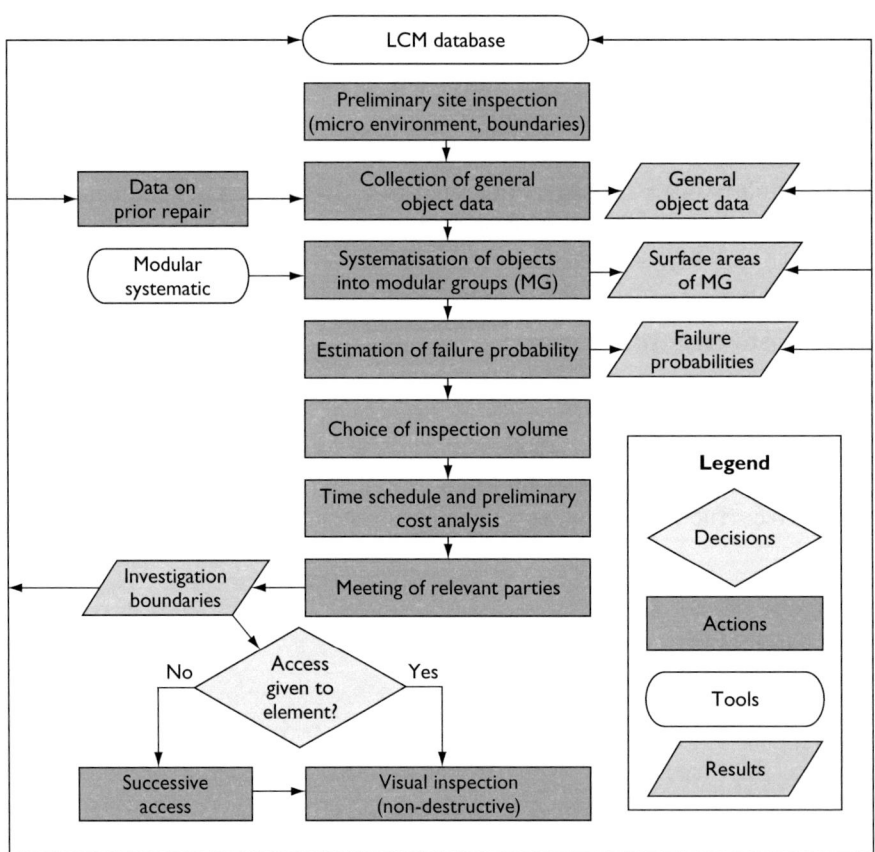

Figure 4.2 Planning of a condition assessment.

4.3.2   First desk study

The first step in the planning process is to get familiar with an object. This is achieved by looking into existing documentations, usually drawings, giving an overview of the object, representing the 'as planned' state. Important questions are:

'What does the object primarily serve for?'
'What are the main paths of loads?'
'Which environmental loads may affect the object?'
'Where are areas of distress to be expected?'
'What consequences could arise from failure events?'

4.3.3   Preliminary site inspection

Preliminary site inspection is suggested for the purpose of gaining a quick overview of a structure especially in respect to details for the in-depth inspection, which may not be obvious from a desk study. This common approach is important for investigators who are not familiar with an object. However, if necessary travelling costs will be high, a preliminary site inspection may not be appropriate. In this case, a first overview is obtained before starting with the actual inspection work, to verify that the inspection effort as planned in the office is suitable.

'Are the data gathered during the first desk study correct?'
   Geometry, materials and, especially, environmental loading identified during the first superficial desk-study have to be verified visually. If the expected conditions do not coincide with the prior assumptions, these need to be documented as a basis for inspection planning.

'What are the micro-environmental loads?'
   Unless the investigators are already familiar with a structure, they should visit the site before starting with the actual investigation work. The main objective is to determine the environmental loads on a micro-level, i.e. at the concrete surface, which is hardly possible by solely performing a desk study. An example for this might be an abutment above a leaching joint far away from traffic.
   The concrete surface would be considered as dry in a desk study, because the abutment is sheltered from rain and is situated far away from traffic. Chloride contact would be regarded as negligible. But chloride-contaminated water from the defective joint will cause severe chloride exposure. A list of such surface areas exposed to environmental loading, which is not derivable from a desk-study, must be prepared on site. The loading will be rated during the later general investigation.

'What are the boundaries for the inspection of the object?'
   The most important aspects to check for are:

*   possibilities of access to components;
*   need for special equipments (lights, etc.);
*   time restrictions for surveys (e.g. only during day- or night-times due to traffic service; working hours of auxiliary personnel, etc.); and
*   appropriate safety measures and equipment during inspection (traffic postings).

4.3.4 Collection of object data

In a second desk study, all information for inspections are gathered. A large quantity of relevant information are contained in design and construction documents. Possible sources for object data may be:

- design documents (drawings, calculations and structural models);
- as-built documents (drawings, specifications);
- inspection reports (photographs, multimedia tools such as digital video);
- operation reports;
- maintenance reports; and
- manufacturers' technical information.

Using these sources, the amount of data to be gathered in cost-intensive field studies can be reduced. The annex provides a default list of data which may be collected and stored for the regarded object. Object data are divided into three categories:

1. administrative data;
2. technical data; and
3. inspection (including repair) data.

The necessary administrative data are usually defined by the owner of an object. Certain data relate to the entire structure. However, as a structure should be divided into various components according to a predefined system ('modular system'), the data should be assigned to these components when stored in a database. A data storing system must account for the fact that certain data will never change, whereas others are variable during the service life. Table 4.3 shows the types, groups and examples of object data.

4.3.5 Prior inspection results

Elements of different structures will behave differently due to the differences in concrete composition, geometrical set-up and micro-climate. In accordance with this concept, it is evident that every structural element must be treated separately. The mathematical equation (model) as such will be the same (e.g. equation for chloride ingress), but the values for the parameters in this equation will be different for elements of different structures.

If no or few data from previous inspections are yet available, default data can be used for predictions. Such default data may be taken from comparable elements of other structures. When using default data, representing a certain type of element in a certain environment, it must however be taken into consideration that the real behaviour of a single element may considerably deviate from the expected results.

If results of previous investigations at the specific structure do exist, in any level (visual, general, structural assessment), they should be extrapolated from appropriate 'service life models' to the current date before performing the inspection (if this has not already been done during earlier inspections to estimate the date for future

Table 4.3 Types, groups and examples of object data

Type	Data group	Example
Constant	Identification	principal and name structure file location in network or road cartographic co-ordinates graphical data
	Administration	authority: design, inspection, maintenance and management dates: construction, modification and demolition contractors and location of contracts
	Technical description	dimensions modular systematic surface areas of homogenous lots graphical data equipment
	Construction	contractors applied construction techniques costs
	Traffic	lanes: number, width and, clearance class for heavy vehicles or military traffic volume
Variable	Inspection	restrictions during inspection checklist for components to be inspected duration of inspection equipments needed inspection data (raw data of tests) causes of damage CR distribution residual service life main conclusion on structure condition costs for inspection date of next inspection
	MR&R	work executed condition of accepted works contractors costs

inspections or MR&R actions). These extrapolations give an indication as to whether a certain inspection (e.g. chloride profiling) should take place once over, because the next condition state is expected to be reached.

4.4 Modular systematic in condition assessment

4.4.1 Requirements for modular systematic

A system to (a) divide concrete structures into homogenous zones and grouping these into lots and (b) quantify surface area per homogenous lot is necessary for a life cycle management system (LMS) as a basis for cost estimation. Such a system is here referred to as the 'Modular Systematic' [7]. The requirements for this system are:

- Comprehensive – the system must be logical to the majority of users.
- Generic – It should be applicable to different types of structures, e.g. bridges, tunnels, buildings, wharfs, etc.
- Variable – It must be suitable for a range of applications, such as CA, life cycle analysis, repair and cost control.

The way in which a structure may be divided into structural items is already defined in national guidelines. As the responsible authorities are already used to these guidelines, existing structures have already been sub-divided and the resulting divisions were already partly incorporated into IT tools. It is hardly intended to change national procedures (nor this would be likely to succeed) when it comes to the Modular Systematic. Nevertheless, most of the current approaches have decisive shortcomings if predictions of future deterioration states are to be integrated. This is because predictions should only be done on macroscopic homogenous zones. Whatever system and level of detail is applied to obtain homogenous zones, existing systems should in any case be extended by two essential steps:

1. Division into surface zones fabricated of *comparable resistance towards degradation* due to environmental action. This means equal material (e.g. concrete composition: mainly equal cement type and water–binder ratio) and geometric resistance (e.g. average concrete cover). The structural capacity (dimensions, reinforcement degree) is not taken into consideration at this stage, as this would lead to an enormous amount of surface zones.
2. These zones are then sub-divided into areas with *comparable environmental loading*. Division into such homogenous zones is not usually obtained by existing systems. The concept of a generic system, applicable to various types of structures obeying the two essential steps mentioned, will be outlined in the following sections.

4.4.2 Modular division

The generic set-up of the proposed framework enables its application to all kinds of structures. This chapter distinguishes six levels of detail which divide the complete object (level 1) in a logical way into modules (level 2), components (level 3), sub-components (level 4), surfaces (level 5) and inner layers (level 6).

Levels 1–4 serve mostly organisational purposes and may be changed or skipped. The most important level for the future working processes is level 5, which represents surface areas with equal resistance and loading with respect to a particular deterioration mechanism. This is because:

- surfaces are the interface between the environment and the structure; and
- the surface area of homogenous zones is required for the prediction of the future condition state, which serves as the basis for the estimation of repair costs.

For these areas one or more potential deterioration mechanisms with possibly observable deterioration indicators will be identified and listed at a later stage. A division into such surface zones should thus in any case be achieved. If desired, level 6 may also be introduced, which divides homogenous lots into inner layers starting at the surface. The division is best organised into a tree, which can be inserted into a database (Figure 4.3).

Level 1: Object   Classification of the complete structure according to its **use**, e.g. bridge, tunnel, building, wharf, etc.

Level 2: Module   The classification criterion is the logical set-up of the structure, which usually yields the order of production (bottom to top). These main modules divide a structure into largest units. A main module is a generic term comprising certain components. A main module can have several, varying functions and is made up of different materials. For the example of a bridge or a wharf, these may be foundation, sub-structure and super-structure. A building may be divided into foundation, floors and roof.

Level 3: Component   A component fulfils a specific function as a unit but can consist of different materials. For each component, the function within the whole object must be identified, as the loss of this function is a consequence which has to be considered in the management process. Such functions may be:

- load bearing (e.g. pillar, abutments, walls);
- functional safety (e.g. guide board, hand rails);
- performance (e.g. expansion joints);
- preservative measures (e.g. roofing);
- protective measure for environment (e.g. noise barrier); and
- aesthetic appearance (e.g. covering of a building).

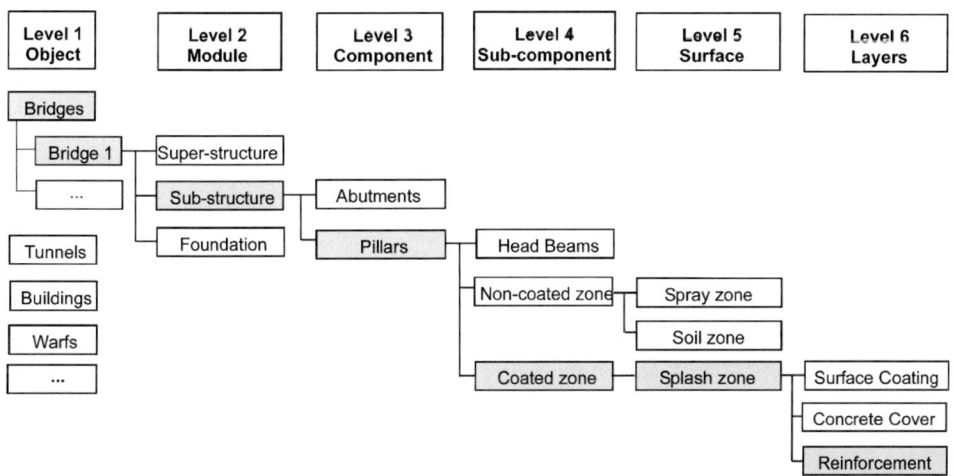

Figure 4.3 Example for 'Modular Systematic'.

Level 4: Sub-component (resistance) The target is to identify sections with equal resistance towards environmental actions at the surface and of the inner layers. Sub-components consist of the same materials (coating, concrete composition) and have a comparable structural design in common (concrete cover, degree of reinforcement). An example is the distinction between the shafts of a row of pillars and the connecting head beams, which may have been produced with different concrete strength class (different water–binder ratio) and concrete cover. If the bottom portion of the pillar shafts (spray zone) is coated, as is a common practice in the road environment, a distinction between coated and non-coated shafts would be required.

Level 5: Surface (environmental loading) Sub-components are divided according to surface zones, which are exposed to the same environmental influences and/or stresses. To identify possible classes of environmental loading, the user is referred to [8] which is an expansion of the concept offered by the European standard EN 206 [9]. In the example of a row of pillars, areas within different exposure zones may be distinguished. In the road environment, pillars may be exposed to the soil zone, splash zone and spray zone. In a marine environment, portions of a pillar may be in the submerged, tidal, splash or atmospheric zone. In these zones, a differing corrosion behaviour of the reinforcement is to be expected.

Level 6: Inner layers Inner layers of a surface are regarded to enable the unique assignment of the location of damage within the concrete. Identification of layers parallel to the surface is aimed at. An example is: surface coating, concrete cover, reinforcement.

4.4.3 Hot spots

'Are any critical zones within macroscopic homogenous groups?'
  The approach of dividing

* modules into components (level 3 → level 4) having equal resistance (surface protection, concrete quality, concrete cover) and
* components into surfaces (level 4 → level 5) subjected to equal loading (static and environmental)

will only lead to homogenous groups, if the two criteria (environmental resistance and load) are constant within the regarded zone. However, 'hot spots' may be found on a micro-level within a regarded zone.

'Are areas with a loading intensity above the average present?'
  An example is a parking garage: Huge costs emerge from MR&R actions due to chloride-induced corrosion. Much higher chloride contents can be found in these areas of the decks when compared to neighbouring ones. If the parking deck is inclined, chloride-containing water may accumulate in low points. The same is true in the road environment, where the chloride salt spray intensity is a function of distance from the chloride source. The environmental loading is thus higher in these 'hot spots' when compared to the average level of the entire zone.

'What is the effect of a combination of deterioration mechanisms?'

In general, a single mechanism can be identified as the dominating cause of deterioration. This is because some mechanisms act mutually exclusive. This is, for example, the case for carbonation and chloride-induced corrosion, because chlorides are brought to the concrete surface dissolved in water, pores will be filled with water, which in turn prevents CO_2 to penetrate. Nevertheless, some mechanisms may also act in a synergistic manner. An example is frost scaling, which reduces the concrete cover and chloride penetration at the same time. In total, a more rapid corrosion induction will take place. If areas with synergistic mechanisms are present, these must be documented separately.

'Are areas with lower deterioration resistance observable?'

Resistance towards ingress of corrosion-inducing media controlled by the material properties (surface coatings, concrete composition) and geometry (concrete cover) can usually be considered as constant at a macro-level within the boundaries of certain sections. But structural stress (flexural moment, shear force) will change as a function of site $M(x)$ and $V(x)$. In the example of a parking deck, cracks may often be found in the edges of plates or over the supports. The resistance towards chloride ingress is thus reduced, compared to uncracked areas. Hence, those areas where degrees of deterioration above the average can be expected, should be distinguished. This requires that the investigator should have a knowledge about deterioration mechanisms and awareness of the static system of the object. In general, 'hot spots' are regarded to be areas:

- with high risk of water ingress (e.g. expansion joints, roots of railings, drainage system, low points of inclined surfaces loaded with corrosive media, etc.);
- where yield points may form in a potential collapse (e.g. end supports, midspan, regions over intermediate supports, construction joints transverse to a post-tensioned cable and duct); and
- weak details (e.g. joints of sealers, especially in the area of kerb lines).

The existence and extent of 'hot spots' is not usually known before the first inspection has taken place, and may remain unknown (because they are not visually accessible) until access has been provided. Such 'hot spots' can and must nevertheless be accounted for in the modular systematic at levels 4 and 5, already during the planning process (Figure 4.4).

In summary, the division of a structure into surface areas serves to answer the questions 'what and where to inspect'.

4.4.4 Checklist

'Which mechanism is relevant?'

The modular division of a structure into surface areas follows the idea to identify and group areas, which are subjected to a main deterioration mechanism. These should be clearly stated and documented, and preferably stored in a database to avoid repeating this type of work over again at later CA.

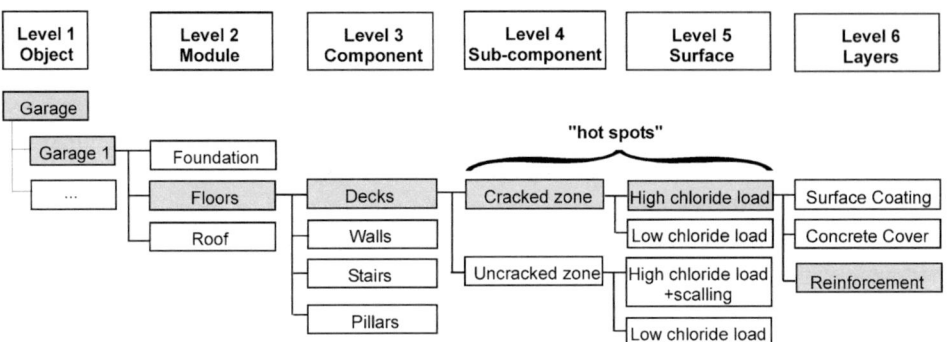

Figure 4.4 Example of integration of hot spots in the modular systematic.

4.4.5 Determination of surface areas

'What is the surface area of zones that will deteriorate *and* may be repaired in the same manner?'

The surface area of homogenous lots is of absolute necessity for various types of calculations in the management process. This is because the conditions of homogenous lots, which correspond to unit MR&R costs, must be related to the surface area involved to calculate the overall cost.

The number of inspections may have to be limited with respect to the number of deterioration indicators and number of investigated areas. It is thus important to evaluate whether or not common mechanisms are the cause of deterioration. If this can be justified by argumentation and/or evidence from the structure, the number of inspections can be reduced by grouping areas and choosing some of these as representative samples for inspection:

1. for every identified surface zone, the surface area must be determined and stored separately, e.g. in a database or a simple spreadsheet; and
2. these individual surface areas are then summed up with all of the comparable surface zones.

Each of these groups will be subjected to a certain portion to in situ investigations. The concept of the grouping assumes that material properties, environmental loading and degradation rating derived from in situ testing can be extrapolated to all surfaces included in the respective group.

4.5 Cost analysis of survey

4.5.1 Principles

Owners, no matter whether they perform inspections themselves or commission a consultant company, always want to know the following before starting with the actual site inspection.

'What will be the price for the survey?'

A cost analysis for the planned survey is essential. Costs items are e.g.:

- personnel ($€/h$);
- equipment ($€/h$);
- travel ($€/km$);
- consumption ($€/item$);
- user costs (e.g. due to influences on traffic caused by inspection; $€/vehicle$).

The extent of sampling can be determined according to Section 4.5.2. This will be the basis for the cost analysis. During the process of CA, the effort and, especially, the degree of technical equipment and thus costs may need to be increased due to actual conditions. This cannot be foreseen during planning. In order to analyse survey costs, various cost aspects are to be considered and prices for these need to be available to the investigator. This is especially important for expensive testing equipment.

'When shall actions take place?'

A detailed time schedule based on the boundary conditions assessed is essential for any inspection and should include targeted dates for:

- meetings of parties concerned (especially if structure is under use);
- on-site inspection;
- evaluation and reporting;
- presentation of results; and
- suggestions for follow up.

4.5.2 Minimisation of inspection costs

'How much surface area of an element needs to be inspected?'
'How many samples should be taken per surface zone?'
'What is the appropriate investigation technique to start with?'

It is generally accepted that important components are inspected more often than others. A series of inspection types (levels) can be distinguished which are fairly similar in most countries. Visual inspections are scheduled at regular intervals. However, for the case of general inspections, 'aim, effort and frequency of inspections need to be adapted to actual structure performance, age, structural type, construction materials and design complexity' [10].

Minimisation of inspection costs during the total service life of a structure can theoretically be achieved by:

1. increasing intervals between inspections;
2. reducing the technical grade of methods and hence the hourly rate for inspection equipment and trained personnel; and
3. reducing the inspection volume (portion of visually inspected area; number of samples or readings).

The main idea of this chapter is to start with basic investigation methods (point 2) and with a low inspection volume (amount of samples, inspected surface portion – point 3), which will be increased or made more sophisticated if intermediate results suggest it, as is current practice in existing international CAPs.

The purpose of investigations is to provide:

- information on the current condition (damage cause and extent); and
- data to calibrate service life models.

Every time inspection data are collected, the uncertainty in the service life models is decreased. However, the choice of the best suitable inspection method depends on the condition state of an element. Two cases can be distinguished:

1. *Element has already been investigated* As the condition does not improve, unless element was repaired, the minimum (previous) condition state is known. The question is: Whether deterioration has reached the next condition level or even one higher? Service life models with updated input parameters using the result of the previous inspections can be utilised to estimate the current condition. The appropriate inspection method must thus be connected to the various condition states, as is achieved by the tables provided for CR.
2. *Element has never been investigated* Service life models with default input parameters (design models) can be utilised to estimate the current condition.

In consequence, irrespective of whether an element has been inspected before or not, the best suitable method can be chosen in dependence on the estimated condition state. A 'first guess at the current condition' may be valuable as most inspection methods will not be of any value if they are applied too early during service life, while others may not bring any additional information and will thus not reduce the uncertainty in the calibrated models if applied at a very late stage. Recent studies on condition indicators (CIs) for the suitability of inspection methods as a function of service life using reliability theory are published, for example, in [11, 12]. These are, however, still far from being widely applied and accepted, but may offer large benefits in the future.

An example: Before accepting a structure, an acceptance concept should be established. This means available information on the 'as-built' condition (depth of concrete cover, migration coefficient of concrete) should be collected to update parameters of a design model (Figure 4.5). This way it can be checked whether the set requirements are fulfilled. Gathering of a chloride profile at a very early stage of service life will not provide much information, as the contact with chlorides was too short. Once chloride profiles have been taken at various times, the degree of information on the penetration behaviour obtained by additional profiles reduces. However, information on the critical chloride content may improve the models. Corrosion sensors may be embedded for this purpose, which indicate corrosion onset at different depths of the concrete cover. These are of course very expensive and only give an indication at a particular position. To obtain information on the activity of the corrosion process on larger surface areas, potential mapping may be used at late stages of the service life once corrosion of a particular portion of the reinforcement will have started.

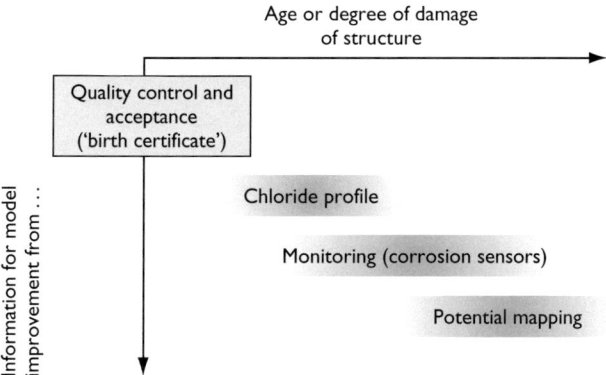

Figure 4.5 Appropriate time and method to obtain additional information for parameter update using Bayesian methods.

Visual signs of corrosion will be detectable once active corrosion leads to cracking of the cover.

An inspector always has the option to 'buy' additional information through an experiment before further decisions (e.g. on strengthening or repairing) are to be made. If the cost of this information is small compared to the potential savings, the inspector should go ahead and perform the experiment. The more the samples are taken, the lower is the scatter and thus the uncertainty of a variable (e.g. chloride penetration depth). More confidence will be given to results obtained by a study with a larger sample size. But every time the number of samples is increased, the degree of additional reduction in scatter becomes smaller and smaller. The appropriate minimum number of samples can be determined using statistical techniques. We are looking for the confidence interval which will cover up the unknown mean value μ of the considered parameter with a defined probability $(1-p)$. The length of the confidence interval is a measure of the precision of the estimation made by sampling. For a given probability of error p, an increase in sample volume will result in a narrower confidence interval. A sample can be used to calculate the empirical mean value $\bar{X}$ as an estimate for μ. Every time we take samples, the results and thus $\bar{X}_i$ will differ from each other. The position of the confidence interval thus depends on the realisations of the samples (Figure 4.6).

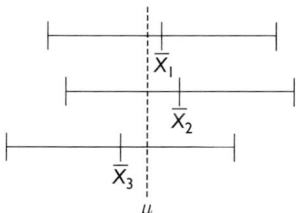

Figure 4.6 Realisations of sample confidence intervals for a given mean value μ.

A large standard variation of the regarded parameter and a high target confidence level both demand a larger sample size. If for simplicity, the investigated parameter is assumed to be normally distributed with unknown mean value μ, the number of samples can be determined as follows:

$$n \geq \left(z_{1-p} \cdot \frac{\sigma}{\delta} \right)^2 \tag{4.1}$$

where
n number of samples
z_{1-p} quantile of order $(1-p)$
p probability that mean value μ is beyond the confidence interval
σ known standard deviation of parameter and
δ targeted interval of confidence.

Table 4.4 shows the important quantile values for the standard normal distribution. Still, sampling is cost-intensive with respect to personnel, equipment, analysis (e.g. in the laboratory) and may be user costs. Therefore, a cost-optimum amount of samples may be determined taking into account the following variables:

- inspection costs

  (a) independent of sample size (e.g. driving costs to reach object, time for preparation and finishing of sampling, reporting, safety measures, access costs)
  (b) unit costs of samples, including costs for personnel and analysis and
  (c) user costs due to inspection (e.g. delay costs)

- consequences of failure for a given limit state (including repair costs)
- failure probabilities as a function of sample size.

The total costs can be expressed as follows:

$$E[C_{total}(t)] = p_i \cdot C_i + p_f \cdot C_f \tag{4.2}$$

where
E expected value
C costs
t time
p probability of an event
i, f inspection, failure (including repair).

Table 4.4 Important quantile values for the standard normal distribution

p	0.10	0.05	0.01	0.001
z_{1-p}	1.28	1.64	2.33	3.09
$z_{1-p/2}$	1.64	1.96	2.58	3.29

In conclusion, the target is to determine a minimum and maximum value for the sample size to obtain an objective basis for the decision. Let us consider an example: A retaining wall situated on both sides along a street which is treated with de-icing salts during the winter time. The wall is fabricated of 200 comparable segments. Coating is considered as adequate if with a probability of 5% two-thirds of the concrete cover d is penetrated by a critical chloride concentration, but not yet at five-sixths of the cover. If immediate action is taken now, the current condition is expected to be conserved for the next 10 years. The current probability of exceeding two-thirds of d is estimated to be $p_1 = 12\%$, but coating is still feasible. The cost of coating is estimated to be € 500 per segment. A first calculation with an 'a priori model' indicates that in 10 years from now, a 'wait' strategy will lead to the state that at approximately $p_2 = 33\%$ of the wall segments the critical chloride threshold value will be exceeded at a depth of two-thirds of the concrete cover d, which will then call for major repair works. Major repair works are assumed to cost € 2500 per segment (about five times the cost necessary for maintenance). In 10 years from nao, repair of the road pavement and other structures along the road is already scheduled. The question is: Perform any maintenance work (coating) now or wait until necessary actions can be combined? The decision problem can be depicted in the form of a decision tree (Figure 4.7), leading to the total costs of the two alternatives:

$$C_1(\text{'wait'}) = 500.000 \cdot p_2 + 0 \cdot (1-p_2) = € 165.000 \tag{4.3}$$

$$C_2(\text{'coating'}) = (500.000 + 100.000) \cdot p_1 + 100.000 \cdot (1-p_1) = € 160.000 \tag{4.4}$$

Immediate coating would currently be the best alternative. However, the inspector knows that additional chloride profiles may reduce the uncertainty in the 'a priori model' and thus the calculated failure probability p_2 for the 'do nothing' strategy. The total

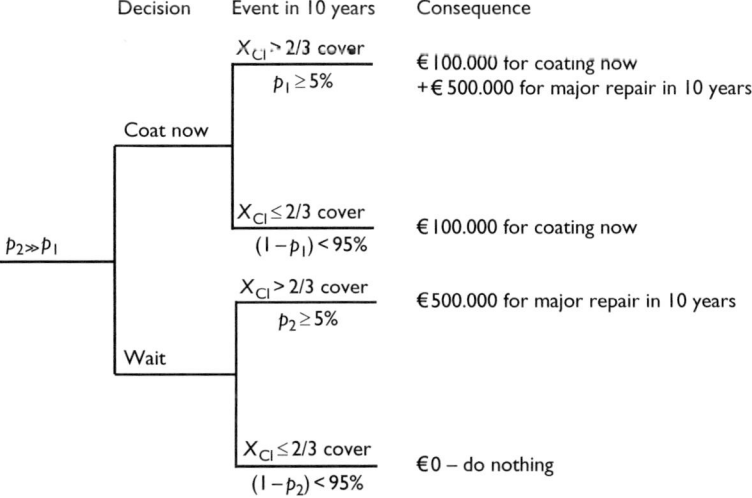

Figure 4.7 Example of decision tree for planning the number of samples *n*.

costs for the 'wait' strategy may thus be lower than according to the current model. But how many profiles should be taken? The minimum sample size n depends on:

1. the acceptable probability p that the real mean value μ of the chloride penetration depth will be larger than the average value of the penetration depth $\bar{X}$ which will be determined by future sampling; and
2. the standard deviation σ of the penetration depth.

From experience, we know that the standard deviation of the penetration depth can be assumed to be $\sigma = 0.75\,\mu$. The mean value μ of the penetration depth is unknown. However, we do not want $\mu \geq 2/3 d_{cover}$, with d_{cover} being the depth of the concrete cover. So as a maximum for the standard deviation, we take $\sigma = 0.75 \cdot 2/3 d_{cover}$. As a criterion for the length of the confidence interval δ, we may regard the remaining safety margin between our limit state criterion (two-thirds of cover) and the reinforcement. Coating is only regarded as an option if at more than 5% of the wall the chloride front has reached two-thirds but not yet five-sixths of the concrete cover. A reasonable size for the confidence interval is thus the margin between those two states, which is one-sixth of d as depicted in Equation (4.5). There is no general rule to set a criterion for the probability p. A common approach is to use $p = 5\%$. The quantile z_{1-p} can then be taken from tables in standard books on statistics.

Example

$$n \geq \left(z_{1-p} \cdot \frac{\sigma}{\delta} \right)^2 = \left(1.64 \cdot \frac{0.75 \cdot 2/3 d_{cover}}{\left(\frac{d_{cover}}{6} \right)} \right)^2 \approx 24$$

(4.5)

The minimum number of samples n from a statistical viewpoint may, however, not always be reasonable from an economic one. A so-called 'pre-posterior analysis' can be performed to solve the problem. 'Posterior' refers to the state that a reliability analysis is performed after sampling, thereby reducing the uncertainty of the model. Thus 'pre-posterior' means that we try to analyse the effect of samples on the improvement of the model, before we actually take these samples. The procedure is as follows:

- simulation of n realisations of the regarded parameter (here penetration depth X) according to 'a priori model';
- stepwise statistical analysis of simulated realisations from a small to a large number of samples $(2, 3, \ldots, n)$, as shown in Figure 4.8;

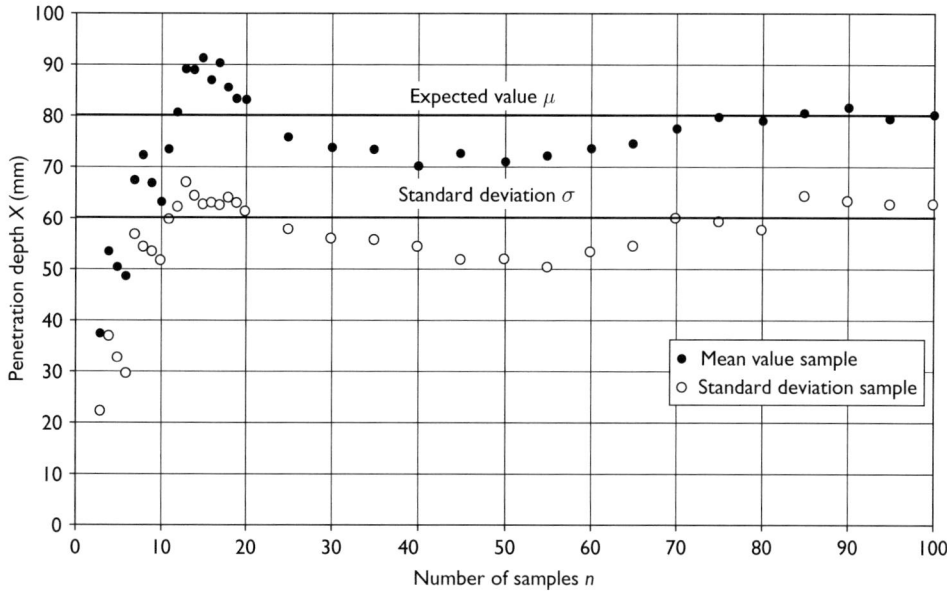

Figure 4.8 Statistical analysis of simulated samples ($\mu = 80$ mm; $\sigma = 0.75 \,\mu$ mm).

• updating of 'a priori model' according to Bayesian methods for every regarded sample number and calculation of probability p for the respective event (Figure 4.9); and
• calculating and summing up of updated repair cost C_R and inspection cost C_i as a function of sample size n (Figure 4.10).

The example demonstrates that dependent on the costs per sample inspections will bring additional benefits by reducing the total risk of major repair works. If costs per sample are below €500, then 40 samples should be taken. If cost per sample is higher, the alternative of immediate coating should be preferred. In these considerations, constant costs for inspection and user costs may play a major role. Let us assume: A sample (chloride profile) shall consist of six depth intervals, which takes about 0.5 h per profile for drilling and will cost around €150 for analysis. Personnel costs will be at €150 per hour, with 8 h of work per day. Equipment costs are around €300 per day (costing is done for complete days). The cost (€) per sample is thus:

$$c_i = 150 \cdot 0.5 + (300/8) \cdot 0.5 + 150 = 244$$
$$\text{personnel} + \text{equipment} + \text{analysis}$$

and sampling is appropriate. However, we may have to add constant costs too. Let us assume fixed reporting costs of 1000 euros. To close down the lane next to the

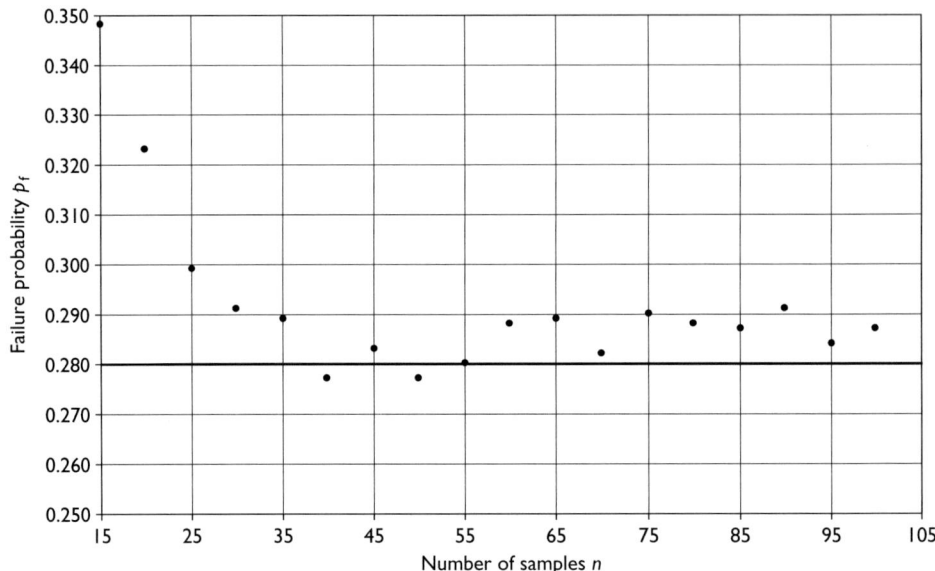

Figure 4.9 Updated failure probability as a function of sample number (for $n \to \infty$ the failure probability $p \to 0.28$ in the example).

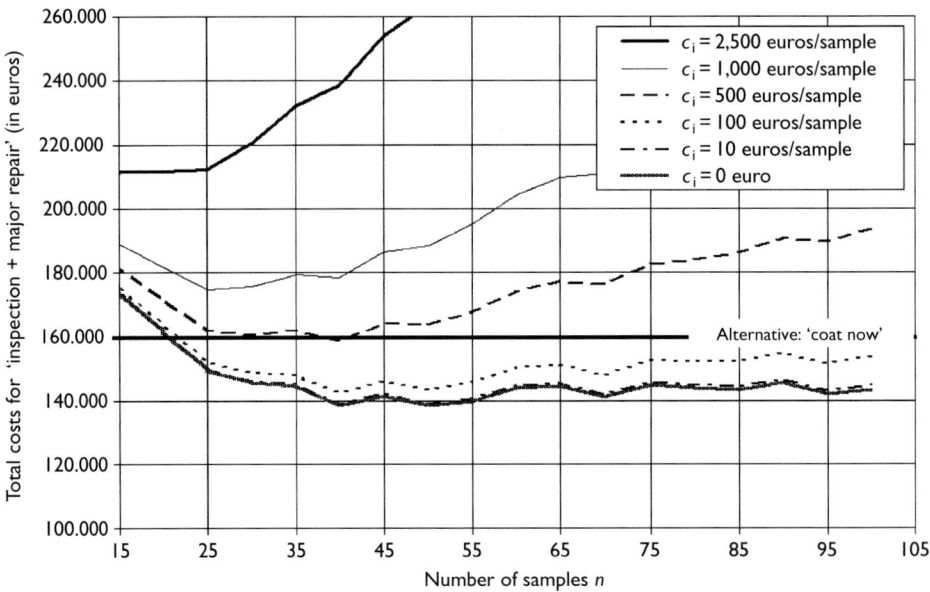

Figure 4.10 Total costs for the alternative 'sampling + wait until major repair' obtained by simulating one set of $n = 100$ realisations of samples.

wall for inspection, additional equipment and personnel is required, which totals to 2000 euros every time a lane is closed. Closing down a lane will affect traffic velocity. For simplicity reasons, it is assumed that every vehicle (8000 vehicles/day) is delayed on average by 5 min, which is taken into account with costs of €5 per vehicle. This totals up to:

$$C_i = 2000 + (150 \cdot 0.5 + (300/8) \cdot 0.5 + 150 + 5 \cdot 8000 \cdot 0.5/24) \cdot n + 1000$$

$$\text{safety} + (\text{personnel} + \text{equipment} + \text{analysis} + \text{user costs}) \cdot \text{samples} + \text{reporting}$$

$$= 3000 + 1077 \cdot n \text{ euros}$$

4.5.3 Safety during inspections

'What safety precautions need to be met?'

The matter of inspection safety is of growing concern, though treated in only a few national documents. Precaution has to be taken especially for inspections:

- under traffic;
- in confined spaces;
- in high altitudes; and
- under water.

Few practical guidelines exist on this topic. The document [13] gives assistance on the safety requirements during inspection. The safety assessment should cover the risk to

- workforce; and
- public (pedestrians, vehicular traffic, waterway and railway travellers).

Special attention must be paid to the topic of confined spaces. In [14] a classification of confined spaces and permit sheets are included for work in these situations. Special attention is paid to diving inspections.

4.5.4 Contracts

'Who will perform the survey under what conditions?'

To carry out a condition survey involves responsibilities and requires competence. If the work is carried out as a consultancy work, contracts have to be drawn up, which must be based on national standards and guides. The contract must define responsibilities (tasks carried out by client or by contractor) and give an unambiguous definition of the task.

4.6   Condition rating

4.6.1   Purpose

'What does condition rating serve for?'

According to [15], there are three factors which trigger repair or replacement actions for structural elements:

1.   function/obsolescence (e.g. inadequate geometry or load capacity due to changes in design standards, traffic volume or roadway class);
2.   vulnerability (e.g. design details, connections, bearings, etc., known to be vulnerable to sudden failure in earthquakes or due to fatigue); and
3.   deterioration and damage (e.g. corrosion of re-bar).

Each of these factors may be reported in the form of a CR, which may be interpreted as a degree for the need for action. The number of different, feasible actions with usually different unit costs determines the number of condition states.

Note: Intensity and extent of damage must always be reported separately, which is not the case in many European countries.

The purpose of rating is to define a CI which will allow the deterioration of a structure and its constituent elements to be tracked with time. The CI will indicate when a specific level of deterioration has been reached in some part of an element. CR obtained at the 'surface' level can be combined to give data on all levels above up to the network level. Combining the CI will enable a CR to be obtained for the object.

Following the allocation of the CI and subsequent CR, a decision will be made upon future action. At any particular CI, there will be a range of repair or remedial maintenance alternatives, including taking no action, which may be appropriate, and others which will not. The CI can thus be used to identify viable maintenance alternatives at each stage in the life of the structure. The decision as to the nature, timing and implementation of maintenance actions will be made by the planning modules of the management process.

A further, more detailed survey will be necessary prior to commencement of repair/remedial work in order to confirm degradation mechanisms and determine the extent of maintenance required. The CI system is not intended to serve this latter function. The CR will be determined by a general inspection to be undertaken 6+ years apart. The CI will be too coarse to detect deterioration on an annual basis and would not be cost-effective.

4.6.2   Condition rating for vulnerability

'What is vulnerability?'

Structures are 'vulnerable' if they may fail suddenly under a specific extreme load, or overload of number of stress cycles. In respect of a specified design life (user requirement), a structure which may fail within a shorter period than would others is considered more vulnerable. For extreme loads, vulnerability is determined by the return period (e.g. seismic overloads, i.e. earthquake). For repetitive loads, vulnerability is determined by the number of bearable repetitions of loads (e.g. fatigue), as shown in Table 4.5.

Table 4.5 Example of condition rating for vulnerability in respect of seismic overloads, fatigue in dependence on return period or service life [12]

Rating	+ Robust	Robust	Vulnerable
Attribute	Exceeds return period or planned service life	Meets return period or planned service life	Below return period or planned service life ('red flag')
Action	None	None	• Elimination of failure (e.g. replacement) • Posting

4.6.3 Condition rating for deterioration

4.6.3.1 Requirements

'What does condition rating for deterioration mechanisms serve for?'

Condition rating transforms quantitative measures for deterioration into a unified, integrated scale. These condition states are used as input for deterioration models according to the Markov chain approach [7, 16], which are of special importance on the network level [17, 18].

'What are integrated condition states?'

Condition states are distinguished by attributes that can be detected and/or measured by inspection methods. Furthermore, each stage in service life should be related to MR&R actions and/or consequences which differ for those of previous or later stages.

To use data collected from single objects on a network level, the data need to be related to limit condition states. Thereby, data is put on a scale. The rating should be related to the viable repair/remedial options. Remedial options are techniques which, if applied now, would appreciably extend the time until the succeeding condition state is reached. Repair options are defined as techniques which would restore the structure to an earlier condition state. The number of steps on a rating scale is thus defined by the number of service life stages with corresponding MR&R actions identifiable until this limit state is reached, plus a failure state.

Quantitative data on conditions are available from many non-destructive evaluation (NDE) and destructive inspection (DI) methods. Data from such measurements can be inserted directly into the database of an LCM system. However, for rating, data have to be interpreted and assigned to condition states. Integrated condition states for deterioration mechanisms may generally be defined by [12]:

- the presence of aggressive agents (chlorides, CO_2);
- activity of a process (processes may be chloride ingress, corrosion, alkali aggregate reaction, etc.); and
- existence of a damage (cracking, spalling, loss of section).

The CR should be based solely on the condition of the structure, i.e. be made without reference to the loading (either structural or environmental) on the structure. An assessment of the condition of a structure in terms of its required continued structural performance lies beyond the scope of a CR system. It is thus implicit that throughout all points, but especially the final point, on a CR scale, the accumulated damage will not degrade structural safety to a significant extent. In the proposal set out in Section 4.6.3.3, it is only at the final stage that a structural strength assessment would be necessary, to be conducted in accordance with, for example, recommendations [19–21].

In conclusion, a generic CR system and the constituting stages should be:

- mutually exclusive (each condition state identifies a distinct stage in service life);
- detectable (states are related to investigation methods);
- an indicator for intensity, not of extent of damage;
- corresponding to maintenance and repair actions;
- utilising currently recognised and established means of inspection (especially non-destructive techniques);
- based on intervals representing approximately equal proportions of the useful life;
- robust;
- independent of deterioration mechanism; and
- presented in a manner such as to assist in formulation of the inspection plan.

4.6.3.2 Shortcomings of current systems

A variety of CR systems (mostly on a scale from 0–4) have been proposed internationally. These have the following features in common:

- the scheme is geared primarily towards ease of inspection;
- guidance on classification is specific to the damage mechanisms; and
- the first transition from condition state 0 to condition state 1 does not occur until corrosion is well established and damage has occurred to the structure.

Current management systems operating with ratings lack information on early deterioration in elements. It is early changes, in particular, which offer opportunities for cost-effective maintenance. The first transition in current systems thus occurs too late for some of the possible preventative maintenance measures to be viable, which is due to the fact that mainly visual CI, available at rather late stages during service life, are regarded. This is not consistent with the aims of a preventative maintenance philosophy. It has therefore been decided to introduce CRs to delineate earlier conditions.

In [15, 17, 18], a classification system is proposed which is based on a more generic system of condition descriptors (Table 4.6). A generic system is particularly desirable for concrete construction in view of the several degradation mechanisms which can occur.

The system in Table 4.6 is geared more strongly to preventative maintenance. At the other end of the degradation scale, however, it is evident that many structures remain in service despite obvious and extensive visual evidence of corrosion in the form of cracking, staining and spalling. Only one classification is allowed for to cover this stage in the life of a structure. Tentative analyses [5] suggest that initiation and propagation phases of deterioration occupy approximately equal portions of the

Table 4.6 Possible generic rating system for degradation

CR	Descriptor	Condition	Current equivalents
0	Protected	By coatings or treatments that block or counteract aggressive agents	0
I	Exposed	When protection is absent or failed	0
2	Vulnerable	When exposure or infiltration of aggressive agents reaches a point to allow damage processes to begin	0
3	Attacked	When damage processes are active	I
4	Damaged	When disruption, discontinuity or losses in elements have occurred	2–4

ultimate life of a non-maintained structure, and hence it is reasonable to allocate equal numbers of stages to each.

4.6.3.3 Proposed system for degradation rating

A 0–6 scale for the assignment of damage is proposed, with states 0–2 representing the initiation phase of damage, state 3 representing the early stage of damage where the consequences of degradation are not visible, and states 4–6 covering various degrees of visible damage. The classifications 0–4 are based upon the proposals in [15, 17, 18]. The definitions given are more detailed than those presented there, however.

0 – Protected	Concrete is protected against agents that can cause deterioration. By definition, maintenance of concrete is not required in this state. Maintenance of protective measures (e.g. coating) may nevertheless be necessary.
1 – Exposed	Concrete does not have protection or protection has failed. Agents have not yet reached a level that may initiate deterioration, if a surface treatment is applied now to prevent agents from reaching initiation levels.
2 – Vulnerable	No damage has occurred up to this point. Maintenance strategies based on retarding ingress of aggressive agents can no longer prevent but only postpone deterioration. However, the structure could be returned to virtually 'as new' condition by suitable maintenance actions.
3 – Attacked	A deterioration process is active but visible damage has not occurred. (For CR 2 and CR 3, the repair alternatives are likely to be the same. They are retained as different inspection techniques are available to determine this condition.)
4 – Damaged	A visible level of deterioration has occurred. The damage is only apparent on close inspection. This may be cracking, but not spalling or delamination. There may remain viable maintenance options that do not entail breakout and recasting.

5 – Service failure	Clearly visible signs of degradation, including cracking, delamination, surface staining and localised spalling are observable. Structural integrity remains intact and safety is not an issue. Investigation techniques may be based upon crack width, which can be related to reinforcement section loss and strength loss. A form of risk assessment may also be undertaken based upon the use and location of structure to determine if a risk to the public is significant. Extensive repair including cut-out and re-casting/patching is the only maintenance option.
6 – Safety risk	Damage of sufficient severity to raise concern over structural adequacy. Calls for a structural assessment and will require specific safety measures for the structure to remain service, possibly under load restriction. Extensive repair including cut-out and re-casting/patching is the only maintenance option.

Every class will be determined by a pre-set range of scores based on recognised inspection techniques. Each class is mutually exclusive and corresponds to a distinct stage of service life when particular repair/remedial actions are viable. Ideally each class should represent an equal proportion of the lifetime of a structure. However, as it is impractical to achieve this, it is deemed sufficient if the duration of each class is greater than the standard time between inspections (taken as 6 years, based on international standards).

4.6.4   Condition rating for functional obsolescence

'Does the object or module still fulfil the functional demands?'

Owners of structures must recognise objects or modules of an individual structure, which today are obsolete with respect to its desired function or will be so at a known time in the future. An additional condition state is defined for those structures with excess of capacity – structures which could meet newer design standards and could accommodate greater traffic loads, for instance (Table 4.7).

Table 4.7 Condition rating for functional adequacy in respect of geometric demands (e.g. deck width or vertical clearance) or load capacity (expressed by, for example, safety margin index) [12]

Rating	+ Adequate	Adequate	Inadequate
Attribute	Exceeds standard or design requirements by 10%	Meets standard or design requirements	Below standard or requirements ('red flag')
Action	None	None	• Elimination of failure (strengthening, replacement) • Posting

4.6.5 Relating condition states to inspection methods

4.6.5.1 General concept

One option of relating condition states and inspection methods to detect these states is that either a single test criterion be used to determine the CR, or that if more than one test could be used, a direct equivalence between the two tests would be given. Different tests would be used for each classification boundary. While this strategy would apparently produce clear-cut decisions, we consider it to be unreliable for a number of reasons:

- Many of the inspection methods do not give sufficiently reliable and unambiguous/precise results to be used as definitive tests for degree of 'damage'. For example, a half-cell potential survey may give different results on a wet or dry day/concrete surface (see literature review in the annex). Moreover, a test value found in one structure may vary from day to day depending on the climatic conditions, although the condition of the structure does not.
- Allowing alternative tests requires a precise linkage between the results from one method and the another. This is not possible; a fixed resistivity value cannot be equated with a fixed half-cell value.
- Different deterioration mechanisms interact. For example, the chloride concentration threshold for initiation of corrosion is lowered in the presence of carbonation. The assessment threshold to be set for chloride concentration should be able to reflect this interaction.
- Limiting the CR to a specific test would impede the flexibility necessary to be applied to structures which have not been inspected under the LIFECON approach.

In order to be robust, the CR system should not rely solely on a single measured value or observation. For example, although a half-cell potential of less than $-350\,\text{mV}$ with respect to a CS reference electrode is commonly cited as a criterion for 95% confidence of active corrosion, many practitioners are wary of basing a decision solely on this criterion. On the other hand, if measurements show chloride concentrations in excess of 1.0%, resistivity measurements show low values, half-cell potentials differ by more than $100\,\text{mV}$ over a distance of $1.0\,\text{m}$ and the minimum half-cell potential is $-340\,\text{mV}$, then most would agree a 99.9% probability of active corrosion. In other words, an 'expert' will look for corroboration from different sources before reaching a conclusion. It is desirable that the CR system represents this process in some way. To meet the need for robustness, a CR system based on cumulative scoring is proposed.

4.6.5.2 Cumulative scoring

In the cumulative approach, the inspection data are allocated a score based on values from a table. This may be referred to an 'expert system', which may even be programmed to an IT tool for automatic rating. An inspection regime that consists of a number of pre-set inspection techniques, to be applied to all structures, is proposed to take these needs into account. A score will be allocated for each individual inspection method. The total score for all tests is then calculated. This cumulative score is then used to allocate the CR. By employing a range of criteria, variations due to different degradation mechanisms or fluctuations in the inspection data may be allowed for. Hence, a single anomalous result will not unduly influence the CR classification. The use of a range of tests will increase the robustness of the inspection and leads to a higher degree of confidence in the final damage index classification. People opposing inspection techniques apart from purely visual inspection may object that a combination of tests to achieve knowledge about a single condition state is a cost-intensive procedure. However, it must be realised that:

- This way of approaching the problem is already well established in the consulting business.
- Test methods which are believed to be less indicative can be given a lower weight (maximum achievable score).
- In the long run, a reliable analysis of the current condition will be the most economical solution by preventing unnecessary and inappropriate spending on MR&R works.
- Dependent on the realisations of measurements, the rating may already be evident after applying a single test method. In this case, further testing is unnecessary, which is indicated by the scoring system.
- The cumulative system can allow for interactions between two deterioration mechanisms: independent 'medium risk' scores under two interacting deterioration mechanisms can be allocated the same CR score as a 'high risk' score under a single mechanism.
- The approach allows the investigator to substitute 'equivalent' test methods once a suitable scoring system is established.

The inspection methods to be used fall into a number of different categories. Due to the different nature of reinforced concrete structures, and to allow operators some leeway, a number of different techniques are given for a number of these categories. The operator conducting the CR can select the most appropriate technique from each generic category for the structure under consideration. Usually, only a single test should be undertaken from within each generic category. If in doubt, other test methods from the same category may nevertheless be applied. Thus, the CR will consist of a number of generic inspection categories with a number of different techniques in each category.

A set of scales is used for each CR boundary. If CR = n has been reached at the previous inspection at the regarded site, then the boundary $n + 1$ should be investigated first. If no previous inspection data exist, then boundary 1–2 should be investigated first (Figure 4.11). To determine an exact CR, all inspection methods should be employed. However, by calculating the cumulative score after each test, it may be possible to eliminate the need for certain tests should the score have passed beyond a boundary. Similarly, it may be possible to eliminate tests if the highest score from that test cannot move the total score beyond the next boundary.

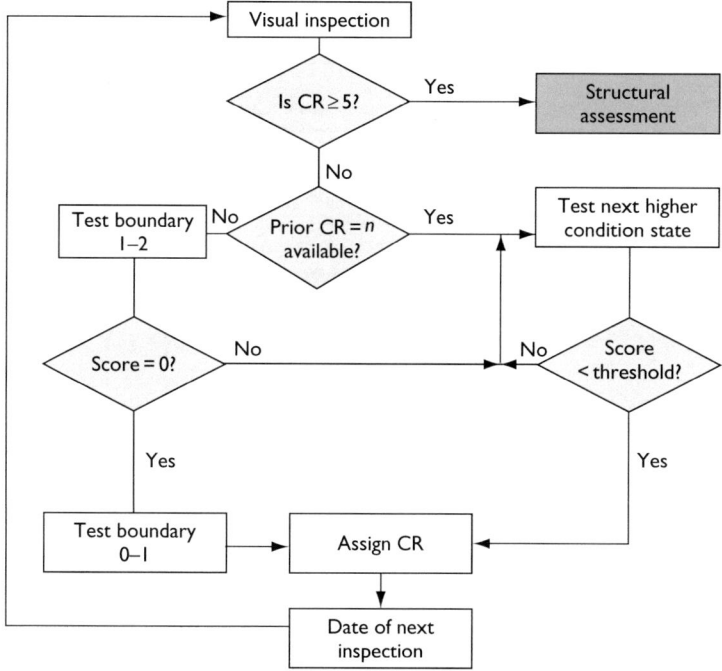

Figure 4.11 Framework of condition rating.

Should the initial visual inspection, indicate a CR value >5, then a more detailed in-depth structural assessment will be necessary. This will require specialist knowledge and will be unique to each structure depending on location, condition, degradation mechanism and level of damage. No uniform set of tests can be prescribed for this part of the CAP.

4.6.5.3 Test methods

4.6.5.3.1 OPTIONS

The domain of CR over which each generic inspection technique is applicable is given in Table 4.8. Shaded beams indicate the most suitable domain for distinction between two condition states. Black lines indicate the potential application domain. If a generic test is required at a specific boundary, a single test from each generic type should be performed. Each test will have a result which will correspond to a score as outlined. Optional techniques for the various inspection targets are tabulated in Table 4.9.

The test methods proposed can be split into a number of specific groups where one technique can be used to measure the performance. The method regarded as most suitable is always given first. More detailed information on the proposed test methods are provided in the 'Inspection Catalogue' in the annex.

Table 4.8 Range of generic test methods with regard to condition indicator for the example of corrosion-induced damages

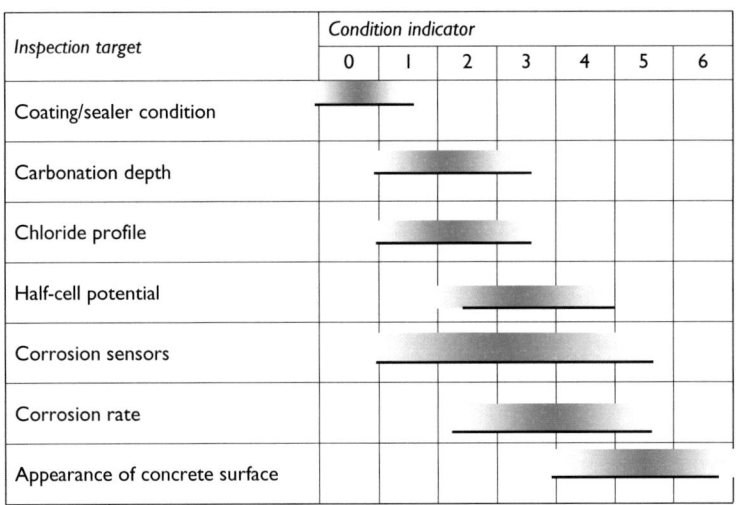

Inspection target	Condition indicator						
	0	1	2	3	4	5	6
Coating/sealer condition							
Carbonation depth							
Chloride profile							
Half-cell potential							
Corrosion sensors							
Corrosion rate							
Appearance of concrete surface							

Table 4.9 Optional inspection techniques for inspection of corrosion

Inspection target	Optional techniques
Coating/sealer condition[1]	– Visual – Resistance
Carbonation depth[1]	– Phenolphthalein indicator on fresh concrete – Pore water pH analysis
Chloride profile[1]	– Dust drillings – Cores – Estimate from chloride diffusion (migration tests)
Half-cell potential[1]	– Portable half cell – Embedded electrodes
Corrosion rate[1]	– Linear polarisation resistance (LPR) – Embedded electrodes – Pulse transient (Destructive opening of cover and measurement of section loss in combination with estimated time for corrosion onset is a viable option for confirmation of each of the above listed techniques.)
Corrosion sensors[1]	– cast in sensors (option for newly built structures) placed in drilled holes – Visual inspection
Appearance of concrete surface	– Sounding – Thermography

Note

1 An alternative approach to determine the coating/sealer condition, carbonation depth and chloride concentration and, finally, corrosion rate is corrosion sensors (galvanic macro cells), which may be embedded in the structure. These will indicate when conditions initiating corrosion have reached set depths. This measurement indicates that (a) the coating has failed and (b) carbonation front or critical corrosion inducing chloride concentration have reached a certain depth. The major benefit compared to carbonation depth measurement and chloride profiling is that the uncertainty about the critical pH value or critical chloride concentration must not be assumed, hereby eliminating an important source of uncertainty. The major disadvantage is the cost per sensor and the fact that a certain measurement period is necessary until conclusions can be drawn.

Table 4.10 Scoring of test methods for boundary 0–1 (coating condition)

Test method	Criterion	Score
Visual	No holes visible	0
	Holes visible	4
Resistance	$>200\,k\Omega$	0
	$<200\,k\Omega$	4

Notes
Average score $0 \rightarrow CI = 0$.
Average score $> 0 \rightarrow CR \geq 1 \rightarrow$ test at next boundary.

4.6.5.3.2 SCORING

Examples of the cumulative score tables are given in Table 4.10. It should be remembered that the system is only thought to provide an objective basis for CR. However, the system will probably work for typical situations. The inspector is nevertheless always asked not to depend solely on the outcome of the scoring system but take into account the specific boundary conditions of the investigated area.

With respect to chloride ingress and carbonation, it is important to find the condition state ($CR = 1$) at which the aggressive media already have penetrated the concrete, but if immediate protection against ingress (PI) is provided now, the onset of deterioration (e.g. corrosion) can still be prevented. If the penetration depth is beyond this, level the onset of deterioration can be retarded but no longer prevented.

The application of a coating to a concrete surface will reduce the access of CO_2, but more importantly may reduce the moisture content and thus may stop the carbonation process. However, the carbonation front is not evenly distributed, but is subjected to scatter. If a sufficient number of carbonation readings are given, the penetration depth can be evaluated statistically. If not, the measurements can be used to determine a mean value, while the standard deviation is assigned from experience. If no more than 5% of the carbonation has reached 75% of the height of the concrete cover, the future depassivation can be eliminated by coating the structure immediately.

In the case of chloride ingress, the simple approach of defining a certain limit for the penetration depth of the critical chloride as a criterion for the feasibility of a coating is inappropriate. This is because chlorides (or other depassivating ions) will redistribute, i.e. proceed with the penetration, even though further chloride supply from the environment is prevented by a coating. Therefore, the whole shape of a chloride profile must be regarded (Figure 4.12). If in the redistributed state, the chloride content exceeds the critical threshold level, the regarded surface area must be assigned to condition state 2, at which a surface coating can no longer prevent depassivation. Sorting of test methods for various boundary levels are shown in Tables 4.11–4.15.

4.6.6 Relating condition states to repair/remedial options

When treating a structure, either repair or remedial action can be undertaken. In repair action, the structure is returned to an earlier CR classification. By remedial action

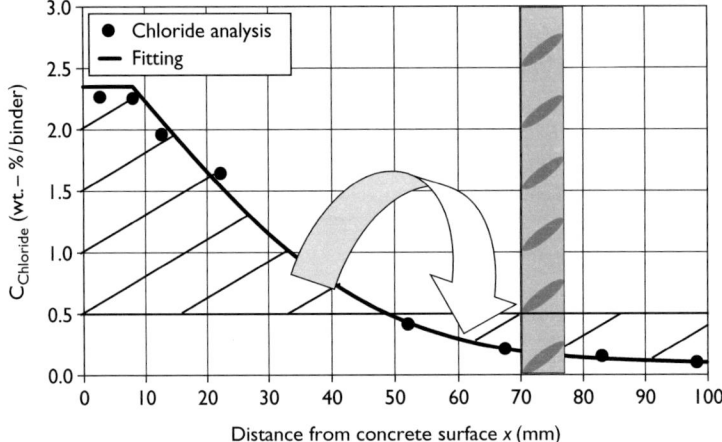

Figure 4.12 Possible redistribution of chlorides after application of a surface coating.

Table 4.11 Scoring of test methods for boundary 1–2

Test	Criterion	Score
Chloride profile	redistribution of chlorides will most likely lead to depassivation	8
	redistribution of chlorides is not critical	0
Carbonation	>5% of penetration front >75% cover	8
	≤5% of penetration front <75%	0
Alkali-aggregate reaction (AAR)	AAR susceptible	8
	Not AAR susceptible	0
Freeze–thaw	Attack visible	8
	No attack visible	0
Visual inspection of concrete surface	Crack	8
	No crack	0

Notes
Total score $0 \rightarrow CR \leq 1$.
Total score $> 0 \rightarrow CR \geq 2 \rightarrow$ test at next boundary.

the onset of the next CR classification is delayed. The CR classification indicates the severity of the damage, but not the extent, neither does it indicate the mechanism. To determine the optimum repair/remedial option, all conditions must be known. As such the CR classification does not on its own indicate what will be the optimum repair/remedial action. However, certain CR classifications will invalidate certain options. The range of repair and remedial actions to be considered are those detailed in [22] (Table 4.16).

Table 4.12 Scoring of test methods for boundary 2–3

Test	Criterion	Score
Visual inspection of concrete surface	Crack/spalling/delamination	12
	No crack	0
AAR	AAR attack (petrographic analysis/map cracking)	12
	AAR susceptible	4
	Not susceptible, no attack	0
Freeze–thaw	Loss of section > 10 mm, delamination of concrete cover	12
	Deterioration up to 10 mm, thin scars in surface, partly loose fine sand grains, scaled cement mortar between grains	4
	Deterioration up to 4 mm, thin scars in surface, partly loose fine sand grains	0
Half-cell potential (absolute)	≤ -700 mV	2
	> -700 mV and ≤ -350 mV	6
	> -350 mV and ≤ -200 mV	4
	> -200 mV	0
Half-cell potential (difference)	Difference ≥ 200 mV in 1 m within segment	8
	Difference ≥ 100 mV in 1 m within segment	6
	Difference < 100 mV in 1 m within segment	0
Chloride concentration at reinforcement	$\geq 1.0\%$ by weight of cement	6
	$\geq 0.4\%$ and $< 1.0\%$	4
	$< 0.4\%$	0
Carbonation	$\geq$ cover	4
	$<$ cover	0
Corrosion rate	$\geq 0.5 \,\mu A\,cm^{-2}$	6
	$\geq 0.1 \,\mu A\,cm^{-2}$ and $< 0.5 \,\mu A\,cm^{-2}$	4
	$< 0.1 \,\mu A\,cm^{-2}$	0

Notes
Total score $<12 \rightarrow CR \leq 2$.
Total score $\geq 12 \rightarrow CR \geq 3 \rightarrow$ test at next boundary.
Half-cell potential gradient is given more weight than absolute half-cell potential and the corrosion current value, as the latter is considered less reliable.

Remedial options are classified into three types as (1) viable, (2) potential and (3) non-viable. Table 4.17 shows under which remedial option each CR classification fall with respect to their condition index. Viable options are those that can be applied to the structure at that CR classification. Non-viable options are those which should not yet (because it would be too early at this stage) or cannot be applied to the structure (because the measure is no longer appropriate) when it is at that CR classification. Potential options may be employed dependent upon the damage mechanism and degree of damage. This must be assessed in the specialist inspection to be conducted post-CR and prior to repair/remedial action.

Once repair action has been conducted and the structure/element has been moved into a lower damage rating classification, the options applicable at that classification may be used to give further protection. Thus two or more options may be used in combination as a repair/remedial strategy.

Table 4.13 Scoring of test methods for boundary 3–4

Test	Criterion	Score
Chloride concentration at reinforcement	$\geq 1.0\%$ by weight of cement	4
	$<1.0\%$ by weight of cement	0
Carbonation	$\geq$ cover	4
	<cover	0
Visual inspection of concrete surface	crack width ≥ 0.3 mm/spalling/delamination	16
	0.3 mm > crack width ≥ 0.1 mm	8
	Crack <0.1 mm	0
Freeze–thaw	Loss of section > 10 mm, delamination of concrete cover	16
	Deterioration up to 10 mm, thin scars in surface, partly loose fine sand grains, scaled cement mortar between grains	4
	Deterioration up to 4 mm, thin scars in surface, partly loose fine sand grains	0
AAR	Attack (petrographic analysis/map cracking)	4
	No attack	0
Corrosion rate	$\geq 1\ \mu A\,cm^{-2}$	8
	$<1\ \mu A\,cm^{-2}$	0

Notes
Total score $\leq 12 \rightarrow$ CR ≤ 3.
Total score $>12 \rightarrow$ CR $\geq 4 \rightarrow$ test at next boundary.

Table 4.14 Scoring of test methods for boundary 4–5

Test	Criterion	Score
Visual inspection of concrete surface	Crack ≥ 0.3 mm/spalling/delamination loss of concrete section from freeze/thaw attack	4
	<0.3 mm	0
E-Modulus	dynamic E-Modulus reduction $\geq 5\%$	8
	dynamic E-Modulus reduction $<5\%$	0

Notes
Score $< 4 \rightarrow$ CR ≤ 4.
Score $= 4 \rightarrow$ CR $\geq 4 \rightarrow$ test at next boundary.

4.6.7 Case study example

The above-explained system for CR is meant to be applied to each area with a comparable level of distress. However, at an inspected surface zone, areas with different intensities of distress, i.e. different CRs, will be found as deterioration usually is a heterogeneous process.

Table 4.15 Scoring of test methods for boundary 5–6

Test	Criterion	Score
Visual inspection of concrete surface	Crack/Spalling/loss of section of concrete and/or rebar: $\geq 5\%$ strength reduction[1]	8
	Crack/Spalling/loss of section of concrete and/or rebar: $<5\%$ strength reduction[1]	0
	Spalling likely	4
	Spalling unlikely	0
	High Risk to traffic[2]	4
	Low Risk to traffic[2]	0
E-Modulus	Dynamic E-Modulus reduction $\geq 5\%$	8
	dynamic E-Modulus reduction $<5\%$	0

Notes
Total score $\leq 8 \rightarrow$ CR ≤ 5.
Total score $>8 \rightarrow$ CR $= 6 \rightarrow$ Consult experts for structural assessment and risk analysis.
1 to be calculated from [17].
2 traffic – vehicles and pedestrians.

Table 4.16 Methods of repair and remedial action

Principle	Definition	Code	Methods
PI	Protection against ingress	1.1	Impregnation
		1.2	Surface coating
		1.3	Locally bandaged cracks
		1.4	Filling cracks
		1.5	Transferring cracks to joints
		1.6	Erecting external panels
		1.7	Applying membranes
MC	Moisture control	2.1	Hydrophobic impregnation
		2.2	Surface coating
		2.3	Sheltering or overcladding
		2.4	Electrochemical treatment
CR	Concrete restoration	3.1	Apply mortar by hand
		3.2	Recasting with concrete
		3.3	Spray concrete or mortar
		3.4	Replacing Elements
SS[1]	Structural strengthening	4.1	Adding or replacing reinforcing bars
		4.2	Installing bonded re-bars in preformed or drilled holes
		4.3	Plate bonding
		4.4	Adding mortar or concrete
		4.5	Injecting cracks, voids or interstices
		4.6	Filling cracks, voids or interstices
		4.7	Pre-stressing
PR	Physical resistance	5.1	Overlays or coatings
		5.2	Impregnation
RC	Resistance to chemicals	6.1	Overlays or coatings
		6.2	Impregnation
RP	Restoring passivity	7.1	Increasing cover to reinforcement with mortar or concrete
		7.2	Replace contaminated or carbonated concrete

Table 4.16 (Continued)

Principle	Definition	Code	Methods
		7.3	Electrochemical realkalisation of carbonated concrete
		7.4	Realkalisation of carbonated concrete by diffusion
		7.5	Electrochemical chloride extraction
IR	Increasing resistivity	8.1	Limiting moisture content by surface treatments
CC	Cathodic control	9.1	Limiting oxygen content by saturation or surface coating
CP	Cathodic protection	10.1	Applying electrical potential
CA	Control of anodic areas	11.1	Painting reinforcement with coatings containing active pigments
		11.2	Painting reinforcement with barrier coatings
		11.3	Applying inhibitors to concrete

Note
1 SS covers only restoration of lost strength and not provision of additional strength above design level.

Table 4.17 Viable (V), Potential (P) and Non-viable (N) remedial options

Condition index	PI	MC	CR	SS	PR	RC	RP	IR	CC	CP	CA
0 Protected	N	N	N	N	N	N	N	N	N	N	N
1 Exposed	V	V	N	N	V	V	N	P	N	N	N
2 Vulnerable	P	P	P	N	P	P	V	V	P	P	V
3 Attacked	P	P	P	N	P	P	V	V	V	V	V
4 Damaged	N	P	V	V	N	N	P	P	P	P	P
5 Service failure	N	N	V	V	N	N	N	N	N	N	N
6 Safety risk	N	N	V	V	N	N	N	N	N	N	N

Example
One of the parking decks of a public garage produced of reinforced concrete has been inspected. The surface is uncoated (condition automatically is CR > 0). A visual examination indicates very few cracks, which seem to be caused by corrosion, as stains are visible. No spalling has yet taken place at all. Partial opening of these areas reveals that pitting corrosion has taken place, but strength reduction can be assumed to be below 5% (CR = 4 in these areas).

In a parking deck, 'hot spots' and areas with 'normal' chloride loading can be identified. As 'hot spots' are considered those areas with potential high chloride loading, e.g. where wheels are situated in parking slots, snow slush is likely to fall off (in bumps, e.g. in entrance area) or chloride-containing water can accumulate (low points). Areas with presumably 'normal' chloride loading are, for example, mid areas of parking slots and wheel tracks. Nevertheless, 'hot spots' may not always be obvious from mere visual inspection. So as a first step, half-cell potential mapping can be performed to obtain a first CI.

A half-cell potential mapping and measurement of cover depth was performed for the entire deck (Figure 4.13). Values in the range of $E_{CSE} = -220$ to $-580\,mV$ were measured. Those areas with values of $E_{CSE} < -350\,mV$ (usually 'normal' areas)

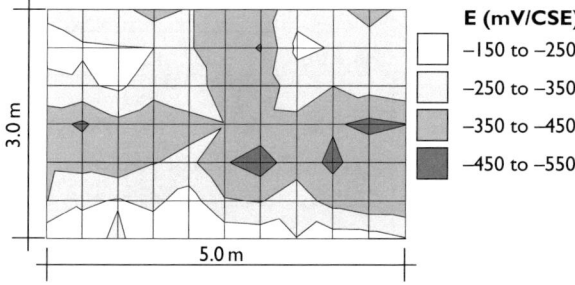

Figure 4.13 Half-cell potential mapping of a single parking slot.

were distinguished from those with values of $-580 \leq E_{CSE} \leq -350\,\text{mV}$ (different scoring according to Table). An example for measurement results and decisions on further inspection using the above given scoring system as a guide is outlined in Figures 4.14–4.16.

All decisions for further inspections and resulting ratings applying Tables 4.10–4.15 are summarised in Figure 4.17. After measuring the half-cell potential, certain measuring points can already be considered as still being in the passive state (CR $\leq$ 2). Here, a chloride profile must be taken at some of the measurement points to distinguish between states 1 and 2. The purpose of this distinction is to determine whether in these areas a surface coating may still prevent future corrosion or will only retard the moment of onset. There is no need to take too many samples as a large extent of detail for rating is not practical having the future repair (and modelling for life cycle analysis) in mind. A couple of samples should be taken. The average result should be considered as representative for the whole area in question, i.e. the whole area is in either state 1 or 2.

At other sites, active corrosion (CR = 3) can already be assumed as certain. Nevertheless, it is recommended to not rely solely on the scoring system, but to verify at a limited number of points that corrosion is really taking place. This should be done by opening up the concrete cover (drilling of holes, coring of concrete or simply chiselling of cover) and visually checking the corrosion state of the re-bar. Moreover, at a few spots, a chloride profile should also be taken in these areas, not so much for future modelling (as corrosion has already started), but more as a confirmation that chloride ingress is actually higher than in areas that are still passive, as would be expected. This information can also be used to obtain a more precise value for the chloride threshold value.

At some measuring points, there will still remain doubt about the condition state (could be 1, 2 or 3), as marked '?' in Figure 4.16. Here the scoring system suggests to proceed with further investigations. The only non-destructive possibility giving instantaneous results (after about 10–15 min) for differentiating between states 2 and 3 is measurement of the corrosion rate. Measuring of all areas in doubt would be very time-consuming. A feasible way is to measure at some points and consider the average of the result as representative for all of the areas in doubt. When deciding on which points to select, the practical procedure for future repair must be kept in mind. As the inspected area is rather small, it is not feasible to have too much detailing

(a)

(m)	0.0	0.5	1.0	1.5	2.0	2.5	3.0
0.0	180	353	375	410	282	212	272
0.5	212	263	385	467	266	242	296
1.0	268	244	420	376	231	247	320
1.5	186	275	355	421	308	250	395
2.0	212	203	324	381	284	340	340
2.5	181	342	441	353	417	421	421
3.0	189	356	505	381	410	455	411
3.5	248	305	386	375	298	228	278
4.0	205	385	478	437	349	257	312
4.5	206	428	380	476	311	274	413
5.0	245	429	391	452	284	286	309

(b)

(m)	0.0	0.5	1.0	1.5	2.0	2.5	3.0
0.0	32	173	-22	128	-70	30	-60
0.5	-32	-51	122	201	-35	-30	-54
1.0	-82	19	176	145	16	-16	-73
1.5	26	-89	-80	113	-77	-3	145
2.0	-31	9	121	-97	24	-90	0
2.5	8	161	117	28	133	-81	-81
3.0	-8	167	149	-28	112	227	133
3.5	-59	-57	-81	-77	-70	29	-50
4.0	1	180	-98	-88	-92	-29	-55
4.5	-1	222	11	165	-37	-17	139
5.0	-39	184	-11	168	2	-12	-23

Figure 4.14 Values of half-cell potential (with respect to copper sulphate electrode) E_{CSE} (mV) (a) and gradient ΔE_{CSE} (mV m^{-1}) (maximum difference from neighbouring cells) of half-cell potential (b) for a single parking slot.

in the CR of the surface area. Measuring points in the area which can already be assigned to repair need not to be investigated any more (see Figure 4.16).

The assumed location of areas with distress and the resulting condition state distribution with corresponding MR&R actions is given in Figure 4.18. The surface portion in each condition state can be plotted against the condition state, hereby obtaining the density distribution p of the discrete condition states (0–6). Integrating the surface portions (summing up from 0 to 6), we obtain the probability distribution P of the discrete ratings (see Figure 4.18). The probability density p is the information needed to built a Markov chain model. In [7, 16], this information is referred to as the condition vector q.

4.6.8　Overall condition rating

In a management system, the overall condition distribution of certain component groups (e.g. pillars) or objects (e.g. bridges) within a network must be known. Therefore all surface areas which are assigned to a certain CR must be summed up. This

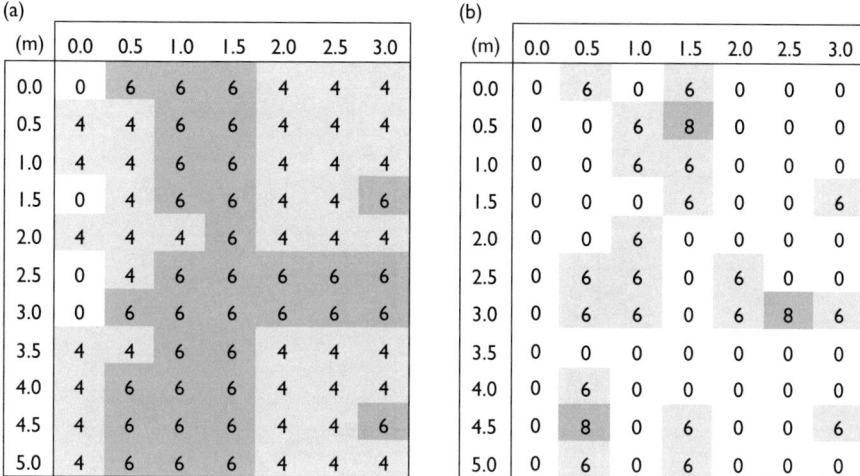

Figure 4.15 Scoring of half-cell potential *E* (a) and gradient of half-cell potential Δ*E* (b).

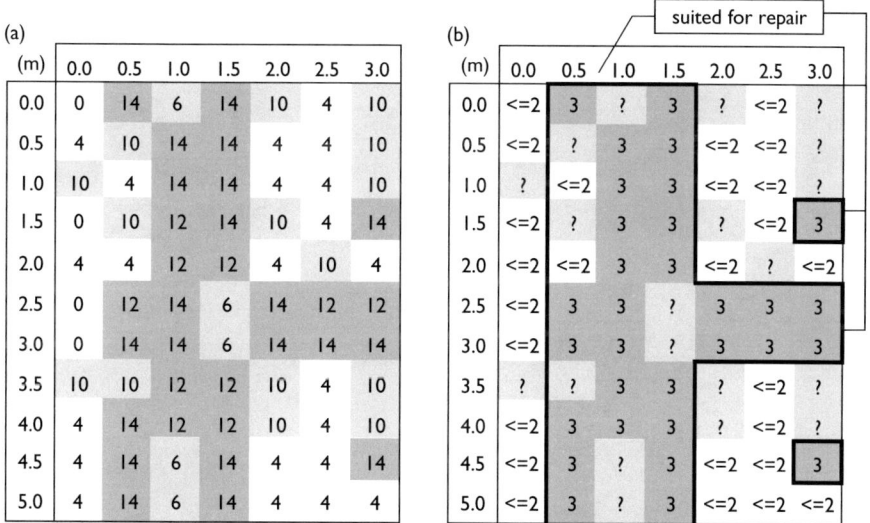

Figure 4.16 Total scoring (scores of *E* and Δ*E*) (a) and resulting condition rating (b). Areas in thick black lines (b) will be repaired as is most practical for such small sites. Areas with a '?' outside the repair area must be investigated in more depth.

is done from bottom (surface level) to top (object level) according to the modular systematic discussed in Section 4.4.

   The above example shows how to proceed for a single surface area. To summarise the condition state of components or an entire object, the surface areas in each condition state must be summed up and expressed as portions of the total surface area. By

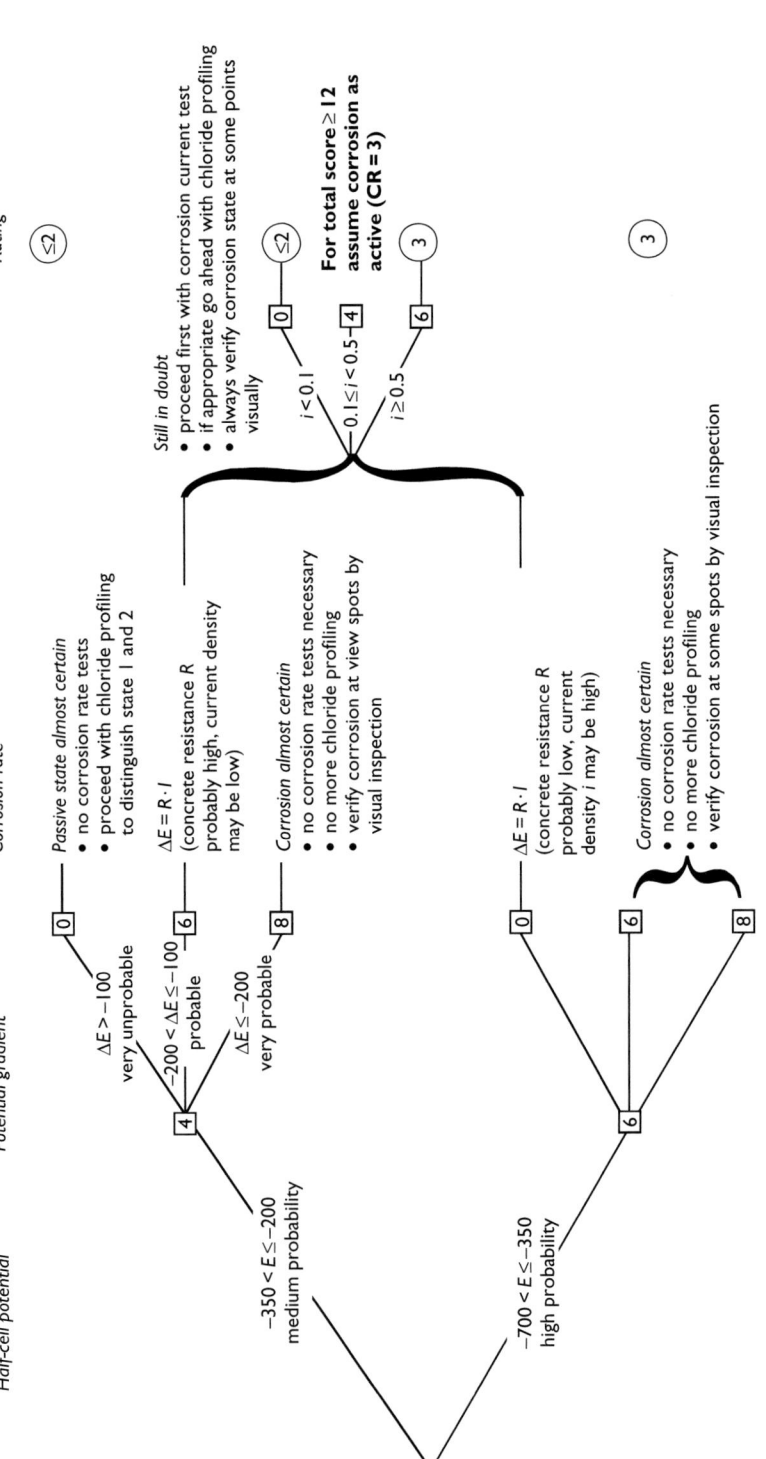

Figure 4.17 Scoring (values in boxes) of inspection results for the rating at condition boundary 2 or 3 (corrosion active or not?).

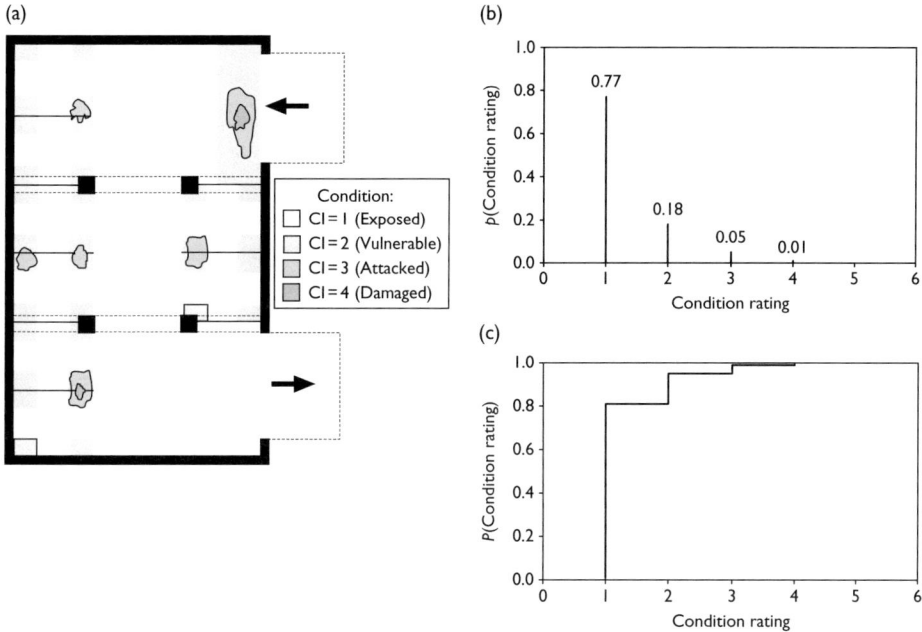

Figure 4.18 Fictive example of condition rating of a parking deck (a), corresponding probability density p (b) and probability distribution P (c) (portion of total surface area).

summing the products of portions per condition state, the corresponding unit costs and the total surface area, the current necessary budget for MR&R actions is obtained:

$$C_R = \sum_{i=0}^{6} p_i \cdot A \cdot c_i \tag{4.6}$$

where
  C_R total repair costs (euros)
  p_i portion of surface area in a condition state i
  i discrete condition states (0–6)
  A total surface area at regarded level (e.g. object level) (m²)
  c_i unit repair costs at each condition state i (depends on chosen measure) (euros m⁻²)

In current CA and management systems, average condition rates are calculated from the condition distributions of entire objects, which is used to prioritise objects for repair on a network level. The disadvantages of the system are that an average condition rate:

1. does not provide information on the expected costs for repair and
2. may be equal for various objects with very different viable types, costs and urgency for MR&R actions.

If we divide the above-stated total repair costs by the total surface area, we obtain the average (mean value) of the unit repair costs, which express the cost intensity of repair measures:

$$\mu(c_R) = \frac{C_R}{A} = \sum_{i=0}^{6} p_i \cdot c_i \tag{4.7}$$

These already give a very much better indication for prioritisation for repair by improving point 1 of the shortcomings. However, unit repair costs may still be equal for many objects in a network of objects (see point 2). This is because surface areas in a very good condition (with low unit costs) can compensate areas in poor condition (with high unit costs). A solution to this problem is to determine additionally the scatter (standard deviation) of unit costs of each object:

$$\sigma(c_r) = \sqrt{\sum_{i=0}^{6} P_i \cdot [c_i - \mu(c_R)]^2} \tag{4.8}$$

If two structures have an equal mean value of unit repair costs $\mu(c_R)$, then the one with higher standard deviation $\sigma(c_R)$ should be prioritised as more portions are already in a more severe state. Still, neither the mean value nor the standard deviation clearly indicates whether a safety risk is already present. Thus, as a third item of information, it should always be stated whether condition state 6 has been reached anywhere within the object ('red flag').

In summary, instead of calculating the average value of the CR, it is proposed to use the following information for prioritisation of repair actions:

1. portion of the object is in condition state 6 (–);
2. average unit repair costs $\mu(c_R)$ (euros m^{-2}); and
3. standard deviation of unit repair costs $\sigma(c_R)$ (euros m^{-2}).

4.7 Visual inspection

4.7.1 Aim and framework

Visual inspection will be performed for all surface groups included in the inspection plan. The major target of visual inspection consists of the identification and documentation of areas with visually detectable distress. The symptoms have to be quantified in such a manner that the necessary repair actions can be based on these in terms of the appropriate type and extent of repair measure. A certain portion of the inspection effort must always be dedicated to inspect the unexpected, which is most cost-effectively done by visual inspections. Therefore, visual inspections are performed at regular intervals, usually between 1 and 3 years.

Visual symptoms may appear rather late in a chain of processes within the line of a damage mechanism, which is thus associated with higher repair costs than earlier stages. Nevertheless, the degree of damage may still increase by far. So the information gathered by visual inspections should also be used to build up service life models or to update already existing ones. This is achieved by rating the inspection results, as described in Section 4.6, and plotting these ratings against time for a comparable set of structural elements. The curves thus obtained are fitted to a so-called 'failure probability matrix' according to the Markov chain approach [7, 16].

The survey is supposed to be done 'hand near', which means that the inspector may place his hand near the surface being inspected. This is because the inspector is also asked to use very basic inspection equipment (e.g. search for hollow areas by sounding with a hammer, see also 'Inspection Catalogue' in the annex). Visual inspections should be performed in regular intervals, e.g. every 6 years, as is commonly the case today. More frequent, superficial visual inspections may be performed without the application of additional access equipment (not hand near) at intervals of, for example, every 3 years. As build-drawings, photos, multimedia documents (digital video), etc., may not be available or may be incomplete, these should be provided during visual inspection if missing or if they no longer represent the current condition.

The elements constituting the framework of a visual inspection, as presented in Figure 4.19, are outlined in the following sections.

4.7.2 Access to the surface

A Prerequisite for any inspection is the access to the surface to be inspected. Direct access may be limited due to the following reasons:

1. confined spaces and
2. item is covered (e.g. asphalt and sealer on top of a deck; re-bar embedded in concrete).

In confined spaces access, for visual inspection can be achieved by endoscopic technique.

For concrete surfaces sheltered by layers on top, these need to be opened up to a certain extent. This should be done step by step with a minimum degree of destruction. The problem with this approach is the fact that the protective layers may still be intact and would be damaged by inspection. If it is obvious that concrete is securely sheltered from environmental loads, there is no need to damage protective layers. This can, for example, be assumed if protective layers have recently been renewed. Visual indicators of worn protective layers could be, for example, pot holes and cracks in top layers (e.g. asphalt).

Even without visual signs, however, protective layers could be ineffective. If no visual indicators of distress are given, an important indicator left for potential failure is the age. Until now no physical models for the service life of protective layers are known to the authors. Only rough estimates can be provided, which may be improved and applied in accordance with the 'Delphi Study' approach. If more data are available, these can be analysed for model building. Once a certain proportion of protective layers is estimated to be worn, visual inspection of concrete surfaces

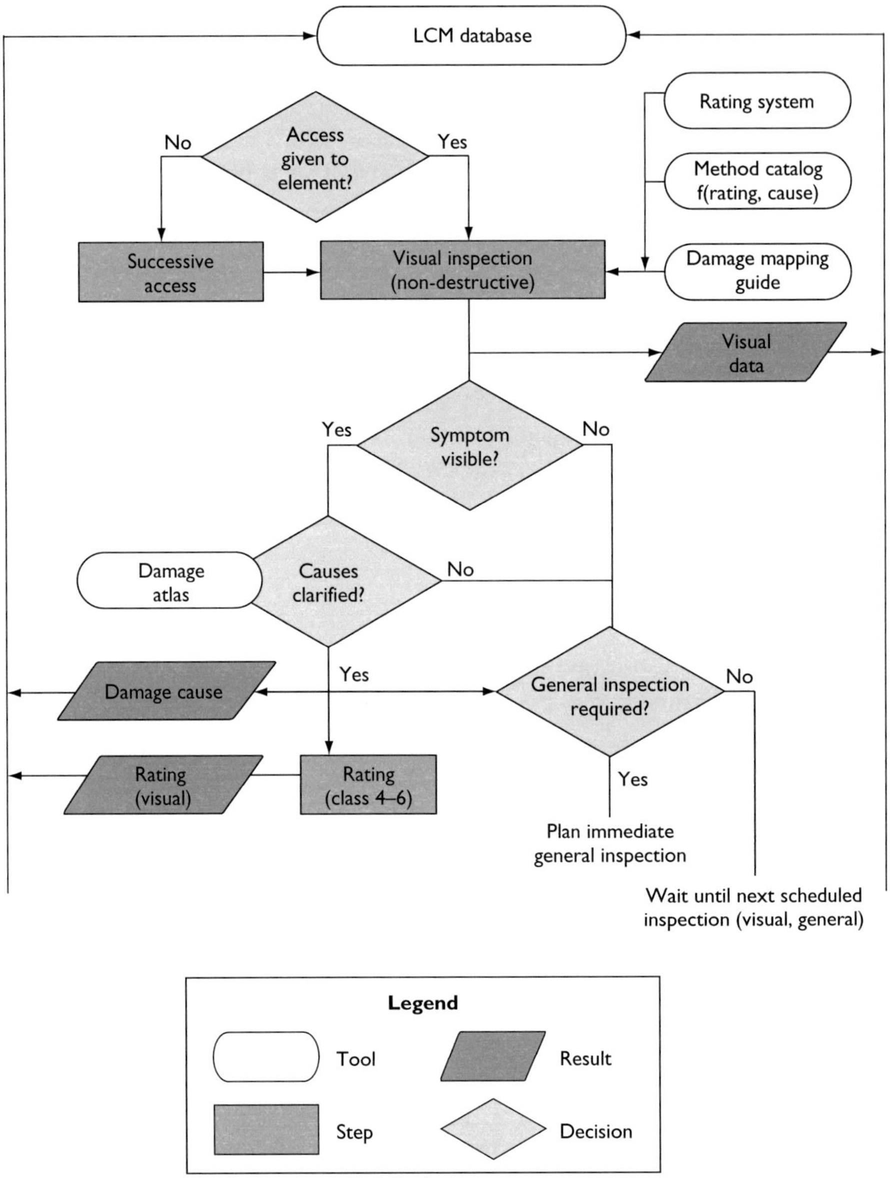

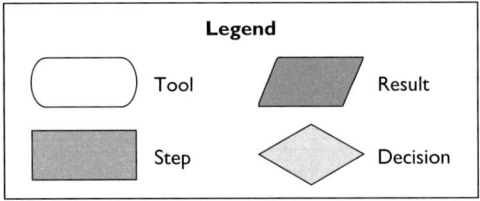

Figure 4.19 Flowchart for visual inspection.

underneath should be scheduled. 'Windows' of appropriate grid size are sawn into the protective layers. This has to be done in such a manner that the protective layers can be repaired afterwards (e.g. overlapping of sealers). The grids should be randomly distributed along the lines of critical areas of protective layers (e.g. joint areas of sealers) and supposedly intact areas.

4.7.3 Performance of visual inspection

4.7.3.1 Essential symptoms

'What to look for?'

Visual inspection is intended to detect the symptoms on concrete surfaces and protective measures for these. Likely symptoms to be detected are in general:

- deflection and movement deviating from the planned state;
- abrasion, cracking, blistering, pealing of protective layers (coatings);
- cracking of concrete;
- loss of concrete section (e.g. spalling, scaling or abrasion);
- water leakage;
- staining and efflorescence; and
- reinforcement corrosion.

Examples of symptoms in dependence on component type are given in Table 4.18. A very detailed 'checklist' for symptoms in dependence on bridge and component type is, for example, given in the Swedish Bridge Inspection Manual [23]. Similar documents exist in most European countries.

4.7.3.2 Damage mapping guide

4.7.3.2.1 GRID SIZE

'What is the appropriate grid size'

Usually surface groups (e.g. the bottom sides of a bridge slab) can be separated into comparable segments according to the fabrication procedure (e.g. fields of a bridge or segments of a tunnel). This is especially true for concrete structures, which are often produced in segments divided by joints. However, irrespective of whether such segments can be identified or not, a grid should be assigned to these surfaces to define the maximum number of potential samples. For visual inspection, the grid size should be chosen in a form appropriate for the expected size of damaged areas; and moreover, the inspector should be able to document visual symptoms (e.g. cracks, spalling, staining) without having to change the position, which is in the range of 2–5 m of grid size, depending on the perspective of the inspector. The number of grids into which a surface is divided represents the maximum number of samples for visual inspection. Only a proportion of this maximum number of grids are chosen for inspection. So the inspector also has to adapt the grid size to the inspection volume. If the inspection volume is low (e.g. 5% of a surface), then a smaller grid size is appropriate to still have an adequate number of grid samples for statistical analysis.

4.7.3.2.2 DATA FORMAT

'What data format is required for the documentation of symptoms?'

The mapping of symptoms found during inspections in general, especially inspections of visual deterioration indicators, requires a systematic procedure if the data

Table 4.18 Examples for symptoms and aspects during visual inspection. (Inspected aspects 1–4 are detailed in Section 4.7.3.2)

Inspected aspect		Symptom
1	Deformations in general	Vertical and horizontal displacements, rotations, settlement, buckling
2	Sealers and paints	abrasion, loss of adhesion, cracks, pealing
3	Components of concrete	Cracks, delamination, buckling, deflection, soaking, efflorescence, corrosion stains, hollow areas, spalling, scaling, dissolution of concrete (chemical attack)
4	Reinforcing and pre-stressing steel	Corrosion of stirrups and main re-bars, corrosion of ducts and pre-stressing reinforcement, exposed tendons, exposed reinforcement and stirrups, broken re-bars, exposed reinforcement and stirrups, broken pre-stressing tendons
5	Concrete foundations	Settlements, overturning, scouring, abrasion and symptoms of Aspects 3 and 4
6	Steel and metal constructions	Cracks, deformation, corrosion
7	Bearings, expansion joints, hinges	blockage of movements, loosening of parts, tightness, corrosion, position, deformation, mechanical damages of steel parts
8	Road surface	wheel tracks in the pavement, cracks, blisters, abrasion, unclean surface
9	Cover elements of walls and ceiling	Cracks, deformations, hollow areas, efflorescence, corrosion (joints)
10	Noise-reducing walls	Cracks, deformations, hollow areas, efflorescence, corrosion (joints)
11	Safety equipment	Loosening of joints, corrosion, deformation, missing components
12	Pipelines	deformation, loosening, corrosion
13	Surrounding landscape	Erosion of riverbed slopes, scour, excessive vegetation on the riverbank slopes or embankment slopes, missing parts or damaged linings of the slopes
14	Structural timber	Moisture, fungus, moss, breakage of timber elements, etc.
15	Stone, brick, plaster	Cracking, spalling of mortar, ingress of moisture, separated bricks and stones, etc.
16	Expansion joints	Mechanical damage of steel parts, deterioration of sealing, leakage, elements, corrosion of steel elements
17	Drainage systems	Corrosion, missing elements, unclean drainage system
18	Cladding	corrosion of fixings
19	Fix points	movement, loss

are to be used for an LMS. The data must be in such a way that repair actions can be based on them. If a damage is detected during the CA, then its cause (mechanism), intensity, extent and location need to be documented. The cause of damage (e.g. cracking caused by chloride-induced corrosion of reinforcement) is essential in order to choose (a) the adequate repair method and (b) the correct service life model for prediction of the future development of the condition state. If in doubt, a 'Damage Atlas' may be consulted to identify the cause, as is provided in the annex. It is very important to distinguish between intensity and extent of CR for degradation. Intensity is a physical measure of the degree of damage (e.g. crack width, penetration depth, etc.). Extent quantifies the amount of surface area or samples damaged at equal intensity. For documentation of symptom location there are, in general, two options.

1. The common approach is to prepare a drawing. If possible, the symptoms are indicated in existing plans in combination with cause, intensity and extent of symptom. Such drawings should be accompanied by photos or other multimedia files (e.g. videos) and verbal descriptions. A 'Damage Mapping Guide' including a list of required data, guide to symbols for drawings and required information for each symptom is provided in the annex. The symptoms are drawn on a map at the scaled location using the symbols appropriate to the structure element. Moreover, the symptoms are labelled using the symptom type number and the defined rating level if applicable. Additional symbols may have to be added to provide more information on the condition state: for example, to indicate whether the crack or joint is well sealed; whether a crack is filled with water or dry; and whether a crack is moving or not. From these kind of documents, digital files may be produced which are stored into a database at the respective site of the modular tree.
2. Alternatively, the type and location of an observed symptom may be indicated with a code, accompanied by the extent.

Which option is preferred is left up to the reader of this chapter. In any case, the cause and total extent of symptoms also have to be stored in a form which enables automatic use of the data during the later management process. Photos and drawings are in general quicker to comprehend and are the common means of communication between engineers. This is why in the existing modern CA systems, photos and drawings are already integrated into digital form. If this option is chosen, an organised system for preparing drawings should be applied. Such files may, however, call for considerable data storing capacities and thus slow down IT systems. The option of using a code system consumes very little data storing capacities but requires considerable efforts during the planning phase, as every item in the modular tree must be given a code. Given today's rather inexpensive IT-systems, the first option of inserting drawings is probably the more user-friendly and in the long run more economical solution. For completeness, coding is nevertheless outlined in the following text.

As applies for the levels for the modular systematic, most countries assign codes to identify the main structural parts or areas of a structure. For LIFECON, a code system was developed, following the set-up of the modular levels. The code system shown in Figure 4.20, which follows the modular systematic, allows the position of every regarded area to be identified.

Level 1: Object   Indicates the *name of the object*, which should be unique in the regarded stock, e.g. Bridge 1.

Level 2: Module   Capital letters (A, B, etc.) indicate the location of a module *in the vertical direction (bottom to top)*. In the case of a bridge, 'A' would be assigned to the foundation, 'B' to the sub-structure and 'C' to the super-structure. In the case of a tunnel, 'A' would be given to the foundation, 'B' to the shell and 'C' would be ignored. In the case of a building, 'A' for foundation, 'B' for floors and 'C' for roof.

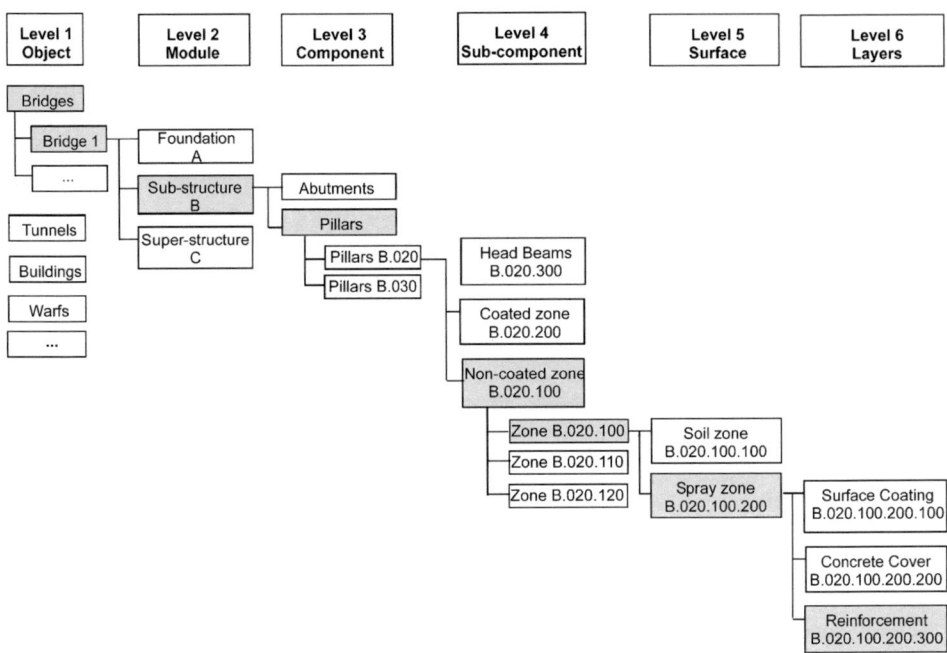

Figure 4.20 Example of the storing of damage sites using a code system according to the modular systematic.

Level 3: Component In axial direction, each module is divided into segments. For tunnels, bridges and retaining walls, the axis is defined by the traffic direction. For buildings, the axis follows the vertical direction. A *three-digit number* added to the letter of the module indicates the *beginning of a new segment in longitudinal direction to the structure axis*. For road structures, the numbering starts at the road joint in longitudinal direction (e.g. with the number A.010) and is increased by increments of ten (e.g. at pole rows or joints, A.020, A.030, etc.). The numbers A.001–A.009 are reserved for segments which are located in front and numbers A.XX1–A.XX9 for those located at the end. *If a component is located in the beginning of a segment with main extensions perpendicular to the axis, the component is given a capital letter with the code of the following segment* (e.g. a row of bridge pillars located at the beginning of the second field is given B.020, those pillars at the beginning of the third field B.030, etc.). *If a component extends over a segment in axial direction of the structure (i.e. the slab of a bridge super-structure or walls in a building), the code is composed of a capital letter and the digits for the beginning and the end of the regarded segment* (e.g. the slab of a bridge super-structure of the second bridge field is called C.020–030). For buildings, this division has to be applied in the vertical direction counting with increasing numbers from bottom to top: A.010 for foundation, B.020 for first ceiling, B.030 for second ceiling, etc. The pillars, walls and stairs in the first storey are given the code B.010–020, the ones in the second floor B.020–030. This system enables the assessment of 99 segments, e.g. bridge fields or storeys and should cover the existing common structures.

Level 4: Sub-component (equal resistance) *Cross sections* (perpendicular to the defined axis) of each segment are looked at. In addition to the above-mentioned number, a *three-digit number for the sub-components*, e.g. B.010.*100*, is added. This number is incremented by steps of 100 from bottom to top. If two or more sub-components of the same type are included in the cross section, they are distinguished by their decimal places, e.g. B.020.*300* and B.020.*310*. Example: Consider a row of three pillars at the beginning of the second bridge field (i.e. at site B.020). The portions of the pillars in the soil and above the splash zone are not coated. The bottom portions above ground level are coated to protect from splash water containing de-icing salts. The various portions of the pillars thus exhibit differing resistance towards the ingress of corrosion-inducing media. The non-coated portions are coded B.020.1XX, and the coated ones are given the code B.020.2XX. To identify the areas of an individual pillar, the decimal places are used. The coated portion of the left pillar is coded B.020.200, the middle one B.020.210 and the right one B.020.220.

Level 5: Surface (equal resistance and environmental loading) Assign, for the differentiation of environmental influence and stress on a surface, a three-digit number to the number code in addition to the already existing number. This number is incremented from bottom to top by steps of 100. Example: The code B.020.110.*100* is assigned to the surface area of the bridge pillar submerged in soil, whereas the upper shaft is exposed to the atmosphere (mist from traffic) and given the code B.020.110.*200*.

Level 6: Layer Attach a three-digit number to each damaged layer. Example: if the reinforcement is of the above-mentioned area is corroded the code is B.020.110.200.*300*.

An application of modular systematic and coding system and an example of detected symptom and coding for documentation are shown in Figure 4.21 and 4.22 respectively.

4.7.3.3 Cause of damage

It is an important principle of rehabilitation that the cause of damage must be controlled or stopped whenever possible. Generally, causes of damage are:

- environmental influence
- load
- wear
- erosion
- accident
- design error
- construction error
- maintenance error.

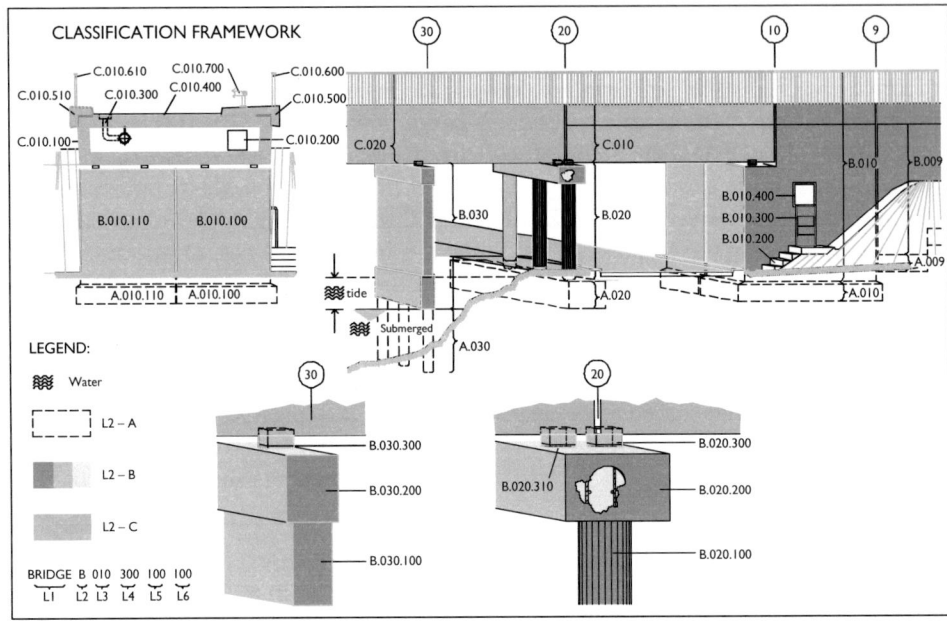

Figure 4.21 Application of modular systematic and coding system.

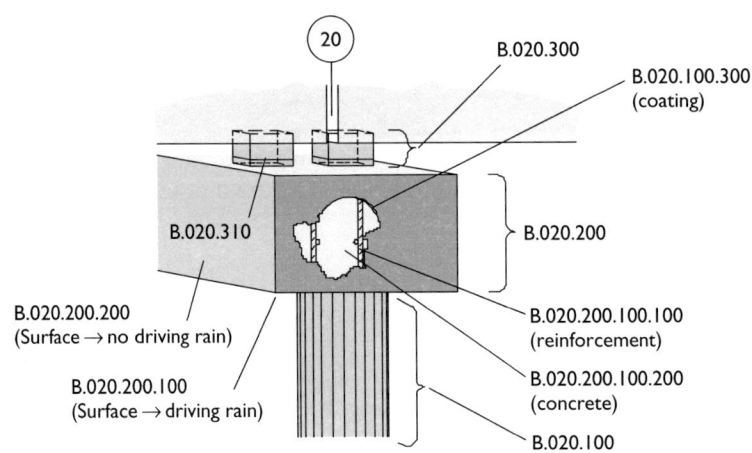

Figure 4.22 Example of detected symptom and coding for documentation.

The exact cause of a visually detected damage cannot be usually identified until sampling takes place. To derive an 'expert system' for the determination of the cause of visually observed damages is a task which would produce a very valuable tool for the future support of experienced inspectors.

Table 4.19 Examples of reference sources for analysis of cause for
visual damage symptoms

Symptom/mechanism	Examples of references
Origin of cracks in concrete	[24–27]
Corrosion	[25, 28, 29]
AAR	[25, 30]
Frost	[25]

Probably, the best structure would be a decision tree guiding the inspector from visual symptoms to further testing ('General Inspections') until a final conclusion on the cause can be drawn. In Table 4.19 and in the following section, only a list of valuable references and a description of the basic mechanisms are provided.

One of the most promising existing expert systems to start off seems to be the software HWYCON, which has been developed under the Strategic Highway Research Program (SHRP) and can be purchased from the Transportation Research Board (TRB). HWYCON is a computer expert system that helps diagnose the cause of distress in concrete pavements and structures, determine appropriate repair and rehabilitation strategies, and select optimum construction materials. The developers acknowledge, however, that the system will be most useful to fresh engineers and inspectors and that the scope of the software is limited. Still bridge decks and sub-structural elements such as columns, piers and parapet walls are included.

4.7.3.4 Damage Atlas

In order to:

- identify the cause for observed symptoms;
- comprehend the environmental aggressiveness classification;
- localise critical areas with respect to deterioration;
- interpret inspection results;
- assign ratings to an observed symptom; and
- suggest repair measures, etc.,

the inspector must have profound knowledge of damage mechanisms leading to observable and measurable symptoms. There is a vast amount of literature existing on the various deterioration mechanisms. The Damage Atlas of the CAP only serves to introduce the user to the basics of each deterioration mechanism according to the following structure:

1. general process description;
2. conditions conducive to mechanism;
3. visual indicators;

4. identifiable service life stages;
5. existing sampling procedures and confirmation tests (detailed description is given in Inspection Catalogue – annex);
6. existing laboratory investigations (detailed description is given in Inspection Catalogue – annex);
7. maintenance and repair actions;
8. existing management procedures for the considered mechanism; and
9. literature references.

The 'Damage Atlas' in the annex is not intended to be comprehensive. More detailed information relating to specific peculiarities has to be looked for in existing literature, using the literature references provided as a starting point.

4.7.4 Need for General Inspection

If the target is to determine data for the modelling of future deterioration, General Inspections are always performed after a preceding visual inspection. This is done to measure the dictating parameters for input to and update of parameters in service life models. For other targets, a general inspection may not always be necessary (e.g. estimation of current value will most likely be based on purely visual inspection). Thus criteria need to be set, which demand for 'General Inspections' (Table 4.20). This criteria need to be set by which to measure the requirement for General Inspections.

Table 4.20 Criteria demanding a 'General Inspection'

Criterion	Example
Rate of deterioration deviates strongly and negatively from expected values	Amount of cracks due to ASR increase at a higher rate as predicted
Cause of deterioration unknown and extent of concern	Severe cracks in the super-structure
Deterioration is likely to be hidden	Corrosion of re-bar in bridge deck covered by defective sealer and asphalt layers is assumed according to service life models or experience
Deterioration of unknown extent	Extent of chloride ingress and corrosion of re-bars are unknown (as are not estimated using 'a priori' design models) but are expected from experience to have progressed considerably
Deterioration of severe extent	Structural reliability has become an issue
Deterioration rate required but unknown	Collection of input data for service life models is necessary (because no model is yet available) or economically beneficial because uncertainties and thus risks will be reduced substantially

4.8 General Inspection

4.8.1 Aim and framework

Visual inspections are not suitable to detect and quantify deterioration at an early stage, as visual symptoms are usually not yet observable. More advanced techniques, preferable non-destructive ones, must be applied. The analogy to the field of medicine is obvious: in certain cases (e.g. a flu), a doctor may evaluate the current condition and cause of the disease from mere visual symptoms and verbal descriptions of the patient. However, in many cases, only a more thorough investigation using advanced techniques, such as X-ray, blood analysis, etc., can give the desired answers. Nobody would expect from a doctor the impossible to evaluate the condition and cause of any disease on the basis only of visual symptoms at hand. In the field of maintenance of concrete structures, this is, however, still the case, with the disadvantage on top that concrete structures do not express their concerns as does a human patient, if not equipped with monitoring devices while still under construction. Both doctors and constructional engineers have to predict the future development of the condition of their 'patients'. Currently, both do so using their expert knowledge and personal experience. The constructional engineer has the benefit that concrete structures are far less complex than human bodies and their behaviour can and thus should be predicted making use of existing models.

The following sections deal with the most important parameters (beyond mere visual symptoms) needed for:

- assessment of the current condition;
- model building for prediction of future condition; and
- the strategy and methods to obtain these parameters

following the flow of Figure 4.23.

4.8.2 Parameters determined during General Inspection

'What parameters can be determined?'

The information needed to perform a rating of the current condition has already been outlined in Section 4.6. Most of the required information is not obtainable by visual inspection. More advanced inspection techniques must be applied as provided in Tables 4.8–4.15 in dependence on the condition states, which are intended to be distinguished. Most of the information needed to describe the current condition can also be used for model building. Apart from these, more specific data can be collected, which can be used for predicting the future condition state. The inspector must thus be familiar with the current possibilities for modelling the deterioration. This aspect must be incorporated into the training of inspectors. Variables on the resistance side may partly already be updated during the 'acceptance'. Some of these resistance variables and all of the stress variables can only be determined after environmental loading has taken place for a sufficient long time. A list of all parameters which may be updated is given in Table 4.21.

It should be borne in mind that predictions become more and more reliable with increasing amounts of information. But updating of all the listed parameters for every

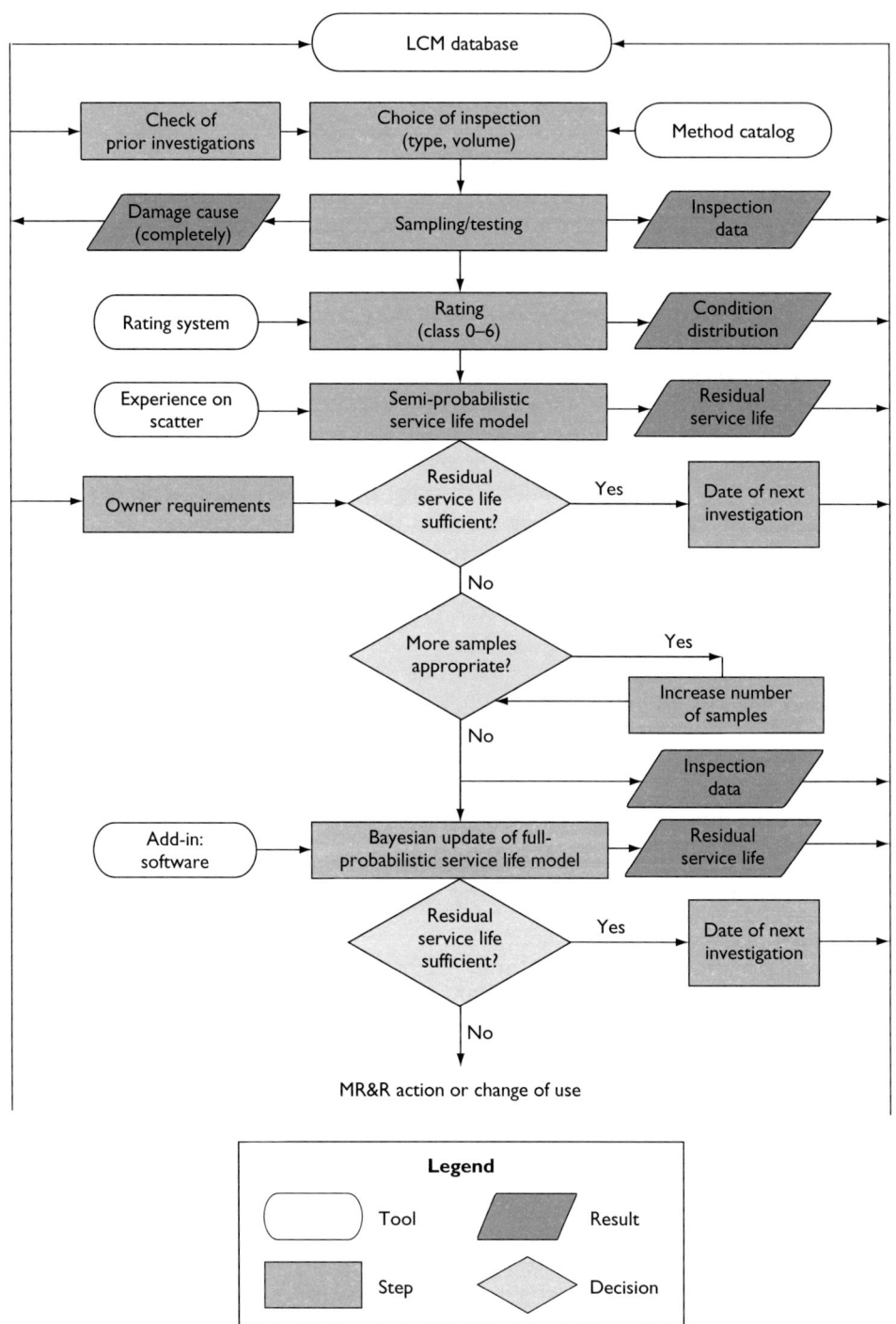

Figure 4.23 Flow of General Inspection.

Table 4.21 Resistance (R) and stress (S) parameters which can be determined through inspection for model updating (for details of test methods see annex)

Mechanism	Parameter[1]
Coating deterioration	• Abrasion depth of coating (S) • Electrical resistivity of coating (R)
Carbonation	• Carbon dioxide concentration in air $\Delta C_S(S)$ should be monitored if concrete surface is subjected to much higher concentrations than are common for atmospheric exposure (e.g. in tunnels) • Concrete cover $d_{cover}(R)$ • Carbonation resistance R_{ACC} (accelerated laboratory test) with separate specimens (cubes or cores) or cores taken from structure (R) • Carbonation depth $X_c(S)$
Chloride ingress	• Concrete cover $d_{cover}(R)$ • Rapid Chloride Migration coefficient $D_{RCM}(R)$; as specimens (cylinders) or cores taken from structure were already subjected to chlorides, a modified test with iodide instead of chloride in the test solution must be applied (Rapid Iodide Migration) • Chloride profiles are taken to determine apparent diffusion coefficient $D_{app}(S)$, concentration $C_{S,\Delta x}(S)$ and depth of convection zone $\Delta x(S)$ • Critical corrosion inducing threshold value $C_{crit}(R)$ is determined with chloride profiles at sites where re-bar is depassivated but has not been active for long periods yet
Re-bar corrosion	• Concrete cover $d_{cover}(R)$ • Electrical concrete resistivity $\rho_0(R)$ in water-saturated state with separate specimens (cubes) or cores taken from structure • Corrosion rate (S) in combination with subsequent visual determination of anodic surface area A_a at inspection site • Corrosion penetration depth (average $P_{AVG}(S)$ and maximum $P_{max}(S)$) at corrosion pits
Structural effect of corrosion	Resistance (R) to corrosion-induced cracking: • tensile strength of concrete f_t • E-modulus of concrete • re-bar diameter d_S Volume expansion (S) leading to corrosion induced cracking: • loss of re-bar section ΔA at spots with cracks that have recently developed for calculation of the section loss necessary to induce cracks Crack growth and time until spalling: • loss of re-bar section $\Delta A(S)$ at spots with larger cracks and spalling Steel specimens can be taken at representative spots to determine (R): • Yield strength of re-bar in corroded state f_y • Ultimate strength of re-bar in corroded state f_u • Strains (λ_y and λ_u) of re-bar in corroded state
AAR	• Monitoring movement of the affected structure component (S) is the preferred method. Instead monitoring the expansion of moist cores from the affected structure (S) can also be performed.
Internal frost damage Frost scaling	• Monitoring of degree of water saturation (S) • Degree of water saturation of recently damaged concrete leads to the critical degree of water saturation (R) • Weight of scaled concrete per area (S) can be monitored to obtain a function of scaling depth against number of frost cycles on site

Note
A parameter is considered as a resistance (R) stress (S) variable if an increase in value will lead to a lower/higher degree of deterioration (failure probability).

individual structure under investigation will cause considerable inspection efforts and is not practical. However, especially owners of large infrastructure networks may investigate a few structures in more depth, which are representative of a large portion of their entire stock, to calibrate existing service life models on an engineering level following the proposed bottom-to-top approach to model building.

4.8.3 Procedure for General Inspection

4.8.3.1 Choice of inspection techniques

Usually, a variety of inspection techniques exist to investigate a certain deterioration mechanism. These serve, in general, for two purposes:

1. identification and verification of mechanism; and
2. provide input to models for the prediction of residual service life.

A useful catalogue to receive method descriptions can be in [31], which is structured according to the:

1. investigation target (e.g. chloride ingress); or
2. denotation of the method.

The structure of each method description is as follows:

1. picture, title and key-word, short description and classification according to
   - technique notation
   - destruction degree – non-destructive, medium destructive, destructive
   - location of testing (laboratory, in situ)
   - evaluation effort
   - handling (degree of education)
   - investigation effort

2. characterisation
   - boundary conditions for application
   - influences
   - applicability for structure investigation
   - target
   - applicability for regular inspection
   - applicability for monitoring
   - versions
   - costs
   - remarks, recommendations, alternative methods

3. application (principle description, peculiarities, qualification of personnel, combination with different methods, criteria and data on accuracy, data format)
4. evaluation
5. producers

6. literature
7. miscellaneous
8. www links.

Techniques and thresholds for evaluation of data are rarely provided when using those methods, which are stated in the 'Inspection Catalogue'. In general, for all inspection techniques, one should have information on:

- the probability of detect damage, if present (can be incorporated for Bayesian update) and
- accuracy of results.

'Which technique is most suitable?'

Few inspection techniques can detect a damage mechanism with a high degree of certainty. Common practice is to use combinations of various techniques.

4.8.3.2 Choice of location for sampling

By following the 'Modular Systematic', an object is divided into segments with equal resistance and equal load classification. Those segments which are subject to the largest risk are prioritised for inspection. If environmental loading can be considered to be equal for the entire surface, samples should be randomly taken from grids of the segment.

Still, within sections, environmental loads may be a function of site, which has to be taken into account. For instance, chloride surface content is a function of distance from chloride source. Although a concrete pillar may be produced with equal concrete composition and cover depth over the entire height, chloride loading will, in general, be more severe close to the source, e.g. traffic or sea level. But geometrical set-up and environment side effects can still lead to quite large deviations from such general rules (lee side may be affected more severely than the weather side, because of air flux conditions, etc.). Specific guidelines for the choice of inspection site in dependence on the regarded deterioration are rather difficult. The inspector must always be aware of the peculiarities of the investigated site. Moreover, the inspector should be familiar with the input parameters to degradation models and be aware of the major influences on these parameters, in order to determine the qualitative profile of the loading in a section, e.g.:

- surface chloride content of vertical structures above the sea level or alongside traffic is higher at the bottom than at the top; and
- driving rain hits a sheltered vertical surface more frequently on the bottom than on the top, hence carbonation should be higher on the top than on the bottom.

Sampling should take place according to these qualitative relationships. Thus, samples should atleast be taken in the areas with most severe, intermediate and lowest loading. This is done to find the function of loading against site, which enables to repair only the areas which are most deteriorated and exclude others, if funding for maintenance is restricted (patch work).

Besides site-dependent relationships for environmental loading, there may be areas which are subject to localised environmental loading (e.g. areas loaded by leaking joints, tyre position of cars with chloride-containing salt slush in a parking deck, etc.). These are here referred to as 'hot spots' and must be included in the sample volume likewise as areas with lesser loading.

The problem is that 'hot spots' and areas with lesser loading are not always obvious from mere visual observation. If non-destructive techniques for mapping of damage indicators exist for a regarded mechanism, these should be applied first to choose sites for later cost-intensive sampling. An example is the mapping of the half-cell potential, as already outlined in Section 4.6.7, which can be used to determine areas where sampling of chloride profiles is appropriate.

4.8.3.3   Choice of grid pattern and number of samples

Dividing a regarded segment into a regular grid pattern is appropriate for most types of inspection techniques. The size and, therefore, the number of grids is an important measure, which may have a large influence on the result of an inspection. Grid pattern defines the number of maximum number of samples. The choice depends on:

1. the scatter of expected results, which depends on the accuracy and resolution of the inspection method and on the type of deterioration (large scatter demands smaller grids, i.e. a large number of samples); and
2. the level of precision of the inspection.

The common approach will be to start with a fairly coarse grid pattern, which can be intensified if intermediate results suggest it. From a statistical viewpoint, the inspector must be aware that there may be correlations between individual points for measurement. An example is the half-cell-potential mapping: a very small grid pattern (less than 0.20 m) will usually not be appropriate, as small anodic areas will influence the measurement results in their vicinity (the resolution of the measurement depends very much on the moisture state of the concrete). On the other hand, anodic areas will provide cathodic protection for others. So the grid should not be too coarse, to still enable the distinction of these areas between two measurement points. A generally accepted grid size for potential mapping is in the range of 0.5–2 m.

4.8.4   Evaluation of residual strength

A proactive concept means that the intended is to provide a framework for early detection and control of deterioration. Guidelines on structural system assessment are not intended here. Nevertheless, a criterion has to be set, which indicates that safety has become an issue. Here, by definition, a loss of 5% bearing capacity of a cross section will immediately call for a 'Structural Assessment' by specialists.

A guideline to check whether this state is reached is provided in the European project CONTECVET [19–21].

4.8.5 Application of semi-probabilistic service life models

Decisions on maintenance actions need to take account of the future development of an object. Future development of deterioration can be predicted by models, which exist on various levels of sophistication. As a first step, simplified engineering models are used on a semi-probabilistic basis. Inspection data is the only input, which combines material resistance and environmental action. Usually, only a very small number of samples will be taken first. These data are used to calculate the mean value of a regarded parameter. In most cases, this small number of samples will not be sufficient to determine the standard deviation with the desired level of confidence. Thus the standard deviation is assumed on the safe side from knowledge of more extensive investigations and theoretical considerations. The output of such calculations is the time until a regarded limit state, which is part of the CR scale, is reached with a pre-defined probability.

The model structure chosen is simple enough to perform such calculations manually (e.g. with a simple pocket calculator). If test results are available already on site (e.g. depth of concrete cover and carbonation depth), service life calculations can be performed right away to decide whether more sampling is appropriate or not.

The results of the calculations on residual service life are compared to the owner's requirements. If demands with the rather pessimistic semi-probabilistic models are satisfied, there is no need for further inspection. Obviously, the considered deterioration process is negligible. The next step is to schedule the next date of inspection.

4.8.6 Date of Next Inspection

When further inspection is not required, the CA ends with the determination of the 'Date of Next Inspection'. The actual component condition should influence the inspection interval in each individual case. In certain cases, continuous monitoring may be required. Factors to consider when choosing inspection intervals are given in [8]. In summary, inspection intervals should be assigned in dependence on risk, being the product of:

- the consequences – safety, functionality, economics, etc.; and
- the failure probability – rate of deterioration, structural set-up.

Consequences can be considered to be constant over time. Failure probability, and therefore risk, is time-dependent. Deterioration rate can be expressed in the form of damage functions. Such damage functions represent the average condition of a regarded surface area or an entire structure against time (Figure 4.24).

After inspection, a life cycle analysis can be performed for the individual structure. Result of such analysis will often be to wait a certain period until the optimal moment for repair is reached. The maximum interval for inspection is the time between the last inspection and the time when repair costs will be at the lowest for the remaining service life. The time for recommended repair is defined by the average condition state (in Figure 4.24 – degree 4) which is considered to be the optimum state for repair action.

In theory, no more inspection would be necessary until the next date of repair, if the prediction of deterioration matches with the real behaviour. But damage functions will always deviate from real behaviour. Safeguards have to be met instead of solely trusting in service life calculations, to ensure that repair actually takes place

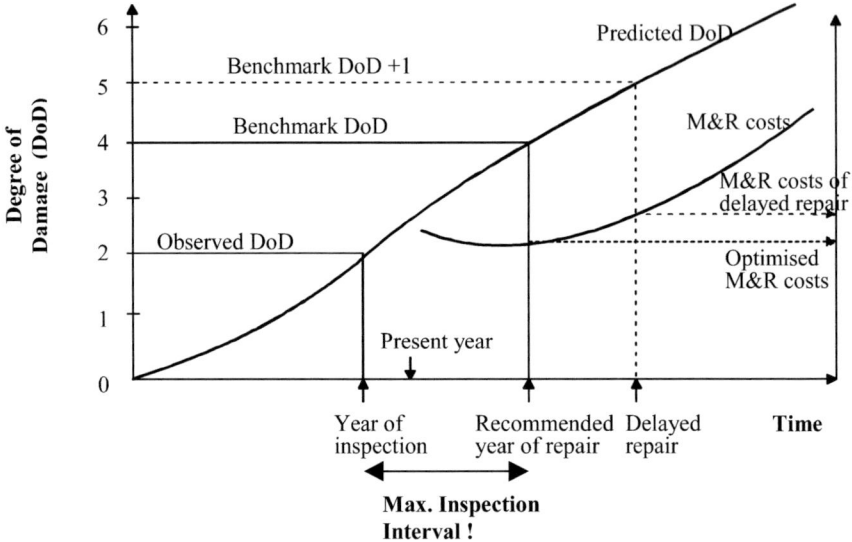

Figure 4.24 Maximum inspection interval in dependence on deterioration rate and repair strategy.

close to the optimal moment in time (Figure 4.24). Based on the above reasoning, it is proposed that the next inspection, when the deterioration rate is determined once over, is planned for the date when the middle of the next condition state is reached. This gives the possibility for the updating of models for a specific structure. The advantage is obvious: for slow degradation development (very flat curves), the risk of failure (passing the optimum point of repair) is low. Therefore, inspections will be scheduled at longer intervals.

For example: say the current condition state of a surface area or an entire structure is 2.5. The next repair is planned for when a condition state of 4 is reached (Figure 4.24). So the next inspection is assigned to the date at which state 3.5 is estimated to be reached.

In summary, the next inspection date is determined as follows:

- Determine the current average CR, which will be in the interval of $i \leq CR \leq i+1$;
- Calibrate a damage function expressed as CR over time (Markov Chain);
- The next inspection should take place once half of the next condition interval is reached, i.e. for $CR = i + 1.5$.

The procedure above is only applied to determine the date of the next general inspection. Visual inspections should be performed at regular intervals (e.g. 3–6 years), as is the current practice. This is necessary because:

- data from visual inspections of entire objects are regularly needed for long-term analysis on a network level [16]; and
- safeguards must be met to detect the unexpected (damage due to accidents, misuse, etc.).

4.9 Repair Inspection

4.9.1 Aim and framework

If a decision is made to perform maintenance works, the scale of such works will be based on partial sampling. The results of such sampling may not always be applicable to an entire surface group. This is because access cannot always be provided to the entire surface area or item group. During maintenance works, such areas will be opened (e.g. asphalt and sealer of bridge deck). The entire component then becomes accessible for inspection. Although this will cause further inspection costs, the advantages are:

* maintenance costs can be reduced, because areas with a lesser degree of deterioration can be detected, which demand for less repair and thus cause lower costs; and
* 'Hidden failure', i.e. areas with more severe deterioration than could be predicted from local inspection, can be detected.

The extent of maintenance work can and should be updated in consequence. Nevertheless, on a network level, decisions have to be based on inspection data before repair works are carried out.

Comment:
As the assumed repair volume during the network planning is subject to uncertainties, these should be taken into account in a probabilistic way!
'Does repair work fulfil the requirements?'
  Subsequent to repair, the quality of works can be checked and used for parameter update.

4.9.2 Reporting

The presentation of results in a unified system is an important issue, but depends very much on the preferences of the administration or consultants performing the CA. Only very general guide can be given. The following items:

* introduction
* main report
* conclusion
* enclosure

form a report and are explained in detail in [2]. The principal groups of information to be included in a report according to [13] are listed in Table 4.22.
  In [32], the following proposal has been made for the contents of an inspection report:

1. administrative and technical data

    * reason for inspection
    * available documents
    * object data

Table 4.22 Principle groups of information to be included in a report according to [13]

Phase/aspect	Necessary information
Preliminary desk study	list of available drawings form of construction type of concrete pre-stressing and reinforcement
Preliminary site inspection	amendments to construction information principal areas and defects requiring site investigation
Site inspection	tests for corrosion concrete material tests results from internal examination of tendon ducts concrete and steel stresses (including temperature conditions)
Structural condition	long-term durability consequences of structural failure
Risk assessment	structural consequences of broken tendons potential collapse mechanisms

2. documentation of damages

    - evaluation of prior investigation data from database
    - data of performed inspections

3. processing of damage data

    - evaluation methods (e.g. statistical methods; assumptions for static calculation)
    - results

4. evaluation

    - description of damage causes
    - textual evaluation of damage
    - rating of damages
    - explanation on data to be inserted to database

5. recommendation on MR&R actions

    - estimation on MR&R costs
    - costs of single actions
    - comparison

6. summary.

4.10 Conclusions

Two essential shortcomings can be identified in today's CA practice:

1. available service life models enabling predictions on future state are not incorporated; and
2. the assumptions as to future deterioration and necessary actions depend very much on the personal experience and preferences of the inspector in charge.

These shortcomings have been tackled by achieving the following:

- The possibilities offered by existing probabilistic service life models and reliability theory are integrated into the framework for CA of concrete structures. Thereby, the best use of scarce resources is made for structure inspections and an objective basis is provided for the decisions on MR&R actions.
- An organised system is provided for collecting, recording, evaluating, rating and storing of data (administrative, technical and inspection data).
- The framework developed follows the idea of adaptation. This means that data are only collected if it is necessary to do so, with appropriate sample volume, and that these data are suitable for the defined purpose of the specific assessment.
- The approach can be applied by users managing assessment projects from the small to the very large.

The CAP in its current form is mainly based on theoretical considerations and has only been tested to a limited extent. Most of the concept is very different and will at first cause more effort for planning of inspections compared to current ways of performing a CA. This increased effort is believed to be more than compensated the benefits of more careful planning. Adaptation to and integration into existing management systems must be tested more in practice in the future to obtain acceptance for this concept.

Acknowledgements

First of all gratitude is expressed to Professor Schießl who chairs the Building Materials Centre at the University of Munich for helping to come up with the developed concept for a CAP.

Special thanks go to Professor John Cairns of Heriot-Watt University (UK) for contributions and fruitful discussions on Section 4.6 on condition rating. Thanks are also go to Christine Kühn of Obermayer Planen+Beraten for contributions on Section 4.4 on modular systematic.

References

[1] International Organization for Standardization. 2000. ISO/DIS No. 13822, *Bases for design of structures – Assessment of existing structures*, ISO/TC 98/SC2. Geneva: International Organization for Standardization.

[2] Norwegian Building Regulations. n.d. *Norwegian Standard 3424 – Condition survey for construction works – Content and execution*, Norwegian Council for Building Standardization, Oslo, unauthorized translation.

[3] LIFECON. 2003. *Instructions on methodology and application of models for the prediction of the residual service life for classified environmental loads and types of structures in europe*, Deliverable D3.2, WP 3, Project G1RD-CT–2000–00378. http://lifecon.vtt.fi/.

[4] LIFECON. 2003. State of the art on condition assessment procedures. Work report for D3.1, WP 3, Project G1RD-CT–2000–00378. http://lifecon.vtt.fi/.

[5] International Organization for Standardization. 2000. ISO/DIS No. 13822, *Bases for design of structures – Assessment of existing structures*, ISO/TC 98/SC2. Geneva: International Organization for Standardization.

[6] Schnütgen, B., and T. Lücken. 2001. *Verfahren der objektbezogenen Schadensanalyse – Schlussbericht*. Bochum: Bundesanstalt für Strassenwesen.

[7] Sarja, A. 2003. *Reliability based methodology for lifetime management of structures*. Lifecon Deliverable D2.1, Project G1RD-CT–2000–00378. http://lifecon.vtt.fi/.

[8] LIFECON. 2003. *Instructions for quantitative classification of environmental degradation loads onto structures*, Deliverable D4.2, WP 4, Project G1RD-CT–2000–00378. http://lifecon.vtt.fi/.

[9] British Standards Institution. 2000. *EN 206–1: Part 1 – Concrete – specification, performance, production and conformity*. London: British Standards Institution.

[10] Organization for Economic Co-operation and Development. 1992. *Road transport research – Bridge management*. Paris: Organization for Economic Co-operation and Development.

[11] Faber, M. H., and J. D. Sorensen. 2000. Indicators for assessment and inspection planning. *Proceedings – Workshop on risk and reliability based inspection planning*, Eidgenössosche Technische Hochschule, Zürich, December 14–15.

[12] Faber, M. H. 2003. *Risk and safety in civil, surveying and environmental engineering*, Lecture notes. Zürich: Swiss Federal Institute of Technology.

[13] British Highways Agency. 1993. BA 50/93: British Highways Agency design manual for roads and bridges, Volume 3, Highway structures inspection and maintenance, Section 1: Inspection, Part 3: Post-tensioned concrete bridges – Planning, organisation and methods for carrying out special inspections. London: British Highways Agency.

[14] British Highways Agency. 1994. BA 63/94: British Highways Agency design manual for roads and bridges, Volume 3, Highway structures inspection and maintenance, Section 1: Inspection, Part 5: Inspection of Highway Structures. London: British Highways Agency.

[15] Hearn, G. 2000. Condition states for highway bridges. *Proceedings of structures congress*, ed. M. Elgaaly. Philadelphia. CD-ROM. Reston, VA: ASCE.

[16] Söderqvist, M.-K., and E. Vesikari. 2003. *Generic technical handbook for a predictive life cycle management system of concrete structures (LMS)*. Lifecon Deliverable D1.1, WP 1, Project G1RD-CT–2000–00378. http://lifecon.vtt.fi/.

[17] Hearn, G. 1998. Integration of bridge management systems and nondestructive evaluation. *Journal of Infrastructure Systems*, June, 49–55.

[18] Hearn, G. 1998. Condition data and bridge management systems. *Structural Engineering International* 3: 221–225.

[19] CONTECVET. 2000. *A validated users manual for assessing the residual service life of concrete structures – Manual for assessing corrosion affected concrete structures*, EC Innovation Programme IN309021. www.ietcc.csic.es/publi_elec/Formulario_Contecvet.html.

[20] CONTECVET. 2000. *A validated users manual for assessing the residual service life of concrete structures – Manual for assessing corrosion of structures affected by ASR*, EC Innovation Programme IN309021.

[21] CONTECVET. 2000. *A validated users manual for assessing the residual service life of concrete structures – Manual for assessing corrosion structures affected by frost*, EC Innovation Programme IN309021.

[22] ENV 1504–9:1997. *Products and systems for the protection and repair of concrete structures – Definitions, requirements, quality control and evaluation of conformity, Part 9: General principles for the use of products and systems*. 2001. Brussels: European Committee for Standardisation.

[23] Swedish national road administration. 1996. *Bridge: Handbook for bridge inspections* (in English), Publication no. 1996:036 (E). Borlänge, Sweden: Swedish National Road Administration.

[24] Jungwirth, D., E. Beyer, and P. Grübl. 1986. *Dauerhafte Betonbauwerke – Substanzerhaltung und Schadensvermeidung in Forschung und Praxis*, Düsseldorf: Beton-Verlag.

[25] Kaetzel, L. J., J. R. Clifton, K. Snyder, and P. Klieger. 1994. *HWYCON: Users guide to the highway concrete (HWYCON) Expert system.* Washington D.C.: Strategic Highway Research Program, National Research Council, SHRP-C–406.

[26] The Concrete Society. 1982. *Non-structural cracks in concrete.* Technical report no. 22, Camberley: The Concrete Society.

[27] The Concrete Society. 1995. *The relevance of cracking.* Technical report no. 44, Camberley: The Concrete Society.

[28] Bundesanstalt für Strassenwesen. 1999. *Handbuch für die Bewertung des Korrosionsschutzes.* Draft. Bergisch Gladbach: Bundesanstalt für Strassenwesen.

[29] Elsener, B. 1996. *Zerstörungsfreie Diagnose der Korrosion von Stahl in Beton. Potentialmessung, Betonwiderstand und Korrosionsgeschwindigkeit.* Zürich: Institut für Werkstoffchemie und Korrosion, Eidgenössische Technische Hochschule.

[30] Reichel, B. 2000. *Empfehlung für die Bauwerksdiagnose und Instandsetzung von Betonbauwerken, die Infolge einer Alkali-Kieselsäure-Reaktion geschädigt sind.* Draft by working group no. 2 'Bauwerksdiagnose und Instandsetzung'. Bundesanstalt für Materialforschung und -prüfung – ZfPBaukompendium. http.//www.bam.de/zfp-kompendium.htm.

[31] Bundesanstalt für Materialforschung und –prüfung – ZfPBaukompendium. http.//www.bam.de/zfp-kompendium.htm.

[32] Schnütgen, B., and T. Lücken. 2001. Verfahren der objektbenzogenen Schadensanalyse – Schußbericht für Bundesanstalt für Strassenwesen, Universität, Bochum.

Chapter 5

Life cycle management process

Marja-Kaarina Söderqvist and Erkki Vesikari

5.1 Maintenance policy in different types of organisations

5.1.1 Principles

Maintenance policy is a target-oriented practice of an organisation for the upkeep of its structures. It is a collection of targets and rules that should be considered in all the maintenance, repair and rehabilitation (MR&R) activities of the organisation. One purpose of a management system is to control in practice that the strategic targets are taken into account at all the decision-making levels related to MR&R, also taking into account the funding constraints [1].

The strategic targets may be categorised as follows:

1. Control of condition

    (a) preservation of assets
    (b) striving to an optimal condition level

2. Control of usability

    (a) safety, security, health and comfort
    (b) functionality and serviceability

3. Control of costs

    (a) life cycle costs: maintenance, repair, rehabilitation, renewal and demolition costs
    (b) operating costs
    (c) user costs in operation

4. Control of ecological efficiency

    (a) energy
    (b) raw materials
    (c) environmental burdens
    (d) wastes
    (e) loss of biodiversity

5.   Control of cultural heritage, acceptance and local compatibility

    (a) construction traditions
    (b) living styles
    (c) working culture in work places
    (d) aesthetics
    (e) architectural styles and trends
    (f) image of assets.

The role of the owner of the structures in MR&R planning is important: the control of the abovementioned requirements should be performed at all decision-making levels of the structural hierarchy – network, object and component/module levels. The strategic targets form and frame the results of a management system. For example, governmental organisations have visions and goals that steer their activities and leave effects on society. The requirements and needs raised from these goals must be taken into account when planning MR&R measures. The important decisions and tasks of an organisation are discussed below.

5.1.1.1   Control of condition

A traditional goal setting for a management system is to keep a steady-state condition of the structures to preserve the asset value of the building stock. This maintenance strategy is typical for a reactive management system.

An optimum condition level for structures is obtained as a result of the network level analysis. This optimum condition level can be considered to be the long-term (LT) goal for the management. The short-term (ST) goal is to define the optimal yearly steps for approaching the optimum condition state at which the MR&R costs are assumed to be minimised. How rapidly the optimal condition level can be achieved depends on the available budget. The financing must be high enough to lift the condition from the 'status quo' level (Figure 5.1).

5.1.1.2   Control of usability

One of the main requirements of a management system is the control of reliability in terms of mechanical loading, functionality and serviceability over time. Generally, the safety of structures is controlled by condition constraints, i.e. by defining the lowest allowable condition states for structures. Once a structure exceeds the allowable condition limit a repair action is triggered by the system.

Another way of controlling the structural safety is to perform capacity and safety analysis for structures that are identified by the system. The results of the analyses may entail actions such as strengthening or renewing of structures.

The functionality and serviceability are generally controlled by condition constraints and by structural serviceability analysis. The lowest tolerable condition states based on the requirements of functionality or serviceability are often decisive as compared to those based on the structural safety. The functionality and serviceability in a predictive life cycle management system are strongly related to the obsolescence of the structures.

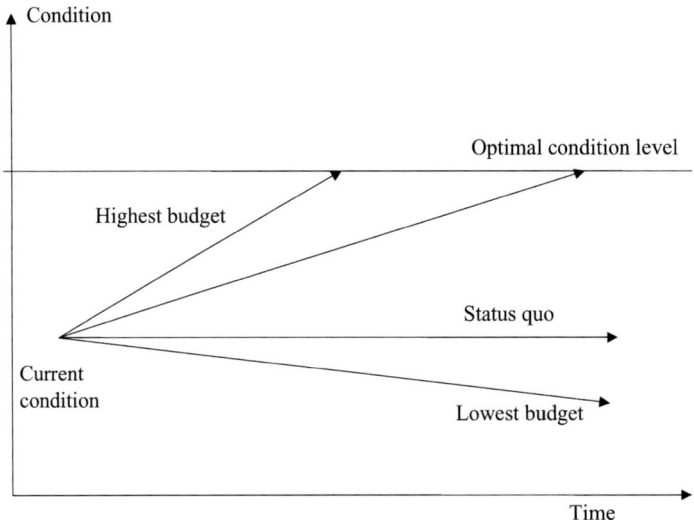

Figure 5.1 The dependency of the future condition level on financing [2].

5.1.1.3 Control of costs

Life cycle costs may be studied with life cycle cost (LCC) analysis. Not only heavy repair actions but also the protective maintenance actions can be considered in the LCC analyses. Necessary elements in the analyses are the degradation, repair action and cost models of structures.

The optimisation of life cycle costs includes optimal timing of actions. This can be done by specific degradation models when the limit condition state is known.

An effective life cycle management system (LMS) also includes the possibility for minimising the delay costs. The delay costs are the result of deferring MR&R actions from the optimal time. The delay costs are often decisive when prioritising actions and projects.

Failure costs can be defined as the product of the probability of failure and the capital consequences of a failure. 'Failure' means either exceeding the serviceability limit state regarding performance requirements or obsolescence or the ultimate limit state. The costs of the consequences of a possible fatal failure would be by far greater than just the repair costs.

A national road administration and other governmental organisations that maintain public infrastructures must consider the user costs, i.e. costs paid by the user when the use of an infrastructure is restricted for some reason. For instance, a weak bridge may cause considerable extra expenses for some users as a result of a longer transport route. A narrow old bridge that causes a bottleneck for traffic results in extra expenses to all road users. Normally, the owner costs form a descending curve and the user costs an ascending curve as a function of increasing degradation of a structure. The minimum socio-economic costs, totalling the owner and user costs, would then lie between the extreme ends of high and low condition, as seen in Figure 5.2. Additional costs due to traffic delays and detours while maintenance action is undertaken must

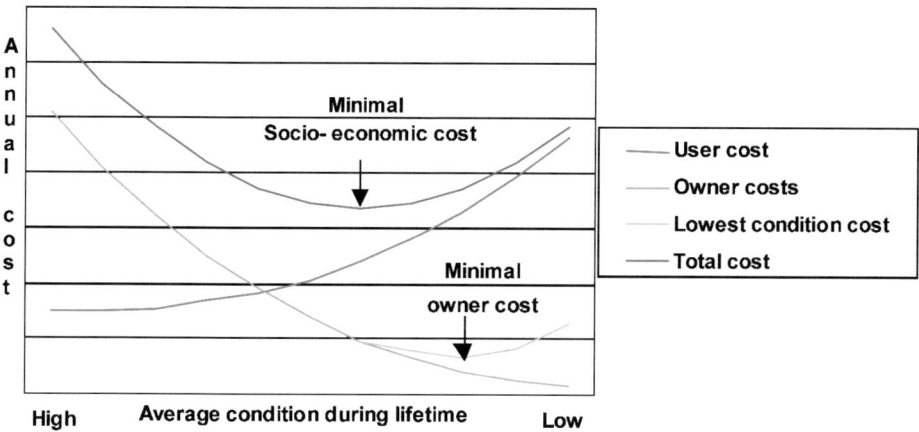

Figure 5.2 Definition of the optimal condition level of concrete structures from a socio-economic point of view [1].

also be considered in the planning of projects. The duration of repair must be as short as possible when the traffic flow is high.

5.1.1.4 Other requirements

The MR&R measures not only cause financial costs but also environmental costs. The environmental costs are the result of consumption of natural raw material resources and energy, emittance of pollutants and loss of biodiversity. The environmental impacts can be considered by life cycle ecology analyses. In the same way as the MR&R costs are totalled in a life cycle cost analysis, the environmental impacts depending on the types and quantities of materials used in repairs can be totalled and used as one attribute in decision-making.

The aesthetic appearance of structures is normally controlled by condition con-straints. In structures where a high aesthetic quality is required the condition constraints may be stricter than in structures where only the structural capacity is the limiting factor.

The cultural value of a building or a structure depends on the national, historical or other social significance of the object. Structures of high cultural value should be maintained with greater care than structures of less importance. Some objects may be defined as national heritage buildings or structures and should be preserved without changing their original appearance.

5.1.2 Owner needs and requirements

Potential interest groups of a life cycle management system (LMS) are owners of the structures, such as

* governmental organisations
* municipal organisations
* private companies.

The main requirements and expectations of an LMS are usually the same, regardless of the abovementioned interest groups. However, there may be differences because of specific duties of organisations in the society.

Potential *governmental organisations* for application of an LMS are the administrations for public roads, railways, waterways, harbours and shipping. These organisations are responsible for the maintenance of bridges, tunnels, canals, dams, pole basements, quays, lighthouses, etc. The governmental administrations usually have a sector responsibility to take care of the traffic needs inside their mandates including the requirements of industrial life and general security. As the MR&R activity of these organisations is financed by public funds they also have a responsibility for using the allocated money cost-effectively. In addition, they have responsibilities and pressures from the society to take care of the ecological, cultural and aesthetic values in their MR&R activity.

Typical needs and requirements of governmental organisations for a computer aided management system for the administration of the organisation are the following:

- need for economic justification of decisions
- objective basis for decisions, based on engineering, economic and ecological grounds
- determination of medium and long-term targets and need for definition of appropriate maintenance strategies
- strategic guidelines for preservation of assets
- optimising MR&R strategies based on engineering and economic grounds
- need for selection of justifiable maintenance decisions within budget constraints
- need for showing value for money in infrastructure provision and maintenance
- need for allocation of funds
- evaluation of whole life costing, including user costs
- implication of lower standards of performance.

For the maintenance engineers and repair designers the needs are the following:

- well-organised condition assessment system and inventory for the structures
- optimisation of MR&R actions for specific components, modules and objects
- guaranteed safety
- safeguarded investments
- correct timing of MR&R actions
- evaluation of MR&R costs
- combination of optimised actions into MR&R projects
- prioritisation of projects
- production of annual repair and reconstruction programmes
- budget control.

Potential *municipal organisations* for adopting an LMS would be organisations for local traffic, water service, drainage, sewage and waste disposal. Typical concrete infrastructures to be maintained by these organisations are bridges, pavements, walls, pipelines, pools, pole basements, etc. The expectations of municipal authorities from a management system are much the same as those of governmental authorities.

However, a little lighter version of a management system might be adequate, as the number of structures is usually less compared to a governmental administration. That is why a separate network level optimisation module may not be necessary in municipal level applications.

Private companies possess a great part of existing concrete infrastructures. Typically such companies are those operating in the production of electricity with nuclear, oil, coal, gas or water power plants. Concrete infrastructures of these companies include containers, reactor buildings, turbine buildings, cooling systems, dams, etc. Companies in the area of electricity distribution might also be interested to set up an LMS if they possess a large electric network with concrete pole basements. Other potential companies would be those operating in winning and refining of natural resources such as oil production, mining and wood processing companies. These companies may possess oil platforms, pipelines, tunnels, silos, basins, reservoirs, basements, etc.

Private companies have an interest in maintain their infrastructures as economically as possible. However, they also have a responsibility for the safety of structures and security of people. Nuclear power plants are obliged to safeguard the radiation security by special regulations. A general requirement of private companies would be that the structures must be robust and maintenance-free enough to guarantee continuous production. Unintended breaks in the production as a result of MR&R actions would be extremely expensive. The requirements of private companies for a management system of their structures could be summarised as follows:

- Safeguard a continuous production.
- Minimise economic and other risks.
- Minimise the MR&R costs of structures.
- Guard the safety of structures (especially in nuclear power plant buildings).
- Ensure a long service life of structures.

5.1.3   User needs at different levels of organisation hierarchy

An LMS provides benefits for all hierarchical levels of an organisation: chief manager, central administration, local administration and repair consultants, as seen in Figure 5.3.

The different levels of an organisation are responsible for decision-making at different levels of the structural hierarchy: network, object, module and component level. Networks mean stocks of bridges, tunnels, quays, etc. Objects refer to single bridges, tunnels, quays, etc. Modules form the main parts of objects such as a superstructure and a substructure of an object. Components are basic elements of structures such as columns, beams, walls and slabs. Usually there is the following correspondence between the level of organisation hierarchy and the interest level in structural hierarchy:

Level of organisation hierarchy	*Level of structural hierarchy*
Chief manager	Network
Central administration	Network, object
Local administration	Local network, object
Repair consultants	Object, module/component

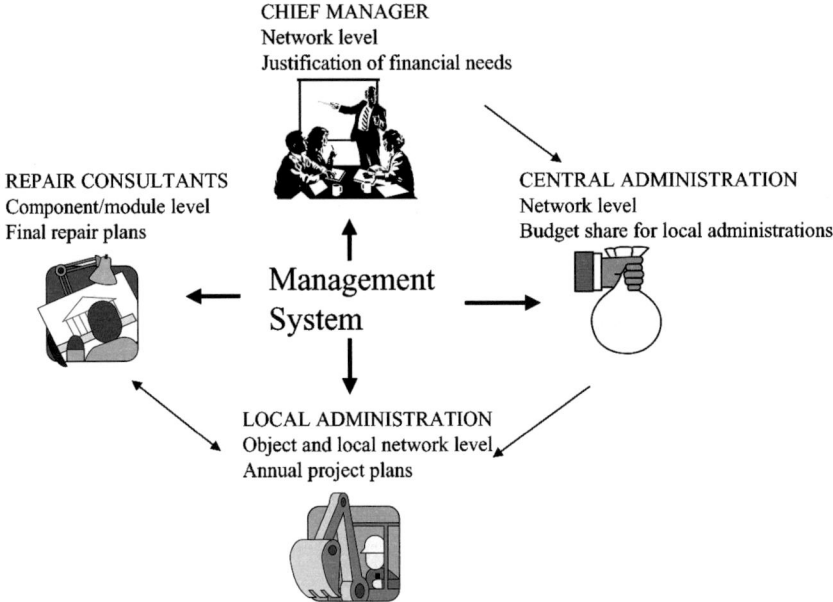

Figure 5.3 Use of a management system in an organisation [3].

If the owner of a building stock is a governmental organisation it will have its annual funding based on the estimates of the state. Politicians decide upon the allocated money for the maintenance of public infrastructure together with all the other needs of the public sector.

An organisation must justify its financial needs to get the money allocated to an appropriate level. It is a task of the chief manager to present and justify the requirements of funding to the minister of a state. For that purpose the chief manager needs statistical scenarios on the future works and costs of MR&R activity. A predictive LMS may offer a valuable tool for producing such scenarios.

When the total budget for all the MR&R activity of the organisation is known, the budget must be allocated to local administrations like districts, departments and units. This is done by the central administration. To do the distribution objectively an analysis on the needs at each local unit is required. As the costs of MR&R works can be evaluated by the management system, a balanced budget share can be implemented.

Another task of the central administration is to define the goals, strategies and general rules to be applied at local level MR&R activities. First of all, the responsibility of an organisation is to maintain the safety of structures and secure the investments of the society. Thus the definition of long- and short-term goals and the control of the realisation of these goals are important tasks of the central administration.

The task of local administrations is to prepare the annual project and resources plans according to the maintenance targets and allocated budgets. To do this, the system must be able to propose MR&R measures to be performed at an optimal time, to produce preliminary plans for projects by combining MR&R measures, to sort and to prioritise projects, to evaluate the costs, to check the budget and to produce the annual project plans. The reached effects of the MR&R measures for the stock of structures have then to be reported to the central administration in the form of reached maintenance goals and targets.

Project planning is completed by maintenance engineers of the organisation itself or repair consultants. The purpose of this work is to prepare detailed project plans and contract documents for individual objects assisted by the system. In addition to all the other helps mentioned above, the system may be equipped with the following analysis services for specific component level planning:

- service life design
- residual service life analysis
- analysis of the structural capacity
- economic and ecological life cycle analysis.

Based on these analyses a repair designer may revise the specifications automatically proposed by the system.

5.2   Tasks and characteristics of the management process on different levels

5.2.1   Network level and object level systems

The process of the life cycle management system consists of two sub-processes: the network level system and the object level system.

The network level system is a tool for administration level operative planning and decision-making for the upkeep of infrastructure. It makes it possible for the administration to optimise the maintenance strategy both technically and financially and to draw up long-term scenarios on financial needs for MR&R activity in the future.

The network level system works with populations of structures while the object level system deals with individual structures and structural components. It uses the results from the network level system to decide on the repair measures in individual repair projects.

The object level system is a practically oriented system which helps the maintainers to plan and execute the MR&R projects based on the inspection and condition assessment data. It specifies and gives optimal timing for MR&R actions at the component level, assists in both preliminary and final planning of projects at the object level, and assists in producing the annual project and resources plans.

The management system can be built up in three versions as a response to the different demands of organisations and companies:

- integrated network and object level life cycle management system (complete version)
- object level life cycle management system (reduced version)
- consultant version.

The complete version of the LMS is meant for wide organisations, e.g. national road administrations or big industry companies, that are responsible for the upkeep of a large network of concrete infrastructures. It consists of both the network level and the object level systems. It is assumed that the owner of the complete version of LMS conducts planned inspection activity on structures and upholds a database. It is also provided that such an organisation will be active in updating the degradation models and the decision trees for the maintenance planning of structures.

The reduced version of the LMS is meant for municipal organisations and companies, which own a limited amount of concrete infrastructures. It consists of only the object level system. However, it enables long-term planning of MR&R projects and drawing up of annual work programmes. The main difference between the users of a complete version and a reduced version is that the building stock of the latter owner is too small to constitute a base for the development of statistical degradation models. So it is assumed that an owner of the reduced version LMS will buy or licence the necessary degradation models and decision trees from a larger organisation. However, it is assumed that the organisation upholds a database and planned inspection of structures.

The consultant version of an LMS is meant for repair consultants who participate in the planning processes of an LMS or otherwise engage in life cycle planning of structures. The system consists only of the modules necessary for the component and object level planning. While participating in the planning processes of an LMS, it is assumed that the consultant has access to the database of an organisation and that the consultant can use the degradation models and decision trees developed by an organisation. A consultant can do the repair planning with the help of the LMS effectively and according to the strategic targets defined by the organisation. If the modules of the system process are used separately from the LMS it is assumed that necessary inspections and condition assessments have been otherwise performed for the treated objects.

5.2.2 Special characteristics of the network and object level systems

5.2.2.1 Network level long-term and short-term analyses

The network level management system consists of two parts: the long-term module to find the ideal optimal condition distribution for the bridge stock and the short-term module to find out how to get the bridge stock from the present condition distribution to the optimal distribution.

The long-term analysis is based on the general idea that the structure stock has an optimal condition distribution. This optimum is intermediate in the following sense: keeping all structures in an excellent state at all times would be excessively expensive and, on the other hand, allowing a severe deterioration of the structure stock would cause expensive major repairs; somewhere in between there is an optimum where the stock can be kept on the same condition level year by year with the smallest possible amount of funding, yet adhering to the safety level and service requirements.

The optimal condition distribution corresponds to a certain optimal set of repair measures. These repair measures would, in the ideal case, be applied to the same extent year by year, although naturally to different individual structures. The set of optimal repair measures and the extent of each, i.e. the optimal repair measure distribution, will ensure the permanency of the optimal condition distribution of the structure stock.

The short-term analysis provides an economically optimal way to reach the long-term optimum condition distribution during the next few years. There are separate short-term solutions for each coming year. Each short-term solution represents a step closer to the long-term optimum.

In reality the long-term optimum will change somewhat year by year due to changes in the variables that influence it. Changes can be expected in repair method costs because of new repair methods and materials. The road policy could change and the level of service standards with it, new improved deterioration modelling could affect the optimum, etc.

The network level system offers the possibility for 'what-if' experiments with respect to the safety and minimum service level policy, repair measure costs, budget limits and other variables. The system will also provide detailed information for future structure designers on the deterioration mechanisms of structure elements and on the life-span cost of different structure types. In addition, the network level results are a key input to the project level system.

5.2.2.2   Object level life cycle cost analysis

The object level system is the key tool for everyday structure repair planning. It helps the repair engineer to plan and schedule the repair projects for individual structures based on the recommendations from the short-term model and the damage data in the database.

The object level system also includes an LCC analysis. This analysis compares the repair measure combinations, recommended by the network level, with each other instead of using the traditional calculation method with its repair measure combination given beforehand. Thus the most advantageous repair measures for an individual structure are found out. This gives the repair engineer the flexibility needed and a possibility of opt for repair measures, which result in the minimum optimal total cost during a structure's life span.

5.3   Essentials of a life cycle management

5.3.1   General

As a prerequisite for building up a predictive, optimising and integrated management system, as described above, certain essentials must be provided. In this chapter these essentials are dealt with in such an extent that is necessary for understanding the later descriptions on the object level and the network level management systems. The essentials focused here are the following:

- database
- inspection and condition assessment system
- system for predicting the future condition of structures
- life cycle cost analysis.

A working management system calls for a database, in which all the data related to the structures are stored. They include material, structural and environmental exposure data together with condition data of the object. The database is a central part of the system being involved in every phase of the management system process.

An equally essential part of an LMS is the inspection and condition assessment system. The inspection and condition assessment system must produce the condition data which is needed in the condition analysis and decision-making processes of the LMS.

A mathematical system for predicting the condition of structures over time is necessary in a predictive management system. A requirement for the mathematical system is that it must be able to predict the condition in a stochastic way over the whole design period, which may include several MR&R actions in an arbitrary sequence. Both degradation models and action effect models are needed in the prediction of condition. As a mathematical framework for the condition analysis the Markov Chain method has proved to be especially suitable.

For optimising the types and timings of actions for various structural components an analysis program for determination of life cycle costs is necessary. An automatic life cycle analysis program integrates the condition analysis based on the Markov Chain and an automatic system for triggering MR&R actions to the LCC analysis framework. An advanced life cycle analysis program would rather be called a 'life cycle planning program' because it is possible to plan MR&R projects automatically by the help of such a program. A decision tree is a necessary add-in in an advanced life cycle planning program. It outputs optimised solutions for MR&R action profiles and selects them taking into account the specific properties, environmental conditions and requirements of structural components.

5.3.2 Database

A computer aided management system is always based on a well-defined data inventory. The data structure of the inventory must be consistent with the system needs. It should allow the input of inspection and condition assessment data, repair data as well as administrative and structural data on all levels of structural hierarchy.

The decision-making in a management system is based on specific component level data including information on materials, structural features, environmental stresses, condition and damages. The condition and damage data are collected during inspections of the structures.

Data in the system database must be such that they contribute to the objectives of the system and are consistent with the data input in the modules. The database also works as an intermediate store and end store for data processing and for the output data from the optimising processes.

The quality of information obtained as an output from the system is directly dependent on the quality of the collected input data. Thus the education and training of inspectors are of great importance. A quality system module for the collected inspection data is recommended.

5.3.2.1 Modular systematics

It is recommended that the database is arranged according to the modular systematics. The modular systematics is needed for:

- dividing the structures into homogenous parts
- providing a system for documentation of damage observations.

The requirements on a modular systematics are:

- common understanding
- applicability to different structure types, e.g. bridges, tunnels, retaining walls, etc.
- suitability for a range of applications like management, monitoring, repair and cost control.

It is important that each structure is divided into homogeneous parts, which are of the same material and are exposed to the same environmental influences. The Markov Chain optimisation algorithm works with populations of structural parts that are similar in their construction and use. For example, in the Finnish Bridge Management System bridge superstructures, substructures, surfacing structures and bridge furnishings are treated separately. A breakdown of building material, structural type and construction technique is also made. In addition there are two environmental categories: salted main roads and other roads.

5.3.2.2 General definition of the structural levels

According to the modular systematics the data of structures are divided into object, module, component and subcomponent level. In principle the data at all levels may be needed in the processes of an LMS but the main focus is addressed to the component level data as the LCC analyses are mainly performed at the component level. However, the data used in the component level analyses may need to be searched on lower levels of the data hierarchy. Also the analyses themselves must sometimes be performed at the subcomponent level or even at the surface level so that the different environmental stresses and materials at different parts of a component can be taken into account.

The complete framework is divided into six levels. The simplicity of the framework makes it easy to use for all kinds of structures. Examples of the levels are explained below.

Level 1: Object Classification of the complete structure, e.g. bridge, tunnel, building, etc.

Level 2: Module The module level is the logical set up of the structure, which yields the order of production. These main modules divide a structure into the largest units. A main module can have several, varying functions and is made up of different materials. A main module is a generic term comprising certain components.

Level 3: Component A component fulfils a certain function as a unit but can consist of different materials. For each component the function within the whole object must be identified.

Level 4: Subcomponent/resistance The subdivision has to be carried out with the target to identify sections with equal resistance at the surface and of the inner layers. The subcomponents consist of the same materials (concrete quality, coating, etc.) and are produced with the same design (concrete cover).

Level 5: Surface/Environmental influences Subcomponents are divided into sections and areas, which are exposed to the same environmental influence and/or stresses.

Level 6: Detail/Material Inner layers of a surface are regarded in depth. An identification of details in perpendicular to the surface is made. A distinction is made for details which consist of the same material.

The above presented modular systematics is designed to be generic and applicable to all concrete, and possibly also other structures. However, for specific groups of structures it is possible to take it only as guidance and redefine the coding system taking into account the real needs of identification and the level of accuracy. For example in some structures the Level 2 module level may not be necessary and thus it may be reasonable to combine Levels 2 and 3. In some other cases it may be possible to merge the Levels 4 and 5, etc.

The recommended code system is illustrated by an example of a bridge, in Figures 5.4 and 5.5.

Capital letters A, B, C, etc. refer to the modules. This alphabetical designation indicates the location of the modules.

The first three-digit number refers to the components, e.g. $A.010$. The number results from the subdivision of the bridge in the transverse direction. For buildings, this division has to be applied in the vertical direction and increased from storey to storey.

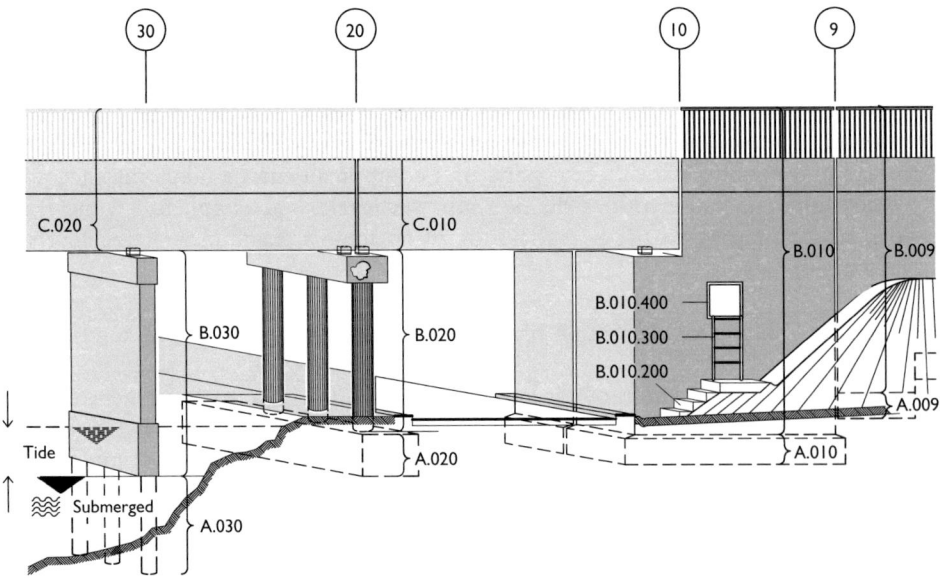

Figure 5.4 Schematic overview of an object [3].

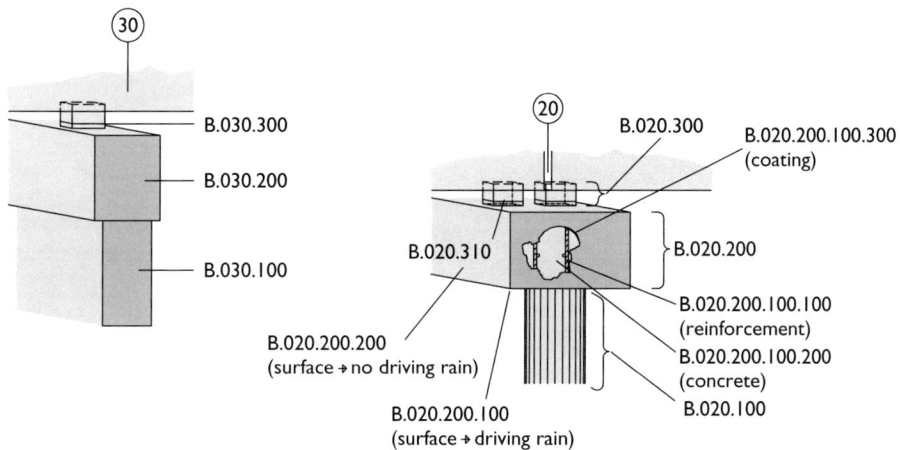

Figure 5.5 Coding for details [3].

The second three-digit number is for the subcomponents, e.g. B.010.*100*, and the third three-digit number is for differentiation of environmental influence and stress on a surface, e.g. A.200.110.*100*. These numbers are incremented from bottom to top by steps of 100.

The forth three-digit number refers to each material, e.g. concrete A.200. 110.100.*100*. For instance, the depth of a reinforcing bar, if necessary, can be indicated in brackets behind the code.

Detailed information on the modular systematics is available in [4].

5.3.2.3   Data organisation

The system database may be the data source for many computer applications. In this connection only the data that is important for an LMS is focused.

The data in the database can principally be divided into two main categories: object oriented data and management system generated data. The object oriented data consists of data describing the object structurally, functionally and administratively. The data generated during the management process includes results from optimisation, condition prediction and life cycle analyses. This provides a possibility to entering feedback data from the executed projects to the system. The feedback data includes data of executed MR&R actions and the costs of these actions. The feedback data is stored in the database for updating the models of the system. An example breakdown broken down into data categories is listed below.

5.3.2.3.1   OBJECT ORIENTED DATA

1.   Administrative data

   object identification data
   object location data, coordinates

information of the owner and upkeeper
historical value
value of the location place in the society's value category

2. Structural data

type of the object
main measures of the object
material, material properties
type and measures of object modules and components
construction year
construction costs + overhead
constructor
designer
design loads
bearing capacity data

3. Data on environment and loadings

- environmental design class
- environmental burdens, climate, etc.
- traffic loads, special heavy loadings
- human loads
- snow and wind loads

4. Inspection and condition data

- condition data on components
- overall condition of the object
- inspection type
- inspection time
- inspection interval
- inspector
- damage data: location, type, severity, damage extent

5. Maintenance, repair and rehabilitation data

repair data: repair urgency class, repair measure and cost recommended by the inspector
fulfilled MR&R measures: previous repairs and their realised costs
planned MR&R measures and costs
unit costs, total costs
year and date of MR&R
unit costs for MR&R actions.

5.3.2.3.2 LMS GENERATED DATA

1. Costs

given budgets
recommended budgets for five coming years

> optimisation budget results
> cost/benefit ratio

2. Condition data

> condition constraints
> present condition state distribution
> predicted condition states for five coming years
> age behaviour models
> environmental degradation models
> condition rating indexes
> target distribution of the condition

3. Maintenance, repair and rehabilitation data

> recommended MR&R measures as optimisation results
> repair and cost models
> work programmes for the next five years
> decision tables.

It is important to keep in mind, when planning and designing the content of the database, that it is expensive to invent, collect and upkeep the data needed for the system. That is why 'nice to know' or unnecessary data should be avoided.

5.3.2.3.3   THE DATA FOR A LIFE CYCLE COST ANALYSIS

The data for an LCC analysis can be divided into two categories:

- material, structural and environmental data
- inspection and condition assessment data.

The data of the data inventory must be consistent with the parameters needed in the LCC and risk analyses. So the data organisation of a database and the parameter values of the analyses should be designed together so that they optimally match with each other taking into account the availability of data and the accuracy requirements of the analyses. An example list of required data is presented in Table 5.1.

There are two types of inspection data: (i) condition assessment data for predictable damage types and (ii) damage data for unpredictable damage types.

The condition assessment data is needed for:

- evaluation of the condition of structures
- prediction of the degradation rate in the specific component
- calibration of degradation models, i.e. adjusting the rate of degradation predicted by models to be consistent with the observed average rate of degradation
- evaluation of the risks in the structure
- repair planning of the structure.

The condition assessment data are given as distributions or average degradation. Some data, such as carbonation and chloride penetration, are only evaluated by

Table 5.1 An example of a bridge component specific data

Data group	Data	Specification	Dimension
Identification data	Code of component		
	Name of component		
Measuring data	Thickness		mm
	Width		m
	Length		m
	Surface area		m²
Burden data	Surface inclination		(H = horizontal/V = vertical)
	Distance from road		m
	Distance from water		m
	Moisture index		(0–1)
	Chloride index	Road	(0–1)
	Chloride index	Crossing road	(0–1)
Material data	Nominal strength		MPa
	water/cement ratio		
	Air content		%
Structural data	Concrete cover		mm
	Width of cracks		mm
	Diameter of steel bar	Nearest to surface	mm
	Diameter of steel bar	Main tensile	mm
	Lever arm		mm
	Phi factor		
	Load factor		
New component data	Nominal strength		MPa
	water/cement ratio		
	Air content		%
	Concrete cover		mm
MR&R data	Is repaired?		0 or 1 (1 = yes, 0 = no)
	Repair system		Code of system
	Is protected by overlayer?		0 or 1 (1 = yes, 0 = no)
	Overlayer system		Code of system
	Is coated?		0 or 1 (1 = Yes, 0 = no)
	Coating system		Code of system

degradation models but the degradation models may be calibrated using the measurement results obtained from special inspections.

In the case of predictable damage types, the specification and timing of repair actions is performed by the system. So an inspector is not necessarily asked to give proposals for repair actions for these degradation types.

There is, however, a group of damage types that are not predictable or occur very seldom and are thus not considered by degradation models. These damage types must be carefully evaluated by the inspector and the inspector will give the first suggestion for the repair measures. The inspector also evaluates the urgency class for the repair. The urgency class has a connection to the maximum allowable delay of the repair from the time of inspection. An inspector evaluates also the extent of the repair area. By these data the system can propose an action, give the timing, and evaluate the costs of the action to counteract the observed damage.

Like the data organisation for material, structural and environmental data, the data organisation of inspection and condition assessment data should also be carefully designed so that it fulfils the demands of the system but does not unreasonably strain the inspection work. Table 5.2 shows a proposal for the data organisation of inspection and condition assessment data.

Detailed information on the condition assessment system is available in LIFECON Deliverable D3.1 [4]. For identification of different damage types the Damage Atlas in the Annex of LIFECON Deliverable D3.1 may be used.

Table 5.2 Data organisation of inspection data

Data group	Data	Specification	Dimension
Condition assessment date	Time of inspection		Date
	Depth of carbonation	Estimated by models	Distribution (%)
	Depth of carbonation	Test	Measured average (mm)
	Depth of critical chloride content	Estimated by models	Distribution (%)
	Depth of critical chloride content	Test	Measured average (mm)
	Frost attack	Depth of scaling	Distribution (%)
	Frost attack	Internal	Distribution (%)
	Corrosion of reinforcement		Distribution (%)
	Width of cracks		mm
	Crack density		m/m^{-2}
	Corrosion of spacer bars at cracks	Estimated by models	Distribution (%)
	Corrosion of spacer bars at cracks	Test	Measured average (mm)
	Corrosion of main bars at cracks	Estimated by models	Distribution (%)
	Corrosion of main bars at cracks	Test	Measured average (mm)
	Frost attack in overlayer		Distribution (%)
	Weathering of coating		Distribution (%)
Damage data	Observed damage	Damage type	Damage code
		Time of observation	Date
		Localisation of damage	(Subcomponent/surface)
		Cause of damage	(Code of damage cause)
		Suggested repair measure	(Code of repair measure)
		Urgency class of repair	(1,2,3)
		Extent of repair	m, m^2

5.3.2.3.4 DATA REQUIREMENTS AT NETWORK LEVEL

Data requirements at network level are much less compared to works programming and project analysis levels. Only a few condition and other parameters are needed:

- the number of structures
- average daily traffic (light and heavy vehicles)
- deterioration models
- maintenance effect models
- maintenance action costs
- user costs
- initial condition distribution (for short-term analysis only)
- condition constraints (optional)
- budget constraint (optional, for entire network).

Number of structures is used to calculate total costs from unit costs. *Traffic data* is used to calculate road user costs, for both roads and bridges. *Deterioration models* describe the process by which structural elements deteriorate and thereby cause increased user costs or increased risk of failure. Deterioration is described using probabilistic models, which determine how many structures, in terms of percent, are in a certain condition state and how many are likely to deteriorate to a worse condition state.

Maintenance effect model describes the improvements that can be expected after a maintenance action is applied. The model structure is similar to that of the deterioration models. The deterioration and repair effect models in the system are probabilistic Markov Chain models.

Maintenance action costs are the average costs of maintenance actions per structure. *User cost models* describe the costs incurred by the user of a structure at any condition state. Increased user costs are usually related to poor conditions, detours and weight restrictions. The key input variables used in the calculation of user costs include average daily traffic, percentage of heavy goods vehicles in traffic mix, vehicle operating costs (unit cost per vehicle), travel time costs (unit cost per vehicle), and accident rates and change of accident rate due to condition variables.

The present (initial) condition distribution is extracted from the database and is used as the initial condition distribution when calculating recommendations for the short-term analysis.

Constraint data includes both condition and budget constraints. Condition constraints describe how many structures, in terms of percent of the network, are allowed to be in any condition class. Each condition variable is considered separately. Budget constraints are used to obey the obvious financial constraints.

The structure of the network level system should be generic. Users can therefore select the number of sub-networks and condition parameters that describe their network, depending upon the data and models.

5.3.3 *Predicting the condition of structures with Markov Chain method*

Markov Chain method is a mathematical framework based on probability calculus and vector algebra. In the application presented here it is used for predicting the future condition of structures over a certain time frame. The condition is presented in the form of condition vectors i.e. frequency distributions based on a predefined set of condition states. The annual changes in the condition state distributions are predicted by matrix multiplications using transition probability matrices.

The Markov Chain is only a mathematical frame work. As such it does not contain any information on the rate of degradation of structures. However, if such data is available in any form it can usually be transferred into transition probabilities of the Markov Chain degradation matrices so that the results of Markov Chain analysis corresponds closely the original information. Markov Chain transition probabilities have also been proved to be suitable for modelling the action effects of various MR&R actions. The action effect models are necessary because the condition analysis must cover the period not only up to the next repair of the structure but over the whole lifetime which may comprise of many MR&R actions of different types.

A general approach for the development of degradation models is presented separately in Sections 3.1–3.4. So the first steps that a developer of a management system has to undergo for providing necessary data on the rate degradation are described in that chapter. The following description on the basics of Markov Chain is aimed to be the next step.

Although the Markov Chain cannot be considered as the only possible solution for the mathematical framework in an LMS it has proved to fulfil the requirements of an advanced system. The main reasons for using Markov Chain mathematics are the following:

- An LMS based on the *reliability theory* is required.
- Continuous presentation of the condition of a structure with analytic stochastic models would be very difficult when the time frame contains several MR&R actions in a deliberate sequence. Certain *concurrent time-dependent degradation processes* which can be emulated by the Markov Chain would be almost impossible to be reproduced by analytic calculation methods.
- The calculation system should serve as the basis for real lifetime design so that a designer would be able to instantly see the effects of the changes in his design on the condition, costs and other possible requirements on a structure. By the Markov Chain method the system can be made automatic enough to render *conversational design*.

The Markov Chain method makes it possible to reproduce the performance (condition) of a structure over the whole treated time frame as a series of sequential annual condition state distributions taking into account the effects of both degradation and various MR&R actions. The calculation tables for condition state vectors can be programmed so that full stochastic condition data can be instantly obtained from the whole treated period of time.

5.3.3.1 *Basics of Markov Chain modelling*

The Markov Chain method evaluates the condition of structures as condition state distributions at each year t. A condition state distribution expresses the relative proportions (= fractions) of structures being at the defined condition states. A condition state distribution is exemplified in Table 5.3.

When studying the condition of structures at the network level the fractions refer to the surface area (sometimes length or other functional unit) of all structures or structural parts belonging to a network of structures. At the object level the fractions refer to the surface area (or other functional unit) of one structure or a structural part. When predicting the condition of structures by the Markov Chain method the condition state vector is interpreted as expressing the probability of a structure or structural part to be at any of the condition states in the future. The sum of all fractions in a condition state vector must always be 1.

The number of condition states is not restricted. In the following examples of the Markov Chain calculus the number of states is assumed to be five consisting of states 0, 1, 2, 3 and 4. The condition state 0 represents the best and 4 the poorest condition. The condition state 3 defines usually the limit state of service life at which the structure should normally be repaired.

The changes in condition states as a result of both degradation and MR&R actions are evaluated by transition probability matrices. The condition state distribution of each year is obtained by multiplying the condition state vector of the previous year by the transition probability matrix. Mathematically the principle is presented in Equation (5.1). By repeated multiplication the condition state distributions can be predicted over time up to several years or even tens of years.

$$W(t) = W(t-1) \times P \tag{5.1}$$

where

$W(t)$ is the condition state distribution of year t and
P transition probability matrix.

There are two kinds of transition probability matrices:

- degradation matrices
- action effect matrices.

Degradation matrices are applied in years when repair actions are not performed, i.e. the changes in the condition state distribution result only from degradation. The action effect matrices predict the condition state distribution, as it will be after the repair action. They are applied only in those years during which repair actions are performed. Accordingly, by the help of the Markov Chain it is possible to reproduce the condition of a structure during the whole time frame as a series of sequential annual condition state distributions. The treated time frame may include various maintenance

Table 5.3 An example of condition state distribution

State	0	1	2	3	4
Fraction	w_0	w_1	w_2	w_3	w_4
Example of fraction	0.25	0.35	0.25	0.1	0.05

and repair actions such as coatings, other predictive maintenance actions, repairs and renewals.

5.3.3.2 Degradation matrices

Usually the form of a degradation matrix is assumed to be as the one presented in Table 5.4. The elements of a transition probability matrix express the probability that a structure which at the beginning of a year was at condition state i (vertical direction) will be at the end of the year at condition state j (horizontal direction).

It has been assumed in the table that within one year the structure either stays at the same condition state where it was at the beginning of that year or drops to the next state, i.e. dropping more than one state in a year is not possible. Accordingly, most of the transition probabilities are 0. Only the diagonal probabilities, i.e. the probabilities that a structure stays at the same condition state and the probabilities next to the right of them expressing the probability that the structure will be transited to the next state during a year, are non-zero elements. The sum of transition probabilities in each row must be 1 $(p_{i,i} + p_{i,i+1} = 1)$.

The transition probabilities of degradation matrices are determined automatically from previously developed degradation model functions by special conversion methods. So the information included in the material, structural and environmental parameters of the model functions is automatically transferred to the transition probabilities of degradation matrices.

The 'drop-from-state' transition probabilities $p_{i,i+1}$ can be deduced from the scaled degradation model functions by derivation of the model function and determination of the average value of the derivative within the interval of the states i and $i+1$.

$$p_{i,i+1} = \mathrm{DoD}'_{i,i+1} = \left(\frac{\partial(\mathrm{DoD}(t))}{\partial t} \right)_{i,i+1} \tag{5.2}$$

where

$p_{i,i+1}$ is transition probability from state i to state $i+1$.
$\mathrm{DoD}(t)$ is a scaled degradation function.
DoD is degree of damage and is considered to be the same as condition state.

The average value of the derivative can be determined either by calculating the value of the derivative in several points within the range $(i, i+1)$ or by determining the value of the derivative in a point that is proved to optimally represent the average.

Table 5.4 Transition probability matrix for degradation (5 state system)

State	0	1	2	3	4
0	p_{00}	p_{01}	0	0	0
1	0	p_{11}	p_{12}	0	0
2	0	0	p_{22}	p_{23}	0
3	0	0	0	p_{33}	p_{34}
4	0	0	0	0	1

The 'remain-in-state' transition probabilities $p_{i,i}$ can be determined by subtracting the corresponding 'drop-from-state' probability from 1.

$$p_{i,i} = 1 - p_{i,i+1} \tag{5.3}$$

At the lower right corner of the matrix the value of the probability element is always 1 as the structures in the highest possible condition state always stay at the same condition state.

The condition state vector after n years is predicted by multiplying the initial condition state vector, $W(0)$, by the transition matrix n times in the row, as shown in the example of Figure 5.6. In this example the limit condition state of service life has been defined to be 3 (DoD = 3). The state 4 is assumed to be a 'terminal state', i.e. an extra state where all structures finally end up. All structures in this case start off in perfect condition, so the initial damage index distribution is (1, 0, 0, 0, 0).

Transition probability matrix

State	0	1	2	3	4
0	0.61	0.39	0	0	0
1	0	0.74	0.26	0	0
2	0	0	0.82	0.18	0
3	0	0	0	0.91	0.09
4	0	0	0	0	1

Year	State 0	1	2	3	4	Average DoD
0	1.000	0.000	0.000	0.000	0.000	0.00
1	0.610	0.390	0.000	0.000	0.000	0.39
2	0.372	0.527	0.101	0.000	0.000	0.73
3	0.227	0.535	0.220	0.018	0.000	1.03
4	0.138	0.484	0.319	0.056	0.002	1.30
5	0.084	0.412	0.388	0.109	0.007	1.54
6	0.052	0.338	0.425	0.169	0.016	1.76
7	0.031	0.270	0.437	0.230	0.032	1.96
8	0.019	0.212	0.428	0.288	0.052	2.14
9	0.012	0.165	0.406	0.339	0.078	2.31
10	0.007	0.126	0.376	0.382	0.109	2.46
11	0.004	0.096	0.341	0.415	0.143	2.60
12	0.003	0.073	0.305	0.439	0.181	2.72
13	0.002	0.055	0.269	0.454	0.220	2.84
14	0.001	0.041	0.235	0.462	0.261	2.94
15	0.001	0.031	0.203	0.463	0.303	3.04
16	0.000	0.023	0.175	0.458	0.344	3.12
17	0.000	0.017	0.149	0.448	0.385	3.20
18	0.000	0.013	0.127	0.434	0.426	3.27
19	0.000	0.010	0.107	0.418	0.465	3.34
20	0.000	0.007	0.091	0.400	0.502	3.40
21	0.000	0.005	0.076	0.380	0.538	3.45
22	0.000	0.004	0.064	0.360	0.573	3.50
23	0.000	0.003	0.053	0.339	0.605	3.55

Figure 5.6 Calculation of sequential condition state distributions by the Markov Chain method.

24	0.000	0.002	0.044	0.318	0.635	3.59
25	0.000	0.002	0.037	0.297	0.664	3.62
26	0.000	0.001	0.031	0.277	0.691	3.66
27	0.000	0.001	0.026	0.258	0.716	3.69
28	0.000	0.001	0.021	0.239	0.739	3.72
29	0.000	0.000	0.018	0.221	0.760	3.74
30	0.000	0.000	0.015	0.205	0.780	3.77
31	0.000	0.000	0.012	0.189	0.799	3.79
32	0.000	0.000	0.010	0.174	0.816	3.81
33	0.000	0.000	0.008	0.160	0.832	3.82
34	0.000	0.000	0.007	0.147	0.846	3.84
35	0.000	0.000	0.006	0.135	0.859	3.85
36	0.000	0.000	0.005	0.124	0.871	3.87
37	0.000	0.000	0.004	0.114	0.883	3.88
38	0.000	0.000	0.003	0.104	0.893	3.89
39	0.000	0.000	0.003	0.095	0.902	3.90
40	0.000	0.000	0.002	0.087	0.911	3.91
41	0.000	0.000	0.002	0.080	0.919	3.92
42	0.000	0.000	0.001	0.073	0.926	3.92
43	0.000	0.000	0.001	0.067	0.932	3.93
44	0.000	0.000	0.001	0.061	0.938	3.94
45	0.000	0.000	0.001	0.055	0.944	3.94
46	0.000	0.000	0.001	0.051	0.949	3.95
47	0.000	0.000	0.001	0.046	0.953	3.95
48	0.000	0.000	0.000	0.042	0.957	3.96
49	0.000	0.000	0.000	0.038	0.961	3.96
50	0.000	0.000	0.000	0.035	0.965	3.96

Figure 5.6 (Continued).

The expectation value of the degree of damage (=expected average DoD) is obtained by multiplying the scale vector $R = (0, 1, 2, 3, 4)$ by the condition state distribution, as shown in Equation (5.4).

$$E(t) = W(t) \times R \qquad (5.4)$$

where

$E(t)$ expectation value for the degree of damage (=average)
R scale vector comprising of the numerical values of condition states.

The probability density functions and the cumulative probability functions for the states 0–4 are depicted in Figures 5.7 and 5.8 according to the calculations in Figure 5.6.

5.3.3.3 *Action effect matrices*

The action effect matrices are built individually for each repair action taking into account the probable changes in the condition of the structure as a result of the action and the risk of failure during repair. Thus the condition state distribution of the structure after a repair action is not necessarily the same as that for a new structure.

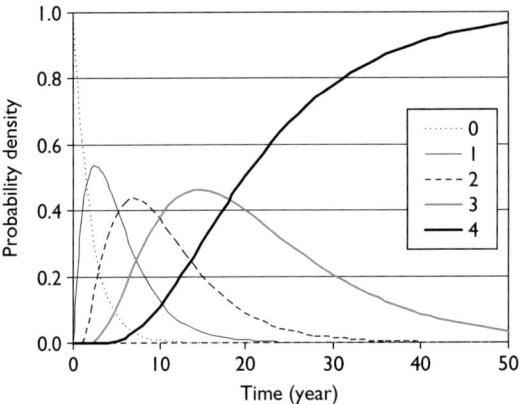

Figure 5.7 Probability density functions for condition states (=degrees of damage) 0–4 calculated by the Markov Chain method.

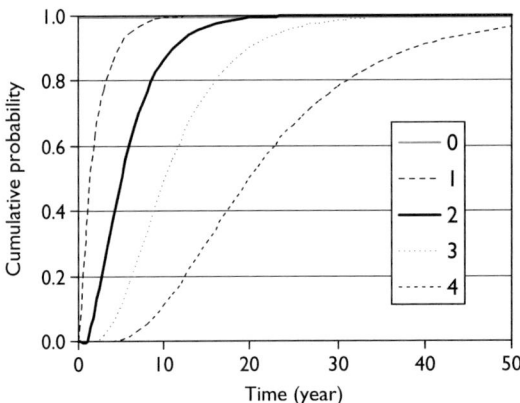

Figure 5.8 Cumulative probability functions for degrees of damage 0–4 determined by the Markov Chain method.

The general appearance of an action effect matrix is as shown in Table 5.5. As it is assumed that the condition state of a structure is always improved or at least remains the same as a result of an MR&R action, all the probability elements above the diagonal are 0. Other elements may have a value between 0 and 1. Again the sum of transition probabilities in each row must be 1. Usually heavy repair actions bring the structures close to the perfect condition so that the elements in the first column of the matrix are near 1 and the others near 0.

Much data lack in this area as very little research work has been done studying the condition-related effects of various repair actions. So there are usually no convertion methods used for action effect matrices as were for degradation matrices. In practice, the transition probabilities of action effect matrices are usually determined based on expert evaluation (Delphi study).

Table 5.5 Transition probability matrix for MR&R action effects
(5 state system)

State	0	1	2	3	4
0	p_{00}	0	0	0	0
1	p_{10}	p_{11}	0	0	0
2	p_{20}	p_{21}	p_{22}	0	0
3	p_{30}	p_{31}	p_{32}	p_{33}	0
4	p_{40}	p_{41}	p_{42}	p_{43}	p_{44}

A typical action effect matrix can be as seen on top of Figure 5.9. The purpose of Figure 5.9 is to visualise the action effects in a Markov Chain process. The calculation table is programmed so that a repair is done every time when signed by 1 in the column at the left side of the figure. The action effects can be readily seen in the condition state distributions and the average DoD curve presented in Figure 5.10.

A repair action may also have an impact on the rate of degradation after the repair. If the rate of degradation is expected to be changed after an MR&R action the degradation matrix is changed respectively.

5.3.3.4   Modelling of the action effects of coatings

When applying coatings and other preventive maintenance measures, the condition state of the structure is not considered to be changed at all but the rate of further degradation is reduced. So no action effect matrix is applied in connection of preventive maintenance actions but the degradation matrix is changed according to the expected rate of degradation. The effects of coatings on the condition of the structure depend on the condition of the coating [5].

Transition probability matrix of repair

State	0	1	2	3	4
0	1	0	0	0	0
1	0.95	0.05	0	0	0
2	0.92	0.05	0.03	0	0
3	0.9	0.05	0.03	0.02	0
4	0.88	0.05	0.03	0.02	0.02

Transition probability matrix of degradation

State	0	1	2	3	4
0	0.61	0.39	0	0	0
1	0	0.74	0.26	0	0
2	0	0	0.82	0.18	0
3	0	0	0	0.91	0.09
4	0	0	0	0	1

Figure 5.9 Action effects in a Markov chain lifetime table.

Repair	Year	State 0	1	2	3	4	Average DoD
	0	1.000	0.000	0.000	0.000	0.000	0.00
	1	0.610	0.390	0.000	0.000	0.000	0.39
	2	0.372	0.527	0.101	0.000	0.000	0.73
	3	0.227	0.535	0.220	0.018	0.000	1.03
	4	0.138	0.484	0.319	0.056	0.002	1.30
	5	0.084	0.412	0.388	0.109	0.007	1.54
	6	0.052	0.338	0.425	0.169	0.016	1.76
	7	0.031	0.270	0.437	0.230	0.032	1.96
	8	0.019	0.212	0.428	0.288	0.052	2.14
	9	0.012	0.165	0.406	0.339	0.078	2.31
	10	0.007	0.126	0.376	0.382	0.109	2.46
	11	0.004	0.096	0.341	0.415	0.143	2.60
	12	0.003	0.073	0.305	0.439	0.181	2.72
	13	0.002	0.055	0.269	0.454	0.220	2.84
	14	0.001	0.041	0.235	0.462	0.261	2.94
I	15	0.902	0.050	0.029	0.014	0.005	0.17
	16	0.550	0.389	0.037	0.018	0.007	0.54
	17	0.336	0.502	0.131	0.023	0.008	0.87
	18	0.205	0.502	0.238	0.045	0.010	1.15
	19	0.125	0.452	0.326	0.084	0.014	1.41
	20	0.076	0.383	0.385	0.135	0.022	1.64
	21	0.046	0.313	0.415	0.192	0.034	1.85
	22	0.028	0.250	0.422	0.249	0.051	2.05
	23	0.017	0.196	0.411	0.303	0.074	2.22
	24	0.011	0.152	0.388	0.349	0.101	2.38
	25	0.006	0.116	0.357	0.388	0.132	2.52
	26	0.004	0.089	0.323	0.417	0.167	2.65
	27	0.002	0.067	0.288	0.438	0.205	2.78
	28	0.001	0.051	0.254	0.450	0.244	2.88
I	29	0.903	0.050	0.028	0.014	0.005	0.17
	30	0.551	0.389	0.036	0.018	0.006	0.54
	31	0.336	0.503	0.131	0.023	0.008	0.86
	32	0.205	0.503	0.238	0.044	0.010	1.15
	33	0.125	0.452	0.326	0.083	0.014	1.41
	34	0.076	0.383	0.385	0.134	0.021	1.64
	35	0.047	0.313	0.415	0.191	0.033	1.85
	36	0.028	0.250	0.422	0.249	0.051	2.04
	37	0.017	0.196	0.411	0.303	0.073	2.22
	38	0.011	0.152	0.388	0.349	0.100	2.38
	39	0.006	0.117	0.358	0.388	0.132	2.52
	40	0.004	0.089	0.324	0.417	0.167	2.65
	41	0.002	0.067	0.288	0.438	0.204	2.77
	42	0.001	0.051	0.254	0.450	0.243	2.88
	43	0.001	0.038	0.221	0.456	0.284	2.98
	44	0.001	0.029	0.191	0.454	0.325	3.07
I	45	0.899	0.050	0.029	0.016	0.007	0.18
	46	0.548	0.388	0.037	0.019	0.008	0.55
	47	0.334	0.501	0.131	0.024	0.010	0.87
	48	0.204	0.501	0.238	0.046	0.012	1.16
	49	0.124	0.450	0.325	0.084	0.016	1.42
	50	0.076	0.382	0.384	0.135	0.024	1.65

Figure 5.9 (Continued).

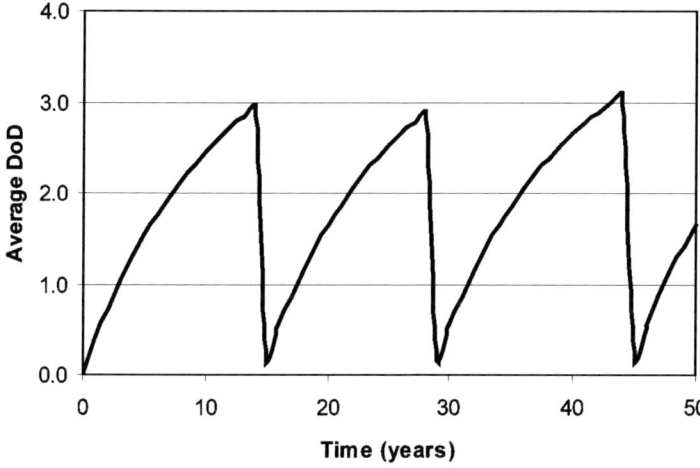

Figure 5.10 The average DoD with time showing the effects of repair on the condition of a structure.

Coatings have both direct and indirect effects on the condition state of a structure. The direct effects are a result of the physical barrier which retards the penetration of aggressive agents, such as CO_2 and chlorides, into the concrete structure. The indirect effects result from the changed moisture content in the structure because of the coating as the moisture content has a remarkable effect on the degradation rate. The model of a degradation matrix, which takes into account the *direct* effects of a coating on the degradation rate of a structure, is presented in Table 5.6.

For more detailed information on the modelling of the condition-related effects of coatings using the Markov Chain method see References [3] and [6]. As the condition and the protection properties of coatings are time-dependent, the condition of the coating is first modelled by the Markov Chain and then the changes in the condition of the structure are determined taking into account the concurrent condition state of the coating. So the transition probabilities of the structure are not any more constant but are dependent on the condition of the coating. Figure 5.11 shows the result of calculation as an example.

Table 5.6 The assumed form of a degradation matrix for a coated structure

State	0	1	2	3	4
0	$1 - p_c \cdot p_{01}$	$p_c \cdot p_{01}$	0	0	0
1	0	$1 - p_c \cdot p_{12}$	$p_c \cdot p_{12}$	0	0
2	0	0	$1 - p_c \cdot p_{23}$	$p_c \cdot p_{23}$	0
3	0	0	0	$1 - p_c \cdot p_{34}$	$p_c \cdot p_{34}$
4	0	0	0	0	1

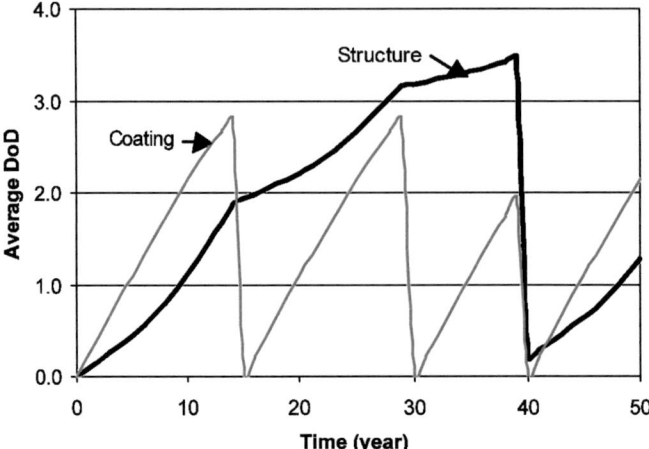

Figure 5.11 Average DoD of the coating and the structure (example).

5.3.4 Combined LCP, LCC and LCE analysis

An advanced LMS, such as Lifecon LMS, is working on the 'life cycle principle'. It means that the profitability of optional maintenance strategies is studied by LCC analyses and the results of these analyses are subject to strategic decision-making. Not only the maintenance costs but also the user costs and environmental costs, i.e. environmental impacts, may be included in the decision-making.

The principles of LCC calculations with predefined MR&R action profiles are well known and described in international standards like ISO 15686-5 [7] and ASTM E 917 [8]. However, the traditional procedure of cost calculation with predefined action profiles could obviously not serve as the basis for an LMS. Rather it is the task of the management system to specify the actions and to define the timings of actions using appropriate degradation models [9]. So the calculation methods for the LCC analyses in an LMS must be more advanced and more automatic than those in a conventional LCC analysis [10–13].

A Markov Chain-based life cycle cost analysis is actually a combination of a life cycle performance (LCP), a life cycle cost (LCC) and a life cycle ecology (LCE) analyses. It integrates the Markov Chain-based condition analysis to a conventional life cycle analysis framework. From the material resources used during the MR&R actions it determines also the life cycle ecological consequences in the form of environmental impacts [6].

5.3.4.1 General principles

The idea of the Markov Chain-based LCC analysis is to combine the Markov Chain lifetime table with a traditional LCC calculation table. The timings of MR&R actions can be defined based on the Markov Chain models and an automatic condition guarding system for triggering actions. The life cycle costs can then be determined using conventional calculation methods. An environmental impact analysis can also be combined into the same composition of analyses.

In a Markov Chain LCC analysis the MR&R actions are timed, based on predefined condition requirements. An action is automatically triggered when the maximum allowable probability for exceeding the predefined limit state is overridden. Every MR&R action causes costs which are summed up. Other costs such as user costs and environmental costs (impacts) can be determined in the same way and integrated in special cost counters attributed to them. The final purpose of the Markov Chain-based LCC analysis is to find the most feasible and economically most effective maintenance strategy to upkeep the structures taking into account MR&R costs, user costs and environmental impacts. It needs an optimisation problem where the user seeks to find the most effective MR&R action profile for each structural part and for the defined period of time.

In the Markov Chain-based LCC analysis the whole MR&R action profile pertaining to the given time frame is reproduced as a series of annual condition state distributions. There are two ways for specification of MR&R actions and definition of timings of them: manual and automatic. In a manual analysis the definition of MR&R actions are done manually. In principle, a designer can specify the necessary actions to be taken at any time during the time frame. However, even in a manual analysis in which the MR&R actions are specified manually, the timing of actions may be based on the condition analysis and automatic triggering of actions. In a fully automatic analysis system (which aims at automatic life cycle planning), the MR&R actions are specified using the decision tree method. The decision tree contains pre-optimised MR&R action profiles for each case. It selects the optimised MR&R action profile based on the material properties, environmental burdens and possible special requirements of the structural part.

A life cycle cost analysis cannot be conducted right away for the whole building or infrastructure if the environmental conditions, materials and structural features in its parts vary. Hence, the building or infrastructure is first divided into components and the LCC analyses are conducted for each component separately. The answers related to the whole building or infrastructure can then be obtained by summing up the analysis results of components.

The LCC analyses can be used both in object level and in network level studies. At the object level the LCC analysis is used for life cycle design of specific components and objects. Specific parameter values of structures (obtained from database) are used in these calculations. The purpose of such analyses is to find out the optimal MR&R action profiles for structural component and to find the optimal project profile for the object.

At the network level the purpose is to use the LCC analysis results for strategic planning of MR&R activities and to make short- and long-term cost scenarios for the future. The structural parts are treated statistically as populations of structural parts. The calculations are conducted using average values of the material, structural and environmental parameters pertaining to the network or a subnetwork of structures. The purpose is to find the optimal maintenance strategy for structures for varying environmental conditions and for varying material and structural properties. Typically, answers for the following questions can be obtained: Is it cost-effective to protect the structures by coatings or other protection methods? Which repair methods should be used? In which condition state should the structure be repaired and in which condition state should the coatings or other protections be renewed to minimise the life cycle costs?

5.3.4.2   Specification of MR&R actions

For both the manual and the automatic analyses methods each MR&R action must be specified. The specification of actions is done by answering the questions presented in Table 5.7.

An *action group* is an MR&R action category composed of similar MR&R systems. For concrete structures the MR&R actions groups may consist of the following:

- coating
- patching of coating
- protection with concrete overlay
- patching of concrete protection
- patching of structure
- repair of structure
- renovation of structure.

Each MR&R action group contains several repair systems or methods. Accordingly, the group of coatings is comprised of several coating systems. The concrete protection group refers to methods in which a layer of shotcrete, conventional concrete or cement mortar is applied on the whole surface of the structure. Cathodic protection methods with a net anode embedded in a layer of concrete on the original structure is also included in this group of actions.

The group of structural repairs refers to major repair actions which improve the condition of the structural part. In concrete structures the structural repairs refer to actions by which the concrete around the reinforcement is renewed. This can be done by removing the concrete around the steel bars and replacing it by mechanical repair methods. Electrochemical methods such as realkalisation and chloride extraction are included in this group as the concrete environment around the reinforcement is renewed by realkalisation or removal of chlorides.

Patching is partial repair of the most attacked areas of the structure. Patching may refer also to partial repair of a coating or other protection. The methods of structural patching are comparable to the structural repair in that they also change

Table 5.7 Definition of actions

1	Is the MR&R action group used during the design period?	Yes/no
2	Which MR&R system?	Code of the MR&R system within the MR&R action group
3	Limit condition state?	Limit state for the action, e.g. 3 or 4
4	Maximum allowable probability for exceeding the limit state?	Probability as %. Exceeding the given percentage will trigger the action.
5	Maximum number of repeated actions?	Number of allowable repetitions of an action before a heavier action.

the environment around the reinforcement. However, this is done only locally and the other parts of the structure remain unchanged. So patching is not considered as staring a new service life but only to extend the on-going service life.

Renovation refers to complete replacement of a component by a new one, so this group consists of methods for renovation. The component can be reconstructed at site or a new prefabricated element can be installed at the place of the old component.

The data related to specific MR&R action systems are presented in Table of MR&R systems. The MR&R systems are arranged in the table according to action groups and they can be referred to by their *code numbers*. For example, in the case of the coating group the code number refers to a specific coating system with defined materials and material thicknesses. In the case of concrete protection group it refers to specific concrete or cathodic protection systems with defined materials, thicknesses and techniques.

The *maximum allowable probability* sets the maximum limit for the probability of exceeding the limit state. In object level studies one can interpret it as expressing the maximum allowable fraction of the surface area of a component to be at the limit state or in still worse condition. In network level studies it means the maximum portion of structures which can be tolerated at the limit state or in still a worse condition. The MR&R actions for structures are automatically triggered when the maximum allowable probability for the defined limit state is exceeded.

Maximum number of repeated actions sets a limit to the number of the same MR&R action that can be taken during the design phase. For instance the number of repairs or recoatings can be limited. In the case of coatings the counter starts from zero every time when the component is repaired and in the case of repairs the repair counter starts from zero when the component is replaced by a new one.

The life of a component is considered to be composed of three phases for which the MR&R actions may be specified independently as follows:

Phase I   Residual service life of the component. All actions of protection and patching are defined until the end of the on-going service life.

Phase II   From the end of the residual service life to the end of the residual life cycle of the component. The repair methods are defined until the end of the life cycle of the component. The patching and protection methods for this period of time can be defined in another way than for the on-going service life. This is necessary as the need of protection may be changed after the repair.

Phase III   From the end of the on-going life cycle to the end of the last life cycle. The methods of renovation are defined. For this period of time the repair methods can be newly defined as also the patching and protection methods.

The division of the life of a component is presented graphically in Figure 5.12. The life of a component can be described as a combination of nested arches which represent the lives of actions.

Several action groups can be selected for the same design phase with appropriate limitations. So it is possible to apply, for example, coating together with structural

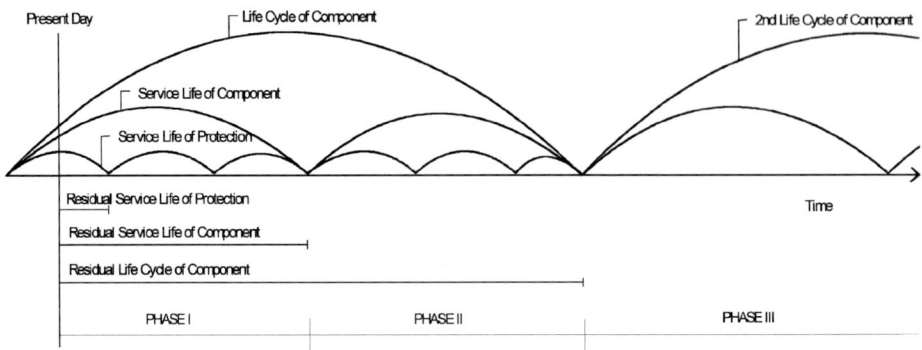

Figure 5.12 Division of the life of a component into phases.

repair or coating and concrete protection together with structural repair. However, in the design phase I no repair is possible and in the design phase II no renovation is possible to select.

As a component can be repaired completely without replacing the whole component by a new one, a new service life of the component is considered to start from the repair. Possibly many consecutive repairs can even be accepted before the component must be replaced. Thus the life cycle of a component is not considered to end until it is completely renovated or replaced by a new one. Accordingly, a structural repair generates a new service life, and a renovation or replacement generates a new life cycle for the component.

5.3.4.3 Specification of MR&R actions by a decision tree

The MR&R actions for a component can be specified automatically by a decision tree. The MR&R action profiles specified by a decision tree have been previously optimised by manually defined LCC analyses and risk analyses. The selection of an MR&R action profile for a particular component is done by the decision tree run during which several decision criteria related to the specific properties, environmental conditions and requirements of the component are evaluated. However, only the types of MR&R actions are defined by the decision tree. The timing of actions is determined by the Markov Chain life cycle table and the automatic triggering of actions.

A decision tree has a 'root' which forks at 'nodes' representing the relevant criteria related to properties of the component, severity of environment and special require-ments of the object and makes with a growing number of nodes an ever-increasing amount of 'branches'. The final branches after the last node are called 'leaves'. The optimal sets of MR&R actions are the results of the tree and are inserted in the leaves of the tree.

An example of a decision tree and its solution is presented in Figure 5.13. The component-specific data is given at the row 'distribution'. The tree is active to find the correct set of MR&R actions corresponding to the given data.

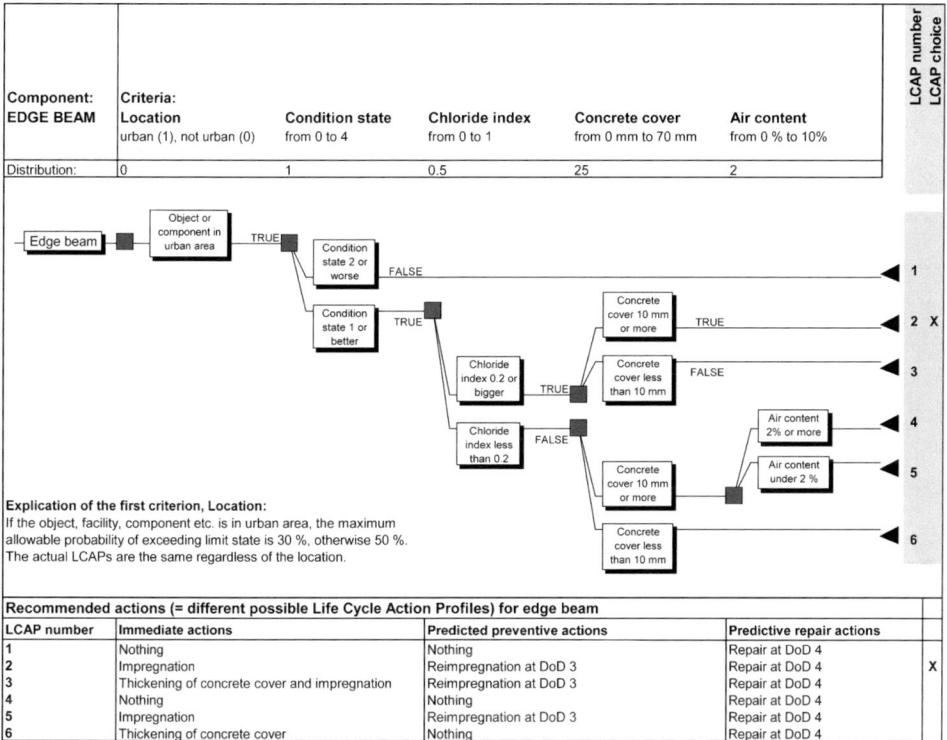

Figure 5.13 Decision tree, illustrative presentation.

In an LCC analysis program the decision tree is usually attached as a subprogram. In a program code of a decision tree the branches are implemented by IF… THEN statements, which can be nested multifold.

Normally the user has no access to the decision tree. However, it is possible to make the computer program so that the user can do some changes in the MR&R specifications of the decision tree.

5.3.4.4   Principles of condition guarding and triggering of actions

In a condition-controlled LCC analysis, the timing of actions is performed automatically. The principle of triggering actions in a Markov Chain life cycle table is presented in Figure 5.14. The sequential annual condition state distributions have been determined by Markov Chain on the left side of the figure. They show the probability of the component at any of the condition states at any time. In the middle of the figure the respective cumulative probabilities which express the probability of exceeding or being equal to any of the condition states are presented. In this example condition state 3 was selected for the limit condition state and 50% as the maximum allowable probability for exceeding the limit condition state. If this criterion is exceeded during a year, a repair action will be performed immediately in the next year. The action

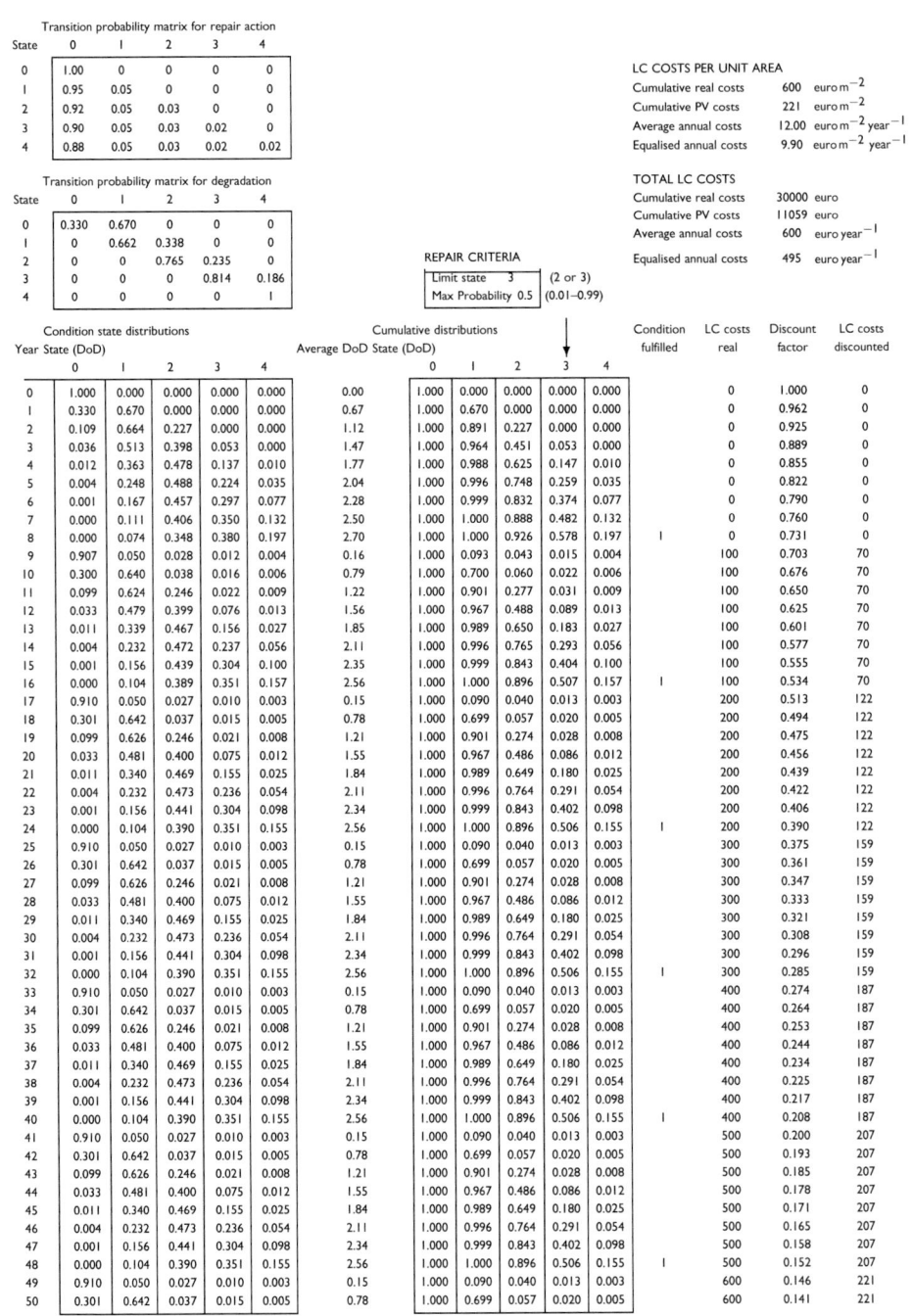

Figure 5.14 Principles for the determination of condition state distributions, triggering of actions and calculation of life cycle costs [6].

effects on the condition state distribution of the structure are obtained by multiplying the condition state distribution of the year by the action effect matrix in the upper left corner. At the same time, the repair costs are added in the cost counters in the right side of the figure. In other years only the increase of degradation is evaluated by the degradation matrix that is situated below the action effect matrix.

Many kinds of maintenance and repair actions can be included in a life cycle of a structure. So Figure 5.14 is inadequate to represent the whole LCC analysis. For instance the degradation of a concrete structure can be retarded by applying an extra layer of concrete or a coating on the structure. However, both the extra layer of concrete and the coating deteriorate themselves. So before evaluation of their effect on the condition of the structure, the condition of the concrete layer and the coating must be first evaluated. In practice three lifetime tables of the form presented in Figure 5.15 are needed:

- Table of coatings;
- Table of extra concrete layer
- Table of the structure.

These tables are connected to each other by rules and formulas, which take into account the mutual condition-related effects, as schematically presented in Figure 5.15.

5.3.4.5 Methods of counting costs

5.3.4.5.1 GENERAL

The costs counted by the cost counters obeying the ISO whole life costing principles [7]. The cost counters get their information from the Markov Chain life cycle table

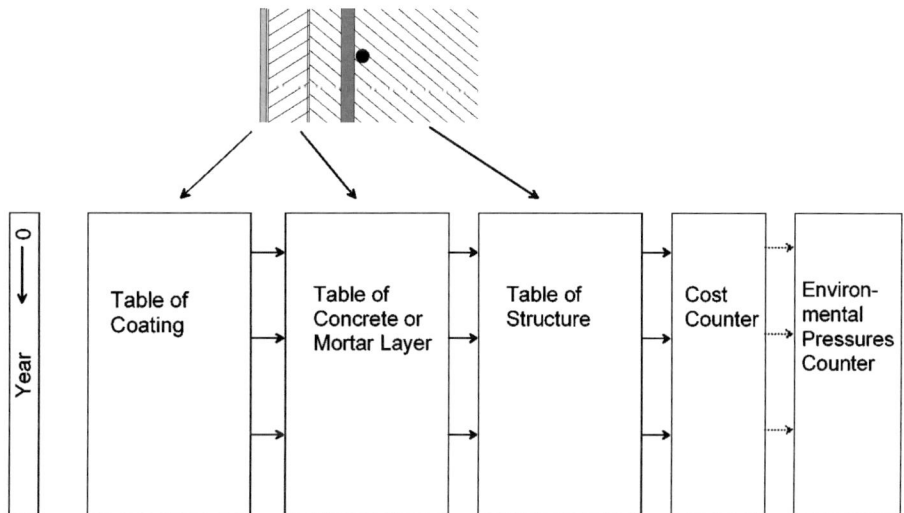

Figure 5.15 Tables of coating, concrete or mortar layer and the structure connected to each other and counters for costs and environmental impacts.

(types and timings of MR&R actions) and the table of the MR&R systems (unit costs for MR&R actions, etc.). The task of the cost counters is to collect and summarise the costs from the total time frame. The costs are understood to cover MR&R costs, user costs and environmental impacts.

5.3.4.5.2 MR&R COSTS

The MR&R costs are comprised of real maintenance costs such as costs of coating, protection, patching, repair, rehabilitation, renovation, etc.

The unit costs of MR&R actions are usually based on statistical data from earlier executed MR&R projects. In some cases the costs depend on the extent of the repair, i.e. the area of repair and the depth of concrete which is replaced from the structure. The unit costs may also depend on the general condition of the structure. Then a single value is not justified for unit costs but a model formula that determines the unit costs as a function of the relevant parameters is applied instead. An example of such a model formula is given in Equation (5.5):

$$\text{UnitCost} = \text{UnitCost}_0 \cdot C_{\text{depth}} \cdot C_{\text{area}} \cdot C_{\text{cond}} \tag{5.5}$$

where

UnitCost	is unit costs of an MR&R action, euro m^{-2}
UnitCost$_0$	unit cost of an MR&R action with respect to the minimum depth and the minimum area of repair, euro m^{-2}
C_{depth}	coefficient depending on the depth of repair
C_{area}	coefficient depending on the area of repair
C_{cond}	coefficient depending on the condition of the structure at the moment of repair

5.3.4.5.3 USER COSTS

In some types of infrastructure, such as bridges, the user costs are included in the decision-making on maintenance strategy. For bridges three kinds of road user costs (RUC) can be identified [2]:

- additional road user costs due to restricted traffic for restricted axle loads and inadequate bridge geometry;
- additional road user costs due to MR&R works (delays);
- risk costs due to failure of a bridge.

In a management system of bridges the additional road user costs due to MR&R actions are of special interest. These costs may result from the following reasons:

- reduced speed (traffic sign)
- diversion
- signal regulation.

The road user costs as a result of the increased travel time (for MR&R works) can be determined by the following formula:

$$\text{RUC} = I \cdot \%_{\text{car}} \cdot \Delta t \cdot \text{TDC}_{\text{car}} + I \cdot \%_{\text{truck}} \cdot \Delta t \cdot \text{TDC}_{\text{truck}} \qquad (5.6)$$

where

RUC is	road user costs, euro day^{-1}
I	average daily traffic (ADT)
$\%_{\text{car}}, \%_{\text{truck}}$	percentage of traffic for cars and trucks
Δt	increased travel time due to the maintenance works (for traffic sign, diversion or signal regulation, h)
$\text{TDC}_{\text{car}}, \text{TDC}_{\text{truck}}$	time-dependent unit costs for cars and trucks, euro h^{-1}.

In the case of diversion the road user costs due to the *increased driving length* must be added to the road user costs. They are determined by the following equation:

$$\text{RUC} = I \cdot \%_{\text{car}} \cdot \Delta L \cdot \text{DDC}_{\text{car}} + I \cdot \%_{\text{truck}} \cdot \Delta L \cdot \text{DDC}_{\text{truck}} \qquad (5.7)$$

where

ΔL	is increased driving length due to the diversion, km
$\text{DDC}_{\text{car}}, \text{DDC}_{\text{truck}}$	driving-dependent unit costs for cars and trucks, euro km^{-1}.

The above-presented equations refer to the road user costs per day. So the total road user costs depend on the total time of the repair work. The total costs per unit area (or other functional unit) can be determined as the product of the user costs per day and the repair time. The repair time may be evaluated based on the production rate of the work (m^2 day^{-1}) for each MR&R action system and the area of repair as follows:

$$t_{\text{r}} = \frac{A}{u_{\text{r}}} \qquad (5.8)$$

where

t_{r}	is repair time, d
A	area of repair, m^2
a_{r}	production rate of the MR&R system applied, m^2 d^{-1}.

This calculation method is not indisputable as in practice several works for several components can be performed at the same time. However, this offers one solution for the problem of addressing user costs for components.

5.3.4.5.4 ENVIRONMENTAL IMPACTS

The purpose of the environmental impact analysis is to provide the decision-makers with comparative data on the environmental impacts of various optional MR&R action profiles. This data is used as one attribute in the optimisation of the maintenance strategy for structures.

As a starting point of the environmental impact analysis it is assumed that the environmental profiles for the used materials are available. The profiles should at least consist of the following variables:

Resources of energy (MJ)

- renewable energy
- non-renewable energy.

Emissions into air (kg), (g) or (mg):

- CO_2
- SO_2
- NO_x
- particles
- CH_4
- non-methane (VOC).

Non-renewable raw materials (kg)

- mineral raw materials.

The results of the environmental profiles are normally given per mass units (kg). So the profiles must be converted into functional units, usually square metres, to know their consumption on the surface of structures. To do this each environmental variable is divided by the coverage ($m^2 kg^{-1}$) of the material.

The emissions related to MR&R actions have many kinds of ecological impacts. The following classification of impacts is normally used:

- climate change
- acidification
- formation of photo-chemical ozone
- ecotoxity
- heavy metals
- cancerous materials
- effect on biodiversity.

The first three environmental impact classes have usually been applied in the analyses of the construction sector.

There are several methods developed for the evaluation of the total environmental impact. However, there is no general agreement on the methods as yet. In the discussion only the Swedish Environmental Priority Strategy (EPS) method is presented [14]. The environmental indicator in this method is Environmental Load Unit (ELU), which is defined in euro as follows:

$$ELU(euro) = 1.557 \cdot CH_4(kg) + 0.191 \cdot CO(kg) + 0.0635 \cdot CO_2(kg)$$
$$+ 0.00707 \cdot particles + 3.4 \cdot (C_xH_y - CH_4)(kg)$$
$$+ 0.395 \cdot NO_x(kg) + 0.0545 \cdot SO_x(kg) \tag{5.9}$$

The principle for calculating the sum of environmental impacts from MR&R actions during the treated time frame is the same as that for life cycle costs. The environmental impacts per functional units like m², m, etc. for each MR&R action are determined in the MR&R system tables. The total impacts for MR&R actions can be determined by multiplying the unit area-based impacts by the total repair area. The impacts of the whole treated time frame are determined in the cost counters by summing up all the impacts of the design period.

5.3.4.5.5 METHODS OF DISCOUNTING

The life cycle costs are determined according to the principles of the standard ISO 15686 Part 5, Whole life costing [7]. Accordingly the costs are determined as *discounted costs* (present value costs).

The total discounted or present value costs are determined by simply summing the MR&R action costs throughout the treated time as presented by Equation (5.10); no discounting is used:

$$C_R = \sum_{i=0}^{t}\sum_{j=1}^{n_i} C_{j,i} \tag{5.10}$$

where

C_R is the total real costs from the treated time frame, euro m⁻²

$C_{j,i}$ is costs of the jth maintenance action in year i, euro m⁻²

n_i number of maintenance action in year i

t number of years in the time frame (length of the span in years).

Discounted or present value (PV) costs refer to maintenance costs discounted to the present day by the discount factor. As the discount factor diminishes with time, the PV costs of actions scheduled near to the start of the time frame are greater than the PV costs of respective actions scheduled later in the time frame. The total PV costs are calculated from the Equation (5.11):

$$C_{PV} = \sum_{i=0}^{t}\sum_{j=1}^{n_i} C_{j,i}\frac{1}{(1+r)^i} \tag{5.11}$$

where

C_{PV} is total PV costs from the treated time frame, euro m⁻²

r discount rate.

To compare different maintenance strategies it is advisable to redistribute the sum of life cycle costs evenly into annual costs. This can be done based on either real costs or present value costs. So two kinds of annual costs are defined:

- average annual costs
- equalised annual costs.

The average annual costs are defined as the total real costs divided by the number of years in the time frame, as in Equation (5.12):

$$A_A = \frac{C_B}{t} \tag{5.12}$$

where
$\quad A_A$ is average annual costs, $euro\,m^{-2}\,year^{-1}$.

The equalised annual costs are determined by multiplying the total PV costs by the annuity factor, an in Equation (5.13):

$$A_E = C_{PV} \cdot \frac{r(1+r)^t}{(1+r)^t - 1} \tag{5.13}$$

where
$\quad A_E$ is equalised annual costs, $euro\,m^{-2}\,year^{-1}$.

The equalised annual costs depend on how the maintenance actions are scheduled within the time frame. Maintenance actions scheduled near to the start of the time frame increase the equalised annual costs more than those scheduled later in the time frame. This feature is emphasised with increasing discount rate.

5.3.4.6   Life cycle cost analysis process

The total life cycle cost analysis process is presented schematically in Figure 5.16. The phases of the analysis are the following:

1. specification of the initial data
2. analysis process
3. presentation of results.

Figure 5.16 also shows schematically the structure of the life cycle analysis program. The program consists of several tables:

1. tables of object- and component-specific data
2. tables of MR&R systems
3. tables for definition of actions
4. Markov Chain life cycle analysis tables
5. tables for counting costs and
6. tables of results.

In the following, the analysis process is described in more detail.

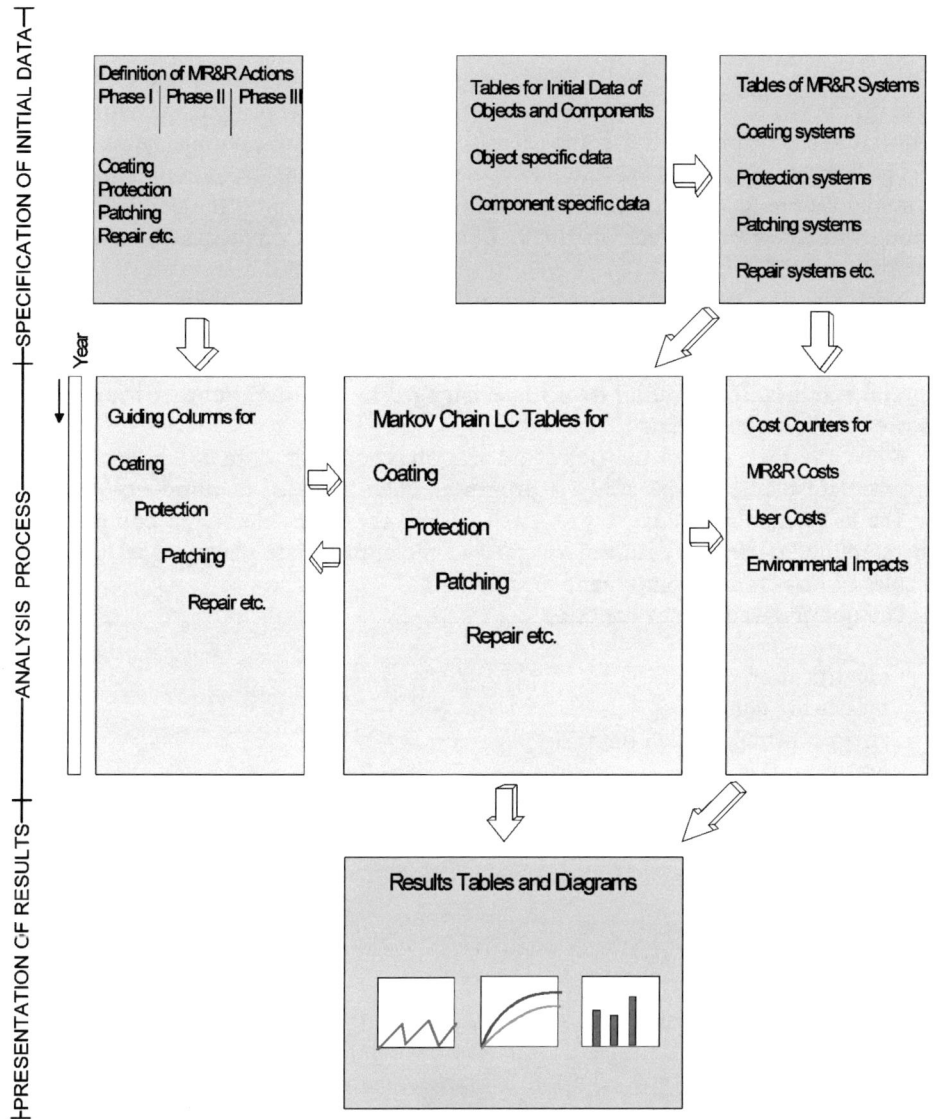

Figure 5.16 General layout of a life cycle cost analysis process.

5.3.4.6.1 SPECIFICATION OF INITIAL DATA

When the LCC analysis computer program is started, the following initial data should be input by the user:

- time frame of the analysis
- discount rate
- object

- component
- MR&R actions (unless not specified automatically by the decision tree).

The time frame (design period) of the analysis is given by the user in years. Usually the time frame is between 50 and 200 years. If it is desired to compare several optional MR&R action profiles for a component the same time frame should be used.

The chosen discount rate should be near the real rate of interest which is the nominal rate minus inflation. If the real rate of interest is used the possible inflation should not have any effect on the results. In industrial countries the real rate of interest in long term has been proved to stay between 2 and 5%.

Next, the user selects the object from the list presented on the screen. The LCC analysis program presents the list of objects according to the current initial data file. The data in the initial data file have been previously gathered from the database by special routines. If the initial data file is changed by the user, another list of objects is presented on the screen.

When the user selects the object all the object-specific data are assigned to the appropriate places of the analysis program. Then a list of components pertaining to the selected object also is presented. When the user selects the component all the component-specific data are assigned to the appropriate places, especially to the Tables of object- and component-specific data.

The object-specific data contain:

- identification data
- measuring data
- environmental burden data
- user cost data, etc.

The component-specific data contain:

- identification data
- measuring data
- structural data
- data on previous MR&R actions
- inspection and condition assessment data, etc.

The Tables of object and component-specific data contain also the default values that are used in the analysis if specific data are not available. The appropriate data in the Tables of object- and component-specific data are then automatically assigned to the Tables of MR&R systems.

The Table of MR&R systems contains all data pertaining to MR&R systems (methods). Each system has a code number in the left column of the table. The row of a specific MR&R system is identified by that code, and the data pertaining to the system are situated in the row indicated by the code number. The table of MR&R systems is more than just a store of data. All models (cost models, degradation models and action effect models) are programmed in the Table of MR&R systems. So, the system table consists of model equations and their parameters.

The MR&R actions are then specified for the selected component as explained Section 5.3.4.3. Manual specification is needed if the optimum MR&R action profile is searched by comparing different optional profiles. Automatic specification by a decision tree is used when the optimum action profiles for each case have already been solved and the decision tree has been provided with the optimum profiles.

5.3.4.6.2 ANALYSIS PROCESS

The principles of the life cycle analysis process are already described. The process has been made automatic so that the user does not have to intervene during the process. The following automatic routines are performed:

- automatic application of object- and component-specific parameter data for degradation, action effect and cost models;
- automatic conversion of degradation models into Markov Chain transition probabilities;
- automatic definition of actions by the decision tree (unless manually defined);
- automatic arrangement of the guiding columns (see. Figure 5.16) according to the specified MR&R action profile;
- automatic determination of the annual condition state distributions in the Markov chain life cycle table;
- automatic timing of actions;
- automatic calculation of life cycle costs, user costs and environmental impacts; and
- automatic presentation of the analysis results in tables and diagrams.

5.3.4.6.3 RESULTS OF LIFE CYCLE COST ANALYSIS

The main results of an LCC analysis can be compacted into a small results table. Table 5.8 shows the life cycle costs calculated per unit area. The annual unit costs are calculated as average annual costs and equalised annual costs.

The discounted component costs are obtained by multiplying the unit cost by the surface area of the component. If, for example, the surface area of the component is $166\,m^2$ and the unit costs are those presented in Table 5.8, the true costs are presented in Table 5.9.

As can be seen from the results in Tables 5.8 and 5.9, the ELU costs calculated based on the EPS method are small compared to both the MR&R costs and user costs.

The design period was in this case 250 years. The condition of the structure changes during this time is as depicted in Figures 5.17 and 5.18.

Table 5.8 An example of the results of life cycle cost analysis, unit costs

Unit costs	MR&R costs	User costs	Total costs	ELU
Cumulative Real Costs (euro m^{-2})	2114	455	2568	1.83
Cumulative PV Costs (euro m^{-2})	98	18	115	
Average Annual Costs (euro m^{-2} year^{-1})	8.46	1.82	10.27	0.01
Equalised Annual Costs (euro m^{-2} year^{-1})	3.91	0.70	4.61	

Table 5.9 An example of the results of life cycle analysis, true component costs

Unit costs	MR&R costs	User costs	Total costs	ELU
Cumulative Real Costs (euro)	350 905	75 460	426 365	303
Cumulative PV Costs (euro)	16 235	2 910	19 146	
Average Annual Costs (euro year^{-1})	1 404	302	1 705	1
Equalised Annual Costs (euro year^{-1})	649	116	766	

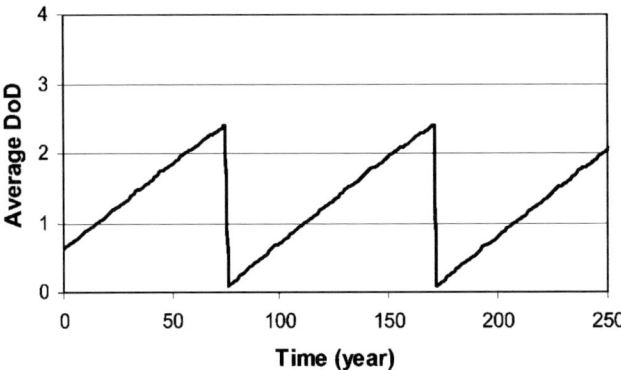

Figure 5.17 Average Degree of Damage as a function of time.

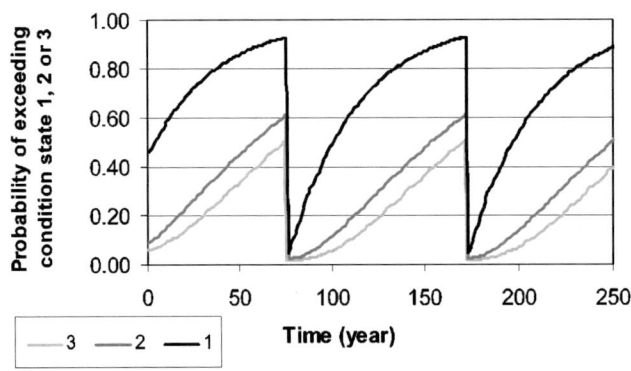

Figure 5.18 Probability of exceeding the condition states 1, 2 and 3 as a function of time.

In this example, the maximum allowable probability of exceeding the condition state 3 (= limitstate) was 50%. From the Figures one can observe that the repair was triggered immediately every time when this limit was exceeded.

The costs can also be presented as a function time. Figure 5.19 shows the cumulative MR&R costs per unit area as real costs and PV costs. The MR&R costs in this case were composed of structural repair cost and coating costs.

Figure 5.20 shows the cumulative MR&R costs and user costs per unit area.

The environmental impact analysis results can be itemised as presented in Figure 5.21.

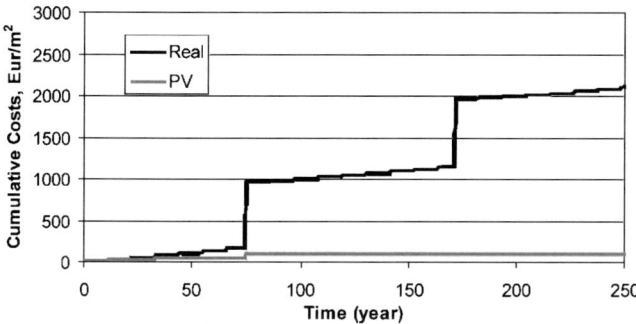

Figure 5.19 MR&R costs per unit area presented cumulatively as a function of time.

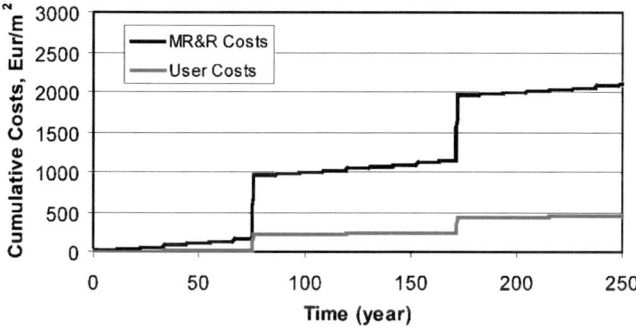

Figure 5.20 MR&R and user costs per unit area as a function of time.

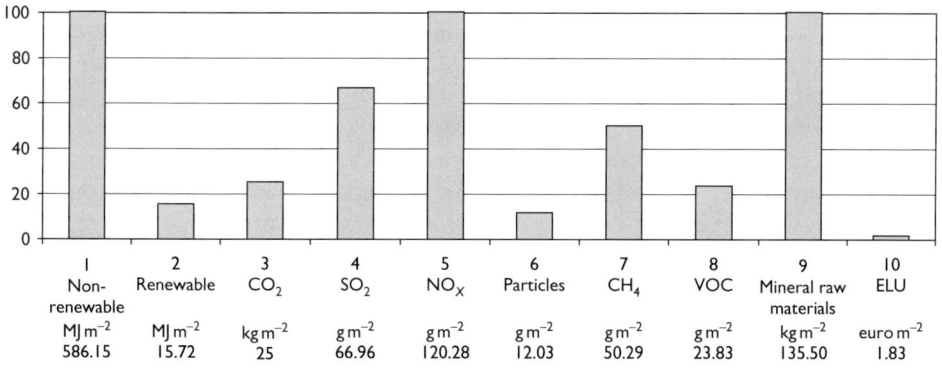

Figure 5.21 The environmental impact analysis itemised for various items of impact.

5.3.4.7 Advanced life cycle analyses programs for object level and network level use

In different variations of the life cycle analysis programs, additional features may be added in the program routine. Such extended analysis programs are those specially designed for the use of the object level and the network level management systems.

5.3.4.7.1 LIFE CYCLE PLANNING PROGRAM FOR THE OBJECT LEVEL MANAGEMENT

In a life cycle planning program for the object level use, all components of an object are analysed one after another and the MR&R actions pertaining to different components of an object are reorganised into 'projects'. By projects we mean here groups of MR&R actions that are scheduled for the same year for the same object. Instead of project planning one could rather call it life cycle planning, as not only is the next project planned but all the projects during the whole life frame are planned at the same time. The planning is done automatically, but the program allows user-defined changes to the plans.

The reason for reorganising the MR&R actions into projects is that the optimal timings for various actions (for various components) will scatter too much. Project planning based only on the optimal timing of actions would result in too many small projects to be executed for the same object. That would be annoying for both maintainers and users. So optimisation in the preliminary project planning is performed from a wider perspective than in the component level optimisation. As a result of proper object level planning in which the single MR&R actions are combined into reasonable groups, economic savings can be achieved by synergy profit.

From many possible ways of combining actions into projects, only one is presented here. It is effective and probably also the fastest method, as it does not require a separate computer run. The combination of actions into projects can be performed already in connection with the first component level runs provided that a reasonable order in the analyses of components is used.

This method of combination is based on the definition of both the minimum and the maximum probability for exceeding the limit state. The timing of an MR&R action is always triggered latest according to the maximum allowable probability. However, to optimise the timings of actions at object level, an MR&R action can be triggered earlier if there is a previously defined action time (for any action in any component of the same object) and if the minimum allowable probability is exceeded (Figure 5.22). The minimum allowable probability is defined in the decision tree for this type project planning.

The specification and timing of actions is performed for each component consecutively in the order of their relative importance. The timings of actions for the first component are defined at their optimal timings corresponding to the maximum probability. However, for the following components the timings of actions may be advanced from their optimal timings provided that any MR&R action (for any of the previously analysed components) was scheduled earlier than the optimal timing and the specified minimum probability is exceeded. The system still guarantees that the higher limit for exceeding the limit state is never overridden.

For the purpose of project planning a new row is added in the MR&R action definitions, and is shown a in Table 5.10.

5.3.4.7.2 PROGRAM FOR COST SCENARIOS AT NETWORK LEVEL

In an analysis program for cost scenarios at the network level, the calculation procedures are essentially the same as those in the object level program. However, the project design as presented above is not performed. The distribution of objects into

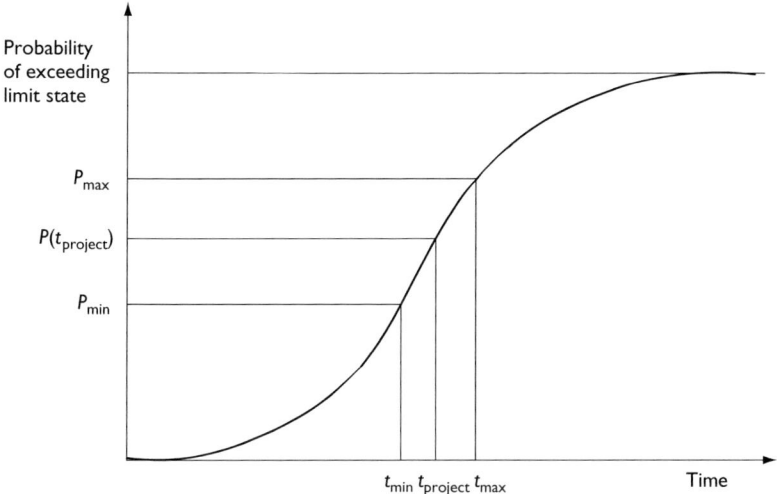

Figure 5.22 Principle of triggering actions.

Table 5.10 Revised table for definition of actions

1	Is the MR&R action group used during the design period?	Yes/no
2	Which MR&R system?	Code of the MR&R system within the MR&R action group
3	Limit condition state?	Limit state for the action, e.g. 3 or 4
4	Minimum allowable probability for exceeding the limit state for accepting the timing of action?	Probability as % (exceeding the given percentage allows timing of the action to equal a previously defined timing of any action for the same object)
5	Maximum allowable probability for exceeding the limit state?	Probability as % (exceeding the given percentage will trigger the action unless not triggered by the previous condition)
6	Maximum number of repeated actions?	Number of allowable repetitions of an action before a heavier action.

components is preferably the same as that in the object level but the surface of components comprises the total surface area of all components in the treated network or subnetwork. The total network is divided into subnetworks according to the decision tree definitions, so that all components of the same type with the same definition of actions can be treated in the same analysis.

Another difference in the network level procedure as compared to the object level procedure is in the mathematical way how the triggering of actions is responded. In an object level analysis the response is that the action is performed and the condition state distribution is completely changed according to the action effect matrix. However, in the network level analysis only the fraction which overrides the maximum allowable probability is considered to be repaired, thus resulting in smaller but more frequent changes in the condition state distribution. The reason for this is that the network level changes in the condition distribution are statistical not individual as at the object level.

5.4 Object level management

5.4.1 General

The main purpose of the object level management system is to assist the organisation in working up MR&R project plans and the annual project and resources plans. Although the object level LMS is aimed mainly at practical assistance of maintenance engineers and repair consultants for laying out MR&R project plans for specific objects, the annual project and resources plans produced by the object level system can be utilised by the whole organisation.

The object level management system is designed to guard, specify and check timing of MR&R actions even at the smallest levels of structural hierarchy. This is a great benefit because it is at these lower hierarchical levels where mistakes and lapses of memory are usual. Without an automatic system that reminds of coatings and repairs to be performed at their optimal time, it would be impossible for a small staff to take care of all MR&R works pertaining to a large building stock. In other words, an object level management system gives an opportunity to sustain a systematic maintenance policy in the upkeep of structures.

Another advantage of the object level LMS is to help the organisation in the readjustment to the new contracting practices such as Design-Build (DB) and Design-Build-Operate-Maintain (DBOM) contracts. In practice, these forms of contracts will reduce the need for expertise among the organisation itself on design, construction and maintenance of structures. On the other hand, if the responsibility for the maintenance of structures is given to private companies by short maintenance contracts in relation to the whole life cycle of structures, one can ask: Who really takes care of the preservation of assets in the long term, and who looks after that the protections and repair actions will be performed at their optimal times? This is probably a worldwide problem in industrial countries, which try to restrict the governmental sector costs. However, if it is possible to prepare in advance long-term MR&R plans for every object using the routines of an LMS then a sustainable upkeep of structures would be possible by a relatively small staff. That would make it possible to brief the maintenance contracts based on carefully prepared plans so that all the MR&R actions will be executed at their optimal times.

5.4.2 Flowchart

The object level LMS is described by the operational flow diagram presented in Figure 5.23. The levels of planning are presented as separated columns in the flow diagram. Their ultimate purposes are explained in Table 5.11.

The boxes in Figure 5.23 refer to the process modules. The boxes plotted with a dotted line may or may not be included to the actual management process as they may also be considered to be separate sectors of MR&R activity. However, from the system point of view the information produced by these modules is vital. The main relationships between modules are marked with arrows. The relationships from the 'updating of models' module are presented with a dotted line for clarity.

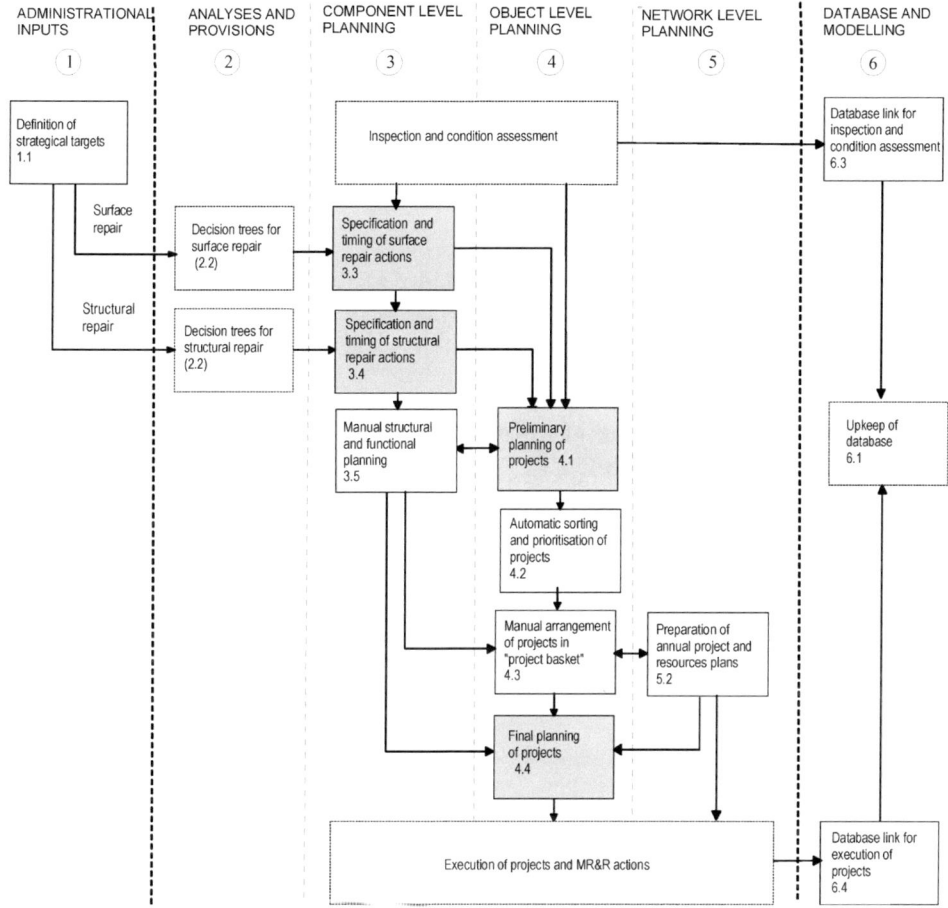

Figure 5.23 Flow diagram of the object level LMS.

Table 5.11 Title and purpose of the planning levels

No.	Title	Purpose
1	Administrational inputs	Define the technical and financial targets for MR&R activity
2	Analyses and provisions	Conducting analyses and drawing up guidelines for actual design
3	Component level planning	Finding optimal MR&R action profiles based on component level LCC analyses
4	Object level planning	Preliminary and final planning of projects
5	Network level planning	Preparation of annual project and resources plans
6	Database and modelling	Upkeep of database and producing the models needed in LCC analyses

The shaded boxes in Figure 5.23 refer to modules in which the Markov Chain-based life cycle cost analysis is used.

For further discussions, the object level management process is divided into the following phases:

- optimisation of the maintenance strategy
- preliminary planning of MR&R projects
- preparation of annual project plans
- final planning of MR&R projects.

The first phase 'optimisation of the maintenance strategy' is shared with the network level system. The practical result of this phase is the construction of decision trees which are used in both network level and object level systems.

5.4.3 Preliminary planning of MR&R projects

The flow diagram for the preliminary planning of projects is presented in Figure 5.24, an extract from Figure 5.23. The modules and their purposes are presented in Table 5.12.

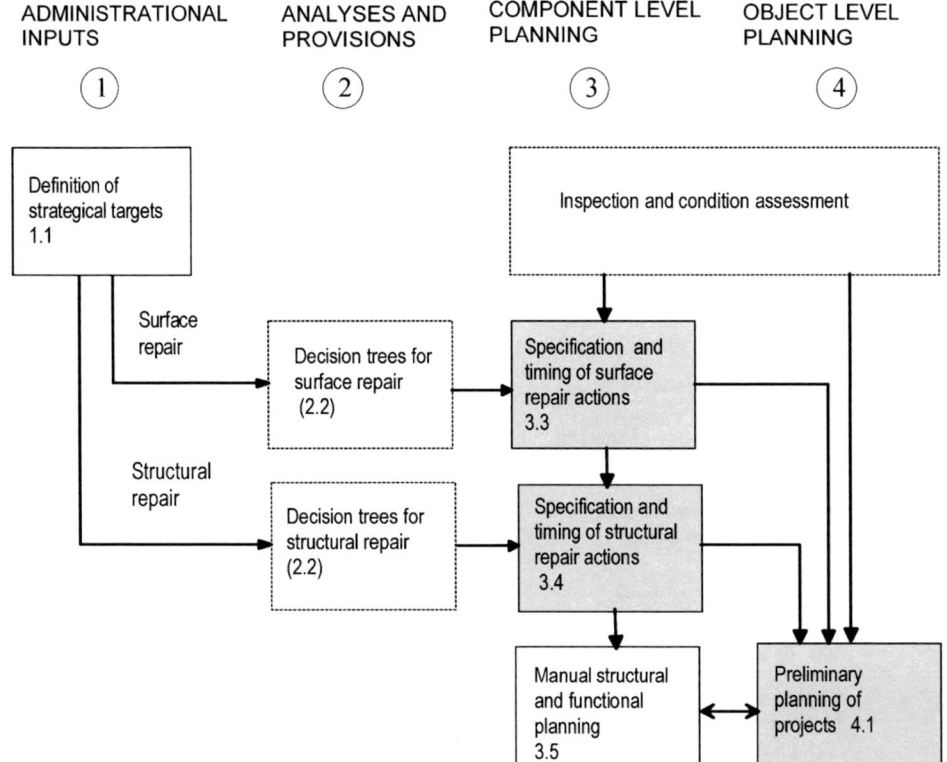

Figure 5.24 Flow diagram for the preliminary planning of projects.

Table 5.12 Modules of preliminary project planning

No.	Module	Purpose
2.2	Provision of decision trees for both 'surface repair' and 'structural repair'	The decision trees define the maintenance strategy (MR&R profiles) for components of different material and structural compositions and in different environmental conditions.
3.3	Specification and timing of 'surface repair' actions	Automatic process which returns the optimal MR&R action profile with optimal timings for surface damage. Surface damage includes damage in the materials of the structure without noticeable consequences in the load bearing capacity.
3.4	Specification and timing of 'structural repair' actions	Automatic process, which returns the optimal MR&R action profile with optimal timings for structural damage. Structural damage consists of damage with remarkable consequences in the load bearing capacity of the structure.
3.5	Manual structural and functional planning	Consists of auxiliary analyses that are conducted for the object to study the structural integrity and the in-service functionality leading to possible repair plans to be considered in the final planning of projects.
4.1	Preliminary planning of projects	Automatic process that combines the component level proposals for MR&R action profiles, rearranges the timing of actions to fit with the project timetable and calculates the project costs.

As can be seen in the flow diagram, there are two branches for specification and timing of the component level repair actions: surface repair and structural repair actions. The specification and timing of actions is in principle the same in both cases as they are based on similar LCC analysis and application of a decision tree. They may also be performed at the same time as a combined analysis and using a combined decision tree.

The key instrument at both the component level and the object level planning is the Markov Chain-based LCC analysis. The shadowed boxes in the flow diagram show the modules in which this LCC analysis method is used.

5.4.3.1 Automatic specification and timing of surface repair actions

In the automatic specification and timing of surface repair actions the definition of actions is performed automatically by the decision tree. The calculation process is fully automatic. The computer program follows the following task list:

- Choose every component of the object in turn.
- Take the component-specific data and insert them into the degradation, repair action and cost models.
- Take the decision tree for definition of MR&R actions and insert the action definitions into the life cycle analysis table.
- Determine the timings and costs for MR&R actions.
- Store the timings and cost data in an output file.

In Table 5.13, results of an automatic analysis are presented as an example. The calculations have been conducted by a spreadsheet program. The MR&R actions for all the concrete components for a bridge have been defined and scheduled. The actions are referred to by code numbers.

Table 5.13 Costs of MR&R actions

DEFINITION OF ACTIONS

Code of component	Code of action	Year	Action group	System	Unit costs (euro m^{-2})	Surface area m^2	Total costs (euro)
16	1	6	1	16	161	646	103866
16	2	6	3	5	295	646	190332
16	3	6	5	13	396	186	73739
16	4	33	1	16	118	646	76450
16	5	33	3	5	217	646	140093
16	6	33	5	13	396	32	12792
16	7	60	1	16	114	646	73751
16	8	60	3	5	209	646	135145
16	9	60	5	13	396	16	6424
16	10	87	1	16	114	646	73815
16	11	87	3	5	209	646	135263
16	12	87	7	18	834	646	539057
16	13	114	1	16	110	646	71222
16	14	114	3	5	202	646	130512
16	15	114	5	13	396	1	390
16	16	141	1	16	111	646	71745
16	17	141	3	5	204	646	131471
16	18	141	5	13	396	4	1644
16	19	168	1	16	112	646	72249
16	20	168	3	5	205	646	132394
16	21	168	5	13	396	7	2849
16	22	195	1	16	113	646	72683
16	23	195	3	5	206	646	133189
16	24	195	7	18	834	646	539057
14	1	1	1	9	20	68	1334
14	2	7	1	9	26	68	1765
14	3	7	6	2	460	68	31313
14	4	13	1	9	17	68	1179
14	5	19	1	9	17	68	1184
14	6	25	1	9	17	68	1189
14	7	31	1	9	18	68	1196
14	8	37	1	9	18	68	1204
14	9	43	1	9	18	68	1215
14	10	49	1	9	18	68	1227
14	11	55	1	9	18	68	1241
14	12	61	1	9	18	68	1256
14	13	67	1	9	19	68	1272
14	14	73	1	9	19	68	1290
14	15	79	1	9	19	68	1309
14	16	85	1	9	20	68	1329
14	17	91	1	9	20	68	1350
9	1	5	1	25	37	11	418
9	2	5	6	4	278	11	3109

9	3	22	I	25	26	II	287
9	4	39	I	25	26	II	290
9	5	56	I	25	26	II	296
9	6	73	I	25	27	II	304
9	7	90	I	25	28	II	314
9	8	107	I	25	29	II	325
9	9	124	I	25	30	II	337
9	10	136	I	25	31	II	344
9	II	136	3	7	135	II	1515
9	12	153	I	25	32	II	356
9	13	170	I	25	33	II	370
9	14	187	I	25	34	II	384
10	I	8	6	3	419	20	8387
II	I	146	6	3	408	20	8154
3	I	I	I	9	20	56	1098
3	2	8	I	9	22	56	1235
3	3	15	I	9	24	56	1317
3	4	22	I	9	25	56	1389
3	5	25	I	9	25	56	1412
3	6	25	6	2	461	56	25839
3	7	32	I	9	17	56	977
3	8	39	I	9	18	56	998
3	9	46	I	9	19	56	1037
3	10	53	I	9	19	56	1091
3	II	60	I	9	21	56	1157
3	12	67	I	9	22	56	1231
3	13	74	I	9	23	56	1310
3	14	81	I	9	25	56	1393

5.4.3.2 Automatic specification and timing of structural repair actions

The specification and timing of structural repair actions are performed in principle in the same way as that for surface repair actions i.e. using an LCC analysis table and decision tree. However, the degradation models, the repair action assortment with action effect models, cost models and the decision trees must be specific for structural damage and structural repair.

The structural models are developed based on the structural design equations in which the degradation models are inserted, as presented in a separate chapter.

As there are common repair actions for both the surface damage and the structural damage, it is recommendable to combine the surface damage analysis and the structural damage analysis so that the action effects of the MR&R could be taken into account to both damage types. That makes the analysis system more complex but on the other hand prevents mistakes in decision-making that two completely separate analyses would inevitably entail.

5.4.3.3 Combination of MR&R actions into projects (project planning)

The purpose of this module is to combine the component level proposals for MR&R action profiles and to rearrange the timings of actions to fit with the project timetable and to calculate the project costs. By combining the MR&R actions into groups of actions the number of projects can be efficiently reduced and financial benefit can be reaped as a result of synergy profit.

Combining separate MR&R actions into projects can be done by special software described in Section 5.3.4.7. The results of such planning process are presented in Table 5.14 and as an example. The order of the actions in Table 5.14 has been resorted so that the actions scheduled for the same year are placed one after another. So the composition of projects is plain to discern. Only actions within the next 55 years are presented.

In this kind of a computer run, the order in which the life cycle planning of components is performed is of much importance. It is advisable that the components, for which the actions are most expensive, such as the bridge deck, will be run first; the expensive actions like renewing the water membrane and repair of the bridge deck will be scheduled at their optimal time. Then the other components also are analysed in the order of their economic importance. The components of least importance will be last in the order. Thus there is a high probability that the actions of the last components will be rescheduled from their optimal timing.

The costs of the MR&R projects up to year 13 are calculated in Table 5.15. The bar graph in Figure 5.25 shows the costs up to 50 years.

Table 5.16 and Figure 5.26 show the corresponding user costs as a result of MR&R actions.

An advanced LCC analysis program enables to evaluate also the delay costs of projects. To determine the delay costs, the timing of projects is delayed by one or several years. The delay costs are obtained by subtracting the delayed project costs from the undelayed project costs. An example of the results of a delay calculation is presented in Table 5.17 and Figure 5.27. The example project is timed originally for the first year. The delay ranges from 1 to 5 years from year 1.

Table 5.14 Table of MR&R actions after combination into projects

DEFINITION OF ACTIONS

Code of component	Code of action	Year	Action group	System	Unit costs (euro m^{-2})	Surface area (m^2)	Total costs (euro)
14	1	1	1	9	20	68	1334
9	1	1	1	25	33	11	368
10	1	1	6	3	366	20	7318
3	1	1	1	9	20	56	1098
18	1	1	6	3	119	156	18582
9	2	1	3	7	145	11	1620
3	2	1	6	2	359	56	20098
9	3	1	6	4	230	11	2575
16	1	6	1	16	161	646	103866
4	1	6	6	2	373	30	11178
16	2	6	3	5	295	646	190332
14	2	6	1	9	25	68	1704
16	3	6	5	13	396	186	73739
14	3	6	6	2	445	68	30237
3	3	6	1	9	17	56	955
14	4	12	1	9	17	68	1176
9	4	12	1	25	25	11	284

3	4	12	1	9	17	56	964
14	5	18	1	9	17	68	1180
3	5	18	1	9	18	56	984
9	5	24	1	25	25	11	285
14	6	24	1	9	17	68	1185
3	6	24	1	9	18	56	1017
14	7	30	1	9	18	68	1192
3	7	30	1	9	19	56	1059
16	4	33	1	16	118	646	76450
16	5	33	3	5	217	646	140093
16	6	33	5	13	396	32	12792
9	6	36	1	25	26	11	286
14	8	36	1	9	18	68	1200
3	8	36	1	9	20	56	1110
14	9	42	1	9	18	68	1211
3	9	42	1	9	21	56	1167
9	7	48	1	25	26	11	287
14	10	48	1	9	18	68	1223
3	10	48	1	9	22	56	1228
14	11	54	1	9	18	68	1237
3	11	54	1	9	23	56	1293
11	1	60	6	3	320	20	6399
16	7	60	1	16	114	646	73751
16	8	60	3	5	209	646	135145
9	8	60	1	25	26	11	289
16	9	60	5	13	396	16	6424
14	12	60	1	9	18	68	1252
3	12	60	1	9	24	56	1360
14	13	66	1	9	19	68	1268
3	13	66	1	9	25	56	1428
9	9	72	1	25	26	11	291
14	14	72	1	9	19	68	1286
3	14	72	1	9	27	56	1496
14	15	78	1	9	19	68	1305
3	15	78	1	9	28	56	1564
9	10	84	1	25	26	11	294
14	16	84	1	9	19	68	1325
3	16	84	1	9	29	56	1630
16	10	87	1	16	114	646	73815
16	11	87	3	5	209	646	135263
16	12	87	7	18	834	646	539057
14	17	90	1	9	20	68	1346
9	11	101	1	25	27	11	301
9	12	114	1	25	27	11	306
16	13	114	1	16	110	646	71222
16	14	114	3	5	202	646	130512
16	15	114	5	13	396	1	390
9	13	131	1	25	28	11	315
22	1	141	6	3	117	90	10518
16	16	141	1	16	111	646	71745
16	17	141	3	5	204	646	131471
16	18	141	5	13	396	4	1644
9	14	148	1	25	29	11	326
9	15	165	1	25	30	11	339

Table 5.15 Combination of MR&R costs for projects

MR&R costs, euro	Discount factor: 1.000 / Year 0	0.962 / 1	0.925 / 2	0.889 / 3	0.855 / 4	0.822 / 5	0.790 / 6	0.760 / 7	0.731 / 8	0.703 / 9	0.676 / 10	0.650 / 11	0.625 / 12	0.601 / 13
Annual costs	0	52992	0	0	0	0	412013	0	0	0	0	0	2424	0
Cumulative costs	0	52992	52992	52992	52992	52992	465005	465005	465005	465005	465005	465005	467429	467429
Annual PV costs	0	50954	0	0	0	0	325620	0	0	0	0	0	1514	0
Cumulative PV costs	0	50954	50954	50954	50954	50954	376574	376574	376574	376574	376574	376574	378088	
Component														
16	0	0	0	0	0	0	367938	0	0	0	0	0	0	0
14	0	1334	0	0	0	0	31942	0	0	0	0	0	1176	0
9	0	4562	0	0	0	0	0	0	0	0	0	0	284	0
10	0	7318	0	0	0	0	0	0	0	0	0	0	0	0
11	0	0	0	0	0	0	0	0	0	0	0	0	0	0
3	0	21197	0	0	0	0	955	0	0	0	0	0	964	0
4	0	0	0	0	0	0	11178	0	0	0	0	0	0	0
17	0	0	0	0	0	0	0	0	0	0	0	0	0	0
18	0	18582	0	0	0	0	0	0	0	0	0	0	0	0
22	0	0	0	0	0	0	0	0	0	0	0	0	0	0

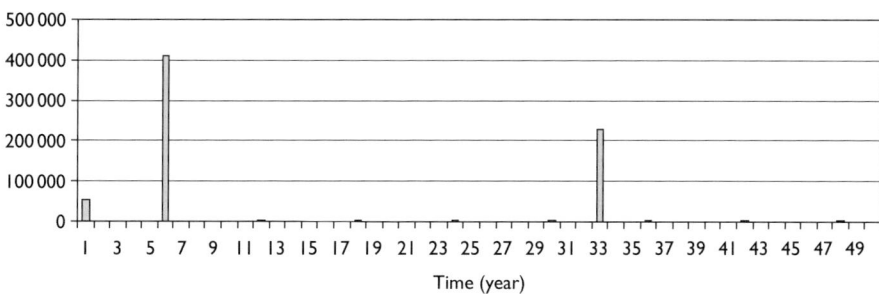

Figure 5.25 The annual project costs presented by a bar graph, Bridge B-43.

5.4.3.4 Consideration of the unpredictable degradation and damage

In the flow chart of the preliminary planning one can find an arrow coming straight from inspection of structures to the module of preliminary planning (see Figure 5.24). This arrow refers to such observations of degradation and damage which call for MR&R actions but which are not considered in the automatic planning of projects. In the automatic planning process only predictable degradation and damage can be considered.

The actions for unpredictable degradation and damage observed during inspections can be divided into three groups based on the way how to respond to these observations:

- actions of routine maintenance
- actions considered by revising the automatically prepared project plans and
- actions that call for separate repair, rehabilitation and renovation plans.

The first option deals with routine maintenance works that are implemented without special planning either by the maintenance staff of the organisation or by a maintenance contractor. These works are executed as quickly as possible using especially reserved financial resources for that purpose. The routine maintenance would include for example the following works:

- straightening of railings
- refurbishment of canals and flumes for drainage and
- service of the mechanical equipments, etc.

The second option consists of degradation and damage that can be repaired by altering the automatically prepared project plans. These observations include:

- leakage of water membranes and joints and
- infrequent degradation that is not considered by degradation models.

Table 5.16 User costs of the planned projects

| User costs, (€) | Discount factor | 1.000 | 0.962 | 0.925 | 0.889 | 0.855 | 0.822 | 0.790 | 0.760 | 0.731 | 0.703 | 0.676 | 0.650 | 0.625 | 0.601 |
	Year	0	1	2	3	4	5	6	7	8	9	10	11	12	13
Annual costs		0	52 524	0	0	0	0	92 142	0	0	0	0	0	459	0
Cumulative costs		0	52 524	52 524	52 524	52 524	52 524	144 665	144 665	144 665	144 665	144 665	144 665	145 124	145 124
Annual PV costs		0	50 504	0	0	0	0	72 821	0	0	0	0	0	287	0
Cumulative PV Costs		0	50 504	50 504	50 504	50 504	50 504	123 325	123 325	123 325	123 325	123 325	123 325	123 611	123 611
Component															
16		0	0	0	0	0	0	80 029	0	0	0	0	0	0	0
14		0	280	0	0	0	0	11 392	0	0	0	0	0	280	0
9		0	1 356	0	0	0	0	0	0	0	0	0	0	149	0
10		0	6 143	0	0	0	0	0	0	0	0	0	0	0	0
11		0	0	0	0	0	0	0	0	0	0	0	0	0	0
3		0	1 294	0	0	0	0	30	0	0	0	0	0	30	0
4		0	0	0	0	0	0	691	0	0	0	0	0	0	0
17		0	0	0	0	0	0	0	0	0	0	0	0	0	0
18		0	43 450	0	0	0	0	0	0	0	0	0	0	0	0
22		0	0	0	0	0	0	0	0	0	0	0	0	0	0

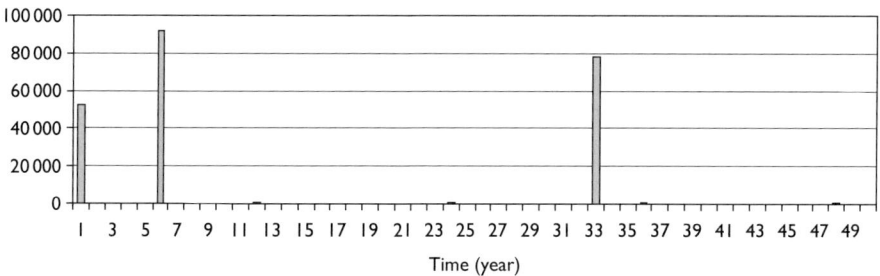

Figure 5.26 User costs of projects during the first 50 years, Bridge B-43.

Table 5.17 Results of the delay cost calculation, Bridge B-43

Delay costs, euro	Year	0	1	2	3	4	5	6
Delay from the timing		0	0	1	2	3	4	5
Timing of the project	1							
Delay costs		0	0	2 324	3 824	4 934	5 859	6 694
MR&R costs		0	52 992	55 316	56 817	57 927	58 851	59 686
Component								
14			1 334	1 573	1 721	1 813	1 872	1 911
9			4 562	4 818	5 032	5 213	5 367	5 498
10			7 318	7 511	7 689	7 852	8 002	8 140
3			21 197	22 832	23 792	24 465	25 026	25 550
18			18 582	18 582	18 582	18 583	18 585	18 587

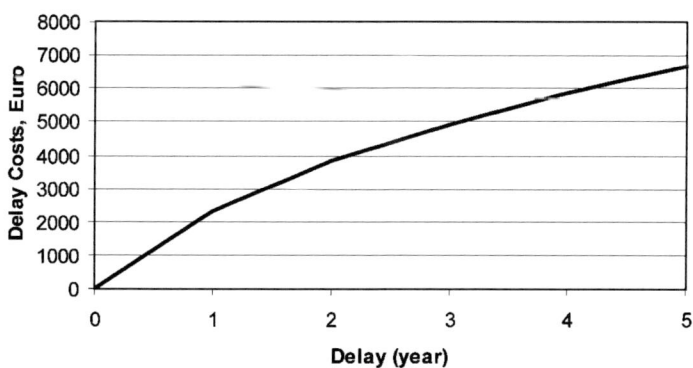

Figure 5.27 Delay costs for the project, Bridge B-43.

The leakage damage observed in the field sets off an alarm. The consequences of water leakage are normally so costly and rapidly progressive that it is not possible to wait until the next automatically scheduled repair. So the repair must be advanced. The timing of the repair can be altered using the methods of manual life cycle planning.

Only by these manual methods, the MR&R action specifications of the component at hand are modified. The rest of the plan remains unchanged.

The infrequent and difficultly predictable degradation modes which are not considered by the degradation models are treated in the same way as the leakage damage. For instance the symptoms of alkali-aggregate-reaction (AAR) may be included in this group as the repair actions for AAR may be similar to those used for other damage types. Then it is possible just to change the automatically prepared project plans so that the repairs and protections are designed especially for the observed damage.

After modifying the plans they are returned and stored among the other project plans. So the modifications will be automatically considered in the further treatment of the projects.

The third group of unpredictable damage consists of serious structural damage types that call for special attention and probably special repair, rehabilitation or renovation projects. The symptoms of such damage types include:

- faulting of supports;
- erosion of subsoil;
- vast deflection;
- collapse, etc.

These serious damage types are studied using the manual methods of structural and functional planning as described in Section 5.4.3.6.

5.4.3.5 Manual structural and functional planning

The purpose of manual structural and functional planning is to study in detail the structural problems observed either during the field inspection or in the automatic preliminary planning. Also the purpose is to study the possible deficiencies in the traffic safety or other performance. All these studies may give rise to repair, rehabilitation or renovation projects which are then planned preliminarily.

5.4.3.5.1 DEFICIENCY IN LOAD BEARING CAPACITY

Observations such as faulting of supports, erosion of subsoil, vast deflection, etc. are serious symptoms that must always be studied by special investigations. It is always important to find out the reason for these symptoms in such a way that the repair is so planned that the symptoms will not recur.

The automatic runs of structural LCC analysis may reveal deficiencies in the load bearing capacity of structures. If such deficiencies occur, it is advisable to carry out a special inspection for these structures to verify that the problem is real and to observe the real extent of damage. In many cases the degradation of the structure may in reality proceed slower than the degradation models would suggest and so no capacity problem exists. On the other hand, if the degradation in structures has really proceeded according to the degradation models or even faster, the load bearing capacity of such structures should be further studied by 'second level' analysis methods.

For second level structural analysis methods, the following data are of importance:

- structural drawings and records on dimensions;
- past and present standards of loading; and
- functional, aesthetic and cultural requirements related to the object.

In the second level study of the structure it is recommended to analyse the structural capacity by the Finite Element Method (FEM) taking into account the possible effects of the observed damage in the materials, possible changes in the regulations of loads, etc.

The real load bearing capacity can also be verified by in-situ loading tests and measurements. These tests, although not extended to the ultimate limit state, usually give enough information of the real structural performance of a structure.

5.4.3.5.2 DEFICIENCY IN TRAFFIC SAFETY OR SERVICEABILITY

The reason for accidents may sometimes be attributed to poor functional condition of structures. So, in the case of bridge accidents it is advisable to check the traffic safety of the bridge by special risk analyses.

Risk of accident or structural failure cannot be treated only as a financial problem if human lives are at risk. However, sometimes the financial consequences resulting from accidents or a structural collapse, which may be by far greater than just the repair or rebuilding costs, can be considered as risk costs.

The risk costs of accidents can be evaluated as the product of the probability of accident and the average accident costs. A simple risk analysis method developed in the PONTIS bridge management system owned by the US Federal Highway Administration serves as an example, and is shown in Equation (5.14) [15].

$$RIC = 365 \cdot V \cdot R \cdot C_a \qquad (5.14)$$

where
  RIC   is risk costs per year
  V predicted Annual Daily Traffic (ADT), cars day^{-1}
  R probability of accident, i.e. risk
  C_a average accident cost, $\text{euro accident}^{-1}$.

The probability of an accident is evaluated from Equation (5.15), which is based on a statistical analysis:

$$R = 200 \cdot 0.3048 \cdot W_r^{-6.5} \cdot \left[1 + 0.5 \cdot \frac{(9 - A_b)}{7} \right] \qquad (5.15)$$

where
  W_r is curb-to-curb width of the roadway, m
  A_b approach alignment rating (typically 2–9).

Using the above formula the probability of failure and the risk costs can be evaluated. For designers there may be rules or instructions on how great an accident risk can be accepted. If this limit is exceeded, the width of the bridge and/or the angle of driving direction must be changed. Thus this kind of simple risk analyses, which uncover deficiencies in the traffic safety, may lead to rehabilitation or renovation projects.

5.4.3.5.3 PREPARATION OF PRELIMINARY PLANS FOR STRUCTURAL REPAIR, REHABILITATION AND RENOVATION

If the risk is not characteristic to a specific component it is treated as an object level problem and a special repair plan is drawn up manually for the object. In the flow chart of the object level LMS, these projects are transferred straight to the 'project basket' in which they are manually prioritised.

5.4.4 Preparation of annual project and resources plans

The importance of annual project resources plans is obvious. These plans form the framework of the future MR&R activity of the organisation. They specify the projects that will be implemented in each year.

The starting point for drawing up an annual project and resources plan is the preliminary project plans, which are automatically prepared for every object. Then the process goes on as presented in Figure 5.28.

The stages, i.e. modules, for preparing annual project and resources plans are presented in Table 5.18

A very important phase in the preparation of annual project and resources plans is the budget check as the annual MR&R costs must be in balance with the budget. The budget check is included in the tasks of the final stage. Before that, the projects must be arranged in the order of urgency.

5.4.4.1 Producing automatic project plans for a group of objects

In order to be able to prioritise the projects they must first be planned. The preliminary project planning is performed in the way presented above. In principle the annual

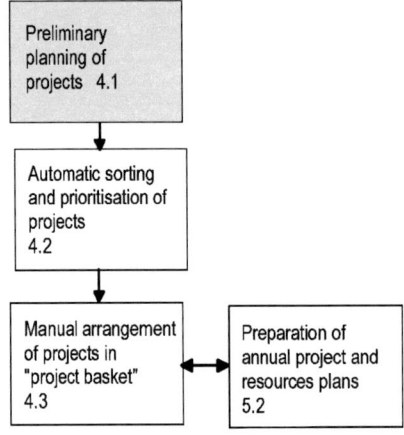

Figure 5.28 Stages for drawing up annual project and resources plans.

Table 5.18 Modules of annual project and resources planning

No.	Module	Purpose
4.1	Producing automatic project plans for all objects	The preliminary project planning is addressed to all objects.
4.2	Automatic sorting and prioritisation of projects	Sorting of projects according to the budget categories and prioritisation of projects in each budget category.
4.3	Manual arrangement of projects in 'project basket'	Allows manual sorting and other arrangement of the project lists.
4.4	Final drawing up of the annual project and resources plans	Finalising the annual project lists with evaluated costs. Includes budget check.

project and resources planning can be addressed to any year in the range of the design period but normally it is addressed to the next coming year or a few of the coming years (1–5 years).

For example, Figures 5.29 and 5.30 show results of project planning for 18 briges. The project costs for year 1 are presented by a bar graph in Figure 5.29. The corresponding delay costs for 1 year delay are presented in Figure 5.30. Note that the costs and delay costs for the bridge B-224 were 0 as there was no project planned for that object in the year 1.

5.4.4.2 Automatic sorting and prioritisation of MR&R projects

5.4.4.2.1 SORTING

In large organisations there may be several budget categories for MR&R projects. The budget categories may be defined as follows:

- routine maintenance;
- normal repair projects;
- rehabilitation projects;
- new investment projects.

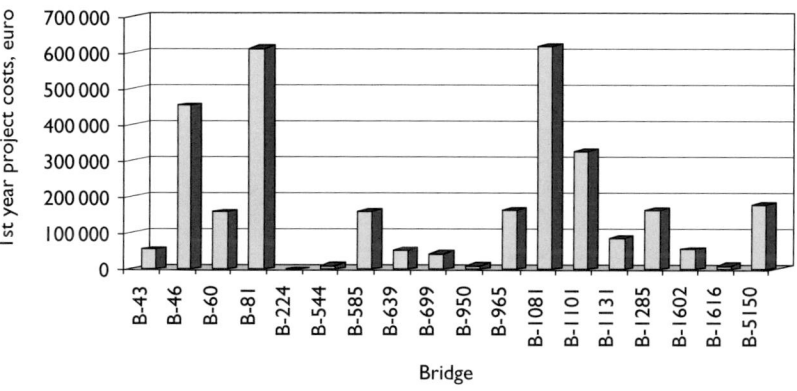

Figure 5.29 Example of project costs for 18 bridges for the 1st year.

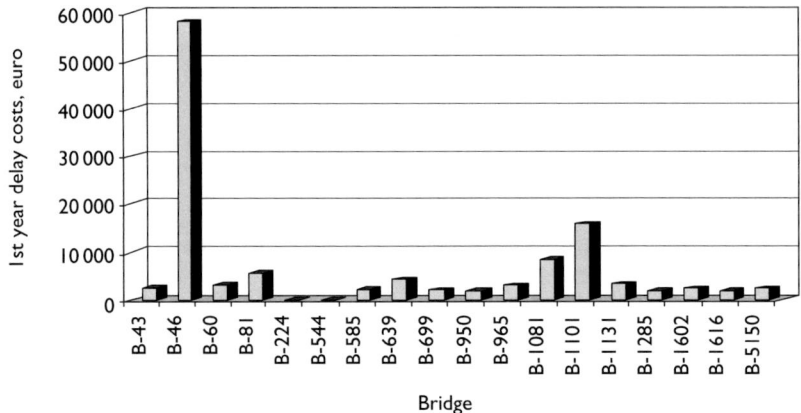

Figure 5.30 Example of one-year delay costs for project costs of 1st year.

The routine maintenance includes smaller maintenance actions which do not require special planning. Normal repair projects contain normal repair actions of components and preventive maintenance actions, such as coatings. The rehabilitation projects incorporate the replacement and modernisation of components as well as some separately defined repair actions. New investment projects are for implementing new structures or buildings.

The purpose of the sorting is to distribute the projects into their correct budget categories. Sorting is performed using automatic database routines.

The determinative factor for the sorting is the heaviest action within an MR&R action group. So, for example, if the heaviest action belongs to the category of rehabilitation, the whole project is placed into that category.

If a project is sorted into the group of new investment, all the repair actions pertaining to that object are ignored, as they automatically will be 'repaired' during the rebuilding. The same applies to rehabilitation projects in which the rehabilitated parts of the object are automatically repaired.

5.4.4.2.2 PRIORITISATION

Sorting is followed by prioritisation procedures in each budget category. The prioritisation is necessary for setting the projects in the order of urgency if not all projects fit into the budget frame. The budget in each budget category sets the limit up to which the projects can be accepted in a prioritised list of projects.

The MR&R action groups are set in the order of urgency by normal database ranking or MADA (Multiple Attribute Decision Aid) [16, 1]. The MADA ranking allows several attributes to be considered at the same time. At least two attributes should obviously be taken into account: one measuring the amount of damage and/or the urgency of repair, and the other evaluating the delay costs of a repair. So the attributes of the MADA ranking could be the following:

Repair need index (RNI) defined based on inspection and condition assessment data
Delay costs of MR&R actions defined based on preliminary project planning.

The following method for the determination of RNI is given as an example [16]. The RNI is determined from the damage data and the urgency class of repair, as given in Equation (5.16):

$$\text{RNI} = Max_i(\text{DoD}_i \times \text{UCL}_i) + \gamma \left[\sum_{j, j \neq j_{\max}} (\text{DoD}_j \times \text{UCL}_j) \right] \quad (5.16)$$

where
  RNI   is Repair need index
  DoD   Degree of degradation of the component
  UCL   Repair urgency class
  γ Reduction factor.

The score of RNI is calculated for every registered damage. The RNI is determined by first counting the most serious damage with greatest product of DoD × UCL. The other scores are counted as multiplied by a reduction factor γ. The value of γ is approximately 0.2.

The delay costs can be determined during the preliminary project planning. As the delay costs correlate well with total project costs and the total project costs correlate with repair needs, reasonable prioritisation can be obtained by using the delay costs as the only ranking parameter. An example of such prioritisation is presented in Table 5.19. As the ranking with delay costs would obviously favour great objects over small objects the ranking parameter was modified a little: the delay costs were divided by the length of the bridge. Thus more objective results are probably obtained.

Table 5.19 Results of prioritisation of 18 bridges

Original order of calculation			Prioritisation		
Bridge	Delay costs	Delay costs per length	Prioritised order	Bridge	Delay costs per length
U-43	2324	39	1	U-46	1614
U-46	58088	1614	2	U-81	473
U-60	3024	126	3	U-1081	357
U-81	5675	473	4	U-1101	327
U-224	0	0	5	U-965	163
U-544	71	1	6	U-699	129
U-585	1950	108	7	U-60	126
U-639	4029	84	8	U-585	108
U-699	1854	129	9	U-1285	93
U-950	1650	69	10	U-1131	92
U-965	2935	163	11	U-639	84
U-1081	8557	357	12	U-950	69
U-1101	15719	327	13	U-1616	55
U-1131	3310	92	14	U-1602	47
U-1285	1681	93	15	U-43	39
U-1602	2258	47	16	U-5150	28
U-1616	1642	55	17	U-544	1
U-5150	2251	28	18	U-224	0

5.4.4.3   Manual arrangement of projects in project basket

Although automatic methods can be used in the timing of projects, the planning process can never be completely automatic. There must be a phase where a maintenance engineer is allowed to intervene in the automatic processes and to make manual changes in the work plans.

The following reasons may cause manual changes in the project and resources plans:

- Manual project plans as a result of structural and functional analyses are not considered in the automatic design procedures.
- Administration level decisions may cause rapid changes in the order of urgency for projects.
- The automatic sorting and prioritisation does not work satisfactorily and calls for corrections.

The necessary changes in the project lists can be implemented using a computer program, which shows all the projects as a list on a monitor screen. This is called the *project basket*. Individual projects can be added or removed or the order of urgency can be changed manually on the screen. The final aim is to render the project lists in the desired form and composition.

5.4.4.4   Final phase in the preparation of annual project and resources plans

When the projects have been arranged in their respective budget categories and within each category in the order of urgency, the final phase in the preparation of annual project and resources plans is to do the budget check. The budget sets the limit to the list of prioritised projects when can be included in the annual project programme of the treated year. The rest of the projects must be postponed to the next year and prioritised with other projects of that year.

The total financial resources for an MR&R activity depend on the budget of the organisation. The resources allocated for MR&R projects are affected on one hand by the financial needs for MR&R activity and on the other hand by the possibilities of funding the MR&R activity. Usually the budget is given as a total allowance for all construction activity, i.e. maintenance, repair, rehabilitation and new construction. Hence, the problem remains how to share the total budget into budget categories.

There is no general rule on how to do the budget share. However, Section 5.4.5.4.1 describes a method which may help solving the problem.

5.4.4.4.1   THE QUALITY FUNCTION DEPLOYMENT METHOD FOR BUDGET SHARE

The Quality Function Deployment (QFD) method takes into account the strategic targets set by the administration of the organisation. For example, the targets and their relative importance are defined as in Table 5.20 [1].

Table 5.20 Strategic targets and their relative weights (sum of the weights is 1)

Target	Weight
Life cycle economy (total agency costs)	0.3
User economy	0.2
Ecology	0.1
Aesthetics	0.1
Mitigation of human risks	0.2
Preservation of cultural values	0.1

The budget share is reasonable to do with the QFD method if there is correlation between the budget categories and the strategic targets. One may consider, for example, that the budget categories are the following:

- normal repair projects
- rehabilitation projects
- new investment projects
- the total budget is, for example, €15 000 000.

The correlation between the targets and the budget categories are evaluated. For instance, the safety aspect is best contributed by heavy repairs such as new investments and rehabilitation. The economic and ecological aspects probably correlate best with the categories of normal repair. The correlation factors are numbers between 0 and 1 expressing the relative importance of the budget category to promote a particular target, which is relative to the other budget categories. The sum of correlation factors must be 1.

The priority index of a budget category, which shows the part of the total budget that is addressed to the budget category, is determined from Equation (5.17):

$$I_i = \sum T_j \cdot K_{ij} \tag{5.17}$$

where
I_i is priority index of ith budget category
T_j weight factor of jth target
K_{ij} correlation between ith budget category and the jth target
i index for budget category and
j index for target.

The budget in each budget category is then determined by Equation (5.18):

$$B_i = I_i \cdot B_{tot} \tag{5.18}$$

where
B_i is the budget of ith budget category and
B_{tot} total budget.

Table 5.21 QFD for budget share

Attributes	Normal repair projects	Rehabilitation projects	New investment projects	Weight
LC economy (total agency costs)	0.3	0.4	0.3	0.3
User economy	0	0.5	0.5	0.2
Ecology	0.6	0.3	0.1	0.1
Aesthetics	0	0.3	0.7	0.1
Mitigation of human risks	0	0.4	0.6	0.2
Preservation of cultural values	0.7	0.3	0	0.1
				Total
Priority	0.22	0.39	0.39	1.00
Budget share, euro	3 300 000	5 850 000	5 850 000 0	15 000 000 0

Thus, the budget share into budget categories is obtained. The budget share in this example is presented in Table 5.21.

5.4.4.4.2 BUDGET CHECK

The aim of this phase is to finalise the annual project programme by checking the budget in each budget category. When the projects are first arranged in the order of urgency, the project costs are cumulatively counted together and the list is cut at the place where the total of project costs exceeds the budget. The projects that are included in the budget frame are chosen for the annual project and resources plan and those staying outside are postponed to the next year. Figure 5.31 shows this principle.

5.4.5 Final planning of MR&R projects

The final planning of projects is performed by maintenance engineers or repair consultants for objects selected for the annual project plan. Although the final planning of projects can be considered to be outside the scope of this book, it is still briefly discussed, as it forms an important part of the management system.

The system assists the final planning of projects by the following add-ins:

- manual life cycle planning
- service life design.

In the final planning a maintenance engineer or a consultant finally decides on the actions on the composition of the project and evaluates the project costs.

Manual life cycle planning   The life cycle planning application is in principle the same as that used in the preliminary planning of projects. However, while the preliminary planning of projects is usually performed automatically by applying the decision tree method, the manual life cycle planning, as the name suggests, is performed manually.

Budget	€ 3 300 000			€ 5 850 000			€ 5 850 000	
Normal repair projects	Total of costs (euro)	Check	Rehabilitation projecs	Total of costs (euro)	Check	New investment projects	Total of costs (euro)	Check
U-1081	86000	OK	L-1534	107 900	OK	U-1285	553 800	OK
O-5004	319900	OK	U-224	367 500	OK	T-295	1 633 300	OK
H-5274	367800	OK	H-1400	764 400	OK	SK-450	4 326 500	OK
H-5275	384800	OK	O-1229	1 422 200	OK	H-1101	7 453 800	
T-673	670400	OK	U-585	1 872 300	OK	T-1711	9 940 400	
T-20	924900	OK	O-1102	2 418 000	OK	H-228	12 503 600	
SK-355	1 198 800	OK	O-293	2 451 800	OK	T-317	14 642 900	
O-201	1 199 200	OK	SK-5044	2 521 400	OK	U-81	18 304 700	
H-1178	1 278 500	OK	SK-324	2 839 500	OK	L-276	21 830 900	
U-639	1 303 300	OK	SK-791	3 040 400	OK	T-1954	25 722 000	
H-143	1 382 300	OK	H-1530	3 527 300	OK	T-1955	25 853 600	
SK-325	1 529 400	OK	O-1286	3 818 500	OK	U-5150	27 752 400	
H-748	1 715 500	OK	SK-698	3 927 200	OK			
L-1567	1 900 900	OK	SK-567	4 366 100	OK			
L-440	2 015 600	OK	SK-602	4 806 600	OK			
U-1101	2 086 300	OK	O-1174	4 873 200	OK			
SK-541	2 306 700	OK	O-1174	5 608 900	OK			
SK-169	2 513 300	OK	U-1131	6 145 300				
L-1900	2 764 700	OK	H-26	6 761 200				
L-26	3 033 800	OK	U-46	7 211 000				
T-1885	3 171 500	OK	O-1099	7 875 200				
H-1440	3 301 700		T-1565	8 465 700				
T-1959	3 412 100		SK-482	8 661 700				
T-41	3 540 900		SK-5042	8 928 600				
SK-663	3 768 000							
U-544	3 846 700							
U-1616	3 895 300							
SK-437	4 119 700							
H-1467	4 385 400							
O-928	4 469 300							
H-714	4 472 400							
L-1493	4 583 400							
U-699	4 785 800							
H-3767	4 995 000							
H-1002	5 016 900							
SK-842	5 260 200							
SK-5043	5 368 600							

Figure 5.31 Principle of the budget check.

The MR&R action profiles to be applied for specific components are more carefully studied in the final planning. Although a designer can start the design from the very beginning, it is usually advantageous to do the automatic decision tree planning first and then modify the plans manually if needed. For instance the designer may take away, add or change MR&R actions or give new timings for them. While doing so the designer can immediately observe the effects on the condition of structures and the respective changes in the life cycle costs [1].

Service life design   Service life design is recommended to be used for the design of new structures and also in the repair of existing structures wherever applicable [9, 1].

5.5   Network level management

The network level LMS is designed for the operative leadership or central administration of an organisation for the prediction of the future MR&R needs and costs and in optimisation of the maintenance strategy both technically and financially.

The network level system is mainly meant for larger organisations in which strategic maintenance planning is important. It is an independent system which can be used with or without the object level system. However, to get full benefit of the network level system it should be integrated with the object level system.

5.5.1   Flowchart

The flow chart of the network level LMS is presented in Figure 5.32. The system consists of the modules presented in Table 5.22.

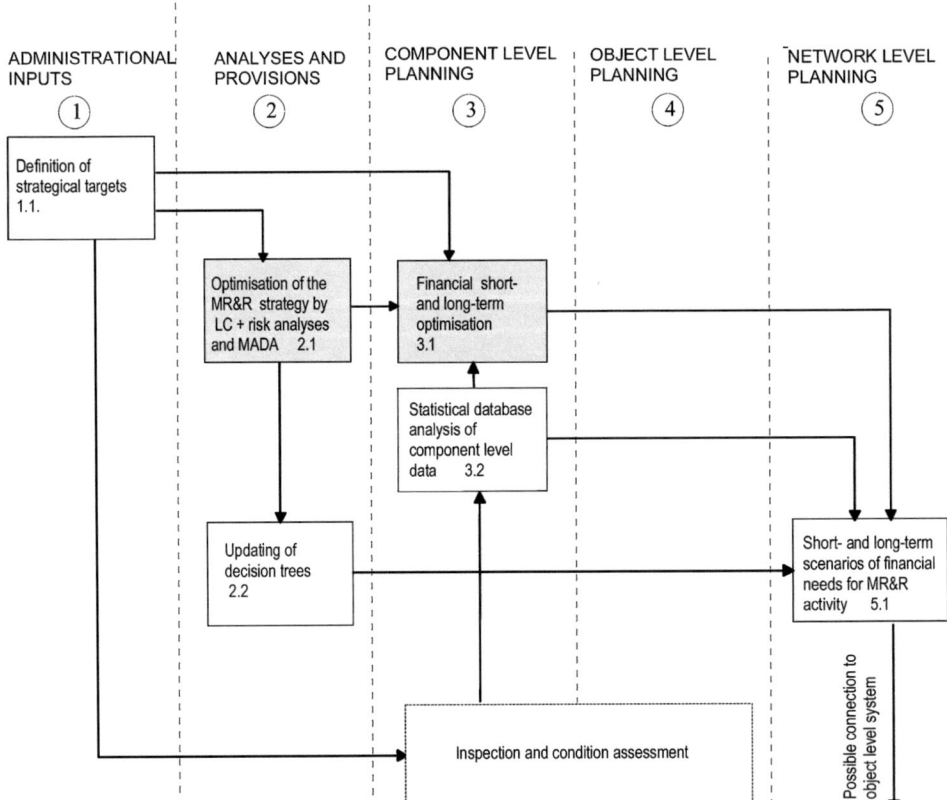

Figure 5.32 Modules of the network level management system.

Table 5.22 Modules of the network level LMS

No.	Module	Purpose
2.1	Optimisation of MR&R strategy by integrated LCC and risk analyses and MADA	Data source for optimisation of MR&R strategy and for updating decision trees.
2.2	Updating decision trees	Defines the maintenance strategy of the organisation for components in different environmental conditions.
3.2	Statistical database analysis of component level data	Returns statistical data on condition and structural and material properties of components for LCC analyses.
3.1	Financial short- and long-term optimisation	Optimisation of MR&R action profiles for components. Returns the annual MR&R costs for the ideal maintenance policy.
5.1	Short- and long-term scenarios of financial needs for MR&R activity	Returns the short- and long-term scenarios by integrating the component level scenarios from 3.2.

For optimisation at the network level it is necessary to divide the objects into smaller parts. It is recommended to use the same breakdown-into-components technique as is used at the object level system. However, at the network level it is possible to use a little coarser breakdown by combining components of the same type and in similar environmental conditions into the same group of structural parts.

At the network level the structural parts are not treated individually but as populations of structural parts in the same structure stock. Hence, although the structures are broken down into structural parts the network level analyses are statistical and cover the whole network of structures.

5.5.2 Optimisation of the technical MR&R strategy

By the optimisation of technical MR&R strategy searching is meant for the optimal MR&R action profiles for the upkeep of structures during a design period. The specification of an MR&R action profile includes the definition of both MR&R methods and condition-related repair criteria to be used for each component. The optimal MR&R action profile may depend on the environmental conditions and/or the technical properties or the condition of the component. By the definition of MR&R methods and the condition-related repair criteria the action profiles are defined as far as it is necessary at this phase. The timing of MR&R actions depends on the rate of degradation, which is specific for each component. Thus the timing of MR&R actions is defined later in connection with the object level planning processes.

If an organisation possesses only a few objects, it is possible to optimise the maintenance strategy for each object separately. However, if the facility stock consists of thousands of objects with tens of components in each, it is impossible to do the optimisation for every object separately. Then the optimisation is done in advance using the network level optimising routines.

The final purpose is to build up a decision tree based on the network level optimisation results. The decision tree takes into account the specific features of the components and makes the decision on the optimal MR&R action profile for them.

So, at the network level it is essential to study the structural parts in various environmental conditions and with varying values of material and structural parameters. The final task is to find the key parameters by which it is possible to specify the maintenance strategy. In a decision tree these parameters must also be equipped with criteria by which the tree branches and makes the decision for each specific case.

Examples of key parameters could be the following [1, 9, 17, 18]:

- chloride stress
- moisture stress
- concrete cover
- frost resistance of concrete.

In the network level analyses the identified key parameters are varied to find the correct criteria for these parameters. Also the condition criteria, i.e. maximum allowable probabilities, are varied to find out the optimal timing of actions. To define the maintenance policy the following questions should be answered:

- Which coatings or other preventive maintenance methods should be used, or is it advisable to leave the structure without any preventive maintenance?
- Which repair methods should be used?
- At which condition state (and allowable probability) should the structures be repaired, and at which condition state (and allowable probability) will the coatings and other preventive maintenance methods be renewed?
- Are there restrictions for the successive application of the same repair or preventive maintenance method?

The optimisation is performed using the LCC analyses, risk analyses and MADA. The following concept may be applied:

1. Conduct LCC analyses

    - using automatic condition guarding and triggering of actions
    - using default (average) values of parameters
    - varying the material, structural and environmental parameters each in turn
    - varying the maximum allowable probability of exceeding the given limit state
    - taking into account the other aspects of design.

2. Conduct the quantitative RAMS analysis.
3. Determine the most appropriate MR&R action profile by MADA by having maintenance costs, user costs, environmental impacts, RAMS, etc. as attributes.

The decision tree reflects the maintenance policy of an organisation. It is up to the organisation itself how it wants to apply the analysis results from the LCC and risk analyses. In some organisations the administration may only want to minimise the MR&R costs. In some other organisations it is desire to combine MR&R costs with the user costs and/or possibly the environmental impacts. The LCC analyses, risk analyses and MADA are just tools for helping the decision-making but usually they offer a good basis for a sustainable maintenance policy.

5.5.3 Statistical database analysis on component level data

To be able to do the long-term and short-term financial optimisation for the network of structures, a division into analysis groups must be first accomplished. The stock of objects is first divided into components. The breakdown into components is preferably the same as that used in the object level planning. Then each component is further divided into MR&R action profile groups according to the decision tree division. The decision tree program itself can be used to categorise the components into MR&R action profile groups. Figure 5.33. shows the principle of grouping.

From each analysis group the following statistical data is determined by exploring the database data:

- total area, and
- statistical data of the parameters used in the LCC analysis.

Usually only the average values of parameters are adequate for the LCC analyses. In some cases it may be relevant to use also the distribution of parameter data to conduct more fine-tuned analyses like sensitivity analyses.

The results of the database exploration are placed into an initial data file which reminds the data file of a single structure. This 'structure' which represents the whole network of structures is divided into 'components'. Each 'component' is divided into subcategories according to decision tree division. For instance in the case of a bridge the edge beams are grouped as 'edge beams 1', 'edge beams 2', etc. each group of edge beams representing different conditions that warrant a different MR&R action profile.

This analysis grouping is used at the network level in both the long-term and short-term optimisation of the maintenance strategy.

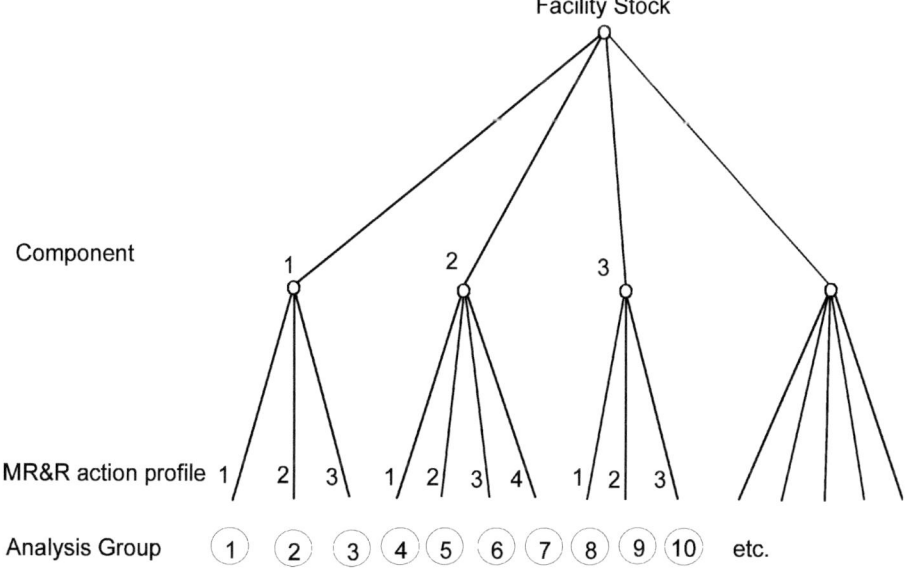

Figure 5.33 Network level division into analysis groups.

5.5.4 Long-term and short-term financial optimisation of the maintenance strategy

Traditionally the financial optimisation of the maintenance strategy includes the long-term and the short-term optimisation. By the long-term optimisation the optimal condition state distribution of each analysis group is identified. By the short-term optimisation the optimal actions and annual condition state distributions are searched having the long-term optimum as the final goal and the annual budgets as constraints. This analysis gives also information on how much funding should at least be allocated for MR&R activity to shift the condition state distribution of the structures towards the optimal distribution.

Two optional methods for the network level optimisation can be applied:

- Linear optimisation method, or
- Life cycle cost analysis method.

The linear optimisation method is the traditional method that is used in several existing management systems. However, there are some limitations in this method as compared to the LCC analysis method, which allows more freedom in the life cycle design of structures. Anyway, both methods are briefly described in this chapter. In simple cases both methods yield the same optimum.

5.5.4.1 Linear optimisation method

In the following example both the long-term and the short-term optimisation as conducted by the linear optimisation method are presented. A more detailed description of the methods is found in the EU RIMES report [2].

The principle of the linear optimisation method in the *long-term optimisation* is presented in Figure 5.34. This example represents a simple case when there is only one damage type defined with five condition states. Such an optimisation problem can be solved by a simple spreadsheet program.

The surface area of structures is considered to be divided into two parts:

- area to which MR&R actions are addressed, and
- area to which no actions are addressed.

The area attributed to different condition states is expressed as relative portions, i.e. fractions, following the principles of the Markov Chain method. The sum of fractions in all condition states must be 1. In principle all the fractions are 'unknown', i.e. they are subject to optimisation. Note, however, that some condition states may be excluded from the optimisation if no MR&R actions are defined for that condition state. In this case repair actions were assigned to condition states 3 and 4.

The absolute areas of repairable and non-repairable structures are obtained by multiplying the relative surface areas by the total surface area of structures (in this example case $10\,000\,m^2$). The MR&R costs at each condition state are obtained by multiplying the surface area with respective unit costs of the MR&R actions. The total costs are obtained by summing up all costs. The 'total cost' is the objective function of optimisation that is to be minimised.

Long term optimisation

Total surface area of structures   10000 m^2

Transition probability matrix for action effects

State	0	1	2	3	4
0	1	0	0	0	0
1	0.99	0.01	0	0	0
2	0.97	0.02	0.01	0	0
3	0.95	0.02	0.02	0.01	0
4	0.93	0.02	0.02	0.02	0.01

Transition probability matrix for degradation

State	0	1	2	3	4
0	0.61	0.39	0	0	0
1	0	0.74	0.26	0	0
2	0	0	0.82	0.18	0
3	0	0	0	0.91	0.09
4	0	0	0	0	1

Structures of no repair actions

State	0	1	2	3	4	Total
Condition distribution	0.087	0.133	0.196	0.399	0.150	0.964
Surface area, m^2	865	1326	1955	3991	1500	9637
Unit costs, Euro/m^2	0.000	0.000	0.000	0.000	0.000	
Cost, Euro	0	0	0	0	0	0

Structures to which repair actions are addressed

State	0	1	2	3	4	Total
Condition distribution	0.000	0.000	0.000	0.000	0.036	0.036
Surface area, m^2	0	0	0	0	363	363
Unit costs, Euro/m^2	0	0	0	80	100	
Cost, Euro	0	0	0	0	36282	36282

Surface area together

State	0	1	2	3	4	Total
Condition distribution	0.087	0.133	0.196	0.399	0.186	1.000
Surface area, m^2	865	1326	1955	3991	1863	
Costs together, Euro	0	0	0	0	36282	36282

Next year

State	0	1	2	3	4	Total
Distribution, no actions	0.053	0.132	0.195	0.398	0.186	0.964
Distribution, repair actions	0.034	0.001	0.001	0.001	0.000	0.036
Distribution, total	0.087	0.133	0.196	0.399	0.186	1.000

Repair actions and unit costs

State	Actions	Costs	
1		0	Euro/m^2
2		0	Euro/m^2
3	C	80	Euro/m^2
4	D	100	Euro/m^2

Figure 5.34 Long-term optimisation by the linear optimisation method.

On top Figure 5.34 the transition probability matrices are presented for degradation and action effects. At the bottom of the figure, condition state distributions of the next year are presented. The next year condition state distribution of non-repairable structures is obtained by multiplying the corresponding distribution of the present year by the degradation matrix. The next year condition state distribution of repairable structures is obtained by multiplying the corresponding distribution of the present year by the action effect matrix. The total condition distribution of the next year is obtained by summing up the distributions of the repaired and non-repaired structures. This distribution is forced to be the same as the original condition state distribution of the treated year, as the long-term condition state distribution is aimed to remain constant (steady state requirement).

Now the cost table is ready for linear optimisation. The optimisation is carried out in the example by using the Excel Solver program. In principle both linear and non-linear problems can be solved by the Solver. When the conditions of linearity are fulfilled the linear option is chosen.

The definition of the optimisation problem is carried out as follows. Minimise the total costs by changing the fractions of non-repairable and repairable structures at each condition state under following conditions:

1. The fractions of both non-repairable and repairable structures at each condition state must be between 0 and 1
2. The sum of all fractions must be 1.
3. The condition state distribution of the next year must be the same as the condition state distribution of the treated year (steady state condition).

By this problem definition, the optimum condition state distribution and the corresponding annual repair action protocol can then be automatically solved.

The results in this case show that the optimum is obtained when the action assigned for condition state 4 is used. Yearly 3.6% of the total stock must be repaired. The total MR&R costs for $10\,000\,m^2$ with the unit costs of $100\,euros\,m^{-2}$ is $36\,282\,euro\,year^{-1}$.

An extra constraint was set for the fraction at condition state 4. Without this extra constraint the optimal steady state would be the case when all the structures are in condition state 4 with no repair actions at all. In this case, no more than 15% of structures are allowed to drop to condition state 4.

The basic formulation in the short-term optimisation closely resembles the long-term optimisation. The costs and condition state transitions within one year are calculated. Repair actions are addressed to part of the structures at each condition state, the rest being left without any actions. An optimisation is performed to find the combination of actions by which the difference between the starting condition distribution and the long-term optimum distribution is minimised. The available annual funding, i.e. budget, works as a constraint.

Only the relative portions of repairable structures can be freely chosen. The relative portions of non-repairable structures are obtained by subtracting the condition distribution of repairable structures from the starting condition distribution of structures.

The total costs cannot be used as the object function of the short-term optimisation as the total costs are limited by the annual funding. So instead, the deviation function that measures the deviation between the starting condition distribution and the long-term optimum distribution is used as an object function of minimisation.

The deviation function can be defined by many means. Examples of such functions are the following:

1. Squared Difference method (sum of the squared differences of fractions at each state);
2. Absolute Difference method (sum of absolute values of fraction differences at each state);
3. Weighted Difference method (sum of absolute values of fraction differences multiplied by the weight coefficients for each state);
4. Cost Difference method (Sum of absolute values of fraction differences multiplied by the average unit costs for each state).

The principle of the short-term optimisation is presented in Figure 5.35 using the Squared Difference method. The optimisation problem is defined as follows. Minimise the deviation function by changing the relative portions of repairable structures at each condition state under following conditions:

1. The fractions of repairable and non-repairable structures at each condition state must be between 0 and 1.
2. The total costs should be equal to the annual budget limit.

The Squared Difference method and the Absolute Difference method favour light repair actions over heavy actions because with a constant budget the difference between the starting distribution and the long-term optimal distribution can be minimised most effectively by making the light repair measures. By choosing the Weighted Difference or the Cost Difference method the heavy repair actions are favoured.

The problem in using the deviation functions as an object function is that the linearity condition is not fulfilled. Then the non-linear modes of the Solver program are used in solving the problem. Because of the non-linearity the result is not always obtained or it can be unreliable. However, if there is only one damage type the result is usually reliable.

The optimum condition state distributions with the given budget are obtained year by year. To obtain the optimum distribution of the year $i + 1$ the optimum distribution of the year i is assigned in the analysis table at the place of the starting condition distribution. So the optimisation can be extended over many years.

5.5.4.2 Life cycle cost analysis method

The long-term and short-term optimisation can be performed also by the LCC analysis method. In this method the Markovian life cycle cost analysis table is used as the framework of calculations. Both the long-term and short-term analyses are performed at the same time and by the same analysis table.

Some changes in the calculation routines are provided as compared to the object level calculations. In the object level analyses when the condition of a structure exceeds the limit state with the maximum allowable probability the whole structure is repaired and the action effects are calculated accordingly. In the network level analysis, where all the building stock is considered, not all the structures are repaired but only the part of structures that exceed the maximum allowable probability. So

Short-term optimisation
Total surface area of structures 10 000 m²
Transition probability matrix for action effects

State	0	1	2	3	4
0	1.000	0.000	0.000	0.000	0.000
1	0.710	0.290	0.000	0.000	0.000
2	0.740	0.000	0.260	0.000	0.000
3	0.810	0.000	0.000	0.190	0.000
4	0.850	0.000	0.000	0.000	0.150

Transition probability matrix for degradation

State	0	1	2	3	4
0	0.963	0.037	0.000	0.000	0.000
1	0.000	0.957	0.043	0.000	0.000
2	0.000	0.000	0.927	0.073	0.000
3	0.000	0.000	0.000	0.880	0.120
4	0.000	0.000	0.000	0.000	1.000

Structures of no repair actions

State	0	1	2	3	4	Total
Condition distribution	0.321	0.212	0.202	0.176	0.052	0.964
Surface area, m²	3209	2124	2023	1758	521	9635
Unit costs, euro m^{-2}	0	0	0	0	0	
Cost euro	0	0	0	0	0	0

Structures to which repair actions are addressed

State	0	1	2	3	4	Total
Condition distribution	0.000	0.000	0.000	0.000	0.037	0.037
Surface area, m²	0	0	0	0	365	365
Unit costs euro m^{-2}	0	0	0	80	100	
Cost, euro	0	0	0	0	36500	36500

Surface area together

State	0	1	2	3	4	Total
Condition distribution	0.321	0.212	0.202	0.176	0.089	1.000
Surface area, m²	3209	2124	2023	1758	886	
Costs together, euro	0	0	0	0	36500	36500

Next year

State	0	1	2	3	4	Total
Distribution, no actions	0.309	0.215	0.197	0.169	0.073	0.964
Distribution, repair actions	0.031	0.000	0.000	0.000	0.005	0.037
Distribution, total	0.340	0.215	0.197	0.169	0.079	1.000
Target distribution	0.087	0.133	0.196	0.399	0.186	1.000
Square of difference	0.06432	0.00682	0.00000	0.05275	0.01159	0.13548

	0	1	2	3	4
Long-term target distribution	0.087	0.133	0.196	0.399	0.186
Present distribution	0.052	0.231	0.245	0.222	0.250
1 Year distribution	0.101	0.223	0.237	0.213	0.226
2 Year distribution	0.148	0.217	0.229	0.205	0.200
3 Year distribution	0.194	0.213	0.222	0.197	0.174
4 Year distribution	0.238	0.211	0.215	0.190	0.146
5 Year distribution	0.280	0.211	0.208	0.182	0.118
6 Year distribution	0.321	0.212	0.202	0.176	0.089
7 Year distribution	0.360	0.215	0.197	0.169	0.059

Figure 5.35 Short-term optimisation by the linear optimisation method.

the fraction of structures that is repaired is smaller than in the object level analysis, but the repair for the exceeding fraction is performed more frequently, actually every year after the first excess of the limit state.

Another difference in the network level programming is that a budget constraint for the annual repair costs may be added. This is done by first determining the maximum repair area that can be repaired by the allocated budget and the current unit costs of the repair. Also the corresponding relative part of the total area, i.e. fraction, is determined. So the fraction of structures that is to be repaired during a year is controlled on one hand by the budget and on the other hand by the difference of the true fraction exceeding the limit state and the maximum allowable fraction for the limit state. The fraction of the repairable structures cannot be greater than the former as the repair costs must keep within the budget frames. So the fraction of the repairable structures is determined from the following Equation (5.19):

$$RepFrac = Min(MaxFracBud; \ TrueFracLS - MaxFracLS) \qquad (5.19)$$

where

RepFrac	is the fraction that is to be repaired during a year
MaxFracBud	the maximum fraction of structures based on the budget
TrueFracLS	the real fraction at the limit state (which may be 3 or 4)
MaxFracLS	the maximum allowable fraction at limit state based on safety, serviceability, aesthetics, etc.

In Figure 5.36 one can see an example of the long-term analysis. In this case there was a criterion set for the condition state 4 that no more than 15% of structures are allowed to drop to condition state 4 without repair. This is the same condition that was set in the example of the linear long-term optimisation. The maximum annual budget was defined to be € 36 500.

In this case the initial condition state distribution was uniform, i.e. 0.2, in each condition state $i(i = 0, 1, 2, 3, 4)$. As one can see from the results the condition state distribution approaches a steady state distribution. After some 30 years the distribution does not change much any more. By comparing the steady state distribution to the long-term optimum distribution obtained by linear optimisation in Section 5.5.4.1 one can see that it is exactly the same.

The final condition state distribution is independent of the initial condition state distribution. No matter whether the initial condition state distribution is (1, 0, 0, 0, 0) or (0, 0, 0, 0, 1), the final steady state distribution is the same with the same repair actions.

The short-term analysis is performed using the same analysis table as that used for the long-term analysis. But more attention is given to the period before the long-term optimum is attained.

The time to reach the long-term steady state depends, of course, on the initial condition state distribution. But it depends also on the budget constraint. The greater the maximum budget per year the sooner the long-term steady state is reached. This fact allows adjusting the length of the short-term period by the budget. The following

Transition probability matrix for action effects

State	0	1	2	3	4
0	1	0	0	0	0
1	0.99	0.01	0	0	0
2	0.97	0.02	0.01	0	0
3	0.95	0.02	0.02	0.01	0
4	0.93	0.02	0.02	0.02	0.01

Transition probability matrix for degradation

State	0	1	2	3	4
0	0.61	0.39	0	0	0
1	0	0.74	0.26	0	0
2	0	0	0.82	0.18	0
3	0	0	0	0.91	0.09
4	0	0	0	0	1

Total area	$10\,000\,m^2$
Repair condition state	4 (3 or 4)
Repair cost	$100\ euro\,m^{-2}$
Rate of interest	4%
Annual max. budget	36 500 euro
Annual max. rep. area	$365\,m^2$
Annual max. rel. rep. area	0.0365

Total costs

Real	PV
1732891	712482

Year	State 0	1	2	3	Criteria (Max %) 15 → 4	3	Average	Repair	Annual costs Real	PV	Cumulative costs Real	PV
0	0.2	0.2	0.2	0.2	0.2	0.400	2.00	0	0	0	0	0	0
1	0.156	0.227	0.217	0.219	0.182	0.401	2.04	1	1	36 500	35 096	36 500	35 096
2	0.125	0.229	0.237	0.239	0.170	0.409	2.10	2	1	31 865	29 461	68 365	64 557
3	0.095	0.219	0.255	0.260	0.172	0.432	2.20	3	1	20 004	17 784	88 369	82 341
4	0.078	0.199	0.266	0.283	0.174	0.457	2.28	4	1	21 682	18 534	110 052	100 875
5	0.070	0.178	0.270	0.306	0.176	0.482	2.34	5	1	23 646	19 435	133 698	120 311
6	0.066	0.160	0.269	0.328	0.178	0.505	2.39	6	1	25 721	20 327	159 418	140 638
7	0.066	0.144	0.262	0.347	0.180	0.527	2.43	7	1	27 800	21 126	187 219	161 764
8	0.068	0.133	0.253	0.364	0.182	0.545	2.46	8	1	29 770	21 752	216 989	183 516
9	0.071	0.126	0.243	0.377	0.183	0.560	2.48	9	1	31 537	22 157	248 525	205 674
10	0.074	0.121	0.233	0.388	0.184	0.572	2.49	10	1	33 046	22 325	281 571	227 998
11	0.077	0.119	0.223	0.395	0.185	0.581	2.49	11	1	34 275	22 264	315 846	250 263
12	0.080	0.119	0.215	0.401	0.186	0.586	2.49	12	1	35 228	22 003	351 074	272 266
13	0.082	0.120	0.208	0.404	0.186	0.590	2.49	13	1	35 928	21 578	387 003	293 844
14	0.084	0.122	0.202	0.406	0.187	0.592	2.49	14	1	36 410	21 026	423 413	314 870
15	0.085	0.123	0.198	0.406	0.187	0.593	2.49	15	1	36 500	20 267	459 913	335 137
16	0.086	0.125	0.195	0.406	0.188	0.594	2.48	16	1	36 500	19 488	496 413	354 625
17	0.086	0.127	0.193	0.405	0.188	0.593	2.48	17	1	36 500	18 738	532 913	373 363
18	0.087	0.128	0.192	0.404	0.188	0.593	2.48	18	1	36 500	18 017	569 413	391 380
19	0.087	0.129	0.192	0.403	0.189	0.592	2.48	19	1	36 500	17 324	605 913	408 705
20	0.087	0.130	0.192	0.402	0.189	0.591	2.48	20	1	36 500	16 658	642 413	425 363
21	0.087	0.131	0.192	0.401	0.189	0.590	2.47	21	1	36 500	16 017	678 913	441 380
22	0.087	0.132	0.192	0.401	0.189	0.589	2.47	22	1	36 500	15 401	715 413	456 782
23	0.087	0.132	0.192	0.400	0.189	0.588	2.47	23	1	36 500	14 809	751 913	471 591
24	0.087	0.132	0.193	0.399	0.189	0.588	2.47	24	1	36 500	14 239	788 413	485 830
25	0.087	0.133	0.193	0.399	0.188	0.587	2.47	25	1	36 500	13 692	824 913	499 522
26	0.087	0.133	0.194	0.398	0.188	0.586	2.47	26	1	36 500	13 165	861 413	512 687
27	0.087	0.133	0.194	0.398	0.188	0.586	2.47	27	1	36 500	12 659	897 913	525 346
28	0.087	0.133	0.194	0.398	0.187	0.585	2.47	28	1	36 500	12 172	934 413	537 518
29	0.087	0.133	0.195	0.398	0.187	0.585	2.46	29	1	36 500	11 704	970 913	549 221
30	0.087	0.133	0.195	0.398	0.187	0.585	2.46	30	1	36 500	11 254	1 007 413	560 475
31	0.087	0.133	0.195	0.398	0.187	0.584	2.46	31	1	36 500	10 821	1 043 913	571 296
32	0.087	0.133	0.196	0.398	0.186	0.584	2.46	32	1	36 500	10 405	1 080 413	581 701
33	0.087	0.133	0.196	0.398	0.186	0.584	2.46	33	1	36 191	9 920	1 116 604	591 620
34	0.087	0.133	0.196	0.398	0.186	0.584	2.46	34	1	36 178	9 535	1 152 782	601 155
35	0.086	0.133	0.196	0.398	0.186	0.585	2.46	35	1	36 187	9 170	1 188 969	610 325
36	0.086	0.133	0.196	0.398	0.186	0.585	2.47	36	1	36 199	8 821	1 225 168	619 146
37	0.086	0.133	0.196	0.399	0.186	0.585	2.47	37	1	36 212	8 484	1 261 380	627 630
38	0.086	0.133	0.196	0.399	0.186	0.585	2.47	38	1	36 225	8 161	1 297 606	635 791
39	0.086	0.133	0.196	0.399	0.186	0.585	2.47	39	1	36 238	7 850	1 333 844	643 641
40	0.086	0.133	0.196	0.399	0.186	0.585	2.47	40	1	36 250	7 550	1 370 094	651 192
41	0.086	0.132	0.196	0.399	0.186	0.585	2.47	41	1	36 260	7 262	1 406 354	658 454
42	0.086	0.132	0.196	0.399	0.186	0.585	2.47	42	1	36 268	6 984	1 442 623	665 438
43	0.086	0.132	0.196	0.399	0.186	0.585	2.47	43	1	36 275	6 717	1 478 898	672 155
44	0.086	0.132	0.196	0.399	0.186	0.585	2.47	44	1	36 280	6 459	1 515 177	678 615
45	0.087	0.132	0.196	0.399	0.186	0.585	2.47	45	1	36 283	6 212	1 551 460	684 826
46	0.087	0.133	0.196	0.399	0.186	0.585	2.47	46	1	36 285	5 973	1 587 745	690 799
47	0.087	0.133	0.196	0.399	0.186	0.585	2.47	47	1	36 286	5 743	1 624 032	696 543
48	0.087	0.133	0.196	0.399	0.186	0.585	2.47	48	1	36 287	5 523	1 660 318	702 066
49	0.087	0.133	0.196	0.399	0.186	0.585	2.47	49	1	36 287	5 310	1 696 605	707 376
50	0.087	0.133	0.195	0.399	0.186	0.585	2.47	50	1	36 286	5 106	1 732 891	712 482

Figure 5.36 Long-term optimisation with LCC analysis table.

linear optimisation can be conducted in the analysis table to control the length of the short-term period:

- Minimise the total costs (real or present value costs)
- by changing the maximum budget
- by subjecting to the constraint that the average condition at the desired time is equal to the average of the long-term optimum distribution.

An example of the short-term optimisation of the budget so that the average condition state of the long-term optimum is attained within 20 years is presented in Figure 5.37. To attain the long-term optimum, the budget must be at least the budget for maintaining the long-term optimum. However, it does not have to be any greater. So, to minimise the MR&R costs in the long term, while the time to attain the long-term optimum distribution is not restricted, the optimal budget and annual work programme for the short-term period is normally the same as that for the long-term period. However, if the average condition state is smaller than that corresponding to the long-term optimum, i.e. the condition is initially better than the long-term optimum, the annual budgets during the short-term period may be smaller than those in the long-term period.

It is also possible to arrange the financial optimisation so that instead of having a constant annual budget the annual budget is a changing variable during the short-term period. This allows more possibilities for the financial planning, especially with capital costs (present value costs). For example, keeping the PV costs as the object function, the optimal financing for the short-term period restricted to 20 years would be that presented in Figure 5.38. During the first 12 years the budget would be around €21 000, because there was a constraint of the average condition which must not be worse than the existing average condition. During the following 5 years the budget would be €60 000, the maximum, and then the budget would rapidly reduce to about €36 300, which is the long-term optimum budget.

In the linear optimisation method there may be alternative MR&R actions assigned to different condition states. In the above presented LCC analysis there is only one MR&R action. This is because the optimisation of methods has already been performed in the first phase of the network level optimisation. If that phase is properly performed, there is no need to have alternative methods neither in the long-term nor in the short-term optimisation.

The examples of long-term and short-term analyses in this Chapter showed the calculation methods only schematically. The real analyses may be more refined and complex.

5.5.4.2.1 ADVANTAGES OF THE LIFE CYCLE COST ANALYSIS METHOD

The greatest advantage of the LCC analysis method over the linear optimisation method in the network level optimisation is that the results with the former are readily consistent with the object level results. It is important to make sure that the MR&R action profiles in the network level analyses and in the object level analyses are the same. If the same MR&R methods are applied, the total annual costs calculated by the network level routines and the object level routines should be approximately the same.

Transition probability matrix for action effects

State	0	1	2	3	4
0	1	0	0	0	0
1	0.99	0.01	0	0	0
2	0.97	0.02	0.01	0	0
3	0.95	0.02	0.02	0.01	0
4	0.93	0.02	0.02	0.02	0.01

Transition probability matrix for degradation

State	0	1	2	3	4
0	0.61	0.39	0	0	0
1	0	0.74	0.26	0	0
2	0	0	0.82	0.18	0
3	0	0	0	0.91	0.09
4	0	0	0	0	1

Total area	$10\,000\,m^2$
Repair condition state	4 (3 or 4)
Repair cost	$100\ euro\,m^{-2}$
Rate of interest	4%
Annual max. budget	58 400 euro
Annual max. rep. area	$584\,m^2$
Annual max. rel. rep. area	0.0584

Total costs

Real	PV
1734540	714081

Year	State 0	1	2	3	Criteria (Max %) 15 → 4	→ 3	Average	Repair	Annual costs Real	PV	Cumulative costs Real	PV
0	0.2	0.2	0.2	0.2	0.2	0.400	2.00	0	0	0	0	0
1	0.169	0.227	0.217	0.219	0.169	0.388	1.99	1	50 000	48 077	50 000	48 077
2	0.120	0.234	0.237	0.239	0.170	0.409	2.10	2	18 500	17 104	68 500	65 181
3	0.092	0.220	0.256	0.260	0.172	0.432	2.20	3	19 895	17 687	88 395	82 868
4	0.076	0.199	0.268	0.283	0.174	0.457	2.28	4	21 684	18 535	110 079	101 403
5	0.068	0.178	0.272	0.307	0.176	0.482	2.34	5	23 649	19 437	133 727	120 841
6	0.066	0.159	0.269	0.328	0.178	0.506	2.39	6	25 743	20 345	159 471	141 186
7	0.066	0.144	0.263	0.348	0.180	0.528	2.43	7	27 846	21 160	187 316	162 346
8	0.068	0.133	0.253	0.364	0.182	0.546	2.46	8	29 831	21 797	217 147	184 144
9	0.071	0.125	0.243	0.378	0.183	0.561	2.48	9	31 607	22 207	248 754	206 350
10	0.074	0.121	0.232	0.388	0.184	0.573	2.49	10	33 118	22 373	281 872	228 723
11	0.077	0.119	0.223	0.396	0.185	0.581	2.49	11	34 342	22 308	316 214	251 031
12	0.080	0.119	0.214	0.401	0.186	0.587	2.49	12	35 288	22 041	351 502	273 072
13	0.082	0.120	0.207	0.404	0.186	0.591	2.49	13	35 979	21 608	387 481	294 680
14	0.084	0.121	0.202	0.406	0.187	0.593	2.49	14	36 451	21 049	423 932	315 730
15	0.085	0.123	0.198	0.406	0.187	0.593	2.49	15	36 743	20 402	460 675	336 132
16	0.086	0.125	0.195	0.406	0.187	0.593	2.48	16	36 897	19 699	497 572	355 831
17	0.087	0.127	0.193	0.405	0.187	0.592	2.48	17	36 948	18 968	534 519	374 799
18	0.087	0.129	0.192	0.405	0.187	0.591	2.47	18	36 928	18 229	571 447	393 028
19	0.088	0.130	0.192	0.403	0.187	0.590	2.47	19	36 864	17 497	608 311	410 525
20	0.088	0.131	0.192	0.402	0.187	0.589	2.47	20	36 777	16 784	645 088	427 309
21	0.088	0.132	0.192	0.402	0.187	0.588	2.47	21	36 681	16 097	681 769	443 406
22	0.087	0.133	0.193	0.401	0.187	0.587	2.47	22	36 587	15 438	718 356	458 844
23	0.087	0.133	0.193	0.400	0.186	0.587	2.47	23	36 502	14 810	754 859	473 654
24	0.087	0.133	0.194	0.400	0.186	0.586	2.46	24	36 430	14 212	791 288	487 866
25	0.087	0.133	0.194	0.399	0.186	0.586	2.46	25	36 371	13 643	827 659	501 510
26	0.087	0.133	0.195	0.399	0.186	0.585	2.46	26	36 326	13 102	863 985	514 612
27	0.087	0.133	0.195	0.399	0.186	0.585	2.46	27	36 293	12 587	900 278	527 199
28	0.087	0.133	0.195	0.399	0.186	0.585	2.46	28	36 271	12 096	936 549	539 295
29	0.087	0.133	0.195	0.399	0.186	0.585	2.47	29	36 258	11 626	972 807	550 921
30	0.087	0.133	0.196	0.399	0.186	0.585	2.47	30	36 251	11 177	1 009 058	562 098
31	0.086	0.133	0.196	0.399	0.186	0.585	2.47	31	36 249	10 746	1 045 306	572 844
32	0.086	0.133	0.196	0.399	0.186	0.585	2.47	32	36 250	10 333	1 081 557	583 177
33	0.086	0.133	0.196	0.399	0.186	0.585	2.47	33	36 254	9 937	1 117 810	593 114
34	0.086	0.133	0.196	0.399	0.186	0.585	2.47	34	36 258	9 556	1 154 069	602 670
35	0.086	0.133	0.196	0.399	0.186	0.585	2.47	35	36 263	9 190	1 190 332	611 860
36	0.086	0.133	0.196	0.399	0.186	0.585	2.47	36	36 268	8 837	1 226 600	620 697
37	0.086	0.133	0.196	0.399	0.186	0.585	2.47	37	36 272	8 498	1 262 872	629 196
38	0.086	0.133	0.196	0.399	0.186	0.585	2.47	38	36 275	8 172	1 299 147	637 368
39	0.086	0.133	0.196	0.399	0.186	0.585	2.47	39	36 278	7 859	1 335 425	645 227
40	0.087	0.133	0.196	0.399	0.186	0.585	2.47	40	36 280	7 557	1 371 706	652 783
41	0.087	0.133	0.196	0.399	0.186	0.585	2.47	41	36 282	7 266	1 407 988	660 050
42	0.087	0.133	0.196	0.399	0.186	0.585	2.47	42	36 283	6 987	1 444 271	667 037
43	0.087	0.133	0.196	0.399	0.186	0.585	2.47	43	36 284	6 719	1 480 555	673 756
44	0.087	0.133	0.196	0.399	0.186	0.585	2.47	44	36 284	6 460	1 516 839	680 216
45	0.087	0.133	0.196	0.399	0.186	0.585	2.47	45	36 284	6 212	1 553 123	686 428
46	0.087	0.133	0.196	0.399	0.186	0.585	2.47	46	36 284	5 973	1 589 407	692 401
47	0.087	0.133	0.196	0.399	0.186	0.585	2.47	47	36 284	5 743	1 625 690	698 144
48	0.087	0.133	0.196	0.399	0.186	0.585	2.47	48	36 284	5 522	1 661 974	703 666
49	0.087	0.133	0.196	0.399	0.186	0.585	2.47	49	36 283	5 310	1 698 257	708 976
50	0.087	0.133	0.196	0.399	0.186	0.585	2.47	50	36 283	5 105	1 734 540	714 081

Figure 5.37 Short-term optimisation arranged so that the average condition state attains the average state of the long-term optimum within 20 years.

Transition probability matrix for action effects

State	0	1	2	3	4
0	1	0	0	0	0
1	0.99	0.01	0	0	0
2	0.97	0.02	0.01	0	0
3	0.95	0.02	0.02	0.01	0
4	0.93	0.02	0.02	0.02	0.01

Transition probability matrix for degradation

State	0	1	2	3	4
0	0.61	0.39	0	0	0
1	0	0.74	0.26	0	0
2	0	0	0.82	0.18	0
3	0	0	0	0.91	0.09
4	0	0	0	0	1

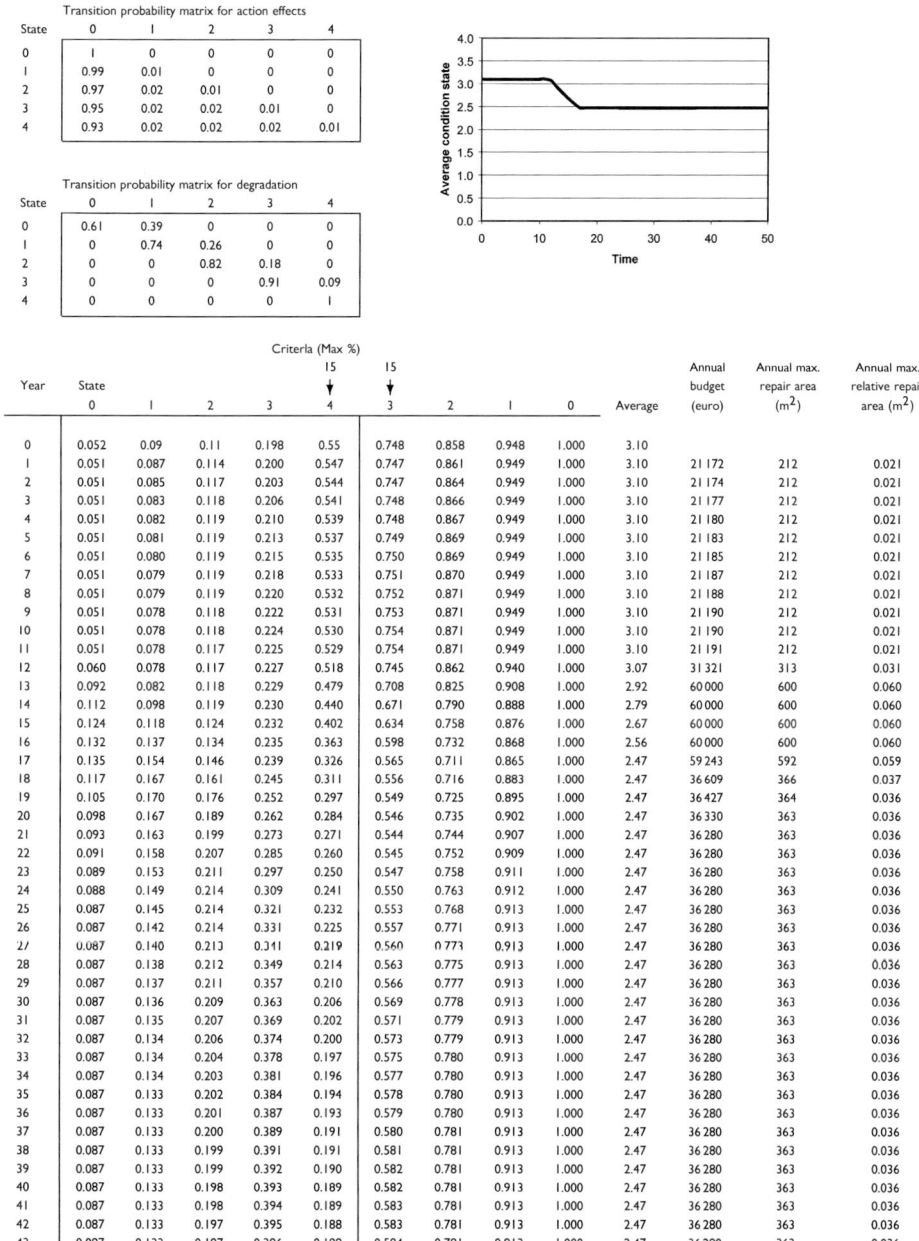

Year	State 0	1	2	3	4 (15↓)	3 (15↓)	2	1	0	Average	Annual budget (euro)	Annual max. repair area (m²)	Annual max. relative repair area (m²)
0	0.052	0.09	0.11	0.198	0.55	0.748	0.858	0.948	1.000	3.10			
1	0.051	0.087	0.114	0.200	0.547	0.747	0.861	0.949	1.000	3.10	21 172	212	0.021
2	0.051	0.085	0.117	0.203	0.544	0.747	0.864	0.949	1.000	3.10	21 174	212	0.021
3	0.051	0.083	0.118	0.206	0.541	0.748	0.866	0.949	1.000	3.10	21 177	212	0.021
4	0.051	0.082	0.119	0.210	0.539	0.748	0.867	0.949	1.000	3.10	21 180	212	0.021
5	0.051	0.081	0.119	0.213	0.537	0.749	0.869	0.949	1.000	3.10	21 183	212	0.021
6	0.051	0.080	0.119	0.215	0.535	0.750	0.869	0.949	1.000	3.10	21 185	212	0.021
7	0.051	0.079	0.119	0.218	0.533	0.751	0.870	0.949	1.000	3.10	21 187	212	0.021
8	0.051	0.079	0.119	0.220	0.532	0.752	0.871	0.949	1.000	3.10	21 188	212	0.021
9	0.051	0.078	0.118	0.222	0.531	0.753	0.871	0.949	1.000	3.10	21 190	212	0.021
10	0.051	0.078	0.118	0.224	0.530	0.754	0.871	0.949	1.000	3.10	21 190	212	0.021
11	0.051	0.078	0.117	0.225	0.529	0.754	0.871	0.949	1.000	3.10	21 191	212	0.021
12	0.060	0.078	0.117	0.227	0.518	0.745	0.862	0.940	1.000	3.07	31 321	313	0.031
13	0.092	0.082	0.118	0.229	0.479	0.708	0.825	0.908	1.000	2.92	60 000	600	0.060
14	0.112	0.098	0.119	0.230	0.440	0.671	0.790	0.888	1.000	2.79	60 000	600	0.060
15	0.124	0.118	0.124	0.232	0.402	0.634	0.758	0.876	1.000	2.67	60 000	600	0.060
16	0.132	0.137	0.134	0.235	0.363	0.598	0.732	0.868	1.000	2.56	60 000	600	0.060
17	0.135	0.154	0.146	0.239	0.326	0.565	0.711	0.865	1.000	2.47	59 243	592	0.059
18	0.117	0.167	0.161	0.245	0.311	0.556	0.716	0.883	1.000	2.47	36 609	366	0.037
19	0.105	0.170	0.176	0.252	0.297	0.549	0.725	0.895	1.000	2.47	36 427	364	0.036
20	0.098	0.167	0.189	0.262	0.284	0.546	0.735	0.902	1.000	2.47	36 330	363	0.036
21	0.093	0.163	0.199	0.273	0.271	0.544	0.744	0.907	1.000	2.47	36 280	363	0.036
22	0.091	0.158	0.207	0.285	0.260	0.545	0.752	0.909	1.000	2.47	36 280	363	0.036
23	0.089	0.153	0.211	0.297	0.250	0.547	0.758	0.911	1.000	2.47	36 280	363	0.036
24	0.088	0.149	0.214	0.309	0.241	0.550	0.763	0.912	1.000	2.47	36 280	363	0.036
25	0.087	0.145	0.214	0.321	0.232	0.553	0.768	0.913	1.000	2.47	36 280	363	0.036
26	0.087	0.142	0.214	0.331	0.225	0.557	0.771	0.913	1.000	2.47	36 280	363	0.036
27	0.087	0.140	0.213	0.341	0.219	0.560	0.773	0.913	1.000	2.47	36 280	363	0.036
28	0.087	0.138	0.212	0.349	0.214	0.563	0.775	0.913	1.000	2.47	36 280	363	0.036
29	0.087	0.137	0.211	0.357	0.210	0.566	0.777	0.913	1.000	2.47	36 280	363	0.036
30	0.087	0.136	0.209	0.363	0.206	0.569	0.778	0.913	1.000	2.47	36 280	363	0.036
31	0.087	0.135	0.207	0.369	0.202	0.571	0.779	0.913	1.000	2.47	36 280	363	0.036
32	0.087	0.134	0.206	0.374	0.200	0.573	0.779	0.913	1.000	2.47	36 280	363	0.036
33	0.087	0.134	0.204	0.378	0.197	0.575	0.780	0.913	1.000	2.47	36 280	363	0.036
34	0.087	0.134	0.203	0.381	0.196	0.577	0.780	0.913	1.000	2.47	36 280	363	0.036
35	0.087	0.133	0.202	0.384	0.194	0.578	0.780	0.913	1.000	2.47	36 280	363	0.036
36	0.087	0.133	0.201	0.387	0.193	0.579	0.780	0.913	1.000	2.47	36 280	363	0.036
37	0.087	0.133	0.200	0.389	0.191	0.580	0.781	0.913	1.000	2.47	36 280	363	0.036
38	0.087	0.133	0.199	0.391	0.191	0.581	0.781	0.913	1.000	2.47	36 280	363	0.036
39	0.087	0.133	0.199	0.392	0.190	0.582	0.781	0.913	1.000	2.47	36 280	363	0.036
40	0.087	0.133	0.198	0.393	0.189	0.582	0.781	0.913	1.000	2.47	36 280	363	0.036
41	0.087	0.133	0.198	0.394	0.189	0.583	0.781	0.913	1.000	2.47	36 280	363	0.036
42	0.087	0.133	0.197	0.395	0.188	0.583	0.781	0.913	1.000	2.47	36 280	363	0.036
43	0.087	0.133	0.197	0.396	0.188	0.584	0.781	0.913	1.000	2.47	36 280	363	0.036
44	0.087	0.133	0.197	0.396	0.188	0.584	0.781	0.913	1.000	2.47	36 280	363	0.036
45	0.087	0.133	0.197	0.397	0.187	0.584	0.781	0.913	1.000	2.47	36 280	363	0.036
46	0.087	0.133	0.196	0.397	0.187	0.584	0.781	0.913	1.000	2.47	36 280	363	0.036
47	0.087	0.133	0.196	0.398	0.187	0.585	0.781	0.913	1.000	2.47	36 280	363	0.036
48	0.087	0.133	0.196	0.398	0.187	0.585	0.781	0.913	1.000	2.47	36 280	363	0.036
49	0.087	0.133	0.196	0.398	0.187	0.585	0.781	0.913	1.000	2.47	36 280	363	0.036
50	0.087	0.133	0.196	0.398	0.187	0.585	0.781	0.913	1.000	2.47	36 280	363	0.036

Figure 5.38 Optimal financing of the short-term period.

From this advantage, it follows that it is normally not necessary to try to harmonise the results of the object level planning with the results of the network level planning. It is, however, advisable to compare these results, not for correcting the object level results but possibly for correcting the initial data for the network level analyses. If there is a difference between the object level and the network level, the reason for it can probably be attributed to inaccurate statistical initial data in the network level analyses, not in the object level planning.

The example of the long-term analysis, Figure 5.34, proves that when the optimal MR&R methods (action profiles) have been defined by the analyses presented in Section 5.5.3, there is actually nothing else to be optimised in the long-term. There is no need to do a special long-term analysis as the 'optimal long-term distribution' is simply the consequence of applying optimal MR&R methods. It is the MR&R methods (action profiles) that are to be optimised, not the distributions.

Some of the problems in the linear optimisation method can be attributed to inaccurate degradation and repair action models. Many time-related processes in the course of a structure's life time cannot be taken into account by the linear optimisation method. The following problems can be noted to cause biased results in the linear optimisation method:

1. The transition probabilities of degradation for a repaired structure are considered to be the same as those of the original structure. In actual practice the rate of degradation of a repaired structure can be totally different from that of the original structure.
2. The retarding effects of coatings and other preventive maintenance methods are impossible to be described correctly by the degradation models as the degradation rate cannot be changed.
3. The degradation of coatings and other protective systems themselves cannot be considered in the models of structures.

The problems related to coatings and other preventive maintenance methods in the linear optimisation method could be partly avoided by considering that coating is not a repair method, as linked to a certain condition state, but rather an integrated part of the structure. Thus the annual costs of coatings are calculated as maintenance costs in the group of non-repairable structures. The effect of a coating on the degradation rate of the structure can be approximately evaluated based on the quality and reapplication frequency of the coating. However, the same level of accuracy as can be achieved by the LCC analysis method cannot be reached by the linear optimisation method.

5.5.5   Long-term and short-term scenarios of financial needs

The last phase in the network level analyses is the production of long-term and short-term scenarios on the financial needs for the MR&R activity. The need of scenarios of this kind is obvious. Every organisation needs them for long-term and short-term planning.

There are two data sources on which the scenarios on the financial needs can be based: (i) the object level LMS and (ii) the network level LMS. One possibility is to aggregate the cost data of the object level LCC planning, which is performed for a longer period of time. The results of such a procedure would look like the example

Table 5.23 MR&R costs counted from the project plans of 18 bridges

MR&R costs, euro	Discount factor 1.000 / Year 0	0.962 / 1	0.925 / 2	0.889 / 3	0.855 / 4	0.822 / 5	0.790 / 6	0.760 / 7	0.731 / 8	0.703 / 9	0.676 / 10	0.650 / 11	0.625 / 12	0.601 / 13
Annual real costs	0	3 139 477	0	1 232 294	576 854	606 941	820 530	19 709	365 299	2 336	202 675	774 476	233 863	1 283 704
Cumulative real costs	0	3 139 477	3 139 477	4 371 771	4 948 625	5 555 566	6 376 096	6 395 805	6 761 104	6 763 440	6 966 115	7 740 591	7 974 453	9 258 157
Annual PV costs	0	3 018 728	0	1 095 505	493 097	498 861	648 477	14 977	266 921	1 641	136 920	503 085	146 070	770 959
Cumulative PV costs	0	3 018 728	3 018 728	4 114 233	4 607 330	5 106 191	5 754 668	5 769 645	6 036 566	6 038 207	6 175 127	6 678 212	6 824 282	7 595 241
Bridge														
B-43	0	52 992	0	0	0	0	412 013	0	0	0	0	0	0	0
B-46	0	451 208	0	0	0	0	0	0	0	0	0	0	2 424	539 057
B-60	0	157 370	0	0	0	0	0	0	0	0	0	12 860	0	0
B-81	0	610 857	0	0	0	0	0	2 128	276	0	0	0	0	221 343
B-224	0	0	0	0	0	0	0	0	0	0	0	222 559	0	0
B-544	0	10 675	0	0	0	33 010	0	0	0	0	0	0	0	0
B-585	0	159 197	0	0	0	0	0	0	0	0	0	0	0	0
B-639	0	52 177	0	0	0	0	0	0	0	0	0	0	0	0
B-699	0	44 383	0	3 660	9 188	0	0	10 448	321 365	0	0	0	0	4 426
B-950	0	9 160	0	615 366	0	570 194	0	0	3 331	1 168	0	0	0	204 443
B-965	0	163 441	0	0	0	1 868	0	1 441	0	0	0	0	229 015	0
B-1081	0	614 083	0	0	0	0	0	2 135	0	0	0	0	0	293 793
B-1101	0	325 832	0	0	0	0	0	2 117	793	0	0	539 057	0	2 130
B-1131	0	84 043	0	0	0	0	0	0	0	0	0	0	0	14 769
B-1285	0	163 634	0	0	0	1 868	0	1 441	0	0	0	0	0	3 743
B-1602	0	52 992	0	0	0	0	408 517	0	0	0	0	0	2 424	0
B-1616	0	9 160	0	613 268	0	0	0	0	4 689	1 168	201 508	0	0	0
B-5150	0	178 271	0	0	567 666	0	0	0	34 845	0	1 167	0	0	0

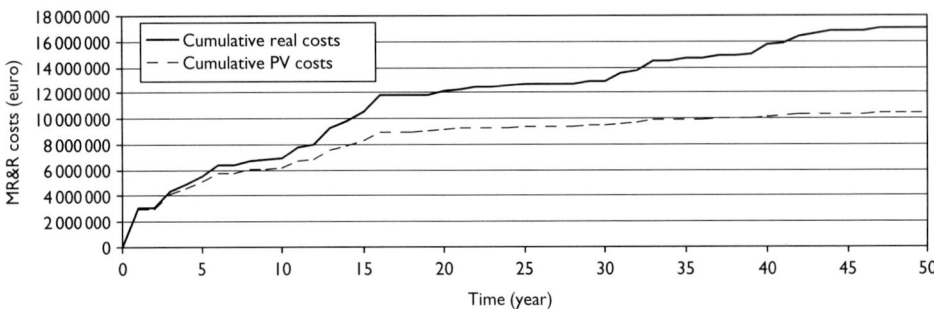

Figure 5.39 Cumulative MR&R costs from 18 bridges.

presented below. In Table 5.23 and in Figure 5.39 the MR&R costs of 18 bridges are aggregated. The corresponding user costs are presented in Table 5.24 and Figure 5.40.

The network level method for drawing up long-term and short-term scenarios on financial needs is based on the aggregation of the data obtained in the long-term and short-term optimisation. By aggregating the cost data as a function of time obtained from the long-term optimisation, long-term scenarios are obtained. By combining the cost data from the short-term optimisation, short-term scenarios can be obtained respectively. Figures 5.41 and 5.42 show examples of cost curves for one analysis group. The final curve representing the whole network of structures is obtained by aggregating the curves for all analysis groups. It is important to note that the surface area by which the unit costs are multiplied in the network level analyses corresponds to the total surface area of components belonging to the analysis group.

As can be observed from Figures 5.41 and 5.42, the calculations performed by the network level analysis method produce smooth curves as opposed to the object level analysis curves, which are rather undulating. The network level curves remain smooth also after aggregation of the analysis groups.

The difference between the long-term and short-term scenarios is that the long-term scenarios express the financial MR&R needs as such while the short-term scenarios take into account the budget limitations and possible financial optimisation performed for analysis groups.

Some corrections have to be made for the network level cost scenarios, however. The corrections are necessary for the following reasons:

1. There is no project planning performed in the network level. So the true costs will probably not be exactly the same as those predicted based on optimal timing of actions.
2. The costs of the unpredictable MR&R actions are not considered in the network level optimisation. So extra costs are expected for these actions.

These corrections can be done based on the comparison of the cost scenarios produced by the object level and network level analyses and the long-term statistics of the

Table 5.24 User costs counted from the project plans of 18 bridges

User Costs, euro	Discount factor													
	1.000	0.962	0.925	0.889	0.855	0.822	0.790	0.760	0.731	0.703	0.676	0.650	0.625	0.601
Year	0	1	2	3	4	5	6	7	8	9	10	11	12	13
Annual real costs	0	5 222 596	0	4 082 971	2 038 872	2 050 750	184 188	2 737	83 772	560	78 250	2 196 456	79 608	2 443 284
Cumulative real costs	0	5 222 596	5 222 596	9 305 567	11 344 439	13 395 189	13 579 377	13 582 114	13 665 886	13 666 445	13 744 695	15 941 151	16 020 760	18 464 044
Annual PV costs	0	5 021 727	0	3 629 747	1 742 836	1 685 567	145 566	2 080	61 211	393	52 863	1 426 776	49 723	1 467 373
Cumulative PV costs	0	5 021 727	5 021 727	8 651 474	10 394 310	12 079 877	12 225 443	12 227 523	12 288 734	12 289 127	12 341 990	13 768 766	13 818 489	15 285 863
Bridge														
B-43	0	52 524	0	0	0	0	92 142	0	0	0	0	0	459	0
B-46	0	96 991	0	0	0	0	0	0	0	0	0	0	0	2 026 170
B-60	0	230 887	0	0	0	0	0	310	0	0	0	597	0	0
B-81	0	2 039 996	0	0	0	0	0	0	9	0	0	0	0	78 606
B-224	0	0	0	0	0	0	0	0	0	0	0	169 689	0	0
B-544	0	6 565	0	0	0	12 391	0	0	0	0	0	0	0	0
B-585	0	62 474	0	0	0	0	0	0	0	0	0	0	0	0
B-639	0	2 801	0	1 011	0	0	0	1 231	79 227	0	0	0	0	845
B-699	0	1 955	0	0	831	2 038 256	0	0	0	0	0	0	0	0
B-950	0	848	0	2 040 980	0	52	0	0	1 019	280	0	0	78 691	78 035
B-965	0	64 878	0	0	0	0	0	288	0	0	0	0	0	0
B-1081	0	2 040 970	0	0	0	0	0	310	25	0	0	0	0	257 942
B-1101	0	284 210	0	0	0	0	0	310	0	0	0	2 026 170	0	310
B-1131	0	51 676	0	0	0	0	0	0	0	0	0	0	0	887
B-1285	0	64 878	0	0	0	52	0	288	0	0	0	0	0	489
B-1602	0	52 524	0	0	0	0	92 046	0	0	0	0	0	0	0
B-1616	0	848	0	2 040 980	0	0	0	0	1 365	280	77 970	0	459	0
B-5150	0	167 574	0	0	2 038 041	0	0	0	2 127	0	280	0	0	0

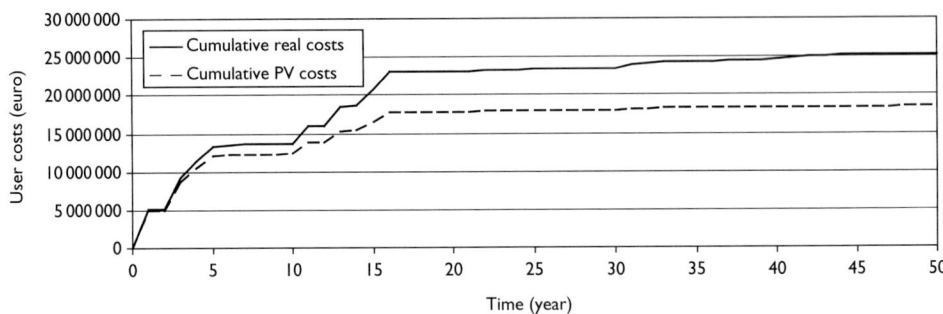

Figure 5.40 Cumulative user costs from 18 bridges.

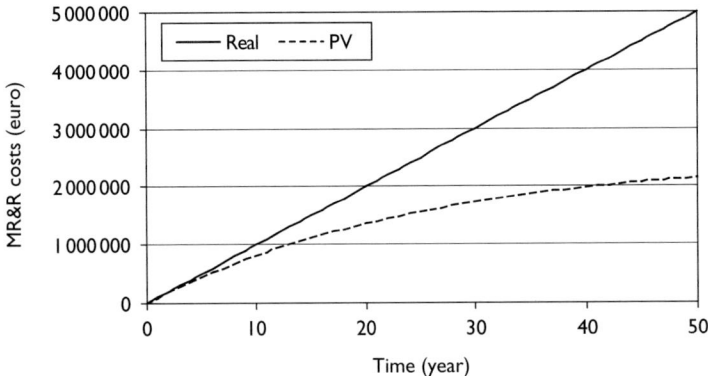

Figure 5.41 Cumulative MR&R cost curve for one analysis group as determined by the long-term optimisation method.

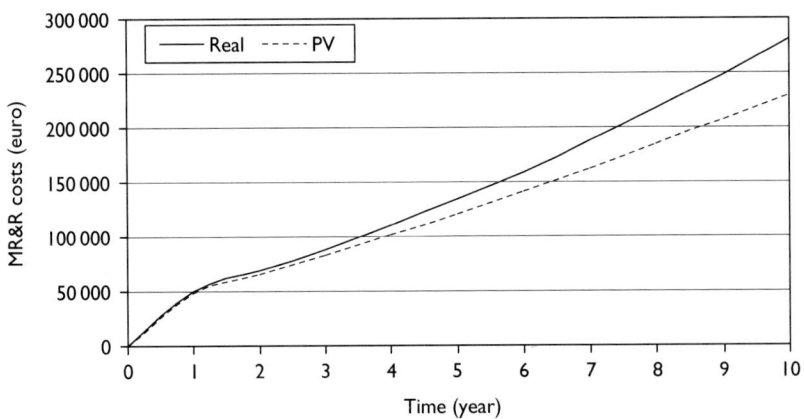

Figure 5.42 Cumulative MR&R cost curve for one analysis group as determined by the short-term optimisation method.

predicted and realised MR&R costs in the past years. Equation (5.20) can be used for evaluating the true annual costs in the future:

$$AC(corrected) = K \cdot AC(predictable, \ uncorrected) + AC(unpredictable) \qquad (5.20)$$

where
    AC   is the annual costs in the cost scenario and
    K   the correction coefficient to take into account the effect of project planning.

Both the correction coefficient K and the annual costs for unpredictable MR&R actions are functions that can be best evaluated after the system has been in use for some years.

5.5.6 Combination of network level LMS with object level LMS

As mentioned earlier the network level system can be used with or without the object level system. However, the full benefit of the network level system can be obtained through the object level system as the strategic planning performed at the network level can only be put into action by assuring that the strategic plans are obeyed at the object level planning too.

Figure 5.43 shows the flowchart of a combined network level and object level management system. Using the same decision trees at both levels of planning is the guarantee that the object level planning is consistent with the network level planning. Also the cost scenarios made at the network level should be in close consistency with the object level scenarios. A checkpoint where this can be verified is presented in Figure 5.43.

Figure 5.43 shows also the process of updating degradation and cost models. At the object level the inspection and condition assessment produces condition data which can be processed at the network level to develop updated degradation models. Likewise the execution of projects produces cost data, which can be studied by suitable analysis methods to produce updated cost models.

5.6 Case: Financial optimisation and lifetime planning in a bridge management system

Erkki Vesikari

5.6.1 Background

The bridge management system consists of both the network level and the object level systems. The object level system includes financial optimisation based on a combined life cycle condition analysis, life cycle cost analysis and life cycle ecology analysis. This system is used in the life cycle planning of maintenance, repair and rehabilitation (MR&R) works for bridges. By life cycle planning one means the MR&R project planning which covers the whole design period of 50–200 years. A design period longer than 100 years is usually applied. The computer programs prepared for the

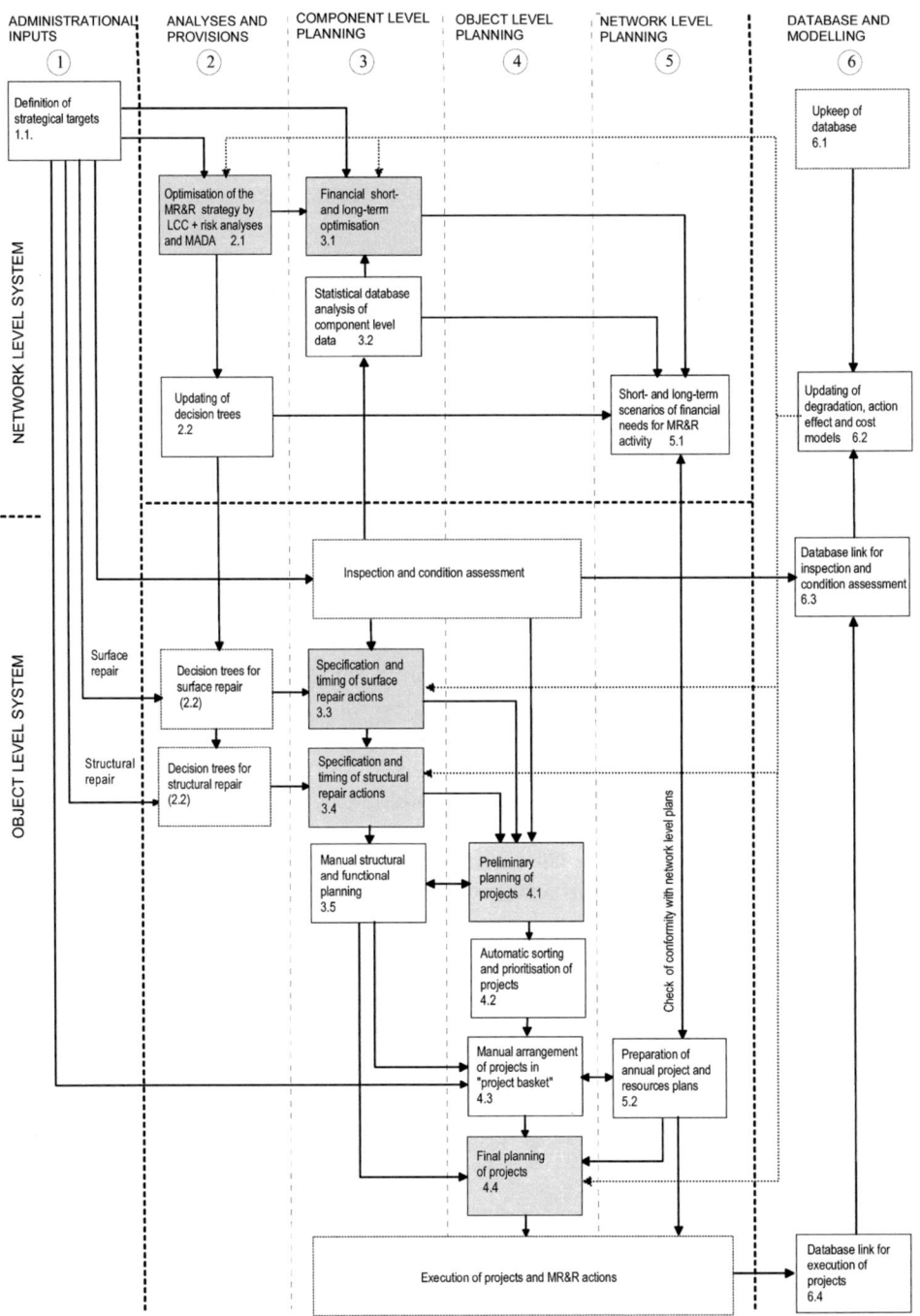

Figure 5.43 Flow diagram of the integrated network and object level LMS.

object level bridge management system include also service life design for new bridge structures.

'Stairs of development', which show schematically the process of development for both the service life design and life cycle planning, are seen in Figure 5.44. The process proceeds by adding new information methods and techniques to the body of systems step by step. At the lower steps the degradation and service life models for concrete structures were developed. The degradation models are a prerequisite for both service life and life cycle design as they together with action effect models give the foundation for the prediction of future condition of structures. The Markov Chain-based condition analysis with automatic triggering of actions is attached to the framework of an LCC analysis by which the life cycle costs can be realistically evaluated. A Markov Chain-based LCC analysis combined with the decision trees by which the optimal MR&R action profiles are defined forms the foundation of an automatic system for life cycle design. At the last step the programs for automatic design of projects are attached to the framework of the bridge management system, giving thus giving the basis for annual project and resources planning and scenarios on MR&R costs.

The first steps of the stairs are explained elsewhere above. In this chapter only the last steps are focussed. The purpose is not to give complete descriptions or instructions of these steps but rather to give by examples of the results an idea how these steps were developed and what was the final outcome of it.

5.6.2 Examples of life cycle cost analysis results

To build up a decision tree for automatic decision-making on MR&R actions, LCC analysis for different structural parts of bridge structures were performed. The purpose of these analyses was to find the optimum MR&R action profile for each component, taking into account the material and structural properties of the structure together with environmental burdens and special requirements pertaining to it. Here the results of an edge beam analysis are presented.

The analysis was performed with the aim of studying various possible maintenance strategies for edge beams of bridges. Usually the edge beams are exposed to rain and heavy chloride burden as de-icing salts are spread on roads. In that case the degradation rate of both the structural concrete and the possible surface treatments on the beam is rapid. On the other hand, the surface treatments may have a great influence on the degradation rate of the structural concrete.

However, if no chlorides are spread on the overpassing road, the chloride burden of the edge beam is low. In the case of low chloride stress the degradation rate of the structural concrete is presumably low and also the surface treatments are long-lasting.

In Tables 5.25 and 5.26 the calculations were performed so that the edge beam was assumed to be repaired every time when the limit condition state of the beam was exceeded by the maximum allowable probability. The edge beam is provided either with a surface treatment or without any treatment. In Table 5.25 the edge beam is exposed to a very high chloride burden. In Table 5.26 the chloride burden is low (no salting).

The tables both have been divided into three parts. The MR&R costs are presented in the upper part A. The user costs are presented in the middle part B and total

Take into use these data, methods and techniques...

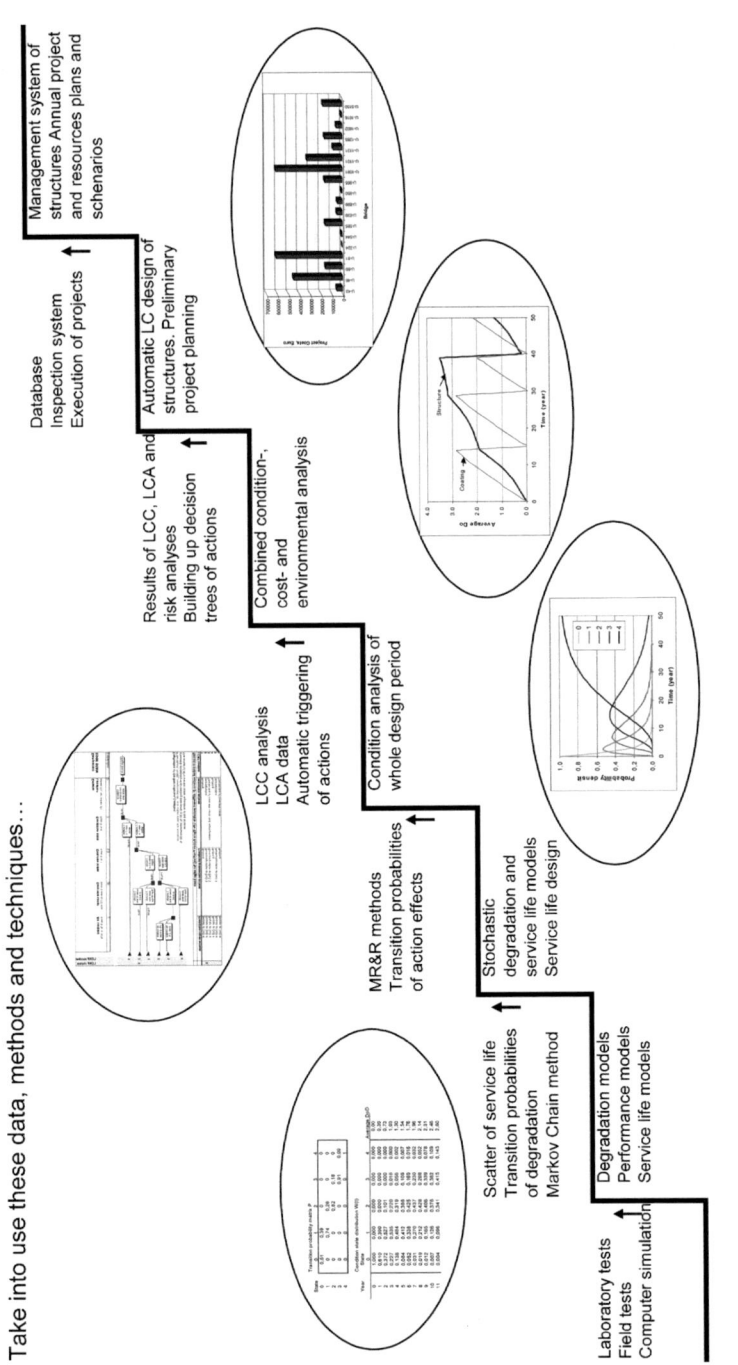

Management system of structures Annual project and resources plans and scenarios

Database
Inspection system
Execution of projects

Automatic LC design of structures. Preliminary project planning

Results of LCC, LCA and risk analyses
Building up decision trees of actions

Combined condition-, cost- and environmental analysis

LCC analysis
LCA data
Automatic triggering of actions

Condition analysis of whole design period

MR&R methods
Transition probabilities of action effects

Stochastic degradation and service life models
Service life design

Scatter of service life
Transition probabilities of degradation
Markov Chain method

Degradation models
Performance models
Service life models

Laboratory tests
Field tests
Computer simulation

...so you can develop these models, tools and systems.

Figure 5.44 'Stairs of development' for an object level bridge management system.

Table 5.25 Average and equalised annual costs of an edge beam with alternative repair methods, high moisture and chloride stress

Repair method Coating	Average annual costs (euro/m/year) Probability of exceeding condition state 3			Equalised annual costs (euro/m/year) Probability of exceeding condition state 3		
	10%	30%	50%	10%	30%	50%
A. MR&R costs						
Shotcrete repair	7.45	9.88	11.78	41.12	31.61	26.60
no coating	9.65	12.33	15.41	30.38	19.24	15.34
silane impregnation	13.15	14.50	17.45	21.19	15.37	11.77
PUR-water membrane						
Repair by casting concrete						
no coating	10.56	13.49	12.27	25.86	13.64	8.12
silane impregnation	12.97	10.74	9.56	18.90	8.13	4.75
PUR-water membrane	10.69	8.47	9.65	16.15	6.34	4.33
Electrochemical removal of chlorides						
no coating	5.76	8.33	10.77	20.91	15.33	11.72
silane impregnation	7.80	10.82	10.28	14.97	10.11	7.30
PUR-water membrane	9.62	9.14	8.78	12.94	8.08	6.06
B. User costs						
Shotcrete repair						
no coating	4.90	4.90	4.90	26.90	15.88	11.05
silane impregnation	5.26	5.26	5.38	18.63	8.87	5.55
PUR-water membrane	5.30	4.78	17.45	11.52	6.07	3.16
Repair by casting concrete						
no coating	4.90	4.90	3.59	11.99	4.94	2.38
silane impregnation	5.44	3.45	2.45	8.26	2.54	1.11
PUR-water membrane	3.68	2.01	1.73	6.17	1.35	0.52
Electrochemical removal of chlorides						
no coating	17.67	17.67	17.67	64.37	32.70	19.07
silane impregnation	18.02	18.17	12.29	40.24	16.53	8.12
PUR-water membrane	18.11	11.00	7.43	27.56	9.28	3.82
C. Total costs (MR&R cost + User costs)						
Shotcrete repair						
no coating	12.35	14.78	16.68	68.02	47.48	37.65
silane impregnation	14.90	17.59	20.79	49.01	28.11	20.89
PUR-water membrane	18.45	19.84	22.24	32.71	21.45	14.93
Repair by casting concrete						
no coating	15.46	18.39	15.87	37.84	18.58	10.50
silane impregnation	18.41	14.19	12.00	27.17	10.67	5.86
PUR-water membrane	14.37	10.49	11.38	22.31	7.69	4.84
Electrochemical removal of chlorides						
no coating	23.42	26.00	28.44	85.29	48.04	30.79
silane impregnation	25.82	28.99	22.57	55.21	26.64	15.42
PUR-water membrane	27.73	20.14	16.21	40.50	17.36	9.88

Table 5.26 Average and equalised annual costs of an edge beam with alternative repair methods, low chloride stress

Repair method Coating	Average annual costs (euro/m/year) Probability of exceeding condition state 3			Equalised annual costs (euro/m/year) Probability of exceeding condition state 3		
	10%	30%	50%	10%	30%	50%
A. MR&R costs						
Shotcrete repair	7.34	8.17	7.01	13.77	6.09	3.12
no coating	8.17	6.41	5.83	10.88	4.25	3.22
silane + acrylic coating	10.76	9.98	8.23	13.60	7.110	5.02
polym.mod.cement coating						
Repair by casting concrete						
no coating	2.79	1.79	1.11	5.68	0.02	0.00
silane + acrylic coating	4.07	2.59	2.57	5.96	2.66	2.66
polym.mod.cement coating	4.78	3.18	3.14	7.77	3.16	3.16
Electrochemical realkalisation						
no coating	2.51	2.00	1.72	3.62	0.85	0.17
silane + acrylic coating	3.85	3.89	3.56	4.92	2.83	2.77
polym.mod.cement coating	4.99	5.11	4.97	6.66	3.63	3.38
B. User costs						
Shotcrete repair						
no coating	4.90	4.25	2.94	9.21	3.16	1.31
silane + acrylic coating	4.42	2.47	1.77	6.07	1.46	0.84
polym.mod.cement coating	5.57	3.67	2.28	7.57	2.40	1.11
Repair by casting concrete						
no coating	1.31	0.65	0.33	2.66	0.01	0.00
silane + acrylic coating	12.97	0.80	0.72	2.28	0.74	0.74
polym.mod.cement coating	10.69	8.47	9.65	16.15	6.34	4.33
Electrochemical realkalisation						
no coating	5.20	2.36	1.42	7.53	1.00	0.14
silane + acrylic coating	3.17	2.13	1.56	5.67	0.80	0.74
polym.mod.cement coating	3.99	2.09	1.55	6.96	0.90	0.60
C. Total costs (MR&R cost + User costs)						
Shotcrete repair						
no coating	12.24	12.41	9.95	22.97	9.26	4.43
silane + acrylic coating	12.58	8.88	7.60	16.95	5.71	4.07
polym.mod.cement coating	16.33	13.66	10.51	21.17	9.52	6.13
Repair by casting concrete						
no coating	4.09	2.44	1.44	8.34	0.03	0.00
silane + acrylic coating	5.65	3.39	3.29	8.23	3.40	3.40
polym.mod.cement coating	6.32	3.93	3.84	10.56	3.75	3.75
Electrochemical realkalisation						
no coating	7.71	4.37	3.14	11.15	1.85	0.32
silane + acrylic coating	7.02	6.02	5.12	10.59	3.63	3.51
polym.mod.cement coating	8.97	7.21	6.52	13.62	4.52	3.99

costs (MR&R costs + user costs) are presented in the lower part C. The costs were calculated as average annual costs and equalised annual costs. The rate of interest for equalised costs was 4%.

In the case of high chloride burden, the analysis results show that the repair by casting concrete is more cost-effective than shotcrete repair, especially when evaluated by equalised costs. By using surface treatments such as silane impregnation or PUR water membraning, the average annual cost can still be lowered. So it is beneficial to use protective surface treatments when the structure is exposed to heavy chloride burden.

Electrochemical removal of chlorides does not seem to be very cost-effective if the user costs are considered in the evaluation, because the process of removal chlorides takes a long time. However, in cases when there is no harm for the traffic, for example in the case of bearing pads, it may be very cost-effective. With increasing deterioration, i.e. with increasing probability for exceeding the limit condition state, the method loses its profitability as the amount of patching work increases.

In the case of low chloride burden, repair by casting concrete is more beneficial than shotcrete repair (as with high chloride burden). However, using surface treatments is not at all so cost-effective as it is with high chloride burden. Both the silane + acrylic coating and the polymer-modified cement coating seem to somewhat increase the average annual costs.

The electrochemical realkalisation is also a cost-effective repair method. It is able to compete with the repair of casting concrete. It can be recommended in such cases when no mechanical removal of concrete is desired.

The purpose of presenting these LCC analysis results is only to demonstrate how the analysis results can be interpreted to get answers to the questions on strategic maintenance. The results of these examples may not be directly applicable to maintenance of edge beams in other countries, as the results are highly dependent on the initial data used. The initial data include the data on unit costs, rate of degradation and protective properties of coatings as well as data on unit costs, rate of degradation and action effects of repair methods.

5.6.3 Decision tree

The final purpose of the LCC analyses integrated with risk analyses and MADA (used in multiple attribute optimisation) is to find the optimal maintenance policy, i.e. the optimised MR&R action profiles, for each combination of material, structural, environmental, etc. parameter values. These optimum MR&R action profiles are arranged into a decision tree, which will then be used in life cycle planning of specific bridge components. These planning routines are designed to be performed fully automatically.

Figure 5.45 shows an example of a decision tree. A decision tree has a 'root', which forks at 'nodes' representing the relevant criteria (related to e.g. material, structural and environmental conditions) and makes with a growing number of nodes an ever-increasing amount of 'branches'. The final branches after the last nodes are called 'leaves'. The optimum MR&R action profiles are the results of the tree and are inserted in the leaves of the tree.

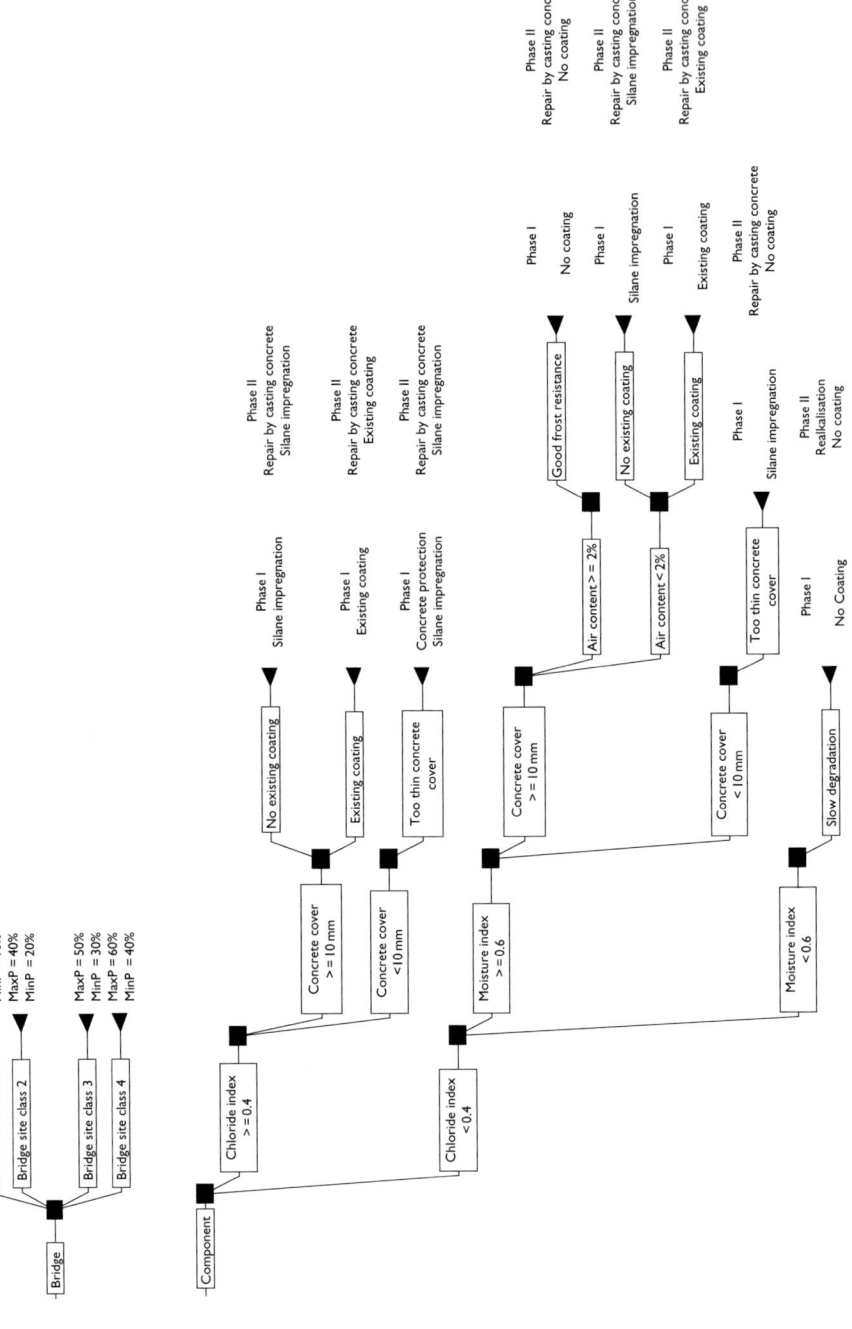

Figure 5.45 Example of a decision tree.

A decision tree can be programmed using normal program code. In normal code language the branches of the decision tree are implemented by nested IF-THEN blocks.

5.6.4 Life cycle planning program of concrete bridge structures

The program for automatic life cycle planning of concrete bridge structures is meant for bridge owners, maintainers, repair engineers and designers of MR&R planning. With the help of the program the condition of bridge structures can be predicted, the required MR&R actions can be specified and scheduled, and the MR&R costs can be evaluated within a defined design period. The costs evaluation includes actual MR&R costs, user costs and environmental impacts.

The program for automatic life cycle planning of bridges is an Excel-based program which uses an initial data file that is prepared by special database routines from the data of the Bridge Register (database). The data of the initial data file comprises of structural, material, environmental and condition assessment data.

The automatic design program uses decision trees which select the optimal MR&R action profiles for structural components based on the specific properties, environmental burdens and condition of the component. Although the preliminary project planning is done fully automatically the program allows also manual intervention for manual corrections and changes to the automatically prepared plans.

In Figure 5.46 the Initial data definition form for life cycle planning is presented. For the planning of a specific bridge, the designer simply selects the object from the

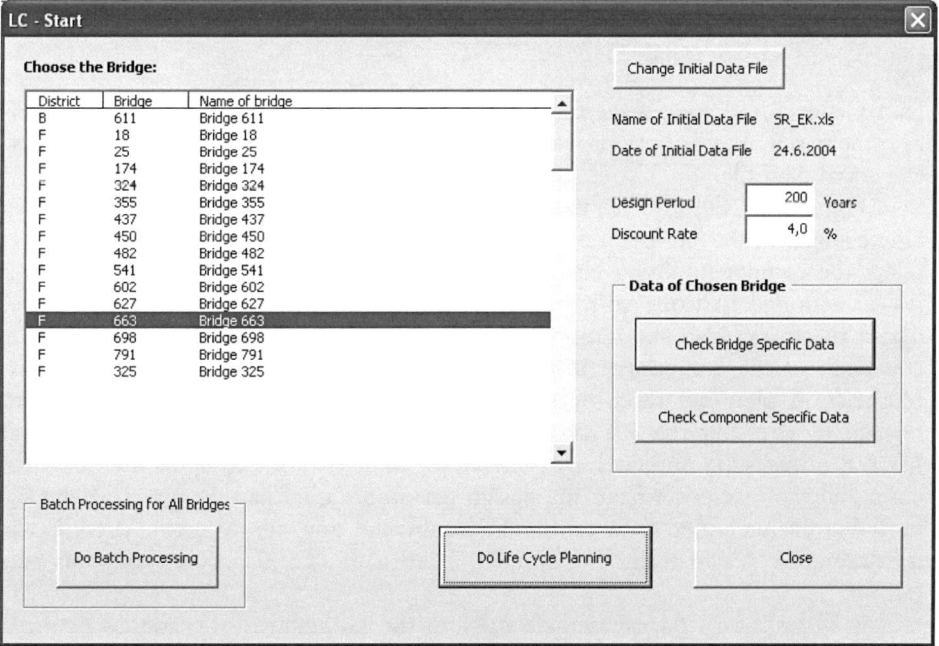

Figure 5.46 Initial data definition form for life cycle planning.

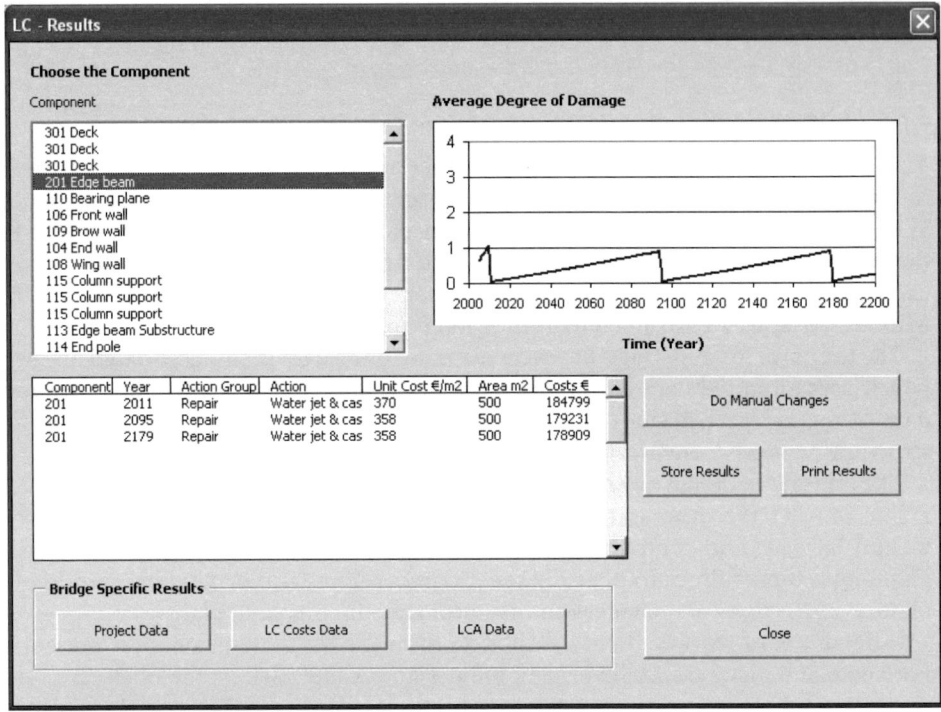

Figure 5.47 Results form for life cycle planning.

list of bridges and presses the button 'Do Life Cycle Planning'. The program allows also the batch process option, which is used for automatic planning of all bridges of the initial data file.

In Figure 5.47 one can see the Results form of the life cycle planning (for a specific bridge). From the list of components in the upper left corner, the designer can select the component for a closer study. The planned MR&R action profile (the list of action definitions with timings) for that component is shown in the middle of the form. Also the figure in the upper right corner that shows the average condition of the component during the design period is updated. In the bottom of the form there are three buttons which report integrated data on the selected bridge. By pressing the left most button costs data and timings of the coming MR&R projects are obtained. By pressing the middle button a report on the MR&R costs and the user costs from the design period are obtained. By pressing the right most button integrated data on the environmental impacts from the MR&R activity during the design frame is obtained. These data can also be obtained as paper output.

If it is desired to do manual changes to the automatically prepared plans, the designer may press the button 'Do Manual Changes'. Then a form presented in

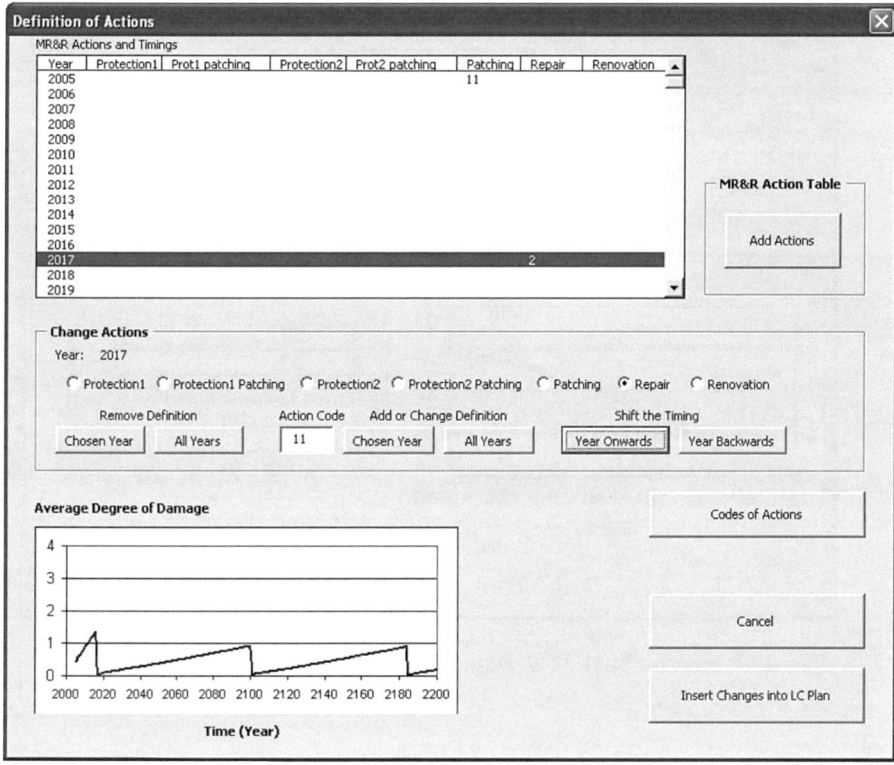

Figure 5.48 Form for Manual life cycle planning.

Figure 5.48 opens. On this form the designer can change the definitions of MR&R actions or the timings of them. The designer can also remove all the previous definitions of actions and define a completely 'own' MR&R action profile with fixed timings of actions. The changes in the design are inserted in the life cycle plan of the bridge by pressing the button 'Insert Changes into LC plan'.

The life cycle planning program for bridges contains also a program package for service life design. With this part of the program a new component can be designed so that it fulfils the performance requirements set for it during its service life. The service life design is based also on the Markov Chain condition analysis, i.e. the degradation models used in the service life design are the same as those used in the life cycle planning.

A 'service Life Design' form is presented in Figure 5.49.

5.6.5 Results of life cycle planning

An example table and graphic schemes of the results of a life cycle planning of life cycle MR&R costs are presented in Table 5.27.

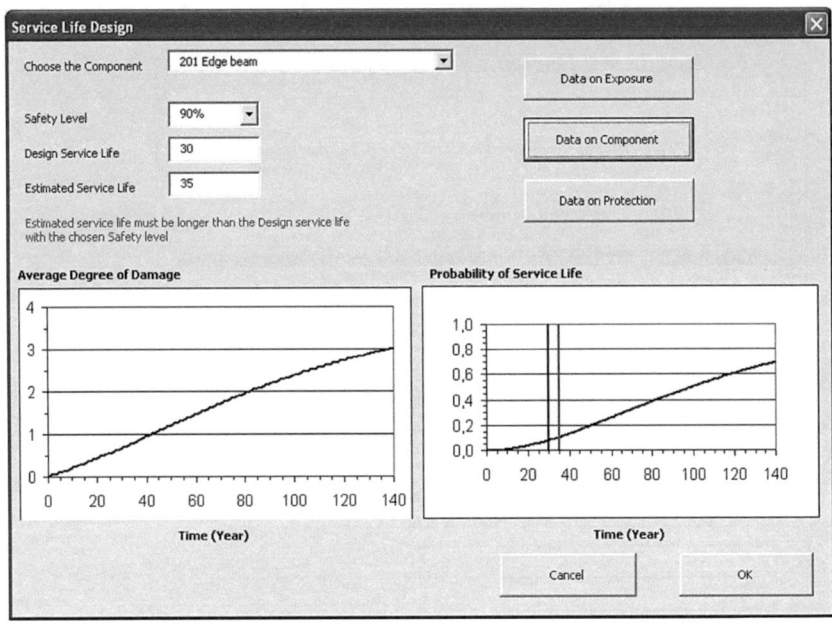

Figure 5.49 Form for Service life design.

Table 5.27 An example of the results of a life cycle planning of MR&R costs

PROJECT COSTS
Date of Analysis   13.4.2005
Bridge   Bridge 663
Bridge _id   8499
Road District   F
Number   663

First Project

		2011	
		Total Costs (euro)	
	Real Costs	Present Value Costs	
MR&R Costs	1 175 374	928 915	
User Costs	4 285	3 386	
Total Costs	1 179 659	932 301	
Delay Costs	19 032	14 463	

Second Project

		2017	
		Total Costs (euro)	
	Real Costs	Present Value Costs	
MR&R Costs	200 404	125 171	
User Costs	792	495	
Total Costs	201 195	125 666	

Next Rehabilitation Project

2039
Total costs (euro)

	Real Costs	Present Value Costs
MR&R Costs	558 971	147 318
User Costs	1 543	407
Total Costs	560 513	147 724

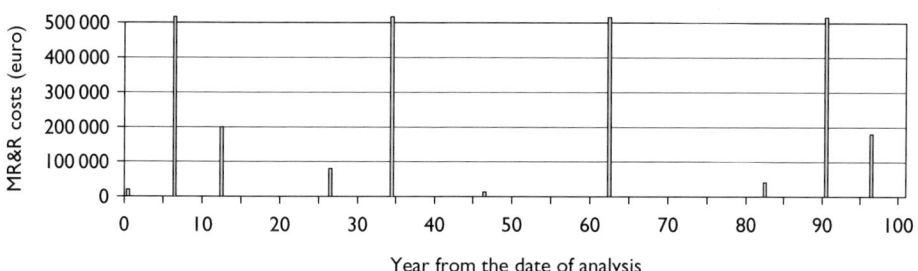

Year from the date of analysis

LIFE CYCLE COSTS

Date of Analysis	13.4.2005
Bridge	Bridge 663
Bridge_id	8499
Design Peiod, Years	200
Discount Rate, %	4
Road District	F
Number	663

MR&R Costs

Cumulative Real Costs	5 747 387 euro
Cumulative PV Costs	1 329 677 euro
Average Unit Costs	28 737 euro/year
Equalised Unit Costs	53 208 euro/year

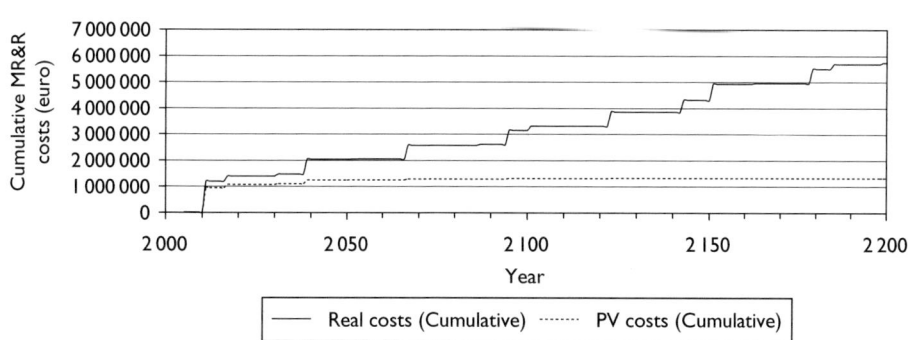

Year

—— Real costs (Cumulative)   ········ PV costs (Cumulative)

User Costs

Cumulative Real Costs	19 336 euro
Cumulative PV Costs	4 630 euro
Average Unit Costs	97 euro/year
Equalised Unit Costs	185 euro/year

Table 5.27 (Continued)

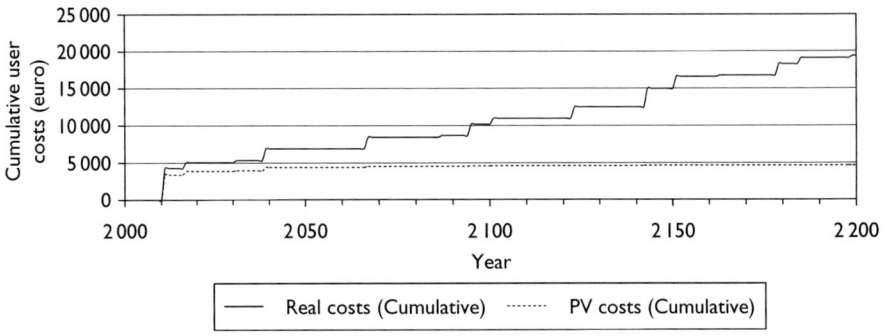

Environmental Impacts
Date of Analysis	13.4.2005
Bridge	Bridge 663
Bridge_id	8499
Design Period, Years	200
Road District	F
Number	663

Non-Renewable Energy	68834	GJ
Renewable Energy	4880.7	GJ
CO_2	6149	1000 kg
SO_2	11454	kg
NO_X	29923	kg
Particles	7151	kg
CH_4	10501	kg
VOC	2390.6	kg
Mineral Raw Materials	51.43	1000 kg
ELU	427445	euro

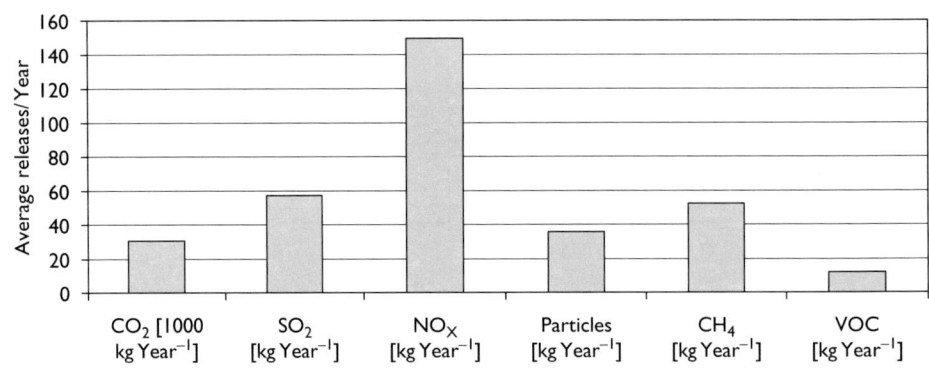

Date	13.4.2005
Bridge	Bridge 663
Bridge_id	8499
Road District	F
Number	663

Code of Component	Year	Code of Action	Action Group	Action	Unit Costs (euro m^{-2})	Area (m^{-2})	Costs (euro)
301	2010	701	Protection1	Membrane waterpro	124	1192	147963
301	2010		Protection2	Milling and concrete	228	1192	271137
301	2010	105	Patching	Patching of deck	43	357	15390
301	2010	701	Protection1	Membrane waterpro	125	10	1201
301	2010		Protection2	Milling and concrete	230	10	2202
301	2010	105	Patching	Patching of deck	48	3	138
301	2010	701	Protection1	Membrane waterpro	125	300	37594
301	2010		Protection2	Milling and concrete	230	300	68891
301	2010	105	Patching	Patching of deck	48	90	4313
201	2010	102	Repair	Water jet & casting	370	500	184799
110	2010	603	Protection1	Maurer service	28	11	313
110	2010		Protection2	Maurer renewal	123	11	1375
106	2010	603	Protection1	Maurer service	0	78	0
109	2010	603	Protection1	Maurer service	0	28	0
104	2010	102	Repair	Water jet & casting	378	211	79683
108	2010	102	Repair	Water jet & casting	372	26	9663
113	2010	102	Repair	Water jet & casting	370	16	5906
114	2010	102	Repair	Water jet & casting	370	7	2765
302	2010	102	Repair	Water jet & casting	387	574	222173
302	2010	102	Repair	Water jet & casting	387	574	222173
303	2010	102	Repair	Water jet & casting	388	80	30992
309	2010	102	Repair	Water jet & casting	387	6	2251
110	2030	603	Protection1	Maurer service	29	11	324
106	2030	603	Protection1	Maurer service	0	78	0
109	2030	603	Protection1	Maurer service	0	28	0
115	2030	102	Repair	Water jet & casting	757	36	27256
115	2030	102	Repair	Water jet & casting	757	46	34827
115	2030	102	Repair	Water jet & casting	757	20	15142
301	2038	701	Protection1	Membrane waterpro	116	1192	137719
301	2038		Protection2	Milling and concrete	212	1192	252366
301	2038	105	Patching	Patching of deck	18	357	6498
301	2038	701	Protection1	Membrane waterpro	121	10	1159

Table 5.27 (Continued)

Code of Component	Year	Code of Action	Action Group	Action	Unit Costs (euro m^{-2})	Area (m^{-2})	Costs (euro)
301	2038		Protection2	Milling and concrete	222	10	2 125
301	2038	105	Patching	Patching of deck	36	3	104
301	2038	701	Protection1	Membrane waterpro	121	300	36 278
301	2038		Protection2	Milling and concrete	222	300	66 479
301	2038	105	Patching	Patching of deck	36	90	3 242
110	2038	119	Repair	Realkalisation	118	11	1 326
106	2038	102	Repair	Water jet & casting	366	78	28 706
110	2050	603	Protection1	Maurer service	25	11	282
106	2050	603	Protection1	Maurer service	0	78	0
109	2050	603	Protection1	Maurer service	0	28	0
109	2050	102	Repair	Water jet & casting	370	28	10 350
301	2066	701	Protection1	Membrane waterpro	114	1192	136 212
301	2066		Protection2	Milling and concrete	209	1192	249 604
301	2066	105	Patching	Patching of deck	15	357	5 405
301	2066	701	Protection1	Membrane waterpro	121	10	1 163
301	2066		Protection2	Milling and concrete	222	10	2 132
301	2066	105	Patching	Patching of deck	38	3	109
301	2066	701	Protection1	Membrane waterpro	121	300	36 396
301	2066		Protection2	Milling and concrete	222	300	66 695
301	2066	105	Patching	Patching of deck	38	90	3 402
110	2066	603	Protection1	Maurer service	25	11	283
106	2066	603	Protection1	Maurer service	0	78	0
109	2066	603	Protection1	Maurer service	0	28	0

References

[1] EN 1990:2002. Eurocode – Basis of structural design. CEN: European Committee for Standardisation. Ref. No. EN 1990:2002 E, pp. 1–87.

[2] RIMES, Road Infrastructure Maintenance Evaluation Study. 1999. Project for EC-DG-VII RTD Programme – Contract No. RO-97-SC 1085/1189, Pavement and structure management system, work package 3, Network level management model, final report.

[3] LIFECON, Life cycle management of concrete infrastructures for improved sustainability. 2003. Project for EC-GP-V-RTD, TRA 1.9 Infrastructures, contract No. G1RD-CT-2000-003788, Deliverable D1.1. Generic technical handbood for a predictive life cycle management system of concrete structures (Lifecon LMS), final report. http://lifecon.vtt.fi/.

[4] LIFECON, Life cycle management of concrete infrastructures for improved sustainability. 2003. Project for EC-GP-V-RTD, TRA 1.9 Infrastructures, Contract No. G1RD-CT-2000-003788, Deliverable D3.1. Prototype of condition assessment protocol, final report. http://lifecon.vtt.fi/.

[5] Vesikari, E. 2002. The effect of coatings on the service life of concrete facades. *Proceedings of the 9th international conference on durability of building materials and components.* Brisbane, Australia, 17–21 March, pp. 1–10.

[6] LIFECON, Life Cycle Management of concrete infrastructures. 2003. Deliverables D1-D15. http://lifecon.vtt.fi/. Deliverable D2.2. Statistical condition management and financial optimisation in lifetime management of structures. Part 1: Markov chain based LCC analysis. Part 2: Reference structure models for prediction of degradation, final report, pp. 1–133. http://lifecon.vtt.fi/.

[7] ISO 15686: *Buildings and constructed assets – Service life planning – Part 5: Whole life costing.*

[8] ASTM E 917. 1994. Measuring life cycle costs of buildings and building systems. American Society for Testing and Materials, pp. 1–12.

[9] Walls, J. and M. R. Smith. 1998. Life cycle cost analysis in pavement design. *Interim technical bulletin.* Federal Highway Administration. FHWA-SA-98-079, pp. 1–107.

[10] Bridge Management Systems. 1987. National co-operative highway research program report 300. Transportation Research Board, pp. 1–74.

[11] Siemes, A., A. Vrouwenvelder, and A. Van Den Beukel. 1985. *Durabiltiy of buildings: A reliability analysis, Heron* 30(3): 1–48.

[12] ASTM E 1765. 1998. Applying Analytical Heirarchy Process (AHP) to multi-attribute decision analysis of investments related to building and building systems. American Society for Testing and Materials, pp. 1–13.

[13] Sarja, A., and E. Vesikari (eds). 1996. Durability design of concrete structures. *RILEM report of TC 130-CSL. RILEM report series 14.* London: E&FN Spon, Chapman & Hall, pp. 1–165.

[14] Systematic approach to Environmental Priority Stages in product development (EPS). 1999. Version 2000 – models and data for the default method. Chalmers University of Technology, Environmental System Analysis. CPM Report, 5.

[15] Thompson, P. D., E. P. Small, M. Johnson, and A. R. Marshall. 1998. The Pontis bridge management system. *Structural engineering international* 4: 303–308.

[16] Söderqvist, M.-K., and M. Veijola. 1999. Finnish project level bridge management system (IBMC99-055). *Proceedings, volume II, F5, TRB international bridge management conference,* Denver, Colorado, April 26–28, pp. 1–7.

[17] Bevc L., I. Peruš, B. Mahut, and K. Grefstad. 2001. *Bridge management systems: Review of existing procedures for optimisation.* BRIME PL97-2220. Deliverable D3.

[18] ENV 1504-9:1996. Products and systems for the protection and repair of concrete structures – Definitions, requirements, quality control and evaluation of conformity. Part 9: General principles for the use of products and systems. European Committee for Standardisation, pp. 1–22.

Chapter 6

Vision for future developments

Asko Sarja

6.1 Description of the framework of lifetime engineering

6.1.1 Background and societal needs

Civil engineering (buildings, civil and industrial infrastructures and environmental engineering), even more than other branches of technique, is strongly tied to the surrounding society [1, 2] (see Figure 1.1).

The aims towards sustainable building are creating for the construction new challenges, which can be fulfilled successfully only through active and innovative changes in the design, products, manufacturing methods and management. In the time period of 10–15 years we can talk on an entirely new generation of building technology. This evolution phase has started in the second part of 1980s and will be implemented into the common practice until 2010–2015.

Lifetime engineering is a new challenge for engineers, who must be educated and trained in this skill. Our intellectual resources must be increasingly directed through renewed education and training to the work towards economically, culturally and ecologically sustainable, healthy, safe and convenient infrastructures.

Working conditions and safety will be an important aspect of the development of construction, maintenance and repair of infrastructures. The well-performing, healthy and safe infrastructures are indirectly increasing working safety especially of traffic, industrial production and power plants as well as their neighbourhood areas.

Future regulations will demand designers to consider not only how the buildings and civil infrastructures can be constructed but also, and equally importantly, how they will be maintained and even demolished and reused. Building owners, users and designers have to be aware of the implications of premature failure of performance of a part of the building in which a business or process operate. The methodology of lifetime engineering expects designers to understand and articulate to the client and user about the risks of loss of performance and propose means of minimising the occurrence of the risks.

6.1.1.1 Business development

Building and civil engineering will be developed to a new level by applying modern technology and knowledge to the Research and Development (R&D) of materials, structures, manufacturing and construction processes, facility management, health monitoring of assets, etc., and this all supported with modern information communications technology (ICT), automation, expert systems, virtual reality, fuzzy systems,

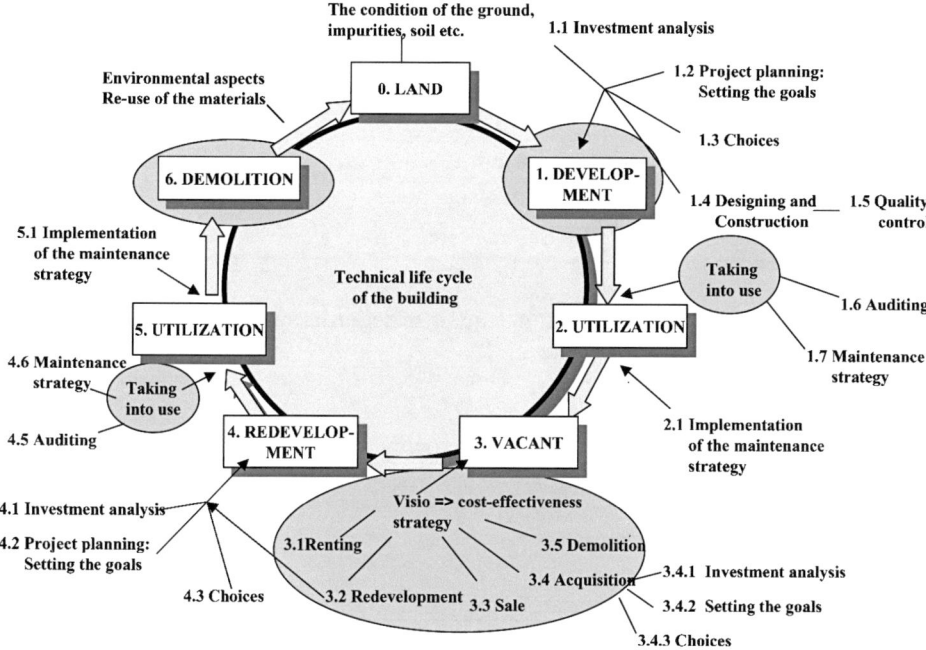

Figure 6.1 Life cycle of a real estate [4].

neural networks, etc. [1, 3]. The current fragmented and sequential building and management process will change to an integrated lifetime-oriented design-construct-asset management process in value added networks of firms and other stakeholders, combined with new goals and requirements.

A special challenge for investors and owners is the management of the life cycle of a facility to guarantee the future value of facilities under continuous changing environment of use. This means that there are several re-development phases during the survival life of a facility, as shown in Figure 6.1 [4].

6.2 Framework of the lifetime engineering

6.2.1 Generic requirements

The integrated lifetime engineering methodology aims at regulating optimisation and guaranteeing lifetime quality, which represents the generic requirements of sustainable construction and the specific requirements of owners and users with technical performance parameters. With the aid of lifetime engineering, we can thus control and optimise the human conditions (functionality, safety, health and comfort), the monetary (financial) economy and the economy of nature (ecology), also taking into consideration the local cultural compatibility (see Figure 2.1). The lifetime economy is expanded into two economical levels: monetary economy and ecology, which means the economy of nature [2].

6.2.2   *Technical and economic components of lifetime quality*

At the practical level of lifetime engineering, the generic requirements can be treated with a number of technical factors. These can be used in analysing the components of lifetime quality [5]. The central life cycle quality components of a facility are:

- functional usability in targeted use
- adaptability in use
- changeability during use
- reliable safety
- health
- durability, which means resistance against degradation loads
- resistance against obsolescence and
- ecological efficiency.

On a practical level, the generic requirements can be classified in different ways, for example [6]:

- mechanical resistance/stability
- safety, in case of fire
- hygiene, health (indoor climate)
- safety in use
- acoustics
- energy use
- durability
- robustness
- lifetime usability
- architecture
- life cycle cost and
- environmental load.

These correspond to the six essential requirements of the EU Construction Product Directive of EU, 1988:

1. mechanical resistance and stability
2. safety, in case of fire
3. hygiene, health and the environment
4. safety in use
5. protection against noise and
6. energy economy and heat retention.

In buildings, the compatibility and easy interchangeability between load-bearing structures, partition structures and building service systems is important. Regarding the life cycle ecology of buildings, the energy efficiency of the building is an important factor. Envelope structures are responsible for most of the energy consumption, and therefore envelopes must be durable and have an effective thermal insulation and safe, static and hygrothermal behaviour. The internal walls have a more moderate length of service life but they have the requirement to cope with relatively high degrees of change, and must therefore possess good changeability and re-usability. In the production phase, it is important to ensure the effective recycling of the production wastes in factories and on site. The final requirement is to recycle the components and materials after demolition. Obsolescence of buildings is either technical or functional, sometimes even aesthetic in nature. Technical and functional obsolescence is usually related to the primary lifetime quality factors of structures. Aesthetic obsolescence is usually architectural in nature [7].

Civil engineering structures like harbours, bridges, dams, off-shore structures, towers, cooling towers, etc. are often massive and their target service life is long. Their repair works under use are difficult. Therefore, their life cycle quality is tied to high durability and easy maintainability during use, saving of materials and selection of environmentally friendly raw materials, minimising and recycling of construction wastes and finally recycling of the materials and components after demolition. Some parts of the civil engineering structures like waterproof membranes and railings have a short or moderate service life and therefore the aspects of easy re-assembly and recycling are most important. Technical or performance-related obsolescence is the dominant reason for demolition of civil engineering structures, which raises the need for careful planning of the whole civil engineering system, e.g. the traffic system, and for selection of relevant and future-oriented design criteria.

6.2.3 *Framework of lifetime engineering*

The framework and content of the lifetime engineering will be as described in Section 2.3.2.

The development of lifetime-oriented integrated processes has to be supported with effective international standardisation work. This work will result in an integrated framework of lifetime engineering, as presented in Figure 6.2.

Efficient maintenance, repair and modernisation of infrastructures means investing more work in saving natural resources and energy, and avoiding environmental

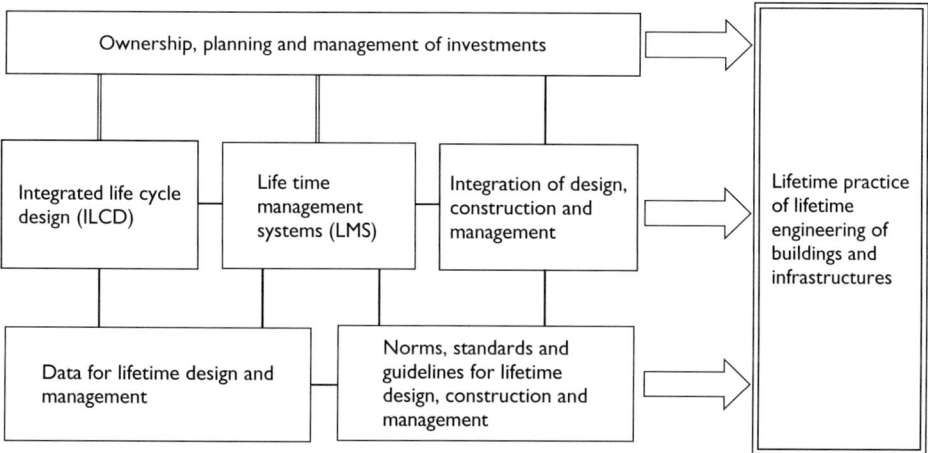

Figure 6.2 Context of the practice of Lifetime Engineering.

burdens and waste. Because maintenance and repair are work-intensive areas of production, this means movement of jobs from new construction to jobs in maintenance and repair. Maintenance and repair works require a high level of skill from the workers, which raises the need for additional training of the workers.

6.3 Process of lifetime engineering

6.3.1 Properties of the lifetime engineering process

Lifetime engineering means a management process over entire life cycle, which is [8]:

- *Predictive* – Future usability, economy, ecology and cultural aspects are evaluated, modelled and used as criteria for selections between alternative solutions and products in all phases.
- *Creative* – Alternative solutions and technologies are created and found at all phases of the process.
- *Optimising* – Comparisons between alternative solutions and products are made with rational methods of applying the criteria, which correspond to the generic criteria on techno-economic and architectural levels.
- *Integrated* – Knowledge and expertise of the disciplines of all stakeholders and of all phases of the life cycle are combined together. This also means effective combination of management with technology in all phases of the lifetime process.

6.3.2 Phases of the integrated lifetime engineering process

The management process of lifetime engineering includes following phases:

- lifetime investment planning;
- integrated lifetime design;
- integrated lifetime responsibility procurement (lifetime contract), or traditional construction contract;
- integrated lifetime management and maintenance planning;
- rehabilitation and modernisation; and
- end-of-life management

    - recovery, reuse
    - recycling and
    - disposal.

6.3.2.1 Investment planning, architectural and technical planning

Lifetime investment planning and decision-making is also called 'Value Management' (Figure 6.3) – a service which maximises the functional value of a project by managing its development from concept to completion and commissioning through the audit (examination) of all decisions against a value system determined by the client. In this phase, the client transmits a clear statement of the value requirements of the

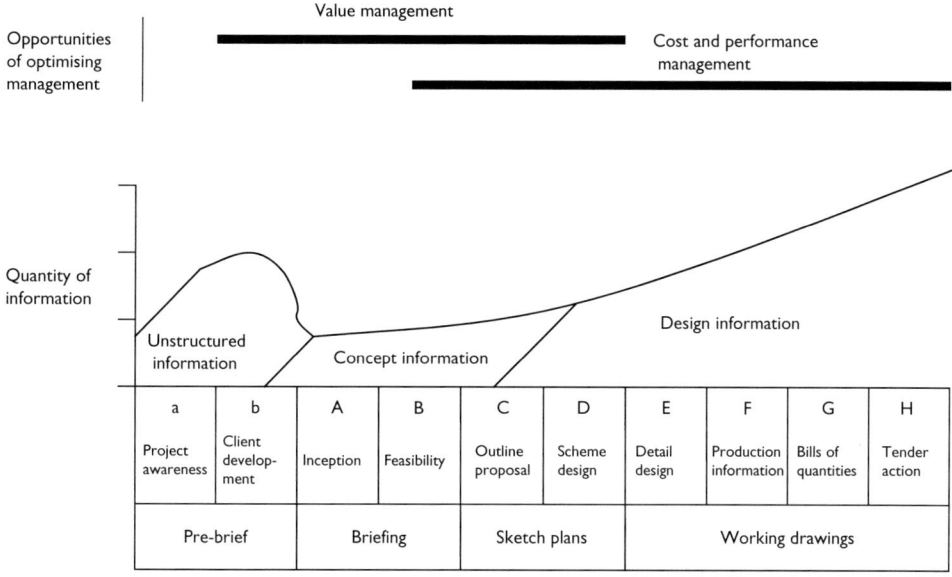

Opportunities of optimising management

Quantity of information

Figure 6.3 Management of lifetime quality during the design process (Modified from Kelly and Male [9]).

project to the designers [9]. The investment planning and decision-making applies value management to audit and optimise:

* the client's use of a facility in relation to its corporate strategy
* the project brief
* the emerging design and
* the production method.

Typical features of value management in lifetime engineering are:

* a proactive and predicting lifetime approach through the use of multi-disciplinary (economics, architecture, structural engineering, building service systems engineering) team-oriented creative process to generate alternatives to the investment solution;
* the use of structured systems method; and
* the relationship of function with value.

   The owner/client defines life cycle objectives like area and functional requirements of a building and its spaces, economy, requirements for use, service life, aesthetic objectives and ecological objectives. Designers in co-operation with the owner/client create alternative investment plans, make a multiple-criteria analysis and decision between alternatives.

The planning process is carried out on following levels:

- Level 5: Concept (Task)

    - represents the first stage wherein the client organisation perceives a problem
    - this problem may be realised through a study of efficiency, safety, markets, profitability, etc.

- Level 4: Spaces

    - the stage where the architect or the whole design team are engaged in the preparation of the brief in conjunction with the client

- Level 3: Technical systems

    - specifications of the performance properties of technical systems (structural system, building service systems)

- Level 2: Modules

    - the stage where the building assumes a structural form

- Level 1: Components

In the concept level, analysis and decision-making happens, which starts from the pre-brief phase and also continues partly in the briefing phase. In the spaces level, the sketching phase is dominant, but technical systems are already included. Finally in the phase of working drawings, the modules and components levels are the central objects of detailed design. Organisationally, this means especially in the first phases of the concept, spaces and technical systems levels, there is a high need of multi-disciplinary planning team, including investor, owner (if different from investor), user (if possible), architect, technical experts of design, operation and MR&R.

In the current management process the optimisation procedure can be adopted in the form of effective computer programs, besides the existing design programs of the stakeholders (Figure 6.4). Software programs contain the optimising methods of multi-attribute analysis and decision-making in different versions for different life cycle phases and management levels, which are presented in the earlier chapters of this book. The differences between these versions are different variables, corresponding to the levels of hierarchy [8].

The components of lifetime quality can be defined differently in each case, but some general checklists can be used. An example of these is the following list:

- Safety, health and comfort

    - Internal air quality (emissions, fungi)
    - Acoustic and visual privacy and convenience
    - Hygrothermal quality of internal conditions
    - Visual quality and aesthetics
    - Working conditions during construction

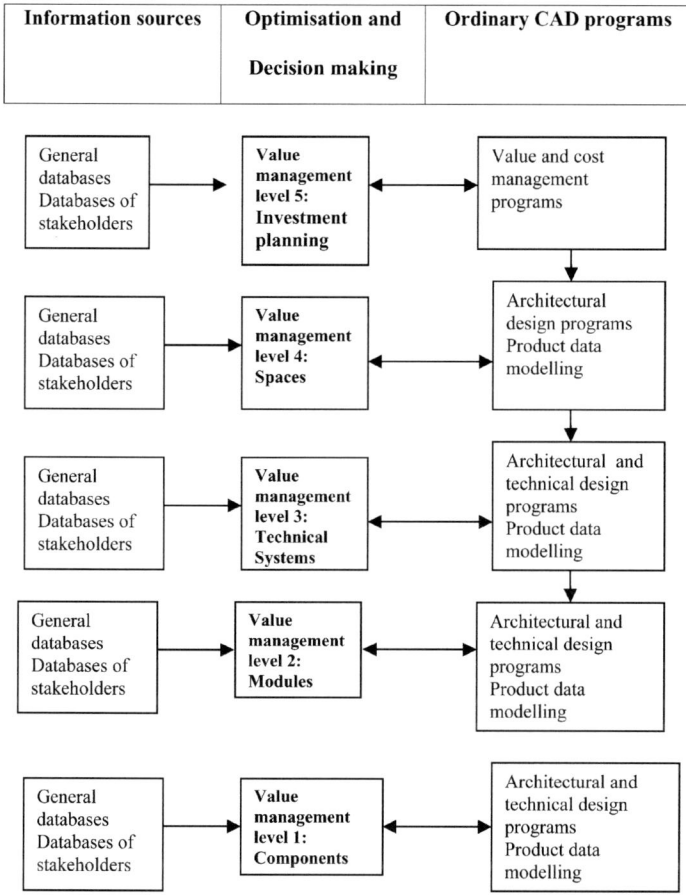

Information sources	Optimisation and Decision making	Ordinary CAD programs

Figure 6.4 Interaction between databases, optimising and decision-making programs, and general managerial, planning and design programs [8].

- Life cycle monetary cost (LCC)
  - Construction cost (40–60% of LCC)
  - Costs during the period of use (50 years: 60–40% of LCC). Following are the costs during design service life:
    - Maintenance cost
    - Repair cost
    - Changing cost
    - Renewal cost
    - Energy cost
      - Recovery + Reuse
      - Recycling
      - Disposal

- Lifetime functionality (usability)

  - Functionality for the first user
  - Flexibility in changes of building services
  - Flexibility in changes of spaces
  - Flexibility in changes of quality level of lifetime quality components of the building
  - Flexibility in changes in performance of structures

- Lifetime maintainability
- Reliability in operation in normal and abnormal conditions

  - Ease
  - Frequency
  - Staff requirements

- Environmental effectiveness of the life cycle (LCE)

  - Consumption of energy in use (heating + lighting) – the most significant factor ($\sim 90\%$)
  - Consumption of energy in production ($\sim 10\%$)
  - Consumption of raw materials: renewal/non-renewal
  - Production of pollutants and disposal into air, soil and water
  - Impact on biodiversity and geodiversity

- Energy Efficiency classification:

  - Class 5: Standard level – heating plus cooling energy economy fits the current standards of each country or region
  - Class 4: Reduced energy level – less than 50% of the current level
  - Class 3: Low energy level – less than 25% of the standard level
  - Class 2: Zero energy level – heating + cooling energy consumption is zero
  - Class 1: Plus energy building – the gain of solar or other natural energy is more than needed for heating and building service systems

- Cultural aspects in harmony with local:

  - building traditions
  - life style
  - business culture
  - aesthetics
  - architectural styles and trends
  - image of the owner and user of a facility.

6.3.2.2   Integrated lifetime design

Central tasks of the lifetime design process after the value engineering (value management) phase are [2]:

- transferring the requirements of owners, users and society (environment included) into functional and technical specifications of materials and structures;
- modular service-life planning and optimisation;

- performance-based design of materials and structures, including service life design (durability); and
- design for the reuse of components and for recycling of materials.

A summary of the integrated lifetime design process is presented in Table 6.1 [2].

During lifetime planning, a modular method is preferred. This allows a systematic allocation and optimisation of the target service life as well as life cycle economy and ecology of different parts of the building [3]. A suited modularisation at the highest level of hierarchy is the following: bearing frame, envelope, foundations, partitions, heating and ventilating services, information, water and sewage system, control services and waste management system. All of these assemblies are specified

Table 6.1 Phases and methods of the integrated lifetime design process [2]

Design phase	Tasks
1. Investment planning	• Set objectives of the building project • Define the study period • Create alternative investment plans • Calculate life cycle costs (monetary economy and ecology) • Calculate cash flows of alternatives • Evaluate benefits of the alternatives • Compare LCCs and make final decision • Define final objectives
2. Analysis of client's and user's needs	• Identify relevant attributes (customer's requirements) • Estimate the rate of importance of each attribute as weight
3. Functional specifications of the buildings	• Transfer the results of needs analysis to demands • Identify relevant functional properties • Define weight of each property
4. Technical performance specifications	• Transfer functional properties and their weights from previous task to demands • Identify technical performance properties • Identify weight of each property
5. Creation and sketching of alternative structural solutions	• Create and sketch alternative solutions for building, its structural systems and building services in co-operation with other designers and project partners
6. Modular life cycle planning and service life optimisation of each alternative	• Define the requirement for design service life of the building • Modulate the building into service life modules of different service life classes • Identify the number of life cycles of each module during the design service life of the building • Identify the design life cycle costs (monetary and ecology) of the modules • Sketch alternative service lifes for the modules • Define optimal service lifes of the modules, based on minimum total costs (monetary cost and ecology)

Table 6.1 (Continued)

Design phase	Tasks
7. Multiple criteria ranking and selection between alternative solutions and products	• Transfer the optimised service life cost of each alternative building concept for previous task • Define multiple attributes from analysed owner's and user's requirements • Evaluate the performance properties of each alternative • Select the alternative for concretisation between the alternatives
8. Detailed design of the selected solution	• Design the structural modules for different performance requirements • Make the synthetic design

during the development or design process in a continuously increasing precision ranging from general performance specifications to detailed designs.

The tasks for each design alternative are the following:

- classification of building modules into target service life classes, following a suitable classification system;
- stating the number of renewals of each module during the design service life of the building;
- calculation of total life cycle monetary costs and costs of the nature (ecology) during the design life cycle of the building; and
- preliminary optimisation of the total life cycle cost varying the value of service life of key modules in each alternative between the allowed values.

The modular scheduling and allocation at the conceptual design phase includes the specification of the alternative structural solutions regarding the target service life and technical performance requirements of each structural module. Based on the specifications, the estimates of lifetime monetary and environmental costs as current values or annual costs are calculated. A model of the modular specification of the technical performance properties is presented in Table 6.2, taking the structural modules of a building as an example. The specification work must be interactive with life cycle optimisation of the central properties to reach the target of optimal design.

Ranking between design alternatives ends the sketch design phase with draft designs. In the sketch design phase, some key products which are connected to the solution of the structural system are considered even if the final selection of products are mainly done at later phases of the design. When applying integrated design procedure, all classes of requirements are systematically taken into account as per their ranking. As a ranking method, Multiattribute Decision Analysis (MADA) is applied. The core properties are mainly calculated quantitatively with numerical equations, but some of the additional properties can be evaluated qualitatively only. These properties are normalised through comparing with a reference alternative. This phase of

Table 6.2 Specification of performance properties for the alternative structural solutions on module levels; an example of a multi-storey apartment building

CENTRAL PERFORMANCE PROPERTIES IN SPECIFICATIONS	STRUCTURAL MODULE									
	Foundations	Bearing frame	Envelope/ Walls	Envelope/ Roof	Envelope/ Ground Floor	Envelope/ Windows	Envelope/ Doors	Partition Floors	Partition walls (including doors)	Bathroom and kitchen
Bearing capacity	•	•								
target service life	•	•	•	•	•	•	•	•	•	•
limits and targets of environmental impact profiles	•	•	•	•	•	•	•	•	•	•
estimated repair intervals		•	•	•	•	•	•	•	•	•
estimated maintenance costs		•	•	•	•	•	•	•	•	•
Target values of thermal insulation			•	•	•	•	•			
Target Values of Sound Insulation								•	•	
Target Values of Sound Insulation and moisture insulation								•	•	
estimated intervals of the renewal of connected installations								•	•	•
estimated intervals of spatial changes in the building									•	

design is usually made under the responsibility of an architect who is supported by structural designers and building service systems designers.

The detailed design generally includes the following phases:

1. ordinary mechanical design
2. durability design against degradation
3. usability design against obsolescence
4. static, dynamic and seismic design and
5. final integrating design.

Mechanical design includes the static, fatigue and dynamic design aspects. This design is traditional and a lot of manuals, guides, norms and standards exist. Therefore, this phase is not discussed further in this context. Detailed descriptions of the integrated lifetime design is presented by Sarja [2]. The procedure of optimising multi-attribute selection and decision-making is presented in Figure 6.5.

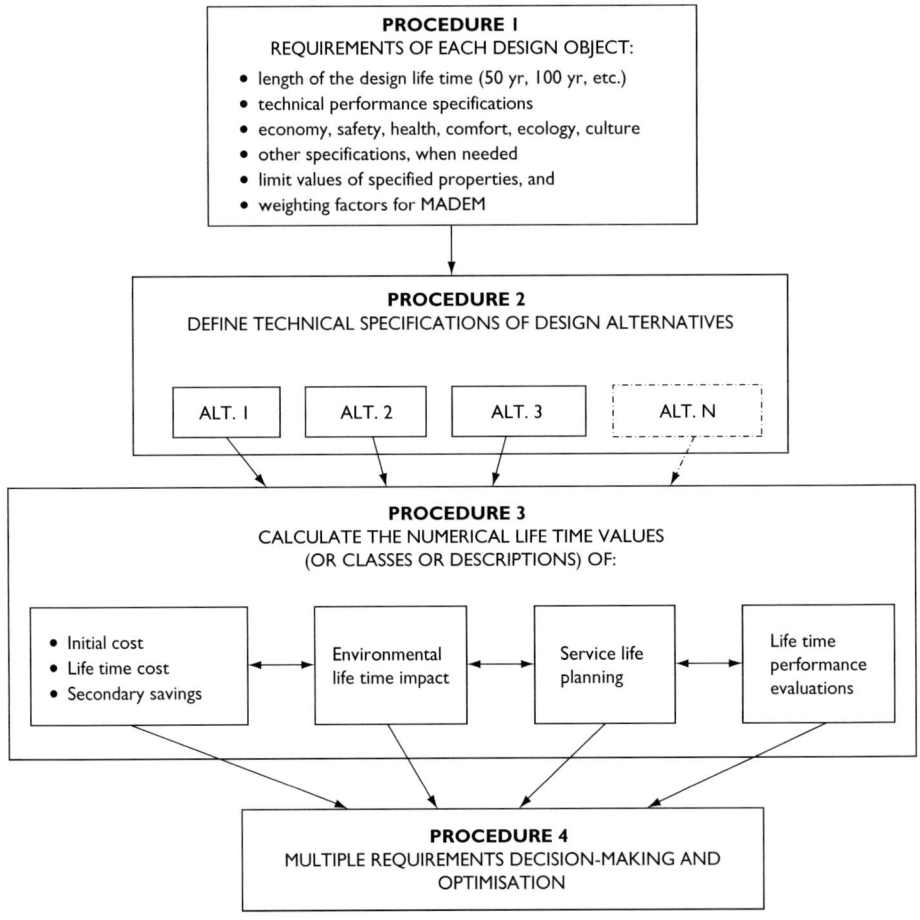

Figure 6.5 The procedure of optimising multi-attribute selection and decision-making.

6.3.3 Procurement and contracting

6.3.3.1 Innovative lifetime procurements

The two most significant recent innovations in public sector project finance are:

- the Private Finance Initiative (PFI) and
- the subsequent evolution of Public Private Partnerships (PPP).

Governments all over Europe and in other continents are turning to the PFI/PPP as an efficient and effective way of delivering services to the public [10].

PFI involves the public and private sectors working together. Traditionally, the public sector procured capital assets by paying for projects up front and in full. In a typical PFI/PPP project, a single, stand-alone, special purpose business, the Project Company, is created by the private sector. This will build a facility, which may be a school, hospital, road, bridge or other asset, which is then operated for a fixed period, known as a 'concession'. The public sector pays for this through a service charge, which will be conditional on the level of service provided. Variations on this are known as 'Design, Build and Operate' (DBO), 'Design, Build, Finance and Operate' (DBFO) and 'Build, Own, Operate and Transfer' (BOOT). Whilst the contractual arrangements vary, they all adopt the PFI/PPP concept.

Because the asset is being built by the operator, who is paid an agreed service charge for providing an asset with a specified functional performance and usually incurs penalties if the functions are not available as, when and to the quality prescribed, the operator has to take a whole-life view of the asset. However, what has happened in practice is that the 'whole life' view has turned out to be the life of the concession plus the time, usually five years, for which the asset must still be serviceable after transfer. Whilst there are no published international standards, there are various other guidance documents related to lifetime procurement [10]. A principal scheme of lifetime procurement is presented in Figure 6.6.

6.3.4 Lifetime management of facilities and their networks

6.3.4.1 Special characteristics of predictive and optimising lifetime management

6.3.4.1.1 HIERARCHY

The decision-making is performed at three levels of structural hierarchy:

1. component and module
2. object
3. network.

The component/module level addresses structural components such as beams and columns and their combinations, i.e. modules. The object level refers to complete

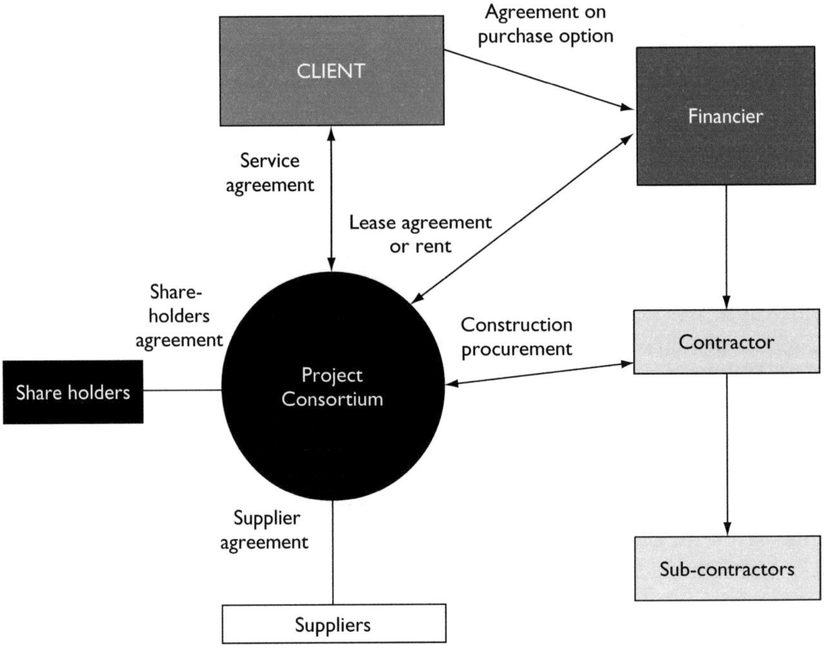

Figure 6.6 A schedule model of a lifetime procurement.

structures or buildings such as bridges and nuclear power plant units. The network level addresses networks of objects such as stocks of bridges or buildings.

6.3.4.1.2 OPTIMISATION OF LIFETIME QUALITY

One of the main objectives in the lifetime management process is the optimisation of lifetime quality with respect to both the network of objects and individual objects. Lifetime quality means the capability of an asset to fulfil the requirements of users, owners and society over a defined design period. Basically, these requirements mean the requirements of sustainable development.

The objective of optimised lifetime quality raises the need for integrated, predictive and optimising management, including management of mechanical performance, serviceability and obsolescence.

The compliance control of various requirements and minimisation of various risks can be done using the respective models. The management of a facility is addressed to the key elements of the management system itself:

- degradation mechanisms
- MR&R methods and life cycle action profiles
- performance or condition of structures
- obsolescence
- adverse events such as accidents and structural failure and
- analysis results and decision-making.

6.3.4.1.3 MODULAR SYSTEMATICS

The database of the management system is arranged according to the modular systematics. The modular systematics is for dividing the facility into homogenous areas and providing a system for documentation of damage observations. The modular systematics is discussed in earlier chapters of this book.

6.3.4.2 Technical development trends of lifetime management

The application of predictive and optimising MR&R strategies for infrastructures raises the need for significant development of condition assessment. Knowledge-based systems combined with Machine-to-Mobile or Machine-to-Machine (M2M) technology are the key issues for this development. Using wireless data as a link between systems, remote devices and individuals can help to update the condition assessment, make it more effective and easier to run. M2M is about enhanced independence of both time and place.

In the management of infrastructures, this M2M System Development means a need for co-operation between:

- owners of the asset stocks (bridges, tunnels, roads, harbours, industrial, office and housing buildings);
- system researchers and developers (Universities, Research Institutes);
- application developers (Mobile phone producers, Mobile operations service firms, IT consulting companies); and
- sensor producers (mechanical, physical and chemical sensors).

Research and development questions for the future are:

- How is the automation platform configured remotely?
- Where is this configuration task done?
- How are the newly added devices notified and the service provider informed?
- How does the service provider get information on the properties of the new and existing devices?
- How are the properties and other data of different kinds of devices described?
- Is it possible to create a universal and generic description of devices?

A schedule of the future management system of infrastructures, combining the predictive and optimising lifetime management system with knowledge-based system engineering and wireless sensor technology is presented in Figure 6.7 [8]. Possibilities for structural health monitoring with wireless sensor networks are discussed, for example, in Markus Krüger and Christian Grosse [11]. A strategy for developing advanced lifetime management facility management systems is presented in Miyamoto [12].

6.3.4.3 Lifetime management of real estates

The maintenance strategy of a real estate is a means for achieving the desired quality level of structural characteristics and the durability of a building's structural parts.

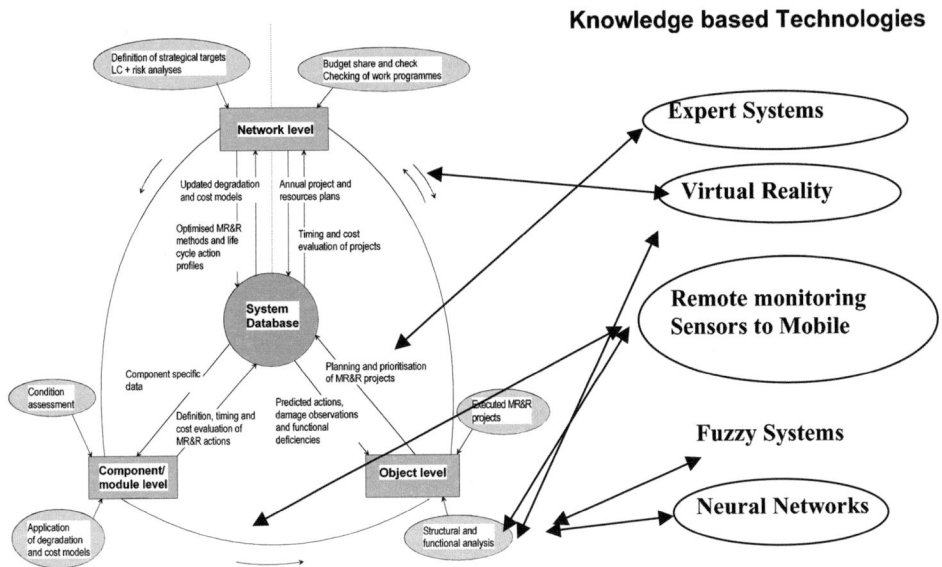

Figure 6.7 A schedule of future management system of infrastructures [8].

The strategy can be defined in terms of accepted cost frames and the chosen risk level. The maintenance strategy also affects the building's residual value. In the management of real estates, the following are the important points [4]:

- In order to be cost-effective, the strategy should be defined and carried out according to the actual need. This need should be determined according to the structural characteristics and according to the significance of the structural item of the building.
- In strategic planning of maintenance, several factors which are important from the viewpoint of use should be taken into account: physical deterioration, economic obsolescence, functional obsolescence, technological obsolescence, social obsolescence, location obsolescence, legal obsolescence, aesthetic and visual obsolescence and environmental obsolescence.
- Maintenance planning should be done so that the life cycle objectives are achieved.
- The maintenance strategy should be done on the one hand according to the needs or restrictions of the users and the usage of the building, but on the other hand according to the needs and objectives of the owners.
- In optimisation of actions, defining the cost–benefit ratio in life cycle and maintenance actions is useful.

Renovation and modernisation, redevelopment phases in the life cycles of buildings, must also be seen as investments, so it is necessary to perform investment analyses about them as well. Since expectations about investments are economic, refurbishment also has to fulfil economic profit expectations. Before starting any investment actions related to a building, the real estate owners should work through their strategic needs and objectives. The needs analysis process should involve as many significant stakeholders as practically possible (owners, managers/executives, facility managers, project managers, staff of employees, tenants, visitors, customers, etc.).

It is better to consider total life cycle costs rather than only initial capital costs. In addition, life cycle costs should be compared to achievable benefits. Total real estate costs can be divided into costs of utilities and real estate costs. Real estate costs are capital costs, taxes and insurance, and maintenance costs, which are the costs of maintaining a building and of repairs:

- Maintenance costs should be budgeted according to the actual need instead of using an annual budgeting method.
- By optimising the characteristics of a building according to the owner's objectives and users' needs, it is easier to make choices related to life cycle costs.
- There are many concepts of value related to real estate and the real estate business. The technical value of a building is the value remaining between its replacement value (either in whole or part) and the decrease in value due to deterioration of characteristics important from the viewpoint of usage.
- Costs can be due to direct or indirect factors. Indirect factors can be related to Indoor Air Quality (IAQ) and healthy buildings.
- In investment situations, the relationship between risks and required rate of return should be considered.

6.3.4.3.1 LIFETIME QUALITY

There are so-called value-generating characteristics of a building. They may be grouped for analytical purposes:

- functional efficiency, the adaptation of the structure to the activities for which it is to be used;
- durability, the physical qualities of the structure, which determine how long the building can continue to render useful service; and
- attractiveness, the aesthetic qualities of the building.

The life cycle quality of a building can be seen from the viewpoint of the owner or the viewpoint of the user or usage. For the owner, in most cases, the building has to be a cost-effective investment. For users, usually, the most important characteristics of a building are its ability to generate added value for the business and its suitability for the intended use.

The Technical Life Cycle Management (TLCM) methodology is a general action description of technical life cycle analysis, strategy development and deployment process (Figure 6.8) [7]. It also describes the roles of different participants in the process:

- The TLCM Methodology is a description of the *interactive* analysis and strategy definition process between the real estate owner, user of a real estate, TLC (Technical Life Cycle) Manager and service provider(s) in which the *technical life cycle analysis, goal-setting and strategy development* are carried out.

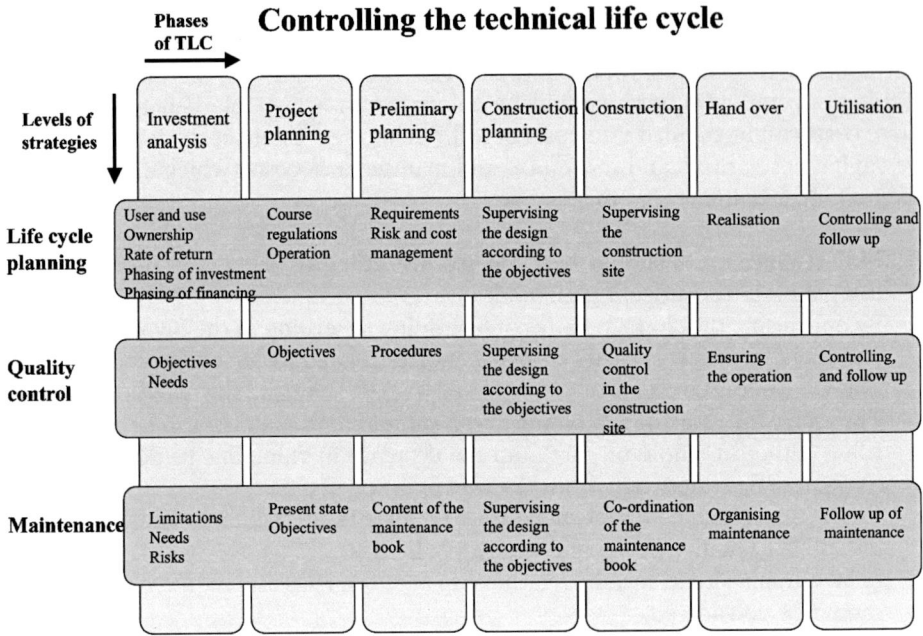

Figure 6.8 Technical Life Cycle Management (TLCM) Methodology [4].

- The methodology has a modular structure. It is systematic, logical and practical in use.
- The methodology describes the tasks of different participants.
- It includes checklists for helping goal-setting and decision-making related to the technical life cycle of an individual real estate.
- It presents connections between phases and task descriptions.
- It presents connections between task descriptions and checklists.
- It provides clear instructions for use and a summary of the theory behind the construction.
- The methodology is a general and comprehensive description, so users can choose phases according to their actual needs.
- The process makes it possible to optimise and set priorities for the building and its life cycle costs. Iteration is possible.
- The process takes into account different stakeholders and their needs.
- Cost, risk and needs analysis are included into the process description:
  - Training is included in the taking-into-use phase.
  - The methodology is easy to use and it is possible to modify it.

The 'Technical Life Cycle Management' concept is principally formed of the

- setting of strategic goals
- making of decisions and

- defining actions that
  - are directed to an individual real estate from the *investment analysis phase until the demolition decision* and
  - affect the *physical characteristics* of the building, which are important from the viewpoint of cost-effective ownership.

The three main starting points for analysing the technical life cycle and determining strategies (as shown in Figure 6.9):

1. Investment analysis for a new building
2. Investment analysis for an old building
3. Technical evaluation of an old owned building to estimate the cost-effectiveness of ownership.

6.3.4.3.2 ASPECTS OF DECISION-MAKING IN A TECHNICAL LIFE CYCLE ANALYSIS

Analysing the technical potential of a building is based on the following factors [4]:

- *Requirements* represent those characteristics of a real estate which are most important from the perspective of cost-effective ownership. Requirements also depend on the planned use of the building. These requirements are such factors

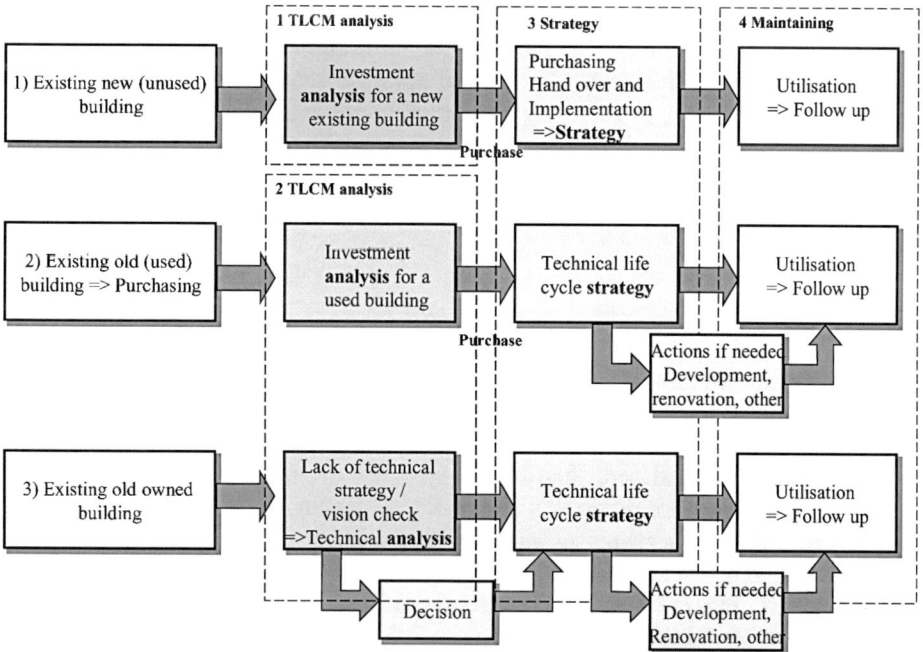

Figure 6.9 The TLCM process consists of two modules: the technical (investment) and potential analysis (TLC-analysis) and the technical life cycle strategy development process [4].

which make investment or ownership profitable and which make it possible for the owner to carry out his *business idea* in the building. The requirements may also depend on user and usage, and what characteristics are critical from the point of view of usage. These are the preliminary objectives from the viewpoint of cost-effective ownership.

- *Limitations* are factors which must be taken into consideration when appraising the cost-effectiveness of ownership. Limitations are connected to risks (in this study, technical risks), expected returns or outcomes and acceptable cost frames during ownership.
- *Expectations* are related to technical, functional and aesthetic quality and the life cycle characteristics of a real estate (life cycle of use or ownership, economic life cycle, technical life cycle, the life cycle of the location) that either exist or are possible to achieve within acceptable risk, cost and return frames. 'Expectations' are hoped for, but achieving them is not critical.
- *Possibilities* is the margin in which the quality and life cycle characteristics can be optimised. 'Possibilities' includes the maintenance required to achieve, in acceptable cost and risk frames, the optimal set of characteristics according to the requirements and expectations of owner and user.

6.3.4.4 End-of-Life Management [13]

6.3.4.4.1 DECONSTRUCTION ASSESSMENT TOOLS

The aim of efficient deconstruction is to reduce the whole duration for dismantling on the site to lower the costs, to improve the working conditions and to assure the required quality of the materials. In order to optimise deconstruction, a methodology for the deconstruction and recycling management for buildings has been developed. To facilitate the task described, a sophisticated computer-aided dismantling and recycling planning system can be used. The methodology for optimisation is based on resource-constrained project scheduling. The structure of this system is illustrated in Figure 6.10.

6.3.4.4.2 AUDIT OF BUILDINGS

The building audit mainly consists of making a detailed description of the building and identifying materials. Based on the documents of the building (construction plans, descriptions, history) detailed data on the composition of the building has to be collected and analysed. Due to the fact that deconstruction normally affects older buildings, reliable information documenting the current state is rarely available. During this audit, indications of substances contained in the building which may influence the quality of the materials need to be collected and analysed. The audit also provides precise information for further investigation on possible pollutant sources and contamination of the building.

The planning system outlined supports the audit by the preparation of bills of materials, which contain details of the materials and the locations of building elements and pollutant sources.

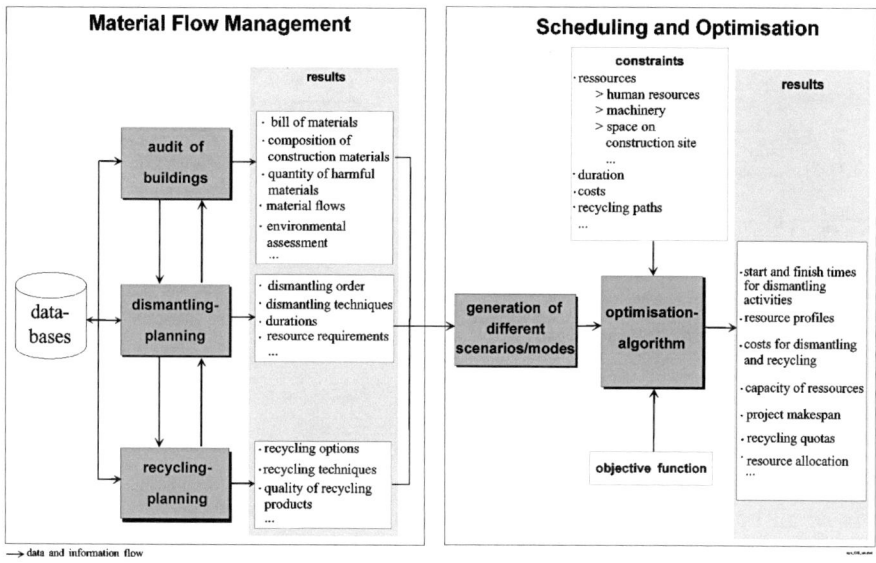

Figure 6.10 Structure of the deconstruction planning system [14].

6.3.4.4.3 DISMANTLING PLANNING

With the available information about the composition of the building combined with the information about the regional framework for waste management, the planning of the dismantling work can be carried out.

On the basis of the bill of materials, appropriate dismantling techniques are selected and aggregated to dismantling activities. The configuration of the dismantling activities comprises the determination of the corresponding construction elements (found in the bill of materials) and the selection of the resources is necessary. Since the aim of the dismantling planning is to do dismantling with minimal costs, dismantling with the aim of preserving building elements intact for later re-use or dismantling due to technical restrictions, etc. the determination of dismantling activities may vary considerably. The dismantling order respecting technological relations as well as security aspects and environmental requirements can be illustrated in the so-called dismantling networks on a modular principle. Figure 6.11 gives an example of a dismantling modules for a residential building.

After determining the dismantling activities and precedence relations, the target of dismantling planning is to find feasible or 'optimal' working schedules. If resources (machines, workers, space on the construction site, budget) are limited, this problem becomes extremely complex.

6.3.4.4.4 RECYCLING AND REUSE PLANNING

The objective of recycling planning is the design of optimal recycling techniques for processing dismantled materials and building components into reusable materials. Depending on the stage of dismantling, the feed can be either a single material or a

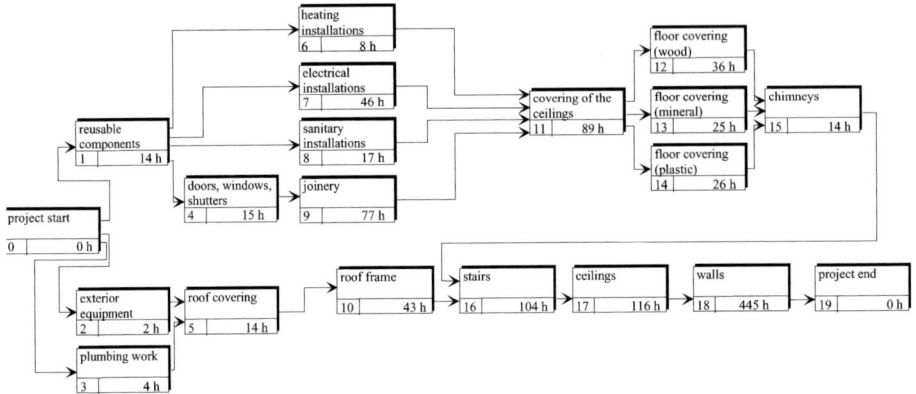

Figure 6.11 Dismantling-network for a residential building [13].

mix of all building materials. For certain individual materials such as metals, glass and minerals or plastics, recycling techniques already exist. In this case, recycling planning is a simple co-ordination. Recycling is difficult, when materials are mixed, when composite materials occur or when pollutants like hydrocarbons or asbestos are present. To obtain materials in an optimal composition for recycling facilities, the available recycling techniques as well as the location of processing facilities have to be considered during dismantling planning. Case studies have shown that direct re-use of elements can be a promising alternative if dismantling is planned well [15].

6.3.4.4.5 OPTIMISATION OF DECONSTRUCTION WORKS

Based on these results, a computer simulation helps to reveal improvement potentials for deconstruction. To show some possible improvements, various simulations and optimisations using the planning tool described above are carried out. Due to the high complexity of the dismantling and recycling planning, a sophisticated mathematical optimisation model is used as a decision support. The model takes into account the interrelations between material flow management (concerning dismantling and recycling) and project management. The consideration of both, material as well as monetary flows during the various planning stages, enables the elaboration of time and cost efficient as well as environmental friendly deconstruction strategies. To evaluate optimal schedules for dismantling, different scenarios might be applied, for instance:

- dismantling of buildings using the possibilities of parallel work as much as possible
- dismantling using mainly manual techniques
- dismantling using partly automated devices and a
- dismantling strategy strictly focused on 'optimal' recycling possibilities according to the material flow analysis.

Computational results for different deconstruction strategies for a building show considerable economic improvement compared with a deconstruction project in

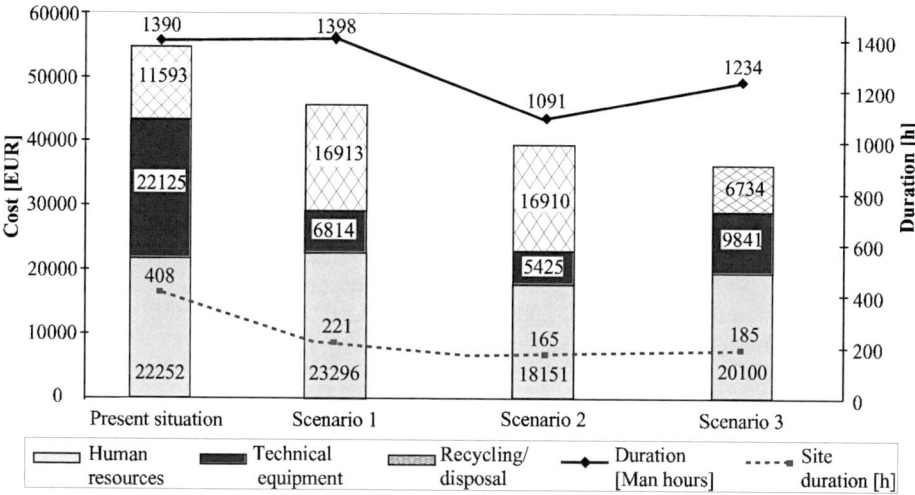

Figure 6.12 Cost and duration of different dismantling strategies for a residential building (Frank Schultmann, 2001 [16]).

practice. As illustrated in Figure 6.12, deconstruction site management can be significantly improved. Optimised dismantling schedules, based on the same framework as in practice, show cost savings up to 50%. In some cases, the dismantling time can be reduced by a factor of 2 applying partly automated devices. Furthermore, a recycling rate of more than 97% can be realised [16].

Based on selected deconstruction strategies, the detailed planning and optimisation of deconstruction work can be done. The complete schedules for two different dismantling scenarios and the corresponding project costs show that an environmental-oriented dismantling strategy imposes a higher effort on the dismantling work. That is, more jobs have to be carried out in order to avoid a mix of hazardous and non-hazardous materials. Nevertheless, environmental-oriented dismantling strategies will not necessarily be disadvantageous from an economic point of view if disposal fees are graded according to the degree of mixed materials.

References

[1] Sarja, A. 1987. Towards the advanced industrialized construction technique of the future. *Betonwerk + Fertigteil-Technik*, 236–9.
[2] Sarja, A. 2002. *Integrated life cycle design of structures*. London: Spon Press.
[3] Sarja, A. 1998. *Open and industrialised building*. London: Spon Press.
[4] Koskelo, T. 2005. *A method for strategic technical life cycle management of real estates.* PhD thesis, Helsinki University of Technology, Doctor Dissertation Series 2005/1.
[5] Sarja, A. 2003. A process towards lifetime engineering in the 5th and 6th Framework Program of EU. *Proceedings of ILCDES 2003 symposium*, Kuopi. Association of Finnish Civil Engineers: Helsinki, 7–15.
[6] Öberg, M. 2005. *Integrated life cycle design, application to Swedish concrete multi-dwelling buildings*. PhD thesis, Lund University of technology, Division of building materials, Report TVBM–1022.

[7] Sarja, A. 2004. Generalised lifetime limit state design of structures. *Proceedings of the 2nd international conference, lifetime-oriented design concepts, ICDLOC*. Ruhr-Universität Bochum, Germany, 51–60.

[8] Sarja, A. 2005. *Generic description of lifetime engineering of buildings, civil and industrial infrastructures*. Manuscript, Deliverable D3.1, Thematic Network Lifetime, EU GROWTH, CONTRACT No: GIRT-CT-2002-05082. http://lifetime.vtt.fi/index.htm.

[9] Kelly, J., and S. Male. 1996. *Value management in design and construction*. London: Spon Press.

[10] Davies, H. 2005. *Review of standards and associated literature on technology and lifetime economy*. Manuscript, Thematic Network Lifetime, EU GROWTH, CONTRACT No: GIRT-CT-2002-05082. http://lifetime.vtt.fi/index.htm.

[11] Krüger, M., and C. U. Grosse. 2004. Structural health monitoring with wireless sensor networks, *Otto Graf Journal*, 15: 77–89.

[12] Miyamoto, A. 2003. Japanese strategy of life cycle management in civil infrastructures systems. *Proceedings of 2nd international symposium ILCDES 2003, Integrated life-time engineering of buildings and civil infrastructures*. Kuopio, Finland. Helsinki: Association of Finnish Civil Engineers.

[13] Scultmann, F. 2004. Manuscript, D3.1, Thematic Network Lifetime, EU GROWTH, CONTRACT No: GIRT-CT-2002-05082. http://lifetime.vtt.fi/index.htm.

[14] Scultmann, F., and O. Rentz. 2002. Scheduling of deconstruction projects under resource constraints. In *Construction management and economics* 20(5): 392–401.

[15] Scultmann, F., and O. Rentz. 1997. Development of a software tool for the optimal dismantling and recycling of buildings. *Proceedings of the symposium computers in the practice of buildings and civil engineering*, Lahti, Finland, September 3–5, pp. 64–68.

[16] Scultmann, F., and O. Rentz. 2001. Environment-oriented project scheduling for the dismantling of buildings. *OR Sektrum* 23(1): 51–78.

Chapter 7

Terms and definitions

Asko Sarja

These terms and definitions are collected and updated from earlier published glossaries [1, 2].

Term	Definition
Life cycle and life time	
Life cycle	The consecutive and inter linked stages of a facility or structure, from the extraction or exploitation of natural resources to the final disposal of all materials as irretrievable wastes or dissipated energy.
Survival life	The time period from the beginning of use until the demolition of a facility.
Working life	The time from the begin of use until the end of the first lifetime period of use, first refurbishment or demolition, usually the equivalent of design life.
Life time	The time from the start of the planning of a building project until a defined point in time, or until the end of its life cycle.
• Life time period	The time from the start of the planning of a facility or construction project until a defined point in time, or between two defined points.
• Design life period	A specified period of time used in design considerations, design calculations and associated decision-making.
• Procurement responsibility period	A specified period defining the responsibility of contractor during the lifetime of a construction project.
• MR&R period	A specified period over which the calculations, optimisations and decision-making in MR&R planning are done.
Serviceability and service life	
Serviceability	Capacity of a structure to perform the service functions for which it is designed and used.
Service life	Period of time after installation during which a facility or its parts meet or exceed the performance requirements.
• Target life	Required service life imposed by general rules, the client or the owner of the structure or its parts.
• Characteristic life	A time period exceeded by the service life with a specified probability, usually with 95% probability.
• Design life	Service life used in the design to provide a required probabilistic safety against the target service life. Design life is calculated dividing the characteristic life by the lifetime safety factor. Design service life has to exceed the target service life.
• Reference service life	Service life forecast for a structure under strictly specified environmental loads and conditions for use as a basis for estimating service life.

(Continued)

Term	Definition
Service life planning	Preparation of the brief for the structure and its parts in order to achieve control of the usability of structures, and to facilitate maintenance and refurbishment in an optimised way.
Service life design	Preparation of the design of the structure and its parts to achieve the desired design life at the defined reliability level.
Reliability and performance	
Performance	Measure by which the structure responses to a certain function.
Performance requirement or performance criterion	Qualitative and quantitative levels of performance required for a critical property of a structure.
Lifetime quality	The capability of the facility to fulfil all the requirements of its owner, user, and society over its specified life time.
Failure	Loss of the ability of a structure or its parts to perform a specified function.
• Durability failure	Exceeding the maximum permitted degradation, or falling below a minimum performance parameter.
Failure probability	The statistical probability of failure occurring.
Risk	Multiplication of the probability of an event such as failure or damage.
Obsolescence	Loss of the ability of an item to perform satisfactorily due to changes in economic, performance, human (safety, health, convenience) or ecological requirements.
Limit state	A specified measure or performance parameter exceeding or falling below a defined value.
• Serviceability limit state	A situation in which conditions of specified service requirement(s) for a structure are no longer being met.
• Ultimate limit state	A state associated with collapse or similar form of failure.
Lifetime safety factor	Coefficient by which the characteristic life is divided to obtain the design life.
Factor method	Modification of reference service life to take account of the specific environmental loads and in-use conditions.
Attribute	A property of an object or its part used in optimisation and selective decision-making between alternatives.
• Multiple attributes	A set of attributes used in optimisation and selective decision-making between alternatives.
Durability	
Durability	The capacity of a structure to maintain minimum performance under actual environmental degrading loads.
Durability limit state	Minimum acceptable state of performance or maximum acceptable state of degradation.
Durability model	Mathematical model for calculating degradation, performance or the service life of a structure.
Performance model	Mathematical model showing performance over time.
Condition	Level of critical properties of a structure or its parts, determining its performance ability.
Condition model	Mathematical model for placing an object, module, component or subcomponent in a specific condition class.

Deterioration	The process of becoming impaired in quality or value.
Degradation	Gradual decrease in performance of a material or structure.
Environmental load	Impact of environment on a structure, including weathering (temperature, temperature changes, moisture, moisture changes, solar effects, etc.), chemical and biological factors.
Degradation load	Any combination of environmental and mechanical loads.
Degradation mechanism	The sequence chemical, physical or mechanical changes that lead to detrimental changes in one or more properties of building materials or structures when exposed to degradation loads.
Degradation model	Mathematical model showing degradation over time.
Management and maintenance Maintenance	Combination of all technical and associated administrative actions during the service life designed to retain a structure in a state in which it can perform its required functions.
Repair	Return of a structure to an acceptable condition by the renewal, replacement or mending of worn, damaged or degraded parts.
Restoration	Actions to bring a structure to its original appearance or state.
Refurbishment or Rehabilitation	Modifications and improvements to an existing structure to bring it up to an acceptable condition.
Renewal	Demolition and rebuilding of an existing object.
M&R	Maintenance, plus repair, restoration, refurbishment and renewal.
Project	Planning and execution of repair, refurbishment, restoration or dismantling of a facility or parts of it.
Life cycle cost	Total cost of a structure throughout its life, including the costs of planning, design, acquisition, operations, maintenance and disposal, less any residual value.
Environmental burden	Any change to the environment which permanently or temporarily results in loss of natural resources or deterioration in the air, water or soil, or loss of biodiversity.
Environmental impact	The consequences for human health, the well-being of flora and fauna, or the future availability of natural resources, attributable to the input and output streams of a system.
Integrated lifetime design of materials and structures	Producing descriptions for structures and their materials, fulfilling the specified requirements of economics, human requirements (safety, health, convenience), ecology (economy of the nature), cultural and social needs, over the life cycle of the structures. Integrated structural design is the synthesis of mechanical design, durability design, physical design and environmental design.
Environmental structural design	That part of the integrated structural design that takes into account environmental aspects during the design process.
Integrated lifetime management	Planning and control procedures in order to optimise economic, human, ecological, cultural and social conditions over the life cycle of a facility.
Hierarchical system System	An integrated entity which functions in a defined way and whose components have defined relationships and rules between them.
Hierarchical system	A system consisting of some value scale, value system or hierarchy.

(Continued)

Term	Definition
Modulated system	A system whose parts or modules are autonomous in terms of performance and internal structure.
Structural system	A system of structural components which fulfil a specified function.
Network	Stock of objects or facilities (e.g. bridges, tunnels, power plants, buildings) under management and maintenance of an owner.
Object	A basic unit of the network serving a specific function.
Module or assembly	A component or set of components designed and manufactured to serve a specific function(s) as part of the system, and whose functional, performance and physical relations with the structural system are specified.
Structural component	A basic unit of the structural system designed and manufactured to serve a specific function(s) as part of a module, and whose functional, performance and physical relations with the structural system are specified.
Subcomponent	Manufactured product forming a part of a component.
Material	A substance that can be used to form products.
Stakeholders	
Stakeholders	Owners, users, designers, contractors, industry sectors, public interest organisations, regional interests and/or government agencies connected with the structure during its life cycle.
Owner	Person or organisation for which a structure is constructed, and/or the person or organisation that has the responsibility for the maintenance and upkeep of the structural, mechanical and electrical systems of the building.
Designer	Person or organisation that prepares a design or arranges for any person under one's control to prepare the design.
Contractor	Person or organisation that undertakes to, or does, carry out or manage construction work. The contractor bids a contract for a new building with information from manufacturers and suppliers. The contractor's representative on the building site is the site supervisor.
Manager	At takeover, the building is administrated by a property manager who engages maintainers to be responsible for proper maintenance inspections or to carry out the necessary maintenance.
Supplier	Person or organisation that supplies structures, parts of structures or services for the construction or maintenance of structures.
User	Person or organisation which occupies a facility.
Dismantler	Person who carries out dismantling work.
Methods	
Allocation	The division of specified financial and physical resources amongst objects, projects and other actions on the network level.
Briefing	Statement of the requirements of a facility.
Service life planning	Preparation of the brief and design for a facility and its parts in order to optimise the required properties of the facility for its owner, and to facilitate maintenance and refurbishment.

Condition assessment	Methodology and methods for quantitative measurement and visual inspection of the properties of an object and its parts, and conclusions drawn from the results regarding to its condition.
Optimisation	Selection between alternative properties of an object or its parts, or of an action taken in order to reach an optimal solution or result.
• Short-term optimisation	Optimisation over a short-time period (usually one or two years).
• Long-term optimisation	Optimisation over a long-term period (usually several or even tens of years).
Decision-making	Methodology for rational choices between alternatives, basing on defined requirements and criteria.

References

[1] Sarja, A. 2002. *Integrated Life Cycle Design of Structures.* 142pp. London: Spon Press. ISBN 0-415-25235-0.

[2] Sarja, A. *Lifecon, Life Cycle Management of Concrete Infrastructures, D2.1.* http://lifecon.vtt.fi/.

Index